MAGNUS, KARRASS, and SOLITAR–Combinatorial Group Theory
MARTIN–Nonlinear Operators and Differential Equations in Banach Spaces
MELZAK–Companion to Concrete Mathematics
MELZAK–Mathematical Ideas, Modeling and Applications (Volume II of Companion to Concrete Mathematics)
MEYER–Introduction to Mathematical Fluid Dynamics
MORSE–Variational Analysis: Critical Externals and Sturmian Extensions
NAYFEH–Perturbation Methods
NAYFEH and MOOK–Nonlinear Oscillations
ODEN and REDDY–An Introduction to the Mathematical Theory of Finite Elements
PAGE–Topological Uniform Structures
PASSMAN–The Algebraic Structure of Group Rings
PRENTER–Splines and Variational Methods
RIBENBOIM–Algebraic Numbers
RICHTMYER and MORTON–Difference Methods for Initial-Value Problems, 2nd Edition
RIVLIN–The Chebyshev Polynomials
RUDIN–Fourier Analysis on Groups
SAMELSON–An Introduction to Linear Algebra
SIEGEL–Topics in Complex Function Theory
Volume 1–Elliptic Functions and Uniformization Theory
Volume 2–Automorphic Functions and Abelian Integrals
Volume 3–Abelian Functions and Modular Functions of Several Variables
STAKGOLD–Green's Functions and Boundary Value Problems
STOKER–Differential Geometry
STOKER–Nonlinear Vibrations in Mechanical and Electrical Systems
STOKER–Water Waves
WHITHAM–Linear and Nonlinear Waves
WOUK–A Course of Applied Functional Analysis

NONLINEAR OSCILLATIONS

NONLINEAR OSCILLATIONS

ALI HASAN NAYFEH

University Distinguished Professor

DEAN T. MOOK

Professor

Department of Engineering Science and Mechanics
Virginia Polytechnic Institute and State University
Blacksburg, Virginia

A WILEY-INTERSCIENCE PUBLICATION

JOHN WILEY & SONS, New York • Chichester • Brisbane • Toronto

Library of Congress Cataloging in Publication Data:

Nayfeh, Ali Hasan, 1933–
Nonlinear oscillations.

(Pure and applied mathematics)
Includes bibliographical references and index.
1. System analysis. 2. Nonlinear theories.
3. Oscillations. I. Mook, Dean T., joint author.
II. Title.

QA402.N34 531 78-27102
ISBN 0-471-03555-6

Printed in the United States of America

10 9 8 7 6 5 4 3 2 1

mms 5-19

To
Our Wives
Samirah and Sally

PREFACE

Recently a large amount of research has been related to nonlinear systems having multidegrees of freedom, but hardly any of this can be found in the many existing books related to this general area. The previously published books emphasized, and some exclusively treated, systems having a single degree of freedom. These include the books of Krylov and Bogoliubov (1947); Minorsky (1947, 1962); Den Hartog (1947); Stoker (1950); McLachlan (1950); Hayashi (1953a, 1964); Timoshenko (1955); Cunningham (1958); Kauderer (1958); Lefschetz (1959); Malkin (1956); Bogoliubov and Mitropolsky (1961); Davis (1962); Struble (1962); Hale (1963); Butenin (1965); Mitropolsky (1965); Friedrichs (1965); Roseau (1966); Andronov, Vitt, and Khaikin (1966); Blaquière (1966); Siljak (1969), and Brauer and Nohel (1969). Exceptions are the books by Evan-Iwanowski (1976) and Hagedorn (1978), which treat multidegree-of-freedom systems. However, a number of recent developments have not been included. The primary purpose of this book is to fill this void.

Because this book is intended for classroom use as well as for a reference to researchers, it is nearly self-contained. Most of the first four chapters, which treat systems having a single degree of freedom, are concerned with introducing basic concepts and analytic methods, although some of the results in Chapter 4 related to multiharmonic excitations cannot be found elsewhere. In the remaining four chapters the concepts and methods are extended to systems having multidegrees of freedom.

This book emphasizes the physical aspects of the systems and consequently serves as a companion to *Perturbation Methods* by A. H. Nayfeh. Here many examples are worked out completely, in many cases the results are graphed, and the explanations are couched in physical terms.

An extensive bibliography is included. We attempted to reference every paper which appeared in an archive journal and related to the material in the book. However omissions are bound to occur, but none is intentional. Many exercises have been included at the end of each chapter except the first. These exercises progress in complexity, and many of them contain intermediate steps to help the reader. In fact, many of them would expand the state of the art if numerical results were computed. Some of these exercises provide further references.

We wish to thank Drs. D. T. Blackstock, M. P. Mortell, and B. R. Seymour for their valuable comments on Chapter 8 and Drs. J. E. Kaiser, Jr., and W. S. Saric for their valuable comments on Chapter 1. A special word of thanks goes to our children Samir (age 7), Tariq (age 10), and Mahir (age 11) Nayfeh and Art Mook (age 16) and to Patty Belcher and Tom Dunyak for their efforts in checking the references. Many of the figures were drawn by Chip Gilbert, Joe Mook, and Fredd Thrasher, and we wish to express our appreciation to them. We wish to thank Janet Bryant for her painstaking typing and retyping of the manuscript. Finally a word of appreciation goes to Indrek Wichman, Jerzy Klimkowski, Helen Reed, Albert Ten, and Ten Liu for proofreading portions of the manuscript.

Ali Hasan Nayfeh
Dean T. Mook

Blacksburg, Virginia
January 1979

CONTENTS

NONLINEAR OSCILLATIONS

CHAPTER 1

Introduction

1.1. Preliminary Remarks

In this chapter we attempt to abstract the entire book. We introduce some of the nonlinear physical phenomena that are discussed in detail in subsequent chapters. The development of many of the results discussed here requires somewhat elaborate algebraic manipulations. Here we describe only the physical phenomena, leaving all the algebra to the subsequent chapters. The descriptions in this chapter are intended to give an overview of the whole book. Thus one might better see how a given topic fits into the overall picture by rereading portions or all of this chapter as one progresses through the rest of the book.

1.2. Conservative Single-Degree-of-Freedom Systems

In Chapter 2, free oscillations of conservative nonlinear systems are considered. Most of these systems are governed by equations having the general form

$$\ddot{u} + f(u) = 0 \tag{1.1}$$

Upon integrating, we obtain

$$\tfrac{1}{2}\dot{u}^2 = h - F(u) \tag{1.2}$$

where $F(u) = \int f\, du$ and h is a constant of integration. Referring to (1.1) and (1.2), we note that $f(u)$ is the (nonlinear) restoring force, $F(u)$ is the potential energy, $\frac{1}{2}\dot{u}^2$ is the kinetic energy, and h (which is determined by the initial conditions) is the total energy level per unit mass.

In the upper portion of Figure 1-1, the undulating line represents the potential energy, while the straight horizontal lines represent total energy levels. Each total energy level corresponds to a different motion, and the vertical distance between a given horizontal line and the undulating line represents the kinetic energy for that motion. Thus motion is possible only in those regions where the potential energy lies below the total energy level.

In the lower portion of Figure 1-1, the variation of $\dot{u}$ with u is shown. Such a graph is called a *phase plane*. For a given set of initial conditions (i.e., for a given

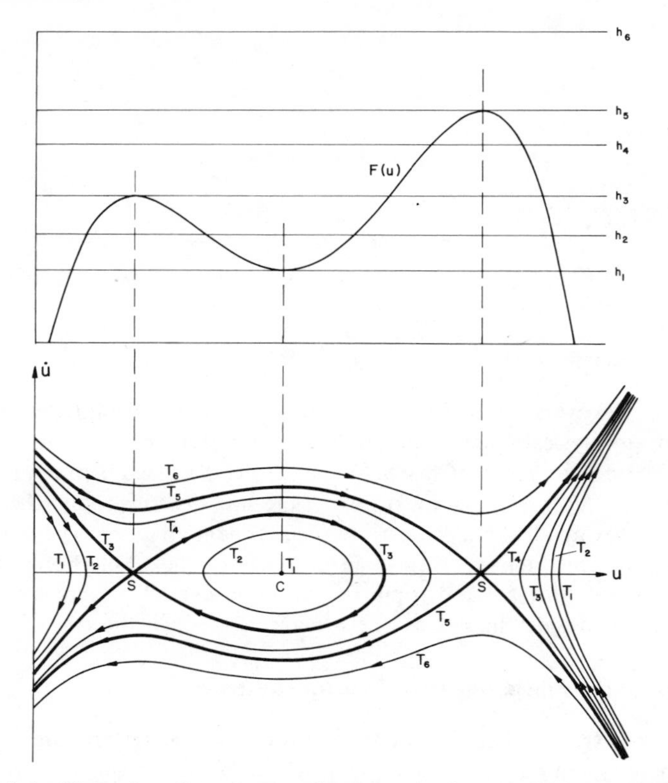

Figure 1-1. Phase plane for a conservative system having a single degree of freedom.

total energy level), the response of the system can be viewed as the motion of a point along a one-parameter (time) curve. Such a curve is called a *trajectory*. The trajectory labeled T_n corresponds to the energy level h_n. The arrows indicate the direction in which the point representing the motion moves as time increases.

The points labeled S are called *saddle points* or *cols*, and the one labeled C is called a *center*. Saddle points and centers correspond to extrema of the potential energy and hence they are *equilibrium points*. Saddle points correspond to maxima while centers correspond to minima of the potential energy. The trajectories that intersect at the saddle points (T_3 and T_5 in Figure 1-1) are called *separatrices*. They are the heavy lines. The point representing the motion moves toward S along two of the separatrices and away from S along the other two. If the representative point is displaced a small distance away from S, there are three possibilities. First, the point can be placed exactly on an inward-bound separatrix, and hence it approaches S as time increases. Second, it can be placed on a closed trajectory, and at times it is far away from S, though it periodically passes close to S. (Here we assume that the equilibrium points are isolated.) Third, it can be placed on an open trajectory, and hence it approaches infinity as time increases. Because the representative point does not stay close to S for all small displacements, the motion is said to be unstable in the neighborhood of a saddle point (i.e., an equilibrium point corresponding to a maximum of the potential energy is unstable).

In the neighborhood of the center, the trajectories are closed, and hence the response is periodic (though not necessarily harmonic). Thus if the motion is displaced slightly from a center, the representative point will always move on a closed trajectory which surrounds the center and stay close to it. (Again we assume that the equilibrium points are isolated.) Thus the motion is said to be stable in the neighborhood of a center (i.e., an equilibrium point corresponding to a minimum of the potential energy is stable). An examination of these closed trajectories shows that the period is a function of the amplitude of the motion. In general, these trajectories do not extend the same distances to the right and the left of the center; thus the midpoint of the motion shifts away from the static center as the amplitude increases. This shift is often called *drift* or *steady-streaming*.

Several analytical methods are introduced and subsequently used to provide approximate expressions for the response. These methods treat small, but finite, periodic motions in the neighborhood of a center. For various examples, the approximate and exact values of the periods are compared.

1.3. Nonconservative Single-Degree-of-Freedom Systems

In Chapter 3, free oscillations of nonconservative systems are introduced. Examples of positive damping due to dry friction (Coulomb damping), viscous effects, form drag, radiation, and hysteresis are presented; examples of negative damping are also included.

In Figure 1-2, a typical phase plane is shown. This one describes the oscillations of a simple pendulum under the action of viscous damping. Depending on the initial conditions, the pendulum may execute several complete revolutions before the oscillatory motion begins. The trajectories spiral into points that correspond to the straight-down position of the pendulum. These points are called *foci*. The straight-up positions correspond to the saddle points in the phase plane. And as in the case of conservative systems, the trajectories that pass through the saddle points are called separatrices.

The concept of a limit cycle is introduced. As an example, we consider Rayleigh's or van der Pol's equation:

$$\ddot{u} + \omega_0^2 u = \epsilon(\dot{u} - \tfrac{1}{3}\dot{u}^3) \tag{1.3}$$

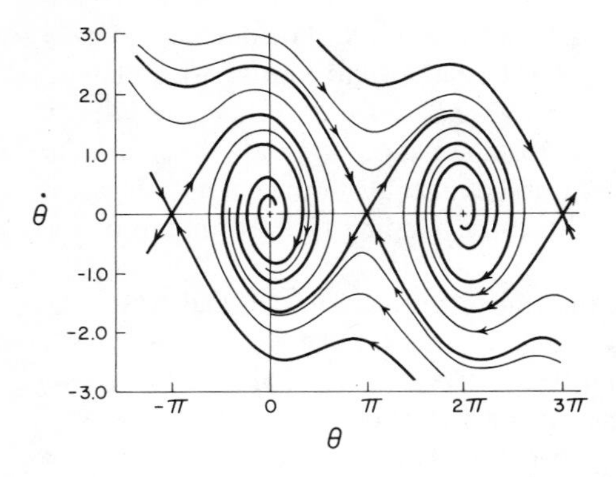

Figure 1-2. Phase plane for a simple pendulum with viscous damping.

We regard the right-hand side of (1.3) as a damping term and note that its influence depends on the amplitude of the motion. When the amplitude of the motion is small, $\frac{1}{3}\dot{u}^3$ is small compared with $\dot{u}$ and the "damping" force has the same sign as the velocity (negative damping); thus the response grows. When the

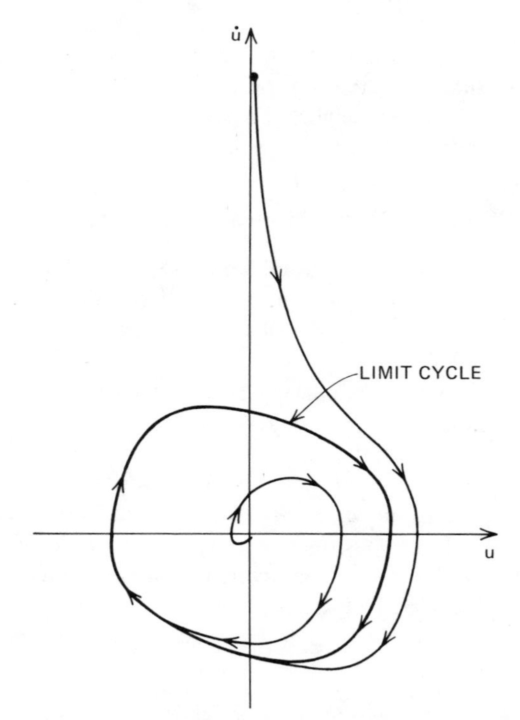

Figure 1-3. Phase plane for van der Pol's equation ($\epsilon = 0.1$).

amplitude is large, $\frac{1}{3}\dot{u}^3$ is large compared with $\dot{u}$ and the damping force has the opposite sign of the velocity (positive damping); thus the motion decays. This behavior of growth when the amplitude is small and decay when the ampitude is large suggests that somewhere in between there exists a motion whose amplitude

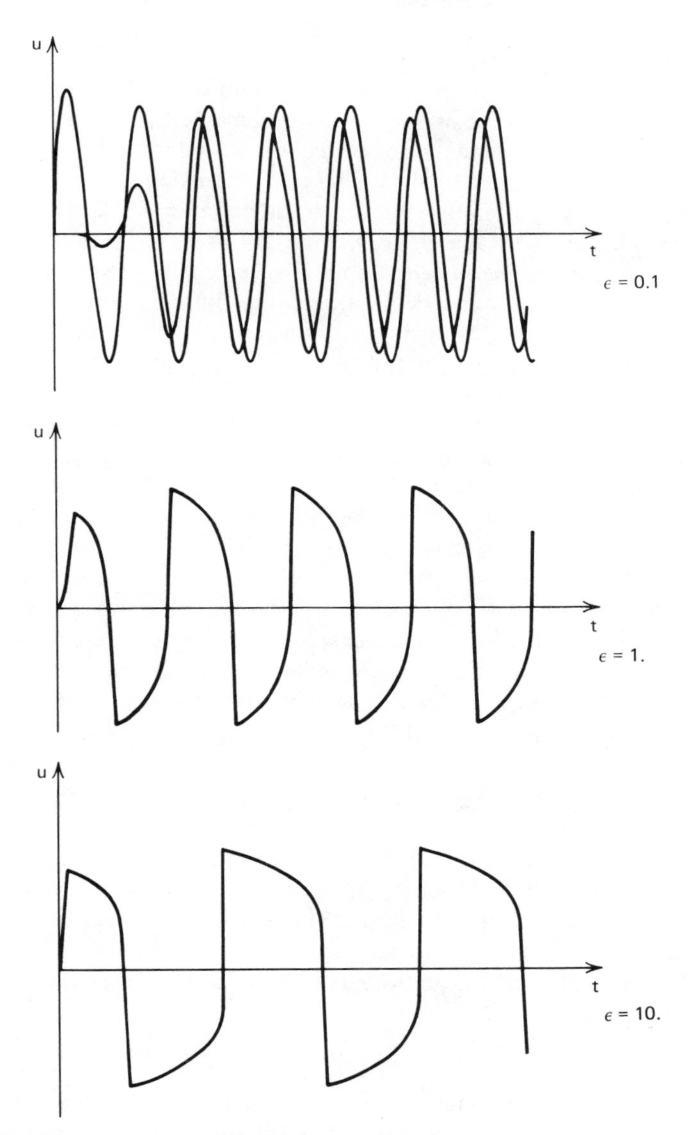

Figure 1-4. Responses of the van der Pol oscillator for various values of ϵ.

neither grows nor decays. This is the case, and the motion is said to approach a *limit cycle*.

In Figure 1-3, a phase plane for the van der Pol equation is shown. There are two trajectories. One begins well outside the limit cycle, while the other begins near the origin. Again the arrows indicate the direction in which the point representing the motion moves. The two trajectories approach the same limit cycle.

The influence of the parameter ϵ on the response is shown in Figure 1-4. The two curves in the top graph correspond to the two trajectories shown in Figure 1-3. We note that, as ϵ increases, the motion becomes jerky; that is, in each cycle there is a period of very rapid motion which is followed by a period of very slow motion. This jerky motion is called a *relaxation oscillation*. Among other examples, this jerky type of motion is characteristic of a beating heart.

A system such as the Rayleigh or van der Pol oscillator is said to be a *self-exciting* or a *self-sustaining* system. Some other examples of self-sustaining systems are found in various other electronic circuits, flutter, supersonic flow past a liquid film, violin strings, a block on a moving belt, Q machines, multimode operation of lasers, ion-sound instability in an arc discharge, and a beam-plasma system.

In Chapter 3, a general discussion of singular points is given, and then various qualitative methods and the analytical methods of multiple scales and averaging are described. The analytical methods treat small, but finite, motions in the neighborhood of a focus or a center. Several examples are worked out, and the analytical results are compared with numerical results.

The comparisons made in the second and third chapters provide confidence for the reader who is not well versed in perturbation methods. Confidence is essential because in the subsequent chapters the analysis predicts many phenomena that are associated only with nonlinear systems and that are in sharp contrast with those associated with linear systems. Some of these phenomena, such as "saturation," are described for the first time in this book.

1.4. Forced Oscillations of Systems Having a Single Degree of Freedom

In Chapter 4, we consider forced oscillations of weakly nonlinear systems having a single degree of freedom. A number of concepts that are associated only with nonlinear systems are introduced. The analytical methods introduced in Chapters 2 and 3 are used for the analysis, and some of the analytical predictions are verified by numerical integration. The problem reduces mathematically to finding the solution of

$$\ddot{u} + \omega_0^2 u = \epsilon f(u, \dot{u}) + E \tag{1.4}$$

where $\epsilon \ll 1$ and E is an externally applied, generalized force called the *excitation*. We distinguish between two types of excitations. The first type of excitation draws on an energy source that is so large the excited system has a negligible effect on it. In this case, $E = E(t)$; that is, E is not a function of the state $(u, \dot{u}, \ddot{u})$ of the excited system. The second type of excitation draws on an energy source that is not large enough to be independent of the response of the excited system. In this case, $E = E(u, \dot{u}, \ddot{u})$. The former is called an *ideal energy source*, while the latter is called a *nonideal energy source*. Both types of energy sources are considered. Thus the response of the system depends on the type of excitation, or energy source, as well as on the natural frequency of the system, the order of nonlinearity, and the type of damping mechanism.

In the next five subsections, we briefly introduce some of the topics treated in detail in Chapter 4.

1.4.1. PRIMARY RESONANCES OF THE DUFFING EQUATION

For an ideal energy source, the response of the system depends on the frequency content of the excitation as well as on the amplitudes and the phases of the different frequency components. In the case of a single-frequency excitation, a cubic nonlinearity, and linear viscous damping, (1.4) becomes

$$\ddot{u} + \omega_0^2 u = -2\epsilon\mu\dot{u} - \epsilon\alpha u^3 + K \cos \Omega t \tag{1.5}$$

where K and Ω are constants. For small amplitudes, the nonlinear term can be neglected and the response of the resulting linear system is

$$u = a \exp(-\epsilon\mu t) \cos [(\omega_0^2 - 4\epsilon^2\mu^2)^{1/2} t + \beta] + K[(\omega_0^2 - \Omega^2)^2 + 4\epsilon^2\mu^2\Omega^2]^{-1/2} \cos(\Omega t + \theta) \tag{1.6}$$

Thus the response of the system consists of two parts: a particular solution and a homogeneous solution (free-oscillation term) having the constants a and β which are determined from the initial conditions. For positive damping (i.e., $\mu > 0$), the free-oscillation term decays with time. The resulting response is called the *steady-state response* and it consists of the particular solution only. Thus the steady-state response has the same frequency as the excitation, but its phase θ is shifted from that of the excitation an amount that depends on the damping and the relative magnitudes of ω_0 and Ω. Moreover it is independent of a and β, and hence it is independent of the initial conditions.

We note from (1.6) that large motions occur when K is large and/or $\Omega \approx \omega_0$. The latter case is called a *primary* or a *main resonance*. When the motions are large, one cannot neglect the nonlinear term $\epsilon\alpha u^3$ in (1.5). When Ω/ω_0 is away from $\frac{1}{3}$, 1, and 3, the response can be written as

$$u = a(t) \cos [\omega_0 t + \beta(t)] + K[(\omega_0^2 - \Omega^2)^2 + 4\epsilon^2\mu^2\Omega^2]^{-1/2} \cos(\Omega t + \theta) + O(\epsilon) \tag{1.7}$$

As $t \to \infty$, $a \to 0$ and to first order the response of the nonlinear system is the same as that of the linear system.

When $\Omega = \omega_0 + \epsilon\sigma$ where $\sigma = O(1)$ and is called the *detuning parameter* or simply the *detuning*, the free-oscillation term cannot be uncoupled from the particular solution. The excitation changes the natural frequency of the system which in turn changes the response of the system to

$$u = a \cos(\Omega t - \gamma) + O(\epsilon) \tag{1.8}$$

where in the steady state a and γ are constants that depend on the amplitude and frequency of the excitation, α, and in some cases the initial conditions. The dependence of the steady-state response on the initial conditions is discussed below. For now we note that, in the presence of damping, this dependence is a nonlinear phenomenon. The amplitude of the response a is related to the amplitude (K) and frequency (σ) of the excitation by the so-called *frequency-response equation*

$$\mu^2 + \left(\sigma - \frac{3\alpha a^2}{8\omega_0}\right)^2 = \frac{K^2}{4\epsilon^2 \omega_0^2 a^2} \tag{1.9}$$

Figure 1-5 shows three representative curves for the cases $\alpha = 0$, $\alpha > 0$, and $\alpha < 0$. Comparing these curves shows that the nonlinearity bends the frequency-response curves to the right when $\alpha > 0$ (hardening nonlinearity) and to the left when $\alpha < 0$ (softening nonlinearity).

The bending of the frequency-response curves leads to multivalued amplitudes and hence to a jump phenomenon. To see the jump phenomenon, let us suppose that an experiment is conducted for $\alpha > 0$ in which the amplitude of the excitation is held constant while the frequency is varied very slowly. We refer to this as a *quasi-stationary process*, and to the excitation as *stationary*. When the experiment is started at an Ω far above ω_0 and Ω is monotonically decreased, the amplitude of the response increases slowly along the curve *AFB* in Figure 1-5*b* until *B* is reached. At that point, any slight decrease in Ω precipitates a spontaneous jump from *B* up to *C*. For further decreases in Ω, the amplitude decreases slowly along the curve from *C* toward *D*. When the experiment is started at an Ω far below ω_0 and Ω is monotonically increased, the amplitude of the response increases slowly along the curve *DCE*. For this process, the amplitude varies smoothly through *C*; there is no downward jump to *B*. The amplitude of the response continues to increase smoothly until *E* is reached. At that point, any further increase in Ω precipitates a spontaneous downward jump from *E* to *F*. For further increases in Ω, the amplitude continues to decrease along the curve from *F* toward *A*.

For $\alpha < 0$, the jumps take place in the opposite directions as shown in Figure 1-5*c*. We emphasize again that the jumps are a consequence of the multivaluedness of the frequency-response curves, which in turn is a consequence of the nonlinearity.

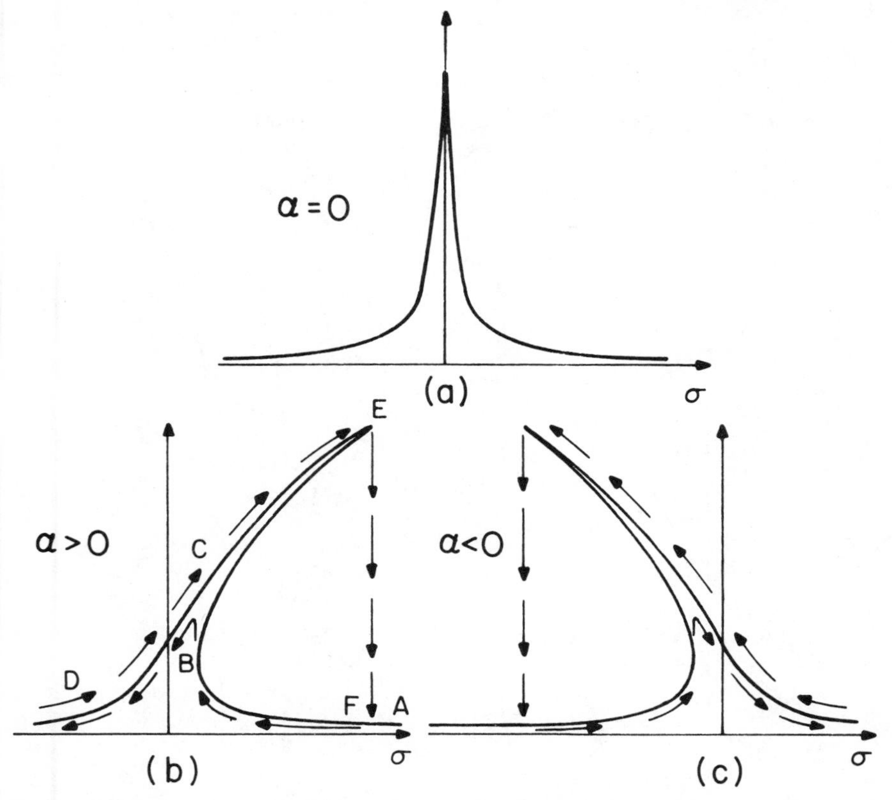

Figure 1-5. Frequency-response curve for the Duffing equation for (*a*) a linear spring, (*b*) a hardening spring, and (*c*) a softening spring.

For frequencies of the excitation in the interval between *BF* and *CE* in Figure 1.5*b*, there are three steady-state solutions for each value of σ. The middle one is a saddle point; hence the response corresponding to it is unstable and unrealizable in any experiment. The other two are stable foci; hence both are realizable. Thus for a given frequency of the excitation, there can be more than one steady-state response. The initial conditions determine which of the possible responses actually develops. This dependence of the steady-state response contrasts sharply with the behavior of positively damped linear systems for which the steady state is independent of initial conditions.

In Figure 1-6, we show a state plane [refer to (1.8) for the meaning of a and γ] for this example when three steady-state solutions exist. Generally the transient response has the same form given by (1.8), but a and γ are functions of time. The trajectories show how the response progresses toward a steady state from any initial conditions. Again the arrows indicate the movement of the

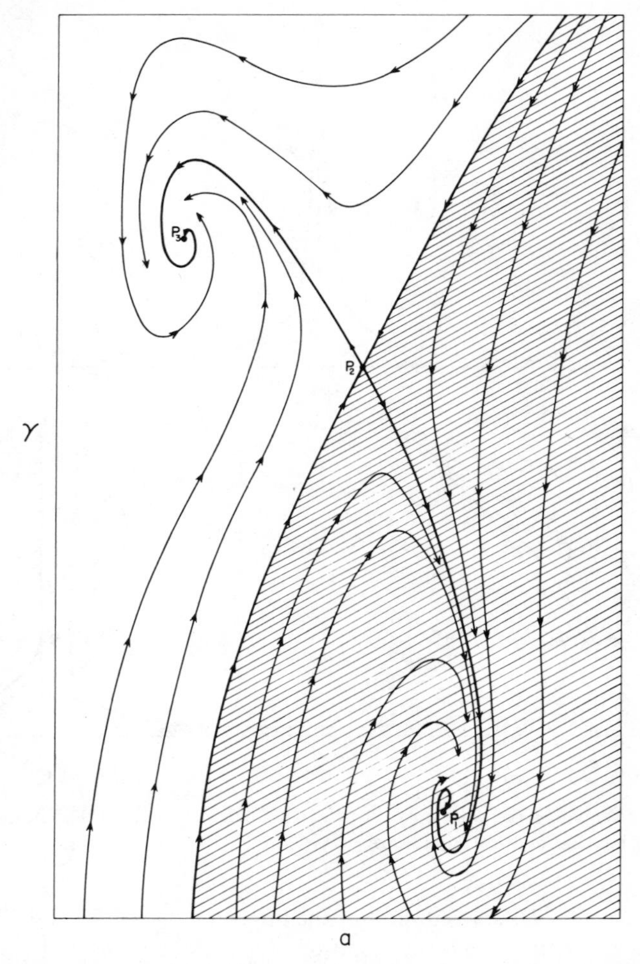

Figure 1-6. State plane for the Duffing equation when three steady-state solutions exist; P_1 is the upperbranch stable focus, P_2 is the saddle point, and P_3 is the lower-branch stable focus.

point representing the motion as time increases. For all the initial conditions lying in the shaded area, the high-amplitude steady state will develop, while for all the initial conditions lying in the unshaded area, the low-amplitude steady state will develop. Thus one says that these areas constitute *domains of attraction* for the possible steady-state responses. We note that the two inward-bound separatrices for the saddle point (the unstable middle-amplitude steady-state)

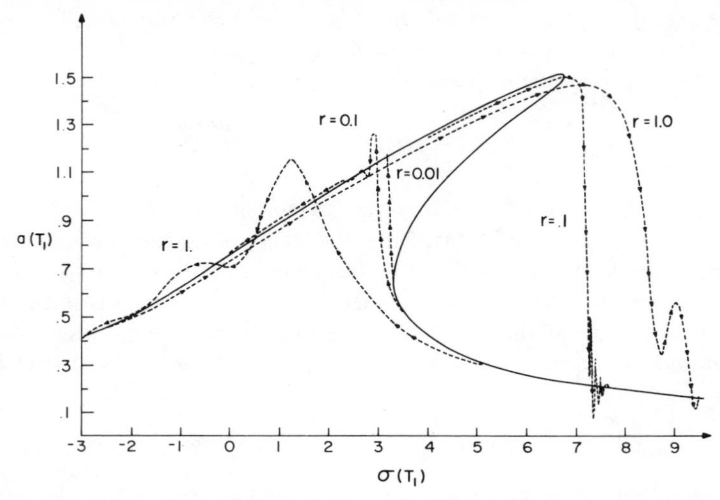

Figure 1-7. Comparison of nonstationary and stationary frequency-response curves.

separate the domains of attraction for the stable steady states. Again we note that, in the presence of damping, the steady-state solution can depend on the initial conditions. This behavior of nonlinear systems contrasts sharply with that of linear systems.

In addition to the quasi-stationary process described above, we consider variations at small, but finite, rates. An excitation whose frequency and/or amplitude vary at a finite rate is said to be *nonstationary*. In this case, the frequency-response curves may develop oscillations and deviate somewhat from the stationary case. The deviations increase as the rates of varying the frequency and amplitude of the excitation increase, as illustrated in Figure 1-7.

1.4.2. SECONDARY RESONANCES OF THE DUFFING EQUATION

Another characteristic of nonlinear systems is the secondary resonance. As an example, when $\Omega = 3\omega_0 + \epsilon\sigma$, the response is given by

$$u = a(t) \cos\left[\tfrac{1}{3}\Omega t - \tfrac{1}{3}\gamma(t)\right] + \frac{K}{\Omega^2 - \omega_0^2} \cos \Omega t + O(\epsilon) \tag{1.10}$$

As $t \to \infty$, there are two possibilities: either $a \to 0$, or $a \to$ a nonzero constant whose value depends on K, σ, and α. The initial conditions determine which possibility represents the actual response. Thus it is possible for the steady-state response to consist of the particular solution, which has the same frequency as the excitation, and a free-oscillation term whose frequency is changed by the nonlinearity to exactly one-third the frequency of the excitation. For this reason, one speaks of this as the *one-third subharmonic resonance*. This subharmonic resonance is a consequence of the nonlinearity.

When $3\Omega = \omega_0 + \epsilon\sigma$, the response is given by

$$u = a(t) \cos [3\Omega t - \gamma(t)] + \frac{K}{\Omega^2 - \omega_0^2} \cos \Omega t + O(\epsilon) \tag{1.11}$$

As $t \to \infty$, a and γ tend to constants that are functions of K, σ, and α, and, in some cases, the initial conditions. Thus the steady-state response consists of a particular solution, which has the same frequency as the excitation, and a free-oscillation term whose frequency is changed by the nonlinearity to exactly three times the frequency of the excitation. For this reason, one speaks of this as a *superharmonic resonance* of order 3. Figure 1-8*a* shows the variation of the

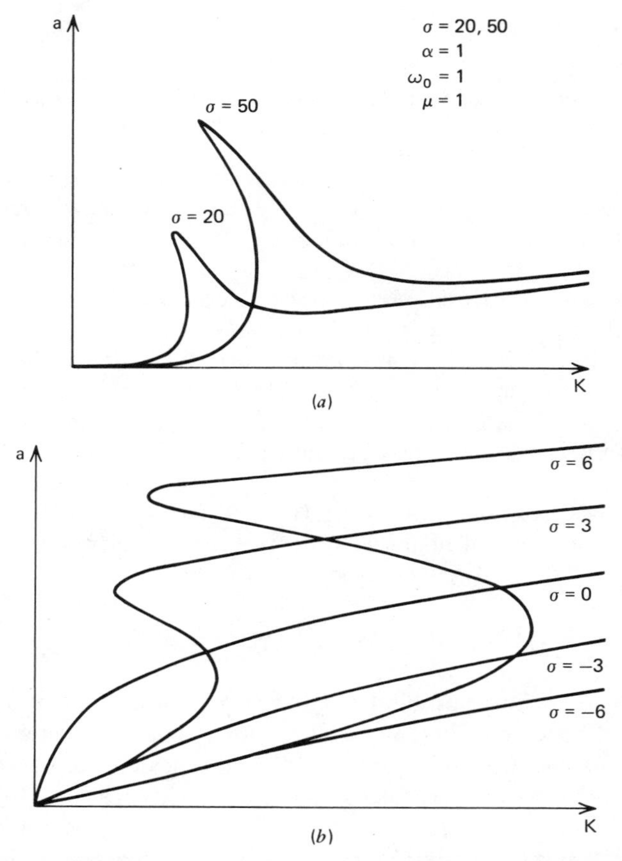

Figure 1-8. Response curves for the Duffing equation: (*a*) superharmonic resonances; (*b*) primary resonances.

steady-state amplitude with the amplitude of the excitation K for a constant σ. If K is very slowly increased from zero, there will be a spontaneous upward jump in a. But after the jump, a does not continue to increase as K increases. This contrasts with the case of primary resonance as shown in Figure 1-8*b*. In the case of superharmonic resonance, the nonlinearity introduces two competing influences. In addition to the direct relationship between the amplitude of the response and the amplitude of the excitation, there is also a relationship between the amplitude of the excitation and the apparent natural frequency of the system. Thus, when K is increased one effect is to increase a, while the other is to detune the system. Right after the jump, the second is stronger.

Figure 1-9 shows the synthesis of a steady-state superharmonic response which occurs when $\omega_0 \simeq 3\Omega$. Figure 1-9*a* is the steady-state free-oscillation term, which differs from zero in spite of the presence of viscous damping in the system. Moreover the amplitude and phase of this steady-state term are influenced by the initial conditions. Figure 1-9*b* is the particular solution, which is, to the approximation being considered, the solution of the linearized governing equation. Figure 1-9*c* is the correct first approximation of the actual response.

The existence of nonlinear phenomena (such as jumps and superharmonic and subharmonic resonances) in nature is well known. As examples, we note that von Kármán observed that certain parts of an airplane can be violently excited

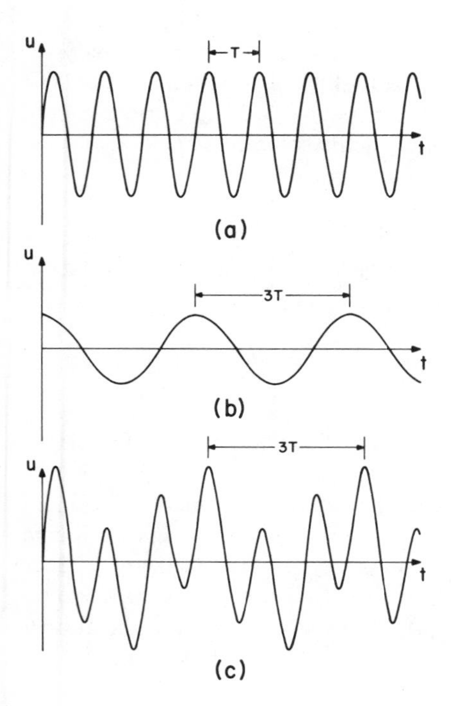

Figure 1-9. Synthesis of the response of the Duffing equation for superharmonic resonance: (*a*) free-oscillation solution; (*b*) particular solution; (*c*) actual response.

by an engine running at an angular speed much larger than their natural frequencies, and that Lefschetz described a commercial airplane in which the propellers induced a subharmonic vibration in the wings which in turn induced a subharmonic vibration in the rudder. The oscillations were violent enough to cause tragic consequences.

1.4.3. SYSTEMS WITH QUADRATIC NONLINEARITIES

All the preceding discussion is for the case of a cubic nonlinearity. When the system has quadratic nonlinearities in addition to the cubic nonlinearity, its response to a sinusoidal excitation in the presence of linear viscous damping is governed by

$$\ddot{u} + \omega_0^2 u = -2\epsilon\mu\dot{u} - \epsilon\alpha_2 u^2 - \epsilon^2\alpha_3 u^3 + K\cos\Omega t \tag{1.12}$$

where K and Ω are constants. When $\alpha_3 = 0$, this system possesses a subharmonic resonance of order $\frac{1}{2}$, and a superharmonic resonance of order 2. That is, the order of the nonlinearity changes the order of the subharmonic and superharmonic resonances. For a subharmonic resonance, the first approximation given by the perturbation analysis predicts unbounded growth of the free-oscillation term under certain conditions. This growth is predicted in the presence of damping and is in sharp contrast with linear systems. Carrying out the expansion to higher order, one finds that, at the point where growth is predicted, the free-oscillation term no longer decays as time increases but grows to a finite value. This result is illustrated in Figure 1-10. For this case, $K = 6$ is the boundary between stable and unstable responses predicted by the perturbation analysis. The graphs in Figure 1-10, which were obtained by numerical integration, clearly show a pronounced change in the character of the solution as K increases beyond 6; Figure 1-10*b* shows the presence of a lower harmonic.

1.4.4. MULTIFREQUENCY EXCITATIONS

There are many interesting phenomena associated with multiharmonic excitations of the form

$$E = \sum_{n=1}^{N} K_n \cos(\Omega_n t + \theta_n), \qquad \Omega_n > \Omega_{n-1} \tag{1.13}$$

where the K_n, Ω_n, and θ_n are constants. For a system with cubic nonlinearities and for $N = 2$, the free-oscillation term will not vanish when $\omega_0 \approx \frac{1}{2}(\Omega_2 \pm \Omega_1)$, $\omega_0 \approx 2\Omega_2 \pm \Omega_1$, and $\omega_0 \approx \Omega_2 \pm 2\Omega_1$ in addition to the primary resonances $\omega_0 \approx \Omega_n$, the subharmonic resonances $\omega_0 \approx \frac{1}{3}\Omega_n$, and the superharmonic resonances $\omega_0 \approx 3\Omega_n$. We note that more than one type of resonance may occur simultaneously such as $\Omega_1 \approx \frac{1}{3}\omega_0$ and $\Omega_2 \approx 3\omega_0$. However, the steady-state response will not be periodic unless $\Omega_n = p_n\Omega_1$ for all $n \geqslant 2$ where p_n is a rational

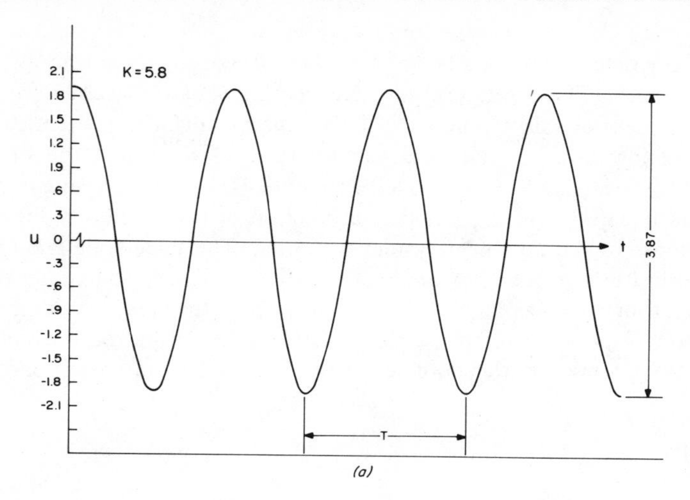

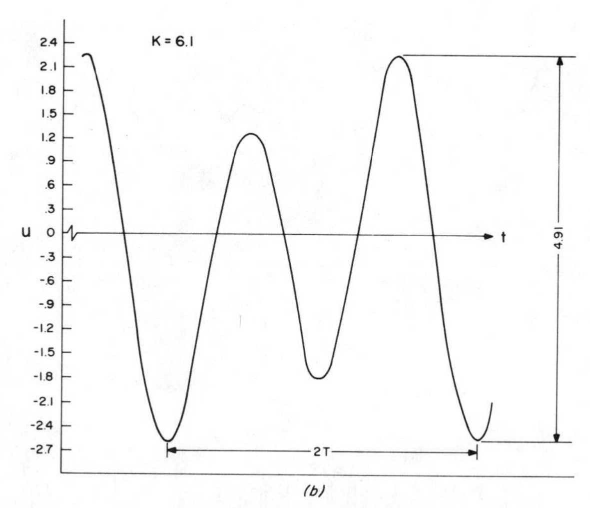

Figure 1-10. Response of a system having quadratic nonlinearities: (*a*) subharmonic response not excited; (*b*) subharmonic response excited.

fraction. When $\Omega_2 = m\Omega_1$ and $\omega_0 \approx \frac{1}{2}(\Omega_2 \pm \Omega_1)$, then $\omega_0 \approx \frac{1}{2}(m \pm 1)\Omega_1$ and an *ultrasubharmonic resonance* is said to exist if m is even.

In considering systems having cubic nonlinearities, we find that when there are three terms in the excitation and $\omega_0 \approx \Omega_3 + \Omega_2 + \Omega_1$, a *combination resonance* occurs which is similar to the superharmonic resonance discussed above.

In this case, however, the response may be aperiodic. This could represent a structural element, such as a beam or a plate, supporting three rotating machines simultaneously. The cubic term in the governing equation accounts for stretching of the neutral axis or midplane. It is common engineering practice to ignore the stretching; however the results in Figure 1-11 show that this practice can dangerously oversimplify the model. In Figure 1-11*a*, the response of the corresponding linear system is plotted as a function of time; while in Figure 1-11*b*, the response of the nonlinear system is plotted. The same scale and excitations are used in both graphs. The amplitude of the response for the nonlinear system is nearly four times as large as the amplitude of the linear system. The nonlinearity is responsible for phase shifts that enable a given set of exciting forces to do more work on the nonlinear system than on the corresponding linear system.

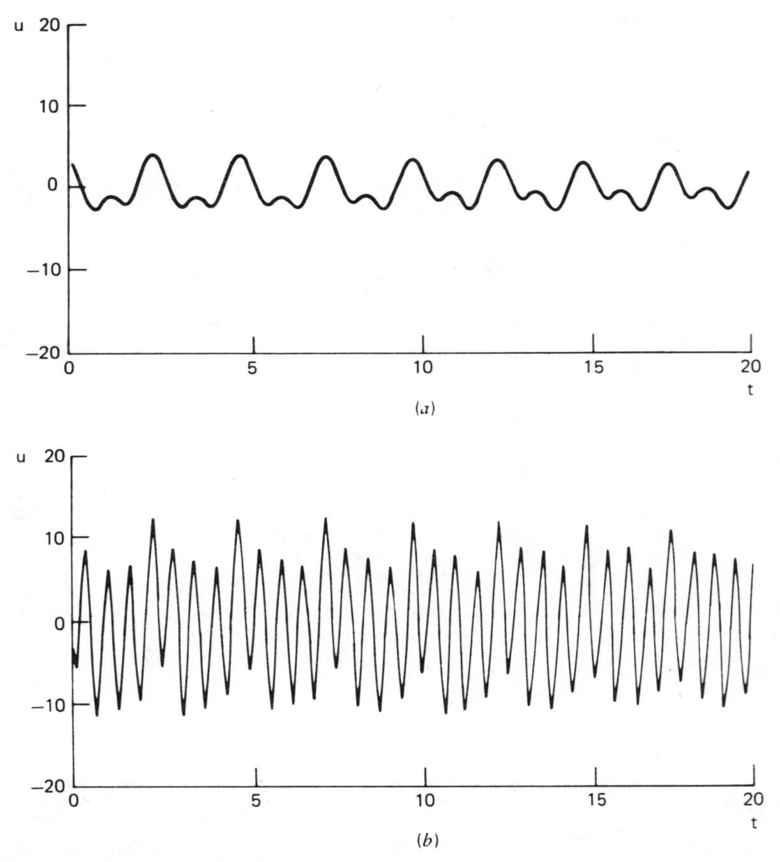

Figure 1-11. Response of a system to a three-frequency excitation: (*a*) linear case; (*b*) nonlinear case.

1.4.5. SELF-SUSTAINING SYSTEMS

Self-sustaining systems are introduced in Chapter 3. Here we determine the response of such a system to a harmonic excitation. As an example, we consider

$$\ddot{u} + \omega_0^2 u = \epsilon(\dot{u} - \tfrac{1}{3}\dot{u}^3) + K \cos \Omega t \tag{1.14}$$

where ω_0, K, and Ω are constants and ϵ is a small parameter. Since the nonlinearity is cubic, in addition to primary resonances, to first-order there are subharmonic resonances of order $\frac{1}{3}$ and superharmonic resonances of order 3. Away from these resonances,

$$u = \left\{\frac{4\eta}{\omega_0^2 + [(4\eta/a_0^2) - \omega_0^2] \exp(-\epsilon\eta t)}\right\}^{1/2} \cos(\omega_0 t + \beta) + \frac{K}{\Omega^2 - \omega_0^2} \cos \Omega t + O(\epsilon) \tag{1.15}$$

where a_0 is the initial amplitude and $\eta = 1 - \frac{1}{2}\Omega^2 K^2(\omega_0^2 - \Omega^2)^{-2}$. Thus as one would expect from experience with linear systems, in the first approximation the response consists of a free-oscillation term (homogeneous solution) and a forced-oscillation term (particular solution). And because the free oscillations develop a limit cycle, one would expect the free-oscillation term to remain permanently and the motion to be essentially the sum of two harmonic terms having the frequencies ω_0 and Ω, which need not be *commensurable*. Such a motion may be aperiodic. However in this case there is a nonlinear interaction between the free- and forced-oscillation terms which can change the character of the damping completely. When the amplitude of the excitation is large enough to make $\eta < 0$ (i.e., $K > \sqrt{2}\,\Omega^{-1}|\omega_0^2 - \Omega^2|$), the free-oscillation term decays with time and the steady-state motion becomes periodic. This process of increasing the amplitude of the excitation until the free-oscillation term decays is called *quenching*. This behavior, which contrasts with those of the previously discussed nonself-sustaining systems, is illustrated in Figure 1-12. Here the parameters of the system are such that the critical value of K is unity. In Figure 1-12*a*, $K = 0.9$ and the motion does not become periodic, while in Figure 1-12*b*, $K = 1.1$ and the motion does become periodic.

It follows from (1.15) that for small K, $\eta > 0$ and in the steady state

$$u = \frac{2\sqrt{\eta}}{\omega_0} \cos(\omega_0 t + \beta) + \frac{K}{\Omega^2 - \omega_0^2} \cos \Omega t + O(\epsilon) \tag{1.16}$$

Thus the steady-state response contains both the forced and natural frequencies. Moreover if K is $O(\epsilon)$, the free-oscillation term is expected to dominate. However as $\Omega \to \omega_0$, the character of the solution is modified drastically and a phenomenon peculiar to self-sustaining systems takes place. As $\Omega \to \omega_0$ the forced response becomes significant, but instead of the persistence of the forced- and free-oscillation solutions independently, the free-oscillation term is entrained by the forced solution. The result is a *synchronization* of the response at the

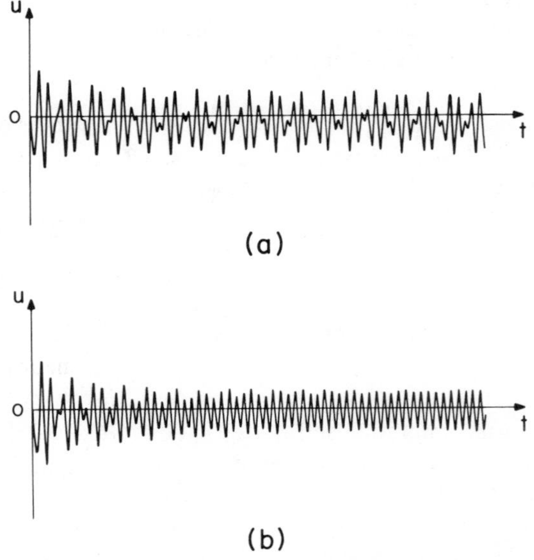

Figure 1-12. Numerical solutions of equation (1.14) ($\omega_0 = 1$, $\Omega = \sqrt{2}$), illustrating the phenomenon of quenching: (*a*) unquenched response; (*b*) quenched response.

frequency of the excitation and, to the first approximation, the steady-state response is given by

$$u = a \cos (\Omega t - \gamma) + O(\epsilon) \tag{1.17}$$

where γ is a constant and a is given by the frequency-response equation

$$4\sigma^2 \rho + \rho(1 - \rho)^2 = \frac{K^2}{4\epsilon^2} \tag{1.18}$$

where $\rho = \frac{1}{4}\omega_0^2 a^2$ and $\epsilon\sigma = \Omega - \omega_0$. In Figure 1-13, ρ is plotted as a function of σ. The dotted curve separates stable from unstable steady-state motions. When the value of ρ is above the dotted curve, the periodic steady-state solution given by (1.17) and shown in Figure 1-14*a* is physically realizable. However when the value of ρ is below the dotted line and there is only one steady-state value, a periodic steady-state solution of the form (1.17) is not physically realizable. In the latter case, the solution (long-time behavior) has the two frequencies ω_0 and Ω, and since they are near each other, a beating phenomenon takes place as shown in Figure 1-14*b*.

Both curves in Figure 1-14 were obtained by numerically integrating (1.14) for $K^2 = 2\epsilon^2$, $\omega_0 = 1$, and $\epsilon = 0.1$. But $\sigma = 0.4$ (i.e., $\Omega = 1.04$) for Figure 1-14*a* and $\sigma = 0.5$ (i.e., $\Omega = 1.05$) for Figure 1-14*b*. The first-order solution predicts that

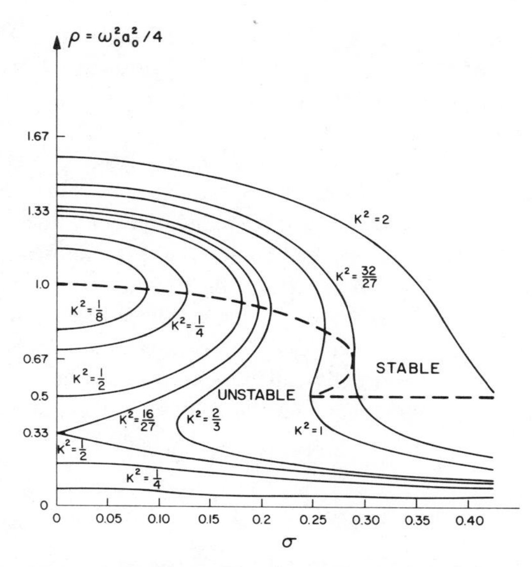

Figure 1-13. Frequency-response curves for primary resonances of the van der Pol oscillator.

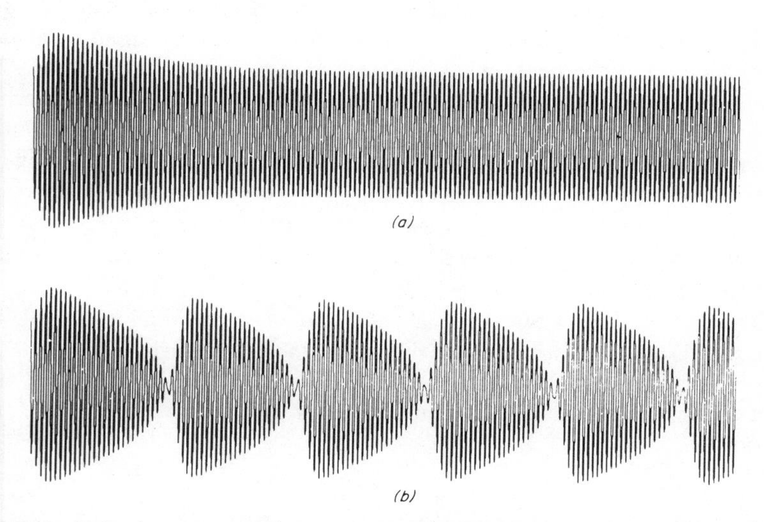

Figure 1-14. Forced response of the van der Pol oscillator when the frequency of the excitation is near the natural frequency, illustrating locking and pulling-out phenomena: (*a*) response below the pull-out frequency; (*b*) response above the pull-out frequency.

the steady-state solution is unstable for $\sigma > \sigma_c = \frac{1}{4}(2K^2\epsilon^{-2} - 1)^{1/2}$. If an experiment is performed with $K^2 = 2\epsilon^2$ and σ is decreased very slowly from a value above $\sigma_c = \frac{1}{4}\sqrt{3}$, initially the response will contain the two frequencies 1 and Ω. As σ is decreased below σ_c, the response becomes periodic with the frequency Ω. In other words, as σ is decreased below σ_c, the free-oscillation term is entrained or *locked* onto the forced-oscillation term. If the experiment is performed by increasing σ very slowly from a value below σ_c, then as σ is increased beyond σ_c, the response will change from a periodic solution having the frequency Ω to an aperiodic solution having the frequencies 1 and Ω. Thus as σ is increased beyond σ_c, the free-oscillation term will be pulled out of the forced-oscillation term. The phenomenon of entrainment of the free-oscillation term by the forced-oscillation term is usually called *locking* while the unlocking is usually called *pulling out*. And one speaks of the frequency associated with σ_c as the *pull-out frequency*.

1.5. Parametrically Excited Systems

In Chapter 5, parametrically excited systems are considered. In contrast with the case of external excitations, which lead to inhomogeneous differential equations with constant or slowly varying coefficients, parametric excitations lead to homogeneous differential equations with rapidly varying coefficients, usually periodic ones. Moreover, in contrast with the case of external excitations for which a small excitation produces a large response only if the frequency of the excitation is close to a linear natural frequency, a small parametric excitation can produce a large response when the frequency of the excitation is away from the linear natural frequencies of the system.

Faraday (1831) seems to have been the first to recognize the phenomenon of parametric resonance. He noted that surface waves in a fluid-filled cylinder under the influence of vertical excitations have twice the period of the excitation. Melde (1859) performed the first serious experiments on parametric resonance. He tied a string between a rigid support and the extremity of the prong of a massive tuning fork of low pitch. For a number of combinations of the mass and tension of the string and the frequency and loudness of the fork, he observed that the string could be made to oscillate laterally, though the exciting force is longitudinal, at one half the frequency of the fork.

The simplest differential equation with periodic coefficients is the Mathieu equation

$$\ddot{u} + (\delta + \epsilon \cos 2t)u = 0 \tag{1.19}$$

where δ and ϵ are constants. This equation governs the response of many physical systems to a sinusoidal parametric excitation. An example is a pendulum consisting of a uniform rod pinned at a point O on a platform that is made to oscillate sinusoidally in the vertical direction as shown in Figure 1-15.

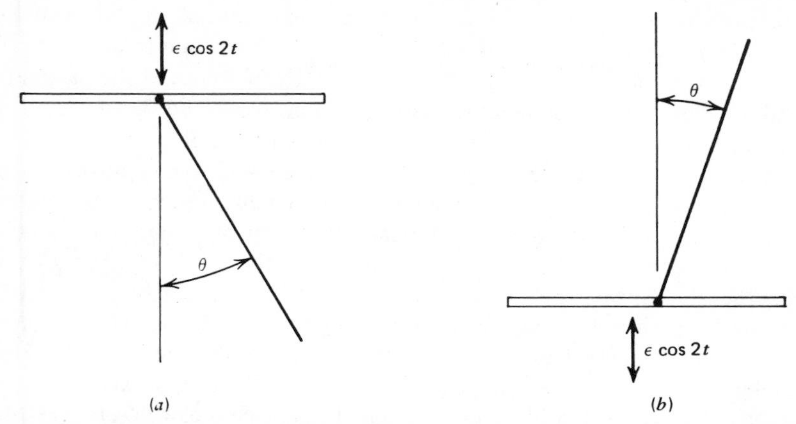

Figure 1-15. Uniform-rod pendulum oscillating in two positions as a result of giving the horizontal platform a harmonic vertical motion.

Using Floquet theory, one can show that (1.19) possesses *normal solutions* having the form

$$u(t) = \exp(\gamma t)\phi(t) \tag{1.20}$$

where γ is called a *characteristic exponent* and $\phi(t) = \phi(t + \pi)$. When the real part of one of the γ's is positive definite, u is unbounded (unstable) with time, while when the real parts of all the γ's are zero or negative, u is bounded (stable) with time. The vanishing of the real parts of the γ's separates stable from unstable motions. The loci of the corresponding values of ϵ and δ are called *transition curves*. They divide the $\epsilon\delta$-plane into regions corresponding to unbounded (unstable) motions and bounded (stable) motions as shown in Figure 1-16. When

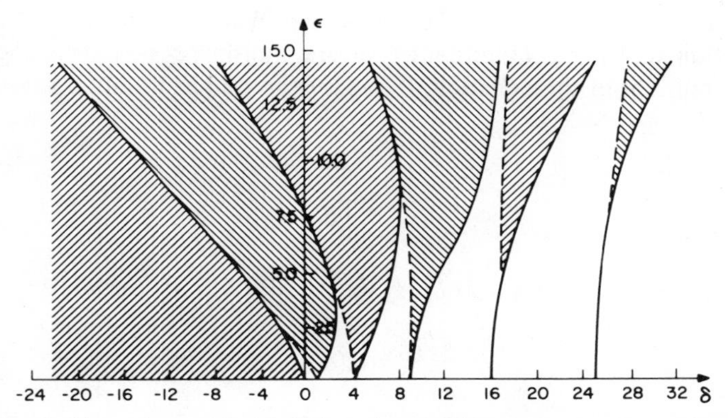

Figure 1-16. Stable and unstable (shaded) regions in the parameter plane for the Mathieu equation.

$\epsilon = 0$, positive values of δ correspond to stable positions of the pendulum (i.e., downward position), while negative values of δ correspond to unstable positions of the pendulum (i.e., upward position). In the presence of the parametric excitation, Figure 1-16 shows that there are values of ϵ and δ for which the downward position is unstable and the upward position is stable.

There are a number of techniques for determining the characteristic exponents and the transition curves separating stable from unstable motions. One method combines Floquet theory with a numerical integration of (1.19). To determine the transition curves by this technique, one divides the $\epsilon\delta$-plane into a grid and checks the solution at each grid point, which is quite a costly procedure. A second technique involves the use of Hill's infinite determinant. When ϵ is small but finite, one can use perturbation methods such as the method of strained parameters, the method of multiple scales, and the method of averaging.

The preceding discussion does not account for dissipation which is present in almost all physical systems. Dissipation has a stabilizing effect on all one-degree-of-freedom systems. Including a linear viscous term, we rewrite (1.19) as

$$\ddot{u} + 2\mu\dot{u} + (\delta + \epsilon \cos 2t)u = 0 \tag{1.21}$$

The transition curves separating stable from unstable solutions of (1.21) are shown in Figure 1-17. Comparing these graphs shows that the presence of the linear viscous term lifts the unstable regions from the δ-axis, rounds the point at the bottom, and narrows the unstable regions.

For a parametric excitation that is periodic but not necessarily sinusoidal, one obtains the following Hill equation:

$$\ddot{u} + [\delta + \epsilon f(t)]\, u = 0 \tag{1.22}$$

where $f(t)$ is periodic in place of the Mathieu equation (1.19). Since Floquet theory is also applicable to this problem, the numerical procedure and the infinite-determinant technique can be used to determine the characteristic exponents and transition curves of this equation. When ϵ is small, one can also use perturbation techniques (Lindstedt-Poincaré, multiple scales, averaging) to analyze the solutions of this equation. If $f(t)$ is expressed in a Fourier series as

$$f(t) = \sum_{n=1}^{\infty} (\alpha_n \cos 2nt + \beta_n \sin 2nt)$$

the transition curves are

$$\begin{aligned} \delta &= n^2 \pm \tfrac{1}{2}\epsilon\sqrt{\alpha_n^2 + \beta_n^2} + O(\epsilon^2) \qquad \text{for } \delta \approx n^2 \text{ and } n \geqslant 1 \\ \delta &= -\tfrac{1}{8}\epsilon^2 \sum_{n=1}^{\infty} \frac{\alpha_n^2 + \beta_n^2}{n} + O(\epsilon^3) \qquad \text{for } \delta \approx 0 \end{aligned} \tag{1.23}$$

Comparing these transition curves with those of the Mathieu equation (i.e., α_n and $\beta_n = 0$ for $n \geqslant 2$), we conclude that the presence of the higher harmonics

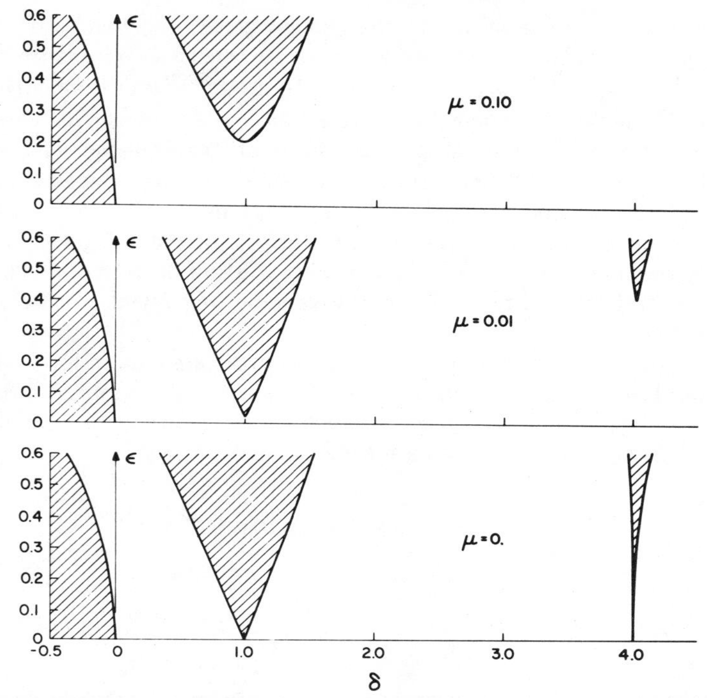

Figure 1-17. Effect of viscous damping on the stability of the solutions of the Mathieu equation. Shaded areas are unstable.

has a destabilizing effect on the transition curves emanating from $\delta = n^2$ where $n \geqslant 1$, but it may have a stabilizing effect on the transition curve emanating from $\delta = 0$.

The problem of a sinusoidal parametric excitation of a system having many degrees of freedom leads mathematically to the following coupled system of differential equations:

$$\ddot{\mathbf{x}} + \{[A] + 2\epsilon [B] \cos \Omega t\}\mathbf{x} = 0 \tag{1.24}$$

where Ω is the frequency of the excitation. The response of such a system depends on the eigenvalues of the matrix $[A]$. For a vibrating system, these eigenvalues are real and positive. If these eigenvalues are distinct, a transformation $\mathbf{x} = [P]\,\mathbf{u}$ can be found such that (1.24) can be rewritten in the form

$$\ddot{u}_n + \omega_n^2 u_n + 2\epsilon \cos \Omega t \sum_{m=1}^{N} f_{nm} u_m = 0 \tag{1.25}$$

In addition to the resonances ($\omega_p \approx \frac{1}{2} m\Omega$, where m is an integer) that occur in the case of a single-degree-of-freedom system, combination resonances of the form $\omega_q \pm \omega_p \approx m\Omega$ might exist in a many-degree-of-freedom system. Moreover, a given mode might be involved in more than one resonance such as $\omega_2 + \omega_1 \approx \Omega$ and $\omega_3 - \omega_2 \approx \Omega$. Figure 1-18 shows the transition curves for a free-clamped column. Comparing Figures 1-16 and 1-18 shows an increase in the number of unstable regions in the case of multi-degree-of-freedom systems. Figure 1-18 also shows that the presence of simultaneous resonances has a destabilizing effect because it decreases the stable regions.

Including linear viscous damping in the analysis of multi-degree-of-freedom systems shows that it may have a destabilizing effect in the case of combination resonances. This contrasts with the always-stabilizing effect of viscosity on simple resonances.

When the eigenvalues of $[A]$ are not distinct, there are cases for which $[A]$ cannot be diagonalized but can be expressed in a *Jordan canonical form*. This occurs in the case of flutter. If all the eigenvalues are distinct except the first pair, one can use a transformation $\mathbf{x} = [P]\,\mathbf{u}$ to rewrite (1.24) as

$$\ddot{u}_1 + \omega_1^2 u_1 + 2\epsilon \cos \omega t \sum_{n=1}^{\infty} f_{1n} u_n = 0 \tag{1.26}$$

$$\ddot{u}_2 + \omega_1^2 u_2 + u_1 + 2\epsilon \cos \omega t \sum_{n=1}^{\infty} f_{2n} u_n = 0 \tag{1.27}$$

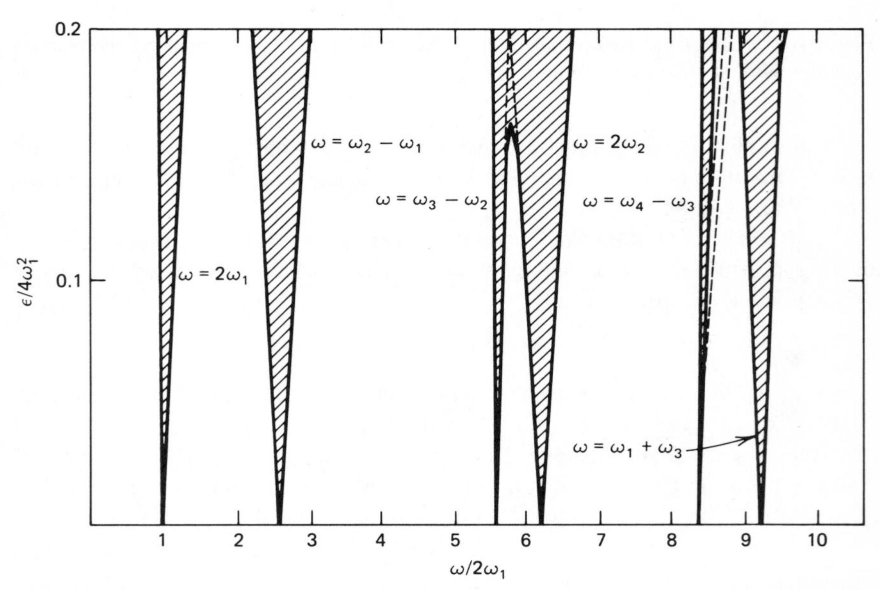

Figure 1-18. Transition curves for the dynamic buckling of a free-fixed column under the influence of a sinusoidal follower force.

$$\ddot{u}_3 + \omega_3^2 u_3 + 2\epsilon \cos \omega t \sum_{n=1}^{\infty} f_{3n} u_n = 0 \tag{1.28}$$

We note that when $\epsilon = 0$ (i.e., in the absence of the parametric excitation), u_2 grows linearly with time and hence the response of the system is unbounded with time. Including the parametric excitation can result in the stabilization of the system depending on the values of ϵ, ω, ω_1, ω_3, and the f's.

Although the linear analysis of parametric excitations is useful in determining the initial growth or decay of the motion, it cannot account for the long-time behavior in the case of growth. Moreover if the initial amplitude is large, the linear analysis may predict a motion that decays to zero in contradiction with the prediction of a nonlinear analysis. The nonlinearity can be the result of damping (form drag and the van der Pol oscillator) and large deformations (Duffing's equation). The latter could represent the lateral vibrations of a column produced by an axial follower force.

Considering a cubic nonlinearity and assuming small viscous damping, we have

$$\ddot{u} + \omega^2 u + 2\epsilon u \cos 2t + \epsilon(2\mu\dot{u} + \alpha u^3) = 0 \tag{1.29}$$

The stability boundaries are shown in Figure 1-19. The line separating Region II from Regions I and III is the stability boundary for the corresponding damped linear system. The boundaries are not influenced by the value of α, the coef-

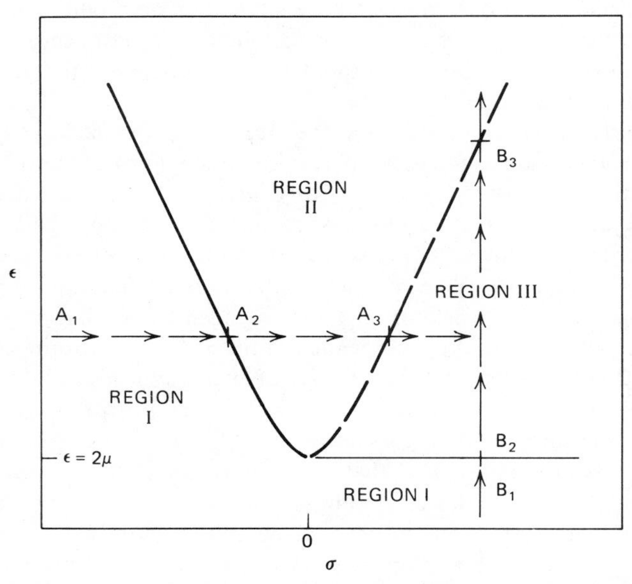

Figure 1-19. Frequency–response curves for the parametrically excited Duffing's equation in the presence of viscous damping.

ficient of the nonlinear term. According to the linear equation (see Figure 1-17), the response to any initial disturbance grows without bound in Region II and decays in Regions I and III. In sharp contrast, the nonlinear equation predicts finite-amplitude motions in both Regions II and III. In Region II, the motion approaches the same steady state regardless of the initial disturbance, but in Region III, for small initial disturbances the motion decays while for large initial disturbances it approaches a finite-amplitude steady state.

1.6. Systems Having Finite Degrees of Freedom

In contrast with the case of systems having a single degree of freedom, available exact solutions of systems having finite degrees of freedom are quite limited. Hence, most of the existing analyses deal with weakly nonlinear systems which are amenable to perturbation analysis. In the case of strongly nonlinear systems, recourse is often made to geometrical methods, numerical analysis, and perturbations about an exact nonlinear solution. Thus, Chapter 6 deals essentially with weakly nonlinear systems having finite degrees of freedom.

In contrast with a single-degree-of-freedom system, which has only a single natural frequency and a single mode of motion, an n-degree-of-freedom system has n natural frequencies $\omega_1, \omega_2, \cdots, \omega_n$ and n corresponding natural modes. All these natural frequencies are assumed to be real and different from zero. The presence of more than one natural frequency and mode produces new physical phenomena such as internal resonances, combinational resonances, saturation, and the nonexistence of periodic responses to a periodic excitation in the presence of positive damping.

New physical phenomena occur in the free oscillations of a system some of whose frequencies are commensurable or nearly commensurable; that is, there exist positive or negative integers $m_1, m_2, m_3, \cdots, m_n$ such that $m_1\omega_1 + m_2\omega_2 + m_3\omega_3 + \cdots + m_n\omega_n \approx 0$. When such a condition exists, we speak of the existence of an *internal resonance*, and conditions might exist for the strong interaction of the modes involved in the internal resonance. For example, consider the motion of a particle of mass m suspended from a linear spring, with a constant k, which is in turn suspended from a fixed platform as shown in Figure 1-20. This system has two modes of oscillation: a pendulumlike mode with the linear natural frequency $\omega_1 = (g/l)^{1/2}$ and a springlike (breathing) mode with the linear natural frequency $\omega_2 = (k/m)^{1/2}$. The parameters k, m, and l can be easily adjusted so that $\omega_2 \approx n\omega_1$ where n is an integer. When $\omega_2 \approx 2\omega_1$, the two modes are strongly coupled and the energy initially imparted to one of them can, in general, be continuously exchanged between them during the ensuing motion as shown in Figure 1-21. This contrasts with the linear solution, which predicts that the two modes are uncoupled. The strong coupling is a consequence of the internal resonance and it decreases as the detuning of this internal resonance increases.

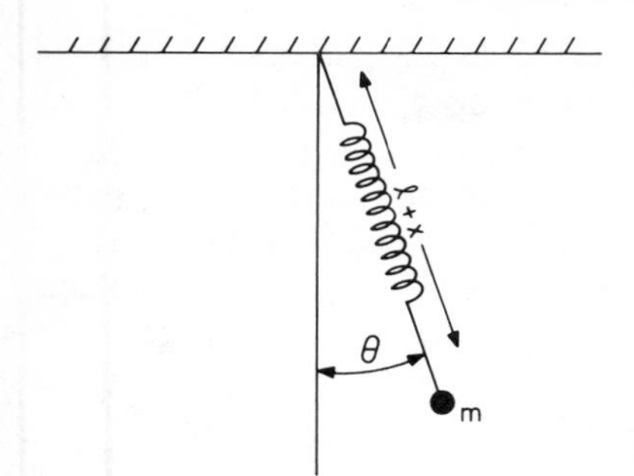

Figure 1-20. Spring pendulum.

If a system having finite degrees of freedom is gyroscopic and possesses an internal resonance, then its free nonlinear oscillations may be unbounded with time even though its linear free oscillations are bounded. This occurs when its first-order Hamiltonian is not positive definite. For a system with two degrees of freedom and $\omega_2 \approx 2\omega_1$, the equations describing the amplitudes and the phases have the form

$$\dot{a}_1 = -\epsilon\Gamma_1 a_1 a_2 \sin\gamma \tag{1.30}$$

$$\dot{a}_2 = \epsilon\Gamma_2 a_1^2 \sin\gamma \tag{1.31}$$

$$\dot{\gamma} = \epsilon f(a_1, a_2, \gamma) \tag{1.32}$$

Eliminating γ from (1.30) and (1.31) and integrating the resulting equation, we have

$$a_2^2 + (\Gamma_2/\Gamma_1)a_1^2 = E \tag{1.33}$$

where E is a constant that is proportional to the Hamiltonian or energy of the first order. If Γ_1 and Γ_2 have the same sign, E is positive definite and a_1 and a_2

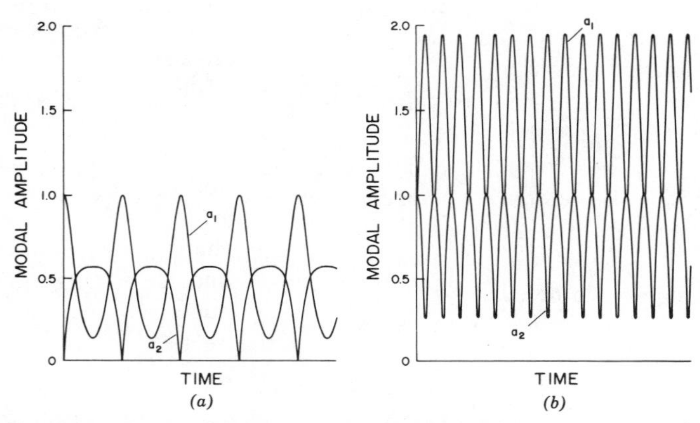

Figure 1-21. Continual exchange of energy in the case of internal resonance.

are bounded for all time. However if Γ_1 and Γ_2 have opposite signs, E is not positive definite and a_1 and a_2 may be unbounded with time depending on the value of the detuning $\sigma = (\omega_2 - 2\omega_1)/\epsilon$.

As in the case of parametric excitations of finite-degree-of-freedom systems, combinational resonances might occur in the forced response of these systems to a single-harmonic excitation of frequency Ω. The type of excited combinational resonance depends on the order of the nonlinearity. For a quadratic nonlinearity, combinational resonances involve to first order two of the linear natural frequencies of the system in the form $\Omega \approx \omega_n \pm \omega_m$. For a cubic nonlinearity, combinational resonances involve to first order two or three modes in one of the following forms: $\Omega \approx \omega_n \pm \omega_m \pm \omega_k$, $\Omega \approx \omega_n \pm 2\omega_m$, $\Omega \approx 2\omega_n \pm \omega_m$, and $\Omega \approx \frac{1}{2}(\omega_n \pm \omega_m)$. If an internal resonance exists in addition to a combinational resonance, a *fractional-harmonic pair* might exist in the response such as $(\frac{1}{3}\Omega, \frac{2}{3}\Omega)$ in the case of quadratic nonlinearities and $(\frac{1}{2}\Omega, \frac{3}{2}\Omega)$ or $(\frac{1}{5}\Omega, \frac{3}{5}\Omega)$ in the case of cubic nonlinearities.

A saturation phenomenon occurs in the forced response of a system with quadratic nonlinearities in the presence of an internal resonance. For example, the forced response of a ship whose motion is restricted to pitch and roll only can be modeled by the following equations:

$$\ddot{u}_1 + \omega_1^2 u_1 = -2\hat{\mu}_1 \dot{u}_1 + 2\alpha_1 u_1 u_2 + F_1 \cos(\Omega t + \tau_1) \tag{1.34}$$

$$\ddot{u}_2 + \omega_2^2 u_2 = -2\hat{\mu}_2 \dot{u}_2 + \alpha_2 u_1^2 + F_2 \cos(\Omega t + \tau_2) \tag{1.35}$$

where u_1 is the roll angle, u_2 is the pitch angle, and the ω_n, $\hat{\mu}_n$, α_n, F_n, and τ_n are constants. For an internal resonance, $\omega_2 \approx 2\omega_1$.

When Ω is near ω_2 and $F_1 = 0$, one expects the u_2-mode to be strongly excited and, in the first approximation, the u_1-mode to be dormant; initially this is so. But the perturbation analysis predicts an upper bound on the amplitude of u_2 and an instability for the trivial solution for u_1 when F_2 increases beyond a critical value. In other words, the u_2-mode becomes *saturated* and the energy "spills over" into the u_1-mode. These results are illustrated in Figure 1-22 where the amplitudes of the two modes, a_1 and a_2, are plotted as functions of F_2. This analysis was verified by numerically integrating (1.34) and (1.35); the small circles and triangles are the numerical data.

These results provide an explanation for a phenomenon first reported by Froude (1863). He wrote that ships having a natural frequency in pitch which is nearly twice the natural frequency in roll (an internal resonance) have undesirable roll characteristics. Thus in accordance with the saturation phenomenon, the ship could be advancing into a head sea, or moving with a following sea, and, if the waves are big enough and at the right frequency, begin to roll violently.

When $F_2 = 0$ and $\Omega \approx \omega_1$, the analytical results also show that for some combinations of the parameters a steady-state response does not exist, in spite of

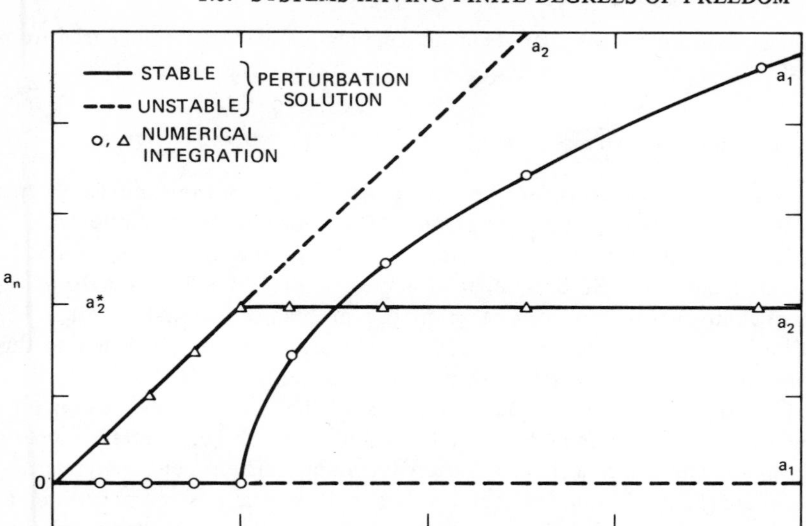

Figure 1-22. Saturation phenomenon: a_1 and a_2 as function of f_2.

the presence of positive damping. Instead of a steady state there is a continual exchange of energy between the two modes. For such a combination of parameters (the combination was predicted by perturbation methods), the numerical results are shown in Figure 1-23. This type of behavior in ships was observed by Robb (1952).

We should note that saturation and the nonexistence of periodic motions under the influence of a periodic excitation in the presence of positive damping are peculiar to systems with quadratic nonlinearities. For systems with cubic

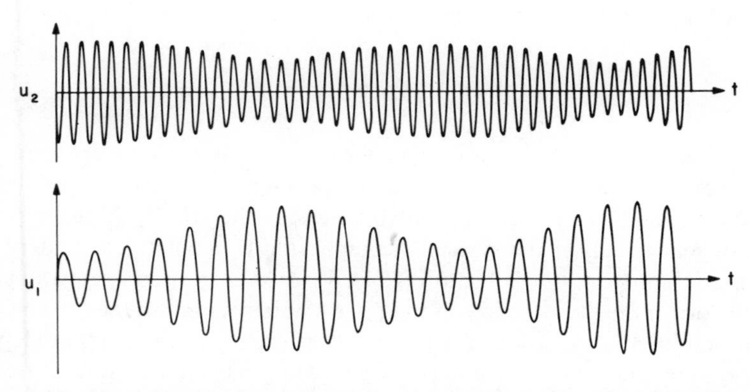

Figure 1-23. Nonexistence of periodic motions in a system with quadratic nonlinearities.

nonlinearities and an internal resonance, energy can be easily transferred from a high-frequency mode to low-frequency modes but not the other way round.

1.7. Continuous Systems

In Chapter 7, we consider the forced oscillations of continuous systems—beams, strings, plates, and membranes. In contrast with the finite-degree-of-freedom systems discussed in Chapters 5 and 6, the systems considered in Chapter 7 have an infinite number of degrees of freedom. The sources of nonlinearities in such systems can be geometric, inertial, or material in nature. Most of Chapter 7 is devoted to geometric nonlinearities and in particular to those arising from midplane stretching in structural elements.

Since exact solutions are generally not available, recourse has been made to approximate analyses including purely analytical techniques, purely numerical techniques, and numerical-perturbation techniques. The purely analytical techniques are applicable to systems with simple geometries, composition, and boundary conditions. Purely numerical techniques may involve the use of finite differences in both space and time, finite differences in time and finite elements in space, and finite elements in both space and time. These purely numerical techniques are especially costly for two and three-dimensional systems. There are two approaches in using numerical-perturbation techniques. One approach assumes the time dependence, uses the method of harmonic balance, and yields nonlinear differential equations describing the spatial behavior. The second approach assumes the spatial variation (such as the linear mode shapes), uses the orthogonality of the mode shapes or the Galérkin procedure, and yields nonlinear coupled second-order ordinary-differential equations describing the temporal behavior. The latter equations are solved by using a perturbation technique such as the method of multiple scales or the method of averaging.

According to the second approach of the numerical-perturbation technique, the deflection $w(\mathbf{r}, t)$ is assumed in the form

$$w(\mathbf{r}, t) = \sum_{n=1}^{\infty} u_n(t)\phi_n(\mathbf{r}) \tag{1.36}$$

where the ϕ_n are the linear natural modes of the system. These are more convenient for interpreting the results than other arbitrarily assumed spatial variations. These mode shapes can be obtained either analytically if the system is simple or numerically if the system is complicated in geometry, boundary conditions, and composition. Substituting (1.36) into the governing equations and using the orthogonality property of the ϕ_n, we obtain an infinite set of nonlinear ordinary-differential equations for the u_n. The form of these equations depends on the system under consideration and the type of nonlinearity. In what follows, we consider linear material properties. For an initially straight beam, a string, an isotropic membrane, and an isotropic plate, these equations have the form

$$\ddot{u}_n + \omega_n^2 u_n = -2\mu_n \dot{u}_n + \sum_{m,p,q=1}^{\infty} \Gamma_{nmpq} u_m u_p u_q + f_n(t) \tag{1.37}$$

where modal damping is assumed, ω_n, μ_n, and Γ_{nmpq} are constants, and $f_n(t)$ is the excitation. For a shell, a laminated plate, and an initially curved beam, these equations have the form

$$\ddot{u}_n + \omega_n^2 u_n = -2\mu_n \dot{u}_n + \sum_{m,p=1}^{\infty} \alpha_{nmp} u_m u_p + \sum_{m,p,q=1}^{\infty} \Gamma_{nmpq} u_m u_p u_q + f_n(t) \tag{1.38}$$

where the α_{nmp} are constants. The interaction of longitudinal and lateral oscillations in a beam is governed by a set of equations having the same form as (1.38).

Most existing analyses of continuous systems are limited to the determination of the amplitude-frequency relationship of a single mode or the steady-state forced response to a single-harmonic excitation. Since many physical phenomena, such as internal resonances, combinational resonances, saturation, and nonexistence of periodic motions, are characteristics of multi-degree-of-freedom systems, we concentrate our discussion on these systems.

As discussed in the preceding section, the response of a system depends on the order of its nonlinearity and its internal resonances. Since the nonlinear vibrations of shells, laminated plates, and buckled beams are governed by differential equations with quadratic nonlinearities, one expects to observe the saturation phenomenon discussed in Section 1.6 as well as the nonexistence of periodic motions when one of the linear frequencies of the system is equal to, or approximately equal to, twice another linear natural frequency (i.e., $\omega_n \approx 2\omega_m$) or when one of the linear natural frequencies is equal to, or approximately equal to, the sum or difference of two other natural frequencies (i.e., $\omega_n \approx \omega_m \mp \omega_k$). The latter case also occurs in the interaction of longitudinal and lateral oscillations in a beam.

We should note that the internal resonances that might occur in a system depend on its geometry, composition, and boundary conditions. In the case of uniform beams, $\omega_3 = 2\omega_2 + \omega_1$ for a hinged-hinged or a free-free beam, $\omega_2 \approx 3\omega_1$ for a clamped-hinged beam, and $\omega_4 \approx \omega_3 + \omega_2 + \omega_1$ for a clamped-clamped beam. However, the interaction terms in (1.37) vanish in the case of hinged-hinged beams and the commensurability of ω_1, ω_2, and ω_3 does not have any effect on the response. Moreover, there is no midplane stretching in the case of free-free beams and the nonlinear terms vanish in (1.37). In the latter case, the nonlinear curvature needs to be included to account for finite-amplitude effects. Strings and membranes have an infinite number of commensurable frequencies. In the case of clamped, homogeneous, isotropic plates,

$\omega_3 \approx 2\omega_2 + \omega_1$ for a circular plate and $\omega_5 \approx \omega_3 + \omega_2 + \omega_1$ and $\omega_4 \approx 2\omega_3 - \omega_1$ for an elliptic plate whose axes are in the ratio of 9 to 10.

Next we consider the forced response of clamped-hinged beams to a harmonic excitation having the form $f_n(t) = F_n \cos \Omega t$. When Ω is near ω_1, the variations of the amplitudes of u_1 and u_2, a_1 and a_2, are shown in Figure 1-24. Although

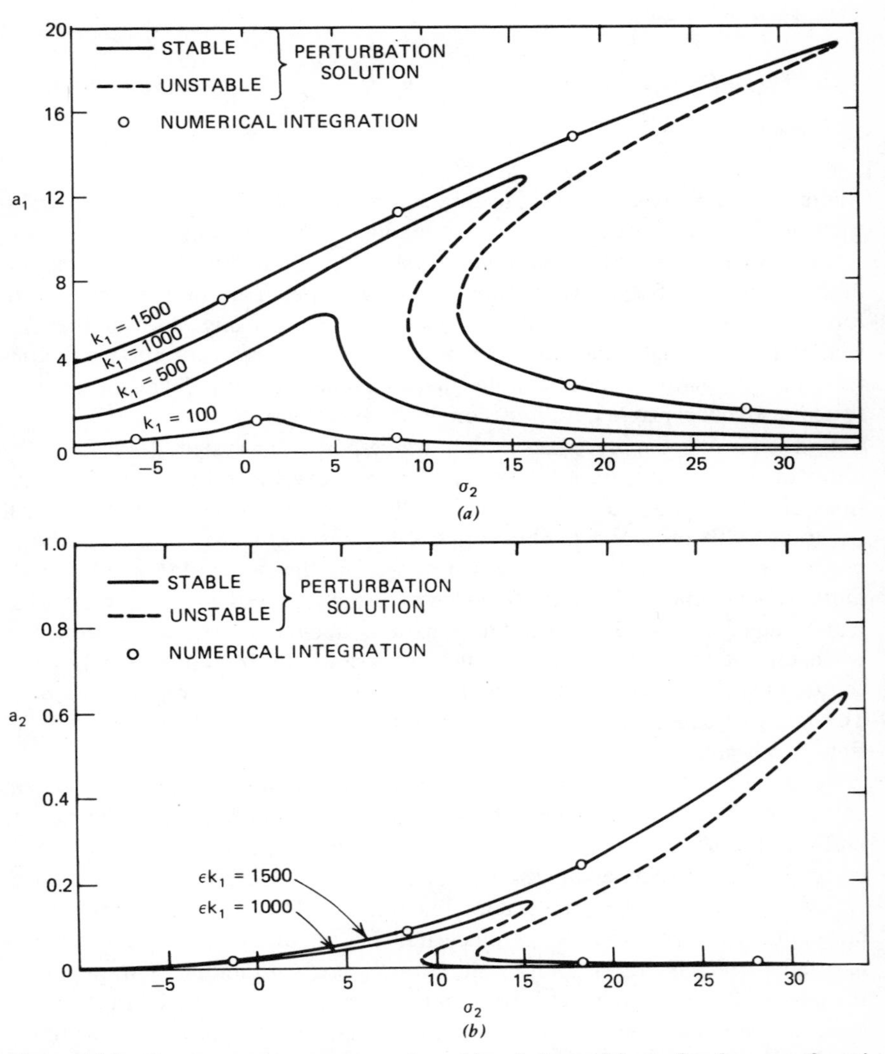

Figure 1-24. Frequency-response curves for a hinged-clamped beam for the case of a primary resonance of the fundamental mode: (*a*) first mode; (*b*) second mode.

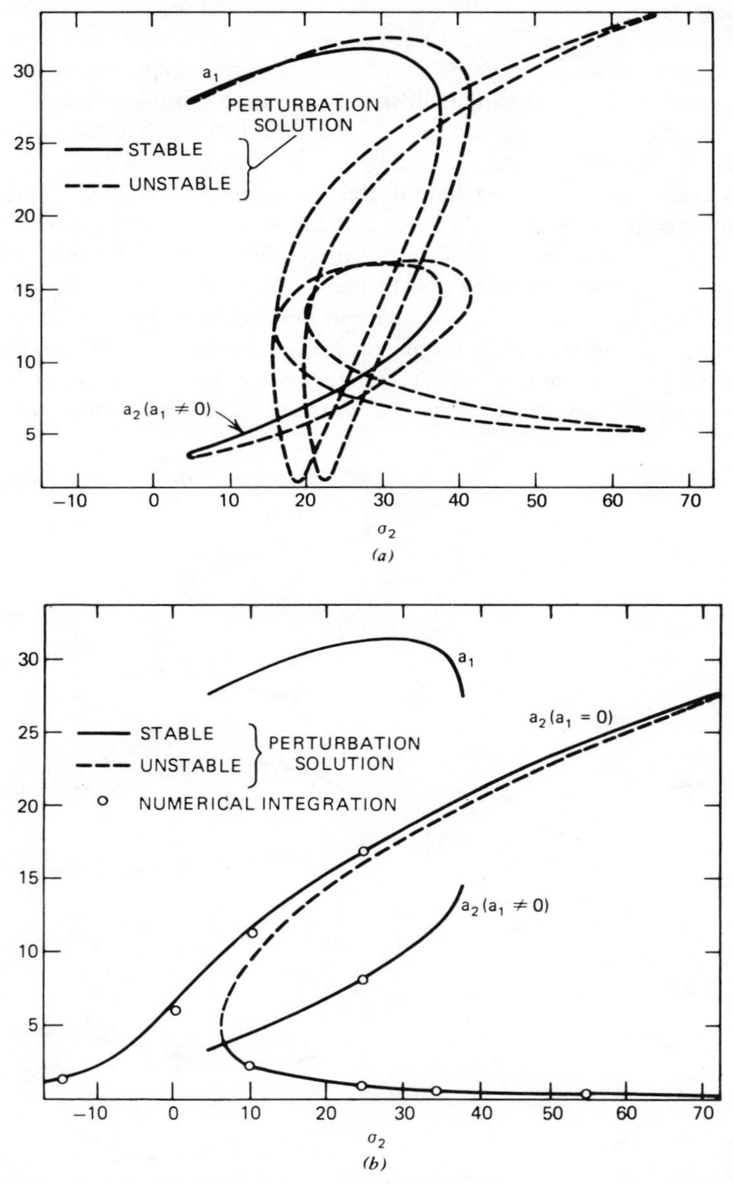

Figure 1-25. Frequency–response curves for a hinged-clamped beam for the case of a primary resonance of the second mode: (*a*) entire solution; (*b*) stable portions of the solution only.

a_2 cannot be zero, typically it is small compared with a_1. This indicates that for all practical purposes the response (deflection) can be described by a single mode, in spite of the presence of an internal resonance. Typically, early investigators considered only one mode in studying finite-amplitude beam vibrations. When Ω is near ω_1 this appears to be justified, but as we shall see next, this is definitely not the case when Ω is near ω_2.

When Ω is near ω_2, the variations of the amplitudes are shown in Figure 1-25. There are two possibilities: either $a_1 = 0$ and $a_2 \neq 0$, or neither a_1 nor a_2 equals zero. Only the stable portion of the solution when $a_1 \neq 0$ is shown in Figure 1-25a; the entire graph is shown in Figure 1-25b. In the latter case, a_1 can be considerably larger than a_2, and once again the deflection can be described by a single term in the expansion for all practical purposes. However, this time the mode is still the fundamental mode, not the second mode, in spite of the fact that the frequency of the excitation is near the second frequency. This possi-

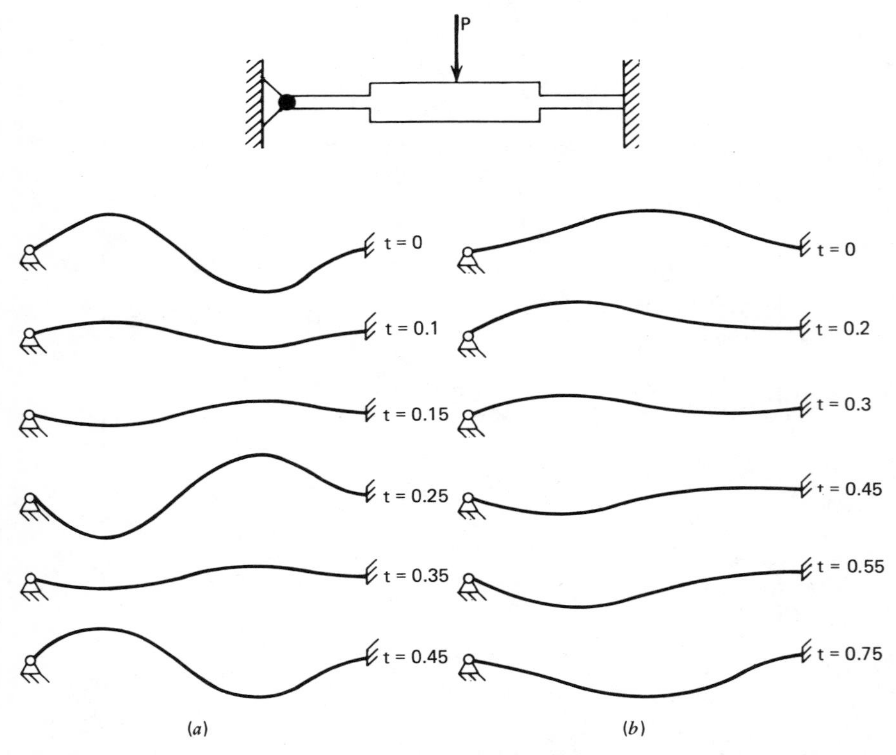

Figure 1-26. Possible steady-state responses of hinged-clamped beam to the same harmonic excitation for the case of a primary resonance of the second mode ($t_n > t_{n-1}$): (a) in the absence of internal resonance; (b) in the presence of internal resonance.

bility cannot be predicted by a linear theory, and it was completely overlooked by all the early investigators, who assumed single-mode expansions for the deflection. Figure 1-26 illustrates the two possibilities. These results show that, through the mechanism of an internal resonance, energy can be passed down to the low mode from the high mode but not from the low to the high in significant quantity.

These figures clearly illustrate the advantage of having an analytical solution. One can easily imagine the difficulty in obtaining these graphs by numerical means alone. In fact for a certain class of problems governed by partial-differential equations, the optimum approach is a combination of numerical and perturbation methods.

Another interesting phenomenon which is a consequence of internal resonance occurs in the stability of planar motions of a string resulting from a harmonic planar force. Experiments show that the response of a string to a plane harmonic excitation is planar provided the response amplitude is smaller than a critical value. Above this critical value, the planar motion becomes unstable and gives way to a nonplanar, *whirling motion*; that is the string begins to whirl like a jump rope. This whirling motion is a direct consequence of the fact that the frequency of the motion in the plane of the excitation is the same as the frequency of the motion in the plane perpendicular to the plane of the excitation. Thus, the two components of motion are strongly coupled.

1.8. Traveling Waves

In contrast with Chapters 6 and 7 which deal with standing waves, Chapter 8 deals with traveling waves. To exhibit the methods and physical phenomena without an elaborate involvement in algebra, we consider the propagation of longitudinal waves along a rod with nonlinear elastic properties and transverse waves along a beam on a nonlinear elastic foundation. These problems are described by the following two partial-differential equations:

$$\frac{\partial^2 u}{\partial t^2} - c^2(e)\frac{\partial^2 u}{\partial x^2} = 0, \qquad e = \frac{\partial u}{\partial x} \tag{1.39}$$

$$\frac{\partial^4 w}{\partial x^4} + \frac{\partial^2 w}{\partial t^2} + w + \alpha w^3 = 0 \tag{1.40}$$

We choose to distinguish dispersive from nondispersive waves by investigating the dispersion relationship $\omega = \omega(\mathbf{k})$ between the frequency and wavenumber of a linear harmonic wave of the form $\exp(i\mathbf{k} \cdot \mathbf{r} - i\omega t)$. The waves are called *dispersive* if the group velocity $\mathbf{c}_g = \partial\omega/\partial\mathbf{k}$ is a function of $\mathbf{k}$ while the waves are called *nondispersive* if $\mathbf{c}_g$ is independent of $\mathbf{k}$. If $\mathbf{c}_g$ is a weak function of $\mathbf{k}$, the waves are called weakly dispersive. Expanding $c^2(e)$ in powers of e, we have

$c^2 = c_0^2(1 + 2E_1 e + \cdots)$, where E_1 is a constant. Then it follows from (1.39) that the linear dispersion relationship for longitudinal waves along a bar is $\omega^2 = c_0^2 k^2$ and hence these waves are nondispersive. On the other hand, it follows from (1.40) that the linear dispersion relationship for transverse waves along a beam on an elastic foundation is $\omega^2 = 1 + k^4$ and hence these waves are dispersive.

Thus waves of different wavelengths travel with the same phase speed if the waves are nondispersive and travel with different phase speeds if the waves are dispersive. In other words, the dispersion tends to sort out the waves based on their phase speeds. If the nonlinearity tends to increase the phase speed with amplitude, then larger waves tend to catch up with smaller waves. The result is a steepening of the waveform with time or propagation distance as shown in Figure 1-27 leading to a shock wave in the bar. Similar effects occur in the propagation of waves in gases. Waves propagating on shallow water also steepen and sometimes break. In the case of dispersive waves, there are two competing effects: a steepening due to the nonlinearity and a spreading due to the dispersion. If the former effect is stronger, the waves focus; otherwise they will disperse.

There are a number of techniques available for the analysis of nondispersive waves traveling in one or two directions in homogeneous as well as heterogeneous media. These include expansions by using the exact characteristics of the problem as the independent variables, the method of renormalization, the method of averaging, and the method of multiple scales. Neglecting viscous effects, one obtains for waves traveling in one direction (simple waves) an equation of the form

$$\frac{\partial f}{\partial x} + f\frac{\partial f}{\partial \xi} = 0 \tag{1.41}$$

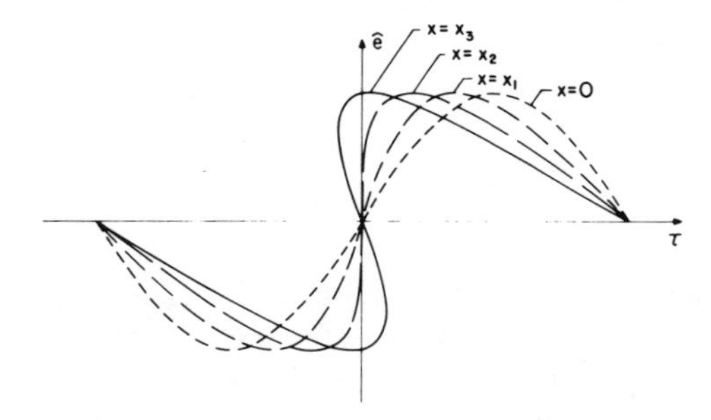

Figure 1-27. Steepening of waveforms propagating along a bar.

Including viscous effects, one obtains a Burgers' equation of the form

$$\frac{\partial f}{\partial x} + f\frac{\partial f}{\partial \xi} = \nu \frac{\partial^2 f}{\partial \xi^2} \tag{1.42}$$

in place of (1.41).

There are also a number of techniques available for analyzing nonlinear dispersive waves. These include the methods of multiple scales and averaging. Seeking a wavepacket solution for a dispersive-wave problem in the form $A(x, t)\phi(y, z) \exp [i(kx - \omega t)]$ + cc, where cc stands for the complex conjugate of the preceding terms, one finds that A is described by one of the following Schrödinger equations:

$$\frac{\partial A}{\partial t} + \omega' \frac{\partial A}{\partial x} - \frac{1}{2} i\omega'' \frac{\partial^2 A}{\partial x^2} = \Gamma_1 A^2 \bar{A} \tag{1.43}$$

$$\frac{\partial A}{\partial x} + k' \frac{\partial A}{\partial t} + \frac{1}{2} ik'' \frac{\partial^2 A}{\partial t^2} = \Gamma_2 A^2 \bar{A} \tag{1.44}$$

where $\omega' = d\omega/dk$, $\omega'' = d^2\omega/dk^2$, $k' = dk/d\omega$, and $k'' = d^2k/d\omega^2$. Here Γ_1 and Γ_2 are known interaction coefficients which depend on the medium. Equations (1.43) and (1.44) possess steady-state solutions, which can be expressed in terms of the Jacobian elliptic functions. These solutions include a bright and a dark soliton, a phase jump, and a plane wave with constant amplitude as special cases.

The preceding solution breaks down in cases of *harmonic resonances* which exist whenever (ω, k) and $(n\omega, nk)$ simultaneously satisfy the dispersion relationship for an integer $n \geqslant 2$. In the case of a beam on an elastic foundation, harmonic resonances exist when

$$\omega^2 = k^4 + 1 \qquad \text{and} \qquad n^2\omega^2 = n^4 k^4 + 1$$

Eliminating ω^2 from these relations yields $k^2 = 1/n$. At or near these critical wavenumbers, the fundamental and its nth harmonic travel with the same phase speed and hence may strongly interact. For example, when $k^2 \approx \frac{1}{3}$, the fundamental and its third harmonic strongly interact. In this case, the deflection has the form

$$w(x, t) = A_1(x, t) \exp [i(k_1 x - \omega_1 t)] + A_3(x, t) \exp [i(k_3 x - \omega_3 t)] + \text{cc} \tag{1.45}$$

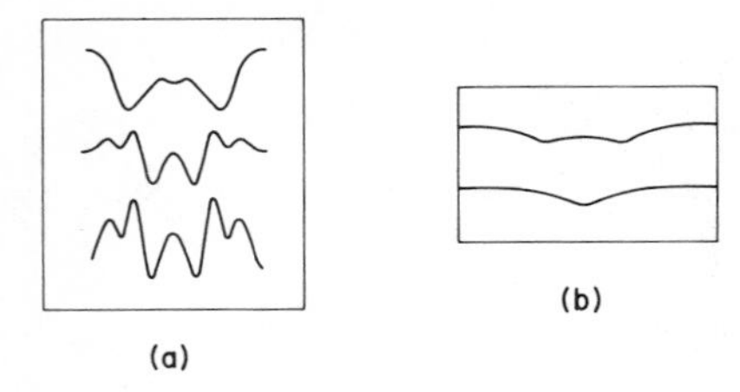

Figure 1-28. Periodic wave profiles that exist in deep water: (*a*) third-harmonic resonance; (*b*) second-harmonic resonance.

where

$$\frac{\partial A_1}{\partial t} + \omega_1' \frac{\partial A_1}{\partial x} - \frac{1}{2} i\omega_1'' \frac{\partial^2 A_1}{\partial x^2} = -\frac{3}{2} i\alpha\omega_1^{-1}(A_1\bar{A}_1 + 2A_3\bar{A}_3)A_1 - \frac{3}{2} i\alpha\omega_1^{-1} A_3\bar{A}_1^2 \exp(i\Gamma) \tag{1.46}$$

$$\frac{\partial A_3}{\partial t} + \omega_3' \frac{\partial A_3}{\partial x} - \frac{1}{2} i\omega_3'' \frac{\partial^2 A_3}{\partial x^2} = -\frac{3}{2} i\alpha\omega_3^{-1}(2A_1\bar{A}_1 + A_3\bar{A}_3)A_3 - \frac{1}{2} i\alpha\omega_3^{-1} A_1^3 \exp(-i\Gamma) \tag{1.47}$$

and $\Gamma = (k_3 - 3k_1)x - (\omega_3 - 3\omega_1)t$. Since $k_3 \approx 3k_1$ and $\omega_3 \approx 3\omega_1$, Γ is a slowly varying function of x and t. Equations (1.46) and (1.47) possess solutions that are stationary and include the nonlinear interaction of the two wavepackets centered at the fundamental and its third harmonic. In the present problem, there are three possible periodic solutions. In the case of waves in deep water they have triple- or quintuple-dimpled profiles as shown in Figure 1-28*a*.

In the case of second-harmonic resonance, the interaction equations have stationary solutions that include solitons and periodic waves. In the case of periodic waves, there are two possible waves. For waves in deep water, they have single- and double-dimpled profiles as shown in Figure 1-28*b*.

CHAPTER 2

Conservative Single-Degree-of-Freedom Systems

In this chapter several examples of conservative, nonlinear systems having one degree of freedom are described. A method for obtaining a qualitative analysis of the free (undamped and unforced) oscillations is presented. Then various methods for obtaining a quantitative analysis are presented. Finally these methods are applied to three specific examples.

2.1. Examples

In this section, we consider a number of conservative systems having a single degree of freedom that are governed by simple nonlinear differential equations having the form

$$\ddot{x} + f(x) = 0 \tag{2.1.1}$$

The examples are chosen to exhibit different sources of nonlinearity.

2.1.1. A SIMPLE PENDULUM

As the first example we consider the motion of a simple pendulum consisting of a mass m attached to a hinged weightless rod of length l as shown in Figure 2-1. The equation describing the motion of the mass is

$$ml\ddot{\theta} + mg \sin \theta = 0$$

or

$$\ddot{\theta} + \omega_0^2 \sin \theta = 0 \tag{2.1.2}$$

where $\omega_0^2 = g/l$. We note that the nonlinearity in this example is due to large motions (it corresponds to large deformations).

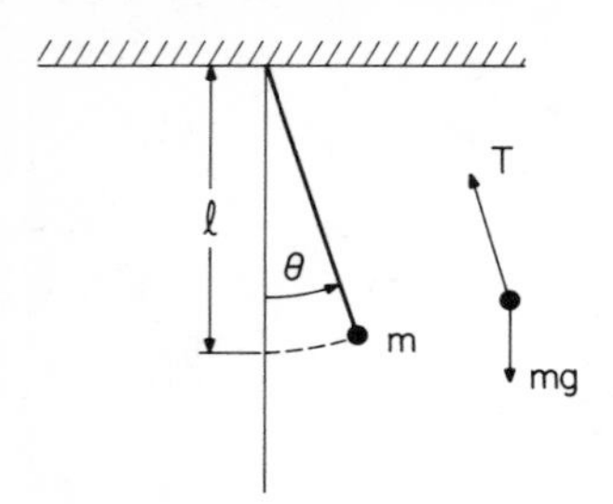

Figure 2-1. Simple pendulum.

2.1.2. A PARTICLE RESTRAINED BY A NONLINEAR SPRING

As the second example we consider the motion of a mass m on a horizontal frictionless plane and restrained by a nonlinear spring as shown in Figure 2-2*a*. If $x(t)$ denotes the position of the mass, then the differential equation describing its motion is

$$m\ddot{x} + f(x) = 0 \tag{2.1.3}$$

where $-f(x)$ is the force exerted by the spring on the mass. For a linear spring, $f(x) = kx$, where k is called the spring constant. For a nonlinear spring, the force is a nonlinear function of the deformation, as shown in Figure 2-2*b*. For a soft spring the nonlinearity decreases the force, while for a hard spring it increases the force. In this section we assume the spring loads and unloads along the same curve and therefore does not exhibit hysteresis, which leads to damping. In this example the nonlinearity is due to the material behavior rather than to large deformations.

2.1.3. A PARTICLE IN A CENTRAL-FORCE FIELD

As the third example we consider the motion of a particle in a plane under the influence of a central-force field as shown in Figure 2-3. In polar coordinates the

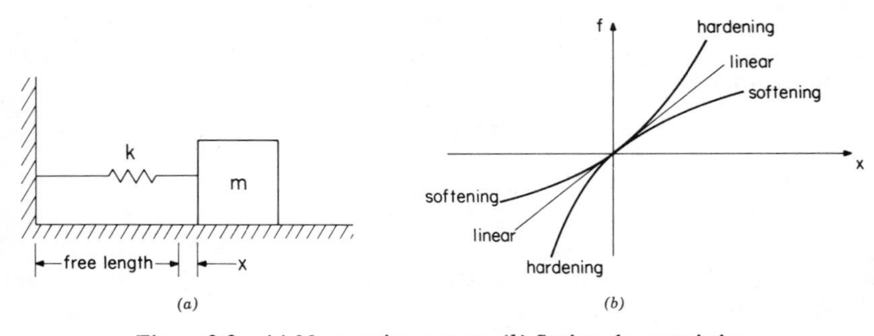

Figure 2-2. (*a*) Mass–spring system. (*b*) Spring characteristics.

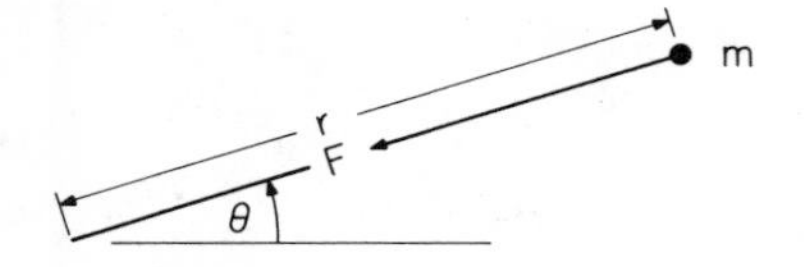

Figure 2-3. Particle in a central-force field.

motion of the particle m is governed by

$$m(\ddot{r} - r\dot{\theta}^2) + mF(r) = 0 \tag{2.1.4}$$

$$m(r\ddot{\theta} + 2\dot{r}\dot{\theta}) = 0 \tag{2.1.5}$$

where m is the mass of the particle if the field is gravitational and m is the charge of the particle if the field is electrical. Equation (2.1.5) has the integral

$$r^2\dot{\theta} = p \tag{2.1.6}$$

where p is a constant; this integral is a statement of conservation of angular momentum. Eliminating $\dot{\theta}$ from (2.1.4) and (2.1.6) yields

$$\ddot{r} - \frac{p^2}{r^3} + F(r) = 0 \tag{2.1.7}$$

Equation (2.1.7) can be put in a simpler form by changing the dependent variable from r to $u = r^{-1}$ and changing the independent variable from t to θ. The derivatives are transformed according to

$$\dot{r} = \frac{dr}{dt} = \frac{dr}{d\theta}\dot{\theta} = -\frac{\dot{\theta}}{u^2}\frac{du}{d\theta} = -p\frac{du}{d\theta} \tag{2.1.8}$$

$$\ddot{r} = -p\frac{d^2u}{d\theta^2}\dot{\theta} = -p^2u^2\frac{d^2u}{d\theta^2} \tag{2.1.9}$$

Hence (2.1.7) becomes

$$\frac{d^2u}{d\theta^2} + u - \frac{1}{p^2u^2}F\left(\frac{1}{u}\right) = 0 \tag{2.1.10}$$

In this example the nonlinearity is due to inertia as well as material properties.

Bond (1974) used the regularizing time transformation $dt/ds = r$ and the Kustaanheimo-Stiefel transformation (Stiefel and Scheifele, 1971) to transform the nonlinear Newtonian differential equations of motion for the two-body problem into four linear harmonic oscillator equations.

2.1.4 A PARTICLE ON A ROTATING CIRCLE

As the fourth example we consider the motion of a mass m moving without friction along a circle of radius R that is rotating with a constant angular velocity Ω about its vertical diameter as shown in Figure 2-4. The forces acting on the

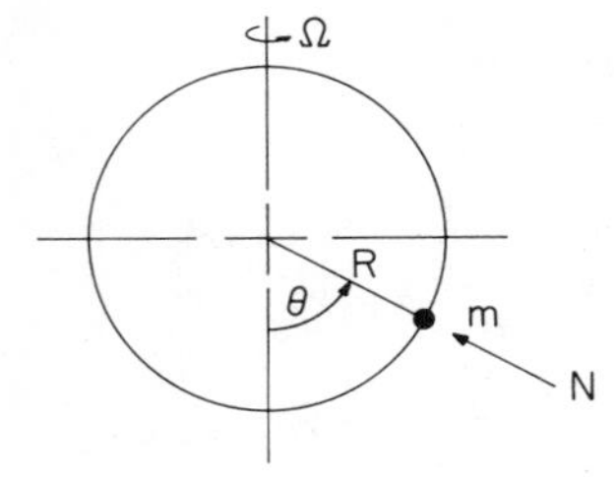

Figure 2-4. Particle moving on a smooth, rotating circular wire.

particle are the gravitational force mg, the centrifugal force $m\Omega^2 R \sin\theta$, and the reaction force N. Taking moments about the center of the circle O and equating their sum to the rate of change of the angular momentum of the particle about O, we obtain

$$mR^2\ddot{\theta} = m\Omega^2 R^2 \sin\theta \cos\theta - mgR \sin\theta \tag{2.1.11}$$

In this example the nonlinearity is due to both inertia and large deformation.

2.2. Qualitative Analysis

The behaviors of the aforementioned physical systems are governed by equations having the form

$$\ddot{u} + f(u) = 0 \tag{2.2.1}$$

In this section we consider a powerful, general method of obtaining many of the distinguishing features of the solutions of this equation.

If follows immediately from (2.2.1) that

$$\int \dot{u}\ddot{u}\, dt + \int f(u)\dot{u}\, dt = h, \quad \int \dot{u}\, d\dot{u} + \int f(u)\, du = h$$

and

$$\tfrac{1}{2}v^2 + F(u) = h \tag{2.2.2}$$

where $v \equiv \dot{u}$ and h is a constant. For a mechanical system, the first term is essentially the kinetic energy; the second term is the potential energy; and the constant h, which is determined from the initial conditions, is the *energy level*. Thus (2.2.2) is a statement of conservation of energy. For a given value of h, the solution (2.2.2) in the uv-plane (called the *phase plane*) is called a *level curve*, or a *curve of constant energy*, or an *integral curve*; the branches of these level curves are called *trajectories*.

As time passes, the point in the phase plane representing the solution moves along a trajectory. The direction or "sense" of the motion of this point can be determined by considering the velocity, $v = \dot{u}$. Clearly u must be increasing if v is positive.

We rewrite (2.2.2) as follows:

$$\tfrac{1}{2}v^2 = h - F(u) \tag{2.2.3}$$

and note that a real solution for v exists if, and only if, $h \geqslant F(u)$ and that the trajectories are symmetric about the u-axis. Moreover we obtain from (2.2.1)

$$\dot{v} = -f(u) \tag{2.2.4}$$

from which it follows immediately that

$$\frac{dv}{du} = -\frac{f(u)}{v} \tag{2.2.5}$$

Thus when the trajectory has a horizontal tangent ($dv/du = 0$), $f(u) = 0$; and when the trajectory has a vertical tangent ($dv/du = \infty$), $v = 0$. As we shall see, the points where either $f(u)$ or v is zero are points of special interest. Also of special interest are the points where v and $f(u)$ are zero simultaneously and thus the slope is indeterminate; these are called *singular points*. Thus singular points correspond to the simultaneous vanishing of the acceleration and the velocity, and hence they are *equilibrium points*. Because the slopes are uniquely determined everywhere except at the singular points, trajectories cannot intersect anywhere except at the singular points. Next we determine the form of the trajectories for various forms of the function $F(u)$.

We begin by considering the case of $F(u)$ being monotonic. In Figure 2-5, the case of $F(u)$ monotonically increasing is shown. We note that each level curve consists of one branch (trajectory) similar in shape to a hyperbola that opens to the left. Clearly the case of $F(u)$ monotonically decreasing would have

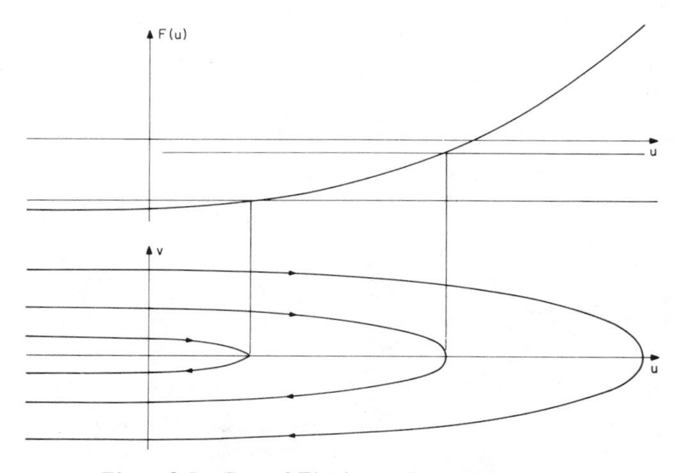

Figure 2-5. Case of $F(u)$ increasing monotonically.

the trajectories opening to the right. We note that, because the trajectories do not close, the motion is not oscillatory in either case.

As a second example we consider the case of $F(u)$ having a maximum as shown in Figure 2-6. When the energy level h is less than h_0, each level curve consists of two branches, which intersect the u-axis and are similar in shape to branches of hyperbolas, one opening to the right and the other opening to the left. When $h > h_0$, each level curve consists also of two branches, but in this case they do not intersect the u-axis. When $h = h_0$, the level curve consists of four branches that meet at the point S, which is a singular point and called a *saddle point*, or *col*. The branches (trajectories) passing through the saddle point are called *separatrices*. None of the other trajectories passes through the point S, and the separatrices are asymptotes to all other trajectories. The equilibrium point S is unstable because any small disturbance will result in a trajectory on which the state of the system deviates more and more from S as $t \to \infty$.

An infinite amount of time is required by a particle to pass along a separatrix from any point in the neighborhood of a saddle point to the saddle point itself. This can be seen as follows. From (2.2.3)

$$\dot{u} = \pm[2h_0 - 2F(u)]^{1/2} \tag{2.2.6}$$

It is convenient to introduce a change of the dependent variable from u to $x = u - u_0$, where u_0 is the location of the saddle point. Thus the expansion of

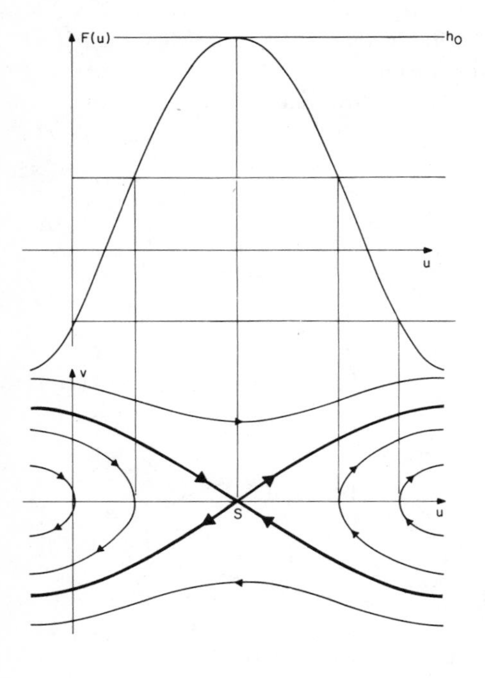

Figure 2-6. Case of $F(u)$ having a maximum.

the function $h_0 - F(u)$ in the neighborhood of the saddle point has the form (see Figure 2-6)

$$h_0 - F(u) = h_0 - F(u_0 + x) = -\tfrac{1}{2}F''(u_0)x^2 + O(x^3) \tag{2.2.7}$$

because $F(u_0) = h_0$ and $F'(u_0) = 0$. Substituting (2.2.7) into (2.2.6) and integrating leads to the following expression for the time required to move from $x_1 = u_1 - u_0$ to $x = u - u_0$:

$$t = -[-F''(u_0)]^{-1/2} \ln\left(\frac{x}{x_1}\right) \tag{2.2.8}$$

We note that $F''(u_0) < 0$ near a col. Thus $x \to 0$ (i.e., $u \to u_0$) as $t \to \infty$.

As a third example we consider the case of $F(u)$ having a minimum as indicated in Figure 2-7. When $h = h_0$, the level curve degenerates into the single singular point C which is called a *center*. When $h < h_0$ there is no real solution, while when $h > h_0$ each level curve consists of a single closed trajectory which need not be an ellipse surrounding the center C. We note that C is stable in the sense of Liapunov (1966) because a small disturbance will result in a closed

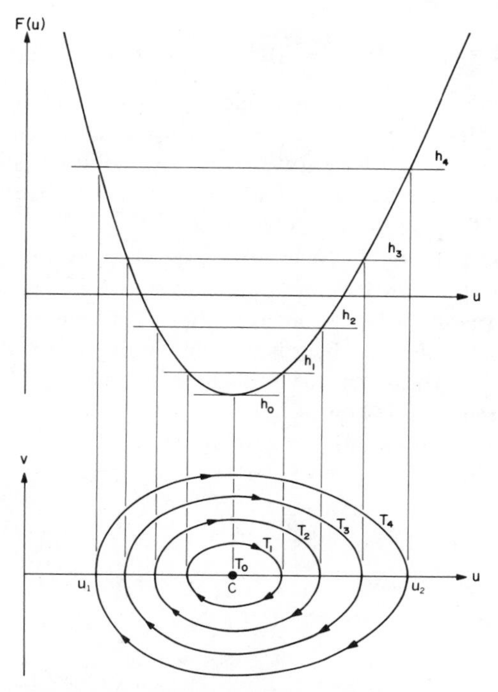

Figure 2-7. Case of $F(u)$ having a minimum.

trajectory that surrounds C along which the state of the system stays close to C. The motions corresponding to the closed curves are periodic but need not be harmonic. Moreover, in contrast with linear systems for which the period is independent of the amplitude (i.e., energy level), the period T of a nonlinear system is a function of h. It can be found from (2.2.6) and Figure 2-7 to be

$$T = 2 \int_{u_1}^{u_2} [2h - 2F(u)]^{-1/2} \, du \tag{2.2.9}$$

Though both centers and cols (saddle points) are singular points, in the neighborhoods of these points the motions produced by small disturbances are quite different as discussed above. Near a singular point, u_0,

$$h - F(u) = -\tfrac{1}{2} F''(u_0) x^2 + O(x^3) \tag{2.2.10}$$

where $x = u - u_0$. If the motion is small, then we may neglect the higher-order terms so that the equation of motion becomes

$$\ddot{x} + F''(u_0) x = 0 \tag{2.2.11}$$

The solution has the form

$$x = c_1 \exp\left[\sqrt{-F''(u_0)}\, t\right] + c_2 \exp\left[-\sqrt{-F''(u_0)}\, t\right] \tag{2.2.12}$$

where c_1 and c_2 are constants. Near a saddle point $F''(u_0)$ is negative, hence one term decays exponentially but the other grows exponentially. On the other hand, near a center $F''(u_0)$ is positive; hence the solution is oscillatory, being described in terms of circular functions. For these reasons the saddle point is called unstable, while the center is called stable.

As a fourth example we consider the case when maximum and minimum points coalesce to form an inflection point as shown in Figure 2-8. Each level curve consists of one branch that opens to the left. The level curve $h = h_0$ passes through the singular point P, which is unstable. It is a *nonelementary* or *degenerate singular point*, which may be thought of as resulting from the fusion or coalescence of a saddle point on the left of P with a center on the right of P. We note that this point corresponds to a cusp in the phase plane; this can be seen by considering the following. At point P,

$$F(u) = h_0, \qquad \frac{dF}{du} = 0, \qquad \text{and} \qquad \frac{d^2F}{du^2} = 0$$

Therefore

$$v = 0 \tag{2.2.13}$$

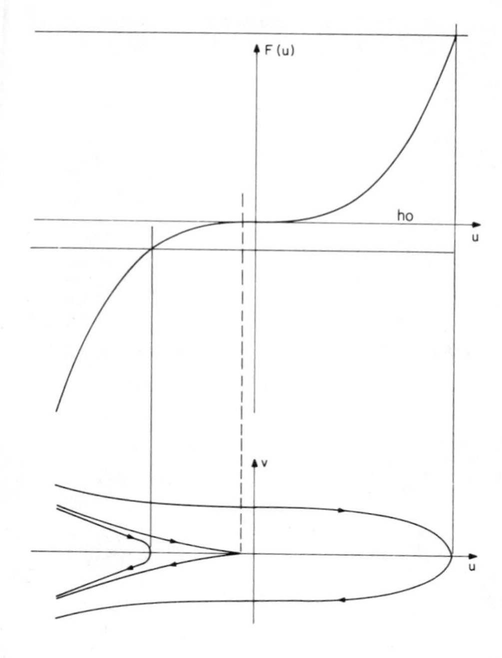

Figure 2-8. Case of $F(u)$ having an inflection point.

Moreover, since

$$\frac{dv}{dt} = v\frac{dv}{du} = -f(u) = -\frac{dF}{du} = 0$$

$$\left(\frac{dv}{du}\right)^2 + v\frac{d^2v}{du^2} = -\frac{df}{du} = -\frac{d^2F}{du^2} = 0 \tag{2.2.14}$$

Because $v = 0$,

$$\frac{dv}{du} = 0 \tag{2.2.15}$$

The preceding three examples constitute an elementary proof of a theorem due to Lagrange and Dirichlet, which states that *if the potential energy has an isolated minimum at an equilibrium point, the equilibrium state is stable.* They also constitute an elementary proof of a converse theorem due to Liapunov, which states that *if the potential energy at an equilibrium point is not a minimum, the equilibrium state is unstable.*

We note that, if the functional form of $F(u)$ or $F'(u)$ is given, one can determine whether a singular point is a saddle point or a center by examining the second derivative. Clearly at a saddle point

$$\frac{d^2F}{du^2} = \frac{df}{du} < 0$$

while at a center

$$\frac{d^2F}{du^2} = \frac{df}{du} > 0$$

As an example, we consider the equation

$$\ddot{u} + (1 - u)(2 - u) = 0 \tag{2.2.16}$$

The singular points are located at

$$u = 1 \quad \text{and} \quad u = 2$$

It follows that

$$\left.\frac{df}{du}\right|_{u=1} = 2u - 3|_{u=1} = -1$$

and

$$\left.\frac{df}{du}\right|_{u=2} = 1$$

Thus $u = 1$ is a saddle point, while $u = 2$ is a center. There are oscillatory solutions in the neighborhood of $u = 2$ but not in the neighborhood of $u = 1$.

When $F(u)$ is more complicated than the cases considered above, the corresponding representations of the solutions in the phase plane are composed of combinations of those presented above. An example is shown in Figure 2-9. When $h = h_0$, the level curve consists of the two centers C_1 and C_2, while when $h = h_3$, the level curve consists of two trajectories (separatrices) meeting at the saddle point S. When $h_0 < h < h_3$, each level curve consists of two closed trajectories, one surrounding the center C_1 and the other surrounding the center C_2. When $h > h_3$, each level curve consists of a single closed trajectory that surrounds the two centers as well as the saddle point. This example illustrates the strong dependence of the state of the system on the initial conditions and the system parameters.

The remainder of this chapter is devoted to finding the solutions, or approximations that exhibit the characteristics of the solutions, in a small but finite neighborhood of a center. From the discussion above one can recognize several features of the motions of nonlinear systems which distinguish them from linear systems. Let us suppose that, in the limit as the amplitude of the motion

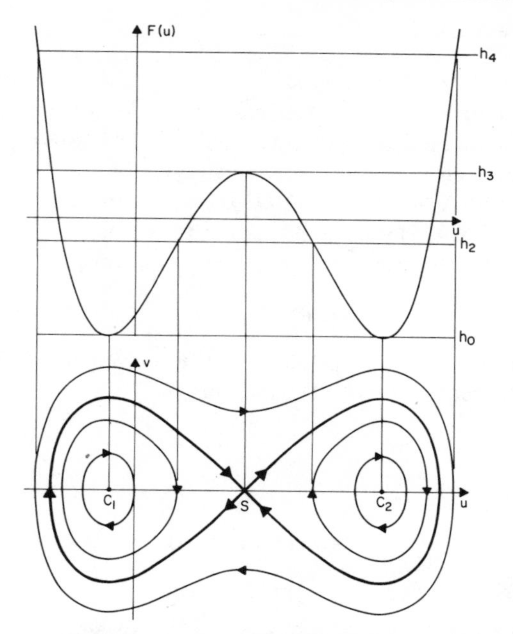

Figure 2-9. Case of $F(u)$ having a maximum and two minima.

vanishes, the solution of the nonlinear equation approaches the solution of the corresponding linear equation. As we shall see, this turns out to be the case. Then a single harmonic will describe the motion in an infinitesimal neighborhood of the center. But as the preceding phase diagrams (Figures 2-7 and 2-9) clearly show, the closed trajectories for large amplitudes are not merely scaled-up versions of those for very small amplitudes. The shape changes noticeably, but the motion is periodic and hence can be represented by a Fourier sine and cosine expansion. Thus one term in the expansion is sufficient to represent the infinitesimal motion accurately; but as the amplitude grows, so does the number of terms required to represent the solution accurately. Consequently one expects higher harmonics in the motion of nonlinear systems. Moreover it appears from (2.2.9) that the period of the motion depends on the amplitude (i.e., it depends on h). Finally we note that the trajectories around a center are not necessarily symmetric with respect to the center. Thus the motion appears to *drift* or *stream* as the amplitude increases; the midpoint of the motion is not the center (equilibrium position). The asymmetry is the result of the presence of even functions of the distance from the center in $f(u)$.

In the next section we present various methods for determining approximate solutions of (2.2.1) in the neighborhood of a center.

2.3. Quantitative Analysis

As the examples in the preceding section illustrate, the motion is oscillatory in the neighborhood of a center. In this section we discuss methods of obtaining approximate expressions describing this oscillatory motion. Numerical methods were used by a number of investigators. Einaudi (1975) used an iterative method; Argyris, Dunne, and Angelopoulos (1973) used a finite-element technique; and Susemihl and Laura (1975) used a collocation technique.

We have been considering systems governed by equations having the form

$$\ddot{u} + f(u) = 0 \tag{2.3.1}$$

where, in general, f is a nonlinear function. It is convenient to shift the origin to the location of the center, $u = u_0$. Thus we let

$$x = u - u_0 \tag{2.3.2}$$

Then (2.3.1) becomes

$$\ddot{x} + f(x + u_0) = 0 \tag{2.3.3}$$

Assuming f can be expanded, we rewrite (2.3.3) as

$$\ddot{x} + \sum_{n=1}^{N} \alpha_n x^n = 0 \tag{2.3.4}$$

where

$$\alpha_n = \frac{1}{n!} f^{(n)}(u_0) \tag{2.3.5}$$

and $f^{(n)}$ denotes the nth derivative with respect to the argument. For a center, $f(u_0) = 0$ and $f'(u_0) > 0$.

The solution describes the response of the system to an initial disturbance. To describe the initial disturbance, one needs to specify both the initial position and the initial velocity, s_0 and v_0, respectively. It is convenient to write the initial conditions in polar form. Thus we introduce an amplitude and a phase according to

$$s_0 = a_0 \cos \beta_0, \qquad v_0 = -a_0 \omega_0 \sin \beta_0 \tag{2.3.6}$$

where

$$\omega_0 = \sqrt{\alpha_1} = [f'(u_0)]^{1/2}$$

and

$$a_0 = \left[s_0^2 + \left(\frac{v_0}{\omega_0} \right)^2 \right]^{1/2}, \qquad \beta_0 = \cos^{-1}\left(\frac{s_0}{a_0} \right) = \sin^{-1}\left(-\frac{v_0}{a_0 \omega_0} \right) \tag{2.3.7}$$

The system governed by the equation obtained from (2.3.4) by deleting all the nonlinear terms is called the corresponding linear system. It plays a key role in the analysis of weakly nonlinear systems. Basically one obtains the response of the nonlinear system by perturbing the response of the corresponding linear system.

There are a number of ways in which this perturbation can be effected. We begin with the so-called straightforward expansion, which is not uniformly valid, and then discuss the details of several modifications of the straightforward procedure which lead to uniformly valid expansions. The present discussion is not meant to be comprehensive; for such a discussion the reader is referred to *Perturbation Methods* by Nayfeh (1973*b*). Here we only discuss the basic concepts of the methods appearing most frequently in the literature.

2.3.1. THE STRAIGHTFORWARD EXPANSION

We seek an expansion that is valid for small- but finite-amplitude motions. It is convenient to introduce a small, dimensionless parameter ϵ which is the order of the amplitude of the motion and can be used as a crutch, or a bookkeeping device, in obtaining the approximate solution.

We assume that the solution of (2.3.4) can be represented by an expansion having the form

$$x(t;\epsilon) = \epsilon x_1(t) + \epsilon^2 x_2(t) + \epsilon^3 x_3(t) + \cdots \tag{2.3.8}$$

Then we substitute (2.3.8) into (2.3.4) and, because the x_n are independent of ϵ, set the coefficient of each power of ϵ equal to zero. This leads to the following set of equations:

Order ϵ

$$\ddot{x}_1 + \omega_0^2 x_1 = 0 \tag{2.3.9}$$

Order ϵ^2

$$\ddot{x}_2 + \omega_0^2 x_2 = -\alpha_2 x_1^2 \tag{2.3.10}$$

Order ϵ^3

$$\ddot{x}_3 + \omega_0^2 x_3 = -2\alpha_2 x_1 x_2 - \alpha_3 x_1^3 \tag{2.3.11}$$

In satisfying the initial conditions, there are the following alternatives:

1. One can substitute the assumed expansion (2.3.8) into the initial conditions (2.3.6) and equate coefficients of like powers of ϵ. The result is

$$x_1(0) = a_0 \cos\beta_0 \quad \text{and} \quad \dot{x}_1(0) = -\omega_0 a_0 \sin\beta_0 \tag{2.3.12}$$

$$x_n(0) = 0 \quad \text{and} \quad \dot{x}_n(0) = 0 \text{ for } n \geqslant 2 \tag{2.3.13}$$

Then one determines the constants of integration in x_1 such that (2.3.12) is satisfied; and one includes the homogeneous solution in the expressions for the

x_n, for $n \geqslant 2$, choosing the constants of integration such that (2.3.13) is satisfied at each step.

2. One can ignore the initial conditions and the homogeneous solutions in all the x_n, for $n \geqslant 2$, until the last step. Then, considering the constants of integration in x_1 to be functions of ϵ, one expands the solution for x_1 in powers of ϵ and chooses the coefficients in the expansion such that (2.3.6) is satisfied.

Initially it may appear that the second alternative is inconsistent because we stipulated that the x_n are independent of ϵ. However, as we demonstrate by an example, the two approaches are equivalent, yielding precisely the same result.

We prefer the second approach because there is much less algebra involved and, in many instances, we are only concerned with steady-state responses, which frequently are independent of the initial conditions.

The general solution of (2.3.9) can be written in the form

$$x_1 = a \cos (\omega_0 t + \beta) \tag{2.3.14}$$

where a and β are constants. Following the first alternative, we let $a = a_0$ and $\beta = \beta_0$ in order to satisfy (2.3.12). Following the second approach, we consider a and β to be functions of ϵ and at this point pay no regard to the initial conditions.

Substituting (2.3.14) into (2.3.10) yields

$$\ddot{x}_2 + \omega_0^2 x_2 = -\alpha_2 a^2 \cos^2 (\omega_0 t + \beta) = -\tfrac{1}{2}\alpha_2 a^2 [1 + \cos (2\omega_0 t + 2\beta)] \tag{2.3.15}$$

where trigonometric identities were used to eliminate all products and powers of the cosines. This is a necessary step in all the subsequent perturbation methods discussed. In accordance with the discussion above, we have two choices for expressing x_2:

$$x_2 = \frac{\alpha_2 a_0^2}{6\omega_0^2} [\cos (2\omega_0 t + 2\beta_0) - 3] + a_2 \cos (\omega_0 t + \beta_2) \tag{2.3.16}$$

or

$$x_2 = \frac{\alpha_2 a^2}{6\omega_0^2} [\cos (2\omega_0 t + 2\beta) - 3] \tag{2.3.17}$$

where a_2 and β_2 are additional constants of integration, independent of ϵ, chosen such that (2.3.13) is satisfied.

Thus following the first alternative, we have

$$x = \epsilon a_0 \cos (\omega_0 t + \beta_0) + \epsilon^2 \left\{ \frac{a_0^2 \alpha_2}{6\omega_0^2} [\cos (2\omega_0 t + 2\beta_0) - 3] + a_2 \cos (\omega_0 t + \beta_2) \right\} + O(\epsilon^3) \tag{2.3.18}$$

Following the second alternative we have

$$x = \epsilon a \cos(\omega_0 t + \beta) + \frac{\epsilon^2 a^2 \alpha_2}{6\omega_0^2}[\cos(2\omega_0 t + 2\beta) - 3] + O(\epsilon^3) \quad (2.3.19)$$

Now into (2.3.19) we put

$$a = \epsilon A_1 + \epsilon^2 A_2 + \cdots, \qquad \beta = B_0 + \epsilon B_1 + \cdots$$

Then

$$\begin{aligned}\epsilon a \cos(\omega_0 t + \beta) &= (\epsilon A_1 + \epsilon^2 A_2 + \cdots)[\cos(\omega_0 t + B_0)\cos(\epsilon B_1 + \cdots) \\ &\quad - \sin(\omega_0 t + B_0)\sin(\epsilon B_1 + \cdots)] = \epsilon A_1 \cos(\omega_0 t + B_0) \\ &\quad + \epsilon^2 [A_2 \cos(\omega_0 t + B_0) - A_1 B_1 \sin(\omega_0 t + B_0)] + O(\epsilon^3) \\ &= \epsilon A_1 \cos(\omega_0 t + B_0) + \epsilon^2 (A_2^2 + A_1^2 B_1^2)^{1/2} \cos(\omega_0 t + \theta_2) \\ &\quad + O(\epsilon^3)\end{aligned}$$

where

$$\theta_2 = B_0 + \tan^{-1}\left(\frac{A_1 B_1}{A_2}\right)$$

We can choose $A_1 = a_0$, $B_0 = \beta_0$, and A_2 and B_1 such that

$$(A_2^2 + A_1^2 B_1^2)^{1/2} = a_2 \quad \text{and} \quad \beta_0 + \tan^{-1}\left(\frac{A_1 B_1}{A_2}\right) = \beta_2$$

Then (2.3.18) and (2.3.19) are equivalent. Thus either alternative can be used in the subsequent schemes and either alternative can be used for higher-order approximations.

Substituting (2.3.14) and (2.3.17) into (2.3.11) yields

$$\begin{aligned}\ddot{x}_3 + \omega_0^2 x_3 &= \frac{\alpha_2^2 a^3}{3\omega_0^2}[3\cos(\omega_0 t + \beta) - \cos(\omega_0 t + \beta)\cos(2\omega_0 t + 2\beta)] - \alpha_3 a^3 \\ &\quad \cdot \cos^3(\omega_0 t + \beta) = \left(\frac{5\alpha_2^2}{6\omega_0^2} - \frac{3\alpha_3}{4}\right) a^3 \cos(\omega_0 t + \beta) - \left(\frac{\alpha_3}{4} + \frac{\alpha_2^2}{6\omega_0^2}\right) \\ &\quad \cdot \cos(3\omega_0 t + 3\beta)\end{aligned} \quad (2.3.20)$$

Any particular solution of (2.3.20) contains the term

$$\left(\frac{10\alpha_2^2 - 9\alpha_3\omega_0^2}{24\omega_0^3}\right) at \sin(\omega_0 t + \beta)$$

If the straightforward procedure is continued, terms containing the factors $t^m \cos(\omega_0 t + \beta)$ and $t^m \sin(\omega_0 t + \beta)$ appear. Terms such as these are called *secular terms*.

Because of secular terms, expansion (2.3.8) is not periodic. Moreover x_3/x_1 and x_3/x_2 grow without bound as t increases; thus x_3 does not always provide a small correction to x_1 and x_2. One says that expansion (2.3.8) is not uniformly valid as t increases.

The discussion in Section 2.2 indicates that one of the features distinguishing nonlinear from linear systems is *frequency–amplitude interaction*. Yet in the procedure used to generate the straightforward expansion, there is no provision for such a relationship. Thus this approach was doomed from the beginning. One modification of the straightforward procedure that does account for the frequency–amplitude interaction is the Lindstedt-Poincaré method, which is discussed next.

2.3.2. THE LINDSTEDT-POINCARÉ METHOD

The idea is to introduce a new independent variable, say $\tau = \omega t$, where initially ω is an unspecified function of ϵ. The new governing equation will contain ω in the coefficient of the second derivative; this permits the frequency and amplitude to interact. One can choose the function ω in such a way as to eliminate the secular terms (i.e., to render the expansion periodic in accordance with the discussion of Section 2.2).

We begin by assuming an expansion for ω:

$$\omega(\epsilon) = \omega_0 + \epsilon\omega_1 + \epsilon^2\omega_2 + \cdots \tag{2.3.21}$$

where ω_1, ω_2, and so on, are unknown constants at this point. Moreover we assume that x can be represented by an expansion having the form

$$x(t;\epsilon) = \epsilon x_1(\tau) + \epsilon^2 x_2(\tau) + \epsilon^3 x_3(\tau) + \cdots \tag{2.3.22}$$

where the x_n are independent of ϵ. Then (2.3.4) becomes

$$(\omega_0 + \epsilon\omega_1 + \epsilon^2\omega_2 + \cdots)^2 \frac{d^2}{d\tau^2}(\epsilon x_1 + \epsilon^2 x_2 + \epsilon^3 x_3 + \cdots) + \sum_{n=1}^{N} \alpha_n(\epsilon x_1 + \epsilon^2 x_2 + \epsilon^3 x_3 + \cdots)^n = 0 \tag{2.3.23}$$

Equating the coefficients of ϵ, ϵ^2, and ϵ^3 to zero and recalling that $\alpha_1 = \omega_0^2$, we obtain

$$\frac{d^2 x_1}{d\tau^2} + x_1 = 0 \tag{2.3.24}$$

$$\omega_0^2\left(\frac{d^2x_2}{d\tau^2}+x_2\right) = -2\omega_0\omega_1\frac{d^2x_1}{d\tau^2} - \alpha_2 x_1^2 \tag{2.3.25}$$

$$\omega_0^2\left(\frac{d^2x_3}{d\tau^3}+x_3\right) = -2\omega_0\omega_1\frac{d^2x_2}{d\tau^2} - 2\alpha_2 x_1 x_2 - (\omega_1^2 + 2\omega_0\omega_2)\frac{d^2x_1}{d\tau^2} - \alpha_3 x_1^3 \tag{2.3.26}$$

We write the general solution of (2.3.24) in the form

$$x_1 = a\cos\phi \tag{2.3.27}$$

where

$$\phi = \tau + \beta \tag{2.3.28}$$

and a and β are constants. Substituting (2.3.27) into (2.3.25) leads to

$$\omega_0^2\left(\frac{d^2x_2}{d\tau^2}+x_2\right) = 2\omega_0\omega_1 a\cos\phi - \tfrac{1}{2}\alpha_2 a^2[1+\cos 2\phi] \tag{2.3.29}$$

Thus we must set $\omega_1 = 0$, or x_2 will contain the secular term $\omega_1\omega_0^{-1}a\tau\sin\phi$. Then disregarding the solution of the homogeneous equation, we write the solution of (2.3.29) as

$$x_2 = -\frac{\alpha_2 a^2}{2\omega_0^2}\left[1 - \tfrac{1}{3}\cos 2\phi\right] \tag{2.3.30}$$

Substituting for x_1 and x_2 into (2.3.26) and recalling that $\omega_1 = 0$, we obtain

$$\omega_0^2\left(\frac{d^2x_3}{d\tau^2}+x_3\right) = 2\left(\omega_0\omega_2 a - \tfrac{3}{8}\alpha_3 a^3 + \tfrac{5}{12}\frac{\alpha_2^2 a^3}{\omega_0^2}\right)\cos\phi - \tfrac{1}{4}\left(\frac{2\alpha_2^2}{3\omega_0^2}+\alpha_3\right)a^3\cos 3\phi \tag{2.3.31}$$

To eliminate the secular term from x_3, we must put

$$\omega_2 = \frac{(9\alpha_3\omega_0^2 - 10\alpha_2^2)a^2}{24\omega_0^3} \tag{2.3.32}$$

Hence from (2.3.2), (2.3.21), (2.3.22), (2.3.27), and (2.3.30), it follows that

$$u = u_0 + \epsilon a\cos(\omega t+\beta) - \frac{\epsilon^2 a^2\alpha_2}{2\alpha_1}\left[1 - \tfrac{1}{3}\cos(2\omega t + 2\beta)\right] + O(\epsilon^3) \tag{2.3.33}$$

where

$$\omega = \sqrt{\alpha_1}\left[1 + \frac{9\alpha_3\alpha_1 - 10\alpha_2^2}{24\alpha_1^2}\epsilon^2 a^2\right] + O(\epsilon^3) \tag{2.3.34}$$

We note that carrying out the expansion to higher order is cumbersome. One seldom has the courage to go beyond third order unless the algebraic manipulations are performed by a computer. Consequently, by using a computer to perform algebraic manipulations, Helleman and Montroll (1974), Montroll and Helleman (1976), Eminhizer, Helleman, and Montroll (1976), Berry (1978), and Helleman (1978) developed a recurrence algorithm by which they solved for the Fourier coefficients of the solution and the frequency corrections rather than solving for the individual u_n.

Imposing the initial conditions (2.3.6), we have

$$a_0 \cos \beta_0 = \epsilon a \cos \beta - \frac{\epsilon^2 a^2 \alpha_2}{2\alpha_1} [1 - \tfrac{1}{3} \cos 2\beta]$$

and (2.3.35)

$$-\omega_0 a_0 \sin \beta_0 = -\epsilon a \omega \sin \beta - \frac{\epsilon^2 a^2 \alpha_2 \omega}{3\alpha_1} \sin 2\beta$$

To solve (2.3.35), we expand a and β in powers of ϵ and equate coefficients of like powers of ϵ. The result is

$$\epsilon a = a_0 + \frac{\alpha_2 a_0^2}{12\alpha_1} (3 \cos \beta_0 + \cos 3\beta_0)$$

$$\beta = \beta_0 - \frac{\alpha_2 a_0}{12\alpha_1} (9 \sin \beta_0 + \sin 3\beta_0) \tag{2.3.36}$$

The resulting solution is in agreement with the solution that can be derived by including the homogeneous solution in x_2 and satisfying the initial conditions at each level of approximation.

In accordance with the qualitative description of the motion given in Section 2.2, we note that the Lindstedt-Poincaré procedure produced (a) a periodic expression describing the motion of the system, (b) a frequency–amplitude relationship (which is a direct consequence of requiring the expression to be periodic), (c) higher harmonics in the higher-order terms of the expression, and (d) a *drift* or *steady-streaming* term $-\frac{1}{2}\epsilon^2 a^2 \alpha_2 \alpha_1^{-1}$.

2.3.3. THE METHOD OF MULTIPLE SCALES

The uniformly valid expansion given by (2.3.33) and (2.3.34) may be viewed as a function of two independent variables rather than a function of one. Namely we may regard x to be a function of t and $\epsilon^2 t$. The underlying idea of the method of multiple scales is to consider the expansion representing the response to be a function of multiple independent variables, or scales, instead of a single variable. The method of multiple scales, though a little more involved, has advantages over the Lindstedt-Poincaré method; for example, it can treat damped systems conveniently.

One begins by introducing new independent variables according to

$$T_n = \epsilon^n t \quad \text{for} \quad n = 0, 1, 2, \cdots \tag{2.3.37}$$

It follows that the derivatives with respect to t become expansions in terms of the partial derivatives with respect to the T_n according to

$$\begin{aligned} \frac{d}{dt} &= \frac{dT_0}{dt}\frac{\partial}{\partial T_0} + \frac{dT_1}{dt}\frac{\partial}{\partial T_1} + \cdots = D_0 + \epsilon D_1 + \cdots \\ \frac{d^2}{dt^2} &= D_0^2 + 2\epsilon D_0 D_1 + \epsilon^2(D_1^2 + 2D_0 D_2) + \cdots \end{aligned} \tag{2.3.38}$$

One assumes that the solution of (2.3.4) can be represented by an expansion having the form

$$x(t;\epsilon) = \epsilon x_1(T_0, T_1, T_2, \cdots) + \epsilon^2 x_2(T_0, T_1, T_2, \cdots) + \epsilon^3 x_3(T_0, T_1, T_2, \cdots) + \cdots \tag{2.3.39}$$

We note that the number of independent time scales needed depends on the order to which the expansion is carried out. If the expansion is carried out to $O(\epsilon^2)$, then T_0 and T_1 are needed. In this section we carry out the expansion to $O(\epsilon^3)$, and hence we need T_0, T_1, and T_2. Substituting (2.3.38) and (2.3.39) into (2.3.4) and equating the coefficients of ϵ, ϵ^2, and ϵ^3 to zero, we obtain

$$D_0^2 x_1 + \omega_0^2 x_1 = 0 \tag{2.3.40}$$

$$D_0^2 x_2 + \omega_0^2 x_2 = -2D_0 D_1 x_1 - \alpha_2 x_1^2 \tag{2.3.41}$$

$$D_0^2 x_3 + \omega_0^2 x_3 = -2D_0 D_1 x_2 - D_1^2 x_1 - 2D_0 D_2 x_1 - 2\alpha_2 x_1 x_2 - \alpha_3 x_1^3 \tag{2.3.42}$$

With this approach it turns out to be convenient to write the solution of (2.3.40) in the form

$$x_1 = A(T_1, T_2) \exp(i\omega_0 T_0) + \bar{A} \exp(-i\omega_0 T_0) \tag{2.3.43}$$

where A is an unknown complex function and $\bar{A}$ is the complex conjugate of A. The governing equations for A are obtained by requiring x_2 and x_3 to be periodic in T_0.

Substituting (2.3.43) into (2.3.41) leads to

$$D_0^2 x_2 + \omega_0^2 x_2 = -2i\omega_0 D_1 A \exp(i\omega_0 T_0) - \alpha_2 [A^2 \exp(2i\omega_0 T_0) + A\bar{A}] + cc \tag{2.3.44}$$

where cc denotes the complex conjugate of the preceding terms. Any particular solution of (2.3.44) has a secular term containing the factor $T_0 \exp(i\omega_0 T_0)$ unless

$$D_1 A = 0 \tag{2.3.45}$$

Therefore A must be independent of T_1. With $D_1 A = 0$, the solution of (2.3.44) is

$$x_2 = \frac{\alpha_2 A^2}{3\omega_0^2} \exp(2i\omega_0 T_0) - \frac{\alpha_2}{\omega_0^2} A\bar{A} + cc \tag{2.3.46}$$

where the solution of the homogeneous equation is not needed as discussed in Section 2.3.1.

Substituting for x_1 and x_2 from (2.3.43) and (2.3.46) into (2.3.42) and recalling that $D_1 A = 0$, we obtain

$$D_0^2 x_3 + \omega_0^2 x_3 = -\left[2i\omega_0 D_2 A - \frac{10\alpha_2^2 - 9\alpha_3\omega_0^2}{3\omega_0^2} A^2\bar{A}\right] \exp(i\omega_0 T_0) - \frac{3\alpha_3\omega_0^2 + 2\alpha_2^2}{3\omega_0^2} A^3 \exp(3i\omega_0 T_0) + cc \tag{2.3.47}$$

To eliminate secular terms from x_3, we must put

$$2i\omega_0 D_2 A + \frac{9\alpha_3\omega_0^2 - 10\alpha_2^2}{3\omega_0^2} A^2\bar{A} = 0 \tag{2.3.48}$$

In solving equations having the form of (2.3.48), we find it convenient to write A in the polar form

$$A = \tfrac{1}{2} a \exp(i\beta) \tag{2.3.49}$$

where a and β are real functions of T_2. Substituting (2.3.49) into (2.3.48) and separating the result into real and imaginary parts, we obtain

$$\omega a' = 0 \quad \text{and} \quad \omega_0 a\beta' + \frac{10\alpha_2^2 - 9\alpha_3\omega_0^2}{24\omega_0^2} a^3 = 0 \tag{2.3.50}$$

where the prime denotes the derivative with respect to T_2. It follows that a is a constant and hence that

$$\beta = \frac{9\alpha_3\omega_0^2 - 10\alpha_2^2}{24\omega_0^3} a^2 T_2 + \beta_0$$

where β_0 is a constant. Returning to (2.3.49), we find that

$$A = \tfrac{1}{2} a \exp\left[i\frac{9\alpha_3\omega_0^2 - 10\alpha_2^2}{24\omega_0^3} \epsilon^2 a^2 t + i\beta_0\right] \tag{2.3.51}$$

where we used the fact that $T_2 = \epsilon^2 t$.

Substituting for x_1 and x_2 from (2.3.43) and (2.3.46) into (2.3.39) and using (2.3.51), we obtain

$$x = \epsilon a \cos(\omega t + \beta_0) - \frac{\epsilon^2 a^2 \alpha_2}{2\alpha_1} [1 - \tfrac{1}{3} \cos(2\omega t + 2\beta_0)] + O(\epsilon^3) \tag{2.3.52}$$

where

$$\omega = \sqrt{\alpha_1} \left[1 + \frac{9\alpha_3 \alpha_1 - 10\alpha_2^2}{24\alpha_1^2} \epsilon^2 a^2\right] + O(\epsilon^3) \tag{2.3.53}$$

in agreement with the solution, (2.3.33) and (2.3.34), obtained in the preceding section by using the Lindstedt-Poincaré procedure.

2.3.4. THE METHOD OF HARMONIC BALANCE

The idea is to express the periodic solution of (2.3.4) in the form

$$x = \sum_{m=0}^{M} A_m \cos(m\omega t + m\beta_0) \tag{2.3.54}$$

Then substituting (2.3.54) into (2.3.4) and equating the coefficient of each of the lowest $M + 1$ harmonics to zero, we obtain a system of $M + 1$ algebraic equations relating ω and the A_m. Usually these equations are solved for $A_0, A_2, A_3, \cdots, A_m$ and ω in terms of A_1. The accuracy of the resulting periodic solution depends on the value of A_1 and the number of harmonics in the assumed solution (2.3.54).

For example, substituting the one-term expansion

$$x = A_1 \cos(\omega t + \beta_0) = A_1 \cos\phi \tag{2.3.55}$$

into (2.3.4) yields

$$-(\omega^2 - \alpha_1)A_1 \cos\phi + \tfrac{1}{2}\alpha_2 A_1^2 [1 + \cos 2\phi] + \tfrac{1}{4}\alpha_3 A_1^3 [3\cos\phi + \cos 3\phi] = 0 \tag{2.3.56}$$

if $N = 3$. Equating the coefficient of $\cos\phi$ to zero, we obtain

$$\omega^2 = \alpha_1 + \tfrac{3}{4}\alpha_3 A_1^2 \tag{2.3.57}$$

which for small A_1 becomes

$$\omega = \sqrt{\alpha_1} [1 + \tfrac{3}{8}\alpha_3 \alpha_1^{-1} A_1^2] \tag{2.3.58}$$

Comparing (2.3.58) with (2.3.34), we conclude that only part of the nonlinear correction to the frequency has been obtained. The reason for the deficiency is that terms $O(A_1^2)$ were neglected in (2.3.56), while terms $O(A_1^3)$ were kept. To obtain the rest of the nonlinear correction, we need to include other terms besides the first harmonic in the expression for x.

Following Mahaffey (1976) and putting

$$x = A_1 \cos\phi + A_0 \tag{2.3.59}$$

in (2.3.4) with $N = 3$, we obtain

$$\alpha_1 A_0 + \alpha_2 A_0^2 + \tfrac{1}{2}\alpha_2 A_1^2 + \alpha_3 A_0^3 + \tfrac{3}{2}\alpha_3 A_0 A_1^2 + [-(\omega^2 - \alpha_1)A_1 + 2\alpha_2 A_0 A_1 + 3\alpha_3 A_0^2 A_1 + \tfrac{3}{4}\alpha_3 A_1^3] \cos\phi + [\tfrac{1}{2}\alpha_2 A_1^2 + \tfrac{3}{2}\alpha_3 A_0 A_1^2] \cos 2\phi + \tfrac{1}{4}\alpha_3 A_1^3 \cos 3\phi = 0 \tag{2.3.60}$$

Equating the constant term and the coefficient of $\cos\phi$ to zero, we have

$$\begin{aligned} \alpha_1 A_0 + \alpha_2 A_0^2 + \tfrac{1}{2}\alpha_2 A_1^2 + \alpha_3 A_0^3 + \tfrac{3}{2}\alpha_3 A_0 A_1^2 &= 0 \\ -(\omega^2 - \alpha_1) + 2\alpha_2 A_0 + 3\alpha_3 A_0^2 + \tfrac{3}{4}\alpha_3 A_1^2 &= 0 \end{aligned} \tag{2.3.61}$$

When A_1 is small, the solutions of (2.3.61) are

$$\begin{aligned} A_0 &= -\tfrac{1}{2}\alpha_2 \alpha_1^{-1} A_1^2 + O(A_1^4) \\ \omega^2 &= \alpha_1 + (\tfrac{3}{4}\alpha_3 - \alpha_2^2 \alpha_1^{-1}) A_1^2 \end{aligned} \tag{2.3.62}$$

Hence

$$\omega = \sqrt{\alpha_1}\left[1 + \frac{3\alpha_3\alpha_1 - 4\alpha_2^2}{8\alpha_1^2} A_1^2\right] \tag{2.3.63}$$

Again, comparing (2.3.63) with (2.3.34), we conclude that the assumption (2.3.59) also produced a solution that does not account for all the nonlinear correction to the frequency to $O(A_1^2)$. Inspecting (2.3.60), we find that we still neglected terms $O(A_1^2)$ while we kept terms $O(A_1^3)$.

Next let us try to include three terms in the solution, that is,

$$x = A_0 + A_1 \cos\phi + A_2 \cos 2\phi \tag{2.3.64}$$

where A_0 and $A_2 << A_1$. Substituting (2.3.64) into (2.3.4) with $N = 3$ yields

$$\begin{aligned} &[(-\omega^2 + \alpha_1)A_1 + \alpha_2 A_1(2A_0 + A_2) + \alpha_3(\tfrac{3}{4}A_1^3 + 3A_0^2 A_1 + 3A_0 A_1 A_2 \\ &\quad + \tfrac{3}{2}A_1 A_2^2)] \cos\phi + [\alpha_1 A_0 + \alpha_2(A_0^2 + \tfrac{1}{2}A_1^2 + \tfrac{1}{2}A_2^2) + \alpha_3(A_0^3 \\ &\quad + \tfrac{3}{2}A_0 A_1^2 + \tfrac{3}{2}A_0 A_2^2 + \tfrac{3}{4}A_1^2 A_2)] + [(-4\omega^2 + \alpha_1)A_2 + \alpha_2(\tfrac{1}{2}A_1^2 + 2A_0 A_2) \\ &\quad + \tfrac{3}{4}\alpha_3(A_2^3 + 4A_0^2 A_2 + 2A_1^2 A_2 + 2A_0 A_1^2)] \cos 2\phi + \text{higher harmonics} = 0 \end{aligned} \tag{2.3.65}$$

Equating the constant term and the coefficients of $\cos\phi$ and $\cos 2\phi$ to zero, we obtain

$$-\omega^2 + \alpha_1 + 2\alpha_2 A_0 + \alpha_2 A_2 + \tfrac{3}{4}\alpha_3 A_1^2 + 3\alpha_3 A_0 A_2 + 3\alpha_3 A_0^2 + \tfrac{3}{2}\alpha_3 A_2^2 = 0 \tag{2.3.66}$$

$$\alpha_1 A_0 + \alpha_2(A_0^2 + \tfrac{1}{2}A_1^2 + \tfrac{1}{2}A_2^2) + \tfrac{3}{2}\alpha_3 A_1^2(A_0 + \tfrac{1}{2}A_2) + \alpha_3(A_0^3 + \tfrac{3}{2}A_0 A_2^2) = 0 \tag{2.3.67}$$

$$-(4\omega^2 - \alpha_1)A_2 + \tfrac{1}{2}\alpha_2 A_1^2 + 2\alpha_2 A_0 A_2 + \tfrac{3}{2}\alpha_3 A_1^2(A_0 + A_2) + 3\alpha_3 A_0^2 A_2 + \tfrac{3}{4}\alpha_3 A_2^3 = 0 \tag{2.3.68}$$

For small A_1, (2.3.66) through (2.3.68) show that $A_0 = O(A_1^2)$ and $A_2 = O(A_1^2)$, and hence

$$A_0 = -\tfrac{1}{2}\alpha_2\alpha_1^{-1}A_1^2 + O(A_1^4) \tag{2.3.69}$$

$$A_2 = \tfrac{1}{6}\alpha_2\alpha_1^{-1}A_1^2 + O(A_1^4) \tag{2.3.70}$$

$$\omega^2 = \alpha_1 + \tfrac{3}{4}\alpha_3 A_1^2 - \tfrac{5}{6}\alpha_2^2\alpha_1^{-1}A_1^2 + O(A_1^4) \tag{2.3.71}$$

Substituting for A_0 and A_2 from (2.3.69) and (2.3.70) into (2.3.64) yields

$$x = A_1\cos\phi - \frac{A_1^2\alpha_2}{2\alpha_1}\left[1 - \tfrac{1}{3}\cos 2\phi\right] + \cdots \tag{2.3.72}$$

Moreover it follows from (2.3.71) that

$$\omega = \sqrt{\alpha_1}\left[1 + \frac{9\alpha_3\alpha_1 - 10\alpha_2^2}{24\alpha_1^2}A_1^2\right] + \cdots \tag{2.3.73}$$

Comparing (2.3.72) and (2.3.73) with (2.3.33) and (2.3.34), we find that they are in full agreement if A_1 is identified with ϵa. Inspecting the coefficients of the higher harmonics in (2.3.65), one finds that they are $O(A_1^4)$, and hence the neglected terms are the order of the error in (2.3.72) and (2.3.73), which is the reason why it is in agreement with the solutions obtained by the Lindstedt-Poincaré procedure and the method of multiple scales.

It is clear from the development above that, to obtain a consistent solution by using the method of harmonic balance, one needs either to know a great deal about the solution *a priori* or to carry enough terms in the solution and check the order of the coefficients of all the neglected harmonics. Otherwise one might obtain an inaccurate approximation (Mahaffey, 1976) such as (2.3.58) or (2.3.63). Therefore we prefer not to use this technique.

Borges, Cesari, and Sanchez (1974) studied the relationship between functional analysis and the method of harmonic balance.

2.3.5. METHODS OF AVERAGING

Foremost among the remaining methods are those based on averaging. These include the Krylov-Bogoliubov method, the Krylov-Bogoliubov-Mitropolsky technique, the generalized method of averaging, averaging using canonical variables, averaging using Lie series and transforms, and averaging using Lagrangians. For a comprehensive treatment, see Chapter 5 of Nayfeh (1973*b*).

Most of the solutions based on averaging start with the method of variation of parameters to transform the dependent variable from x to a and β where

$$\begin{aligned} x &= a(t) \cos [\omega_0 t + \beta(t)] \\ \dot{x} &= -\omega_0 a(t) \cos [\omega_0 t + \beta(t)] \end{aligned} \tag{2.3.74}$$

and $\omega_0 = \sqrt{\alpha_1}$. Then it follows from (2.3.4) that the equations governing a and β are

$$\begin{aligned} \dot{a} &= \omega_0^{-1} \sin \phi \, [\alpha_2 a^2 \cos \phi + \alpha_3 a^3 \cos^3 \phi] \\ \dot{\beta} &= \omega_0^{-1} \cos \phi \, [\alpha_2 a \cos^2 \phi + \alpha_3 a^2 \cos^3 \phi] \end{aligned} \tag{2.3.75}$$

where $\phi = \omega_0 t + \beta(t)$.

Using the Krylov-Bogoliubov first approximation, one averages the right-hand sides of (2.3.75) over ϕ from 0 to 2π, assuming a and β to be constants. The result is

$$\dot{a} = 0, \qquad \dot{\beta} = \tfrac{3}{8} \omega_0^{-1} \alpha_3 a^2 \tag{2.3.76}$$

which when combined with (2.3.74) yields

$$x = a \cos (\omega t + \beta_0) \tag{2.3.77}$$

where β_0 is a constant and

$$\omega = \sqrt{\alpha_1} \, [1 + \tfrac{3}{8} \alpha_3 \alpha_1^{-1} a^2] \tag{2.3.78}$$

Comparison of (2.3.78) with (2.3.34) shows that the first approximation does not account for all the nonlinear correction to the frequency. Hence one must use a technique that is valid to second order rather than first order.

To obtain a consistent approximate solution to (2.3.75), one needs to employ the generalized method of averaging. The resulting solution will be in agreement with those obtained in Sections 2.3.2 and 2.3.3 by using the Lindstedt-Poincaré technique and the method of multiple scales.

It should be noted that using the methods of averaging correctly leads to valid results. On the other hand, averaging in an ad hoc manner may lead to an incorrect answer. For example, Mahaffey (1976) wrote (2.3.4) when $N = 3$ in the form

$$\ddot{x} + (\alpha_1 + \alpha_2 x + \alpha_3 x^2) x = 0 \tag{2.3.79}$$

Then he interpreted the quantity inside the parentheses as the square of the nonlinear frequency, that is

$$\omega = \sqrt{\alpha_1}\left(1 + \frac{\alpha_2}{\alpha_1}x + \frac{\alpha_3}{\alpha_1}x^2\right)^{1/2} \tag{2.3.80}$$

Hence

$$\omega = \sqrt{\alpha_1}\left[1 + \frac{\alpha_2}{2\alpha_1}x + \left(\frac{\alpha_3}{2\alpha_1} - \frac{\alpha_2^2}{8\alpha_1^2}\right)x^2\right] + \cdots \tag{2.3.81}$$

Assuming $x = a\cos\phi$ in (2.3.81) and averaging over ϕ from 0 to 2π, he obtained

$$\omega = \sqrt{\alpha_1}\left[1 + \left(\frac{4\alpha_3\alpha_1 - \alpha_2^2}{16\alpha_1^2}\right)a^2\right] \tag{2.3.82}$$

Comparison of (2.3.82) with (2.3.34) shows that the nonlinear correction to the frequency is totally incorrect. Therefore one must be careful in using an ad hoc technique, for one may obtain a solution that may be believable but nevertheless totally incorrect. In this book we only use consistent methods.

2.4. Applications

In this section we apply the general results obtained in the preceding section to some specific examples.

2.4.1. THE MOTION OF A SIMPLE PENDULUM

The equation describing the motion of a simple pendulum was derived in Section 2.1.1 as

$$\ddot{\theta} + \omega_0^2 \sin\theta = 0 \tag{2.4.1}$$

where $\omega_0^2 = g/l$. A first integral is

$$\dot{\theta}^2 = 2[h - F(\theta)] \tag{2.4.2}$$

where $F(\theta) = -\omega_0^2\cos\theta$ and h, the energy level, depends on the initial conditions. Let us take

$$2h = \dot{\theta}_0^2 - 2\omega_0^2\cos\theta_0 \tag{2.4.3}$$

Since $F(\theta)$ has the minima $-\omega_0^2$ at even multiples of π, the level curve $h = -\omega_0^2$ consists of an infinite number of discrete centers located along the θ-axis at even multiples of π. The centers correspond to the stable equilibrium position of the pendulum. Moreover since $F(\theta)$ has the maxima ω_0^2 at odd multiples of π, the level curve $h = \omega_0^2$ consists of the two separatrices shown in Figure 2-10 that meet at an infinite number of saddle points located along the θ-axis at odd multiples of π. The saddle points correspond to the unstable equilibrium position (inverted pendulum). It follows from (2.4.2) that the equation describing the separatrices is

$$\dot{\theta}^2 = 4\omega_0^2 \cos^2 \frac{\theta}{2}$$

or

$$\dot{\theta} = \pm 2\omega_0 \cos \frac{\theta}{2} \tag{2.4.4}$$

When $-\omega_0^2 < h < \omega_0^2$, the level curves consist of an infinite number of closed trajectories each of which surrounds one of the centers; they correspond to periodic motions about an equilibrium position of the pendulum. When $h > \omega_0^2$, a level curve consists of two wavy trajectories outside the separatrices which correspond to rotating or spinning motions of the pendulum.

Rearranging (2.4.2), we can obtain

$$t = \pm \int_{\theta_0}^{\theta} [\dot{\theta}_0^2 + 2\omega_0^2(\cos\theta - \cos\theta_0)]^{-1/2}\, d\theta \tag{2.4.5}$$

For convenience, let us regard the motion as one started in the vertical position ($\theta_0 = 0$) with the angular velocity $\dot{\theta}_0$. Then one can rewrite (2.4.5) as

$$t = \pm \frac{1}{|\dot{\theta}_0|} \int_0^{\theta} \frac{d\theta}{(1 - \kappa^2 \sin^2 \frac{1}{2}\theta)^{1/2}} \tag{2.4.6}$$

where $\kappa = 2\omega_0/|\dot{\theta}_0|$.

The character of the motion varies according to the value of κ. If $\kappa < 1$ (i.e., $|\dot{\theta}_0| > 2\omega_0$), the integrand is always real and the value of θ increases indefinitely. In this case $h > \omega_0^2$ according to (2.4.3), the motion is unbounded, and the pendulum undergoes a spinning, rather than an oscillatory, motion. The separatrices in Figure 2-10 are between the trajectories representing this motion and the θ-axis. If $\kappa = 1$, the integrand is real and approaches ∞ as θ approaches π. Thus the motion carries the pendulum from straight down to straight up. However θ approaches π asymptotically as t becomes infinite (see Section 2.2). In this case $h = \omega_0^2$ according to (2.4.3), and the trajectories representing the motion are the separatrices. If $\kappa > 1$, the integrand is only real if

$$|\theta| \leqslant \theta_m = 2 \sin^{-1} \frac{|\dot{\theta}_0|}{2\omega_0} \tag{2.4.7}$$

Thus the pendulum oscillates between $\pm\theta_m$. In this case $-\omega_0^2 < h < \omega_0^2$, and the closed trajectories represent this motion.

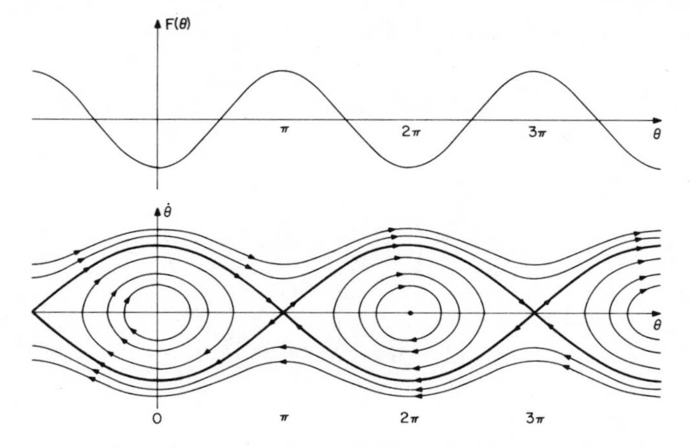

Figure 2-10. Phase plane for a simple pendulum.

The value $\kappa = 1$ is often called a *bifurcation value* because it separates values of κ for which the trajectories vary qualitatively (from open to closed).

In the case of oscillatory motion, the integral (2.4.6) from 0 to θ_m, where we must use the positive sign, yields one fourth the period. Thus the period T is

$$T = \frac{4}{|\dot{\theta}_0|} \int_0^{\theta_m} \frac{d\theta}{(1 - \kappa^2 \sin^2 \frac{1}{2}\theta)^{1/2}} \tag{2.4.8}$$

This expression can be put into a more convenient form by letting

$$\kappa \sin \tfrac{1}{2}\theta = \sin \phi$$

It follows that

$$\phi = \tfrac{1}{2}\pi \quad \text{when} \quad \theta = \theta_m \quad \text{and} \quad d\theta = \frac{2 \cos \phi \, d\phi}{\kappa(1 - k^2 \sin^2 \phi)^{1/2}} \tag{2.4.9}$$

where $k = |\dot{\theta}_0|/2\omega_0 = \sin(\frac{1}{2}\theta_m)$. Then the period becomes

$$T = \frac{4}{\omega_0} \int_0^{\frac{1}{2}\pi} \frac{d\phi}{(1 - k^2 \sin^2 \phi)^{1/2}} \tag{2.4.10}$$

This exact expression for the period is in terms of the elliptic function of the first kind.

In this case one can readily obtain an approximate value for the period by expanding the integrand and integrating term by term, that is,

$$T = \frac{4}{\omega_0} \int_0^{\frac{1}{2}\pi} [1 + \tfrac{1}{2} k^2 \sin^2 \phi + \tfrac{3}{8} k^4 \sin^4 \phi + \cdots] \, d\phi$$

Hence

$$T_a = \frac{2\pi}{\omega_0} \left(1 + \frac{1}{4} k^2 + \frac{9}{64} k^4 + \cdots\right)$$

But

$$k = \sin (\tfrac{1}{2} \theta_m) = \tfrac{1}{2} \theta_m - \tfrac{1}{48} \theta_m^3 + \cdots$$

Hence

$$T_a = \frac{2\pi}{\omega_0} \left(1 + \frac{1}{16} \theta_m^2 + \frac{11}{3072} \theta_m^4 + \cdots\right) \tag{2.4.11}$$

An approximation to the periodic orbit surrounding the origin can also be obtained by using the Lindstedt-Poincaré technique. To this end we expand $\sin \theta$ in (2.4.1) about $\theta = 0$ and obtain

$$\ddot{\theta} + \omega_0^2 (\theta - \tfrac{1}{6} \theta^3 + \cdots) = 0 \tag{2.4.12}$$

We let $\tau = \omega t$ and expand θ and ω as

$$\begin{aligned} \theta &= \epsilon \theta_1(\tau) + \epsilon^3 \theta_3(\tau) + \cdots \\ \omega &= \omega_0 + \epsilon^2 \omega_2 + \cdots \end{aligned} \tag{2.4.13}$$

where ϵ is a small dimensionless parameter characterizing the amplitude of the motion. The term $\epsilon^2 \theta_2$ is missing from (2.4.13) because the nonlinearity appears at $O(\epsilon^3)$. The term $\epsilon \omega_1$ is missing because the frequency is independent of the sign of ϵ (the amplitude of the motion). Substituting (2.4.13) into (2.4.12) and equating coefficients of like powers of ϵ, we obtain

$$\omega_0^2 (\theta_1'' + \theta_1) = 0 \tag{2.4.14}$$

$$\omega_0^2 (\theta_3'' + \theta_3) + 2\omega_0 \omega_2 \theta_1'' - \tfrac{1}{6} \omega_0^2 \theta_1^3 = 0 \tag{2.4.15}$$

The solution of (2.4.14) is

$$\theta_1 = a \cos (\tau + \beta) \tag{2.4.16}$$

where a and β are constants. Hence (2.4.15) becomes

$$\omega_0^2 (\theta_3'' + \theta_3) = (2\omega_0 \omega_2 a + \tfrac{1}{8} \omega_0^2 a^3) \cos (\tau + \beta) + \tfrac{1}{24} a^3 \omega_0^2 \cos (3\tau + 3\beta) \tag{2.4.17}$$

Eliminating the term that produces secular terms in (2.4.17) gives $\omega_2 = -\frac{1}{16}\omega_0 a^2$. Hence a first approximation to θ is

$$\theta = \epsilon a \cos\left[\omega_0(1 - \tfrac{1}{16}\epsilon^2 a^2)t + \beta\right] + O(\epsilon^3) \tag{2.4.18}$$

To compare the exact and approximate solutions, we set $\epsilon a = \theta_m$ so that from (2.4.18) a first approximation to the period T_a is given by

$$\omega_0 T_a = 2\pi(1 - \tfrac{1}{16}\theta_m^2)^{-1} \approx 2\pi(1 + \tfrac{1}{16}\theta_m^2) \tag{2.4.19}$$

in agreement with (2.4.11). Thus we have seen that the perturbation solution and the exact solution yield the same results for small values of θ_m. Table 2-1 shows that T_a/T gets closer and closer to unity as $\theta_m \to 0$.

TABLE 2-1. The Ratio of the Approximate Period to the Exact Period for Various Amplitudes of the Motion of a Simple Pendulum

θ_m	0	10°	20°	30°	40°	50°	60°	70°	80°	90°
T_a/T	1.	1.0000	1.0000	0.9997	0.9992	0.9979	0.9956	0.9920	0.9862	0.9778

2.4.2. MOTION OF A CURRENT-CARRYING CONDUCTOR

As an example of a single-degree-of-freedom system in which the restoring force depends on a parameter in addition to the coordinate, we consider the motion of a current-carrying wire with mass m in the field of an infinite current-carrying conductor and restrained by linear elastic springs as shown in Figure 2-11. The differential equation describing the motion of the wire is

$$m\frac{d^2\tilde{x}^2}{d\tilde{t}^2} + k\tilde{x} - \frac{2i_1 i_2 l}{b - \tilde{x}} = 0 \tag{2.4.20}$$

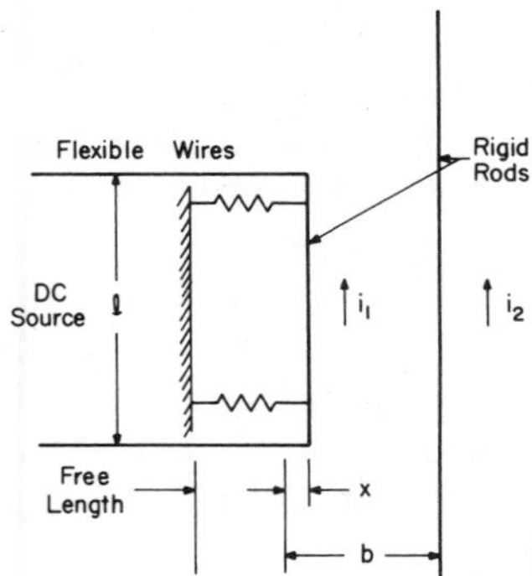

Figure 2-11. Current-carrying wire in the field of an infinite current-carrying conductor.

where $k\tilde{x}$ is the restoring force due to the springs and $2i_1 i_2 l/(b - \tilde{x})$ is the attraction force between the conductors due to the magnetic fields produced by the currents. Equation (2.4.20) can be rewritten as

$$\frac{d^2 x}{dt^2} + x - \frac{\Lambda}{1 - x} = 0 \tag{2.4.21}$$

where $x = \tilde{x}/b$, $t = \omega_0 \tilde{t}$, $\omega_0^2 = k/m$, and $\Lambda = 2i_1 i_2 l/kb^2$. A first integral of (2.4.21) is

$$\tfrac{1}{2}\dot{x}^2 + \tfrac{1}{2}x^2 + \Lambda \ln|1 - x| = h \tag{2.4.22}$$

We let

$$v = \dot{x} \text{ and } F(x) = \tfrac{1}{2}x^2 + \Lambda \ln|1 - x|$$

The function $F(x)$ has singular points at x_i, where

$$F'(x_i) = x_i - \Lambda(1 - x_i)^{-1} = 0$$

or

$$x_1, x_2 = \tfrac{1}{2} \pm (\tfrac{1}{4} - \Lambda)^{1/2} \tag{2.4.23}$$

The nature of the singular points can be determined by examining $F''(x_i)$. Since

$$F''(x_i) = 1 - \Lambda(1 - x_i)^{-2} = 1 - \Lambda^{-1}x_i^2$$

then

$$\begin{aligned} F''(x_2) &= 2\Lambda^{-1}(\tfrac{1}{4} - \Lambda)^{1/2}[\tfrac{1}{2} - (\tfrac{1}{4} - \Lambda)^{1/2}] \\ F''(x_1) &= -2\Lambda^{-1}(\tfrac{1}{4} - \Lambda)^{1/2}[\tfrac{1}{2} + (\tfrac{1}{4} - \Lambda)^{1/2}] \end{aligned} \tag{2.4.24}$$

It follows from (2.4.23) and (2.4.24) that there are five cases to be considered: $\Lambda < 0$, $\Lambda = 0$, $0 < \Lambda < \frac{1}{4}$, $\Lambda = \frac{1}{4}$, and $\Lambda > \frac{1}{4}$. These are taken up individually.

When $\Lambda < 0$, $F(x)$ has minima at x_1 and at x_2 (here we take $x_1 > x_2$); hence both points are centers. Because $x_1 > 1$, there can be motion in the neighborhood of x_1 only if the bracket for the spring-supported conductor extends around the fixed conductor. Because $|v| \to \infty$ as $x \to 1$, the moving conductor remains on the same side of the fixed conductor (this is true in all the cases considered). The possible motions are represented in Figure 2-12. The energy levels are labeled h_n and the corresponding trajectories T_n. When $h < h_1$, no motion is possible. When $h_1 < h < h_3$, a level curve consists of a single branch, which encircles the center at x_2. For $h > h_3$ a level curve consists of two branches, one encircling the center at x_2 and the other encircling the center at x_1.

When $\Lambda = 0$, the electromagnetic force vanishes, and there is only one singu-

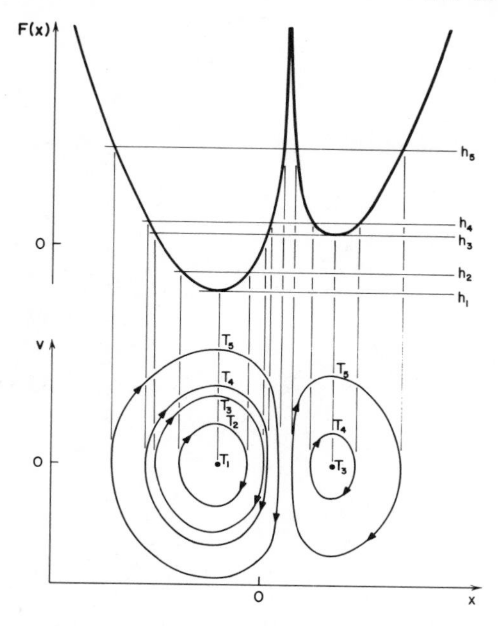

Figure 2-12. Phase plane for a current-carrying wire in the field of an infinite current-carrying conductor when $\Lambda < 0$.

lar point, a minimum at $x = 0$. Thus $x = 0$ is a center. The possible motions in the neighborhood of this point are represented in Figure 2-13*a*.

When $0 < \Lambda < \frac{1}{4}$, $F(x)$ has a maximum at x_1 and a minimum at x_2; hence x_1 is a saddle point and x_2 is a center. Both points lie on the left-hand side of the fixed conductor. The possible motions on the left-hand side of the fixed conductor are represented in Figure 2-13*b*. When $h_2 < h < h_4$, a level curve consists of two branches, a closed branch encircling the center and an open branch on the right-hand side of the saddle point. Both branches have the same label in Figure 2-13*b*. When $h > h_4$, the level curves are open and pass on the left-hand side of both singular points. When $h = h_4$, the level curve consists of a separatrix whose equation is

$$\frac{1}{2}v^2 + \frac{1}{2}x^2 + \Lambda \ln|1 - x| = h_1 = \frac{1}{2}x_1^2 + \Lambda \ln|1 - x_1| \qquad (2.4.25)$$

When $h < h_2$, the level curves are open but pass on the right-hand side of both singular points. When $h = h_2$, the level curve consists of the center x_2 and a branch that is similar in shape to a hyperbola that opens to the right. Thus the motion can be bounded only if $h_2 \leqslant h \leqslant h_4$, and then only if the initial condi-

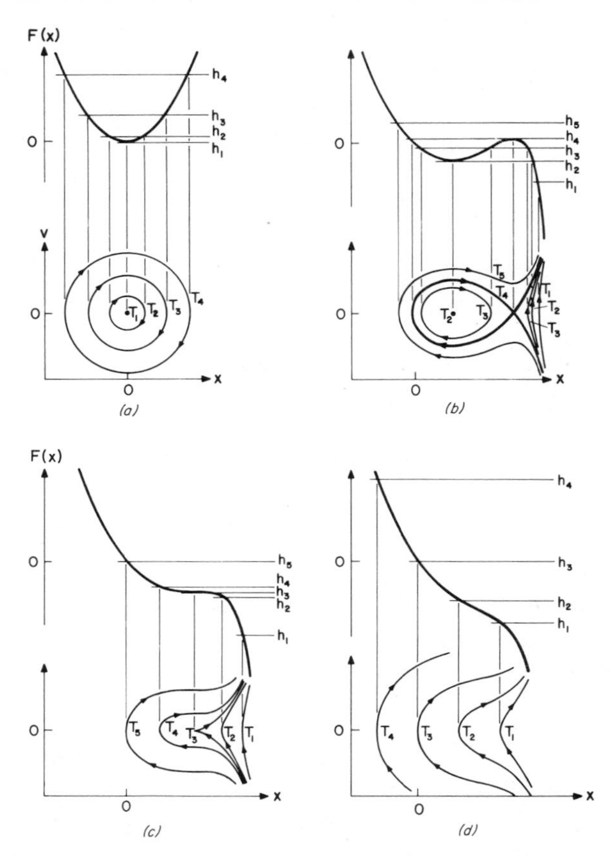

Figure 2-13. Phase plane for a current-carrying conductor in the field of another: (*a*) $\Lambda = 0$; (*b*) $0 < \Lambda < \frac{1}{4}$; (*c*) $\Lambda = \frac{1}{4}$; (*d*) $\Lambda > \frac{1}{4}$.

tions place the representative point on the closed branch of the level curve near the center.

As Λ increases, the center and the saddle point approach each other. They coalesce at $x = \frac{1}{2}$ when $\Lambda = \frac{1}{4}$. Thus at $x = \frac{1}{2}$, $F(x)$ has an inflection point. The possible motions are represented in Figure 2-13*c*. We note that there is a cusp on one of the trajectories and that bounded motion does not exist.

Finally when $\Lambda > \frac{1}{4}$, there are no singular points and hence there is no bounded motion. This is represented in Figure 2-13*d*.

The scale in the four parts of Figure 2-13 is uniform, but not the same as the scale in Figure 2-12. Figure 2-12 covers a much larger area in the phase plane than Figure 2-13. We note that $\Lambda = 0$ and $\frac{1}{4}$ are bifurcation values because they separate values of Λ for which the phase plane varies qualitatively.

If $x = x_0$ and $\dot{x} = 0$ at $t = 0$, then $h = \frac{1}{2}x_0^2 + \Lambda \ln |1 - x_0|$ according to (2.4.22). Moreover an exact solution of (2.4.22) can be obtained by separating variables. The result is

$$t = \int_{x_0}^{x} \left(x_0^2 - x^2 + 2\Lambda \ln \frac{1 - x_0}{1 - x}\right)^{-1/2} dx \tag{2.4.26}$$

For one of the closed trajectories shown in Figure 2-13, the period T is

$$T = 2\int_{x_{01}}^{x_{02}} \left(x_{01}^2 - x^2 + 2\Lambda \ln \frac{1 - x_{01}}{1 - x}\right)^{-1/2} dx \tag{2.4.27}$$

where x_{01} and x_{02} are the abscissas of the points of intersection of the closed trajectory with the x-axis.

To obtain an approximate solution for x, we let $x = x_2 + u$ in (2.4.21) and expand the resulting equation in a Taylor series about $u = 0$. The result is

$$\ddot{u} + \alpha_1 u + \alpha_2 u^2 + \alpha_3 u^3 + \cdots = 0 \tag{2.4.28}$$

where

$$\alpha_1 = 1 - \Lambda(1 - x_2)^{-2}, \qquad \alpha_2 = -\Lambda(1 - x_2)^{-3}, \qquad \alpha_3 = -\Lambda(1 - x_2)^{-4} \tag{2.4.29}$$

Since (2.4.28) is the same as (2.3.4), then according to (2.3.33) and (2.3.34) it has the following approximate solution:

$$x = x_2 + \epsilon a \cos(\omega t + \beta) - \frac{\epsilon^2 a^2 \alpha_2}{2\alpha_1}\left[1 - \tfrac{1}{3}\cos(2\omega t + 2\beta)\right] + O(\epsilon^3) \tag{2.4.30}$$

where

$$\omega = \sqrt{\alpha_1}\left[1 + \frac{9\alpha_3\alpha_1 - 10\alpha_2^2}{24\alpha_1^2}\epsilon^2 a^2\right] + O(\epsilon^3) \tag{2.4.31}$$

and a and β are constants of integration.

To compare the approximate period $T_a = 2\pi/\omega$ with the exact period T given by (2.4.27), we set $\beta = 0$ so that $x_{01} = x_2 + \epsilon a - \frac{1}{3}\alpha_2\alpha_1^{-1}\epsilon^2 a^2 + O(\epsilon^3)$. Table 2-2 shows the variation of T_a/T with ϵa for $\Lambda = \frac{1}{8}$. We note that $T_a/T \to 1$ as $\epsilon a \to 0$ as expected.

TABLE 2-2. The Ratio of the Approximate Period to the Exact Period for Various Amplitudes of the Motion of a Current-Carrying Conductor

ϵa	0.2	0.3	0.4	0.5	0.6
T_a/T	1.000	0.998	0.993	0.977	0.924

2.4.3. MOTION OF A PARTICLE ON A ROTATING PARABOLA

As an example of a single-degree-of-freedom conservative system that is described by an equation different from (2.2.1), we consider the motion of a ring of mass m sliding freely on the wire described by the parabola $z = px^2$ which rotates with a constant angular velocity Ω about the z-axis as shown in Figure 2-14.

It is convenient to write the equation of motion of the ring by using an Euler-Lagrange formulation. For a conservative, holonomic (constraints are integrable) system, we express the kinetic and potential energies T and V in terms of what are usually called generalized coordinates $\mathbf{q}$, where $\mathbf{q}$ is a vector whose elements are the independent coordinates needed to describe the system under consideration. Then we form the Lagrangian L as

$$L(\mathbf{q}, \dot{\mathbf{q}}, t) = T(\mathbf{q}, \dot{\mathbf{q}}, t) - V(\mathbf{q}, \dot{\mathbf{q}}, t) \tag{2.4.32}$$

Applying Hamilton's principle leads to the following Euler-Lagrange equations:

$$\frac{d}{dt}\left(\frac{\partial L}{\partial \dot{\mathbf{q}}}\right) - \frac{\partial L}{\partial \mathbf{q}} = 0 \tag{2.4.33}$$

For the present problem,

$$T = \tfrac{1}{2} m(\dot{x}^2 + \Omega^2 x^2 + \dot{z}^2), \qquad V = mgz \tag{2.4.34}$$

Using the constraint $z = px^2$, we rewrite these energies as

$$T = \tfrac{1}{2} m\,[(1 + 4p^2x^2)\dot{x}^2 + \Omega^2 x^2], \qquad V = mgpx^2 \tag{2.4.35}$$

Since the kinetic energy is not a quadratic function of the velocity, the system is usually called a *nonnatural system* (e.g., Meirovitch, 1970, p. 77). Substituting for T and V into (2.4.32) yields

$$L = \tfrac{1}{2} m\,[(1 + 4p^2x^2)\dot{x}^2 + \Omega^2 x^2] - mgpx^2 \tag{2.4.36}$$

Substituting for L into (2.4.33) and letting $q = x$, we find that the equation describing the motion of the ring is

$$(1 + 4p^2x^2)\ddot{x} + \Lambda x + 4p^2\dot{x}^2 x = 0 \tag{2.4.37}$$

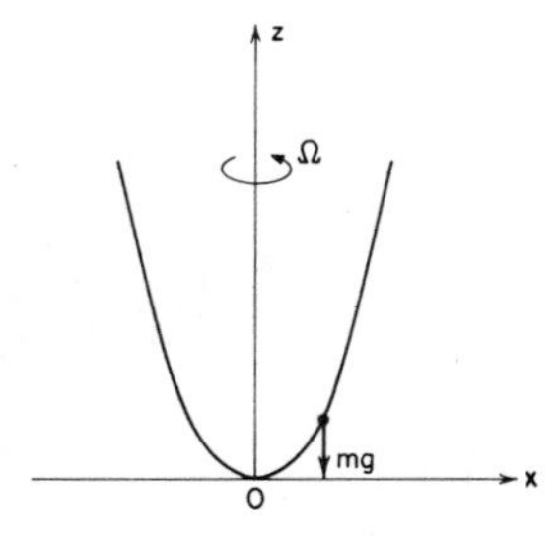

Figure 2-14. Particle on a rotating parabola.

where

$$\Lambda = 2gp - \Omega^2 \tag{2.4.38}$$

The equations describing the motion in the phase plane are obtained next. Let

$$\dot{x} = v, \qquad \dot{v} = -\frac{\Lambda x + 4p^2 x v^2}{1 + 4p^2 x^2} \tag{2.4.39}$$

Eliminating t from these equations, we obtain

$$\frac{dv}{dx} = -\frac{\Lambda x + 4p^2 x v^2}{v(1 + 4p^2 x^2)} \tag{2.4.40}$$

which can be rewritten as

$$\tfrac{1}{2}(1 + 4p^2 x^2)\, d(v^2) + 4p^2 x v^2\, dx + \Lambda x\, dx = 0 \tag{2.4.41}$$

Equation (2.4.41) has the integral

$$(1 + 4p^2 x^2) v^2 + \Lambda x^2 = h \tag{2.4.42}$$

where h is a constant. Equation (2.4.42) shows that $T + V$ is not a constant for this system; this is a consequence of the system being nonnatural. The integral (2.4.42) is called the *Jacobi integral.*

For a general holonomic, conservative system described by a Lagrangian that does not depend explicitly on t, the Jacobi integral can be obtained as follows:

$$\frac{dL}{dt} = \frac{\partial L}{\partial \mathbf{q}} \dot{\mathbf{q}} + \frac{\partial L}{\partial \dot{\mathbf{q}}} \ddot{\mathbf{q}} \tag{2.4.43}$$

But

$$\frac{\partial L}{\partial \mathbf{q}} = \frac{d}{dt}\left(\frac{\partial L}{\partial \dot{\mathbf{q}}}\right) \tag{2.4.44}$$

according to (2.4.33). Hence (2.4.43) can be rewritten as

$$\frac{dL}{dt} = \frac{d}{dt}\left(\frac{\partial L}{\partial \dot{\mathbf{q}}}\right) \dot{\mathbf{q}} + \frac{\partial L}{\partial \dot{\mathbf{q}}} \ddot{\mathbf{q}} = \frac{d}{dt}\left(\dot{\mathbf{q}} \frac{\partial L}{\partial \dot{\mathbf{q}}}\right) \tag{2.4.45}$$

Therefore

$$\frac{d}{dt}\left[\dot{\mathbf{q}} \frac{\partial L}{\partial \dot{\mathbf{q}}} - L\right] = 0 \tag{2.4.46}$$

or

$$\dot{\mathbf{q}} \frac{\partial L}{\partial \dot{\mathbf{q}}} - L = h \tag{2.4.47}$$

where h is a constant. Substituting for L from (2.4.36) into (2.4.47), we obtain an integral that is equivalent to (2.4.42), which we obtained by direct integration. Rearranging (2.4.42) can produce

$$v^2 = \frac{h - \Lambda x^2}{1 + 4p^2 x^2} \tag{2.4.48}$$

and it follows that

$$\frac{dv}{dx} = \pm \frac{(\Lambda + 4p^2 h)x}{(h - \Lambda x^2)^{1/2}(1 + 4p^2 x^2)^{3/2}} \tag{2.4.49}$$

Next we consider the influence of the parameter Λ on the character of the solutions. We consider three cases: $\Lambda > 0$, $\Lambda < 0$, and $\Lambda = 0$.

When $\Lambda > 0$, it follows form (2.4.48) that v^2 decreases from the value h at $x = 0$ to zero at $x^2 = h/\Lambda$. For $x^2 > h/\Lambda$, there is no real solution for v. Thus the motion, which is bounded, is represented by closed trajectories surrounding the origin, which is a center as shown in Figure 2-15*a*. We note that as Λ decreases the trajectories become more oblong as shown in Figure 2-15*b*.

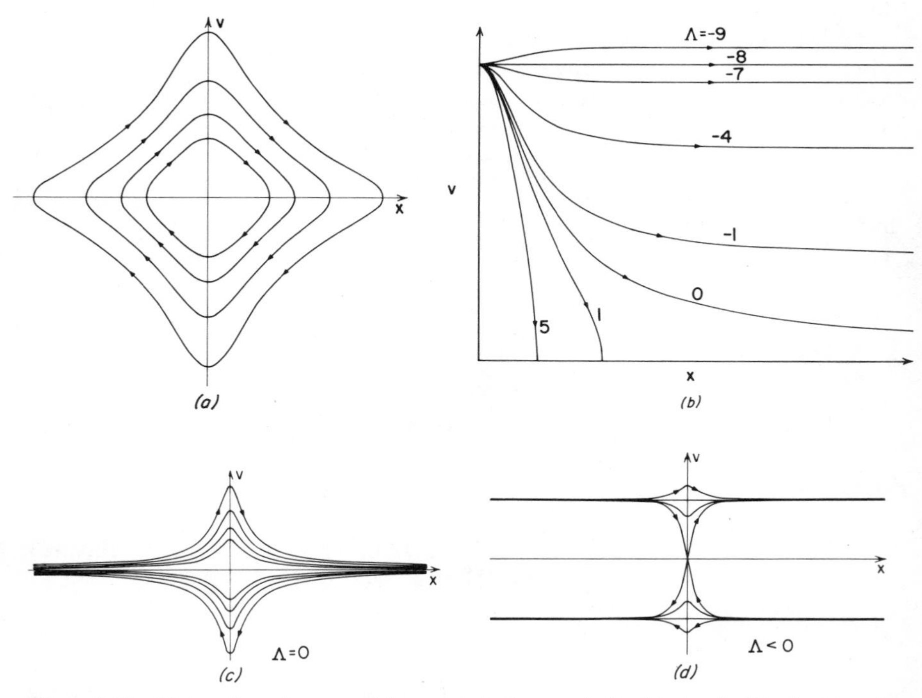

Figure 2-15. Phase plane for a particle on a rotating parabola: (*a*) $\Lambda = 1$; (*b*) $-9 \leqslant \Lambda \leqslant 5$; (*c*) $\Lambda = 0$; (*d*) $\Lambda < 0$.

When $\Lambda < 0$, v^2 approaches $|\Lambda|/4p^2$ and dv/dx approaches zero as x approaches infinity. The values of v^2 at large x are independent of the value of h, and consequently all trajectories approach the same asymptote. The motion, which is unbounded, is represented by open trajectories, and the origin is a saddle point as shown in Figure 2-15*d*. The velocity approaches the same asymptotes (which depend on the value of p and Λ) regardless of the value of h.

As Λ decreases toward zero, the trajectories in Figure 2-15*a*, become more oblong; and as Λ increases toward zero, the asymptote of all the trajectories in Figure 2-15*d* approaches zero. The limiting case when $\Lambda = 0$ marks the boundary between the bounded and the unbounded motions; for this reason, $\Lambda = 0$ is called the bifurcation value. In Figure 2-15*c* the trajectories are shown for the bifurcation value of Λ (zero) and various values of h. In this case the velocity approaches zero regardless of the value of h.

The discussion above shows that periodic motion exists when $\Lambda > 0$. We consider this periodic motion next. Proceeding as before, we manipulate (2.4.48) to obtain

$$t = \pm \int_0^x \left(\frac{1 + 4p^2 x^2}{h - \Lambda x^2} \right)^{1/2} dx \tag{2.4.50}$$

where we assumed that $t = 0$ when $x = 0$. Changing the variable according to $x = (h/\Lambda)^{1/2} \cos\theta$ leads to

$$t = \Lambda^{-1/2} \int_\theta^{\frac{1}{2}\pi} \left(1 + \frac{4hp^2}{\Lambda} \cos^2\theta\right)^{1/2} d\theta$$

The time required for x to change from 0 to $(h/\Lambda)^{1/2}$ is one fourth of the period T. Hence

$$T = 4\Lambda^{-1/2} \int_0^{\frac{1}{2}\pi} \left(1 + \frac{4hp^2}{\Lambda} \cos^2\theta\right)^{1/2} d\theta \tag{2.4.51}$$

We can rewrite (2.4.51) in terms of the elliptic function of the second kind as

$$T = 4\Lambda^{-1}(\Lambda + 4hp^2)^{1/2} \int_0^{\frac{1}{2}\pi} (1 - k^2 \sin^2\theta)^{1/2}\, d\theta \tag{2.4.52}$$

where $k^2 = 4hp^2/(\Lambda + 4hp^2)$.

An approximate value for T can be obtained by first expanding the integrand in (2.4.51) and then integrating term by term. The result is

$$T = 4\Lambda^{-1/2} \int_0^{\frac{1}{2}\pi} \left[1 + \frac{2hp^2}{\Lambda} \cos^2\theta + \cdots\right] d\theta \approx T_a = 2\pi\Lambda^{-1/2} \left(1 + \frac{hp^2}{\Lambda}\right) \tag{2.4.53}$$

An approximate expression for the period can also be obtained by the Lindstedt-Poincaré method. Thus we let $\tau = \omega t$ and expand x and ω as

$$x = \epsilon x_1(\tau) + \epsilon^3 x_2(\tau) + \cdots \tag{2.4.54}$$

$$\omega = \omega_0 + \epsilon^2 \omega_2 + \cdots \tag{2.4.55}$$

In (2.4.54) the term proportional to ϵ^2 is missing because the nonlinearity is cubic rather than quadratic; and in (2.4.55) the term proportional to ϵ is missing because the frequency must be independent of the sign of ϵ, as discussed in Section 2.2.

Substituting (2.4.54) and (2.4.55) into (2.4.37) and equating coefficients of like powers of ϵ, we obtain

$$\omega_0^2 x_1'' + \Lambda x_1 = 0 \tag{2.4.56}$$

$$\omega_0^2 x_3'' + \Lambda x_3 = -2\omega_0\omega_2 x_1'' - 4p^2\omega_0^2 x_1^2 x_1'' - 4p^2\omega_0^2 x_1'^2 x_1 \tag{2.4.57}$$

We choose $\omega_0 = \sqrt{\Lambda}$ so that the solution of (2.4.56) becomes

$$x_1 = a\cos(\tau + \beta) \tag{2.4.58}$$

where a and β are constants. Hence (2.4.57) becomes

$$\omega_0^2(x_3'' + x_3) = (2\omega_0\omega_2 a + 2p^2\omega_0^2 a^3)\cos(\tau + \beta) + 2p^2\omega_0^2 a^3 \cos[3(\tau + \beta)] \tag{2.4.59}$$

To eliminate secular terms from x_3, we must put

$$\omega_2 = -p^2\omega_0 a^2 \tag{2.4.60}$$

Hence

$$x = \epsilon a \cos[\omega_0(1 - \epsilon^2 p^2 a^2)t + \beta] + O(\epsilon^3) \tag{2.4.61}$$

To compare this result with the exact result, we put the amplitude of the motion $\epsilon a = (h/\Lambda)^{1/2}$. Then it follows that the approximate expression for the period is

$$T_a = 2\pi\Lambda^{-1/2}\left(1 - \frac{hp^2}{\Lambda}\right)^{-1} \approx 2\pi\Lambda^{-1/2}\left(1 + \frac{hp^2}{\Lambda}\right) \tag{2.4.62}$$

We note that (2.4.62) agrees with (2.4.53). Table 2-3 shows that the agreement between the approximate and exact values for the period improves as hp^2/Λ decreases.

TABLE 2-3. The Ratio of the Exact Period to the Approximate Period for Various Values of hp^2/Λ for the Motion of a Particle on a Rotating Parabola

hp^2/Λ	0.	0.05	0.1	0.15	0.25	0.4
T/T_a	1.	0.995	0.994	0.988	0.973	0.946

Exercises

2.1. For each of the following systems, (1) sketch the solution trajectories in the phase plane, and (2) indicate on the sketch the singular points and their types, as well as the separatrices:

(a) $\ddot{u} + u = 0$

(b) $\ddot{u} + u - u^3 = 0$

(c) $\ddot{u} - u + u^3 = 0$

(d) $\ddot{u} + u + u^3 = 0$

(e) $\ddot{u} - u - u^3 = 0$

(f) $\ddot{u} + u^3 = 0$

(g) $\ddot{u} + u - \dfrac{\lambda}{a - u} = 0$

2.2. Determine a two-term expansion for the frequency–amplitude relationship for the systems governed by the following equations:

(a) $\ddot{u} + \omega_0^2 u(1 + u^2)^{-1} = 0 \qquad (\omega = \omega_0 - \frac{3}{8}\omega_0 a^2)$

(b) $\ddot{u} + \omega_0^2 u + \alpha u^5 = 0 \qquad (\omega = \omega_0 + \frac{5}{16}\omega_0^{-1}\alpha a^4)$

(c) $\ddot{u} - u + u^3 = 0 \qquad (\omega^2 = 2 - 3a^2)$

(d) $\ddot{u} + \omega_0^2 u + \alpha u^2 \ddot{u} = 0 \qquad (\omega = \omega_0 - \frac{3}{8}\omega_0 \alpha a^2)$

2.3. The relativistic motion of a particle having a mass m_0 at rest and attached to a linear spring with stiffness k on a smooth horizontal plane is

$$\frac{d}{dt}\left[\frac{m_0 \dot{u}}{(1 - \dot{u}^2/c^2)^{1/2}}\right] + ku = 0$$

where c is the speed of light. Determine a two-term expansion for the relationship between the frequency and the amplitude.

2.4. Determine a two-term expansion for the relationship between the frequency and the amplitude for a system governed by

$$\ddot{u} + \omega_0^2 u + u|u| = 0$$

2.5. Consider the system shown in Figure 2-16*a*.

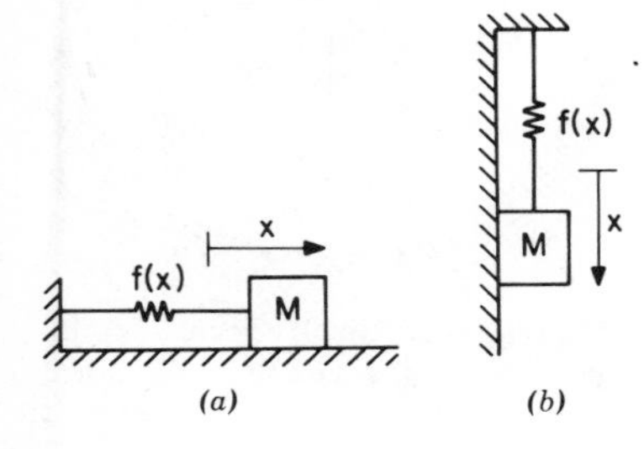

Figure 2-16. Mass restrained by a nonlinear spring: (*a*) in the absence of gravity force; (*b*) in the presence of gravity force.

(a) Determine a two-term expansion for the relationship between the frequency and the amplitude for this system where the spring force $f(x) = k_1 x + k_3 x^3$ with x being the spring deformation.

(b) Suppose that the same system is rotated 90° as shown in Figure 2-16*b*. Compare the relationship between the frequency and the amplitude for this configuration with that for part (a).

2.6. A rigid rod slides back and forth on the smooth walls as shown in Figure 2-17. Show that its motion is governed by

$$\ddot{\theta} + \frac{g(R^2 - l^2)^{1/2}}{R^2 - \frac{2}{3}l^2} \sin\theta = 0$$

What is the linear natural frequency? What effect does increasing l have on the nonlinear natural frequency? Obtain a two-term expression relating the period to the amplitude of the motion.

2.7. The small cylinder rolls without slip on the circular surface (Figure 2-18).

(a) Show that the governing equation for θ is

$$\ddot{\theta} + \frac{2g}{3(R - r)} \sin\theta = 0$$

(b) What is the minimum value of $\dot{\theta}$ at $\theta = 0$ for which the cylinder will make a complete revolution? (Note that the normal force at point A must be equal to or greater than zero.)

(c) How long does it take the cylinder to make a complete revolution in part (b)?

2.8. Reconsider the motion of a particle on a rotating parabola that was discussed in Section 2.4.3. However now assume that the wire is weightless and that its angular velocity Ω is changing with the position of the mass along the wire. There is no outside influence acting on the wire.

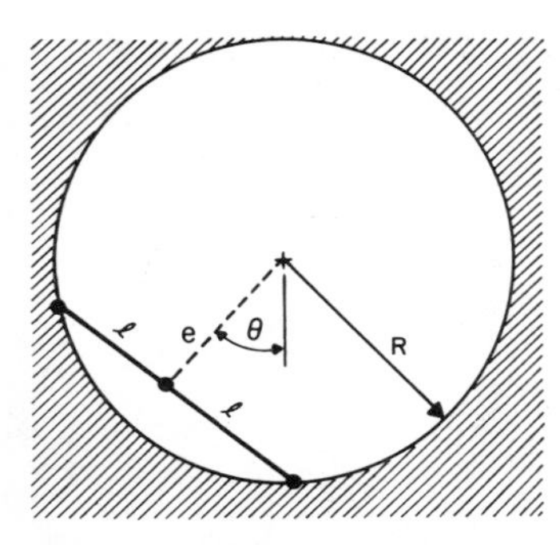

Figure 2-17. Rigid rod slides on the smooth walls of a circular cylindrical surface.

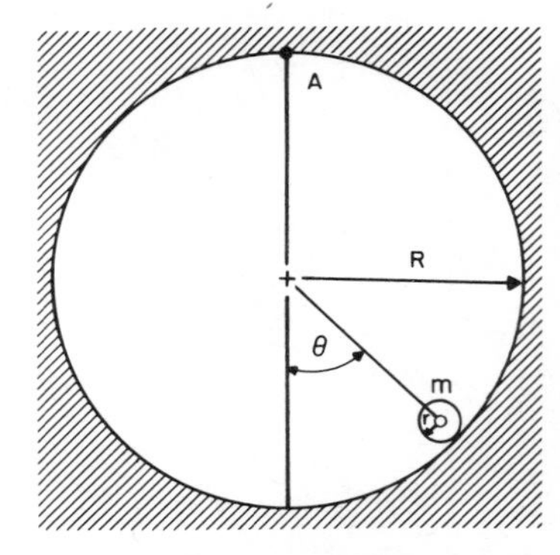

Figure 2-18. Small circular cylinder rolls on a circular cylindrical surface.

(a) Show that the equations of motion are

$$2\Omega\dot{x} + \dot{\Omega}x = 0$$

and

$$(1 + 4p^2x^2)\ddot{x} + 4p^2x\dot{x}^2 + (2pg - \Omega^2)x = 0$$

(b) Show that

$$\Omega x^2 = \sqrt{H}$$

where $\sqrt{H}$ is a constant of integration (essentially this is a statement of conservation of angular momentum) and that the governing equation for x can be written in the form

$$(1 + 4p^2x^2)\ddot{x} + 4p^2x\dot{x}^2 + \left(2pg - \frac{H}{x^4}\right)x = 0$$

(c) In a manner similar to that used in Section 2.4.3, discuss the motion of the mass along the parabola. Show that the motion is always bounded in this system, in contrast with Section 2.4.3.

(d) For $p = 1$, $g = 32.2$, $h = 1000$, and $H = 12$, plot the trajectories in the phase plane.

2.9. The rigid frame (Figure 2-19) is forced to rotate at the fixed rate Ω. While the frame rotates, the simple pendulum oscillates.

(a) Show that the equation governing θ is

$$\ddot{\theta} + (1 - \Lambda \cos \theta) \sin \theta = 0$$

where $\Lambda = (\Omega^2 r/g)$ and the new independent variable is $\tau = (g/r)^{1/2}t$. Compare this with the motion of a particle on a rotating circle considered in Section 2.1.4.

(b) Show that this equation can be integrated to yield the following:

$$\tfrac{1}{2}\dot{\theta}^2 = \tfrac{1}{2}\dot{\theta}_0^2 - F(\theta)$$

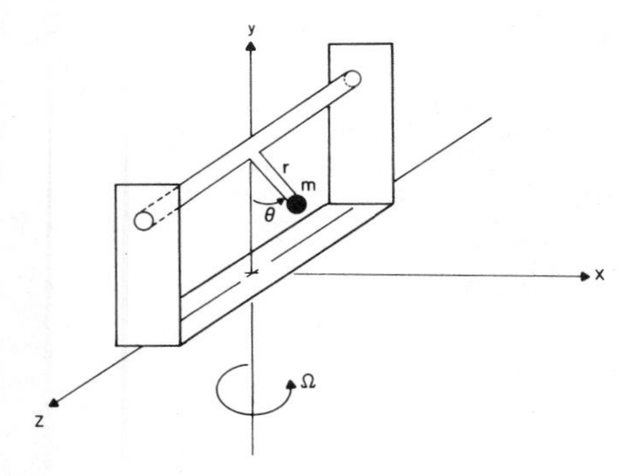

Figure 2-19. Simple pendulum attached to a rotating rigid frame.

where $\dot{\theta}_0$ is the speed at $\theta = 0$ and

$$F(\theta) = 1 - \tfrac{1}{2}\Lambda - (1 - \tfrac{1}{2}\Lambda \cos\theta)\cos\theta$$

Sketch the motion in the phase plane for $\Lambda < 1$, $\Lambda = 1$, and $\Lambda > 1$. Discuss the changes in the characteristics of the motion as Λ increases. What is the significance of $\Lambda = 1$?

(c) Assuming $\Lambda > 1$, obtain a two-term, approximate relationship between the amplitude and the frequency for small but finite motions.

(d) Assuming $\Lambda < 1$, obtain a two-term, approximate relationship between the amplitude and the frequency for small but finite motions.

2.10. In the preceding problem, suppose that the rigid frame is free to rotate (i.e., it is not driven at constant Ω). Thus Ω becomes a function of time.

(a) Neglecting the mass of the frame, show that the equations governing θ and Ω can be written in the following form:

$$(r\sin\theta)^2\Omega = \sqrt{H}$$

$$\ddot{\theta} + \left(\frac{g}{r} - \Omega^2\cos\theta\right)\sin\theta = 0$$

where H is a constant of integration ($\sqrt{H}$ is, essentially, the angular momentum).

(b) These equations can be combined to yield

$$\ddot{\theta} + \frac{g}{r}(\sin\theta - \Lambda\cot\theta) = 0$$

where $\Lambda = H/gr$. Show that essentially only one singular point exists, regardless of the value of Λ, and that it is always a center. For $\Lambda = 1$, what is the value of θ at the center?

(c) Find the two-term, approximate amplitude–frequency relationship for small but finite oscillations around this center.

2.11. Consider the system shown in Figure 2-20.

(a) Show that the equation of motion is

$$\left(m_1 + \frac{m_2x^2}{l^2 - x^2}\right)\ddot{x} + \frac{m_2l^2x\dot{x}^2}{(l^2 - x^2)^2} + kx + m_2g\frac{x}{(l^2 - x^2)^{1/2}} = 0$$

(b) Let $R = m_2/m_1$ and $u = x/l$. Then expanding for $|u| \ll 1$, obtain

$$(1 + Ru^2)\ddot{u} + Ru\dot{u}^2 + \omega_0^2u + \frac{Rg}{2l}u^3 + \cdots = 0$$

where

$$\omega_0^2 = \frac{k}{m_1} + \frac{Rg}{l}$$

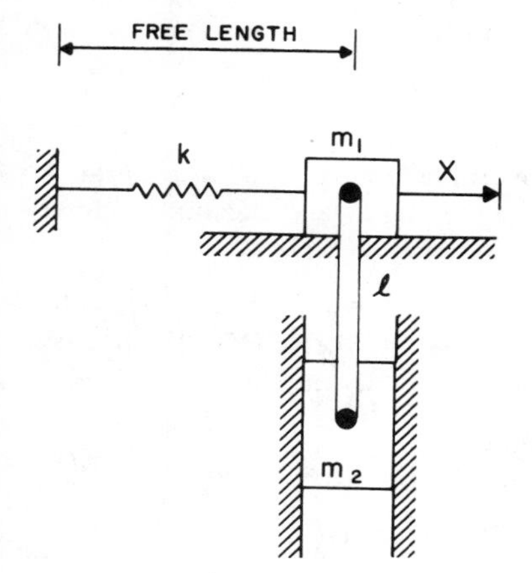

Figure 2-20. Exercise 2.11.

(c) Obtain a two-term, approximate relationship between the amplitude and the frequency of the motion.

2.12. Consider the system shown in Figure 2-21.

(a) Show that the equation governing y is given by

$$m_2 g - \frac{2m_1 g y}{(l^2 + y^2)^{1/2}} = m_2 \ddot{y} + \frac{2m_1 y}{l^2 + y^2}\left(y\ddot{y} + \frac{l^2 \dot{y}^2}{l^2 + y^2}\right)$$

(b) Show that the equilibrium value of y is

$$y_e = \frac{l m_2}{(4m_1^2 - m_2^2)^{1/2}} \qquad (\text{assume } 2m_1 > m_2)$$

(c) Let $u = y/l$ and $R = m_1/m_2$ and obtain

$$\frac{g}{l} - \frac{2Rgu}{l(1 + u^2)^{1/2}} = u + \frac{2Ru}{1 + u^2}\left(u\ddot{u} + \frac{\dot{u}^2}{1 + u^2}\right)$$

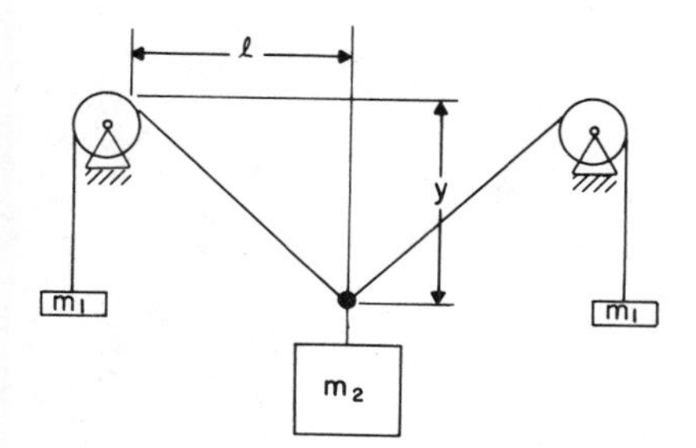

Figure 2-21. Exercise 2.12.

(d) Let

$$u = \frac{y_e}{l} + \eta$$

and expand for small values of η. Use this governing equation for η to obtain a two-term, approximate relationship between the frequency and the amplitude of the motion.

2.13. A rigid rod is rigidly attached to the axle as shown in Figure 2-22. The wheels roll without slip as the pendulum swings back and forth. Only the ball on the end of the pendulum has appreciable mass, and it may be considered a particle.

(a) Show that the equation governing θ is

$$(l^2 + r^2 - 2rl \cos \theta)\ddot{\theta} + rl \sin \theta \dot{\theta}^2 + gl \sin \theta = 0$$

(b) For small but finite motions, determine a two-term, approximate frequency–amplitude relationship.

2.14. Consider the same system as in the preceding exercise, except now there is a linear spring attached to the axle as shown in Figure 2-23. The spring force is zero when θ is zero.

(a) Show that the governing equation for θ is

$$m(l^2 + r^2 - 2rl \cos \theta)\ddot{\theta} + mrl \sin \theta \dot{\theta}^2 + mgl \sin \theta + kr^2\theta = 0$$

(b) For small but finite motions, obtain a two-term, approximate frequency–amplitude relationship.

2.15. Consider a simple pendulum that makes repeated inelastic impacts with a wall as shown in Figure 2-24 (Meissner, 1932). While the mass is not in contact with the wall, the governing equation is

$$\ddot{\theta} + \frac{g}{l} \sin \theta = 0$$

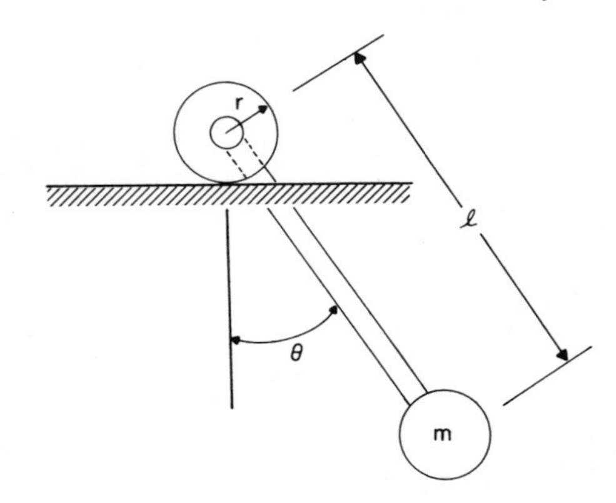

Figure 2-22. Pendulum attached to rolling wheels.

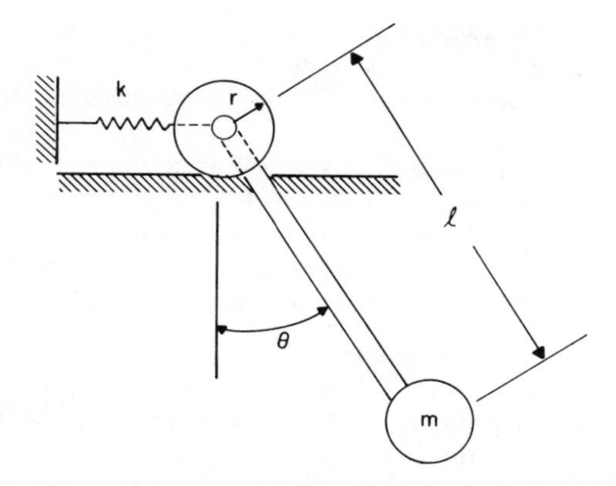

Figure 2-23. Pendulum attached to rolling wheels that are restrained by a spring.

This equation can be rewritten in a convenient form by changing the independent variable: Let

$$\tau = \omega t \quad \text{where} \quad \omega^2 = \frac{g}{l}$$

Then

$$\theta'' + \sin \theta = 0$$

For small angles, this equation becomes

$$\theta'' + \theta = 0$$

(a) Show that, in the case of no impacts at all, the trajectories in the phase plane are a family of circles.

(b) Sketch the family of trajectories when

$$(1)\ \alpha > 0, \qquad (2)\ \alpha = 0, \qquad \text{and} \qquad (3)\ \alpha < 0$$

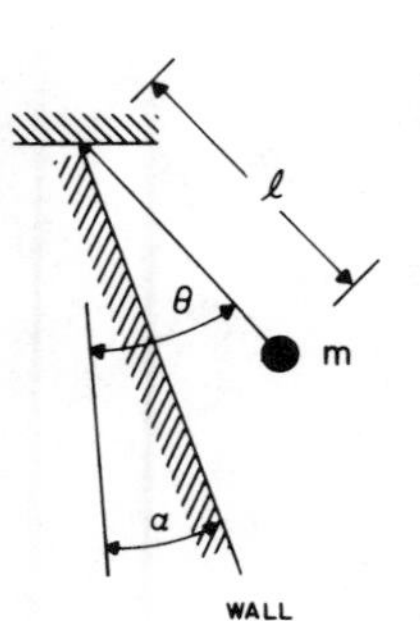

Figure 2-24. Simple pendulum that makes inelastic impacts with a wall.

Hint: Idealizing the situation, we can model the impact by a sudden change in velocity while the position remains fixed. The impact causes the direction of the velocity to change and, because the impact is inelastic, the magnitude to decrease.

(c) Show that the rest position is reached in a finite length of time when $\alpha > 0$ and not at all (infinite length of time) if $\alpha \leqslant 0$. Hint: Show that the time required for the representative point to traverse a trajectory between impacts is equal to the central angle of the trajectory in the phase plane. Then examine the behavior of the central angle as the time increases.

2.16. Consider the structure shown in Figure 2-25. The mass m moves in the horizontal direction only. Using this model to represent a column, we demonstrate how one can study its static stability by determining the nature of the singular point at $x = 0$ of the dynamic equations. This "dynamic" approach is simpler to use, and the arguments are more satisfying than the "static" approach. Vito (1974) analyzed the stability of vibrations of a particle in a plane constrained by identical springs.

(a) Neglecting the weight of all but the mass, show that the governing equation for the motion of m is

$$m\ddot{x} + \left(k_1 - \frac{2P}{l}\right) x + \left(k_3 - \frac{P}{l^3}\right) x^3 + \cdots = 0$$

where the spring force is given by

$$F_{\text{spring}} = k_1 x + k_3 x^3 + \cdots$$

This equation can be put in the general form

$$\ddot{x} + \alpha_1 x + \alpha_3 x^3 + \cdots = 0$$

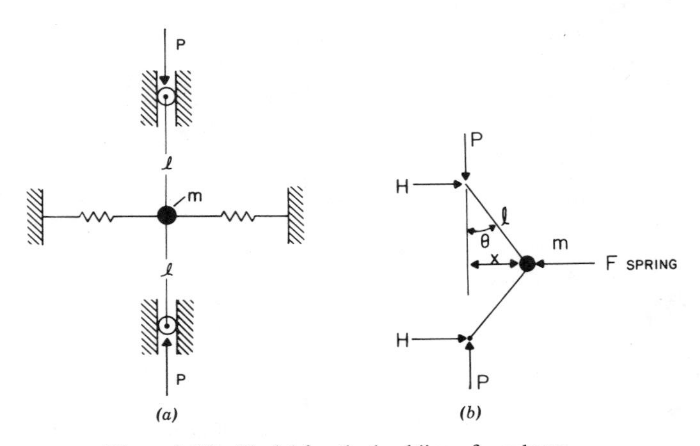

Figure 2-25. Model for the buckling of a column.

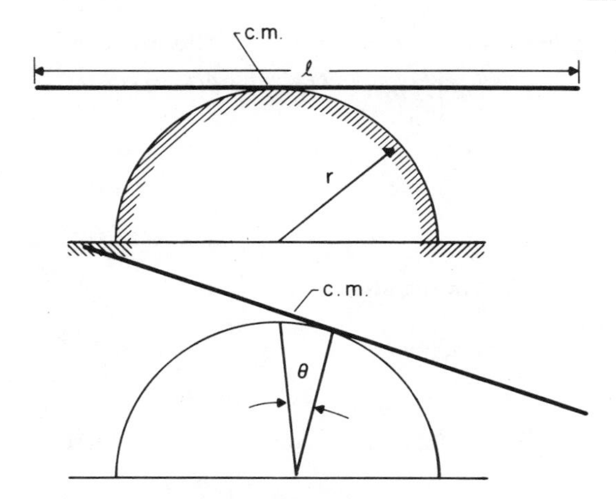

Figure 2-26. Rigid rod rocks on a circular surface.

(b) Sketch the potential energies and the phase planes for $\alpha_1 < 0$ and $\alpha_1 > 0$. These sketches describe the behavior of the system when it is disturbed from its equilibrium position at $x = 0$. What is the critical value of P (i.e., the buckling load which causes a large response to a small initial disturbance)?

2.17. The rigid rod (Figure 2-26) rocks back and forth on the circular surface without slipping.

(a) Show that the equation governing θ is (Gaylord, 1959)

$$\left(\tfrac{1}{12}l^2 + r^2\theta^2\right)\ddot{\theta} + r^2\theta\dot{\theta}^2 + gr\theta\cos\theta = 0$$

(b) Obtain the two-term, approximate relationship between the amplitude and the frequency of the motion.

2.18. Consider the system shown in Figure 2-27.

(a) Show that the equation of motion is

$$m\ddot{x} + kx(x^2 + l^2)^{-1/2}[(x^2 + l^2)^{1/2} - \hat{l}] = 0 \tag{1}$$

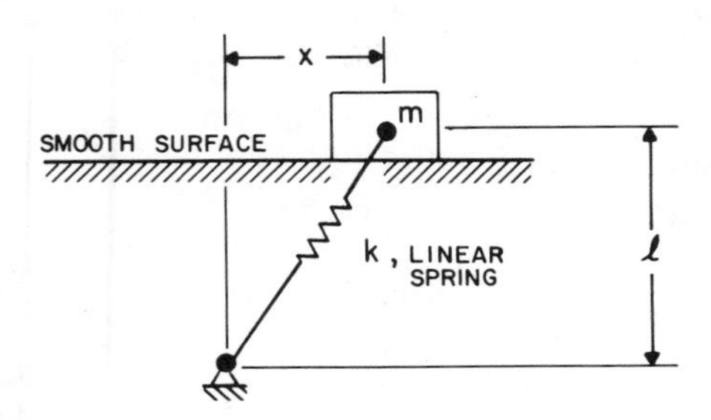

Figure 2-27. Mass slides on a smooth surface while restrained by a linear spring.

Then show that this equation can be put in the following convenient form:

$$\ddot{u} + 2u(1 + u^2)^{-1/2}[(u^2 + 1)^{1/2} - L] = 0 \tag{2}$$

where

$$u = \frac{x}{l} \quad \text{and} \quad L = \frac{\hat{l}}{l}$$

The new independent variable is $\tau = \omega t$ where $\omega^2 = k/2m$.

(b) Sketch the motion in the phase plane for $L < 1$, $L = 1$, and $L > 1$. What is the significance of $L = 1$?

(c) Expanding (2) for small u (i.e., for x small compared with l), show that

$$\ddot{u} + u^3 + \cdots = 0 \quad \text{when} \quad L = 1 \tag{3}$$

(d) Using (3), obtain the following exact relationship between u and τ:

$$\tau = \sqrt{2} \int_{-u_0}^{u} \frac{du}{[u_0^4 - u^4]^{1/2}} \tag{4}$$

where the motion begins from rest at $u = -u_0$. This expression can be put in a more convenient form by letting

$$u = -u_0 \cos \phi \tag{5}$$

Using (5), show that (4) becomes

$$\tau = \frac{1}{u_0} \int_0^{\theta} \frac{d\phi}{(1 - \frac{1}{2} \sin^2 \phi)^{1/2}} = F(\tfrac{1}{2}, \theta) \tag{6}$$

where F is called the elliptic integral of the first kind and is a tabulated function. Show that the period of the motion is $T \approx 7.416/u_0$.

(e) Equation (6) defines ϕ as a function of τ. One writes

$$\phi = am(\tau)$$

(and one says amplitude of τ). Then it follows that

$$u = -u_0 \cos \phi = -u_0 \cos [am(\tau)] = -u_0 cn(\tau)$$

The function cn is one of Jacobi's elliptic functions. Plot u as a function of τ and compare this graph with a plot of the cosine function, giving both the same frequency and amplitude. Plot only one half of the cycle.

2.19. Consider the behavior of a system governed by

$$u'' + u^3 = 0$$

(a) Use the method of harmonic balance and show that

$$u = u_0 \cos (\omega\tau + \beta)$$

where $\omega = (\sqrt{3}/2)u_0$. Hence $T = 7.255/u_0$. Compare this result with that obtained in part (d) of Exercise 2.18. What is the percent error?

(b) By expanding the integrand and integrating term by term, refer to (6) in Exercise 2.18, obtain an approximate expression for the period. Show that three terms are needed in the expansion before one can obtain greater accuracy than that given in part (a) above.

Number of Terms in Expansion	1	2	3	4	5	6
Tu_0	6.283	7.069	7.289	7.366	7.396	7.407

2.20. Use the method of harmonic balance to show that the first term in the nonlinear frequency of the system

$$\ddot{u} + \alpha u^5 = 0$$

is $\omega = (5\alpha/8)^{1/2} a^2$, where a is the amplitude of oscillation.

2.21. Apply the method of equivalent linearization (e.g., Caughey, 1963; Blaquiére, 1966; Iwan 1969, 1973; Patula and Iwan, 1972; van der Werff, 1973, 1975; Srirangarajan, Srinivasan, and Dasarathy, 1974; Dasarathy, 1975; Chou and Sinha, 1975; Iwan and Miller, 1977) to

$$u'' + u^3 = 0 \tag{1}$$

that is, replace (1) by

$$x'' + \lambda x = 0$$

(the "equivalent" linear equation). Then the integrated square of the error over a time interval T is given by

$$e = \int_0^T (\lambda x - x^3)^2 \, d\tau \equiv \langle(\lambda x - x^3)^2\rangle$$

Minimize this error with respect to λ and obtain

$$\lambda = \langle x^4 \rangle / \langle x^2 \rangle$$

Next write the solution of the linear equation in the form

$$x = u_0 \cos(\omega\tau + \beta)$$

where u_0 and β are constants of integration and

$$\omega = \sqrt{\lambda}$$

Choosing T to be the period of x, show that

$$\omega = \frac{\sqrt{3}}{2} u_0$$

in agreement with the results of Exercise 2.19.

2.22. Again consider

$$u'' + u^3 = 0 \tag{1}$$

Now let the approximate form of u be

$$u = u_0 \cos(\omega\tau + \beta) \tag{2}$$

where u_0 and β are arbitrary constants and ω is to be chosen in such a way as to optimize the approximation. Substituting (2) into (1), show that the "residue" is

$$R = (\tfrac{3}{4}u_0^3 - \omega^2 u_0)\cos(\omega\tau + \beta) + \tfrac{1}{4}u_0^3 \cos(3\omega\tau + 3\beta)$$

and that the average of R^2 over the interval $2\pi/\omega$ is

$$\langle R^2 \rangle = \tfrac{1}{2}(\tfrac{3}{4}u_0^3 - \omega^2 u_0)^2 + \tfrac{1}{32}u_0^6$$

(a) Show that if $\langle R^2 \rangle$ is minimized with respect to u_0 (assuming u_0 is not zero),

$$\omega = \tfrac{1}{2}(6 \pm \sqrt{6})^{1/2} u_0$$

The root with the positive sign maximizes $\langle R^2 \rangle$ and must be discarded. The other root is not in agreement with the result obtained by the methods of harmonic balance and equivalent linearization.

(b) Instead of minimizing $\langle R^2 \rangle$ with respect to u_0, show that if $\langle R^2 \rangle$ is minimized with respect to ω,

$$\omega = \frac{\sqrt{3}}{2} u_0$$

(c) Instead of minimizing $\langle R^2 \rangle$, show that if R is made orthogonal to the assumed solution (i.e., in this case $\langle u_0 R \cos(\omega\tau + \phi) \rangle = 0$)

$$\omega = \frac{\sqrt{3}}{2} u_0$$

This is the Galérkin procedure.

2.23. Is the method of multiple scales appropriate for solving $u'' + u^3 = 0$? Explain your answer.

2.24. Consider the system of Exercise 2.18 when the free length of the spring is $\tfrac{1}{2}l$ (i.e., $L = \tfrac{1}{2}$).

(a) Show that the equation of motion becomes

$$u'' + u + \tfrac{1}{2}u^3 + \cdots = 0$$

Assuming the motion begins from rest at $u = -u_0$, show that

$$\tau = 2\int_{-u_0}^{u} \frac{du}{[(u_0^2 - u^2)(4 + u_0^2 + u^2)]^{1/2}}$$

(b) Let $u = -u_0 \cos\theta$ and put this expression in the form

$$T = \frac{2}{(4 + 2u_0^2)^{1/2}} \int_0^{\theta} \frac{d\theta}{(1 - k^2 \sin^2\theta)^{1/2}}$$

where $k^2 = u_0^2/(4 + 2u_0^2)$.

(c) Make a plot of the period as a function of u_0.

(d) For u_0 much less than unity, expand the integrand, then integrate term by term, and obtain the following two-term expression for the period:

$$T = 2\pi\left(1 - \frac{3u_0^2}{16} + \cdots\right)$$

Compare this result with (2.3.34).

2.25. (a) Consider a system governed by the nonlinear equation

$$\ddot{x} + \alpha_1 x + \alpha_3 x^3 = 0$$

Using the method of equivalent linearization, show that it can be replaced by

$$\ddot{x} + \lambda x = 0$$

where

$$\lambda = \frac{\alpha_1\langle x^2\rangle + \alpha_3\langle x^4\rangle}{\langle x^2\rangle}$$

Let $x = a\cos(\omega t + \beta)$, where a and β are constants of integration and $\omega = \sqrt{\lambda}$. Perform the averaging over $2\pi/\omega$ and obtain

$$\lambda = \alpha_1 + \tfrac{3}{4}\alpha_3 a^2$$

in agreement with (2.3.34).

(b) Now consider a system governed by the nonlinear equation

$$\ddot{x} + \alpha_1 x + \alpha_2 x^2 + \alpha_3 x^3 = 0$$

Using the method of equivalent linearization, replace it by

$$\ddot{x} + \lambda x = 0$$

where

$$\lambda = \frac{\alpha_1\langle x^2\rangle + \alpha_2\langle x^3\rangle + \alpha_3\langle x^4\rangle}{\langle x^2\rangle}$$

Show that in this case, one also obtains

$$\lambda = \alpha_1 + \tfrac{3}{4}\alpha_3 a^2$$

in disagreement with (2.3.34).

(c) Can you make any conclusions about the method of equivalent linearization?

2.26. Show that the results obtained by using the Galérkin procedure depend on the assumed form of the solution for the following governing equation:

$$\ddot{x} + \alpha_1 x + \alpha_2 x^2 + \alpha_3 x^3 = 0$$

(a) Let

$$x = a \cos(\omega t + \beta)$$

and show that

$$\omega^2 = \alpha_1 + \tfrac{3}{4}\alpha_3 a^2$$

(b) Let

$$x = a \cos(\omega t + \beta) + B$$

and show that

$$\begin{aligned} R = {} & [-a\omega^2 + \alpha_1 a + 2\alpha_2 aB + \tfrac{3}{4}\alpha_3 a^3 + 3\alpha_3 aB^2] \cos(\omega t + \beta) + \alpha_1 B \\ & + \alpha_2(\tfrac{1}{2}a^2 + B^2) + \alpha_3(B^3 + \tfrac{3}{2}a^2 B) + [\tfrac{1}{2}\alpha_2 a^2 + \tfrac{3}{2}\alpha_3 a^2 B] \cos(2\omega t + \beta) \\ & + \tfrac{1}{4}\alpha_3 a^3 \cos(3\omega t + 3\beta) \end{aligned}$$

(c) Using the condition that

$$\langle aR \cos(\omega t + \beta) \rangle = 0$$

over the interval $T = 2\pi/\omega$, obtain

$$-a\omega^2 + \alpha_1 a + 2\alpha_2 aB + \tfrac{3}{4}\alpha_3 a^3 + 3\alpha_3 aB^2 = 0$$

(d) Using the condition that

$$\langle BR \rangle = 0$$

over the interval $T = 2\pi/\omega$, obtain

$$\alpha_1 B + \alpha_2(\tfrac{1}{2}a^2 + B^2) + \alpha_3(B^3 + \tfrac{3}{2}a^2 B) = 0$$

Then show that

$$B = -\tfrac{1}{2}\alpha_2 \alpha_1^{-1} a^2 + \cdots$$

and

$$\omega^2 = \alpha_1 + \tfrac{3}{4}\alpha_3 a^2 - \alpha_2^2 \alpha_1^{-1} a^2 + \cdots$$

(e) Let

$$x = a \cos(\omega t + \beta) + B + C \cos(2\omega t + 2\beta)$$

and obtain

$$\omega^2 = \alpha_1 + \tfrac{3}{4}\alpha_3 a^2 - \tfrac{5}{6}\alpha_2^2 \alpha_1^{-1} a^2$$

(f) Compare these results with (2.3.34) and draw conclusions about the use of the Galérkin procedure.

2.27. Consider the system governed by the following equation of motion:

$$\ddot{u} - u + u^4 = 0$$

(a) Show that $u = 1$ is a center. Then let $v = u - 1$ and obtain

$$\ddot{v} + 3v + 6v^2 + 4v^3 + v^4 = 0$$

(b) Use either the Lindstedt-Poincaré method or the method of multiple scales to obtain a two-term frequency–amplitude relationship.

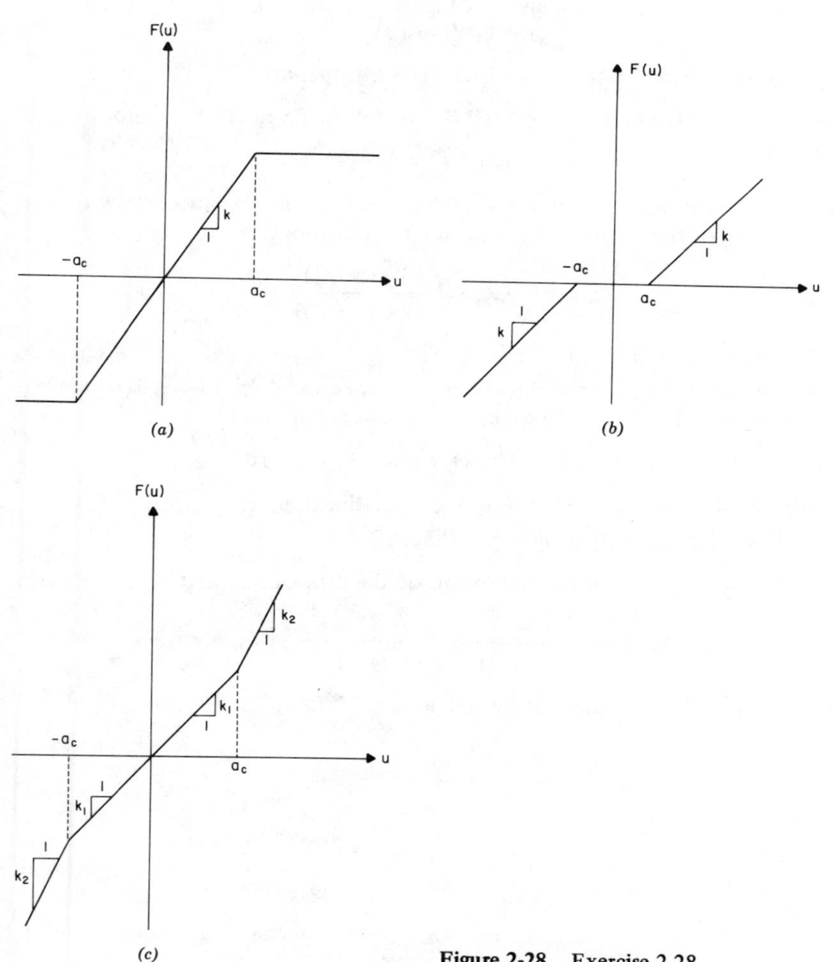

Figure 2-28. Exercise 2.28.

2.28. Consider the free oscillations of a system governed by

$$\ddot{u} + F(u) = 0$$

where $F(u)$ is defined in Figure 2-28 for three different cases. Show that in the first approximation

(a) $$\omega_0^2 = \frac{2k}{\pi}\left[\sin^{-1}\left(\frac{a_c}{a}\right) + \frac{a_c}{a}\left(1 - \frac{a_c^2}{a^2}\right)^{1/2}\right]$$

(b) $$\omega_0^2 = k - \frac{2k}{\pi}\left[\sin^{-1}\left(\frac{a_c}{a}\right) + \frac{a_c}{a}\left(1 - \frac{a_c^2}{a^2}\right)^{1/2}\right]$$

(c) $$\omega_0^2 = k_2 - \frac{2}{\pi}(k_2 - k_1)\left[\sin^{-1}\left(\frac{a_c}{a}\right) + \frac{a_c}{a}\left(1 - \frac{a_c^2}{a^2}\right)^{1/2}\right]$$

[Many more cases were treated by Šiljak (1969, Appendix F).]

2.29. Consider the system governed by the following equation of motion:

$$\ddot{u} + ku^n = 0$$

where n is an odd integer. Using the method of harmonic balance, show that an approximation of the frequency–amplitude relationship is

$$\omega^2 \approx \frac{2k}{\sqrt{\pi}} a^{n-1} \frac{\Gamma[\frac{1}{2}(n+2)]}{\Gamma[\frac{1}{2}(n+3)]}$$

where Γ is the gamma function.

2.30. Consider the motion of a system governed by $\ddot{u} - u + u^n = 0$, where n is a positive integer. Let $u = 1 + v$, expand for small ϵ, and obtain

$$\ddot{v} + (n-1)v + \tfrac{1}{2}n(n-1)v^2 + \tfrac{1}{6}n(n-1)(n-2)v^3 + \cdots = 0$$

Determine a two-term frequency–amplitude relationship.

2.31. Consider the system shown in Figure 2-29.

(a) Show that the equation of motion of the disk can be written in the form

$$\ddot{\theta} + \omega^2\left[1 - \frac{f}{[2r(r+l)(1-\cos\theta) + l^2]^{1/2}}\right]\sin\theta = 0$$

where $\omega^2 = 2k(r+l)/mr$ and f is the free length of the spring.

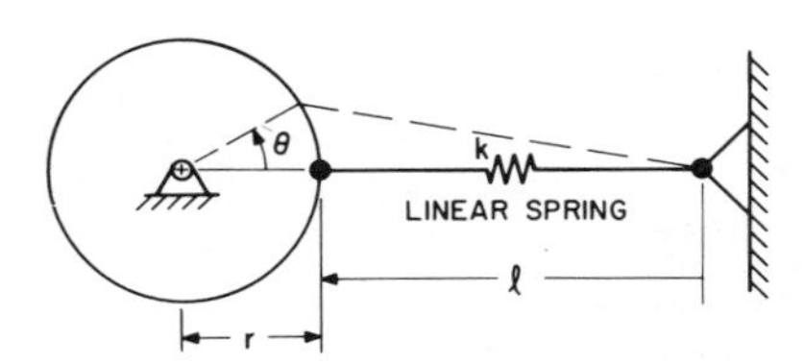

Figure 2-29. Particle on a wheel restrained by a spring.

(b) Sketch the potential energy as a function of r for

(i) $f \leqslant l$
(ii) $l < f < l + 2r$
(iii) $f \geqslant l + 2r$

Show the equilibrium positions and indicate whether they are stable or unstable.
(c) For $f > l$ obtain a two-term expression relating the amplitude and the frequency of small oscillations around the equilibrium position.

2.32. The cylinder rolls back and forth without slip as shown in Figure 2-30.

(a) Show that the equation of motion can be written in the form

$$\ddot{x} + \omega^2 [1 - f(1 + x^2)^{-1/2}]x = 0$$

where $\omega^2 = 2k/3M$ and f is the free length of the spring. All lengths were made dimensionless with respect to the radius r.
(b) Sketch the potential energy as a function of x for

(i) $1 \leqslant f$
(ii) $1 > f$

Show the equilibrium positions and indicate whether they are stable or unstable.
(c) For $f = \sqrt{2}$, obtain a two-term frequency–amplitude relationship for small oscillations around the equilibrium position.

2.33. Consider the system governed by (Baker, Moore, and Spiegel, 1971):

$$\dddot{u} - \delta(1 - u^2)\dot{u} = -\epsilon[\ddot{u} + (1 - \delta)\dot{u}]$$

where $\epsilon \ll 1$ and $\delta = O(1)$

(a) When $\epsilon = 0$, show that

$$\dot{u}^2 + 2\delta(\tfrac{1}{12}u^4 - \tfrac{1}{2}u^2 + bu) = E$$

Show that there are three equilibrium positions if $|b| < \frac{2}{3}$ and only one if $|b| > \frac{2}{3}$. Sketch the trajectories in the phase plane for the cases

(i) $b = 0$
(ii) $0 < b < \frac{2}{3}$
(iii) $b > \frac{2}{3}$

(b) Determine a first-order approximate solution for small ϵ.

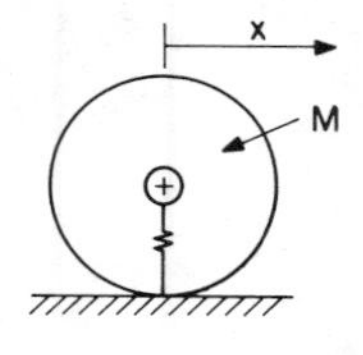

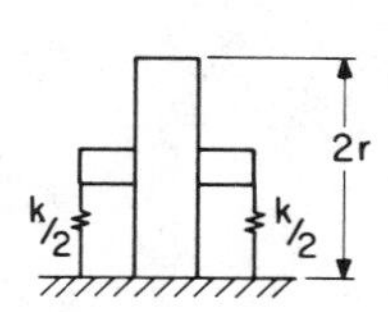

Figure 2-30. Cylinder rolls without slip.

2.34. Consider the motion of the simple pendulum of Section 2.4.1.

(a) Use the Lindstedt-Poincaré technique to show that

$$\theta = \epsilon a \cos(\omega t + \beta) - \frac{\epsilon^3 a^3}{192} \cos(3\omega t + 3\beta) + O(\epsilon^5)$$

$$\omega = \omega_0 \left(1 - \frac{\epsilon^2 a^2}{16} + \frac{\epsilon^4 a^4}{1024}\right) + O(\epsilon^6)$$

(b) Use the initial conditions $\theta(0) = 0$ and $\dot{\theta}(0) = \dot{\theta}_0$ to show that

$$\beta = \tfrac{1}{2}\pi, \qquad \dot{\theta}_0 = -\epsilon a \omega \left(1 + \frac{\epsilon^2 a^2}{64}\right) + \cdots$$

(c) Use (a) to eliminate ω from (b) and obtain

$$\frac{\dot{\theta}_0}{\omega_0} = 2k = -\epsilon a + \frac{3\epsilon^3 a^3}{64} + \cdots$$

Then show that

$$\epsilon a = -2k - \tfrac{3}{8} k^3 + \cdots$$

(d) Determine the period of oscillation and compare it with (2.4.6).

CHAPTER 3

Nonconservative Single-Degree-of-Freedom Systems

In this chapter we consider the damped unforced oscillations of systems having a single degree of freedom. We begin with a discussion of several damping mechanisms. For a comprehensive discussion we refer the reader to the book of Lazan (1968). Next we discuss methods for obtaining qualitative and quantitative analyses. Then we consider systems with slowly varying coefficients, and finally we consider relaxation oscillations.

Forces that are functions of the velocity are called damping forces. When the damping force, or simply the damping, causes the amplitude of the unforced motion to decrease, it is called *positive damping*. When the damping causes the amplitude to increase, it is called *negative damping*. In this chapter we consider both types of damping mechanisms.

3.1. Damping Mechanisms

3.1.1. COULOMB DAMPING

When the contact surface between two solids is dry, the friction force opposing their relative motion is called *Coulomb damping*. When an external force is applied to move the block in Figure 3-1a from rest, a friction force which opposes the impending motion develops. The magnitude of the friction force f increases until a critical value is reached, and then the block moves. After the motion begins, the magnitude of f decreases as long as $|\dot{x}| < |\dot{x}_m|$ and then increases when $|\dot{x}|$ becomes larger than $|\dot{x}_m|$ as shown schematically in Figure 3-1b. The critical value is usually expressed as $\mu_s N$, where μ_s is the so-called static coefficient of friction and N is the normal force between the block and the surface; in this case, it is mg. In many applications, the Coulomb force is approximated by a constant. Thus referring to Figure 3-1a, one writes the equa-

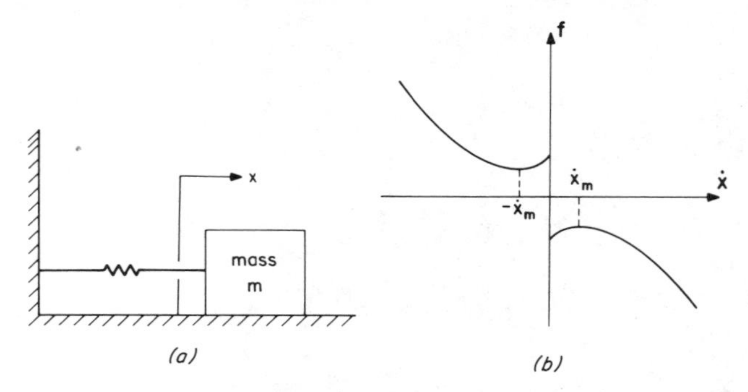

Figure 3-1. (*a*) Spring-mass system. (*b*) Friction force as a function of velocity.

tion of motion as

$$m\ddot{x} + F(x) = f = \begin{cases} \mu_d mg \text{ when } \dot{x} < 0 \\ -\mu_d mg \text{ when } \dot{x} > 0 \end{cases} \tag{3.1.1}$$

where μ_d is the so-called dynamic, or kinetic, coefficient of friction and $F(x)$ is the negative of the restoring force of the spring.

3.1.2. LINEAR DAMPING

When the contact surface in Figure 3-1*a* is covered with a thin liquid film so that the two surfaces do not touch, it is usual to assume that the friction force is proportional to the velocity gradient (i.e., $f \propto \dot{x}/h$, where $\dot{x}$ is the relative velocity and h is the thickness of the film) and opposes the motion (Newton). Thus referring to Figure 3-1*a*, one writes the equation of motion in the form

$$m\ddot{x} + c\dot{x} + F(x) = 0 \tag{3.1.2}$$

where c is a positive constant that is a function of the fluid properties and the condition of the surfaces. Mahalingam (1975) investigated the combined influence of Coulomb and linear damping on the response of vibratory systems.

Another example of the drag force being proportional to the velocity occurs when an immersed body moves through a fluid at very low Reynolds numbers (Stokes flow). The case when the Reynolds number is large is discussed next.

3.1.3. NONLINEAR DAMPING

When an immersed body moves through a fluid at high Reynolds numbers, the flow separates and the drag force is very nearly proportional to the square of the velocity. Thus one writes the equation of motion in the form

$$m\ddot{x} + F(x) = -c|\dot{x}|\dot{x} \tag{3.1.3}$$

where c is a positive constant that is a function of the body geometry and the fluid properties. For moderate Reynolds numbers, the damping force lies between the linear and quadratic forms. Consequently some researchers have represented the damping force as $-c|\dot{x}|^{\alpha}\dot{x}$, where $0 < \alpha < 1$. Since these models of damping are not analytic, other researchers have used damping forms such as $-cf(x)\dot{x}$ or $-cg(\dot{x})\dot{x}$, where $f(x)$ and $g(\dot{x})$ are even analytic functions of x and $\dot{x}$, respectively. Hemp (1972) proposed a combination of Coulomb and quadratic damping for a runaway escapement mechanism.

3.1.4. HYSTERETIC DAMPING

Let us consider the system shown in Figure 3-2*a*, which is a simple example illustrating hysteretic damping. The mass m lies on a smooth surface, and the restoring mechanism consists of an elastic spring (not necessarily linear) in parallel with a linear elastic spring and a "Coulomb damper" in series. The functions f_1 and f_2 give the forces in the springs.

Let us suppose that the motion is started from rest by moving the mass m to the right. The restoring force in the top element is always given by $f_1(x)$, where x is the position of the mass m. However the restoring force in the bottom element depends on the path traveled by m. If $x \leqslant x_s$, where $f_2(x_s) = f_s$ is the critical friction force in the damper, the elongation in the bottom spring is x and the restoring force is kx, where k is the spring constant. When $x \geqslant x_s$, the

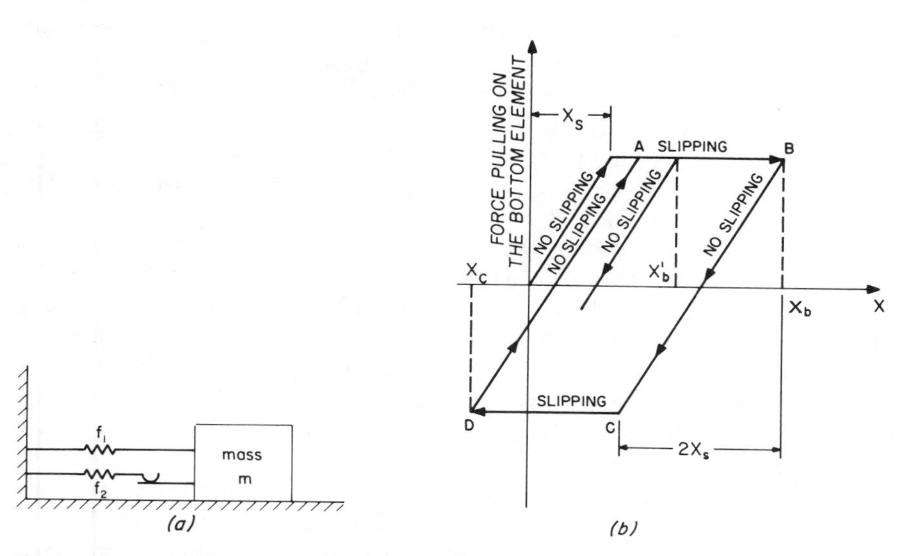

Figure 3-2. (*a*) Simple example of hysteretic damping. (*b*) Typical loading diagram for the bottom element in the restoring mechanism of Figure 3-2*a*.

damper slips, the elongation in the spring remains x_s, and the restoring force remains $f_s = kx_s$ as shown in Figure 3-2*b*.

Let us assume that the motion reverses its direction at $x = x_b > x_s$. Initially the damper does not slip and the restoring force in the bottom spring decreases from kx_s to $k(x + x_s - x_b)$ along the line *BC*. As the mass m reaches the location $x = x_c = x_b - 2x_s$, the force in the bottom spring reaches the critical friction value f_s, but now it is compressive. As x decreases beyond x_c, the damper slips and the restoring force in the bottom element remains $-kx_s$. If the motion reverses its direction at $x = x_d$, the damper does not slip initially and the restoring force in the bottom element is given by $k(x - x_s - x_d)$ along the line *DA*. As x reaches x_a, slipping occurs and the restoring force in this element remains kx_s. If the motion reverses its direction again at x'_b, initially no slipping occurs and the restoring force in the bottom element starts decreasing as shown in Figure 3-2*b*. The area enclosed in the diagram as the load is cycled equals the energy dissipated.

The hysteresis described by the above example is of the hardening type and is linear. There are many structures that exhibit a hardening behavior under cyclic loading. These include riveted and bolted structures (Iwan and Furuike, 1973), externally reinforced masonry walls (Tso, Pollner, and Heidebrecht, 1974), and reinforced concrete shear walls and beam-column connections (Shiga, Shibata, and Takahashi, 1974; Townsend and Hanson, 1974). Miller (1977) presented a physical model, which produces a form of hardening hysteresis, and used it to determine the steady-state response of single- and multidegree-of-freedom systems. Caughey and Vijayaraghavan (1976) treated a one-degree-of-freedom system with linear hysteretic damping. Systems with bilinear hysteretic damping models were analyzed by Jacobsen (1952); Goodman and Klumpp (1956); Thomson (1957); Berg and DaPeppo (1960); Caughey (1960b, c); Tanabashi and Kaneta (1962); Iwan (1965); Jong (1969); Tso and Asmis (1970); Drew (1974); and Karasudhi, Tan, and Lee (1974).

In addition to the structural systems discussed above, there are composites of ductile materials that are assembled in such a way as to slip or yield gradually. These exhibit a softening hysteretic behavior under cyclic loading; that is, their hysteresis loops are generally composed of smooth curves with rounded knees. Such systems were treated by a number of investigators including Jennings (1964), Iwan (1967, 1968a, b), and Jennings and Husid (1968). Distributed-element models for hysteresis were used by Pisarenko (1948), Iwan (1966, 1970), and Jong and Chen (1971).

Other interesting examples of hysteresis can be found in nonlinear aerodynamics. In Figure 3-3, from Atta, Kandil, Mook, and Nayfeh (1977), the pitching moment and normal-force coefficients are plotted as functions of angle of attack. For this plot a rectangular wing is initially in a steady state at an angle of attack of 11°. Then the angle of attack suddenly starts to increase. When the

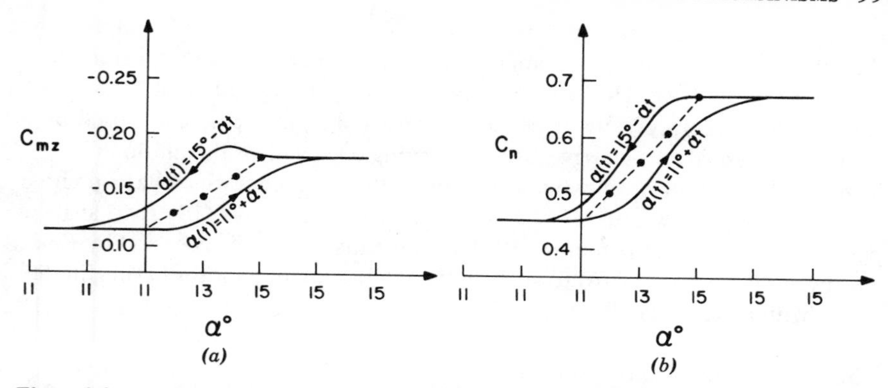

Figure 3-3. (*a*) Pitching-moment and (*b*) normal-force coefficients as functions of angle of attack.

angle of attack reaches 15°, it stops changing and a new steady state develops. Then the angle of attack suddenly begins to decrease. It continues to decrease until it reaches 11°. Then it stops decreasing and a new steady state develops.

In Figure 3-4 from Thrasher, Mook, Kandil, and Nayfeh (1977), a rectangular wing at an angle of attack undergoes a harmonic yawing motion around an axis

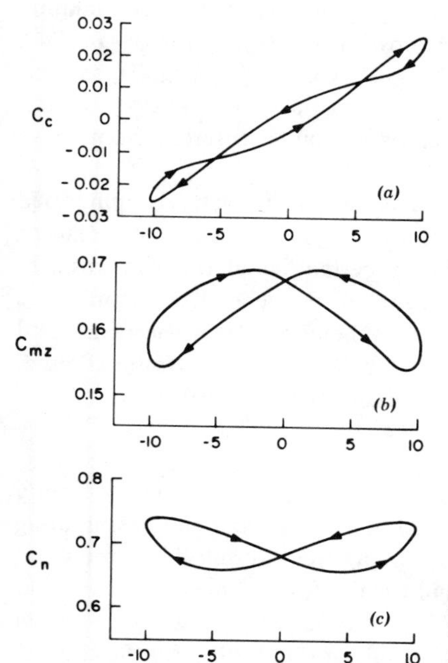

Figure 3-4. (*a*) Rolling-moment, (*b*) pitching-moment, and (*c*) normal-force coefficients as functions of yaw angle.

through the leading edge and perpendicular to the wing surface. The plots show the rolling-moment, pitching-moment, and normal-force coefficients as functions of the angle of yaw. The loads are calculated by using potential theory and allowing for viscous effects by separated vortex sheets. The area enclosed by a loading path equals the energy dissipated during one cycle of the motion.

Davidenkov (1938) proposed an analytic hysteretic model which has a pointed hysteretic loop. It was used by Mozer and Evan-Iwanowski (1972) to analyze parametrically excited columns with hysteretic material properties and by Rajac and Evan-Iwanowski (1976) to analyze the interaction of a motor having limited power with a dissipative foundation.

3.1.5. MATERIAL DAMPING

When a real material is deformed, certain internal mechanisms are responsible for a dissipation of energy. Several models have been proposed for these mechanisms; among them are analogies to springs and dashpots in series (Maxwell fluid) and springs and dashpots in parallel (Kelvin and Voight solid). For more details of these models of viscoelastic materials and other more complicated models we refer the reader to the books of Flügge (1967) and Lazan (1968). Besides the investigators mentioned in Chapter 7 who treated the nonlinear vibrations of viscoelastic beams, plates, and shells, a number of other investigators studied the nonlinear vibrations of viscoelastic systems. These include Maiboroda and Morgunov (1972); Maiboroda, Koltunov, and Morgunov (1972); Movlyankulov (1974); Kravchuk, Morgunov, and Troyanovskii (1974); Karimov (1974); Ibragimov (1975); and Nambudiripad and Neis (1976). Caughey (1960a) showed that the variations of the amplitude and the phase of a wave train along a semi-infinite rod exhibiting weak bilinear hysteresis are quite different from those along a linear viscoelastic rod.

It should be noted that viscous damping, including viscoelastic damping used to simulate the internal damping mechanism of physical systems, may lead to the prediction of some peculiar instabilities in certain circulatory systems subjected to nonconservative forces which depend on the generalized coordinates (Ziegler, 1952, 1953, 1956, 1968; Bolotin, 1963; Leipholz, 1964; Herrmann and Jong, 1965, 1966; Nemat-Nasser and Herrmann, 1966; Nemat-Nasser, Prasad, and Herrman, 1966; Herrmann, 1967; Nemat-Nasser, 1967, 1970).

3.1.6. RADIATION DAMPING

One mechanism of radiation damping involves the transfer of energy from a moving body to a surrounding, unbounded fluid. As examples, a pulsating bubble in a compressible fluid does work on the fluid during the motion and hence causes the kinetic energy of the fluid (in the form of pressure waves that radiate outward) to increase; similarly, a body vibrating at the interface between two liquids creates surface waves that radiate outward. In both cases the amount of fluid set into motion continually increases. The effect is to damp the motion of the body. To illustrate this concept, we consider in detail the symmetric pulsations of a spherical bubble in an infinite, slightly compressible liquid.

We assume that the bubble is initially in equilibrium with the liquid. The initial radius of the bubble is R_0, the initial density of the gas inside the bubble is ρ_{0g}, the initial pressure in both the gas and the liquid is p_0, and the initial density of the liquid is ρ_0.

The linearized equations describing the motion of the liquid are

Conservation of Mass

$$\frac{1}{r^2}\frac{\partial}{\partial r}(r^2 u) + \frac{1}{\rho_0}\frac{\partial \rho}{\partial t} = 0 \tag{3.1.4}$$

Conservation of Radial Momentum

$$\frac{\partial u}{\partial t} = -\frac{1}{\rho_0}\frac{\partial p}{\partial r} \tag{3.1.5}$$

Equation of State

$$dp = c^2 d\rho \tag{3.1.6}$$

where u, ρ, p, and c are the radial component of the velocity, the density, the pressure, and the speed of sound in the liquid, respectively. The problem formulation is completed by the specification of the boundary conditions. The linearized form of these boundary conditions at the bubble surface is

$$p = p_g \quad \text{and} \quad u = \epsilon R_0 \dot{\eta} \quad \text{at } r = R_0 \tag{3.1.7}$$

where

$$R = R_0[1 + \epsilon\eta(t)] \tag{3.1.8}$$

is the instantaneous radius of the bubble. Since disturbances decay away from the bubble,

$$u \to 0 \quad \text{and} \quad p \to p_0 \quad \text{as} \quad r \to \infty \tag{3.1.9}$$

Equations (3.1.4) through (3.1.6) can be combined to yield the wave equation

$$\frac{1}{r^2}\frac{\partial}{\partial r}\left(r^2 \frac{\partial p}{\partial r}\right) = \frac{1}{c^2}\frac{\partial^2 p}{\partial t^2} \tag{3.1.10}$$

The solution corresponding to an outward-propagating wave is

$$p = p_0 + \frac{f(r - ct)}{r} \tag{3.1.11}$$

Hence

$$\frac{\partial}{\partial t}(rp) = -c\frac{\partial}{\partial r}(rp) \tag{3.1.12}$$

To determine the equation describing η, we integrate (3.1.5) from $r = R_0$ to $r = \infty$, use (3.1.7) and (3.1.9), and obtain

$$\int_\infty^{R_0} \frac{\partial u}{\partial t}\,dr + \frac{p_g - p_0}{\rho_0} = 0 \tag{3.1.13}$$

To evaluate the integral in (3.1.13), we replace $\partial u/\partial t$ by $(1/r^2)(\partial/\partial t)(r^2 u)$, integrate the result by parts, and obtain

$$\int_\infty^{R_0} \frac{\partial u}{\partial t}\,dr = \int_\infty^{R_0} \frac{1}{r^2}\frac{\partial}{\partial t}(r^2 u)\,dr = -r\frac{\partial u}{\partial t}\bigg|_\infty^{R_0} + \int_\infty^{R_0} \frac{1}{r}\frac{\partial^2}{\partial r\,\partial t}(r^2 u)\,dr \tag{3.1.14}$$

But from (3.1.4) and (3.1.6) one can obtain

$$\frac{1}{r}\frac{\partial^2}{\partial r\,\partial t}(r^2 u) = -\frac{r}{\rho_0 c^2}\frac{\partial^2 p}{\partial t^2} = -\frac{1}{\rho_0 c^2}\frac{\partial^2}{\partial t^2}(rp) \tag{3.1.15}$$

Using (3.1.12) we express (3.1.15) as

$$\frac{1}{r}\frac{\partial^2}{\partial r\,\partial t}(r^2 u) = \frac{1}{\rho_0 c}\frac{\partial^2}{\partial t\,\partial r}(rp) = \frac{1}{\rho_0 c}\frac{\partial}{\partial r}\left(r\frac{\partial p}{\partial t}\right) \tag{3.1.16}$$

Combining (3.1.14) and (3.1.16) and performing the integration, we obtain

$$\int_\infty^{R_0} \frac{\partial u}{\partial t}\,dr = -r\frac{\partial u}{\partial t}\bigg|_\infty^{R_0} + \frac{1}{\rho_0 c}r\frac{\partial p}{\partial t}\bigg|_\infty^{R_0} \tag{3.1.17}$$

Using (3.1.7) through (3.1.9) in (3.1.17) and combining the resulting equation with (3.1.13), we obtain

$$-\epsilon R_0^2 \ddot{\eta} + \frac{R_0}{\rho_0 c}\frac{\partial p_g}{\partial t} + \frac{p_g - p_0}{\rho_0} = 0 \tag{3.1.18}$$

Conservation of mass in the bubble gives

$$\rho_g R^3 = \rho_{0g} R_0^3 \tag{3.1.19}$$

For an isentropic motion of the gas inside the bubble

$$\frac{p_g}{p_0} = \left(\frac{\rho_g}{\rho_{0g}}\right)^\gamma \tag{3.1.20}$$

where γ is the ratio of the specific heats of the gas. Thus from (3.1.8) it follows that

$$\frac{p_g}{p_0} = \left(\frac{R_0}{R}\right)^{3\gamma} = (1 + \epsilon\eta)^{-3\gamma} \approx 1 - 3\epsilon\gamma\eta \tag{3.1.21}$$

Substituting (3.1.21) into (3.1.18) yields

$$\ddot{\eta} + \frac{3\gamma p_0}{\rho_0 c R_0}\dot{\eta} + \frac{3\gamma p_0}{\rho_0 R_0^2}\eta = 0 \tag{3.1.22}$$

The second term in (3.1.22) is a damping term which is due to the radiation of energy from the bubble outward via the liquid.

For this damping mechanism there is actually conservation of energy if one regards the vibrating body and the surrounding fluid to be a single dynamic system.

3.1.7. NEGATIVE DAMPING

In all the examples described above, the damping is positive. Here we consider a van der Pol oscillator in detail. This is a simple example of a system having a dampinglike mechanism which acts to increase the energy (i.e., which is negative) when the amplitude of the motion is small and to decrease the energy when the amplitude is large. As a result, the system reaches a so-called *limit cycle* which is independent of the initial conditions.

The van der Pol oscillator is an electrical circuit consisting of two dc power sources, resistors, inductance coils, a capacitor, and a triode as shown in Figure 3-5. The triode has three main elements: a cathode (filament A) heated by one of the dc sources so that it emits electrons; an anode (plate P, positively charged), which attracts the electrons emitted by the cathode; and a grid (course mesh G), which controls the flow of electrons from the anode to the cathode. The control is accomplished by changing the potential of the grid by a mutual inductance.

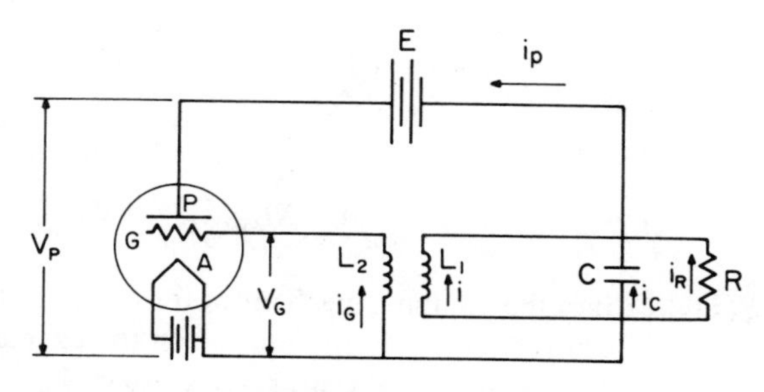

Figure 3-5. Circuit diagram for the van der Pol oscillator.

To derive the equation governing the current i in the inductance coil of the oscillator, we apply Kirchhoff's laws:

$$L_1 \frac{di}{dt} + M \frac{di_G}{dt} = Ri_R \tag{3.1.23}$$

$$L_1 \frac{di}{dt} + M \frac{di_G}{dt} = \frac{q_c}{C} \tag{3.1.24}$$

$$V_G = L_2 \frac{di_G}{dt} + M \frac{di}{dt} \tag{3.1.25}$$

$$V_p = E - Ri_R \tag{3.1.26}$$

$$i_p = i + i_R + i_c \tag{3.1.27}$$

In addition,

$$i_c = \frac{dq_c}{dt} \tag{3.1.28}$$

where q_c is the charge on the capacitor.

The current in the grid i_G is assumed to be negligible. Moreover it was shown experimentally, and it can be shown theoretically, that

$$i_p = \phi(V_G + \Delta V_p) \tag{3.1.29}$$

where Δ is a constant. The reciprocal of Δ is called the amplification factor, and the function ϕ is called the characteristic of the vacuum tube.

Using these assumptions one can combine (3.1.23) through (3.1.29) to obtain

$$L_1 C \frac{d^2 i}{dt^2} + \frac{L_1}{R} \frac{di}{dt} + i = \phi \left[\Delta E + (M - \Delta L_1) \frac{di}{dt} \right] \tag{3.1.30}$$

To simplify (3.1.30), we let

$$x = i - \phi(\Delta E)$$

and find

$$L_1 C \ddot{x} + x + F(\dot{x}) = 0 \tag{3.1.31}$$

where

$$F(\dot{x}) = \frac{L_1}{R} \dot{x} - \phi[\Delta E + (M - \Delta L_1)\dot{x}] + \phi(\Delta E) \tag{3.1.32}$$

The function $F(\dot{x})$ describes the damping mechanism for the circuit.

The character of the damping mechanism depends on the characteristic of the triode ϕ. Typically the function ϕ has the properties shown in Figure 3-6. The

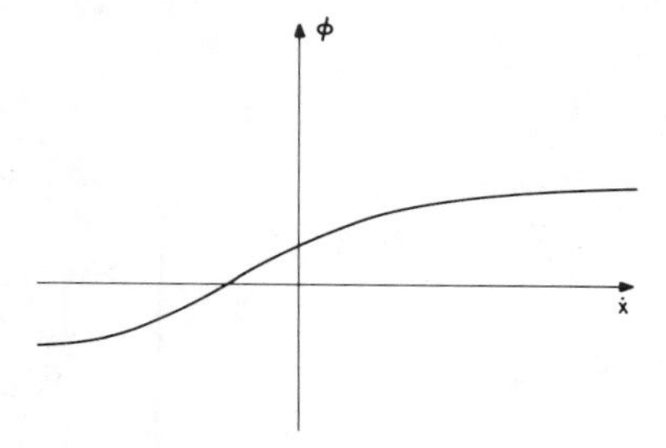

Figure 3-6. The characteristic of the triode.

saturation exhibited by the triode is the result of the plate current being limited by the rate of production of electrons.

When the amplitude of the motion is small ($|\dot{x}| << 1$), we may write

$$F(\dot{x}) = F'(0)\dot{x} = [L_1R^{-1} - (M - \Delta L_1)\phi'(\Delta E)]\,\dot{x} \tag{3.1.33}$$

We note from Figure 3-6 that $\phi'(\Delta E)$ is positive. Thus if one can make $(M - \Delta L_1) > 0$ and $(M - \Delta L_1)\phi'(\Delta E) > L_1/R$, then one can make the damping negative. It is possible to build such a circuit.

When the amplitude of the motion is large, ϕ is nearly a positive constant if $\dot{x}$ is positive, and ϕ is nearly a negative constant if $\dot{x}$ is negative. Now the damping is positive. Thus the character of the damping mechanism changes from negative to positive as the amplitude of the motion increases.

To approximate the function $F(\dot{x})$ in a region around the point $\dot{x} = 0$ which includes the transition from negative to positive damping, one can represent $F(\dot{x})$ by a cubic function as

$$F(\dot{x}) = -\alpha_1\dot{x} + \alpha_3\dot{x}^3 \tag{3.1.34}$$

where α_1 and α_3 are positive constants. Substituting (3.1.34) into (3.1.31) and letting $x = u\sqrt{\alpha_1/\alpha_3}$, we obtain

$$\ddot{u} + \omega_0^2 u - \epsilon(\dot{u} - \dot{u}^3) = 0 \tag{3.1.35}$$

where $\epsilon = \alpha_1\omega_0^2$ and $\omega_0^2 = (L_1C)^{-1}$. Equation (3.1.35) is often called Rayleigh's equation. Differentiating (3.1.35) with respect to t yields

$$\ddot{v} + \omega_0^2 v - \epsilon(1 - v^2)\dot{v} = 0, \quad v = \sqrt{3}\dot{u} \tag{3.1.36}$$

which is often called van der Pol's (1922) equation.

Next we describe a mechanical system that exhibits negative damping. Specifically we consider the system shown in Figure 3-7*a*. It consists of a block of mass m resting on a rough belt which moves with a constant speed $\dot{x}_0$ and connected to a spring attached to a rigid support. If x is the displacement of the block from

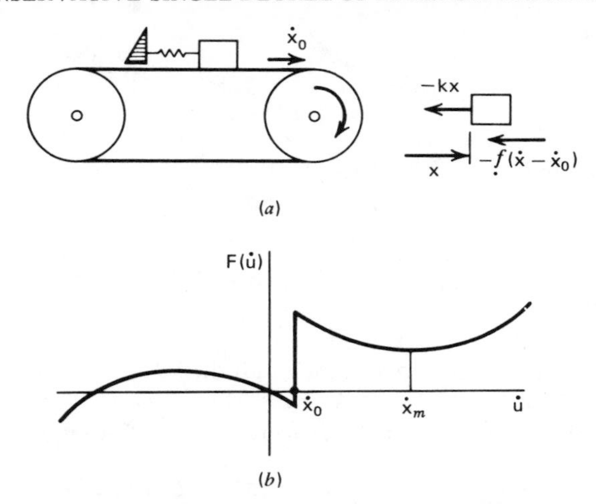

Figure 3-7. (*a*) Mechanical system capable of executing self-sustained oscillations. (*b*) The characteristic of the mechanical system.

the free-length position of the spring, then

$$m\ddot{x} + kx - f(\dot{x} - \dot{x}_0) = 0 \tag{3.1.37}$$

where f is the Coulomb friction force between the block and the belt and is shown in Figure 3-1*b*.

We introduce a new variable u defined by

$$u = x - k^{-1} f(-\dot{x}_0) \tag{3.1.38}$$

Then (3.1.37) becomes

$$\ddot{u} + \omega_0^2 u + F(\dot{u}) = 0 \tag{3.1.39}$$

where $\omega_0^2 = k/m$ and

$$F(\dot{u}) = m^{-1} [f(-\dot{x}_0) - f(\dot{u} - \dot{x}_0)] \tag{3.1.40}$$

If $\dot{x}_0$ is not too large, $F(\dot{u})$ has the form shown in Figure 3-7*b*. We note that the slope of $F(\dot{u})$ at the origin is negative if $\dot{x}_0 < |\dot{x}_m|$. If $\dot{x}_0$ is large, then the slope of $F(\dot{u})$ at the origin will be positive. Thus negative damping occurs only for values of $\dot{x}_0 < |\dot{x}_m|$, where $|\dot{x}_m|$ corresponds to the relative extrema of the curves in Figure 3-1*b*.

We note that dry friction can be used to exhibit negative damping in many other mechanical systems. Rayleigh used arguments similar to the above to explain the production of oscillations in a violin string resulting from drawing a bow in one direction across the string. Dry friction also produces self-excited oscillations in a pendulum swinging from a rotating shaft, and it causes whirling of a shaft in a loose bearing. Dry friction can also be used to explain the chattering of the brake shoes against the wheels of a railroad car when the brakes are applied.

Self-excited oscillations resulting from a form of negative damping occur in many other physical systems. Lamb (1964) used a van der Pol model to describe the multimode operation of lasers, Lashinsky (1969) used van der Pol models to describe mode locking and frequency pulling in Q machines, Keen and Fletcher (1970) used a van der Pol model to describe suppression of the ion sound instability in an arc discharge, and DeNeef and Lashinsky (1973) used a van der Pol model for unstable waves on a beam–plasma system. A van der Pol model was also used to describe the effect of a beam modulation on standing waves on an electron beam-produced plasma in which the endplate potential reflects the electron beam back into itself (Nakamura, 1971).

Self-excited oscillations also occur in supersonic flutter of plates and shells (Fung, 1963; Dowell, 1975, Section 7.6) and in oil film journal bearings (Jain and Srinivasan, 1975). They also occur whenever a supersonic gas flows over a thin liquid film (Nayfeh and Saric, 1971a; Saric, Nayfeh, and Lekoudis, 1976).

3.2. Qualitative Analysis

The equations governing the systems discussed in the previous section have the form

$$\ddot{x} = f(x, \dot{x}) \tag{3.2.1}$$

where in general f is a nonlinear function of both x and $\dot{x}$. It is convenient to replace (3.2.1) by a system of first-order equations by introducing two new dependent variables:

$$x_1 = x \quad \text{and} \quad x_2 = \dot{x}$$

Then, in place of (3.2.1) we can write

$$\begin{aligned} \dot{x}_1 &= x_2 \\ \dot{x}_2 &= f(x_1, x_2) \end{aligned} \tag{3.2.2}$$

In subsequent chapters we consider more general systems of equations having the form

$$\dot{x}_i = X_i(x_1, x_2, x_3, \ldots, x_n) \tag{3.2.3}$$

for $i = 1, 2, \ldots, n$. Clearly (3.2.2) is a special case of (3.2.3). Here we begin the discussion with some general observations which apply to (3.2.3) for all n. Then we limit the discussion to the case $n = 2$.

It is convenient to introduce the matrix notation

$$\mathbf{x} = \begin{Bmatrix} x_1 \\ x_2 \\ \cdot \\ \cdot \\ \cdot \\ x_n \end{Bmatrix} \text{ and } \mathbf{X} = \begin{Bmatrix} X_1 \\ X_2 \\ \cdot \\ \cdot \\ \cdot \\ X_n \end{Bmatrix} \tag{3.2.4}$$

so that (3.2.3) becomes simply

$$\dot{\mathbf{x}} = \mathbf{X}(\mathbf{x}) \tag{3.2.5}$$

We assume that the vector function $\mathbf{X}$ has bounded first partial derivatives in the region D. It follows from the mean-value theorem for functions of several variables that in D there exists a constant M for which

$$|\mathbf{X}(\mathbf{x}) - \mathbf{X}(\mathbf{y})| \leqslant M|\mathbf{x} - \mathbf{y}| \tag{3.2.6}$$

Here $|\ |$ indicates the norm of the vectors, which is defined by

$$|\mathbf{x} - \mathbf{y}| = \sum_{i=1}^{n} |x_i - y_i| \tag{3.2.7}$$

and

$$|\mathbf{X}(\mathbf{x}) - \mathbf{X}(\mathbf{y})| = \sum_{i=1}^{n} |X_i(x_1, x_2, \ldots, x_n) - X_i(y_1, y_2, \ldots, y_n)| \tag{3.2.8}$$

Vector functions $\mathbf{X}$ that satisfy (3.2.6) are called *Lipschitz functions*. The significance of the vector function $\mathbf{X}$ being a Lipschitz function is that there exists a unique solution to the initial-value problem defined by (3.2.5) and a set of initial conditions having the form

$$\mathbf{x} = \mathbf{c} \quad \text{at } t = 0 \tag{3.2.9}$$

(e.g., Coddington and Levinson, 1955; Struble, 1962).

At any instant, we may regard the solution vector as a point in space. For $n = 1$, 2, or 3, this space can easily be identified with the ordinary physical space, but for n greater than 3 it is an abstract space. The point representing the solution is often called the state of the system, or simply the *state*. The totality of all possible states, corresponding to all initial conditions and all times in the range under consideration, forms the *state space*. In two dimensions this space is simply a plane. In the special case when the equations have the form (3.2.2), the state space is called the *phase plane*, as in Chapter 2.

It follows that we may view the response of any system to an initial disturbance as the motion of a single point through the state space. The path followed by this point representing the solution is a curve defined in terms of the parameter t. As in Chapter 2, it is called the *trajectory*. The portion of the trajectory corresponding to $t < 0$ is called the *negative half-trajectory*, while the portion corresponding to $t \geqslant 0$ is called the *positive half-trajectory*.

It may happen that the initial conditions, $\mathbf{c}$, are such that

$$X_i(\mathbf{c}) = 0 \qquad \text{for } i = 1, 2, \ldots, n \tag{3.2.10}$$

Then it follows that for all t

$$\dot{x}_i = 0 \qquad \text{for } i = 1, 2, \ldots, n \tag{3.2.11}$$

In this case the trajectory consists of a single point, $\mathbf{c}$. Such a trajectory is called a *singular trajectory*, or a *singular point*, and the corresponding solution is called a *singular solution*. Points that are not singular are called *regular*.

The state speed is defined by

$$v = \text{mag}(\dot{\mathbf{x}}) = \left[\sum_{i=1}^{n} \dot{x}_i^2\right]^{1/2} = \left[\sum_{i=1}^{n} X_i^2\right]^{1/2} \tag{3.2.12}$$

Thus the state speed is zero if, and only if, each $\dot{x}_i$ is zero. It follows that the state speed is zero at a singular point.

One can determine the components e_i of the unit vectors which are tangent to the trajectories from

$$e_i = \frac{\dot{x}_i}{\left[\sum_{j=1}^{n} \dot{x}_j^2\right]^{1/2}} \tag{3.2.13}$$

At every regular point, the direction field given by (3.2.13) is definite, and hence only one trajectory can pass through such a point. But at a singular point the direction field is not definite, and more than one trajectory can pass through such a point. There are cases for which all the X_n are zero at a point, but the direction field is definite. This occurs when the X_n have a common factor. Such cases are not included in this book.

If trajectories pass through a singular point, as in the case of the saddle point considered in Chapter 2, an infinite amount of time is required for the state to reach the singular point from any other point. This is proved in Chapter 2 and is indicated here by the fact that the system is both at rest and in equilibrium at a singular point. In other words, if the initial conditions start the motion at such a singular point, the state of the system never changes; or if the initial conditions start the motion on a trajectory going into a singular point, but not at the point itself, the state of the system continually changes but never reaches the singular point.

A nontrivial trajectory corresponds to a periodic motion if, and only if, it does not pass through a singular point and it is closed. This follows from the fact that the representative point, having begun its motion at an arbitrary point on the closed trajectory, will return to its initial point in a finite period of time (the period of oscillation) because the state speed is nonzero at every point of the trajectory.

One can obtain a good qualitative representation of the motion in the entire state space if one knows the character of the motion in the neighborhoods of the singular points. In the remainder of Section 3.2, we restrict our attention to systems for which $n = 2$. First we obtain the character of various singular points and in the process classify them; then we construct trajectories for the entire phase plane; and finally we study the stability of the motion in the neighborhood of a singular point. For a topological description of the singular points of systems with $n \geqslant 3$, we refer the reader to the book of Blaquière (1966) and the monograph of Tondl, Fiala, and Šklíba (1970).

3.2.1. A STUDY OF THE SINGULAR POINTS

As discussed above, the singular points of (3.2.5) are the solutions of (3.2.10). To study the behavior of the solution near one of these points, we shift the origin to this point by introducing the transformation

$$y_1 = x_1 - x_{10}, \qquad y_2 = x_2 - x_{20} \tag{3.2.14}$$

in (3.2.3) to obtain

$$\begin{aligned} \dot{y}_1 &= X_1(x_{10} + y_1, x_{20} + y_2) \\ \dot{y}_2 &= X_2(x_{10} + y_1, x_{20} + y_2) \end{aligned} \tag{3.2.15}$$

If X_1 and X_2 have bounded first partial derivatives at x_{10} and x_{20}, we rewrite (3.2.15) as

$$\begin{aligned} \dot{y}_1 &= X_1(x_{10}, x_{20}) + a_{11}y_1 + a_{12}y_2 + R_1(y_1, y_2) \\ \dot{y}_2 &= X_2(x_{10}, x_{20}) + a_{21}y_1 + a_{22}y_2 + R_2(y_1, y_2) \end{aligned} \tag{3.2.16}$$

where $a_{ij} = \partial X_i / \partial x_j(x_{10}, x_{20})$ and $R_j(y_1, y_2) = o(r)$ as $r = \sqrt{y_1^2 + y_2^2} \to 0$. Since $X_j(x_{10}, x_{20}) = 0$ for $j = 1$ and 2 and since R_1 and R_2 are small compared with $\sqrt{y_1^2 + y_2^2}$, one expects the character of the trajectories near the origin to be given by the linearized form of (3.2.16); namely

$$\dot{\mathbf{y}} = [A]\,\mathbf{y} \tag{3.2.17}$$

where

$$\mathbf{y} = \begin{Bmatrix} y_1 \\ y_2 \end{Bmatrix} \quad \text{and} \quad [A] = \begin{bmatrix} a_{11} & a_{12} \\ a_{21} & a_{22} \end{bmatrix} \tag{3.2.18}$$

This is so provided that $|[A]| = a_{11}a_{22} - a_{12}a_{21} \neq 0$ (i.e., the origin is an isolated singularity). The condition $|[A]| \neq 0$ is violated when the development of X_1 and X_2 begins with terms of higher order than the first. In the latter case, singularities of higher order and of types which differ from those obtainable from (3.2.17) can occur; examples of these can be found in Section 2.4.2 when $\Lambda = \frac{1}{4}$ and Section 2.4.3 when $\Lambda = 0$.

To study the singular points of (3.2.17), we find it convenient to introduce a linear transformation from $\mathbf{y}$ to $\mathbf{u}$ by using the nonsingular constant matrix $[P]$ according to

$$\mathbf{y} = [P]\,\mathbf{u} \tag{3.2.19}$$

This linear transformation preserves the topological features of the system (3.2.17). Under this transformation the origin is mapped into the origin, straight lines are mapped into straight lines, and parallel lines are mapped into parallel lines, with the spacing between them remaining proportional.

Substituting (3.2.19) into (3.2.17) and premultiplying the result by $[P]^{-1}$, we obtain

$$\dot{\mathbf{u}} = [B]\,\mathbf{u} \tag{3.2.20}$$

where

$$[B] = [P]^{-1}\,[A]\,[P] \tag{3.2.21}$$

Matrices $[A]$ and $[B]$, related as in (3.2.21), are said to be *similar matrices*. The significance of this is that these matrices have the same eigenvalues. This is shown next. The eigenvalues of $[B]$ are given by

$$\det\left([B] - \lambda[I]\right) = 0 \tag{3.2.22}$$

Using (3.2.21) and observing that

$$[P]^{-1}\,[I]\,[P] = [I]$$

we can rewrite (3.2.22) as

$$\det\left([P]^{-1}\left([A] - \lambda[I]\right)[P]\right) = 0$$

or, from the properties of determinants,

$$\det\left([P]^{-1}\right)\det\left([A] - \lambda[I]\right)\det\left([P]\right) = 0$$

and because

$$\det([P]^{-1})\det([P]) = 1$$

it is also true that

$$\det([A] - \lambda[I]) = 0 \tag{3.2.23}$$

We can choose $[P]$ in such a way as to make $[B]$ have the simplest possible form, the so-called *Jordan canonical form*. If the eigenvalues are distinct,

$$[B] = \begin{bmatrix} \lambda_1 & 0 \\ 0 & \lambda_2 \end{bmatrix} \tag{3.2.24}$$

If the eigenvalues are not distinct, $[B]$ may have the form (3.2.24) where $\lambda_1 = \lambda_2$, or it may have the following form:

$$[B] = \begin{bmatrix} \lambda & 1 \\ 0 & \lambda \end{bmatrix} \tag{3.2.25}$$

Generally, the transformation (3.2.19) defines each component of $\mathbf{y}$ as a combination of both components of $\mathbf{u}$. Consequently $\mathbf{u}$ may not be a convenient form for studying the response quantitatively. However $\mathbf{u}$ can be very convenient for studying the response qualitatively, which we do in this section.

The eigenvalues of $[A]$ are the solutions of

$$\begin{vmatrix} a_{11} - \lambda & a_{12} \\ a_{21} & a_{22} - \lambda \end{vmatrix} = 0$$

or

$$\lambda^2 - (a_{11} + a_{22})\lambda + a_{11}a_{22} - a_{12}a_{21} = 0 \tag{3.2.26}$$

Hence

$$\lambda_{1,2} = \tfrac{1}{2}p \pm (\tfrac{1}{4}p^2 - q)^{1/2} \tag{3.2.27}$$

where $p = a_{11} + a_{22}$, the trace of $[A]$; and $q = a_{11}a_{22} - a_{12}a_{21}$, the determinant of $[A]$. Thus the eigenvalues are real and distinct if $p^2 > 4q$, real and equal if $p^2 = 4q$, and complex conjugates if $p^2 < 4q$. These cases are discussed in order below.

The Case of Distinct Real Roots. In this case the Jordan canonical form is the following diagonal form:

$$[B] = \begin{bmatrix} \lambda_1 & 0 \\ 0 & \lambda_2 \end{bmatrix} \tag{3.2.28}$$

Hence (3.2.20) becomes

$$\dot{u}_1 = \lambda_1 u_1, \qquad \dot{u}_2 = \lambda_2 u_2 \tag{3.2.29}$$

whose solutions are

$$u_1 = u_{10} \exp(\lambda_1 t), \qquad u_2 = u_{20} \exp(\lambda_2 t) \tag{3.2.30}$$

Eliminating t from (3.2.30) gives

$$u_2 = u_{20}\left(\frac{u_1}{u_{10}}\right)^{\alpha}, \qquad \alpha = \frac{\lambda_2}{\lambda_1} \tag{3.2.31}$$

if $u_{10} \neq 0$. If $u_{10} = 0$, the half-trajectories coincide with the u_2-axis.

The behavior of the trajectories passing through the origin depends on the sign of α. If α is positive (i.e., λ_1 and λ_2 have the same sign), the origin is called a *node*, or *nodal point*, and the arrangement of the trajectories is shown in Figures 3-8*a* and 3-8*b*. When $\alpha > 1$, the trajectories are tangent to the u_1-axis, and when $\alpha < 1$, the trajectories are tangent to the u_2-axis at the origin. When λ_1 and λ_2 are positive, the representative point moves away from the origin as t increases, and the origin is called an unstable node. Figure 3-8 shows the arrangement of the trajectories for negative λ_1 and λ_2.

If λ_1 and λ_2 have different signs, the origin is called a *saddle point*; it is an unstable point. The arrangement of the trajectories is shown in Figure 3-9. It is clear that the u_1- and u_2-axes are integral curves.

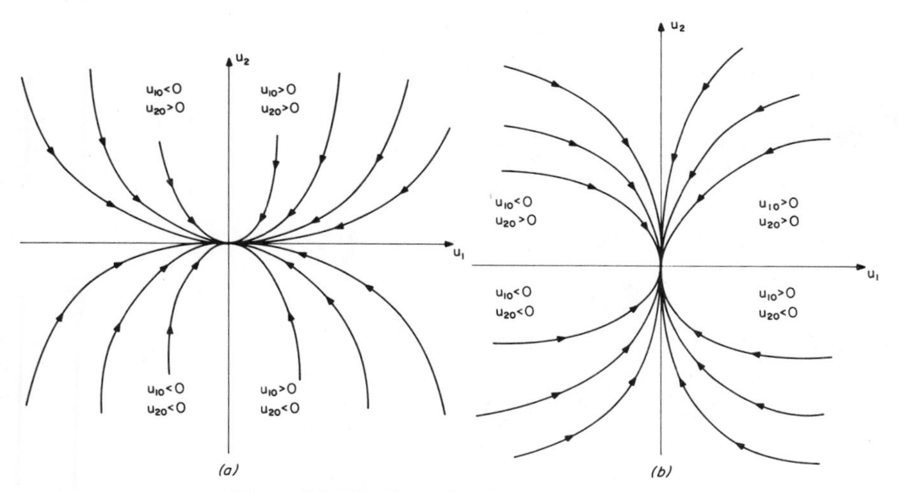

Figure 3-8. Stable nodes: (*a*) $\alpha > 1$; (*b*) $\alpha < 1$.

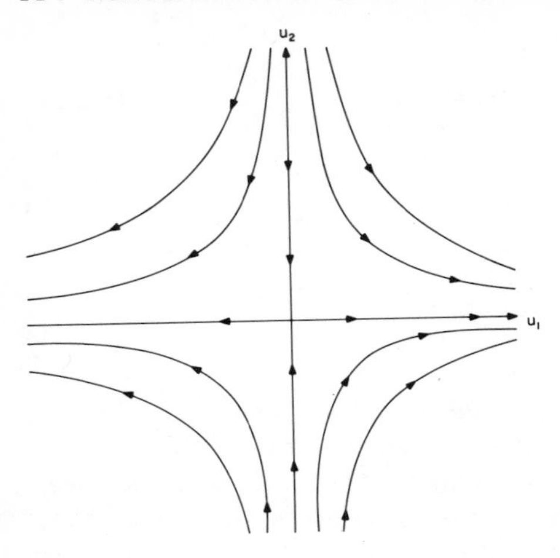

Figure 3-9. Saddle point.

The Case of Equal Real Roots. In this case two Jordan forms are possible, namely

$$[B] = \begin{bmatrix} \lambda & 0 \\ 0 & \lambda \end{bmatrix} \quad \text{or} \quad [B] = \begin{bmatrix} \lambda & 1 \\ 0 & \lambda \end{bmatrix} \tag{3.2.32}$$

When $[B]$ has the first form, (3.2.20) becomes

$$\dot{u} = \lambda u_1, \qquad \dot{u}_2 = \lambda u_2 \tag{3.2.33}$$

Hence

$$u_1 = u_{10} \exp(\lambda t), \qquad u_2 = u_{20} \exp(\lambda t), \qquad \frac{u_1}{u_2} = \frac{u_{10}}{u_{20}} \tag{3.2.34}$$

and the origin is called a node. The node is stable if $\lambda < 0$ and unstable if $\lambda > 0$. The arrangement of the trajectories when $\lambda < 0$ is shown in Figure 3-10.

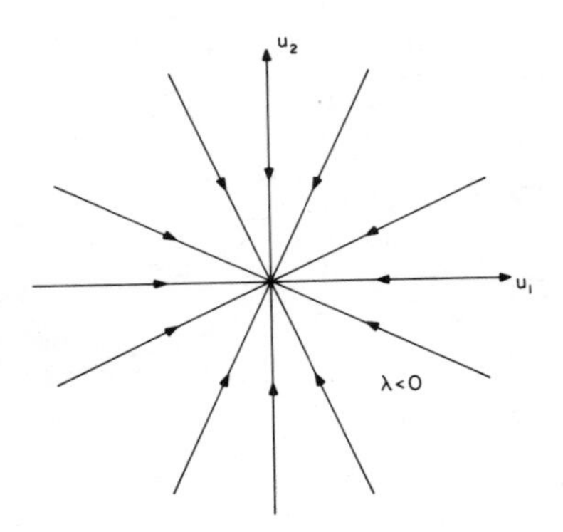

Figure 3-10. Stable node when the eigenvalues are equal and the Jordan form is diagonal.

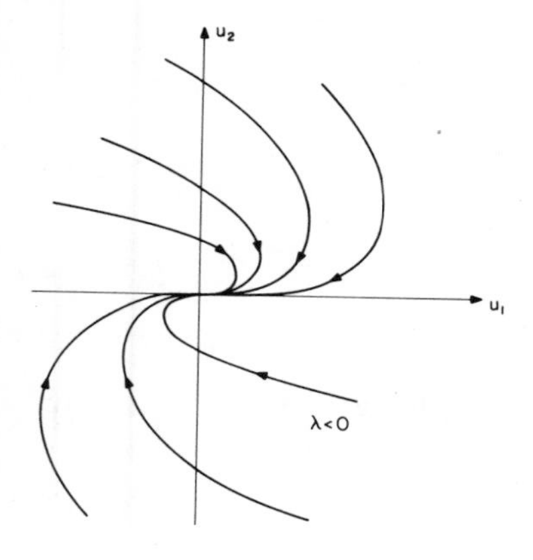

Figure 3-11. Stable node when the eigenvalues are equal and the Jordan form is not diagonal.

When $[B]$ has the second Jordan form, (3.2.20) leads to

$$\dot{u}_1 = \lambda u_1 + u_2, \qquad \dot{u}_2 = \lambda u_2 \tag{3.2.35}$$

with the solution

$$u_2 = u_{20} \exp(\lambda t), \qquad u_1 = (u_{10} + u_{20} t) \exp(\lambda t) \tag{3.2.36}$$

The origin is also called a node in this case. The node is stable if $\lambda < 0$ and unstable if $\lambda \geqslant 0$. The arrangement of the trajectories when $\lambda < 0$ is shown in Figure 3-11. The half-trajectories corresponding to $u_{20} = 0$ coincide with the u_1-axis. From (3.2.36)

$$\frac{u_2}{u_1} = \frac{u_{20}}{u_{10} + u_{20} t} \to \frac{1}{t} \quad \text{as} \quad t \to \infty \tag{3.2.37}$$

for $u_{20} \neq 0$.

It follows from (3.2.36) that u_2 cannot change sign. Thus any one trajectory must lie entirely in either the upper half plane or the lower half plane. Moreover, as t approaches infinity u_1 and u_2 have the same sign and u_2/u_1 approaches zero. Thus all trajectories in the lower half plane approach the origin from the left with zero slope, while all trajectories in the upper half plane approach the origin from the right with zero slope. This is illustrated in Figure 3-11.

The Case of Complex Roots. In this case the Jordan canonical form is

$$[B] = \begin{bmatrix} \lambda & 0 \\ 0 & \bar{\lambda} \end{bmatrix} \tag{3.2.38}$$

so that (3.2.20) can be rewritten as

$$\dot{u}_1 = \lambda u_1, \qquad \dot{u}_2 = \bar{\lambda} u_2 \tag{3.2.39}$$

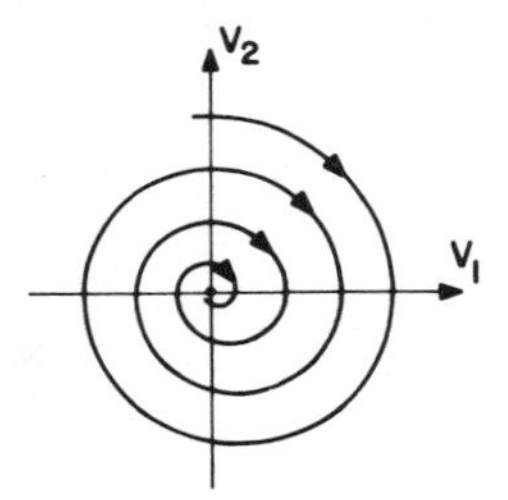

Figure 3-12. Stable focus.

where the overbar denotes the complex conjugate. Hence $u_2 = \bar{u}_1$. The solution of (3.2.39) is

$$u_1 = u_{10} \exp(\lambda_r t + i\lambda_i t) \tag{3.2.40}$$

where λ_r and λ_i are the real and imaginary parts of λ. In this case u_1 and u_2 and $[P]$ are complex; however y_1 and y_2 are real.

Letting $u_1 = v_1 + iv_2$ in (3.2.40), where v_1 and v_2 are real and separating real and imaginary parts, we obtain

$$v_1 = a \exp(\lambda_r t) \cos(\lambda_i t + \beta), \qquad v_2 = a \exp(\lambda_r t) \sin(\lambda_i t + \beta) \tag{3.2.41}$$

where $u_{10} = a \exp(i\beta)$. In this case the origin is called a *focal point*, or *focus*, when $\lambda_r \neq 0$, and it is called a *center* when $\lambda_r = 0$. The focus is stable if $\lambda_r < 0$ and unstable if $\lambda_r > 0$.

In a similar way u_2 can be represented as the sum of a real and an imaginary part. A possible arrangement for u_1 is shown in Figure 3-12 for $\lambda_r < 0$. The corresponding arrangement for u_2, which is not shown, is similar.

The preceding discussion shows that the character of the linear system (3.2.17) depends on the eigenvalues of $[A]$ and hence on the parameters p and q according to (3.2.27). Thus one can divide the pq-plane into regions characterizing different singular points as shown in Figure 3-13. When $q > 0$, λ_1 and λ_2 are complex or real having the same sign depending on whether $\Delta = q - \frac{1}{4}p^2$ is positive or negative. Hence $\Delta > 0$ and $p > 0$ correspond to unstable foci, while $\Delta > 0$ and $p < 0$ correspond to stable foci. Moreover $\Delta < 0$ and $p > 0$ correspond to unstable nodes. The curve $p^2 = 4q$, which corresponds to repeated eigenvalues, separates the nodes from the foci; while the positive q-axis, which separates the stable from the unstable foci, corresponds to centers.

Again we emphasize that usually it is y_1 and y_2 which are convenient for describing the state of the system (i.e., it is y_1 and y_2 which can readily be associated with observed data), not u_1 and u_2. But u_1 and u_2 are more convenient to use for the analysis; consequently we obtained a qualitative representation of the state space using u_1 and u_2. Though the y_1y_2-plane and the u_1u_2-plane are not the same, they are similar and the results above also provide a true, albeit qualitative, picture of the state plane in terms of y_1 and y_2.

There exist a number of techniques for graphically constructing the trajectories of systems governed by two first-order ordinary-differential equations (see, for example, Butenin, 1965; Andronov, Vitt, and Khaikin, 1966). In this book we describe the method of isoclines and Liénard's method.

3.2.2. THE METHOD OF ISOCLINES

If the equations describing the motion of the system are

$$\dot{x}_1 = X_1(x_1, x_2) \quad \text{and} \quad \dot{x}_2 = X_2(x_1, x_2) \tag{3.2.42}$$

the direction field is given by

$$\frac{dx_2}{dx_1} = \frac{X_2(x_1, x_2)}{X_1(x_1, x_2)} = \psi(x_1, x_2) \tag{3.2.43}$$

The curve $\psi(x_1, x_2) = c$ for a fixed c is called an *isocline*; it is the locus of all state points for which the slopes of the trajectories are the same.

This method consists of constructing a family of isoclines in the state plane as shown in Figure 3-14. The points of intersection of the isoclines are singular points. If a trajectory is to be initiated at a point A_1 on the isocline corresponding to $c = c_1$, we draw two straight lines passing through A_1 and having the slopes c_1 corresponding to the isocline on which A_1 lies and c_2 corresponding to the adjacent isocline. We extend these lines until they meet the isocline c_2 at a_2 and b_2 and take the point A_2, lying halfway between a_2 and b_2 on the isocline c_2, as the next point on the desired trajectory. Then we draw two straight lines passing through A_2 with the slopes c_2 and c_3. The third point on the trajectory

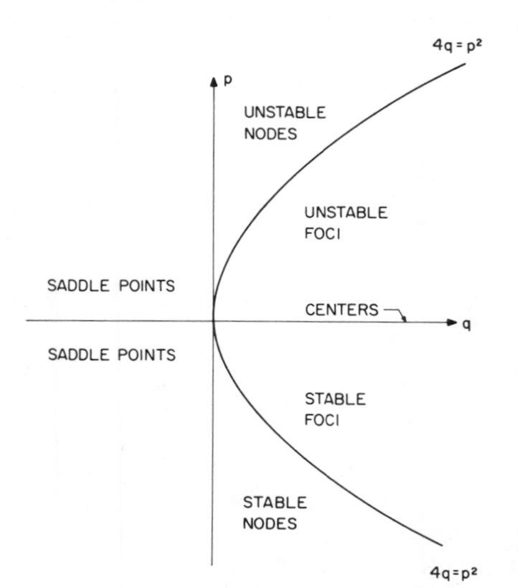

Figure 3-13. Singular points of equation (3.2.17).

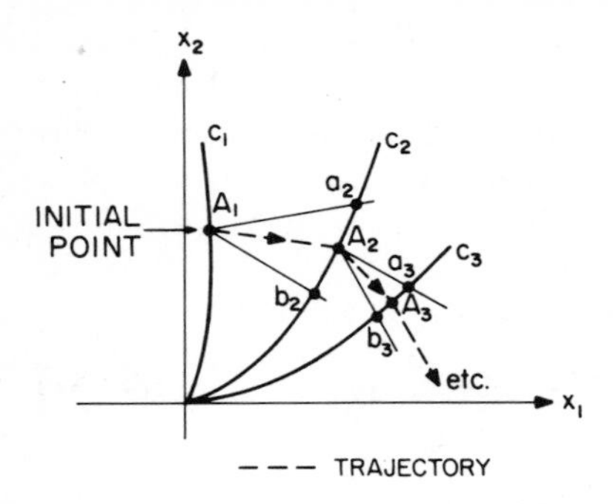

Figure 3-14. Method of isoclines.

A_3 is the point on c_3 halfway between a_3 and b_3. By repeating the process, we determine an approximate but fairly detailed portrait of the trajectory. Obviously the more dense the family of isoclines is, the more accurate the trajectory is.

3.2.3. LIÉNARD'S METHOD

A simple method was devised by Liénard (1928) for constructing the integral curves of equations having the form

$$\ddot{x} + \tilde{\phi}(\dot{x}) + \omega_0^2 x = 0 \tag{3.2.44}$$

We let $\tau = \omega_0 t$ and transform (3.2.44) into

$$x'' + \phi(x') + x = 0 \tag{3.2.45}$$

where primes indicate derivatives with respect to τ. Moreover we let $x_1 = x$ and $x_2 = x'$ and obtain the following differential equation for the trajectories:

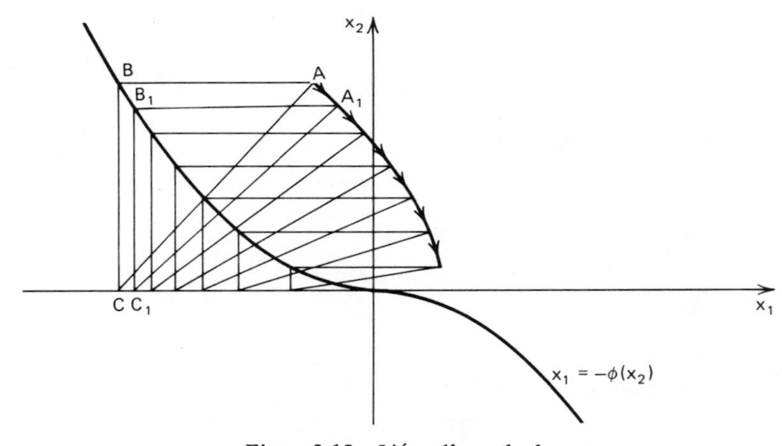

Figure 3-15. Liénard's method.

$$\frac{dx_2}{dx_1} = -\frac{\phi(x_2) + x_1}{x_2} \tag{3.2.46}$$

To draw the trajectories, we first plot the curve $x_1 = -\phi(x_2)$ in the phase plane as shown in Figure 3-15. To initiate the trajectory passing through point A, we draw a line parallel to the x_1-axis until it intersects the curve $x_1 = -\phi(x_2)$. We denote the point of intersection by B and construct a line which is parallel to the x_2-axis from B to C. Then the line CA is perpendicular to the direction field at A because the slope of CA is $BC/AB = x_2/[x_1 + \phi(x_2)]$. Hence we draw a line from A perpendicular to AC and approximate the integral curve by a short segment AA_1 along the direction field. Then, starting with A_1, we repeat the process.

3.3. Approximate Solutions

In this section we determine approximate solutions of the equations governing the oscillations of systems having a single degree of freedom. Different types of damping are considered. In particular, we obtain approximate solutions of equations having the form

$$\ddot{u} + \omega_0^2 u = \epsilon f(u, \dot{u}) \tag{3.3.1}$$

where ϵ is a small dimensionless parameter and $f(u, \dot{u})$ is a general nonlinear function of u and $\dot{u}$. Klotter (1955) treated systems having quadratic damping and arbitrary restoring forces. Rasmussen (1970, 1973, 1977), Soudack and Barkham (1971), Christopher and Brocklehurst (1974), and Beshai and Dokainish (1975) devised approximate solutions for oscillators with strong nonlinear forces and small damping. Ludeke and Wagner (1968) treated a generalized Duffing equation with large damping. Popov and Paltov (1960), Mendelson (1970), and Arya, Bojadziev, and Farooqui (1975) used the method of averaging to determine an approximate solution (Exercise 3.29) for

$$\ddot{u} + 2\mu\dot{u} + \omega_0^2 u = \epsilon f(u, \dot{u})$$

when $\mu = O(1)$. Tondl (1973b) investigated some properties of nonlinear systems and used his results to identify their damping characteristics. Cap (1974) devised a method based on the method of averaging to determine approximate solutions for systems governed by equations having the form (Exercise 3.30)

$$\ddot{u} + f(u) = \epsilon F(u, \dot{u})$$

in terms of the solutions of $\ddot{u} + f(u) = 0$.

Since ϵ is small, (3.3.1) is weakly nonlinear, and a number of perturbation methods are available for the determination of approximate solutions of this equation (see, for example, Nayfeh, 1973b). In this section we use the methods of multiple scales and averaging to determine first-order expansions, which are valid as t increases, and use these expansions to investigate the effects of the various types of damping.

3.3.1. THE METHOD OF MULTIPLE SCALES

In using this technique, we introduce different time scales according to (2.3.37) and expand the time derivatives according to (2.3.38). We consider u to be a function of the various new scales, instead of t, and assume that u can be represented by an expansion having the form

$$u = u_0(T_0, T_1, T_2, \ldots) + \epsilon u_1(T_0, T_1, T_2, \ldots) + \cdots \tag{3.3.2}$$

Substituting (2.3.37), (2.3.38), and (3.3.2) into (3.3.1) and equating coefficients of like powers of ϵ, we obtain

$$D_0^2 u_0 + \omega_0^2 u_0 = 0 \tag{3.3.3}$$

$$D_0^2 u_1 + \omega_0^2 u_1 = -2D_0 D_1 u_0 + f(u_0, D_0 u_0) \tag{3.3.4}$$

$$D_0^2 u_n + \omega_0^2 u_n = F(u_0, u_1, \ldots, u_{n-1}) \quad \text{for } n \geqslant 2 \tag{3.3.5}$$

It is convenient to write the general solution of (3.3.3) in the complex form

$$u_0 = A(T_1, T_2, \ldots) \exp(i\omega_0 T_0) + \bar{A}(T_1, T_2, \ldots) \exp(-i\omega_0 T_0) \tag{3.3.6}$$

The function A is still arbitrary at this level of approximation; it is determined by eliminating the secular terms (invoking the so-called *solvability conditions*) at the higher levels of approximation.

Substituting for u_0 into (3.3.4), we have

$$\begin{aligned} D_0^2 u_1 + \omega_0^2 u_1 = & -2i\omega_0 D_1 A \exp(i\omega_0 T_0) + 2i\omega_0 D_1 \bar{A} \exp(-i\omega_0 T_0) \\ & + f[A \exp(i\omega_0 T_0) + \bar{A} \exp(-i\omega_0 T_0), i\omega_0 A \exp(i\omega_0 T_0) \\ & - i\omega_0 \bar{A} \exp(-i\omega_0 T_0)] \end{aligned} \tag{3.3.7}$$

Depending on the function A, all particular solutions of (3.3.7) contain terms proportional to $T_0 \exp(\pm i\omega_0 T_0)$ (these are the so-called *secular terms*). Thus ϵu_1 can dominate u_0 for large t, resulting in a nonuniform expansion. We choose the function A so that secular terms are eliminated from u_0 and thereby obtain a uniformly valid expansion. To this end we expand $f[u_0, D_0 u_0]$ in a Fourier series as

$$f = \sum_{n=-\infty}^{\infty} f_n(A, \bar{A}) \exp(in\omega_0 T_0) \tag{3.3.8}$$

where

$$f_n(A, \bar{A}) = \frac{\omega_0}{2\pi} \int_0^{2\pi/\omega_0} f \exp(-in\omega_0 T_0)\, dT_0 \tag{3.3.9}$$

Hence the condition for the elimination of secular terms is

$$2iD_1A = \frac{1}{2\pi}\int_0^{2\pi/\omega_0} f \exp(-i\omega_0 T_0)\, dT_0 \tag{3.3.10}$$

For a first approximation we consider A to be a function of T_1 only and end the solution here. To solve (3.3.10) we find it convenient to express $A(T_1)$ in its polar form as

$$A(T_1) = \tfrac{1}{2}a(T_1)\exp[i\beta(T_1)] \tag{3.3.11}$$

so that we rewrite (3.3.6) as

$$u_0 = a(T_1)\cos\phi, \qquad \phi = \omega_0 T_0 + \beta(T_1) \tag{3.3.12}$$

Substituting (3.3.11) into (3.3.10) we have

$$i(a' + ia\beta') = \frac{1}{2\pi\omega_0}\int_0^{2\pi} f(a\cos\phi, -\omega_0 a\sin\phi)\exp(-i\phi)\, d\phi$$

Separating real and imaginary parts we obtain

$$a' = -\frac{1}{2\pi\omega_0}\int_0^{2\pi} \sin\phi\, f(a\cos\phi, -\omega_0 a\sin\phi)\, d\phi \tag{3.3.13}$$

$$\beta' = -\frac{1}{2\pi\omega_0 a}\int_0^{2\pi} \cos\phi\, f(a\cos\phi, -\omega_0 a\sin\phi)\, d\phi \tag{3.3.14}$$

Therefore a first approximation to the solution of (3.3.1) is

$$u = a(T_1)\cos[\omega_0 T_0 + \beta(T_1)] + O(\epsilon) \tag{3.3.15}$$

where a and β are given by (3.3.13) and (3.3.14).

3.3.2. THE METHOD OF AVERAGING

When $\epsilon = 0$, the solution of (3.3.1) can be written as

$$u = a\cos(\omega_0 t + \beta) = a\cos\phi \tag{3.3.16}$$

where a and β are constants. When $\epsilon \neq 0$, the solution of (3.3.1) can still be expressed in the form (3.3.16) provided that a and β are considered to be functions of t rather than constants. Thus (3.3.16) can be viewed as a transformation from the dependent variable $u(t)$ into the dependent variables $a(t)$ and $\beta(t)$. Since (3.3.1) and (3.3.16) constitute two equations for the three variables u, a, and β, we are at liberty to impose an additional condition. It is convenient to require the velocity to have the same form as the case when $\epsilon = 0$; that is,

$$\dot{u} = -\omega_0 a\sin\phi \tag{3.3.17}$$

To determine the equations describing $a(t)$ and $\beta(t)$, we differentiate (3.3.16) with respect to t and obtain

$$\dot{u} = -\omega_0 a \sin \phi + \dot{a} \cos \phi - a\dot{\beta} \sin \phi \tag{3.3.18}$$

Comparing (3.3.17) and (3.3.18), we find that

$$\dot{a} \cos \phi - a\dot{\beta} \sin \phi = 0 \tag{3.3.19}$$

Differentiating (3.3.17) with respect to t, we have

$$\ddot{u} = -\omega_0^2 a \cos \phi - \omega_0 \dot{a} \sin \phi - \omega_0 a\dot{\beta} \cos \phi \tag{3.3.20}$$

Substituting for $\dot{u}$ and $\ddot{u}$ in (3.3.1) yields

$$\omega_0 \dot{a} \sin \phi + \omega_0 a\dot{\beta} \cos \phi = -\epsilon f(a \cos \phi, -\omega_0 a \sin \phi) \tag{3.3.21}$$

Solving (3.3.19) and (3.3.21) for $\dot{a}$ and $\dot{\beta}$, we obtain

$$\dot{a} = -\frac{\epsilon}{\omega_0} \sin \phi \, f(a \cos \phi, -\omega_0 a \sin \phi) \tag{3.3.22}$$

$$\dot{\beta} = -\frac{\epsilon}{\omega_0 a} \cos \phi \, f(a \cos \phi, -\omega_0 a \sin \phi) \tag{3.3.23}$$

Equations (3.3.16), (3.3.22), and (3.3.23) are exactly equivalent to (3.3.1) because no approximations have been made yet.

For small ϵ, $\dot{a}$ and $\dot{\beta}$ are small; hence a and β vary much more slowly with t than $\phi = \omega_0 t + \beta$. In other words, a and β hardly change during the period of oscillation $2\pi/\omega_0$ of $\sin \phi$ and $\cos \phi$. This enables us to average out the variations in ϕ in (3.3.22) and (3.3.23). Averaging these equations over the period $2\pi/\omega_0$ and considering a, β, $\dot{a}$, and $\dot{\beta}$ to be constants while performing the averaging, we obtain the following equations describing the slow variations of a and β:

$$\dot{a} = -\frac{\epsilon}{2\pi\omega_0} \int_0^{2\pi} \sin \phi \, f(a \cos \phi, -\omega_0 a \sin \phi) \, d\phi \tag{3.3.24}$$

$$\dot{\beta} = -\frac{\epsilon}{2\pi\omega_0 a} \int_0^{2\pi} \cos \phi \, f(a \cos \phi, -\omega_0 a \sin \phi) \, d\phi \tag{3.3.25}$$

in agreement with (3.3.13) and (3.3.14) obtained in the previous section by using the method of multiple scales.

Next we use these results to analyze the effects of the different types of damping on the response of systems having a single degree of freedom.

3.3.3. DAMPING DUE TO FRICTION

In this section we consider the effects of linear damping, quadratic damping, constant damping, and hysteretic damping.

Linear Damping. In this case the equation of motion is

$$\ddot{u} + \omega_0^2 u = -2\epsilon\mu\dot{u} \tag{3.3.26}$$

Hence $f = -2\mu\dot{u}$, and (3.3.24) and (3.3.25) become

$$\dot{a} = -\frac{\epsilon\mu a}{\pi}\int_0^{2\pi} \sin^2\phi \, d\phi = -\epsilon\mu a \tag{3.3.27}$$

$$\dot{\beta} = -\frac{\epsilon\mu}{\pi}\int_0^{2\pi} \sin\phi \cos\phi \, d\phi = 0 \tag{3.3.28}$$

Solving these equations yields

$$a = a_0 \exp(-\epsilon\mu t), \qquad \beta = \beta_0 \tag{3.3.29}$$

where a_0 and β_0 are constants. Hence (3.3.16) becomes

$$u = a_0 \exp(-\epsilon\mu t)\cos(\omega_0 t + \beta_0) + O(\epsilon) \tag{3.3.30}$$

To this approximation the frequency is not affected by the viscosity, while the amplitude decays exponentially with time.

In Figure 3-16 the exact solution of (3.3.26) (solid line) and the amplitude given by (3.3.29) (dotted line) are plotted.

Quadratic Damping. In this case the equation describing the motion has the form

$$\ddot{u} + \omega_0^2 u = -\epsilon\dot{u}|\dot{u}| \tag{3.3.31}$$

Then $f = -\dot{u}|\dot{u}|$, and (3.3.24) and (3.3.25) become

$$\dot{a} = -\frac{\epsilon a^2 \omega_0}{2\pi}\int_0^{2\pi} \sin^2\phi|\sin\phi| \, d\phi$$

$$\dot{\beta} = -\frac{\epsilon a \omega_0}{2\pi}\int_0^{2\pi} \sin\phi\cos\phi|\sin\phi| \, d\phi$$

To perform the integration above, we break each integral into two parts—one with the limits 0 and π and the other with the limits π and 2π. That is

$$\dot{a} = -\frac{\epsilon a^2\omega_0}{2\pi}\left[\int_0^{\pi} \sin^3\phi \, d\phi - \int_{\pi}^{2\pi} \sin^3\phi \, d\phi\right] = -\frac{4}{3\pi}\epsilon a^2\omega_0 \tag{3.3.32}$$

$$\dot{\beta} = 0 \tag{3.3.33}$$

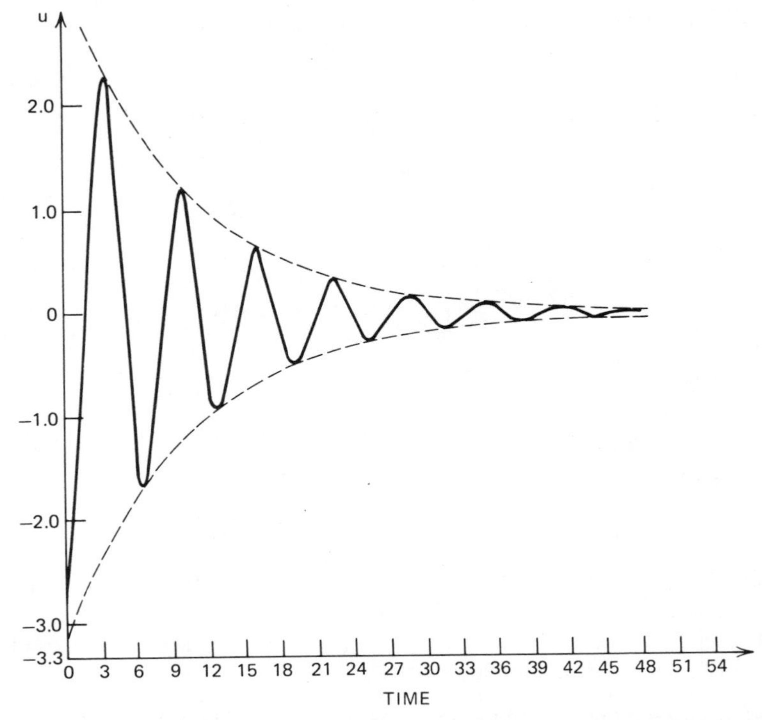

Figure 3-16. Linear damping: (– – –) equation (3.3.29); (——) numerical integration.

The solution of (3.3.33) is $\beta = \beta_0$, while the solution of (3.3.32) is

$$\frac{1}{a} - \frac{1}{a_0} = \frac{4\epsilon\omega_0}{3\pi} t$$

or

$$a = \frac{a_0}{1 + \dfrac{4\epsilon\omega_0 a_0}{3\pi} t} \tag{3.3.34}$$

where a_0 is a constant. Hence (3.3.16) becomes

$$u = \frac{a_0}{1 + \dfrac{4\epsilon\omega_0 a_0}{3\pi} t} \cos(\omega_0 t + \beta_0) + O(\epsilon) \tag{3.3.35}$$

As in the case of linear damping, the frequency is not affected by the damping to this order. However the amplitude decays algebraically rather than exponentially with time.

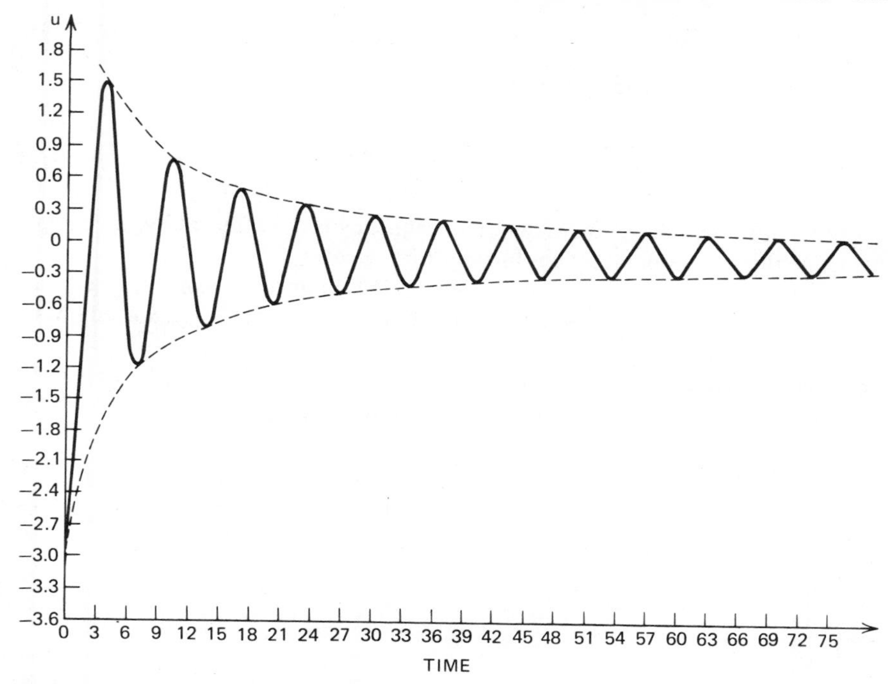

Figure 3-17. Quadratic damping: (– – –) equation (3.3.34); (——) numerical integration.

In Figure 3-17 the exact solution of (3.3.31) (obtained by numerical integration) and the amplitude as given by (3.3.34) are plotted (solid and dotted lines, respectively).

It follows from (3.3.27) and (3.3.32) that initially the rate at which the amplitude of the response decreases is proportional to the amplitude of the initial disturbance for linear damping and to the square of the amplitude of the initial disturbance for quadratic damping. Thus when the amplitude of the initial disturbance is large, one expects the initial decay to be slower for linear damping than for quadratic damping. If the initial disturbance is small, one expects the opposite to be true. A glance at Figures 3-16 and 3-17 will confirm this.

Coulomb Damping. In this case

$$\ddot{u} + \omega_0^2 u = f = \begin{cases} -\mu & \text{when } \dot{u} > 0 \\ \mu & \text{when } \dot{u} < 0 \end{cases} \tag{3.3.36}$$

Substituting for f into (3.3.24) and (3.3.25) and splitting the integration as in the previous case, we obtain

$$\dot{a} = -\frac{\epsilon\mu}{2\pi\omega_0}\left[\int_0^{\pi} \sin\phi \, d\phi - \int_{\pi}^{2\pi} \sin\phi \, d\phi\right] = -\frac{2\epsilon\mu}{\pi\omega_0} \tag{3.3.37}$$

$$\dot{\beta} = 0 \tag{3.3.38}$$

The solutions of these equations are

$$a = a_0 - \frac{2\epsilon\mu}{\pi\omega_0} t, \qquad \beta = \beta_0 \tag{3.3.39}$$

Hence to this level of approximation, the frequency is not affected by the damping, while the amplitude decreases linearly with time.

In Figure 3-18*a* the exact solution of (3.3.36) (obtained by numerical integration) and the amplitude given by (3.3.39) are plotted (the solid and dotted lines, respectively).

The phase plane can be obtained exactly by Liénard's method (Meissner, 1932). The motion can be described by the differential equation

$$\ddot{x} + \phi(\dot{x}) + x = 0$$

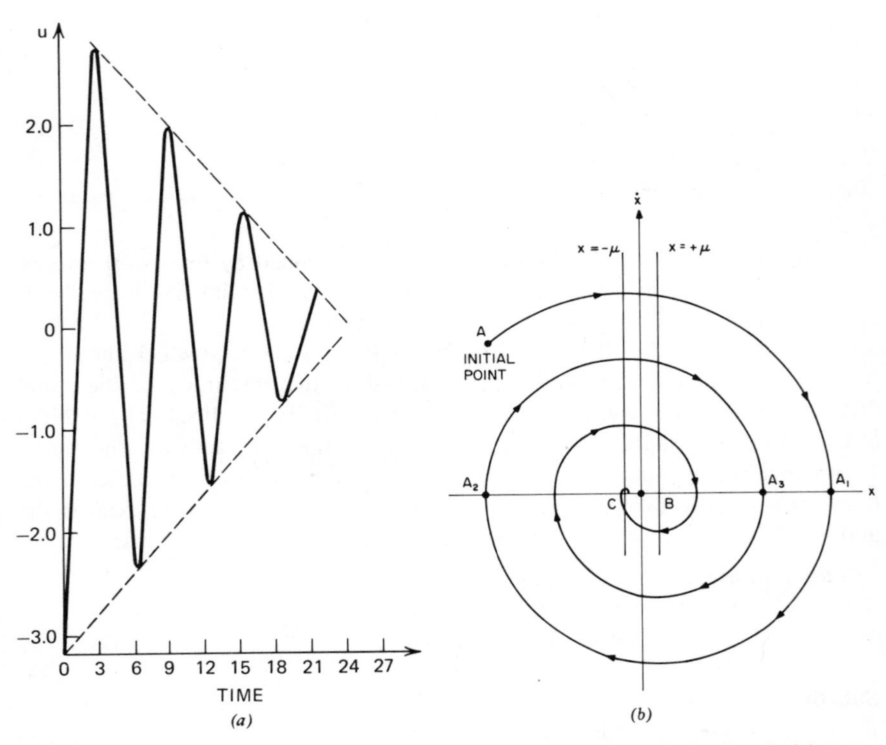

Figure 3-18. (*a*) Coulomb damping: (– – –) equation (3.3.39); (——) numerical integration. (*b*) Liénard's method applied to Coulomb damping.

where

$$\phi(\dot{x}) = \begin{cases} \mu & \text{for } \dot{x} > 0 \\ -\mu & \text{for } \dot{x} < 0 \end{cases}$$

We draw the lines $x = \pm\mu$ in the state plane as shown in Figure 3-18b. The trajectories consist of a series of circular arcs having the centers C and B, depending on whether the representative point is in the upper or lower half-plane. Thus if a trajectory is initiated at A, we draw a circle clockwise with center C and radius CA until it meets the x-axis at A_1. Then we switch the center to B and draw half a circle clockwise with a radius BA_1 which meets the x-axis again at A_2. We switch the center to C and draw half a circle clockwise with a radius CA_2 which meets the x-axis at A_3. We continue the process until the representative point intersects the x-axis between C and B. The motion ceases at such a point because the maximum possible friction force exceeds the force in the spring.

Hysteretic Damping. As an example we let the springs be linear and denote the constant of the top spring by k and that of the bottom spring by ϵm in Figure 3-2b. Then if we neglect other forms of damping, the equation of motion becomes

$$\ddot{x} + \omega_0^2 x = \epsilon f \tag{3.3.40}$$

where $\omega_0^2 = k/m$ and

$$-f = \begin{cases} x + x_s - x_b & x_b \geqslant x \geqslant x_c \\ -x_s & x_c \geqslant x \geqslant x_d \\ x - x_s - x_d & x_a \geqslant x \geqslant x_d \\ x_s & x_b \geqslant x \geqslant x_a \end{cases} \tag{3.3.41}$$

where

$$x_c = x_b - 2x_s \quad \text{and} \quad x_a = x_d + 2x_s$$

Substituting for f from (3.4.41) into (3.3.24) and (3.3.25), we obtain

$$\dot{a} = \frac{\epsilon}{2\pi\omega_0}\left[\int_{x_b}^{x_c} (x + x_s - x_b) \sin\phi \, d\phi(x) - \int_{x_c}^{x_d} x_s \sin\phi \, d\phi(x) + \int_{x_d}^{x_a} (x - x_s - x_d) \sin\phi \, d\phi(x) + \int_{x_a}^{x_b} x_s \sin\phi \, d\phi(x)\right] \tag{3.3.42}$$

$$\dot{\beta} = \frac{\epsilon}{2\pi\omega_0}\left[\int_{x_b}^{x_c} (x + x_s - x_b)\cos\phi\, d\phi(x) - \int_{x_c}^{x_d} x_s \cos\phi\, d\phi(x)\right.$$

$$\left. + \int_{x_d}^{x_a} (x - x_s - x_d)\cos\phi\, d\phi(x) + \int_{x_a}^{x_b} x_s\cos\phi\, d\phi(x)\right] \qquad (3.3.43)$$

where the integrations over the cycle *BCDAB* of Figure 3-2*b* have been broken into four parts over the segments *BC*, *CD*, *DA*, and *AB*.

To perform the integrations in (3.3.42) and (3.3.43), we change the integration variable from x to ϕ. To accomplish this, we note that the period in the variable ϕ is 2π; and since the motion is periodic, we set $\phi = 0$ at point B so that $\phi = \pi$ at point D. Since $x = a\cos\phi$,

$$\begin{aligned} x_b &= a, \qquad x_c = x_b - 2x_s = a\cos\phi_1 \\ x_d &= -a, \qquad x_a = x_d + 2x_s = a\cos\phi_2 \end{aligned} \qquad (3.3.44)$$

where

$$\phi_1 = \cos^{-1}\left(\frac{a - 2x_s}{a}\right) \quad \text{and} \quad \phi_2 = \cos^{-1}\left(\frac{2x_s - a}{a}\right) \qquad (3.3.45)$$

Then (3.3.42) and (3.3.43) become

$$\dot{a} = \frac{\epsilon}{2\pi\omega_0}\left[\int_0^{\phi_1} (a\cos\phi + x_s - a)\sin\phi\, d\phi - x_s\int_{\phi_1}^{\pi} \sin\phi\, d\phi\right.$$

$$\left. + \int_{\pi}^{\phi_2} (a\cos\phi - x_s + a)\sin\phi\, d\phi + x_s\int_{\phi_2}^{2\pi} \sin\phi\, d\phi\right] \qquad (3.3.46)$$

$$\dot{\beta} = \frac{\epsilon}{2\pi\omega_0 a}\left[\int_0^{\phi_1} (a\cos\phi + x_s - a)\cos\phi\, d\phi - x_s\int_{\phi_1}^{\pi} \cos\phi\, d\phi\right.$$

$$\left. + \int_{\pi}^{\phi_2} (a\cos\phi - x_s + a)\cos\phi\, d\phi + x_s\int_{\phi_2}^{2\pi} \cos\phi\, d\phi\right] \qquad (3.3.47)$$

Performing the integrations in (3.3.46) and (3.3.47) and using (3.3.45), we obtain

$$\dot{a} = \frac{2\epsilon x_s}{\pi\omega_0 a}(x_s - a) \qquad (3.3.48)$$

$$\dot{\beta} = \frac{\epsilon}{\pi\omega_0}\left[\frac{1}{2}\cos^{-1}\left(\frac{a - 2x_s}{a}\right) - \left(1 - \frac{2x_s}{a}\right)\left(\frac{x_s}{a} - \frac{x_s^2}{a^2}\right)^{1/2}\right] \qquad (3.3.49)$$

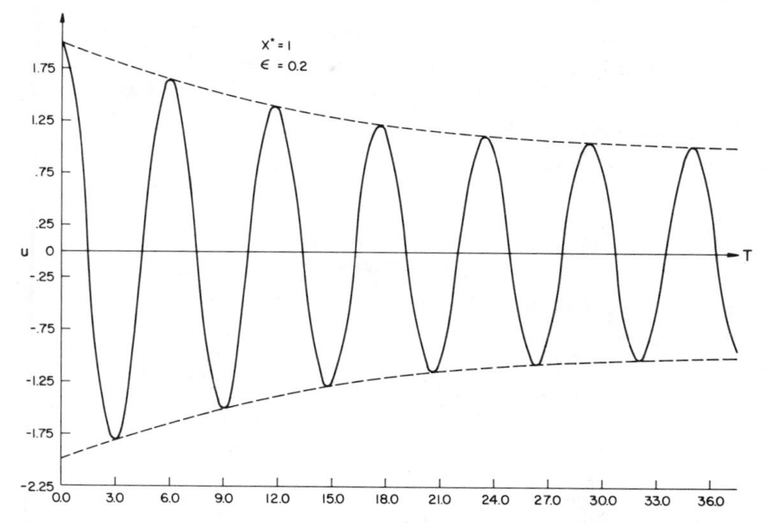

Figure 3-19. Hysteretic damping: (– – –) equation (3.3.50); (——) numerical integration.

The solution of (3.3.48) is

$$a + x_s \ln (a - x_s) = -\frac{2\epsilon x_s}{\pi\omega_0} t + c \qquad (3.3.50)$$

where c is a constant of integration. Thus a is given implicitly in terms of t. We cannot integrate (3.3.49) as we were able to do for a; however we note that, as $t \to \infty$, $a \to x_s$ and $\dot{\beta} \to \epsilon/2\omega_0$, which agrees with the exact solution when the hysteretic mechanism is not activated.

In Figure 3-19 the numerical solution of (3.3.40) is compared with the asymptotic result (3.3.50).

3.3.4. NEGATIVE DAMPING

For a comprehensive treatment of self-excited mechanical oscillations, we refer the reader to the monographs of Tondl (1970b, 1976b). Tondl (1968) and El-Owaidy (1974) studied the perturbations of a class of self-excited oscillators, while Nguyen (1976b) analyzed some properties of the generalized van der Pol equation. George, Gunderson, and Hahn (1975) studied sustained small oscillations in nonlinear control systems, while St. Hilaire (1976) studied the response of a self-excited structure. Warncke (1973) studied the vibrations of a rigid rotor running in a sliding bearing, while Kelzon and Yakovlev (1974) experimentally investigated self-excited vibrations of a high-speed rotor. Tarantovich and Kohnkin (1975) treated a two-frequency system with a stiff excitation. Klotter (1955) and Klotter and Kreyszig (1957, 1960) studied equations of the form

$$\ddot{u} - \mu \dot{u}^2 f(u) \operatorname{sgn} \dot{u} + g(u) = 0 \qquad (3.3.51)$$

Gumpert (1974) included the effects of friction when he investigated the existence of nonrelaxation cycles for self-excited vibrations. Gyozo (1974) investigated the elimination of self-excitations by the use of Lanchester-type dampers.

For the Rayleigh oscillator

$$f = \dot{u} - \dot{u}^3 \tag{3.3.52}$$

so that (3.3.1) becomes

$$\ddot{u} + \omega_0^2 u = \epsilon(\dot{u} - \dot{u}^3) \tag{3.3.53}$$

and (3.3.24) and (3.3.25) become

$$\dot{a} = \frac{\epsilon a}{2\pi} \int_0^{2\pi} (\sin^2 \phi - \omega_0^2 a^2 \sin^4 \phi)\, d\phi = \tfrac{1}{2} \epsilon a(1 - \tfrac{3}{4} \omega_0^2 a^2) \tag{3.3.54}$$

$$\dot{\beta} = \frac{\epsilon}{2\pi} \int_0^{2\pi} (1 - \omega_0^2 a^2 \sin^2 \phi) \sin \phi \cos \phi\, d\phi = 0 \tag{3.3.55}$$

The solution of (3.3.55) is $\beta = \beta_0$, while the solution of (3.3.54) can be obtained by separation of variables. The result is

$$a^2 = \frac{a_0^2}{\tfrac{3}{4} \omega_0^2 a_0^2 + (1 - \tfrac{3}{4} \omega_0^2 a_0^2) \exp(-\epsilon t)} \tag{3.3.56}$$

where a_0 is the initial amplitude.

Equation (3.3.56) shows that the amplitude of oscillation tends to $a_s = 2/\sqrt{3}\,\omega_0$, irrespective of the magnitude of the initial amplitude as long as it is different from zero. Oscillators of this type are called *self-sustained oscillators*. Equation (3.3.54) shows that when $a < a_s$, $\dot{a} > 0$, and hence a tends to increase; while when $a > a_s$, $\dot{a} < 0$, and a tends to decrease. The value $a = a_s$ is a stable amplitude.

In Figure 3-20 the numerical solutions of (3.3.53) are compared with the

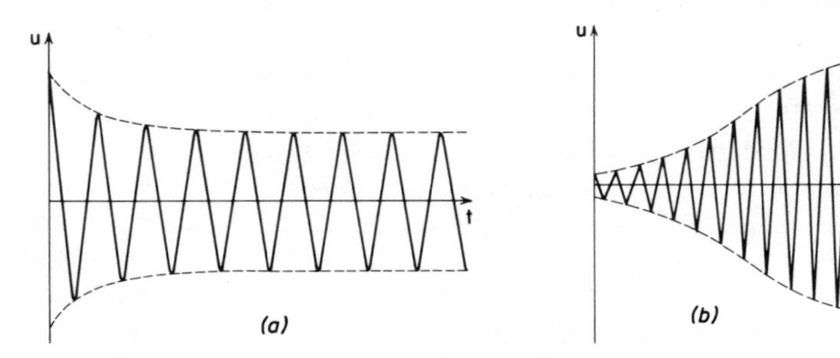

Figure 3-20. Rayleigh oscillator: (*a*) large initial disturbance; (*b*) small initial disturbance; (– – –) equation (3.3.56); (——) numerical integration.

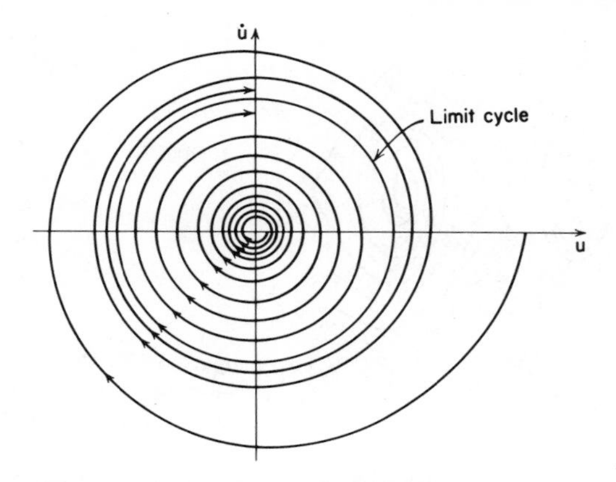

Figure 3-21. Phase plane for the Rayleigh oscillator.

asymptotic results (3.3.56) for $a_0 > a_s$ and $a_0 < a_s$, respectively. In Figure 3-20*a* the initial amplitude is greater than the amplitude of the limit cycle; hence the damping is positive initially and the amplitude decays until it reaches the limit cycle. In contrast the initial amplitude in Figure 3-20*b* is less than that of the limit cycle; hence the damping is negative initially and the amplitude increases until it reaches the limit cycle. The corresponding trajectories in the phase plane are shown in Figure 3-21.

When ϵ is very large, systems such as the one being considered here exhibit rather jerky motions called relaxation oscillations. These are discussed in Section 3.5.

3.3.5. EXAMPLES OF POSITIVELY DAMPED SYSTEMS HAVING NONLINEAR RESTORING FORCES

For a pendulum with viscous damping the governing equation has the form

$$\ddot{\theta} + 2\hat{\mu}\dot{\theta} + \omega_0^2 \sin\theta = 0 \tag{3.3.57}$$

and the exact phase plane, obtained by numerical methods, is shown in Figure 3-22. The singular points of (3.3.57) are θ = integral multiples of π, as in the undamped case. The even multiples of π are stable foci, while the odd multiples are saddle points.

We seek an approximate solution of (3.3.57) that is uniformly valid near $\theta = 0$ and that accounts for nonlinear effects. To this end we expand $\sin\theta$ and retain only the first two terms; consequently (3.3.57) becomes

$$\ddot{\theta} + 2\hat{\mu}\dot{\theta} + \omega_0^2(\theta - \tfrac{1}{6}\theta^3) = 0 \tag{3.3.58}$$

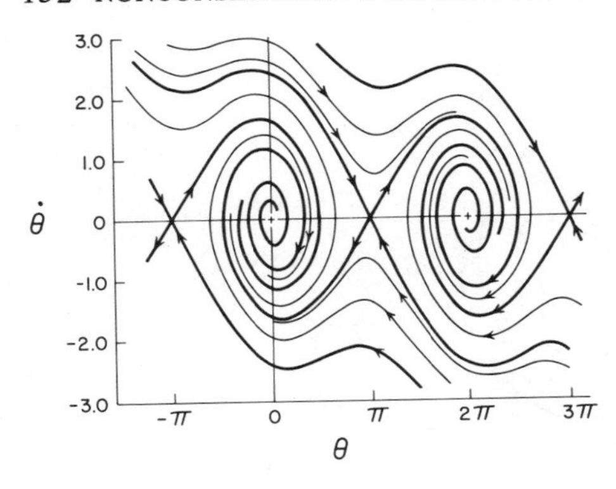

Figure 3-22. Phase plane for a simple pendulum with viscous damping.

We are primarily concerned with lightly damped motions. Consequently we let

$$\hat{\mu} = \epsilon^2 \mu \tag{3.3.59}$$

where ϵ is a measure of the amplitude of the motion. Now the nonlinear and damping terms will interact at the same level of approximation.

Following the method of multiple scales we assume

$$\theta(t;\epsilon) = \epsilon\theta_1(T_0, T_1, T_2) + \epsilon^2\theta_2(T_0, T_1, T_2) + \epsilon^3\theta_3(T_0, T_1, T_2) + \cdots \tag{3.3.60}$$

Substituting (3.3.59) and (3.3.60) into (3.3.58) and equating coefficients of like powers of ϵ, we obtain

$$D_0^2\theta_1 + \omega_0^2\theta_1 = 0 \tag{3.3.61}$$

$$D_0^2\theta_2 + \omega_0^2\theta_2 = -2D_0D_1\theta_1 \tag{3.3.62}$$

$$D_0^2\theta_3 + \omega_0^2\theta_3 = -2D_0D_1\theta_2 - 2D_0D_2\theta_1 - D_1^2\theta_1 - 2\mu D_0\theta_1 + \tfrac{1}{6}\theta_1^3 \tag{3.3.63}$$

We can write the solution of (3.3.61) in the form

$$\theta_1 = A_1(T_1, T_2)\exp(i\omega_0 T_0) + cc \tag{3.3.64}$$

where cc stands for the complex conjugate of the preceding terms. Substituting (3.3.64) into (3.3.62) yields

$$D_0\theta_2 + \omega_0^2\theta_2 = -2i\omega_0 D_1 A_1 \exp(i\omega_0 T_0) + cc \tag{3.3.65}$$

To eliminate secular terms from θ_2, we must put

$$D_1 A_1 \equiv 0 \tag{3.3.66}$$

Thus $A_1 = A_1(T_2)$. It follows that

$$\theta_2 = A_2(T_1, T_2) \exp(i\omega_0 T_0) + cc \tag{3.3.67}$$

Substituting (3.3.64) and (3.3.67) into (3.3.63), we obtain

$$\begin{aligned} D_0^2\theta_3 + \omega_0^2\theta_3 = &-2i\omega_0 D_1 A_2 \exp(i\omega_0 T_0) - 2i\omega_0 D_2 A_1 \exp(i\omega_0 T_0) \\ &- 2i\omega_0 \mu A_1 \exp(i\omega_0 T_0) + \tfrac{1}{6}A_1^3 \exp(3i\omega_0 T_0) \\ &+ \tfrac{1}{2}A_1^2\bar{A}_1 \exp(i\omega_0 T_0) + cc \end{aligned} \tag{3.3.68}$$

To eliminate secular terms from θ_3 we must put

$$-2i\omega_0 D_1 A_2 - 2i\omega_0 D_2 A_1 - 2i\omega_0 \mu A_1 + \tfrac{1}{2}A_1^2\bar{A}_1 = 0 \tag{3.3.69}$$

Since $A_1 = A_1(T_2)$, it follows from (3.3.69) that $A_2 \propto T_1$, and hence θ_2/θ_1 is unbounded as $T_1 \to \infty$ unless $D_1 A_2 = 0$ and

$$2i\omega_0(D_2 A_1 + \mu A_1) - \tfrac{1}{2}A_1^2\bar{A}_1 = 0 \tag{3.3.70}$$

Consequently θ_1 and θ_2 have exactly the same form, and we may drop θ_2.

It is convenient to introduce the polar notation

$$A_1 = \tfrac{1}{2}a \exp(i\beta) \tag{3.3.71}$$

where a and β are real functions of T_2. Then substituting (3.3.71) into (3.3.70) and separating the result into real and imaginary parts, we obtain

$$a' + \mu a = 0, \qquad \beta' + \frac{a^2}{16\omega_0} = 0 \tag{3.3.72}$$

where primes denote derivatives with respect to T_2. Thus

$$a = a_0 \exp(-\mu T_2), \qquad \beta = -\frac{a_0^2}{32\omega_0\mu} \exp(-2\mu T_2) + \beta_0 \tag{3.3.73}$$

where a_0 and β_0 are constants of integration. In terms of the original variables and parameters, we obtain

$$\theta = \epsilon a_0 e^{-\hat{\mu}t} \cos\left[\omega_0 t - \frac{\epsilon^2 a_0^2}{32\omega_0\hat{\mu}} e^{-2\hat{\mu}t} + \beta_0\right] + O(\epsilon^3)$$

for the first approximation.

When the motion is started from rest by giving the pendulum an initial displacement, this result becomes

$$\theta = \theta_0 e^{-\hat{\mu}t} \cos\left[\omega_0 t - \frac{\theta_0^2}{32\omega_0\hat{\mu}}(e^{-2\hat{\mu}t} - 1)\right] + O(\epsilon^2) \tag{3.3.74}$$

where $\theta_0 = \epsilon a_0$, the initial displacement. Thus in the first approximation, the amplitude decays as in the linear case; but unlike the linear case, here the frequency is a function of the amplitude. However we note that the dependence of the frequency on the amplitude vanishes rapidly after the motion begins.

It is essential to make the nonlinearity and damping interact at the same order [recall (3.3.59)]. If the damping term had been lower in order than the nonlinear term, the solution would have predicted a slight perturbation of the linear damped solution, not the nonlinear frequency-amplitude dependence obtained above.

As a rule in this book, we do not consider these highly damped systems. Instead we focus on the lightly damped systems, which strongly exhibit nonlinear effects.

For a pendulum with quadratic damping, the governing equation has the form

$$\ddot{\theta} + 2\hat{\mu}\dot{\theta}|\dot{\theta}| + \omega_0^2 \sin\theta = 0 \tag{3.3.75}$$

We can obtain an exact expression for the trajectories in the phase plane. We note that, as before, the singular points of (3.3.75) are θ = integral multiples of π; even multiples are stable foci, while odd multiples are saddle points. We let $\dot{\theta} = v$ so that (3.3.75) can be rewritten as

$$\dot{v} = -2\hat{\mu}v|v| - \omega_0^2 \sin\theta \tag{3.3.76}$$

It follows that

$$v\frac{dv}{d\theta} = -2\hat{\mu}v|v| - \omega_0^2 \sin\theta \tag{3.3.77}$$

Hence,

$$\frac{1}{2}\frac{dv^2}{d\theta} \pm 2\hat{\mu}v^2 = -\omega_0^2 \sin\theta \tag{3.3.78}$$

which we can integrate to obtain v^2 as a function of θ. The sign is chosen according to (3.3.77). It follows that

$$v = \pm\left[c\, e^{\mp 4\hat{\mu}\theta} + \frac{2\omega_0^2}{1 + 16\hat{\mu}^2}\cos\theta \mp \frac{8\omega_0^2\hat{\mu}}{1 + 16\hat{\mu}^2}\sin\theta\right]^{1/2} \tag{3.3.79}$$

where c is a constant of integration. The signs change where the trajectories cross the θ-axis; at each crossing a new constant of integration must be chosen to insure the continuity of the trajectory. The exact trajectories in the phase plane, constructed by using (3.3.79), are shown in Figure 3-23.

Next we obtain a uniformly valid approximate solution of (3.3.75) by using the method of multiple scales. Again we suppose the motion is started from rest by giving the pendulum an initial displacement, and we assume

$$\theta(t;\epsilon) = \epsilon\theta_1(T_0, T_1, T_2) + \epsilon^2\theta_2(T_0, T_1, T_2) + \epsilon^3\theta_3(T_0, T_1, T_2) + \cdots \tag{3.3.80}$$

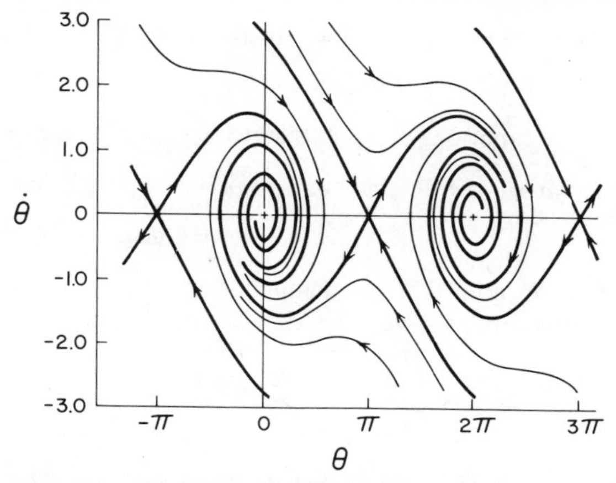

Figure 3-23. Phase plane for a simple pendulum with quadratic damping.

where ϵ is a measure of the amplitude. In this case we must put

$$\hat{\mu} = \epsilon\mu \tag{3.3.81}$$

so that the nonlinear and damping terms will first appear at the same order. Again we expand $\sin\theta$ and retain only the first two terms.

It follows from (3.3.75) that

$$D_0^2\theta_1 + \omega_0^2\theta_1 = 0 \tag{3.3.82}$$

$$D_0^2\theta_2 + \omega_0^2\theta_2 = -2D_0D_1\theta_1 \tag{3.3.83}$$

$$D_0^2\theta_3 + \omega_0^2\theta_3 = -2D_0D_1\theta_2 - 2D_0D_2\theta_1 - D_1^2\theta_1 - 2\mu D_0\theta_1|D_0\theta_1| + \tfrac{1}{6}\theta_1^3 \tag{3.3.84}$$

Proceeding as before, we express the solution of (3.3.82) in the form

$$\theta_1 = A(T_1, T_2)\exp(i\omega_0 T_0) + cc \tag{3.3.85}$$

Then we find that secular terms are eliminated from θ_2 if A is a function of T_2 only. As in the case of a pendulum with viscous damping, we may drop θ_2. It follows that

$$\begin{aligned} D_0^2\theta_3 + \omega_0^2\theta_3 = {}& -2i\omega_0[A'\exp(i\omega_0 T_0) - \bar{A}'\exp(-i\omega_0 T_0)] \\ & - 2\mu[i\omega_0 A\exp(i\omega_0 T_0) - i\omega_0\bar{A}\exp(-i\omega_0 T_0)]\,|i\omega_0 A \\ & \cdot\exp(i\omega_0 T_0) - i\omega_0\bar{A}\exp(i\omega_0 T_0)| + \tfrac{1}{6}A^3\exp(3i\omega_0 T_0) \\ & + \tfrac{1}{2}A^2\bar{A}\exp(i\omega_0 T_0) + \tfrac{1}{2}\bar{A}^2 A\exp(-i\omega_0 T_0) \\ & + \tfrac{1}{6}\bar{A}^3\exp(-3i\omega T_0) \end{aligned} \tag{3.3.86}$$

In this case it is convenient to introduce the polar notation

$$A = \tfrac{1}{2} a \exp(i\beta) \tag{3.3.87}$$

where a and β are real functions of T_2. Introducing (3.3.87) into (3.3.86) leads to

$$D_0^2\theta_3 + \omega_0^2\theta_3 = 2\omega_0 a' \sin\phi + 2\omega_0 a\beta' \cos\phi + \tfrac{1}{8}a^3 \cos\phi + \tfrac{1}{24}a^3 \cos 3\phi + 2\mu\omega_0^2 a^2 \sin\phi|\sin\phi| \tag{3.3.88}$$

where $\phi = \omega_0 t + \beta$. We note that the damping term is periodic in ϕ and can be expanded in a Fourier series (as we did earlier in this section); that is

$$\sin\phi|\sin\phi| = \sum_{n=1}^{\infty} f_n \sin n\phi \tag{3.3.89}$$

where in particular $f_1 = 8/3\pi$. Thus secular terms are eliminated from θ_3 if

$$a' + \frac{8\mu\omega_0}{3\pi} a^2 = 0, \qquad \beta' + \frac{a^2}{16\omega_0} = 0 \tag{3.3.90}$$

It follows that

$$a = \frac{3\pi a_0}{3\pi + 8\mu\omega_0 a_0 T_2}, \qquad \beta = \frac{9\pi^2 a_0}{128\mu\omega_0^2(3\pi + 8\mu\omega_0 a_0 T_2)} + \beta_0 \tag{3.3.91}$$

where a_0 and β_0 are constants of integration.

Rewriting the solution in terms of the original variables and using the initial conditions, we obtain

$$\theta = \frac{3\pi\theta_0}{3\pi + 8\hat{\mu}\omega_0\theta_0 t} \cos\left[\omega_0 t + \frac{9\pi^2\theta_0}{128\omega_0^2\hat{\mu}(3\pi + 8\hat{\mu}\omega_0\theta_0 t)} - \frac{3\pi\theta_0}{128\omega_0^2\hat{\mu}}\right] + O(\epsilon^3) \tag{3.3.92}$$

when $\theta_0 = \epsilon a_0$, the initial displacement. Again the amplitude decays as in the linear case; but unlike the linear case here the frequency is a function of the amplitude.

3.4. Nonstationary Vibrations

In this section we consider nonlinear systems having components that are time dependent. These components may be masses, lengths, rigidities, coefficients of friction, material properties, the spin rate of the circle in Section 2.1.4 and the parabola in Section 2.4.3, etc. In analyzing the response of such systems by using the method of averaging, one uses the method of variation of parameters as in Section 3.3.2 to arrive at the following equations describing the amplitudes and the phases:

$$\dot{\mathbf{x}} = \epsilon \mathbf{f}(\mathbf{x}, t; \epsilon) \tag{3.4.1}$$

where ϵ is a small dimensionless real parameter and $\mathbf{x}$ and $\mathbf{f}$ are real vectors. Since ϵ is small, $\mathbf{x}$ is a slowly varying function of t, and to the first approximation one is tempted to average (3.4.1) and obtain

$$\dot{\boldsymbol{\xi}} = \epsilon \mathbf{f}_0(\boldsymbol{\xi}) \tag{3.4.2}$$

where

$$\mathbf{f}_0(\boldsymbol{\xi}) = \lim_{T \to \infty} \left[\frac{1}{T} \int_t^{t+T} \mathbf{f}(\boldsymbol{\xi}, s; 0)\, ds \right] \tag{3.4.3}$$

Krylov and Bogoliubov (1947) and Mitropolsky (1965) gave sufficient conditions for the solutions of (3.4.1) to remain close to the solutions of (3.4.2) satisfying $\boldsymbol{\xi}(t_0) = \mathbf{x}(t_0)$, over time intervals which are $O(\epsilon^{-1})$. Hale (1969) generalized these results to systems governed by

$$\dot{\mathbf{x}} = \epsilon \mathbf{f}(\mathbf{x}, \mathbf{y}, t; \epsilon) \quad \text{and} \quad \dot{\mathbf{y}} = [A]\mathbf{y} + \epsilon \mathbf{g}(\mathbf{x}, \mathbf{y}, t; \epsilon) \tag{3.4.4}$$

where $\mathbf{x}$ and $\mathbf{y}$ are real vectors and $[A]$ is a matrix which may be constant or periodic in t. These results were generalized further by Volosov (1961) to systems governed by

$$\dot{\mathbf{x}} = \epsilon \mathbf{f}(\mathbf{x}, \mathbf{y}; \epsilon) \quad \text{and} \quad \dot{\mathbf{y}} = \mathbf{g}(\mathbf{x}, \mathbf{y}; \epsilon) \tag{3.4.5}$$

The above systems depend on one time scale. Mitropolsky (1965) obtained an asymptotic solution valid for $t = O(\epsilon^{-1})$ for systems governed by

$$\dot{\mathbf{x}} = \epsilon \mathbf{f}(\mathbf{x}, t, \epsilon t; \epsilon) \tag{3.4.6}$$

while Sethna and Balachandra (1974a) obtained a solution valid for finite time for systems with several slow time scales. Sethna (1967a) obtained an asymptotic solution valid for all t for systems governed by

$$\dot{\mathbf{x}} = \epsilon \mathbf{f}(\mathbf{x}, t) + \epsilon \mathbf{g}(\mathbf{x}, \epsilon t) \tag{3.4.7}$$

while Roseau (1969) generalized these results to systems governed by

$$\dot{\mathbf{x}} = \epsilon \mathbf{f}(\mathbf{x}, \mathbf{y}, t, \epsilon t; \epsilon) \quad \text{and} \quad \dot{\mathbf{y}} = [A]y + \epsilon \mathbf{g}(\mathbf{x}, \mathbf{y}, t, \epsilon t; \epsilon) \tag{3.4.8}$$

where $\mathbf{f}$ and $\mathbf{g}$ are almost periodic in t and periodic in ϵt. Sethna (1969, 1973) removed the restriction of almost periodicity and obtained results valid for all t provided that $\mathbf{f}$ and $\mathbf{g}$ are bounded in t for all t. Balachandra and Sethna (1975) proved under certain hypotheses the existence of bounded solutions of

$$\dot{\mathbf{x}} = \epsilon \mathbf{f}(\mathbf{x}, \mathbf{y}, t, \epsilon t; \epsilon) \quad \text{and} \quad \dot{\mathbf{y}} = \mathbf{g}(\mathbf{x}, \mathbf{y}, t, \epsilon t; \epsilon) \tag{3.4.9}$$

which approach certain special solutions of a derived averaged system of equations as $\epsilon \to 0$.

In this section we consider a single-degree-of-freedom system having slowly varying coefficients. Thus we consider equations having the form

$$\ddot{u} + \omega_0^2(\tau)u = \epsilon f(u, \dot{u}, \tau) \tag{3.4.10}$$

where $\tau = \epsilon t$ and ϵ is a small, positive, dimensionless parameter. Since ϵ is small, τ is a slow scale. Such systems were treated by Kuzmak (1959), Nayfeh (1969), Meyer (1976), Kaper (1976), and Rubenfeld (1977). Systems whose material properties exhibit time-dependent characteristics were studied by Paria (1968).

The presence of the slow scale in (3.4.10) suggests the use of the method of multiple scales. Thus we seek an expansion for the solution in terms of the two scales τ and ϕ, where

$$\frac{d\phi}{dt} = \omega_0(\tau) \tag{3.4.11}$$

In terms of these variables, the time derivatives become

$$\begin{aligned} \frac{d}{dt} &= \omega_0 \frac{\partial}{\partial \phi} + \epsilon \frac{\partial}{\partial \tau} \\ \frac{d^2}{dt^2} &= \omega_0^2 \frac{\partial^2}{\partial \phi^2} + \epsilon \left(2\omega_0 \frac{\partial^2}{\partial \tau \partial \phi} + \omega_0' \frac{\partial}{\partial \phi}\right) + \epsilon^2 \frac{\partial^2}{\partial \tau^2} \end{aligned} \tag{3.4.12}$$

Hence (3.4.10) becomes

$$\omega_0^2 \frac{\partial^2 u}{\partial \phi^2} + \epsilon \left(2\omega_0 \frac{\partial^2 u}{\partial \tau \partial \phi} + \omega_0' \frac{\partial u}{\partial \phi}\right) + \epsilon^2 \frac{\partial^2 u}{\partial \tau^2} + \omega_0^2 u = \epsilon f\left[u, \omega_0 \frac{\partial u}{\partial \phi} + \epsilon \frac{\partial u}{\partial \tau}, \tau\right] \tag{3.4.13}$$

We expand u as

$$u = u_0(\phi, \tau) + \epsilon u_1(\phi, \tau) + \cdots \tag{3.4.14}$$

Substituting (3.4.14) into (3.4.13) and equating coefficients of like powers of ϵ, we obtain

$$\omega_0^2 \left(\frac{\partial^2 u_0}{\partial \phi^2} + u_0\right) = 0 \tag{3.4.15}$$

$$\omega_0^2 \left(\frac{\partial^2 u_1}{\partial \phi^2} + u_1\right) = -2\omega_0 \frac{\partial^2 u_0}{\partial \tau \partial \phi} - \omega_0' \frac{\partial u_0}{\partial \phi} + f\left(u_0, \omega_0 \frac{\partial u_0}{\partial \phi}, \tau\right) \tag{3.4.16}$$

The general solution of (3.4.15) can be written in the complex form

$$u_0 = A(\tau) \exp(i\phi) + cc \tag{3.4.17}$$

Hence, (3.4.16) becomes

$$\omega_0^2\left(\frac{\partial^2 u_1}{\partial\phi^2}+u_1\right) = -i(2\omega_0 A' + \omega_0' A)\exp(i\phi) + i(2\omega_0\bar{A}' + \omega_0'\bar{A})\exp(-i\phi)$$
$$+ f\{A\exp(i\phi) + \bar{A}\exp(-i\phi), i\omega_0[A\exp(i\phi) - \bar{A}\exp(-i\phi)], \tau\} \quad (3.4.18)$$

Eliminating the terms in (3.4.18) that produce secular terms in u_1 yields

$$2i\omega_0 A' + i\omega_0' A = \frac{1}{2\pi}\int_0^{2\pi} f(A, \bar{A}, \phi, \tau)\exp(-i\phi)\,d\phi \quad (3.4.19)$$

Letting $A = \frac{1}{2}a\exp(i\beta)$ in (3.4.19), where a and β are real, and then separating real and imaginary parts, we obtain

$$a' = -\frac{\omega_0'}{2\omega_0}a - \frac{1}{2\pi\omega_0}\int_0^{2\pi}\sin\psi\, f(a\cos\psi, -\omega_0 a\sin\psi, \tau)\,d\psi \quad (3.4.20)$$

$$\beta' = -\frac{1}{2\pi\omega_0 a}\int_0^{2\pi}\cos\psi\, f(a\cos\psi, -\omega_0 a\sin\psi, \tau)\,d\psi \quad (3.4.21)$$

where $\psi = \phi + \beta$. From (3.4.14) and (3.4.17) we can write a first approximation to u as

$$u = a(\tau)\cos[\phi + \beta(\tau)] + O(\epsilon) \quad (3.4.22)$$

where a and β are obtained from (3.4.20) and (3.4.21). The frequency of oscillation is given by

$$\omega = \frac{d}{dt}(\phi + \beta) = \omega_0 + \epsilon\beta' \quad (3.4.23)$$

When ω_0 is a constant and f is independent of τ, (3.4.20) and (3.4.21) reduce to (3.3.13) and (3.3.14) obtained for the case of systems having constant parameters. Next we specialize these formulas for conservative systems and for systems restrained by linear elastic forces and resisted by nonlinear friction forces.

3.4.1. CONSERVATIVE SYSTEMS

To allow for the variation of mass, charge, and length, we take the equation of motion in the form

$$\frac{d}{dt}[m(\tau)\dot{u}] + k(\tau)u = \epsilon g(u, \tau) \quad (3.4.24)$$

Hence

$$\ddot{u} + \omega_0^2 u = \epsilon f(u, \dot{u}, \tau) = -\frac{\epsilon}{m}[m'\dot{u} - g(u, \tau)] \quad (3.4.25)$$

where $\omega_0^2 = k/m$. Substituting for f into (3.4.20) yields

$$a' = -\frac{1}{2}\left(\frac{\omega_0'}{\omega_0} + \frac{m'}{m}\right)a \tag{3.4.26}$$

whose solution is

$$a = a_0\left[\frac{\omega_0(0)\,m(0)}{\omega_0(\tau)\,m(\tau)}\right]^{1/2} = a_0\left[\frac{m(0)\,k(0)}{m(\tau)\,k(\tau)}\right]^{1/4} \tag{3.4.27}$$

where $a(0) = a_0$ is a constant. Substituting for f into (3.4.21) gives

$$\beta' = -\frac{1}{2\pi\omega_0 ma}\int_0^{2\pi} \cos\psi\, g(a\cos\psi, \tau)\, d\psi \tag{3.4.28}$$

Combining (3.4.23) and (3.4.28), we obtain the following expression for the frequency:

$$\omega = \omega_0(\tau) - \frac{\epsilon}{2\pi\omega_0 ma}\int_0^{2\pi} \cos\psi\, g(a\cos\psi, \tau)\, d\psi \tag{3.4.29}$$

Thus the oscillations described by (3.4.24) are nearly sinusoidal, with an amplitude that varies inversely with the fourth root of $m(\tau)\,k(\tau)$ and a frequency that is given by (3.4.29).

We apply these results to the motion of a particle having a slowly varying mass $m(\tau)$ and restrained by a nonlinear cubic spring whose parameters vary slowly with time. The equation describing the motion can be written as

$$\frac{d}{dt}\,[m(\tau)\,\dot{x}] + k(\tau)\,x + \alpha(\tau)\,x^3 = 0 \tag{3.4.30}$$

Letting $x = \sqrt{\epsilon}\,u$ and rearranging terms, we rewrite (3.4.30) as

$$\frac{d}{dt}\,[m(\tau)\dot{u}] + k(\tau)u = \epsilon g(u, \tau) = -\epsilon\alpha(\tau)u^3 \tag{3.4.31}$$

Hence the amplitude of oscillation is given by (3.4.27), and from (3.4.29) the frequency of oscillation is

$$\omega = \sqrt{\frac{k(\tau)}{m(\tau)}} + \frac{3\epsilon a_0^2\alpha(\tau)\sqrt{m(0)\,k(0)}}{8m(\tau)\,k(\tau)} + O(\epsilon^2) \tag{3.4.32}$$

As a second example we consider the oscillation of a simple pendulum whose mass m is constant but whose length varies slowly with time. The equation describing the motion is

$$\frac{d}{dt}\,[l^2(\tau)\,\dot{\theta}] + gl(\tau)\sin\theta = 0 \tag{3.4.33}$$

We let $\theta = \sqrt{\epsilon}\, u$ in (3.4.33), expand $\sin \theta$, and obtain

$$\frac{d}{dt}\,[l^2(\tau)\dot{u}] + gl(\tau)u = \tfrac{1}{6}\,\epsilon gl(\tau)u^3 + O(\epsilon^2) \tag{3.4.34}$$

Comparing (3.4.31) with (3.4.34), we conclude that the solution of the latter can be obtained from the solution of the former if we identify $m(\tau)$ with $l^2(\tau)$, $k(\tau)$ with $gl(\tau)$, and α with $-\frac{1}{6}gl(\tau)$. Hence the amplitude and frequency of the pendulum are given by

$$\begin{aligned} a &= a_0\left[\frac{l(0)}{l(\tau)}\right]^{3/4} \\ \omega &= \sqrt{\frac{g}{l(\tau)}}\left\{1 - \tfrac{1}{16}\,\epsilon a_0^2\left[\frac{l(0)}{l(\tau)}\right]^{3/2}\right\} + O(\epsilon^2) \end{aligned} \tag{3.4.35}$$

3.4.2. SYSTEMS WITH NONLINEAR DAMPING ONLY

In this case the equation describing the oscillations has the form

$$\frac{d}{dt}\,[m(\tau)\dot{u}] + k(\tau)u = \epsilon g(\dot{u}, \tau) \tag{3.4.36}$$

Hence

$$\ddot{u} + \omega_0^2 u = \epsilon f(u, \dot{u}, \tau) = -\frac{\epsilon}{m}\,[m'\dot{u} - g(\dot{u}, \tau)] \tag{3.4.37}$$

Substituting for f into (3.4.20) and (3.4.21) yields

$$a' = -\tfrac{1}{2}\left(\frac{\omega_0'}{\omega_0} + \frac{m'}{m}\right)a - \frac{1}{2\pi m\omega_0}\int_0^{2\pi} \sin\psi\, g(-\omega_0 a \sin\psi, \tau)\, d\psi \tag{3.4.38}$$

$$\beta' = 0 \tag{3.4.39}$$

Hence to the first approximation, the oscillations are sinusoidal with the frequency $\omega_0(\tau)$ and an amplitude given by (3.4.38).

We apply these results to the case of quadratic damping, that is, to a system governed by

$$\frac{d}{dt}\,[m(\tau)\dot{u}] + k(\tau)u = \epsilon g(\dot{u}, \tau) = -\epsilon\alpha(\tau)\dot{u}|\dot{u}| \tag{3.4.40}$$

Substituting for g into (3.4.38) and carrying out the indicated integration, we obtain

$$a' = -\frac{1}{2}\left(\frac{\omega_0'}{\omega_0} + \frac{m'}{m}\right)a - \frac{4\alpha\omega_0}{3\pi m}\,a^2 \tag{3.4.41}$$

Equation (3.4.41) has the integral

$$\frac{1}{a\sqrt{\omega_0(\tau)\,m(\tau)}} - \frac{1}{a_0\sqrt{\omega_0(0)\,m(0)}} = \frac{4}{3\pi}\int_0^\tau \frac{\alpha(\tau)\,\omega_0^{1/2}(\tau)}{m^{3/2}(\tau)}\,d\tau$$

or

$$a = a_0\left[\frac{m(0)\,k(0)}{m(\tau)\,k(\tau)}\right]^{1/4}\left\{1 + \frac{4a_0}{3\pi}\int_0^\tau \alpha(\tau)\left[\frac{m(0)\,k(0)\,k(\tau)}{m^7(\tau)}\right]^{1/4} d\tau\right\}^{-1} \tag{3.4.42}$$

where a_0 is the initial amplitude. When k and m are constants, (3.4.42) reduces to (3.3.34).

3.5. Relaxation Oscillations

In this section we return to Rayleigh's equation

$$\ddot{u} + u - \epsilon(\dot{u} - \dot{u}^3) = 0 \tag{3.5.1}$$

We have already studied the behavior of the solutions of this equation for small ϵ. We found that these solutions are approximately sinusoidal with slowly varying amplitudes and phases and that they always approach a limit cycle as $t \to \infty$, irrespective of the initial conditions. There are electrical and mechanical systems of interest for which ϵ is large; van der Pol (1922) mentions that there are cases for which ϵ is approximately 10^5. Here we consider the case of large ϵ.

Figure 3-24 shows the trajectories in the phase plane for the solutions of

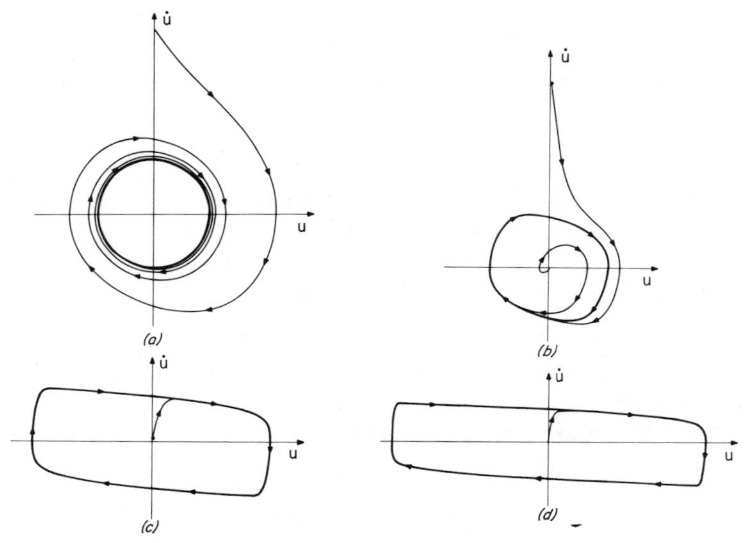

Figure 3-24. Phase planes for Rayleigh's equation; (*a*) $\epsilon = 0.01$; (*b*) $\epsilon = 0.1$; (*c*) $\epsilon = 1.0$; (*d*) $\epsilon = 10.0$.

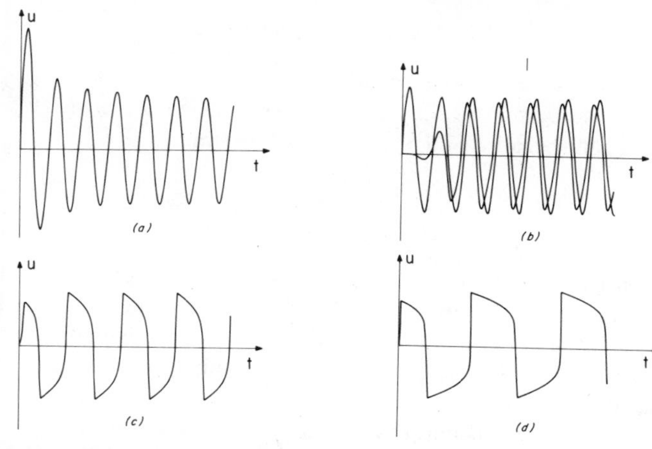

Figure 3-25. Solutions of Rayleigh's equation: (*a*) $\epsilon = 0.01$; (*b*) $\epsilon = 0.1$; (*c*) $\epsilon = 1.0$; (*d*) $\epsilon = 10.0$.

(3.5.1) for four values of ϵ, namely $\epsilon = 0.01$, 0.1, 1, and 10. For small ϵ, the resulting limit cycle (i.e., closed curve) is nearly a circle, and the corresponding motion is nearly a harmonic motion with a definite amplitude as shown in Figure 3-25*a*. As ϵ increases, the limit cycle in the phase plane deviates more and more from a circle, and the corresponding motion deviates more and more from a simple harmonic motion. We note from Figure 3-25 that the distortion of the limit cycle (steady-state motion) from a sinusoidal form increases markedly with increasing ϵ. When $\epsilon = 10$, the corresponding motion is jerky and consists of slowly varying stretches followed by abrupt changes. Such oscillations are often called *relaxation oscillations* (van der Pol and van der Mark, 1928). For the van der Pol circuit the energy is being stored in the capacitor during the slowly varying part of the motion, while during the abrupt changes the energy is being suddenly released.

To analyze the behavior of these relaxation oscillations, we introduce the change of variable

$$\xi = \epsilon^{-1}u, \qquad \dot{u} = v \tag{3.5.2}$$

Then (3.5.1) can be replaced by the following pair of first-order ordinary-differential equations:

$$\epsilon^{-1}\dot{v} = -\xi + v - v^3 \tag{3.5.3}$$

$$\dot{\xi} = \epsilon^{-1}v \tag{3.5.4}$$

The set of equations (3.5.3) and (3.5.4) is a special case of the set

$$\epsilon^{-1}\dot{\mathbf{x}} = \mathbf{f}(\mathbf{x}, \mathbf{y}, t; \epsilon^{-1}) \tag{3.5.5}$$

$$\dot{\mathbf{y}} = \mathbf{g}(\mathbf{x}, \mathbf{y}, t; \epsilon^{-1}) \tag{3.5.6}$$

where **x**, **y**, **f**, and **g** are real-valued vectors.

When $\epsilon \to \infty$, the system (3.5.5) and (3.5.6) is called a *singularly perturbed system* of equations because (3.5.5) and (3.5.6) become

$$\mathbf{f}(\mathbf{x}, \mathbf{y}, t) = 0 \tag{3.5.7}$$

$$\dot{\mathbf{y}} = \mathbf{g}(\mathbf{x}, \mathbf{y}, t) \tag{3.5.8}$$

whose solution cannot satisfy in general all the initial and boundary conditions because (3.5.5) is reduced from a differential to an algebraic equation. Thus the solutions of (3.5.7) and (3.5.8) cannot be expected to approximate the solutions of (3.5.5) and (3.5.6) for all t. Consequently a number of techniques have been developed to determine uniform solutions for such singularly perturbed systems. Foremost among these techniques are asymptotic methods, especially the method of matched asymptotic equations. Dorodnicyn (1947) and Cartwright and Littlewood (1947) applied this technique to the van der Pol equation in the phase plane, while Cole (1968, Section 2.6) also applied it to the van der Pol equation but in the physical plane. In addition to the method of matched asymptotic expansions, a number of other techniques have been developed. They involve idealizing the system under consideration by representing the solution as a combination of continuous segments and discontinuous or quasi-discontinuous segments (Andronov, Vitt, and Khaikin, 1966; Minorsky, 1962). Numerical solutions of the van der Pol equation for large ϵ were obtained by Yanagiwara (1960), Krogdahl (1960), and Urabe (1963).

Recently asymptotic methods have been the most widely used to study relaxation oscillations. Besides the aforementioned studies, systems governed by a second-order differential equation were studied by Levinson and Smith (1942), Graffi (1942), Corbeiller (1931), LaSalle (1949), Stoker (1950), Caprioli (1954), and Coddington and Levinson (1955). Anh (1973) and Tondl (1970b) studied mechanical relaxation oscillations. Flatto and Levinson (1955) treated the case in which **f** and **g** are periodic in t, while Hale and Seifert (1961) and Hale (1963) treated the case in which **f** and **g** are almost periodic in t. Balachandra (1973, 1975) obtained periodic solutions of singularly perturbed equations arising from gyroscopic systems and obtained new results for (3.5.5) and (3.5.6).

In the remainder of this section we restrict our attention to (3.5.3) and (3.5.4). Eliminating t from these equations, we obtain the following equation for the trajectories in the phase plane:

$$\epsilon^{-2}\frac{dv}{d\xi} = \frac{v - v^3 - \xi}{v} \tag{3.5.9}$$

The curve $\xi = v - v^3$ is shown by the solid line in Figure 3-26a. Along this curve the slopes of the trajectories are zero, and because ϵ is very large, the magnitude of the slope is very large away from this line. Around this line there is a bound-

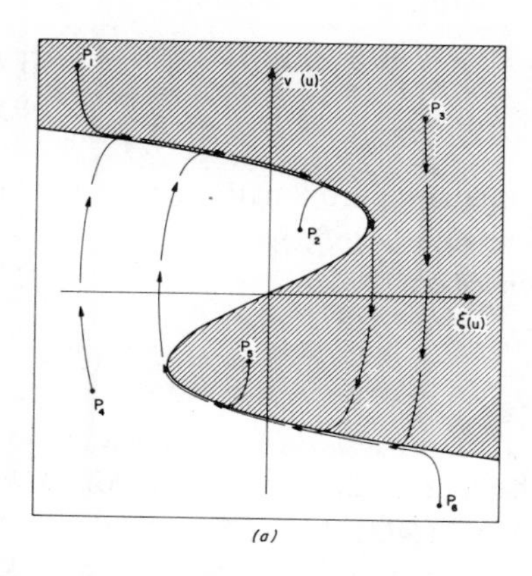

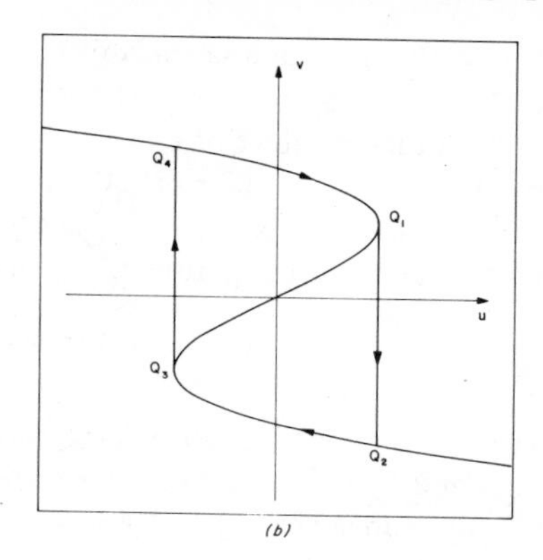

Figure 3-26. Relaxation oscillations of Rayleigh's equation: (*a*) motion of representative points; (*b*) limit cycle.

ary layer inside which the slopes of the trajectories change very rapidly. Also shown in Figure 3-26*a* are six trajectories, corresponding to six different starting points. In the shaded regions the representative point moves downward, slightly to the right when v is positive and slightly to the left when v is negative. In the unshaded region the representative point moves upward, slightly to the right when v is positive and slightly to the left when v is negative.

This construction clearly indicates the existence of a limit cycle as shown in Figure 3-26*b*. As $\epsilon \to \infty$, the limit cycle approaches $Q_1Q_2Q_3Q_4Q_1$. Thus it consists of the two vertical segments Q_1Q_2 and Q_3Q_4 and the segments Q_4Q_1 and Q_2Q_3.

In Figures 3-27 two phase planes are shown. Each phase plane was constructed

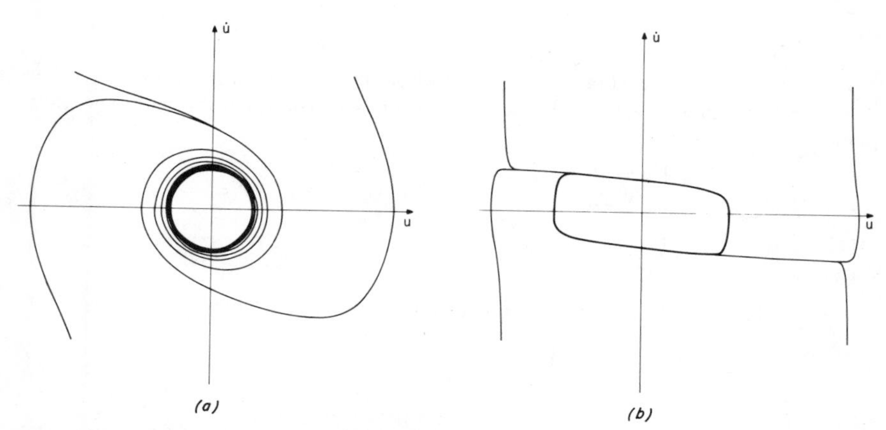

Figure 3-27. Phase planes for Rayleigh's equation: (*a*) $\epsilon = 0.01$; (*b*) $\epsilon = 10.0$.

numerically, and each is for a different value of ϵ. Several different trajectories are shown in each figure. The trajectories apparently form limit cycles very rapidly for large values of ϵ.

Next we follow Stoker (1950) and determine an approximation of the period of the limit cycle as $\epsilon \to \infty$. Referring to Figure 3-26*b*, we note that a first approximation to the period of the limit cycle as $\epsilon \to \infty$ is

$$T = \epsilon \oint \frac{d\xi}{v} \tag{3.5.10}$$

Since as $\epsilon \to \infty$ the segments Q_1Q_2 and Q_3Q_4 are vertical, $d\xi \approx 0$ along these segments and hence the times needed to traverse these segments are zero. Moreover since the times needed to traverse the segments Q_2Q_3 and Q_4Q_1 are the same

$$T = 2\epsilon \int_{Q_4}^{Q_1} \frac{d\xi}{v} = 2\epsilon \int_{Q_4}^{Q_1} \frac{dv - 3v^2\, dv}{v} \tag{3.5.11}$$

The point Q_1 corresponds to $d\xi/dv = 0$ or $1 - 3v^2 = 0$. Hence $v = 1/\sqrt{3}$ and $\xi = 2/3\sqrt{3}$. Moreover Q_3 corresponds to $\xi = -2/3\sqrt{3}$, and hence Q_4 corresponds to $v = 2/\sqrt{3}$. Then (3.5.11) becomes

$$T = 2\epsilon \int_{2/\sqrt{3}}^{1/\sqrt{3}} \left(\frac{dv}{v} - 3v\, dv\right) = 2\epsilon\left(\frac{3}{2} - \ln 2\right) \approx 1.614\epsilon \tag{3.5.12}$$

Using the method of matched asymptotic expansions, one can find the following expansion for the period (Dorodnicyn, 1947):

$$T = 1.6137\epsilon + 7.0143\epsilon^{-1/3} - \tfrac{1}{3}\epsilon^{-1} \ln \epsilon - 1.3246\epsilon^{-1} + O(\epsilon^{-4/3}) \tag{3.5.13}$$

The last two terms in (3.5.13) were in error in the original paper of Dorodnicyn, and they were corrected by Urabe (1963).

Exercises

3.1. Determine the singular points and their types for the following equations, and for each case sketch the trajectories and the separatrices in the phase plane:

(a) $\ddot{u} + 2\mu\dot{u} + u + u^3 = 0, \quad \mu > 0$

(b) $\ddot{u} + 2\mu\dot{u} + u - u^3 = 0$

(c) $\ddot{u} + 2\mu\dot{u} - u + u^3 = 0$

(d) $\ddot{u} + 2\mu\dot{u} - u - u^3 = 0$

3.2. Determine the singular points and their types for the system

$$\dot{x} = x^2 - y$$
$$\dot{y} = x - y$$

Sketch the trajectories and the separatrices in the state plane.

3.3. Determine the singular points and their types of the system

$$\dot{x} = x^2 + y^2 - 5$$
$$\dot{y} = xy - 2$$

Sketch the trajectories and the separatrices in the state plane.

3.4. Consider the Rayleigh equation

$$\ddot{x} - \epsilon(\dot{x} - \tfrac{1}{3}\dot{x}^3) + x = 0$$

Let $u = \dot{x}$ and show that u is governed by the van der Pol equation

$$\ddot{u} - \epsilon(1 - u^2)\dot{u} + u = 0$$

3.5. Consider van der Pol's equation in the following form:

$$\ddot{u} + \epsilon(\beta\dot{u}^2 - 1)\dot{u} + u = 0$$

Use Liénard's method to construct two trajectories in the phase plane—one starting far outside the limit cycle and the other starting near the origin. This construction provides another, rather convincing, argument for the existence of a limit cycle.

3.6. Consider the system governed by

$$\ddot{u} + \mu \sin \dot{u} + u = 0$$

Using Liénard's method, construct several trajectories. Show that more than one limit cycle exists. Some limit cycles are stable while others are unstable. How can one determine the stability of the various limit cycles by examining the trajectories in the phase plane? Indicate which limit cycles are stable and which are unstable in this example (see Figure 3-28).

3.7. Consider the following system of equations:

$$\dot{x}_1 = -\mu x_1 + k \sin x_2$$
$$\dot{x}_2 = \sigma - \alpha x_1^2 + \frac{k}{x_1} \cos x_2$$

Locate the singular points of this system. (Hint: Obtain σ as a function of x_1.) Show that the maximum value of x_1 is given by k/μ and occurs when $\sigma = \alpha k^2/\mu^2$. Show your results by plotting x_1 and x_2 as functions of σ for

(a) $\alpha = 1$, $k = 1$, and $\mu = \frac{1}{2}$
(b) $\alpha = 0$, $k = 1$, and $\mu = \frac{1}{2}$
(c) $\alpha = -1$, $k = 1$, and $\mu = \frac{1}{2}$

For $\alpha = 1$, $k = 1$, and $\mu = \frac{1}{2}$, determine the nature of the singular points (i.e., focus, col, etc.) when $\sigma = 0$, 3, and 4. Sketch the trajectories in the state plane for each case.

3.8. In studying the primary-resonance response of the van der Pol oscillator with delayed amplitude limiting, Nayfeh (1968) encountered the following sys-

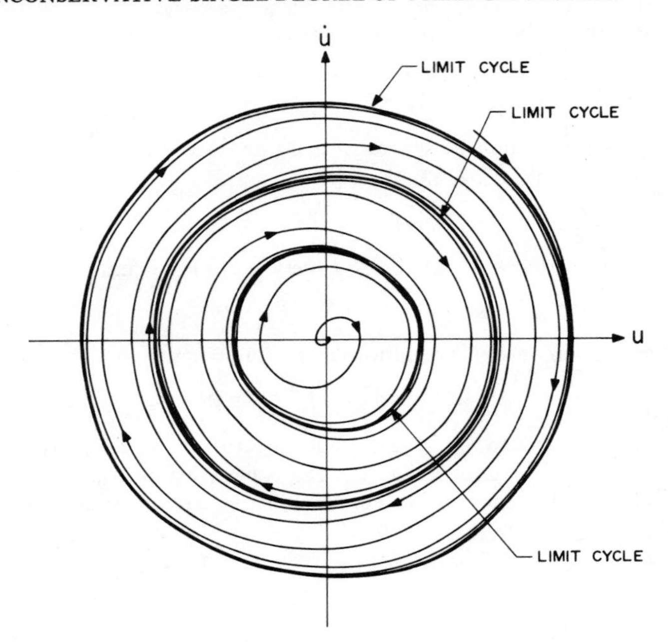

Figure 3-28. Exercise 3.6.

tem of equations:

$$\dot{x}_1 = x_1(1 - x_1^2) + f\cos x_2$$

$$\dot{x}_2 = \sigma + \nu x_1^2 - \frac{f}{x_1}\sin x_2$$

(a) Show that the x_1-coordinate of a singular point is a solution of

$$\rho[(1 - \rho)^2 + (\sigma + \nu\rho)^2] = f^2$$

where

$$\rho = x_1^2$$

(b) Using $\nu = -0.15$, plot the locus of the singular points in the $\rho\sigma$-plane for $f^2 = 1, \frac{1}{3}, \frac{4}{27}$, and $\frac{1}{10}$. What is the significance of the value $\frac{4}{27}$?

(c) Show that the interior points of the ellipse defined by

$$(1 - \rho)(1 - 3\rho) + (\sigma + \nu\rho)(\sigma + 3\nu\rho) = 0$$

are saddle points and hence unstable. Also show that the exterior points are nodes if $D \geqslant 0$ and foci if $D < 0$, where

$$D = 4[(1 - 3\nu^2)\rho^2 - 4\nu\rho\sigma - \sigma^2]$$

(d) Finally, show that the exterior points are stable if $\rho > \frac{1}{2}$ and unstable if $\rho < \frac{1}{2}$.

3.9. Consider a pendulum with a dry, rusty hinge. Assume that friction has constant magnitude so that the governing equation has the form

$$\ddot{\theta} + 2\mu \operatorname{sgn} \dot{\theta} + \omega^2 \sin \theta = 0$$

Determine a first approximation for θ which accounts for both damping and nonlinearity. Ottl (1975) determined the transient motion of an oscillator with Coulomb damping.

3.10. Consider the system defined by (Hayes, 1953)

$$\ddot{\theta} + 2\mu\dot{\theta} + \omega^2 \sin \theta = f$$

where f is a constant (pendulum with linear damping and constant torque). When $f = 0$, let $x_1 = \theta$ and $x_2 = \dot{x}_1$. Then study the motion near the singular point defined by $x_1 = 0$ and $x_2 = 0$, using the discussion of Section 3.2.1. Then compare your results with (3.3.74). Repeat for $x_1 = \pi$. When $f \neq 0$, determine the equilibrium points and discuss the nonlinear motion around these points.

3.11. Consider the system defined by

$$\ddot{\theta} + 2\mu\dot{\theta}|\dot{\theta}| + \omega^2 \sin \theta = 0$$

(pendulum with quadratic damping). Let $x_1 = \theta$ and $x_2 = \dot{x}_1$. Then study the motion near the singular point defined by $x_1 = 0$ and $x_2 = 0$, using the discussion of Section 3.2.1. Then compare your results with (3.3.92). Explain the differences. Compare these results with those of Exercise 3.10 above.

3.12. Consider Rayleigh's equation

$$\ddot{u} + \omega_0^2 u - \epsilon(\dot{u} - \dot{u}^3) = 0 \tag{1}$$

Let $x_1 = u$ and $x_2 = \dot{x}_1$. Study the singular point(s), using the discussion of Section 3.2. Then, instead of (1), consider the system

$$\dot{a} = \tfrac{1}{2}\epsilon a(1 - \tfrac{3}{4}\omega_0^2 a^2), \qquad \dot{\beta} = 0 \tag{2}$$

which is equivalent to (1) when ϵ is small according to Section 3.3.4.

3.13. The motion of a particle in the restricted three-body problem is governed by

$$\ddot{x} - 2\dot{y} - x = -\frac{m(x - 1 + m)}{d_1^3} - \frac{(1 - m)(x + m)}{d_2^3}$$

$$\ddot{y} + 2\dot{x} - y = -\frac{my}{d_1^3} - \frac{(1 - m)y}{d_2^3}$$

where m is a reduced mass and

$$d_1^2 = (1 - m - x)^2 + y^2, \qquad d_2^2 = (m + x)^2 + y^2$$

(a) Show that there are five equilibrium points of this system; two of them are $(\frac{1}{2} - m, \pm\frac{1}{2}\sqrt{3})$, while the other three are the solutions of

$$x - \frac{m}{(1-m-x)^2} - \frac{1-m}{(m+x)^2} = 0$$

These are usually called the *libration*, or *Lagrange's*, *points.*

(b) Show that the first two are linearly stable when $m < m_c$ and unstable when $m \geqslant m_c$, where $m_c = \frac{1}{2}(1 - \frac{1}{9}\sqrt{69})$.

(c) Show that the remaining three points are always unstable.

3.14. In analyzing the effect of two-to-one resonances on the nonlinear stability of the triangular points in the restricted three-body problem and in analyzing second-harmonic resonances in the problem of capillary-gravity waves, Simmons (1969), McGoldrick (1970b), and Nayfeh (1971b, 1973c) encountered the following set of equations:

$$\dot{a}_1 = J_1 a_1 a_2 \sin\gamma$$

$$\dot{a}_2 = J_2 a_1^2 \sin\gamma$$

$$a_2\dot{\gamma} = \sigma a_2 + (J_2 a_1^2 + 2J_1 a_2^2)\cos\gamma$$

where J_1, J_2, and σ are constants.

(a) Show that the singular points of this system are given by

$$\gamma = n\pi, \qquad \sigma a_2 + (J_2 a_1^2 + 2J_1 a_2^2)\cos n\pi = 0$$

(b) Show that these singular points are stable if $J_1 J_2 < 0$ and may be unstable if $J_1 J_2 > 0$.

3.15. Consider a simple pendulum with a dashpot as shown in Figure 3-29.

(a) Show that the equation of motion is

$$ml_1\ddot{\theta} = -mg\sin\theta - \hat{\mu}ml_1\dot{\theta}\cos^2(\beta - \theta) \tag{1}$$

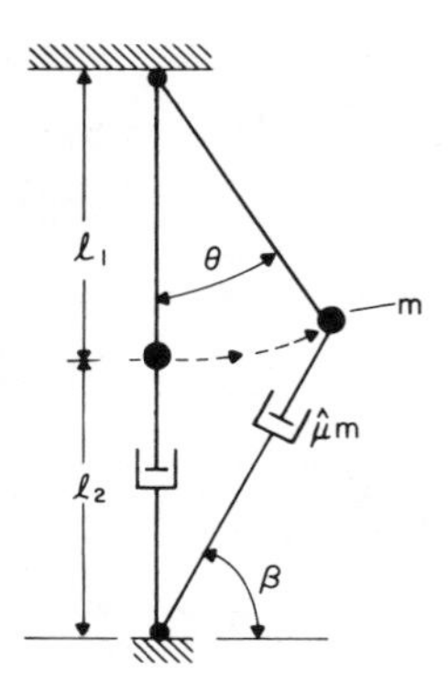

Figure 3-29. A simple pendulum with a dashpot.

Then show that (1) can also be written as

$$\ddot{\theta} + \omega^2 \sin\theta + \frac{\hat{\mu}(l_1 + l_2)^2 \sin^2\theta}{l_2^2 + 2l_1(l_1 + l_2)(1 - \cos\theta)}\dot{\theta} = 0 \tag{2}$$

(b) Expanding and retaining through the cubic terms, show that (2) becomes

$$\ddot{\theta} + \omega^2(1 - \tfrac{1}{6}\theta^2)\theta + 2\mu\theta^2\dot{\theta} = 0 \tag{3}$$

where

$$2\mu = \frac{\hat{\mu}(l_1 + l_2)^2}{l_2^2} \tag{4}$$

Using (3), obtain the following first approximation for θ when the amplitude of the motion is small but finite:

$$\theta = \frac{a_0}{\sqrt{1 + \frac{1}{2}\mu a_0^2 t}} \cos\left\{\omega\left[t - \frac{\ln(1 + \frac{1}{2}\mu a_0^2 t)}{8\mu}\right] + \beta_0\right\} \tag{5}$$

where a_0 and β_0 are constants of integration. Note that μ is not small and that in this case the frequency is affected by the damping in the first approximation. As a check, show that in the limit as $\mu \to 0$ equation (5) reduces to equation (2.4.18).

3.16. The free oscillations of a van der Pol oscillator with delayed amplitude limiting are governed by

$$\ddot{v} + \omega_0^2 v = 2\epsilon[(1 - z)\dot{v} - \dot{z}v]$$

$$\tau\dot{z} + z = v^2$$

(a) Use the method of multiple scales and show that (Nayfeh, 1967)

$$v = a\cos(\omega_0 t + \theta) + O(\epsilon)$$

$$z = b\exp(-t/\tau) + \tfrac{1}{2}a^2 + \tfrac{1}{2}a^2(1 + 4\omega_0^2\tau^2)^{-1/2}$$

$$\cdot \cos(2\omega_0 t + 2\theta - \arctan 2\omega_0\tau) + O(\epsilon)$$

where

$$\dot{a} = \epsilon a(1 - \tfrac{1}{4}\alpha_r a^2), \qquad \alpha_r = (3 + 8\omega_0^2\tau^2)(1 + 4\omega_0^2\tau^2)^{-1}$$

$$\dot{\theta} = -\tfrac{1}{4}\epsilon\alpha_i a^2, \qquad \alpha_i = -2\omega_0\tau(1 + 4\omega_0^2\tau^2)^{-1}$$

$$\dot{b} = 2(1 + 2\omega_0^2\tau^2)(1 + 4\omega_0^2\tau^2)^{-1}a^2 b$$

(b) Solve for a, θ, and b.

(c) Determine the steady-state motions and their stability. Free oscillations of other third-order systems were studied by Dasarathy and Srinivasan (1969), Srirangarajan and Srinivasan (1973, 1974), Srirangarajan and Dasarathy (1975), and Joshi, Srirangarajan, and Srinivasan (1976). Baker, Moore, and Spiegel (1971) analyzed a third-order nonlinear system as a model for nonlinear instability. Nayfeh (1968) and Tondl (1968, 1974) determined the forced response of two third-order systems.

3.17. The motion of a particle restrained by a linear spring and under the combined influence of Coulomb and viscous damping is governed by

$$\ddot{u} + \omega_0^2 u + \epsilon(\mu_0 \operatorname{sgn} \dot{u} + 2\mu_1 \dot{u}) = 0$$

where μ_0 and $\mu_1 > 0$ and $\epsilon \ll 1$.

(a) Show that

$$u = a \cos(\omega_0 t + \beta) + O(\epsilon)$$

where

$$\dot{a} = -\epsilon\left(\frac{2\mu_0}{\pi\omega_0} + \mu_1 a\right)$$

and

$$\dot{\beta} = 0$$

(b) Show that

$$a = \left(a_0 + \frac{2\mu_0}{\pi\omega_0\mu_1}\right)\exp(-\epsilon\mu_1 t) - \frac{2\mu_0}{\pi\omega_0\mu_1}$$

where a_0 is the initial amplitude.

(c) Explain how the motion decays. When does it stop?

3.18. The motion of a particle restrained by a linear spring and under the combined influence of Coulomb and square damping is governed by

$$\ddot{u} + \omega_0^2 u + \epsilon(\mu_0 \operatorname{sgn} \dot{u} + \mu_2 \dot{u}|\dot{u}|) = 0$$

where μ_0 and $\mu_2 > 0$ and $\epsilon \ll 1$.

(a) Show that

$$u = a \cos(\omega_0 t + \beta) + O(\epsilon)$$

where

$$\dot{a} = -\epsilon\left(\frac{2\mu_0}{\pi\omega_0} + \frac{4}{3\pi}\mu_2\omega_0 a^2\right)$$

and

$$\dot{\beta} = 0$$

(b) Show that

$$a = \frac{1}{\omega_0} \sqrt{\frac{3\mu_0}{2\mu_2}} \tan \left(c - \frac{\epsilon}{\pi} \sqrt{\frac{8\mu_2\mu_0}{3}}\, t \right)$$

where c is a constant of integration.

(c) Explain how the motion decays. When does it stop?

3.19. The motion of a particle under the combined influence of viscous and square damping is governed by

$$\ddot{u} + \omega_0^2 u + \epsilon(2\mu_1 \dot{u} + \mu_2 \dot{u}|\dot{u}|) = 0$$

where μ_1 and $\mu_2 > 0$ and $\epsilon \ll 1$.

(a) Show that

$$u = a \cos(\omega_0 t + \beta) + O(\epsilon)$$

where

$$\dot{a} = -\epsilon\left(\mu_1 a + \frac{4}{3\pi}\mu_2 \omega_0 a^2\right)$$

and

$$\dot{\beta} = 0$$

(b) Show that

$$a = a_0 \left\{ \exp(\epsilon\mu_1 t) + a_0 \frac{4\mu_2\omega_0}{3\pi\mu_1} [\exp(\epsilon\mu_1 t) - 1] \right\}^{-1}$$

where a_0 is the initial amplitude.

3.20. Consider the free oscillations of a system governed by

$$\ddot{u} + \omega_0^2 u + \epsilon(2\mu_1 \dot{u} - \mu_0 \operatorname{sgn} \dot{u}) = 0$$

where $\epsilon \ll 1$ and μ_1 and $\mu_0 > 0$.

(a) Show that

$$u = a \cos(\omega_0 t + \beta) + O(\epsilon)$$

where

$$\dot{a} = \epsilon\left(\frac{2\mu_0}{\pi\omega_0} - \mu_1 a\right) \quad \text{and} \quad \dot{\beta} = 0$$

(b) Show that

$$a = \left(a_0 - \frac{2\mu_0}{\pi\omega_0\mu_1}\right) \exp(-\epsilon\mu_1 t) + \frac{2\mu_0}{\pi\omega_0\mu_1}$$

where a_0 is the initial amplitude.

(c) Discuss the limit cycle(s), if any.

3.21. Consider the motion of a system governed by

$$\ddot{u} + \omega_0^2 u + \epsilon(\mu_2 \dot{u}|\dot{u}| - \mu_0 \operatorname{sgn} \dot{u}) = 0$$

where $\epsilon \ll 1$ and μ_2 and $\mu_0 > 0$.

(a) Show that

$$u = a \cos(\omega_0 t + \beta) + O(\epsilon)$$

where

$$\dot{a} = \frac{2\epsilon}{\pi}\left(\frac{\mu_0}{\omega_0} - \frac{2\mu_2\omega_0}{3} a^2\right) \quad \text{and} \quad \dot{\beta} = 0$$

(b) Show that a can be written in the form

$$a = \sqrt{\frac{3\mu_0}{2\mu_2}\left\{\frac{1}{\omega_0} + \frac{2}{c \exp\left[\frac{\epsilon}{\pi}\sqrt{32\mu_0\mu_2/3}\, t\right] - \omega_0}\right\}}$$

where c is a constant of integration.

(c) Discuss the limit cycle(s), if any.

3.22. Consider the motion of a system governed by

$$\ddot{u} + \omega_0^2 u + \epsilon(\mu_2 \dot{u}|\dot{u}| - 2\mu_1 \dot{u}) + \epsilon\alpha u^3 = 0$$

where $\epsilon \ll 1$ and μ_2 and $\mu_1 > 0$. Klotter (1955) studied the free oscillations of a system having quadratic damping and arbitrary restoring forces.

(a) Show that

$$u = a \cos(\omega_0 t + \beta) + O(\epsilon)$$

where

$$\dot{a} = \epsilon\left(\mu_1 a - \frac{4}{3\pi}\mu_2\omega_0 a^2\right)$$

and

$$\dot{\beta} = \frac{3\epsilon\alpha}{8\omega_0} a^2$$

(b) Show that a can be written in the form

$$a = \frac{3\pi\mu_1}{4\mu_2\omega_0}[1 + C \exp(-\epsilon\mu_1 t)]^{-1}$$

where C is a constant of integration.

(c) Discuss the limit cycle(s). What influence does $\epsilon\alpha u^3$ have?

3.23. Consider the motion of a system governed by

$$\ddot{u} + \omega_0^2 u + \epsilon(2\mu_1 \dot{u} + \mu_2 \dot{u}|\dot{u}| + \mu_3 \dot{u}^3) = 0$$

where $\epsilon \ll 1$. Baum (1972) determined an approximate solution when $\mu_2 = 0$.

(a) Show that, to the first approximation, the amplitude is governed by

$$\dot{a} = -\epsilon \left(\mu_1 a + \frac{4}{3\pi} \mu_2 \omega_0 a^2 + \frac{3}{8} \mu_3 \omega_0^2 a^3\right)$$

(Note that μ_3 must be positive for a realistic system.)

(b) Determine the stationary motions and their stability as a function of the magnitudes and the signs of μ_1 and μ_2.

3.24. A clock pendulum excited by pulses is governed by

$$J\ddot{x} + kx + \lambda \dot{x} - \tfrac{1}{2} I(\dot{x} - |\dot{x}|)\, \delta(x - x_0) = 0$$

where J, k, λ, and I are constants and $\delta(x - x_0)$ is Dirac's delta function.

(a) Show that to the first approximation

$$x = a \cos \psi, \qquad \psi = \omega t + \phi$$

where

$$\omega^2 = \frac{k}{J}$$

and

$$\dot{a} = -\frac{\lambda a}{2J\pi} \int_0^{2\pi} \sin^2 \psi \, d\psi + \frac{Ia}{4J\pi} \int_0^{2\pi} \delta(a \cos \psi - x_0)$$

$$\cdot \sin \psi (\sin \psi + |\sin \psi|)\, d\psi = -\frac{\lambda a}{2J} + \frac{Ia}{2J\pi} \int_0^{\pi} \delta(a \cos \psi - x_0) \sin^2 \psi \, d\psi$$

$$\dot{\phi} = -\frac{\lambda}{2J\pi} \int_0^{2\pi} \cos \psi \sin \psi \, d\psi + \frac{I}{2J\pi} \int_0^{\pi} \delta(a \cos \psi - x_0) \sin \psi \cos \psi \, d\psi$$

(b) If $a \geqslant x_0 \geqslant 0$, show that

$$\dot{a} = -\frac{\lambda a}{2J} + \frac{I}{2J\pi} \sin \psi_a$$

$$\dot{\phi} = \frac{I}{2J\pi a} \cos \psi_a$$

where ψ_a is the root of $a \cos \psi - x_0 = 0$. Hence

$$\dot{a} = -\frac{\lambda a}{2J} + \frac{I}{2J\pi} \left(1 - \frac{x_0^2}{a^2}\right)^{1/2}$$

$$\dot{\phi} = \frac{I x_0}{2J\pi a^2}$$

Hint:

$$\int_0^\pi \delta(a \cos \psi - x_0) \sin^2 \psi \, d\psi = \frac{1}{a} \int_{-a}^{a} \delta(z - x_0) \sin [\psi(z)] \, dz = \frac{1}{a} \sin \psi_a$$

(c) If $x_0 > a$, show that

$$\dot{a} = -\frac{\lambda a}{2J}, \qquad \dot{\phi} = 0$$

(d) Show that finite-amplitude stationary oscillations exist only when $a \geqslant x_0$ and $x_0 \leqslant I/2\pi\lambda$.

3.25. A number of problems involving convection phenomena, such as the formation of Bénard cells, the Rayleigh-Taylor instability, and various aperiodic plasma instabilities, lead to a model equation of the form

$$\ddot{u} + \mu \dot{u} - \alpha_1 u + \alpha_2 u^3 = 0$$

where μ, α_1, and α_2 are positive. Determine an approximate solution to this equation and show that $u = \sqrt{\alpha_1/\alpha_2}$ is a saturation value (Cap and Lashinsky, 1973).

3.26. Use Liénard's construction to describe the behavior of the system governed by

$$\ddot{u} + \omega_0^2 u + \epsilon(2\mu_1 \dot{u} - \mu_0 \operatorname{sgn} \dot{u}) = 0$$

where $\epsilon \gg 1$ and μ_1 and $\mu_0 > 0$. Show that the period of the motion $\rightarrow \infty$ as $\epsilon \rightarrow \infty$. Describe the motion when the signs of μ_1 and μ_0 are reversed.

3.27. Use Liénard's construction to describe the behavior of the system governed by

$$\ddot{u} + \omega_0^2 u + \epsilon(\mu_2 \dot{u}|\dot{u}| - 2\mu_1 \dot{u}) = 0$$

where $\epsilon \gg 1$ and μ_2 and $\mu_1 > 0$. What is the period of motion? Describe the motion when the signs of μ_2 and μ_1 are reversed.

3.28. Use Liénard's construction to describe the behavior of the system governed by

$$\ddot{u} + \omega_0^2 u + \epsilon(-2\mu_1 \dot{u} - \mu_2 \dot{u}|\dot{u}| + \mu_3 \dot{u}^3) = 0$$

where $\epsilon \gg 1$ where μ_1, μ_2, and $\mu_3 > 0$. What is the period of the motion? Baum (1972) analyzed the case $\mu_2 = 0$.

3.29. Consider the response of a system governed by

$$\ddot{u} + 2\mu \dot{u} + u + \epsilon u^3 = 0$$

where $\mu = O(1)$ and $1 - \mu^2 > 0$.

(a) Seek a solution in the form (Popov and Paltov, 1960; Mendelson, 1970; Arya, Bojadziev, and Farooqui, 1975)

$$u = a \cos \psi + \epsilon u_1(a, \psi) + \cdots$$
$$\dot{a} = -\mu a + \epsilon \xi_1(a) + \cdots = \xi(a)$$
$$\dot{\psi} = \omega_0 + \epsilon \omega_1(a) + \cdots = \omega(a)$$

where $\omega_0^2 = 1 - \mu^2$. Show that the time derivatives are transformed into

$$\frac{d}{dt} = \omega \frac{\partial}{\partial \psi} + \xi \frac{\partial}{\partial a}$$

$$\frac{d^2}{dt^2} = \omega^2 \frac{\partial^2}{\partial \psi^2} + 2\omega\xi \frac{\partial^2}{\partial \psi \partial a} + \xi^2 \frac{\partial^2}{\partial a^2} + \xi \frac{d\xi}{da} \frac{\partial}{\partial a} + \xi \frac{d\omega}{da} \frac{\partial}{\partial \psi}$$

(b) Show that u_1 is governed by

$$\omega_0^2 \frac{\partial^2 u_1}{\partial \psi^2} - 2\mu\omega_0 a \frac{\partial^2 u_1}{\partial \psi \partial a} + \mu^2 a^2 \frac{\partial^2 u_1}{\partial a^2} + 2\mu\omega_0 \frac{\partial u_1}{\partial \psi} - \mu^2 a \frac{\partial u_1}{\partial a} + (\omega_0^2 + \mu^2) u_1$$

$$= 2\omega_0 \omega_1 a \cos \psi + 2\omega_0 \xi_1 \sin \psi - \mu a^2 \frac{d\omega_1}{da} \sin \psi + \mu \left(a \frac{d\xi_1}{da} - \xi_1 \right) \cos \psi$$

$$- \tfrac{3}{4} a^3 \cos \psi - \tfrac{1}{4} a^3 \cos 3\psi$$

(c) Show that elimination of secular terms leads to

$$2\omega_0 \omega_1 a + \mu a \frac{d\xi_1}{da} - \mu \xi_1 - \tfrac{3}{4} a^3 = 0$$

$$2\omega_0 \xi_1 - \mu a^2 \frac{d\omega_1}{da} = 0$$

Hence

$$\xi_1 = \tfrac{3}{8} \mu a^3 \quad \text{and} \quad \omega_1 = \tfrac{3}{8} \omega_0 a^2$$

(d) Show that

$$\dot{a} = -\gamma a + \tfrac{3}{8} \epsilon \mu a^3$$
$$\dot{\psi} = \omega_0 + \tfrac{3}{8} \epsilon \omega_0 a^2$$

Hence

$$a = a_0 e^{-\mu t} [1 + \tfrac{3}{8} \epsilon a_0^2 (e^{-2\mu t} - 1)]^{-1/2}$$

$$\psi = \psi_0 + \omega_0 t - \frac{\omega_0}{2\mu} \ln [1 + \tfrac{3}{8} \epsilon a_0^2 (e^{-2\mu t} - 1)]$$

3.30. The free oscillation of a single-degree-of-freedom system is governed by

$$\ddot{u} + \omega^2 \sin u = \epsilon f(u, \dot{u}), \qquad \epsilon \ll 1$$

(a) When $\epsilon = 0$, show that

$$u = 2 \sin^{-1} [k \operatorname{sn} (\psi, k)]$$

where sn is the Jacobi elliptic sine function, $\psi = \omega t + \phi$, k is the modulus of the elliptic function (amplitude of the oscillation), and ϕ is the phase.

(b) When $\epsilon \neq 0$, seek a solution as in (a) but with time-varying k and ϕ. Moreover impose the condition

$$\dot{u} = 2k\omega \operatorname{cn}(\psi, k)$$

where cn is the Jacobi elliptic cosine function. Show that (Cap, 1974)

$$\dot{k} = \frac{\epsilon \operatorname{cn}}{2\omega} f[2 \sin^{-1}(k \operatorname{sn}), 2k\omega \operatorname{cn}]$$

$$\dot{\phi} = -\frac{\epsilon}{2\omega k \operatorname{dn}}\left(\operatorname{sn} + k\frac{d \operatorname{sn}}{dk}\right) f[2 \sin^{-1}(k \operatorname{sn}) + 2k\omega \operatorname{cn}]$$

where $\operatorname{dn}^2 = 1 - k^2 \operatorname{sn}^2$.

(c) Determine a first approximate solution by averaging the equations for $\dot{k}$ and $\dot{\phi}$ over the period $4K$ of sn, cn, and dn. Note that

$$K = \int_0^{1/2\pi} \frac{d\theta}{\sqrt{1 - k^2 \sin^2 \theta}}$$

3.31. In analyzing the motion of a particle constrained to move on a circular path within a body that is spinning and coning, Mingori and Harrison (1974) encountered the equation

$$v\frac{dv}{du} + \mu_1(v - 1) - \mu_2\mu_3 \sin u - \tfrac{1}{2}\mu_3^2 \sin 2u = 0$$

where μ_1, μ_2, and μ_3 are constants.

(a) Determine the equations describing the singular points.

(b) When $\mu_1 = 0.1$ and $\mu_2 = 2.0$, show that bifurcation occurs at $\mu_3 = 0.0502$, 0.3, and 2.265.

(c) Calculate the singular points and their types as μ_3 is varied from zero past 2.265.

(d) Sketch the trajectories in the phase plane.

3.32. Consider the system govened by

$$\ddot{u} + f(u, \tau) = 0 \tag{1}$$

where $d\tau/dt = \epsilon$, a very small quantity. The variable τ appears explicity in the coefficients of the various functions of u; thus (1) describes a system that has slowly varying parameters.

(a) Following the generalized method of multiple scales (Nayfeh, 1973b, Section 6.4), assume

$$u(t) = u_0(\tau, \eta) + \epsilon u_1(\tau, \eta) + \cdots \tag{2}$$

where $d\eta/dt = g(\tau)$, a function to be determined as part of the solution. Here η is the fast scale and τ is the slow scale. We note that, in contrast with the derivative-expansion version, here the derivative of η with respect to t is a slowly varying function of time.

Show that

$$\frac{d^2}{dt^2} = g^2 \frac{\partial^2}{\partial \eta^2} + \epsilon\left(g' \frac{\partial}{\partial \eta} + 2g \frac{\partial^2}{\partial \tau\, \partial \eta}\right) + O(\epsilon^2) \tag{3}$$

Substituting (2) and (3) into (1), expanding f around $u = u_0$, and equating coefficients of like powers of ϵ, show that

$$g^2\left(\frac{\partial u_0}{\partial \eta}\right)^2 + F(u_0, \tau) = c_1(\tau) \tag{4}$$

$$g^2 \frac{\partial^2 u_1}{\partial \eta^2} + \frac{\partial f(u_0, \tau)}{\partial u_0} u_1 = -g' \frac{\partial u_0}{\partial \eta} - 2g \frac{\partial^2 u_0}{\partial \eta\, \partial \tau} \tag{5}$$

where

$$F(u_0, \tau) = 2\int f(u_0, \tau)\, du_0$$

and c_1 is an arbitrary function. In terms of the variable η, (4) describes periodic motions around a center.

Multiplying (5) by $\partial u_0/\partial \eta$ and integrating, obtain

$$g^2\left(\frac{\partial u_1}{\partial \eta}\frac{\partial u_0}{\partial \eta} - u_1 \frac{\partial^2 u_0}{\partial \eta^2}\right)\Bigg|_0^T = -\int_0^T \frac{\partial}{\partial \tau}\left[g\left(\frac{\partial u_0}{\partial \eta}\right)^2\right] d\eta \tag{6}$$

where T is the period of motion in terms of the fast scale; T is independent of τ. The left-hand side vanishes if u_1 is periodic; thus the condition of periodicity (i.e., the solvability condition which eliminates secular terms) is

$$g(\tau)\int_0^T \left(\frac{\partial u_0}{\partial \eta}\right)^2 d\eta = 2c \tag{7}$$

where c is an arbitrary constant.

Using (4), show that (7) can be rewritten as follows:

$$\int_{y_1}^{y_2} \sqrt{c_1 - F(u_0, \tau)}\, du_0 = c \tag{8}$$

where y_1 and y_2 are the zeros of $c_1 - F$.

(b) Consider a linear oscillator having a time-dependent, restoring-force coefficient

$$f(u_0, \tau) = k(\tau)\, u_0 \tag{9}$$

Show that (4) and (9) lead to

$$u_0 = \sqrt{\frac{c_1(\tau)}{k(\tau)}} \sin \phi \tag{10}$$

where

$$\eta = \frac{g(\tau)}{\sqrt{k(\tau)}} \left(\phi + \frac{\pi}{2} \right) \tag{11}$$

Hence the period (which is independent of τ) is given by

$$T = \frac{2\pi g(\tau)}{\sqrt{k(\tau)}} \tag{12}$$

If we let $T = 2\pi$, then it follows that

$$g(\tau) = \sqrt{k(\tau)} \tag{13}$$

Show that (8) and (9) lead to

$$\frac{c_1(\tau)}{\sqrt{k(\tau)}} = \frac{2c}{\pi}, \text{ a constant} \tag{14}$$

Finally combine (14), (13), (11), and (10) to obtain

$$u_0 = -\sqrt{\frac{2c}{\pi}}\, k^{-1/4} \cos \eta \tag{15}$$

Recalling that

$$\frac{d\eta}{dt} = g(\tau)$$

we can rewrite (15) as follows:

$$u_0 = \frac{a}{[k(\tau)]^{1/4}} \cos \left[\frac{1}{\epsilon} \int_0^{\epsilon t} g(\tau)\, d\tau + \eta_0 \right]$$

where a and η_0 are constants. Compare this result with that in Section 3.4.1.

CHAPTER 4

Forced Oscillations of Systems Having a Single Degree of Freedom

In the previous chapters we considered systems having one degree of freedom that were initially disturbed and then allowed to respond with no further external excitation. In contrast, we now consider systems having one degree of freedom that are continuously excited. In this book two types of excitations are considered: (1) the excitation appears as an inhomogeneous term in the equations governing the motion of the system, and (2) the excitation appears as a variable (i.e., time-dependent) coefficient in the governing equations. The second type, which is called a *parametric excitation*, is considered in the following chapter. The first type, which is called an *external excitation*, is considered in this chapter.

Here we consider systems governed by

$$\ddot{u} + \omega_0^2 u = \epsilon f(u, \dot{u}) + E$$

where ϵ is a small parameter, f is a nonlinear function of u and $\dot{u}$, and E is an externally applied force called the excitation. We distinguish between two types of excitations. First, the excitation draws on an energy source that is assumed to be unlimited or so large that the excited system has a negligible effect on it. In this case $E = E(t)$; that is, E is not a function of the state of the system u, $\dot{u}$, or $\ddot{u}$. Such sources are said to be *ideal sources of energy*. Second, the excitation draws on an energy source that is limited so that the excited system has an appreciable effect on it. In this case $E = E(t, u, \dot{u}, \ddot{u})$; that is, E is a function of the state of the system. Such sources are said to be *nonideal sources of energy*. Systems are classified as ideal or nonideal according to the energy source.

In Sections 4.1 through 4.4 we treat ideal systems and take the excitation to be the sum of N terms, each of which is harmonic:

$$E(t) = \sum_{n=1}^{N} K_n \cos(\Omega_n t + \theta_n)$$

If the K_n (amplitudes), Ω_n (frequencies), and θ_n are constants, the excitation is said to be stationary; otherwise it is said to be nonstationary. Perturbation methods lend themselves to the analysis of nonstationary systems when the amplitudes and frequencies are slowly varying functions of time.

In Sections 4.1 through 4.3 we treat stationary excitations. Systems having cubic nonlinearities are treated in Section 4.1, those having quadratic and cubic nonlinearities are treated in Section 4.2, and self-excited systems are treated in Section 4.3. Generally the discussion treats only one-term (harmonic) excitations; however multifrequency excitations are treated in Sections 4.1.5–4.1.7, 4.2.4, and 4.3.5. In these subsections, the frequencies are assumed to be distinct and away from each other. The case in which two or more frequencies are close to each other is best treated by the approach used in Section 4.4, where nonstationary excitations are considered. This is so because, if

$$E(t) = K_1 \cos(\Omega_1 t + \theta_1) + K_2 \cos(\Omega_2 t + \theta_2)$$

then we can write

$$E(t) = K(t) \cos [\Omega_1(t) + \theta(t)]$$

where

$$K^2 = (K_1 + K_2 \cos\beta)^2 + K_2^2 \sin^2\beta = K_1^2 + K_2^2 + 2K_1K_2 \cos\beta$$

$$\theta = \theta_1 + \arctan\left(\frac{K_2 \sin\beta}{K_1 + K_2 \cos\beta}\right)$$

$$\beta = (\Omega_2 - \Omega_1)t + \theta_2 - \theta_1$$

Thus, if $\Omega_1 \simeq \Omega_2$, the excitation may be considered a monofrequency excitation with slowly varying amplitude and frequency.

For a detailed treatment of nonstationary excitations we refer the reader to the book by Mitropolsky (1965). Anderson (1974a, b) analyzed a system subjected to a step-function excitation; Arya, Bojadziev, and Farooqui (1975) analyzed a system subjected to a slowly varying excitation; and Srirangarajan and Srinivasan (1973) analyzed a system subjected to a pulse excitation. Helfenstein (1950) and Hsu (1960) obtained exact solutions for the Duffing equation when $E(t)$ is a Jacobian elliptic function. In a series of papers, Loud (1957, 1968, 1969) studied the response of nonlinear systems to large-amplitude excitations.

In Section 4.5 we treat nonideal systems.

4.1. Systems with Cubic Nonlinearities

Instead of treating general systems for which the algebra is involved, we treat simple systems that exhibit the essential ideas. Thus we consider the forced oscillations of a particle attached to a nonlinear spring under the influence of slight viscous damping so that the equation of motion has the form

$$\ddot{u} + \omega_0^2 u = -2\epsilon\mu\dot{u} - \epsilon\alpha u^3 + E(t) \tag{4.1.1}$$

where μ is positive and α can be either a positive (*hard spring*) or a negative (*soft spring*) constant. As mentioned in the introduction, we assume that

$$E(t) = K \cos \Omega t \tag{4.1.2}$$

except in Sections 4.1.5–4.1.7, where $E(t)$ is a multifrequency excitation. Primary resonances (i.e., $\Omega \approx \omega_0$) are considered in the next section, and other resonances are considered in Sections 4.1.2 through 4.1.4. Besides the books listed in the preface, there are many studies dealing with primary and secondary resonances in single-degree-of-freedom systems under the influence of monofrequency excitations (e.g., Klotter, 1953a, b; Klotter and Pinney, 1953; Sethna, 1954; Caughey, 1954; Loud, 1955, 1965; Shen, 1959; Lamb, 1960; Plotnikova, 1962, 1963b; Kononenko, 1964; Newland, 1965; Moser, 1965; Aks and Carhart, 1970; Ness, 1971; Bykov and Chinkaraev, 1972; Stanišíc and Euler, 1973; van Dooren, 1973a; Ablowitz, Funk, and Newell, 1973; Chao and Sikarskie, 1974; Dobias, 1974; Samoilenko and Momot, 1974; Varga and Aks, 1974; Nocilla and Riganti, 1974; Ovcharova and Goloskokov, 1975; Hsieh, 1975; Cheshankov, 1975; Anderson, 1975a, c; Bastin and Delchambre, 1975; Plakhtienko, 1975; Beshai and Dokainish, 1975; Eminhizer, Helleman, and Montroll, 1976; Mishra and Singh, 1976). Tondl (1970a, 1973a, b) analyzed the primary responses of general systems with various types of damping. He intended the results to be used for the identification of the damping character and for finding the most suitable function expressing the damping force from experimental observations.

4.1.1. PRIMARY RESONANCES, $\Omega \approx \omega_0$

Instead of using the frequency of the excitation Ω as a parameter, we introduce a detuning parameter σ, which quantitatively describes the nearness of Ω to ω_0. This has the advantage of helping one to recognize the terms in the governing equation for u_1 that lead to secular, and nearly secular (*small divisor*), terms. Accordingly we write

$$\Omega = \omega_0 + \epsilon\sigma \tag{4.1.3}$$

where $\sigma = O(1)$. The linear undamped theory will predict unbounded oscillations when $\sigma = 0$ irrespective of how small the excitation is. In the actual system these large oscillations are limited by the damping and the nonlinearity. Thus to obtain a uniformly valid approximate solution of this problem, we need to order the excitation so that it will appear when the damping and the nonlinearity appear. To accomplish this, we set $K = \epsilon k$. We note that this scheme for ordering the terms is consistent with our primitive notions of primary resonance; namely we anticipate that in a lightly damped system a small-amplitude excitation produces a relatively large-amplitude response.

An approximate solution of the problem can be obtained by a number of perturbation techniques. Here we use the method of multiple scales. Accordingly we express the solution in terms of different time scales as

$$u(t;\epsilon) = u_0(T_0, T_1) + \epsilon u_1(T_0, T_1) + \cdots \tag{4.1.4}$$

where $T_0 = t$ and $T_1 = \epsilon t$. We also express the excitation in terms of T_0 and T_1 as

$$E(t) = \epsilon k \cos(\omega_0 T_0 + \sigma T_1) \tag{4.1.5}$$

Substituting (4.1.4) and (4.1.5) into (4.1.1) and equating the coefficients of ϵ^0 and ϵ on both sides, we obtain

$$D_0^2 u_0 + \omega_0^2 u_0 = 0 \tag{4.1.6}$$

$$D_0^2 u_1 + \omega_0^2 u_1 = -2D_0 D_1 u_0 - 2\mu D_0 u_0 - \alpha u_0^3 + k \cos(\omega_0 T_0 + \sigma T_1) \tag{4.1.7}$$

We note that, as a result of the ordering, the excitation, damping, and nonlinear terms appear in (4.1.7).

The general solution of (4.1.6) can be written as

$$u_0 = A(T_1) \exp(i\omega_0 T_0) + \bar{A}(T_1) \exp(-i\omega_0 T_0) \tag{4.1.8}$$

where $A(T_1)$ is an undetermined function at this point; it will be determined by eliminating the secular terms from u_1. Substituting u_0 into (4.1.7) and expressing $\cos(\omega_0 T_0 + \sigma T_1)$ in complex form, we have

$$\begin{aligned} D_0^2 u_1 + \omega_0^2 u_1 = &- [2i\omega_0(A' + \mu A) + 3\alpha A^2 \bar{A}] \exp(i\omega_0 T_0) \\ &- \alpha A^3 \exp(3i\omega_0 T_0) + \tfrac{1}{2} k \exp[i(\omega_0 T_0 + \sigma T_1)] + cc \end{aligned} \tag{4.1.9}$$

where cc stands for the complex conjugate of the preceding terms. Secular terms will be eliminated from the particular solution of (4.1.9) if we choose A to be a solution of

$$2i\omega_0(A' + \mu A) + 3\alpha A^2 \bar{A} - \tfrac{1}{2} k \exp(i\sigma T_1) = 0 \tag{4.1.10}$$

To solve (4.1.10), we write A in the polar form

$$A = \tfrac{1}{2} a \exp(i\beta) \tag{4.1.11}$$

where a and β are real. Then we separate the result into its real and imaginary parts and obtain

$$\begin{aligned} a' &= -\mu a + \frac{1}{2} \frac{k}{\omega_0} \sin(\sigma T_1 - \beta) \\ a\beta' &= \frac{3}{8} \frac{\alpha}{\omega_0} a^3 - \frac{1}{2} \frac{k}{\omega_0} \cos(\sigma T_1 - \beta) \end{aligned} \tag{4.1.12}$$

Substituting (4.1.11) into (4.1.8) and substituting that result into (4.1.4), we obtain the first approximation

$$u = a \cos (\omega_0 t + \beta) + O(\epsilon) \tag{4.1.13}$$

where a and β are given by (4.1.12).

Equations (4.1.12) can be transformed into an autonomous system (i.e., one in which T_1 does not appear explicitly) by letting

$$\gamma = \sigma T_1 - \beta \tag{4.1.14}$$

The result is

$$\begin{aligned} a' &= -\mu a + \tfrac{1}{2} \frac{k}{\omega_0} \sin \gamma \\ a\gamma' &= \sigma a - \tfrac{3}{8} \frac{\alpha}{\omega_0} a^3 + \tfrac{1}{2} \frac{k}{\omega_0} \cos \gamma \end{aligned} \tag{4.1.15}$$

The system of equations (4.1.15) have the general form of the equations discussed in Section 3.2. To determine the character of the solutions, we first locate the singular points and then examine the motion in their neighborhoods. Because the amplitude and phase are not changing at a singular point, the response of the system is said to be a steady-state motion. The nature of the trajectories in the neighborhoods of singular points shows whether a small perturbation in the steady-state motion decays or grows; that is, they illustrate the stability of the steady-state motion.

Steady-State Motions. Steady-state motions occur when $a' = \gamma' = 0$, which corresponds to the singular points of (4.1.15); that is, they correspond to the solutions of

$$\begin{aligned} \mu a &= \tfrac{1}{2} \frac{k}{\omega_0} \sin \gamma \\ a\sigma - \tfrac{3}{8} \frac{\alpha}{\omega_0} a^3 &= -\tfrac{1}{2} \frac{k}{\omega_0} \cos \gamma \end{aligned} \tag{4.1.16}$$

Squaring and adding these equations, we obtain

$$\left[\mu^2 + \left(\sigma - \tfrac{3}{8} \frac{\alpha}{\omega_0} a^2\right)^2\right] a^2 = \frac{k^2}{4\omega_0^2} \tag{4.1.17}$$

Equation (4.1.17) is an implicit equation for the amplitude of the response a as a function of the detuning parameter σ (i.e., the frequency of the excitation) and the amplitude of the excitation k; it is called the *frequency-response equation*.

Substituting (4.1.14) and (4.1.3) into (4.1.13), we find that the first approximation to the steady-state solution is given by

$$u = a \cos (\omega_0 t + \epsilon \sigma t - \gamma) + O(\epsilon) = a \cos (\Omega t - \gamma) + O(\epsilon) \tag{4.1.18}$$

where a and γ are constants. Hence the steady-state response is exactly tuned to the frequency of the excitation. However the phase of the response is shifted, in general, from that of the excitation by $-\gamma$.

The plot of a as a function of σ for given μ and k is called a *frequency-response curve*. Each point on this curve corresponds to a singular point in a different state plane; there is one state plane for each combination of parameters. Later an example of a state plane is presented. To draw such a curve, one can solve a cubic equation for a^2 as a function of σ, or one can solve for σ in terms of a. The latter approach, which is easier, gives

$$\sigma = \frac{3}{8}\frac{\alpha}{\omega_0}a^2 \pm \left(\frac{k^2}{4\omega_0^2 a^2} - \mu^2\right)^{1/2} \tag{4.1.19}$$

Figure 4-1 shows a comparison of the linear ($\alpha = 0$) and nonlinear ($\alpha > 0$) response curves. Equation (4.1.19) indicates that the peak amplitude, which is given by $a_p = k/(2\omega_0\mu)$, is independent of the value of α. The linear results are symmetric to this order of approximation and represent the solution in a very narrow band around the resonant frequency (recall that $\Omega = \omega_0 + \epsilon\sigma$; so the frequency scale σ is greatly expanded). The effect of the nonlinearity is to bend the amplitude curve and distort the phase curve. In both cases multivalued

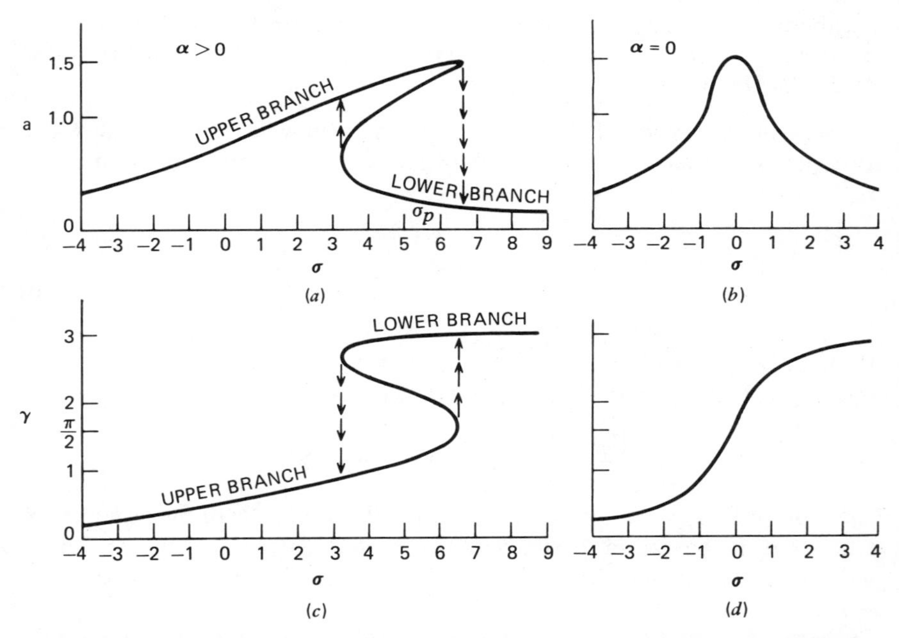

Figure 4-1. Comparison of linear and nonlinear response curves: (*a*) and (*b*) amplitudes; (*c*) and (*d*) phases.

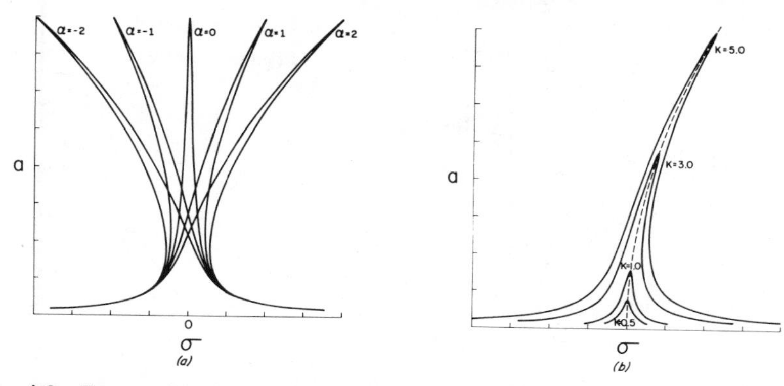

Figure 4-2. Frequency-response curves for primary resonances of the Duffing equation: (*a*) effect of nonlinearity; (*b*) effect of amplitude of excitation.

regions are formed. As we shall discuss later, the multivaluedness is responsible for a jump phenomenon; arrows indicate the jumps.

Figure 4-2*a* shows that the nonlinearity bends the frequency-response curve away from the linear curve ($\alpha = 0$), to the right for hard springs (i.e., $\alpha > 0$) and to the left for soft springs (i.e., $\alpha < 0$). Figure 4-2*b* shows the variation of the frequency-response curves with the amplitude of the excitation for a hard spring. As the amplitude of the excitation increases, the frequency-response curves bend away from the $\sigma = 0$ axis. The locus of the peak amplitudes is the parabola $\sigma = \frac{3}{8}(\alpha/\omega_0)a^2$, which is shown by a dotted line in Figure 4-2*b*. It is often called the *backbone curve*. We note that, depending on the value of k, some of the frequency-response curves are multivalued while others are single-valued.

Figure 4-3 shows the influence of the damping coefficient μ on the response curves. In the absence of damping, the peak amplitude is infinite, and the frequency-response curve consists of two branches having as their asymptote the

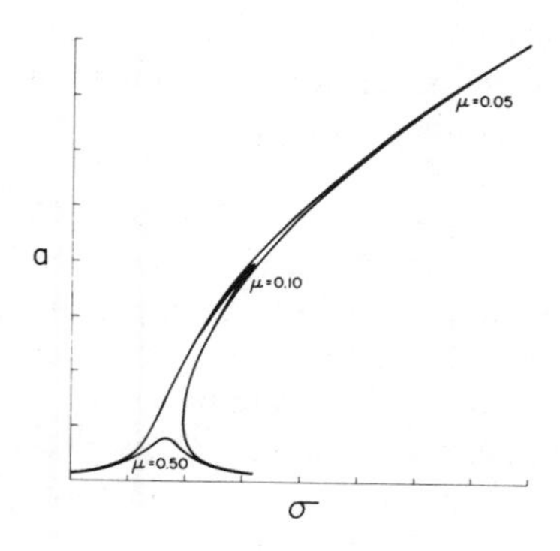

Figure 4-3. Effect of damping on the response of the Duffing equation to a primary-resonance excitation.

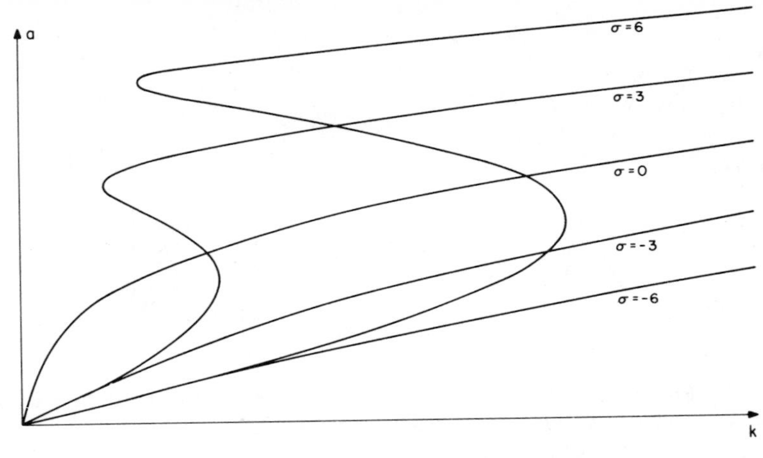

Figure 4-4. Amplitude of the response as a function of amplitude of the excitation for several detunings.

curve $\sigma = \frac{3}{8}(\alpha/\omega_0)a^2$. When $\mu = 0$, $\gamma = n\pi$, where n is an integer according to (4.1.16). Hence (4.1.18) shows that the response is either in phase or 180° out of phase with the excitation. However in the presence of damping, the peak amplitude is finite. Moreover the first of (4.1.16) shows that $\gamma = \sin^{-1}(2\mu a\omega_0/k)$, and hence the damping alters the phase shift of the response.

Figure 4-4 shows the variation of the amplitude of the response with the amplitude of the excitation for several values of σ. The values of α and μ are the same for all curves. These curves were obtained directly from (4.1.17). We note that, depending on the value of σ, some curves are multivalued while others are single-valued.

Jump Phenomena. The multivaluedness of the response curves due to the nonlinearity has a significance from the physical point of view because it leads to jump phenomena. To explain this, we imagine that an experiment is performed in which the amplitude of the excitation is held fixed, the frequency of the excitation (i.e., σ) is very slowly varied up and down through the linear natural frequency, and the amplitude of the harmonic response is observed. The experiment is started at a frequency corresponding to point 1 on the curve in Figure 4-5*a*. As the frequency is reduced, σ decreases and a slowly increases through point 2 until point 3 is reached. As σ is decreased further, a jump from point 3 to point 4 takes place with an accompanying increase in a and a large shift in γ, after which a decreases slowly with decreasing σ. If the experiment is started at point 5 and σ is increased, a increases slowly through point 4 until point 6 is reached. As σ is increased further, a jump from point 6 to point 2 takes place with an accompanying decrease in a and a large shift in γ, after which a decreases

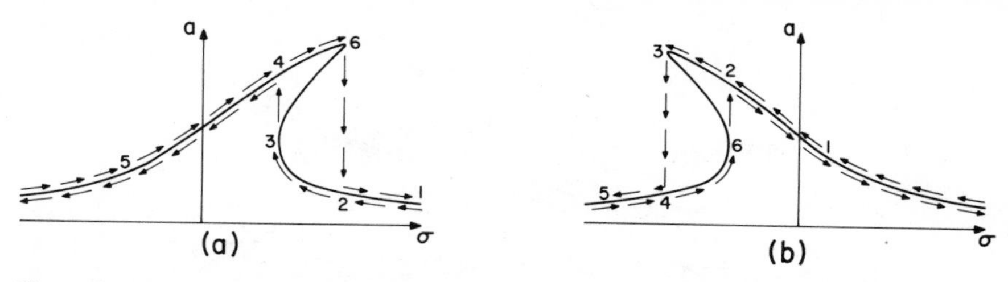

Figure 4-5. Jump phenomena for primary resonance of the Duffing equation: (*a*) $\alpha > 0$; (*b*) $\alpha < 0$.

slowly with increasing σ. The maximum amplitude corresponding to point 6 is attainable only when approached from a lower frequency. The portion of the response curve between points 3 and 6 is unstable and hence cannot be produced experimentally. The stability is discussed below.

For a soft spring, if the experiment is started at point 1 in Figure 4-5*b* and σ is slowly decreased, a jump from point 3 to point 4 takes place. On the other hand, if the experiment is started at point 5 and σ is increased, a jump from point 6 to point 2 takes place. Thus the jump phenomenon is a nonlinear phenomenon which takes place for soft as well as hard springs. As the frequency is decreased, the response amplitude jumps to a lower amplitude for a soft spring and to a higher amplitude for a hard spring. As σ is increased, the opposite takes place.

If the experiment is performed with the frequency of the excitation Ω held fixed while the amplitude of the excitation is varied slowly, a similar jump phenomenon can be observed. Suppose that the experiment is started at point 1 in Figure 4-6. As k is increased, a slowly increases through point 2 to point 3. As k is increased further, a jump takes place from point 3 to point 4, with an accompanying increase in a and a large shift in γ, after which a increases slowly with k. If the process is reversed, a decreases slowly as k decreases from point 5 to point 6. As k is decreased further, a jump from point 6 to point 2 takes place, with an accompanying decrease in a and a large shift in γ, after which a decreases slowly with decreasing k.

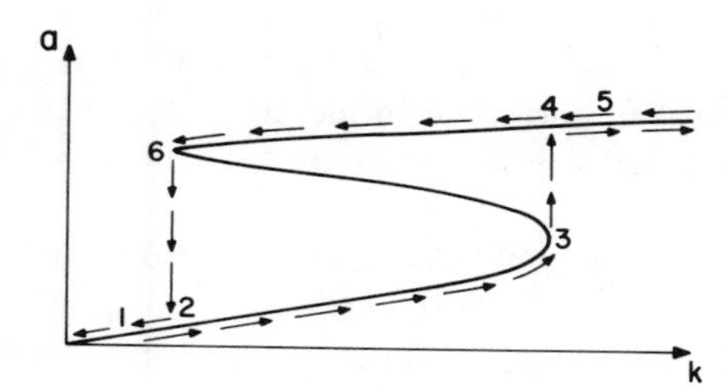

Figure 4-6. Jump phenomena for primary resonance of the Duffing equation.

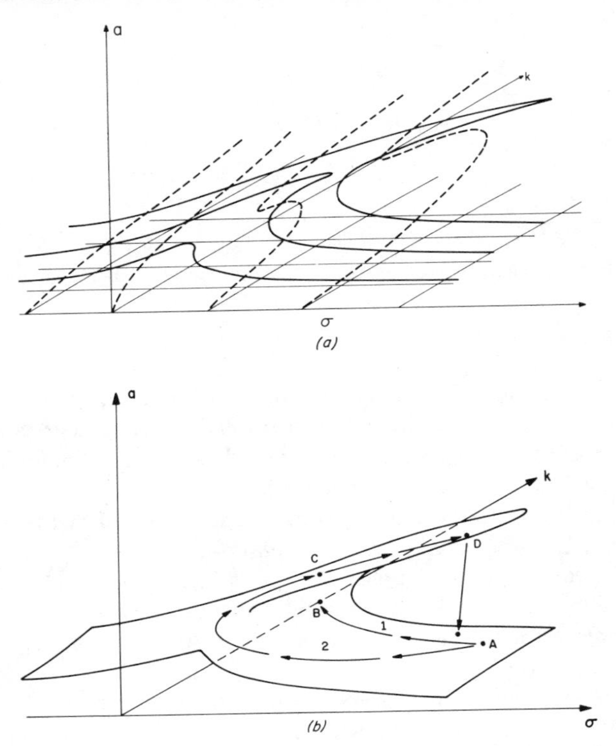

Figure 4-7. Amplitude of the response as a function of amplitude and frequency of the excitation: (*a*) catastrophe surfaces; (*b*) a schematic of a cusp.

When two stable steady-state solutions exist, the initial conditions determine which of these represents the actual response of the system. Thus in contrast with linear systems, the steady-state solution of a nonlinear system can depend on the initial conditions. This point is illustrated later when we discuss the state plane for a case in which two stable solutions exist.

Another way of viewing the jump phenomenon involves the use of *catastrophe theory*. For a given μ one can regard (4.1.17) as the equation of a surface $[a = a(\sigma, k)]$. In Figure 4-7*a*, the intersections of several planes of constant k with this surface are shown by the solid lines, while the intersections with several planes of constant σ are shown by the broken lines. The former are the curves shown in Figure 4-2*b*, and the latter are the lines shown in Figure 4-4. This type of surface is called a *cusp*.

In Figure 4-7*b* a schematic view of a cusp is shown. If a quasi-steady process is started at point A and follows path 1 (only σ changes), there will be a spontaneous jump from point B to point C (a catastrophe). On the other hand if a process is started at point A and follows path 2 (requiring both k and σ to change), the amplitude will increase smoothly to its value at point C. If the process continues from point C to point D, a spontaneous downward jump (a catastrophe) occurs. With this approach all possible responses and quasi-steady processes can be represented by a single surface and viewed simultaneously. For a comprehensive treatment and more references on catastrophe theory, we refer the reader to the book of Lu (1976).

The jump phenomenon is a result of the nonlinear phase-amplitude interaction indicated in the second equation of (4.1.16).

Stability of Steady-State Motions. The stability of the different portions of the response curves can be determined either by investigating the nature of the singular points of (4.1.15) as in Section 3.2 or by superposing a perturbation $v(t)$ on the steady-state solution given by (4.1.18). In the latter case one lets $u = a\cos(\Omega t - \gamma) + v(t)$ in (4.1.1), uses (4.1.2), and obtains

$$\begin{aligned}\ddot{v} + \omega_0^2 v + 2\epsilon\mu\dot{v} + 3\epsilon\alpha a^2 v\cos^2(\Omega t - \gamma) + [(\omega_0^2 - \Omega^2 + \tfrac{3}{4}\epsilon\alpha a^2)a\cos(\Omega t - \gamma) \\ - 2\epsilon\mu a\Omega\sin(\Omega t - \gamma) + \tfrac{1}{4}\epsilon\alpha a^3\cos(3\Omega t - 3\gamma) - \epsilon k\cos\Omega t] \\ + 3\epsilon a v^2\cos(\Omega t - \gamma) + \epsilon v^3 = 0\end{aligned} \tag{4.1.20}$$

Neglecting the term $\cos(3\Omega t - 3\gamma)$, one can easily show that the term in the square brackets vanishes on account of (4.1.3) and (4.1.16). Then (4.1.20) becomes

$$\ddot{v} + \omega_0^2 v + 2\epsilon\mu\dot{v} + 3\epsilon\alpha a^2 v\cos^2(\Omega t - \gamma) + 3\epsilon a v^2\cos(\Omega t - \gamma) + \epsilon v^3 = 0 \tag{4.1.21}$$

Thus the stability of the steady-state motion is transformed into the stability of the solutions of (4.1.21), which is an equation with variable coefficients. Equations with variable coefficients are discussed in Chapter 5.

In this chapter we determine the stability of the steady-state motion by investigating the nature of the singular points of (4.1.15). To accomplish this, we let

$$\begin{aligned}a &= a_0 + a_1 \\ \gamma &= \gamma_0 + \gamma_1\end{aligned} \tag{4.1.22}$$

Substituting (4.1.22) into (4.1.15), expanding for small a_1 and γ_1, noting that a_0 and γ_0 satisfy (4.1.16), and keeping linear terms in a_1 and γ_1, we obtain

$$\begin{aligned}a_1' &= -\mu a_1 + \left(\frac{k}{2\omega_0}\cos\gamma_0\right)\gamma_1 \\ \gamma_1' &= -\left(\frac{3\alpha a_0}{4\omega_0} + \frac{k}{2\omega_0 a_0^2}\cos\gamma_0\right)a_1 - \left(\frac{k}{2\omega_0 a_0}\sin\gamma_0\right)\gamma_1\end{aligned} \tag{4.1.23}$$

Thus the stability of the steady-state motions depends on the eigenvalues of the coefficient matrix on the right-hand sides of (4.1.23).

Using (4.1.16) one can obtain the following eigenvalue equation:

$$\begin{vmatrix} -\mu - \lambda & -a_0\left(\sigma - \dfrac{3\alpha a_0^2}{8\omega_0}\right) \\ \dfrac{1}{a_0}\left(\sigma - \dfrac{9\alpha a_0^2}{8\omega_0}\right) & -\mu - \lambda \end{vmatrix} = 0$$

Expanding this determinant yields

$$\lambda^2 + 2\mu\lambda + \mu^2 + \left(\sigma - \frac{3\alpha a_0^2}{8\omega_0}\right)\left(\sigma - \frac{9\alpha a_0^2}{8\omega_0}\right) = 0$$

Hence the steady-state motions are unstable when

$$\Gamma = \left(\sigma - \frac{3\alpha a_0^2}{8\omega_0}\right)\left(\sigma - \frac{9\alpha a_0^2}{8\omega_0}\right) + \mu^2 < 0 \tag{4.1.24}$$

and are otherwise stable. The condition (4.1.24) corresponds to the portion between points 3 and 6 in Figure 4-5 because points 3 and 6 correspond to $\Gamma = 0$, which is the locus of the vertical tangents to the frequency-response curves. This can be shown by differentiation of (4.1.17) implicitly with respect to a^2 and setting $d\sigma/da^2 = 0$.

The preceding analysis determines the linear or local stability of the steady-state solutions. The stability of motions in the large can be determined theoretically by the use of the Liapunov method. This method depends on the existence of a so-called Liapunov function, which in practice can be found for very few problems (e.g., LaSalle and Lefschetz, 1961; Szego, 1966; Hahn, 1967).

In the case of linear systems and in the presence of positive damping, the steady-state forced response is independent of the initial conditions. In nonlinear systems the initial conditions play a crucial role. When more than one stable steady-state solution exist the initial conditions determine which steady-state solution is physically realized by the system. It turns out that there are instances in which a small change in the initial conditions produces a large change in the response of the system. To illustrate this point we used (4.1.15) to calculate several trajectories in the state plane corresponding to σ_p in Figure 4-1. The trajectories are plotted in Figure 4-8. Points P_1 and P_3 are stable foci, and point P_2 is a saddle point. All initial conditions in the shaded area lead to the steady-state solution on the upper branch P_1, while all initial conditions in the unshaded area lead to the lower branch P_3. The arrows indicate the direction of the motion of the representative point. Thus all the shaded area constitutes the *domain of attraction* of point P_1, and all the unshaded area constitutes the

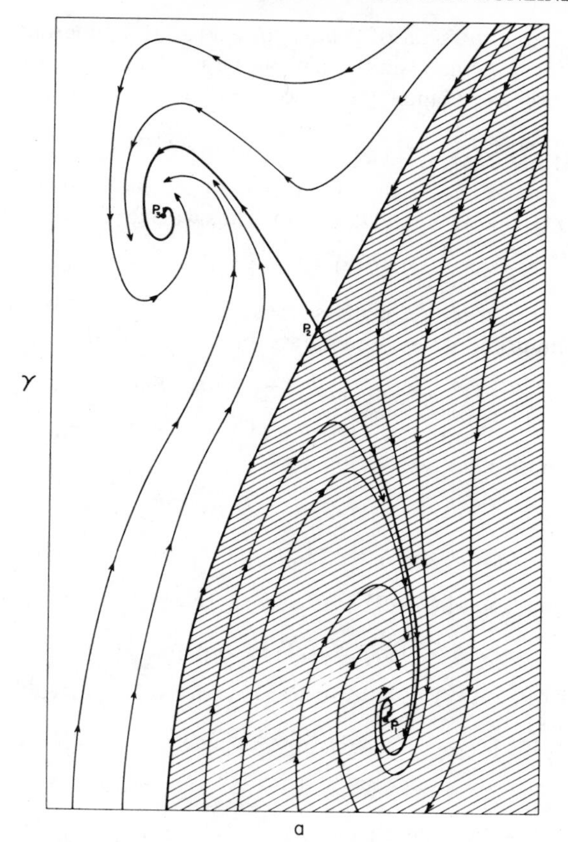

Figure 4-8. State plane for the Duffing equation when three steady-state solutions exist.

domain of attraction of point P_3. Domains of attraction were studied by a number of investigators (e.g., Loud and Sethna, 1966; Sethna, 1967b; Tondl, 1970a, 1973a).

In summary, the question of the stability to small disturbances can be settled with relative ease as in Section 3.2 because the analysis is linear. However to find the stability in the large and to determine the effects of changes in the initial conditions and the system parameters, one has to use a state plane as used above or formulate some form of integrals of motion or energy levels or Liapunov functions (e.g., Struble, 1962; Brauer and Nohel, 1969; Roseau, 1966; Leipholz, 1970). The use of state planes is most suited to systems governed by two first-order differential equations as in this section. Although the other approach is not limited by the degrees of freedom of the systems, it is limited by the ability to find an integral of motion or a Liapunov function. For canonical systems some form of the Hamiltonian can be used, but for general systems the integrals of motion and Liapunov functions are known for very few problems.

4.1.2. NONRESONANT HARD EXCITATIONS

When Ω is away from ω_0, the effect of the excitation will be small unless its amplitude is *hard*; that is, unless $K = O(1)$. Thus we express the excitation as

$$E(t) = K \cos \Omega T_0 \tag{4.1.25}$$

As in the case of primary resonances, we seek an approximate solution by using the method of multiple scales. We express the solution in the form

$$u(t; \epsilon) = u_0(T_0, T_1) + \epsilon u_1(T_0, T_1) + \cdots \tag{4.1.26}$$

Substituting (4.1.26) into (4.1.1), using (4.1.25), and equating the coefficients of ϵ^0 and ϵ on both sides, we obtain

$$D_0^2 u_0 + \omega_0^2 u_0 = K \cos \Omega T_0 \tag{4.1.27}$$

$$D_0^2 u_1 + \omega_0^2 u_1 = -2D_0 D_1 u_0 - 2\mu D_0 u_0 - \alpha u_0^3 \tag{4.1.28}$$

The general solution of (4.1.27) can be written as

$$u_0 = A(T_1) \exp(i\omega_0 T_0) + \Lambda \exp(i\Omega T_0) + cc \tag{4.1.29}$$

where $\Lambda = \frac{1}{2} K(\omega_0^2 - \Omega^2)^{-1}$. Substituting u_0 into (4.1.28) yields

$$\begin{aligned} D_0^2 u_1 + \omega_0^2 u_1 = &- [2i\omega_0(A' + \mu A) + 6\alpha A \Lambda^2 + 3\alpha A^2 \bar{A}] \exp(i\omega_0 T_0) \\ &- \alpha \{A^3 \exp(3i\omega_0 T_0) + \Lambda^3 \exp(3i\Omega T_0) \\ &+ 3A^2 \Lambda \exp[i(2\omega_0 + \Omega)T_0] + 3\bar{A}^2 \Lambda \exp[i(\Omega - 2\omega_0)T_0] \\ &+ 3A\Lambda^2 \exp[i(\omega_0 + 2\Omega)T_0] + 3A\Lambda^2 \exp[i(\omega_0 - 2\Omega)T_0]\} \\ &- \Lambda[2i\mu\Omega + 3\alpha\Lambda^2 + 6\alpha A\bar{A}] \exp(i\Omega T_0) + cc \end{aligned} \tag{4.1.30}$$

In addition to the terms that are proportional to $\exp(\pm i\omega_0 T_0)$, secular or nearly secular (small divisor) terms may occur whenever $\Omega = O(\epsilon)$ or whenever there is a *secondary resonance*, that is, whenever $\omega_0 \approx (m\omega_0 + n\Omega)$, where m and n are integers such that $|m| + |n| = 3$. This occurs whenever $\Omega \approx \frac{1}{3}\omega_0$ or $\Omega \approx 3\omega_0$; the first case is called *superharmonic resonance*, and the second is called *subharmonic resonance*. Thus in eliminating terms that produce secular terms, we need to distinghish four cases: (a) Ω is away from 0, $\frac{1}{3}\omega_0$, and $3\omega_0$; (b) $\Omega \approx 0$; (c) $\Omega \approx \frac{1}{3}\omega_0$; and (d) $\Omega \approx 3\omega_0$. The first case is discussed in this section, the second case is a special case of the problem discussed in Exercise 4.37, the third case (superharmonic resonance) is discussed in the next section, and the fourth case (subharmonic resonance) is discussed in Section 4.1.4.

In the nonresonant case secular terms are eliminated if

$$2i\omega_0(A' + \mu A) + 6\alpha\Lambda^2 A + 3\alpha A^2 \bar{A} = 0 \tag{4.1.31}$$

Letting $A = \frac{1}{2}a \exp(i\beta)$ in (4.1.31), where a and β are real, and separating real and imaginary parts, we obtain

$$\begin{aligned} a' &= -\mu a \\ \omega_0 a\beta' &= 3\alpha(\Lambda^2 + \tfrac{1}{8}a^2)a \end{aligned} \tag{4.1.32}$$

Therefore for the first approximation

$$u = a\cos(\omega_0 t + \beta) + K(\omega_0^2 - \Omega^2)^{-1}\cos\Omega t + O(\epsilon) \tag{4.1.33}$$

where a and β are given by (4.1.32). The general solution for a is $a = a_0 \exp(-\mu T_1)$, where a_0 is a constant. Thus the free-oscillation (homogeneous) solution decays with time so that the steady-state response consists of the forced (particular) solution only, as in the linear case. While the free-oscillation term is decaying, however, its frequency is a function of the amplitude of the particular solution.

4.1.3. SUPERHARMONIC RESONANCES, $\Omega \approx \frac{1}{3}\omega_0$

Besides the books listed in the preface, there are a number of studies that treat superharmonic and higher-harmonic resonances in single-degree-of-freedom systems under the influence of monofrequency excitations (e.g., Atkinson, 1957; Szemplińska-Stupnicka, 1968; and Maezawa and Furukawa, 1973).

In this case we express the nearness of Ω to $\frac{1}{3}\omega_0$ by introducing the detuning parameter σ according to

$$3\Omega = \omega_0 + \epsilon\sigma \tag{4.1.34}$$

In addition to the terms proportional to $\exp(\pm i\omega_0 T_0)$ in (4.1.30), there is another term that produces a secular term in u_1. This is $-\alpha\Lambda^3 \exp(\pm 3i\Omega T_0)$. To eliminate the secular terms, we express $3\Omega T_0$ in terms of $\omega_0 T_0$ according to

$$3\Omega T_0 = (\omega_0 + \epsilon\sigma)T_0 = \omega_0 T_0 + \sigma\epsilon T_0 = \omega_0 T_0 + \sigma T_1 \tag{4.1.35}$$

Using (4.1.35), we find that the secular terms in u_1 are eliminated if

$$2i\omega_0(A' + \mu A) + 6\alpha\Lambda^2 A + 3\alpha A^2\bar{A} + \alpha\Lambda^3 \exp(i\sigma T_1) = 0 \tag{4.1.36}$$

Letting $A = \frac{1}{2}a\exp(i\beta)$ in (4.1.36), where a and β are real, and separating real and imaginary parts, we have

$$\begin{aligned} a' &= -\mu a - \frac{\alpha\Lambda^3}{\omega_0}\sin(\sigma T_1 - \beta) \\ a\beta' &= \frac{3\alpha}{\omega_0}(\Lambda^2 + \tfrac{1}{8}a^2)a + \frac{\alpha\Lambda^3}{\omega_0}\cos(\sigma T_1 - \beta) \end{aligned} \tag{4.1.37}$$

Equations (4.1.37) can be transformed into an autonomous system by setting

$$\gamma = \sigma T_1 - \beta \tag{4.1.38}$$

and thereby obtaining

$$a' = -\mu a - \frac{\alpha\Lambda^3}{\omega_0} \sin \gamma$$
$$a\gamma' = \left(\sigma - \frac{3\alpha\Lambda^2}{\omega_0}\right) a - \frac{3\alpha}{8\omega_0} a^3 - \frac{\alpha\Lambda^3}{\omega_0} \cos \gamma \tag{4.1.39}$$

Therefore for the first approximation

$$u = a \cos (3\Omega t - \gamma) + K(\omega_0^2 - \Omega^2)^{-1} \cos \Omega t + O(\epsilon) \tag{4.1.40}$$

where a and γ are given by (4.1.39).

The steady-state motions correspond to $a' = \gamma' = 0$; that is, they correspond to the solutions of

$$-\mu a = \frac{\alpha\Lambda^3}{\omega_0} \sin \gamma$$
$$\left(\sigma - 3\frac{\alpha\Lambda^2}{\omega_0}\right) a - \frac{3\alpha}{8\omega_0} a^3 = \frac{\alpha\Lambda^3}{\omega_0} \cos \gamma \tag{4.1.41}$$

Squaring and adding these equations leads to the frequency-response equation

$$\left[\mu^2 + \left(\sigma - 3\frac{\alpha\Lambda^2}{\omega_0} - \frac{3\alpha}{8\omega_0} a^2\right)^2\right] a^2 = \frac{\alpha^2\Lambda^6}{\omega_0^2} \tag{4.1.42}$$

Solving this equation for σ in terms of a yields

$$\sigma = 3\frac{\alpha\Lambda^2}{\omega_0} + \frac{3\alpha}{8\omega_0} a^2 \pm \left(\frac{\alpha^2\Lambda^6}{\omega_0^2 a^2} - \mu^2\right)^{1/2} \tag{4.1.43}$$

Therefore when $\Omega \approx \frac{1}{3}\omega_0$, the free-oscillation term does not decay to zero in spite of the presence of damping and in contrast with the linear case. Moreover the nonlinearity adjusts the frequency of the free-oscillation term to exactly three times the frequency of the excitation so that the response is periodic. Since the frequency of the generated free-oscillation term is three times the frequency of the excitation, such resonances are called *superharmonic resonances*, or *overtones*. In Figure 4-9 the three curves show how the response is formed from the particular solution and the free-oscillation term [recall (4.1.40)].

In this case the peak amplitude of the free-oscillation term is given by

$$a_p = \frac{\alpha\Lambda^3}{\mu\omega_0}$$

In contrast with the case of primary resonance, a_p is a function of α, the coefficient of the nonlinear term. The corresponding value of the detuning is given by

$$\sigma_p = \frac{3\alpha\Lambda^2}{\omega_0}\left(1 + \frac{\alpha^2\Lambda^4}{8\mu^2\omega_0^2}\right)$$

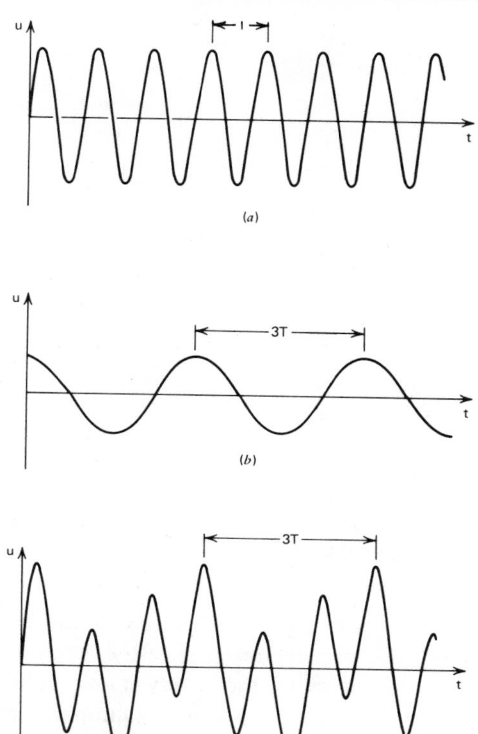

Figure 4-9. Synthesis of the response of the Duffing equation to superharmonic excitation: (*a*) free-oscillation solution; (*b*) particular solution; (*c*) actual response.

It follows from (4.1.42) that when the amplitude of the free-oscillation term is small, it is proportional to Λ^3; thus the response approaches that of the linear system as Λ vanishes.

In Figure 4-10 various frequency–response curves are shown. These curves show the influences of varying α, Λ, and μ. Here, as in the case of primary resonance, the bending of the frequency–response curves is responsible for a jump phenomenon. There is symmetry about the $\sigma = 0$ line when the sign of α is changed [symmetry is indicated by (4.1.43)].

In Figure 4-11 the amplitude of the free-oscillation term is plotted as a function of the amplitude of the excitation. The broken lines show the unstable portions of the curve, and the arrows indicate the jumps that occur as the amplitude of the excitation is increased (upward jump) and decreased (downward jump). By comparing Figures 4-4 and 4-11, we can discover important differences between the character of the two responses. As the amplitude of the excitation decreases, the amplitude of the response vanishes much more rapidly

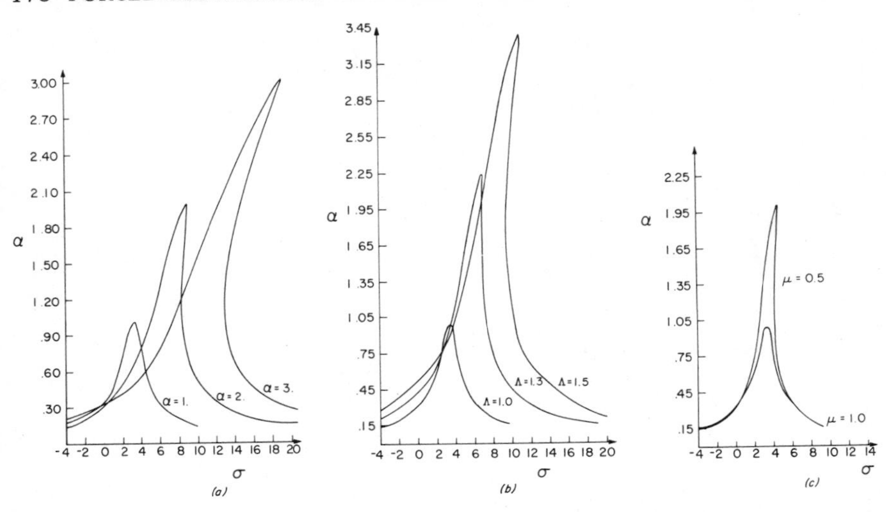

Figure 4-10. Superharmonic frequency–response curves for the Duffing equation: (*a*) effect of nonlinearity; (*b*) effect of amplitude of the excitation; (*c*) effect of damping.

in Figure 4-11 than it does in Figure 4-4. This can also be seen by comparing (4.1.41) with (4.1.16); for primary resonances the amplitude of the excitation appears in the equations governing the steady-state solution, while for superharmonic resonances the cube appears. After the jump in Figure 4-11, the amplitude of the response decreases as the amplitude of the excitation increases until point E is reached, while in Figure 4-4 the amplitude of the response increases monotonically. This can also be seen by comparing (4.1.41) with (4.1.16); in the second equation of (4.1.41) there is an extra term multiplying the amplitude a, $-3\alpha\Lambda^2/\omega_0$. As Λ is increased for a fixed σ, the effect is to decrease the apparent detuning. The effect of decreasing σ can be seen in Figure 4-4. Thus

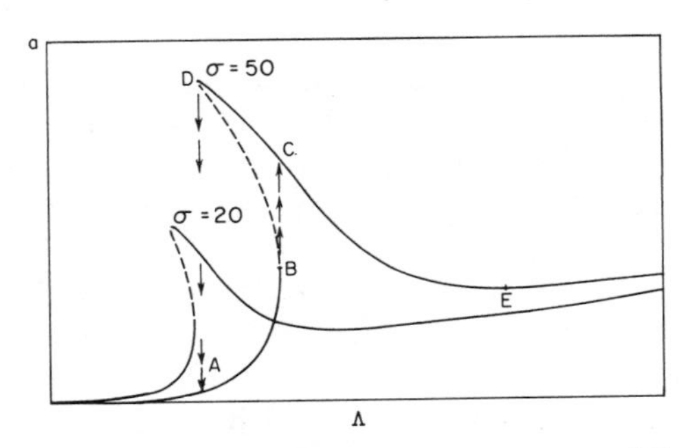

Figure 4-11. Jump phenomenon in the superharmonic response of the Duffing equation.

when Λ increases, there are two influences competing simultaneously: one tends to increase the amplitude of the response while the other tends to decrease the amplitude of the response.

4.1.4. SUBHARMONIC RESONANCES, $\Omega \approx 3\omega_0$

Subharmonic and ultrasubharmonic resonances in conservative systems having a single degree of freedom were studied by Cartwright and Littlewood (1947); Reuter (1949); Levenson (1949, 1968); Obi (1950); Stoker (1950); Hayashi (1953b); Gambill and Hale (1956); Rosenberg (1958); Hsu (1959); Kronauer and Musa (1966b); Tomás and Tondl (1967); Proskuriakov (1971); Loud (1972); Fu (1974); Yamamoto, Yasuda, and Nagasaka (1976); and Prosperetti (1976).

To analyze subharmonic resonances for (4.1.1) and (4.1.2), we introduce the detuning parameter σ according to

$$\Omega = 3\omega_0 + \epsilon\sigma \tag{4.1.44}$$

In addition to the terms proportional to $\exp(\pm i\omega_0 T_0)$, the term proportional to $\exp[\pm i(\Omega - 2\omega_0)T_0]$ produces a secular term in u_1. We express $(\Omega - 2\omega_0)T_0$ as

$$(\Omega - 2\omega_0)T_0 = \omega_0 T_0 + \epsilon\sigma T_0 = \omega_0 T_0 + \sigma T_1 \tag{4.1.45}$$

Therefore to eliminate the terms in (4.1.30) that produce secular terms in u_1, we put

$$2i\omega_0(A' + \mu A) + 6\alpha\Lambda^2 A + 3\alpha A^2\bar{A} + 3\alpha\Lambda\bar{A}^2 \exp(i\sigma T_1) = 0 \tag{4.1.46}$$

Letting $A = \frac{1}{2}a\exp(i\beta)$ in (4.1.46), where a and β are real, and separating real and imaginary parts, we obtain

$$\begin{aligned} a' &= -\mu a - \frac{3\alpha\Lambda}{4\omega_0}a^2 \sin(\sigma T_1 - 3\beta) \\ a\beta' &= \frac{3\alpha}{\omega_0}(\Lambda^2 + \tfrac{1}{8}a^2)a + \frac{3\alpha\Lambda}{4\omega_0}a^2\cos(\sigma T_1 - 3\beta) \end{aligned} \tag{4.1.47}$$

To transform (4.1.47) into an autonomous system, we let

$$\gamma = \sigma T_1 - 3\beta \tag{4.1.48}$$

and obtain

$$\begin{aligned} a' &= -\mu a - \frac{3\alpha\Lambda}{4\omega_0}a^2 \sin\gamma \\ a\gamma' &= \left(\sigma - \frac{9\alpha\Lambda^2}{\omega_0}\right)a - \frac{9\alpha}{8\omega_0}a^3 - \frac{9\alpha\Lambda}{4\omega_0}a^2\cos\gamma \end{aligned} \tag{4.1.49}$$

Therefore for the first approximation

$$u = a \cos\left[\tfrac{1}{3}(\Omega t - \gamma)\right] + K(\omega_0^2 - \Omega^2)^{-1} \cos \Omega t + O(\epsilon) \tag{4.1.50}$$

where a and γ are given by (4.1.49).

The steady-state motions correspond to the solutions of

$$\begin{aligned} -\mu a &= \frac{3\alpha\Lambda}{4\omega_0} a^2 \sin \gamma \\ \left(\sigma - \frac{9\alpha\Lambda^2}{\omega_0}\right) a - \frac{9\alpha}{8\omega_0} a^3 &= \frac{9\alpha\Lambda}{4\omega_0} a^2 \cos \gamma \end{aligned} \tag{4.1.51}$$

Eliminating γ from these equations leads to the frequency-response equation

$$\left[9\mu^2 + \left(\sigma - \frac{9\alpha\Lambda^2}{\omega_0} - \frac{9\alpha}{8\omega_0} a^2\right)^2\right] a^2 = \frac{81\alpha^2\Lambda^2}{16\omega_0^2} a^4 \tag{4.1.52}$$

Equation (4.1.52) shows that either $a = 0$ or

$$9\mu^2 + \left(\sigma - \frac{9\alpha\Lambda^2}{\omega_0} - \frac{9\alpha}{8\omega_0} a^2\right)^2 = \frac{81\alpha^2\Lambda^2}{16\omega_0^2} a^2 \tag{4.1.53}$$

which is quadratic in a^2. Its solution is

$$a^2 = p \pm (p^2 - q)^{1/2} \tag{4.1.54}$$

where

$$p = \frac{8\omega_0\sigma}{9\alpha} - 6\Lambda^2 \quad \text{and} \quad q = \frac{64\omega_0^2}{81\alpha^2}\left[9\mu^2 + \left(\sigma - \frac{9\alpha\Lambda^2}{\omega_0}\right)^2\right] \tag{4.1.55}$$

We note that q is always positive, and thus nontrivial free-oscillation amplitudes occur only when $p > 0$ and $p^2 \geqslant q$. These conditions demand that

$$\Lambda^2 < \frac{4\omega_0\sigma}{27\alpha} \quad \text{and} \quad \frac{\alpha\Lambda^2}{\omega_0}\left(\sigma - \frac{63\alpha\Lambda^2}{8\omega_0}\right) - 2\mu^2 \geqslant 0 \tag{4.1.56}$$

It follows that α and σ must have the same sign.

It follows from (4.1.56) that, for a given Λ, nontrivial solutions can exist only if

$$\alpha\sigma \geqslant \frac{2\mu^2\omega_0}{\Lambda^2} + \frac{63\alpha^2\Lambda^2}{8\omega_0} \tag{4.1.57}$$

while for a given σ, nontrivial solutions can exist only if

$$\frac{\sigma}{\mu} - \left(\frac{\sigma^2}{\mu^2} - 63\right)^{1/2} \leqslant \frac{63\alpha\Lambda^2}{4\omega_0\mu} \leqslant \frac{\sigma}{\mu} + \left(\frac{\sigma^2}{\mu^2} - 63\right)^{1/2} \tag{4.1.58}$$

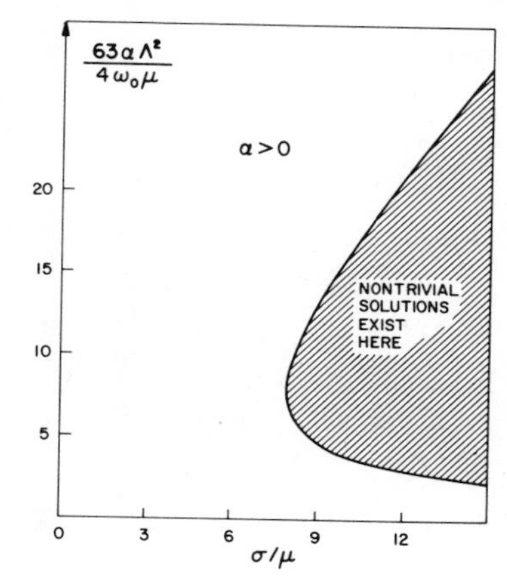

Figure 4-12. Regions where subharmonic responses exist.

In the $\Lambda\sigma$-plane the boundary of the region where nontrivial solutions can exist is given by

$$\frac{63\alpha\Lambda^2}{4\omega_0\mu} = \frac{\sigma}{\mu} \pm \left(\frac{\sigma^2}{\mu^2} - 63\right)^{1/2}$$

which is shown in Figure 4-12 for $\alpha > 0$.

When these conditions hold, it is possible for the system to respond in such a way that the free-oscillation term does not decay to zero in spite of the presence of damping and in contrast with the linear solution. Moreover in the steady state, the nonlinearity adjusts the frequency of the free-oscillation term to one third the frequency of the excitation so that the response is periodic. Since the frequency of the free-oscillation term is one third that of the excitation, such resonances are called *subharmonic resonances*, or *frequency demultiplication*. Several frequency-response curves are shown in Figure 4-13*a*, and the amplitude of the free-oscillation term is plotted as a function of the amplitude of the excitation in Figure 4-13*b*.

We note that although the frequency of the excitation is three times the natural frequency of the system, the response is quite large. For example, certain parts of an airplane can be violently excited by an engine running at an angular speed that is much larger than their natural frequencies (von Kármán, 1940). Lefschetz (1956) described a commercial airplane in which the propellers induced a subharmonic vibration of order $\frac{1}{2}$ in the wings which in turn induced a subharmonic of order $\frac{1}{4}$ in the rudder. The oscillations were so violent that the airplane broke up.

We note that there is no jump phenomenon in this case. In the regions where two stable solutions exist ($a = 0$ and $a \neq 0$), the initial conditions determine

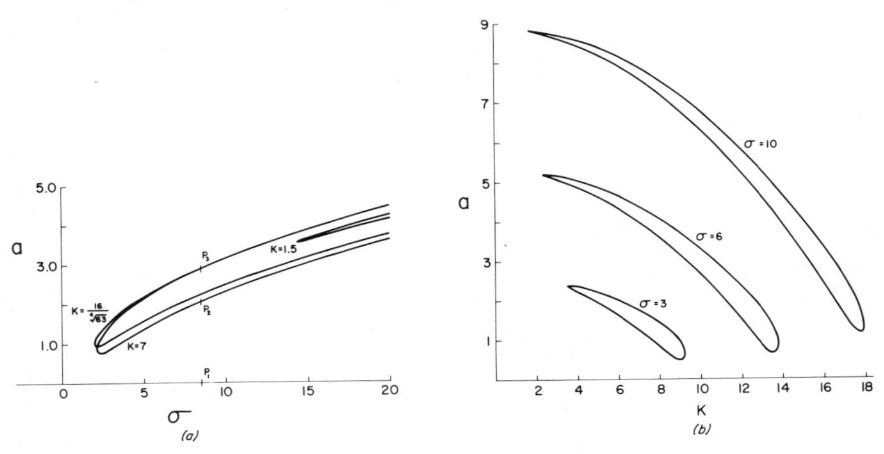

Figure 4-13. Subharmonic response for the Duffing equation; amplitude of the free-oscillation term versus (*a*) detuning and (*b*) amplitude of the excitation.

which solution represents the response. One could construct a figure similar to Figure 4-9, showing the synthesis of the response. In this case the high-frequency component is the particular solution and the low-frequency component is the free-oscillation term.

A possible state plane for the free-oscillation term is shown in Figure 4-14. The

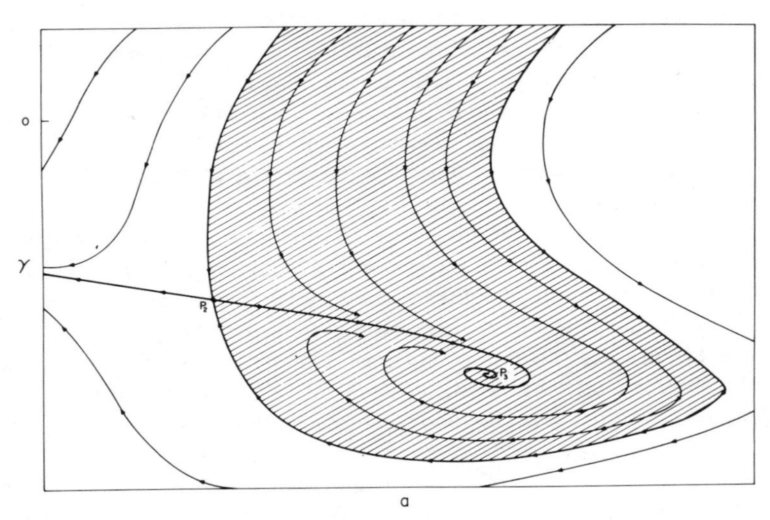

Figure 4-14. State plane for the subharmonic response of the Duffing equation.

trajectories were calculated by using (4.1.49). Point P_2 is a saddle point and point P_3 is a stable focus. These points correspond to those labeled in Figure 4-13. The initial conditions must fall in the shaded area if the subharmonic is to appear in the first approximation of the steady-state response. All initial conditions outside this area lead to the trivial solution for the free-oscillation term.

4.1.5. COMBINATION RESONANCES FOR TWO-TERM EXCITATIONS

In this section we consider excitations that consist of two terms having the form

$$E(t) = K_1 \cos(\Omega_1 t + \theta_1) + K_2 \cos(\Omega_2 t + \theta_2) \tag{4.1.59}$$

where K_n, Ω_n, and θ_n are constants. Moreover we assume that $K_n = O(1)$, and we exclude the primary-resonance cases $\omega_0 \approx \Omega_n$ for $n = 1$ and 2. We assume an expansion of the form

$$u(t;\epsilon) = u_0(T_0, T_1) + \epsilon u_1(T_0, T_1) + \cdots \tag{4.1.60}$$

in (4.1.1), use (4.1.59), equate the coefficients of ϵ^0 and ϵ on both sides, and obtain

$$D_0^2 u_0 + \omega_0^2 u_0 = K_1 \cos(\Omega_1 T_0 + \theta_1) + K_2 \cos(\Omega_2 T_0 + \theta_2) \tag{4.1.61}$$

$$D_0^2 u_1 + \omega_0^2 u_1 = -2D_0 D_1 u_0 - 2\mu D_0 u_0 - \alpha u_0^3 \tag{4.1.62}$$

The general solution of (4.1.61) can be written in the form

$$u_0 = A(T_1)\exp(i\omega_0 T_0) + \Lambda_1 \exp(i\Omega_1 T_0) + \Lambda_2 \exp(i\Omega_2 T_0) + cc \tag{4.1.63}$$

where

$$\Lambda_n = \tfrac{1}{2} K_n (\omega_0^2 - \Omega_n^2)^{-1} \exp(i\theta_n)$$

Substituting u_0 into (4.1.62) gives

$$\begin{aligned} D_0^2 u_1 + \omega_0^2 u_1 = &-[2i\omega_0(A' + \mu A) + 3\alpha(A\bar{A} + 2\Lambda_1\bar{\Lambda}_1 + 2\Lambda_2\bar{\Lambda}_2)A] \\ &\cdot \exp(i\omega_0 T_0) - [2i\Omega_1\mu + 3\alpha(2A\bar{A} + \Lambda_1\bar{\Lambda}_1 + 2\Lambda_2\bar{\Lambda}_2)]\Lambda_1 \\ &\cdot \exp(i\Omega_1 T_0) - [2i\Omega_2\mu + 3\alpha(2A\bar{A} + 2\Lambda_1\bar{\Lambda}_1 + \Lambda_2\bar{\Lambda}_2)]\Lambda_2 \\ &\cdot \exp(i\Omega_2 T_0) - \alpha A^3 \exp(3i\omega_0 T_0) - \alpha\Lambda_1^3 \exp(3i\Omega_1 T_0) \\ &- \alpha\Lambda_2^3 \exp(3i\Omega_2 T_0) - 3\alpha A^2\Lambda_1 \exp[i(2\omega_0 + \Omega_1)T_0] \\ &- 3\alpha A^2\Lambda_2 \exp[i(2\omega_0 + \Omega_2)T_0] - 3\alpha A^2\bar{\Lambda}_1 \\ &\cdot \exp[i(2\omega_0 - \Omega_1)T_0] - 3\alpha A^2\bar{\Lambda}_2 \exp[i(2\omega_0 - \Omega_2)T_0] \\ &- 3\alpha A\Lambda_1^2 \exp[i(\omega_0 + 2\Omega_1)T_0] - 3\alpha A\Lambda_2^2 \end{aligned}$$

$$
\begin{aligned}
&\cdot \exp\left[i(\omega_0 + 2\Omega_2)T_0\right] - 3\alpha A\overline{\Lambda}_1^2 \exp\left[i(\omega_0 - 2\Omega_1)T_0\right] \\
&- 3\alpha A\overline{\Lambda}_2^2 \exp\left[i(\omega_0 - 2\Omega_2)T_0\right] - 6\alpha A\Lambda_1\Lambda_2 \\
&\cdot \exp\left[i(\omega_0 + \Omega_1 + \Omega_2)T_0\right] - 6\alpha A\overline{\Lambda}_1\overline{\Lambda}_2 \\
&\cdot \exp\left[i(\omega_0 - \Omega_1 - \Omega_2)T_0\right] - 6\alpha A\overline{\Lambda}_1\Lambda_2 \\
&\cdot \exp\left[i(\omega_0 - \Omega_1 + \Omega_2)T_0\right] - 6\alpha A\Lambda_1\overline{\Lambda}_2 \\
&\cdot \exp\left[i(\omega_0 + \Omega_1 - \Omega_2)T_0\right] - 3\alpha\Lambda_1^2\Lambda_2 \exp\left[i(2\Omega_1 + \Omega_2)T_0\right] \\
&- 3\alpha\Lambda_1^2\overline{\Lambda}_2 \exp\left[i(2\Omega_1 - \Omega_2)T_0\right] - 3\alpha\Lambda_1\Lambda_2^2 \\
&\cdot \exp\left[i(\Omega_1 + 2\Omega_2)T_0\right] - 3\alpha\overline{\Lambda}_1\Lambda_2^2 \exp\left[i(2\Omega_2 - \Omega_1)T_0\right] + cc
\end{aligned} \tag{4.1.64}
$$

Equation (4.1.64) exhibits a number of resonant combinations some of which we encountered earlier in cases of monofrequency excitations and some of which are characteristic of multifrequency excitations. These combinations are

$\omega_0 \approx 3\Omega_n$	superharmonic resonance
$\omega_0 \approx \frac{1}{3}\Omega_n$	subharmonic resonance
$\omega_0 \approx \lvert \pm 2\Omega_m \pm \Omega_n \rvert$	combination resonance
$\omega_0 \approx \frac{1}{2}(\Omega_m \pm \Omega_n)$	combination resonance

where $m = 1$ and 2 and $n = 1$ and 2. For excitations with three or more frequencies, the resonant combination $\omega_0 \approx \lvert \pm\Omega_m \pm \Omega_n \pm \Omega_k \rvert$ might occur.

We note that for a multifrequency excitation, more than one resonant condition might occur simultaneously; that is, both superharmonic and subharmonic resonances can occur simultaneously or both superharmonic and combination resonances can occur simultaneously, etc. For a two-frequency excitation, at most two resonances can occur simultaneously. If these frequencies are denoted by Ω_1 and Ω_2, where $\Omega_2 > \Omega_1$, the only secondary resonances that can occur are

$$
\begin{aligned}
&\omega_0 \approx 3\Omega_1 \quad \text{or} \quad 3\Omega_2 \\
&\omega_0 \approx \tfrac{1}{3}\Omega_1 \quad \text{or} \quad \tfrac{1}{3}\Omega_2 \\
&\omega_0 \approx \Omega_2 \pm 2\Omega_1 \quad \text{or} \quad 2\Omega_1 - \Omega_2 \\
&\omega_0 \approx 2\Omega_2 \pm \Omega_1 \\
&\omega_0 \approx \tfrac{1}{2}(\Omega_2 \pm \Omega_1)
\end{aligned}
$$

Inspection of these resonances shows that more than one of them occur simultaneously if

(a) $\Omega_2 \approx 9\Omega_1 \approx 3\omega_0$
(b) $\Omega_2 \approx \Omega_1 \approx 3\omega_0$
(c) $\Omega_2 \approx \Omega_1 \approx \frac{1}{3}\omega_0$
(d) $\Omega_2 \approx 5\Omega_1 \approx \frac{5}{3}\omega_0$
(e) $\Omega_2 \approx 7\Omega_1 \approx \frac{7}{3}\omega_0$
(f) $\Omega_2 \approx 2\Omega_1 \approx \frac{2}{3}\omega_0$
(g) $\Omega_2 \approx \frac{7}{3}\Omega_1 \approx 7\omega_0$
(h) $\Omega_2 \approx \frac{5}{3}\Omega_1 \approx 5\omega_0$

where we note that cases (b) and (c) in which $\Omega_2 \approx \Omega_1$ can be best treated by considering the excitation to be a monofrequency nonstationary excitation, as discussed at the beginning of this chapter.

Since the individual primary, superharmonic, and subharmonic resonances were discussed at length in Sections 4.1.1 through 4.1.4, we devote the rest of this section to individual combination resonances and to a case in which subharmonic and superharmonic resonances occur simultaneously, that is, case (a) above. Combination resonances in one-degree-of-freedom systems were treated by Tomáš and Tondl (1967); Tondl (1972); Yamamoto, Yasuda, and Nakamura (1974a, b, c); Efstathiades (1974); Tiwari and Subramanian (1976); and Mojaddidy, Mook, Nayfeh (1977).

In the remainder of this section we consider the case in which $\omega_0 \approx 2\Omega_1 + \Omega_2$. We introduce a detuning parameter σ according to

$$\omega_0 = 2\Omega_1 + \Omega_2 - \epsilon\sigma \tag{4.1.65}$$

and express $(2\Omega_1 + \Omega_2)T_0$ as

$$(2\Omega_1 + \Omega_2)T_0 = \omega_0 T_0 + \epsilon\sigma T_0 = \omega_0 T_0 + \sigma T_1 \tag{4.1.66}$$

Then the secular terms will be eliminated if

$$2i\omega_0(A' + \mu A) + \alpha(3A\bar{A} + 6\Lambda_1\bar{\Lambda}_1 + 6\Lambda_2\bar{\Lambda}_2)A + 3\alpha\Lambda_1^2\Lambda_2 \exp(i\sigma T_1) = 0 \tag{4.1.67}$$

Letting $A = \frac{1}{2}a \exp(i\beta)$ in (4.1.67), where a and β are real, using the definition of Λ_n, and separating real and imaginary parts, we obtain

$$a' = -\mu a - \alpha\Gamma_1 \sin\gamma \tag{4.1.68}$$

$$a\beta' = \alpha\Gamma_2 a + \frac{3\alpha}{8\omega_0}a^3 + \alpha\Gamma_1 \cos\gamma \tag{4.1.69}$$

where

$$\Gamma_1 = \tfrac{3}{8} K_1^2 K_2 \omega_0^{-1} (\omega_0^2 - \Omega_1^2)^{-2} (\omega_0^2 - \Omega_2^2)^{-1}$$
$$\Gamma_2 = \tfrac{3}{4} \omega_0^{-1} [K_1^2 (\omega_0^2 - \Omega_1^2)^{-2} + K_2^2 (\omega_0^2 - \Omega_2^2)^{-2}] \tag{4.1.70}$$
$$\gamma = \sigma T_1 - \beta + 2\theta_1 + \theta_2$$

Eliminating β from (4.1.69) and (4.1.70) gives

$$a\gamma' = (\sigma - \alpha\Gamma_2) a - \frac{3\alpha}{8\omega_0} a^3 - \alpha\Gamma_1 \cos\gamma \tag{4.1.71}$$

Therefore for the first approximation

$$u = a \cos [(2\Omega_1 + \Omega_2) t - \gamma + 2\theta_1 + \theta_2] + K_1 (\omega_0^2 - \Omega_1^2)^{-1} \cos (\Omega_1 t + \theta_1)$$
$$+ K_2 (\omega_0^2 - \Omega_2^2)^{-1} \cos (\Omega_2 t + \theta_2) + O(\epsilon) \tag{4.1.72}$$

For steady-state solutions of (4.1.68) and (4.1.71), $a' = \gamma' = 0$ so that a and γ are the solutions of

$$-\mu a = \alpha\Gamma_1 \sin\gamma$$
$$(\sigma - \alpha\Gamma_2) a - \frac{3\alpha}{8\omega_0} a^3 = \alpha\Gamma_1 \cos\gamma \tag{4.1.73}$$

Eliminating γ from these equations leads to the frequency-response equation

$$\left[\mu^2 + \left(\sigma - \alpha\Gamma_2 - \frac{3\alpha}{8\omega_0} a^2\right)^2\right] a^2 = \alpha^2 \Gamma_1^2 \tag{4.1.74}$$

It follows that the peak amplitude a_p is given by

$$a_p = |\alpha| \Gamma_1 / \mu \tag{4.1.75}$$

and occurs when

$$\sigma = \alpha\Gamma_2 + \frac{3\alpha a^2}{8\omega_0} = \alpha\Gamma_2 + \frac{3\alpha^3 \Gamma_1^2}{8\omega_0 \mu^2}$$

We note that the peak amplitude is independent of Γ_2, but the frequency at which it occurs is not. Equation (4.1.72) shows that unless Ω_2 and Ω_1 are *commensurable* (i.e., unless there exist integers m and n such that $m\Omega_1 + n\Omega_2 = 0$), the motion cannot become periodic.

In Figure 4-15, several frequency–response curves are shown. These figures illustrate the effects on the amplitude of varying Γ_1, Γ_2, α, and μ, respectively. The bending of the frequency–response curves produces a jump phenomenon.

Equation (4.1.74) shows that a is always different from zero. Consequently, in spite of the presence of damping, the free-oscillation term, tuned by the nonlinearity to exactly the combination frequency $2\Omega_1 + \Omega_2$, is part of the steady-

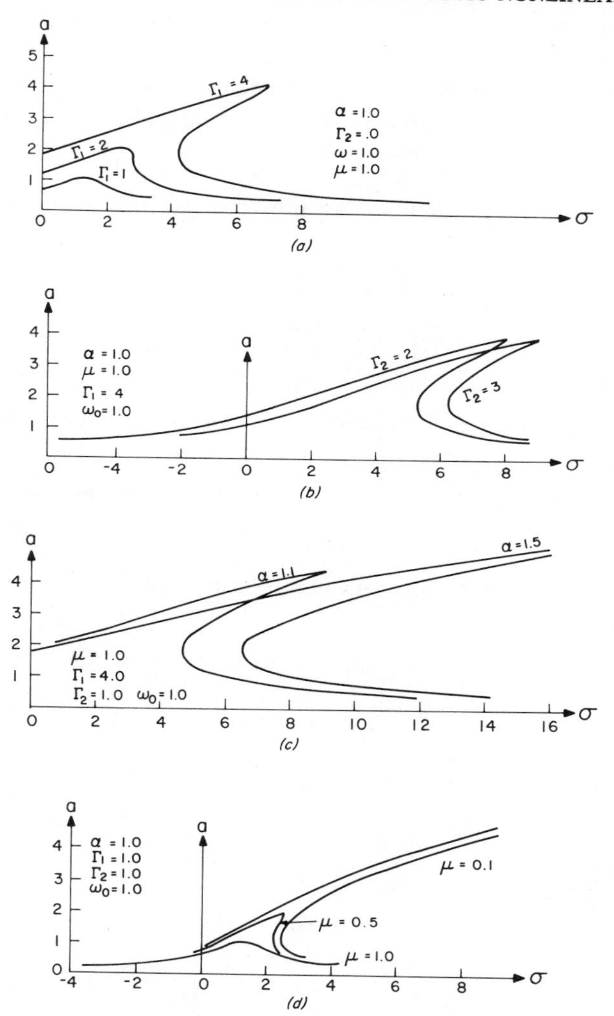

Figure 4-15. Amplitude of the free-oscillation term as a function of the detuning for a combination resonance of the Duffing equation: (*a*) effect of Γ_1; (*b*) effect of Γ_2; (*c*) effect of α; (*d*) effect of μ.

state motion for all conditions, in contrast with the subharmonic-resonance case (Section 4.1.4).

The results for the resonant case $\omega_0 \approx \Omega_2 + 2\Omega_1$ can be obtained from the above results by simply interchanging the subscripts 1 and 2. Moreover the results for the resonant case $\omega_0 \approx 2\Omega_1 - \Omega_2$ can be obtained from the above by simply changing the sign of Ω_2.

4.1.6. SIMULTANEOUS RESONANCES: THE CASE IN WHICH $\omega_0 \approx 3\Omega_1$ AND $\omega_0 \approx \frac{1}{3}\Omega_2$

In this section we consider a case of double resonance in which subharmonic and superharmonic resonances exist simultaneously, that is, the case $\omega_0 \approx 3\Omega_1 \approx \frac{1}{3}\Omega_2$. Other cases of simultaneous resonance can be treated in a similar fashion.

To analyze the case in question, we introduce the two detuning parameters σ_1 and σ_2 according to

$$3\Omega_1 = \omega_0 + \epsilon\sigma_1 \quad \text{and} \quad \Omega_2 = 3\omega_0 + \epsilon\sigma_2 \tag{4.1.76}$$

Then we express $\Omega_1 T_0$ and $\Omega_2 T_0$ as

$$\Omega_1 T_0 = \tfrac{1}{3}\omega_0 T_0 + \tfrac{1}{3}\sigma_1 T_1 \quad \text{and} \quad \Omega_2 T_0 = 3\omega_0 T_0 + \sigma_2 T_1 \tag{4.1.77}$$

Using (4.1.77) in eliminating the terms in (4.1.64) that produce secular terms in u_1, we obtain

$$2i\omega_0(A' + \mu A) + 3\alpha(A\bar{A} + 2\Lambda_1\bar{\Lambda}_1 + 2\Lambda_2\bar{\Lambda}_2)A + \alpha\Lambda_1^3 \exp(i\sigma_1 T_1) + 3\alpha\Lambda_2\bar{A}^2 \exp(i\sigma_2 T_1) = 0 \tag{4.1.78}$$

Letting $A = \frac{1}{2}a\exp(i\beta)$, where a and β are real, in (4.1.78) and separating real and imaginary parts, we have

$$a' + \mu a + \alpha\Gamma_1 \sin(\sigma_1 T_1 + 3\theta_1 - \beta) + \alpha\Gamma_2 a^2 \sin(\sigma_2 T_1 + \theta_2 - 3\beta) = 0 \tag{4.1.79}$$

$$-a\beta' + \frac{3\alpha a^3}{8\omega_0} + \alpha\Gamma_3 a + \alpha\Gamma_1 \cos(\sigma_1 T_1 + 3\theta_1 - \beta) + \alpha\Gamma_2 a^2 \cos(\sigma_2 T_1 + \theta_2 - 3\beta) = 0 \tag{4.1.80}$$

where

$$\Gamma_1 = \frac{K_1^3}{8\omega_0(\omega_0^2 - \Omega_1^2)^3}, \quad \Gamma_2 = \frac{3K_2}{8\omega_0(\omega_0^2 - \Omega_2^2)}, \quad \text{and}$$

$$\Gamma_3 = \frac{3}{4\omega_0}\left[\frac{K_1^2}{(\omega_0^2 - \Omega_1^2)^2} + \frac{K_2^2}{(\omega_0^2 - \Omega_2^2)^2}\right]$$

Recalling (4.1.76), one can rewrite these as follows:

$$\Gamma_1 = \frac{729K_1^3}{4096\omega_0^7} + O(\epsilon), \quad \Gamma_2 = -\frac{3K_2}{64\omega_0^3} + O(\epsilon), \quad \text{and}$$

$$\Gamma_3 = \frac{3}{256\omega_0^5}(81K_1^2 + K_2^2) + O(\epsilon)$$

We note that Γ_1, Γ_2, and Γ_3 are not independent. They are functions of K_1 and K_2 only, which are the true independent parameters characterizing the amplitudes of the excitation.

Inspection of (4.1.79) reveals that steady-state motions (i.e., $a' = 0$) exist if, and only if, both $\sigma_1 T_1 - \beta$ and $\sigma_2 T_1 - 3\beta$ are constants. That is,

$$\sigma_1 = \beta' \quad \text{and} \quad \sigma_2 = 3\beta'$$

Therefore steady-state motions exist only when $\sigma_2 = 3\sigma_1 = 3\sigma$. It follows from (4.1.76) that $\Omega_2 = 9\Omega_1$. That is, steady-state motions occur only when the excitation is periodic.

When steady-state motions occur, (4.1.79) and (4.1.80) show that they correspond to the solutions of

$$\mu a + \alpha\Gamma_1 \sin(\gamma + 3\theta_1) + \alpha\Gamma_2 a^2 \sin(3\gamma + \theta_2) = 0 \tag{4.1.81}$$

$$-a\sigma + \frac{3\alpha a^3}{8\omega_0} + \alpha\Gamma_3 a + \alpha\Gamma_1 \cos(\gamma + 3\theta_1) + \alpha\Gamma_2 a^2 \cos(3\gamma + \theta_2) = 0 \tag{4.1.82}$$

where $\gamma = \sigma T_1 - \beta$. Letting $A = \frac{1}{2} a \exp(i\beta)$ in (4.1.63) and substituting the result into (4.1.60), we obtain

$$u = K_1(\omega_0^2 - \Omega_1^2)^{-1} \cos(\Omega_1 t + \theta_1) + a \cos(3\Omega_1 t - \gamma)$$
$$+ K_2(\omega_0^2 - 9\Omega_1^2)^{-1} \cos(9\Omega_1 t + \theta_2) + O(\epsilon) \tag{4.1.83}$$

Therefore steady-state motions, if they exist, are periodic. The effect of the nonlinearity is to adjust the frequency of the free-oscillation term to be perfectly commensurable with the frequencies of the excitation.

In Figure 4-16*a* a typical frequency-response graph is shown. The various solid and dotted branches indicate stable and unstable solutions, respectively. We note that there are as many as seven branches for a given σ and that these branches do not intersect at some large value of σ; that is, the curves do not

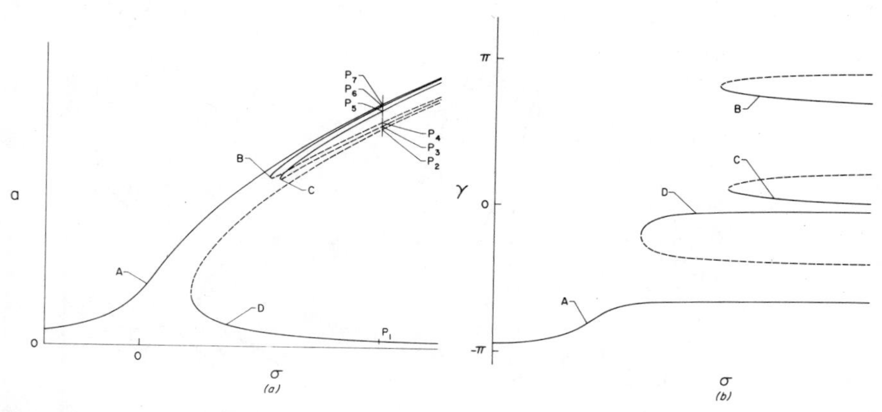

Figure 4-16. (*a*) Amplitude vs. detuning and (*b*) phase vs. detuning for simultaneous resonances of the Duffing equation.

close. When two branches are close together in this figure, one must examine the phases, which are quite different, to recognize the existence of all the distinct branches, see Figure 4-16*b*.

Comparing Figure 4-16 with Figures 4-10 and 4-13 shows that the simultaneous presence of superharmonic and subharmonic resonances changes the character of the frequency–response curves in a number of ways. The two branches A and D resemble those for a superharmonic resonance in the absence of damping. The remaining two curves B and C resemble those for a subharmonic resonance. For the subharmonic resonance, there are two possibilities: either a single trivial solution or three solutions one of which is trivial and another of which is unstable. For the superharmonic resonance there are also two possibilities: a stable nontrivial solution or three nontrivial solutions one of which is unstable. In contrast, for the simultaneous resonances there are four possibilities: (a) a nontrivial stable solution, (b) three nontrivial solutions one of which is unstable, (c) five nontrivial solutions two of which are unstable, and (d) seven nontrivial solutions three of which are unstable. In the case of more than one stable solution the initial conditions determine which one is physically realized, as illustrated in the state plane shown in Figure 4-17. The points P_1, P_5, P_6, and P_7 are stable foci, while the points P_2, P_3, and P_4 are saddle points. If the initial conditions fall in the shaded area 5, 6, or 7, the motion ends up at points P_5, P_6, or P_7, respectively. Otherwise, the motion ends up at point P_1. Figure 4-17 shows that small changes in the system parameters or the initial conditions might drastically change the response of the system.

Figure 4-18 shows the influence of the relative phase $(\theta_1 - \theta_2)$ of the two excitations on the response of the system. This relative phase has only a slight

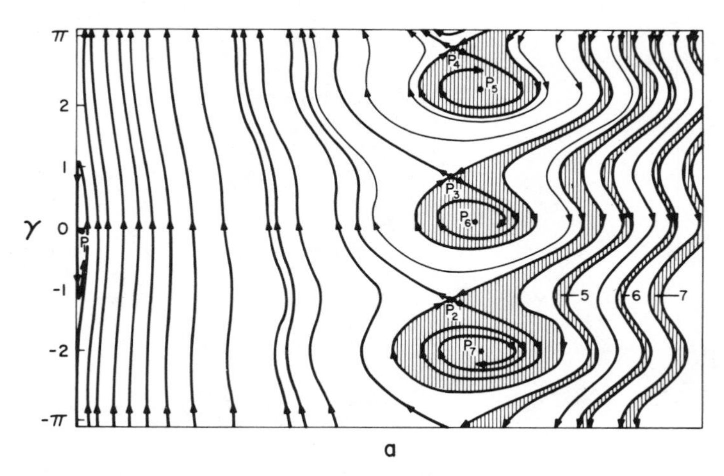

Figure 4-17. State plane for the free-oscillation term for simultaneous resonances of the Duffing equation.

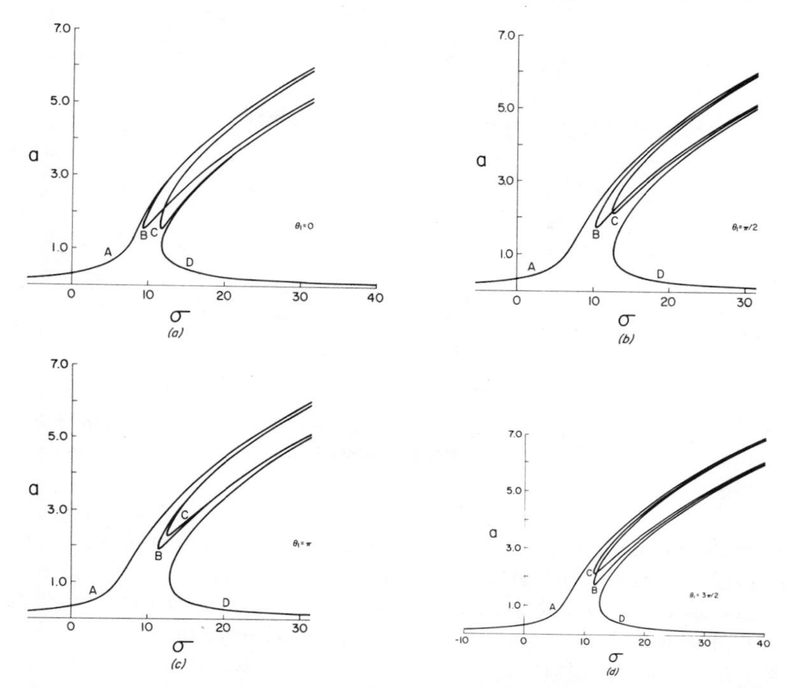

Figure 4-18. Influence of the relative phase of the excitations on the response of the Duffing equation to simultaneous subharmonic and superharmonic excitations: (*a*) $\theta_1 = 0$; (*b*) $\theta_1 = \frac{1}{2}\pi$; (*c*) $\theta_1 = \pi$; (*d*) $\theta_1 = \frac{3}{2}\pi$.

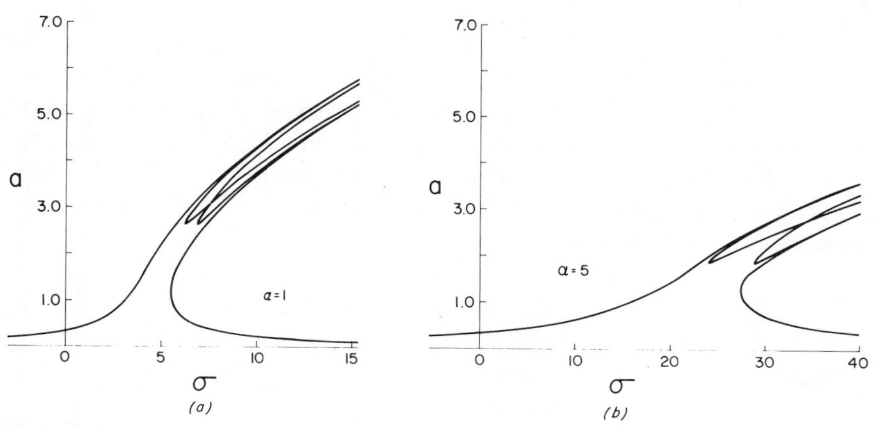

Figure 4-19. Effect of nonlinearity on the response of the Duffing equation to simultaneous subharmonic and superharmonic excitations: (*a*) $\alpha = 1$; (*b*) $\alpha = 5$.

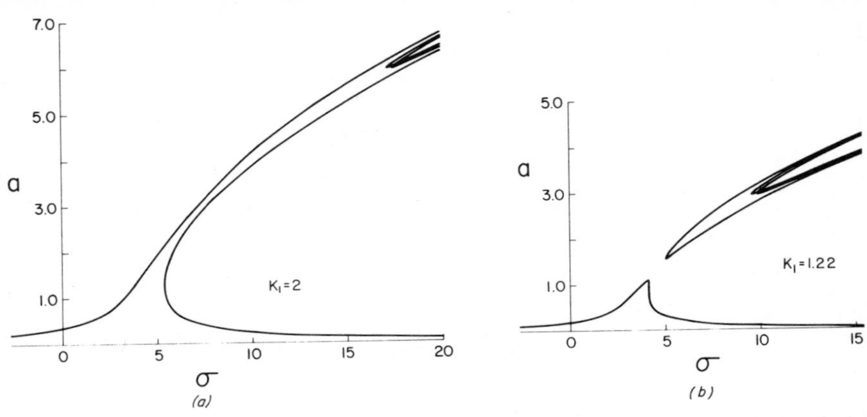

Figure 4-20. Influence of the amplitudes of the excitations on the response of the Duffing equation to simultaneous subharmonic and superharmonic excitations: (*a*) $K_1 = 2$; (*b*) $K_1 = 1.22$.

effect on the superharmonic-resonance type of curves (branches A and D), but it has a significant effect on the subharmonic-resonance type of curves (B and C). As θ_1 increases from 0 to 2π, curves B and C are shifted to the right, while curve C is shifted upward and curve B is shifted downward until they are nested when $\theta_1 = \pi$. As θ_1 increases further, the upward and downward shift of curves B and C continues; in addition they shift to the left. When $\theta_1 \to 2\pi$, the frequency-response curves are the same as the case $\theta_1 = 0$, except that curves B and C are interchanged.

Figure 4-19 shows the influence of the nonlinearity (i.e., the influence of α). As α increases, all curves bend and shift to the right. The shift to the right is due to an apparent increase in the natural frequency with increasing nonlinearity. Moreover an increase in α results in the separation of the various curves.

Figure 4-20 shows the influence of K_1, the amplitude of the subharmonic excitation. As K_1 decreases, the superharmonic-resonance type of curve separates into two curves. Figure 4-20*b* shows that when $K_1 = 1.22$, one of these curves has qualitatively the character of the superharmonic-resonance curves of Figure 4-10, while the second curve has qualitatively the character of the subharmonic-resonance curves of Figure 4-13*a*.

4.1.7. AN EXAMPLE OF A COMBINATION RESONANCE FOR A THREE-TERM EXCITATION

Here we consider systems governed by equations having the form

$$\ddot{u} + \omega^2 u = \epsilon(-\alpha u^3 - 2\mu\dot{u}) + \sum_{n=1}^{3} K_n \cos(\Omega_n + \theta_n) \tag{4.1.84}$$

where the Ω_n, K_n, and θ_n are constants. Equations of this type arise in the study of transverse vibrations of plates and beams. In this context $\ddot{u}$ represents the inertia, $2\epsilon\mu\dot{u}$ represents the damping force, $\omega^2 u$ represents the restoring force due to bending, and the cubic term represents the restoring force due to stretching of the midplane or neutral axis. For the case above the structure supports three harmonic loads simultaneously. However we note that the frequencies Ω_n are not necessarily commensurable, and hence the load is not necessarily periodic. Because the equations are cubic, three loads can interact nonlinearly to produce several interesting resonances (Mojaddidy, Mook, and Nayfeh, 1977).

There are many possible cases that can produce some sort of resonant response. Here we consider the following:

$$\Omega_1 + \Omega_2 + \Omega_3 = \omega + \epsilon\sigma \tag{4.1.85}$$

We use the method of multiple scales and obtain from (4.1.84)

$$u_0 = A(T_1) \exp(i\omega T_0) + \sum_{n=1}^{3} \Lambda_n \exp(i\Omega_n T_0 + i\theta_n) + cc \tag{4.1.86}$$

where

$$\Lambda_n = \tfrac{1}{2} K_n(\omega^2 - \Omega_n^2)$$

Then one finds, after some manipulation, that secular terms are eliminated from u_1 if

$$2i\omega(A' + \mu A) + \omega\alpha\Gamma_1 A + 3\alpha A^2\bar{A} + \omega\alpha\Gamma_2 \exp[i(\sigma T_1 + \theta_1 + \theta_2 + \theta_3)] = 0 \tag{4.1.87}$$

where

$$\Gamma_1 = 6\omega^{-1}(\Lambda_1^2 + \Lambda_2^2 + \Lambda_3^2) \quad \text{and} \quad \Gamma_2 = 6\omega^{-1}\Lambda_1\Lambda_2\Lambda_3$$

The steady state of the homogeneous term corresponds to the solution of

$$\mu a + \alpha\Gamma_2 \sin\gamma = 0 \quad \text{and} \quad a\left(\sigma - \tfrac{1}{2}\alpha\Gamma_1 - \frac{3\alpha a^2}{8\omega}\right) - \alpha\Gamma_2 \cos\gamma = 0 \tag{4.1.88}$$

where

$$A = \tfrac{1}{2} a \exp(i\beta) \quad \text{and} \quad \gamma = \sigma T_1 - \beta + \theta_1 + \theta_2 + \theta_3 \tag{4.1.89}$$

The frequency-response equation follows immediately:

$$\sigma = \tfrac{1}{2}\alpha\Gamma_1 + \frac{3\alpha a^2}{8\omega} \pm \left(\frac{\alpha^2\Gamma_2^2}{a^2} - \mu^2\right)^{1/2} \tag{4.1.90}$$

Comparing (4.1.90) with (4.1.43), we see the similarity between simple superharmonic resonance and the present example.

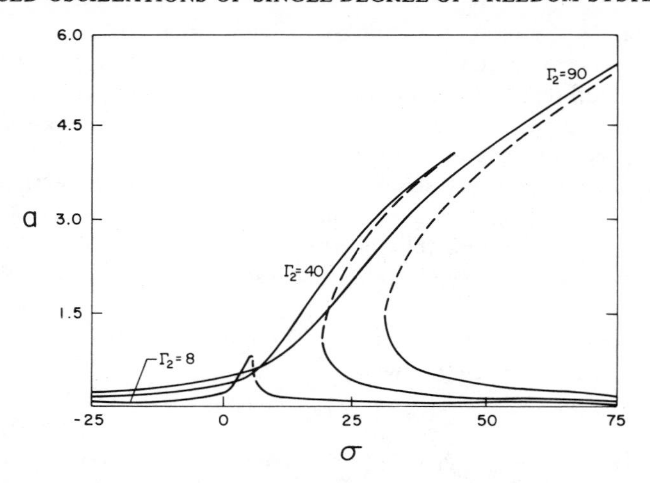

Figure 4-21. Frequency–response curves for a combination resonance involving three-term excitation.

Because σ must be real, the peak amplitude is given by

$$a_p = \frac{\alpha\Gamma_2}{\mu} \tag{4.1.91}$$

and it occurs when

$$\sigma_p = \tfrac{1}{2}\alpha\Gamma_1 + \frac{3\alpha^3\Gamma_2^2}{8\omega\mu^2} \tag{4.1.92}$$

In Figure 4-21, a is plotted as a function of σ. Again we note the presence of a jump phenomenon.

It follows from (4.1.89) that

$$\omega_0 T_0 + \beta = \omega_0 T_0 + \sigma T_1 - \gamma + \theta_1 + \theta_2 + \theta_3 = (\Omega_1 + \Omega_2 + \Omega_3) T_0 + \theta_1 + \theta_2 + \theta_3 - \gamma$$

Thus for the first approximation

$$u = a \cos\left[(\Omega_1 + \Omega_2 + \Omega_3)t + \theta_1 + \theta_2 + \theta_3 - \gamma\right] + \sum_{n=1}^{3} \frac{K_n}{\omega^2 - \Omega_n^2} \cos(\Omega_n t + \theta_n) + O(\epsilon) \tag{4.1.93}$$

where a and γ are given by (4.1.88). We note again that the Ω_n are not necessarily commensurable, and hence u is not necessarily periodic.

In Figure 4-22, u is plotted as a function of time. In Figure 4-22*a*, α is zero and hence this corresponds to the solution of the linearized problem. In Figure 4-22*b*, α is unity, the other parameters are the same, and the scales in the two drawings are the same. We note that when α is nonzero, the maximum a is approximately four times as large as in the case when α is zero. The presence of a

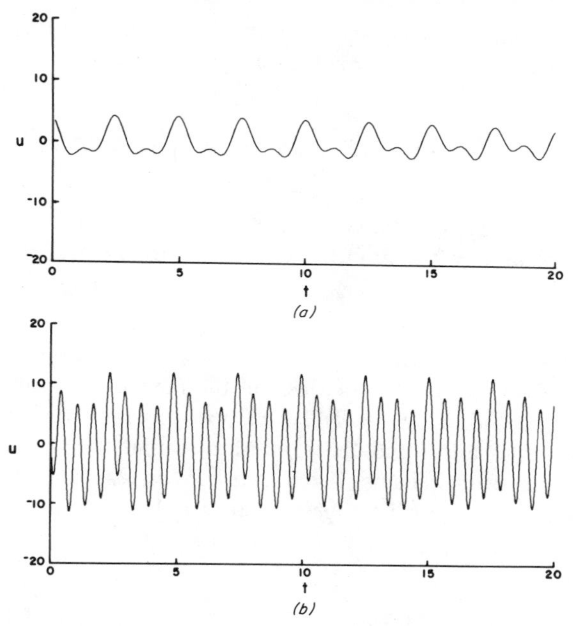

Figure 4-22. Response of the Duffing equation to a three-frequency excitation: (*a*) linear case; (*b*) nonlinear case.

strong high-frequency component in the second case is clearly evident. These results were verified by integrating (4.1.84) numerically.

Including the cubic term in the equation has the effect of stiffening the restoring force and significantly altering the phase. The latter is responsible for the drastic increase in amplitude; it makes the forces act closely in phase with the velocity and hence increases the rate at which they do work.

It is the practice in structural design to consider the structure safe from a resonant response if all the frequencies of the loads are below the fundamental frequency of the structure. Such a practice is based on linear models of the structural elements. However as large amplitudes of the response develop, the midplane stretches. The term which accounts for the stretching is cubic, and ultimately the equation governing the response can be put in the form of (4.1.84). Thus the present example clearly illustrates the need to consider nonlinear effects in the models of structural elements.

4.2 Systems with Quadratic and Cubic Nonlinearities

In this section we consider systems governed by equations having the form

$$\ddot{x} + 2\hat{\mu}\dot{x} + \omega_0^2 x + \alpha_2 x^2 + \alpha_3 x^3 = E(t) \tag{4.2.1}$$

The damping is taken to be linear, for simplicity. As in the previous case we assume that the excitation is a linear combination of harmonic functions, that is,

$$E(t) = \sum_{j=1}^{N} \tilde{K}_j \cos(\Omega_j t + \theta_j) \tag{4.2.2}$$

where $\tilde{K}_j$, Ω_j, and θ_j are constants. In the next three sections we consider monofrequency excitations, while multifrequency excitations are treated in Section 4.2.4.

4.2.1. PRIMARY RESONANCES

We consider $E(t) = \tilde{K} \cos \Omega t$ in this and the next two sections. For the primary resonance to occur, $\omega_0 \approx \Omega$. To analyze this case we need to order the damping, the nonlinearity, and the excitation so that they appear at the same time in the perturbation scheme. If we let $x = \epsilon \cos \omega_0 t$, then the nonlinearity generates a term proportional to $\cos \omega_0 t$ at $O(\epsilon^3)$. Therefore if we let $x = \epsilon u$, we need to order $\hat{\mu}\dot{x}$ as $\epsilon^2 \mu \dot{x}$ and $\tilde{K} \cos \Omega t$ as $\epsilon^3 k \cos \Omega t$ so that the governing equation becomes

$$\ddot{u} + \omega_0^2 u = -2\epsilon^2 \mu \dot{u} - \epsilon \alpha_2 u^2 - \epsilon^2 \alpha_3 u^3 + \epsilon^2 k \cos \Omega t \tag{4.2.3}$$

In studying the forced response of a spherical bubble to a harmonic excitation, Lauterborn (1970) solved numerically an equation similar to (4.2.3). He found large responses when $\Omega/\omega_0 = 1, \frac{1}{2}, \frac{1}{3}, \frac{1}{4}, \frac{2}{1}, \frac{3}{1}, \frac{2}{3}, \frac{3}{2}, \ldots$.

We seek an approximate solution to this equation by letting

$$u(t;\epsilon) = u_0(T_0, T_1, T_2) + \epsilon u_1(T_0, T_1, T_2) + \epsilon^2 u_2(T_0, T_1, T_2) + \cdots \tag{4.2.4}$$

Since the excitation is $O(\epsilon^2)$, $\Omega - \omega_0$ is assumed to be $O(\epsilon^2)$ for consistency. Hence we put

$$\Omega = \omega_0 + \epsilon^2 \sigma \tag{4.2.5}$$

Substituting (4.2.4) and (4.2.5) into (4.2.3) and equating the coefficients of ϵ^0, ϵ, and ϵ^2 on both sides, we obtain

$$D_0^2 u_0 + \omega_0^2 u_0 = 0 \tag{4.2.6}$$

$$D_0^2 u_1 + \omega_0^2 u_1 = -2D_0 D_1 u_0 - \alpha_2 u_0^2 \tag{4.2.7}$$

$$D_0^2 u_2 + \omega_0^2 u_2 = -2D_0 D_1 u_1 - 2D_0 D_2 u_0 - D_1^2 u_0 - 2\mu D_0 u_0 - 2\alpha_2 u_0 u_1 - \alpha_3 u_0^3 + k \cos(\omega_0 T_0 + \sigma T_2) \tag{4.2.8}$$

The general solution of (4.2.6) can be written in the form

$$u_0 = A(T_1, T_2) \exp(i\omega_0 T_0) + \bar{A}(T_1, T_2) \exp(-i\omega_0 T_0) \tag{4.2.9}$$

Substituting u_0 into (4.2.7) yields

$$D_0^2 u_1 + \omega_0^2 u_1 = -2i\omega_0 D_1 A \exp(i\omega_0 T_0) - \alpha_2 [A^2 \exp(2i\omega_0 T_0) + A\bar{A}] + cc \tag{4.2.10}$$

Eliminating the terms in (4.2.10) that produce secular terms in u_1 yields $D_1 A = 0$, or $A = A(T_2)$. Hence the solution of (4.2.10) becomes

$$u_1 = \frac{\alpha_2}{\omega_0^2} [-2A\bar{A} + \tfrac{1}{3} A^2 \exp(2i\omega_0 T_0) + \tfrac{1}{3} \bar{A}^2 \exp(-2i\omega_0 T_0)] \tag{4.2.11}$$

Substituting u_0 and u_1 into (4.2.8) gives

$$D_0^2 u_2 + \omega_0^2 u_2 = -\left[2i\omega_0(A' + \mu A) + \left(3\alpha_3 - \frac{10\alpha_2^2}{3\omega_0^2}\right) A^2\bar{A} - \tfrac{1}{2} k \exp(i\sigma T_2)\right] \exp(i\omega_0 T_0) + cc + NST \tag{4.2.12}$$

where the prime denotes the derivatives with respect to T_2 and NST stands for terms proportional to $\exp(\pm 3i\omega_0 T_0)$. Secular terms will be eliminated from u_2 if

$$2i\omega_0(A' + \mu A) + \left(3\alpha_3 - \frac{10\alpha_2^2}{3\omega_0^2}\right) A^2\bar{A} - \tfrac{1}{2} k \exp(i\sigma T_2) = 0 \tag{4.2.13}$$

Letting $A = \frac{1}{2} a \exp(i\beta)$ in (4.2.13) and separating real and imaginary parts, we have

$$a' = -\mu a + \frac{k}{2\omega_0} \sin\gamma \tag{4.2.14}$$

$$a\beta' = \frac{9\alpha_3\omega_0^2 - 10\alpha_2^2}{24\omega_0^3} a^3 - \frac{k}{2\omega_0} \cos\gamma \tag{4.2.15}$$

where

$$\gamma = \sigma T_2 - \beta \tag{4.2.16}$$

Eliminating β from (4.2.15) and (4.2.16) yields

$$a\gamma' = a\sigma - \frac{9\alpha_3\omega_0^2 - 10\alpha_2^2}{24\omega_0^3} a^3 + \frac{k}{2\omega_0} \cos\gamma \tag{4.2.17}$$

Therefore to the second approximation

$$u = a\cos(\Omega t - \gamma) + \tfrac{1}{2}\epsilon\alpha_2\omega_0^{-2} a^2 [-1 + \tfrac{1}{3}\cos[2\Omega t - 2\gamma)] + O(\epsilon^2) \tag{4.2.18}$$

where a and γ are defined by (4.2.14) and (4.2.17).

Since (4.2.14) and (4.2.17) have the same form as (4.1.15), the discussion in Section 4.1.1 applies to this case provided that we identify α with $\alpha_3 - \frac{10}{9}\alpha_2^2\omega_0^{-2}$. When $\alpha_3 = 0$, $\alpha < 0$, and the quadratic nonlinearity has a softening effect which tends to bend the frequency-response curves to lower frequencies, irrespective of the sign of α_2. Thus unless $\alpha_3 > \frac{10}{9}\alpha_2^2\omega_0^{-2}$, the nonlinearity has a softening effect. When $\alpha_3 = \frac{10}{9}\alpha_2^2\omega_0^{-2}$, the nonlinearity has no effect on the response to this order because the effects of the quadratic and cubic nonlinearities cancel each other.

Even-power, nonlinear terms are responsible for drift or steady streaming, as we discussed before. In this case we note the presence of the drift term $-\frac{1}{2}\epsilon\alpha_2\omega_0^{-2}a^2$, which indicates that the oscillatory motion is not centered at $u = 0$.

4.2.2. SUPERHARMONIC RESONANCES

To examine subharmonic and superharmonic resonances, we need to order the excitation so that it appears at the same time as the free-oscillation part of the solution; that is, the excitation should appear in the lowest-order perturbation equation. Thus if $x = \epsilon u$, $\widetilde{K} = \epsilon K$. Moreover to analyze subharmonic and superharmonic resonances generated by the quadratic nonlinearity, we need to order the damping so that it appears in the same perturbation equation that generates these resonances. In this case these are the result of $x^2 = \epsilon^2 u^2$. Hence we let $\hat{\mu}\dot{x} = \epsilon^2 \mu\dot{u}$ so that (4.2.1) for a monofrequency excitation becomes

$$\ddot{u} + \omega_0^2 u = -2\epsilon\mu\dot{u} - \epsilon\alpha_2 u^2 - \epsilon^2\alpha_3 u^3 + K\cos\Omega t \tag{4.2.19}$$

where Ω is assumed to be away from ω_0 in this section.

We seek a first approximation to the solution of (4.2.19) in the form

$$u = u_0(T_0, T_1) + \epsilon u_1(T_0, T_1) + \cdots \tag{4.2.20}$$

Substituting (4.2.20) into (4.2.19) and equating the coefficients of ϵ^0 and ϵ on both sides, we obtain

$$D_0^2 u_0 + \omega_0^2 u_0 = K\cos\Omega T_0 \tag{4.2.21}$$

$$D_0^2 u_1 + \omega_0^2 u_1 = -2D_0 D_1 u_0 - 2\mu D_0 u_0 - \alpha_2 u_0^2 \tag{4.2.22}$$

The general solution of (4.2.21) is taken in the form

$$u_0 = A(T_1)\exp(i\omega_0 T_0) + \Lambda\exp(i\Omega T_0) + cc \tag{4.2.23}$$

where $\Lambda = \frac{1}{2}K(\omega_0^2 - \Omega^2)^{-1}$. Then (4.2.22) becomes

$$\begin{aligned} D_0^2 u_1 + \omega_0^2 u_1 = &-2i\omega_0(A' + \mu A)\exp(i\omega_0 T_0) - 2i\mu\Lambda\Omega\exp(i\Omega T_0) \\ &- \alpha_2\{A^2\exp(2i\omega_0 T_0) + \Lambda^2\exp(2i\Omega T_0) + A\bar{A} + \Lambda^2 \\ &+ 2\bar{A}\Lambda\exp[i(\Omega - \omega_0)T_0] + 2A\Lambda\exp[i(\Omega + \omega_0)T_0]\} + cc \end{aligned} \tag{4.2.24}$$

Equation (4.2.24) shows that secondary resonances occur whenever $\omega_0 \approx 2\Omega$ or $\omega_0 \approx \frac{1}{2}\Omega$; the first corresponds to a superharmonic resonance, while the second corresponds to a subharmonic resonance. In this section we treat superharmonic resonances, and in the following section we take up subharmonic resonances.

For superharmonic resonances the particular solution and the homogeneous solution for u_0 interact in u_1. Thus we introduce the detuning parameter σ defined by

$$2\Omega = \omega_0 + \epsilon\sigma \quad \text{and} \quad 2\Omega T_0 = \omega_0 T_0 + \sigma T_1 \tag{4.2.25}$$

Then eliminating the terms in (4.2.24) that produce secular terms in u_1, we have

$$2i\omega_0(A' + \mu A) + \alpha_2 \Lambda^2 \exp(i\sigma T_1) = 0 \tag{4.2.26}$$

whose solution is

$$A = a_0 \exp(-\mu T_1) + \frac{i\alpha_2 \Lambda_2^2}{2\omega_0(\mu + i\sigma)} \exp(i\sigma T_1) \tag{4.2.27}$$

where a_0 is a complex constant of integration. As $t \to \infty$, $T_1 \to \infty$, and

$$A \to \frac{i\alpha_2 \Lambda^2}{2\omega_0(\mu + i\sigma)} \exp(i\sigma T_1) \tag{4.2.28}$$

Therefore to the first approximation the steady-state solution is

$$u = \frac{K}{\omega_0^2 - \Omega^2} \cos \Omega t - \frac{\alpha_2 K^2}{4\omega_0(\omega_0^2 - \Omega^2)^2 (\mu^2 + \sigma^2)^{1/2}} \sin(2\Omega t - \gamma) + O(\epsilon) \tag{4.2.29}$$

where $\gamma = \tan^{-1}(\sigma/\mu)$. Thus a steady-state superharmonic term exists for all conditions and the response is periodic, as in the case of cubic nonlinearity. We note that the second term is proportional to the square of the amplitude of the excitation K. Thus the second term vanishes more rapidly than the first term as K vanishes, and this solution approaches the solution of the linear problem.

4.2.3. SUBHARMONIC RESONANCES

Subharmonic resonances in systems with quadratic nonlinearities were studied by Eller and Flynn (1969), Neppiras (1969), Lauterborn (1970), Safar (1970), Nayfeh and Saric (1973), Ashwell and Chauhan (1973), and Eller (1974). To analyze these subharmonic resonances to first order, we let

$$\Omega = 2\omega_0 + \epsilon\sigma \tag{4.2.30}$$

so that

$$(\Omega - \omega_0) T_0 = \omega_0 T_0 + \sigma T_1 \tag{4.2.31}$$

Then eliminating the terms in (4.2.24) that produce secular terms in u_1 yields

$$i\omega_0(A' + \mu A) + \alpha_2 \bar{A} \Lambda \exp(i\sigma T_1) = 0 \tag{4.2.32}$$

We let $A = B \exp(\frac{1}{2} i\sigma T_1)$ so that (4.2.32) becomes

$$i\omega_0(B' + \tfrac{1}{2} i\sigma B + \mu B) + \alpha_2 \bar{B}\Lambda = 0 \tag{4.2.33}$$

Putting $B = B_r + iB_i$ in (4.2.33), where B_r and B_i are real, and separating real and imaginary parts, we obtain

$$B_r' + \mu B_r - \left(\tfrac{1}{2}\sigma + \frac{\alpha_2 \Lambda}{\omega_0}\right) B_i = 0 \quad \text{and} \quad B_i' + \mu B_i + \left(\tfrac{1}{2}\sigma - \frac{\alpha_2 \Lambda}{\omega_0}\right) B_r = 0 \tag{4.2.34}$$

These are linear equations having constant coefficients; so the solution can be expressed in the form

$$B_r = b_r \exp(\lambda T_1) \quad \text{and} \quad B_i = b_i \exp(\lambda T_1) \tag{4.2.35}$$

where b_r, b_i, and λ are constants. Upon substituting (4.2.35) into (4.2.34), one finds that in order to have a nontrivial solution

$$\lambda = -\mu \pm \left(\frac{\alpha_2^2 \Lambda^2}{\omega_0^2} - \frac{\sigma^2}{4}\right)^{1/2} \tag{4.2.36}$$

Hence we find that

1. if $\sigma^2 > 4\alpha_2^2\Lambda^2/\omega_0^2$, B oscillates and decays
2. if $4\alpha_2^2\Lambda^2/\omega_0^2 > \sigma^2 > 4(\alpha_2^2\Lambda^2/\omega_0^2 - \mu^2)$, B decays without oscillating
3. if $4(\alpha_2^2\Lambda^2/\omega_0^2 - \mu^2) > \sigma^2$, B grows without bound (4.2.37)

Actually the motion does not grow without bound. As the amplitude increases, the problem is no longer weakly nonlinear and the present analysis is not valid. However the present analysis clearly points out the conditions when one should expect to see a marked change in the solution develop. This is illustrated in Figure 4-23.

In Figures 4-23 the steady-state results obtained by integrating (4.2.19) numerically are shown for the case $\sigma = 0$ and α_2 and ω_0 are unity. For Figure 4-23*a* $K = 5.8$, while in Figure 4-23*b* $K = 6.1$. It follows from (4.2.37) that, to within an error of $O(\epsilon)$, $K = 6$ is the boundary between growing and decaying free-oscillation terms. The presence of a lower harmonic (i.e., the free-oscillation term) is obvious in Figure 4-23*b*. We note that the frequency of the free-oscillation term is adjusted to exactly one half that of the excitation. Also we note the drift in Figure 4-23*b*, as one would expect because of the quadratic term in the equation of motion.

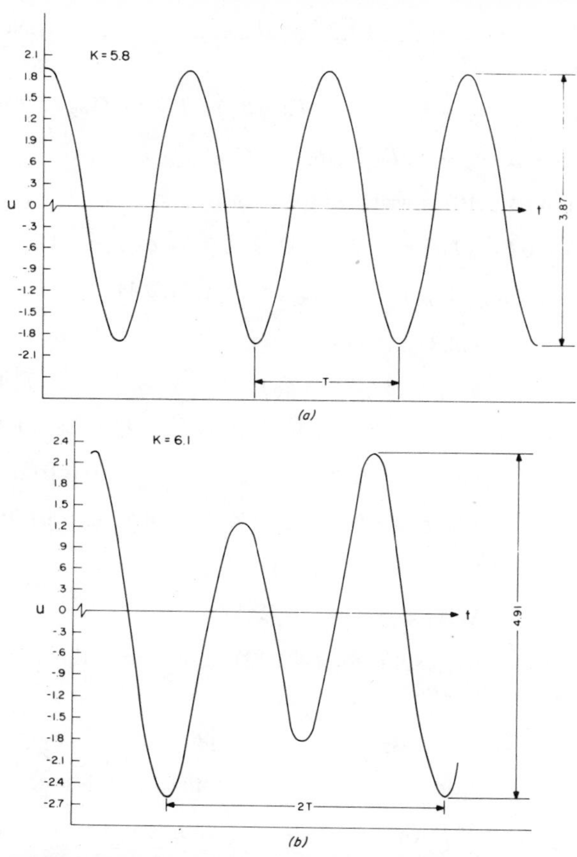

Figure 4-23. Response of a system having quadratic nonlinearities: (*a*) subharmonic response not excited; (*b*) subharmonic response excited.

4.2.4. COMBINATION RESONANCES

In this section we consider multifrequency excitations, using the same ordering as in the preceding two sections. Thus we consider first-order approximate solutions of

$$\ddot{u} + \omega_0^2 u = -2\epsilon\mu\dot{u} - \epsilon\alpha_2 u^2 + K_1 \cos(\Omega_1 t + \theta_1) + K_2 \cos(\Omega_2 t + \theta_2) \quad (4.2.38)$$

where $\Omega_2 > \Omega_1$.

We exclude primary resonances and seek a first approximation to the solution of (4.2.38) in the form

$$u(t;\epsilon) = u_0(T_0, T_1) + \epsilon u_1(T_0, T_1) + \cdots \quad (4.2.39)$$

Substituting this expansion into (4.2.38) and equating the coefficients of ϵ^0 and ϵ on both sides, we obtain

$$D_0^2 u_0 + \omega_0^2 u_0 = K_1 \cos(\Omega_1 T_0 + \theta_1) + K_2 \cos(\Omega_2 T_0 + \theta_2) \quad (4.2.40)$$

$$D_0^2 u_1 + \omega_0^2 u_1 = -2D_0 D_1 u_0 - 2\mu D_0 u_0 - \alpha_2 u_0^2 \quad (4.2.41)$$

The solution of (4.2.40) is written in the form

$$u_0 = A(T_1)\exp(i\omega_0 T_0) + \Lambda_1 \exp(i\Omega_1 T_0) + \Lambda_2 \exp(i\Omega_2 T_0) + cc \quad (4.2.42)$$

where $\Lambda_m = \frac{1}{2}K_m(\omega_0^2 - \Omega_m^2)^{-1}\exp(i\theta_m)$. Then (4.2.41) becomes

$$\begin{aligned} D_0^2 u_1 + \omega_0^2 u_1 = & -2i\omega_0(A' + \mu A)\exp(i\omega_0 T_0) - 2i\mu\Omega_1\Lambda_1 \exp(i\Omega_1 T_0) \\ & - 2i\mu\Omega_2\Lambda_2 \exp(i\Omega_2 T_0) - \alpha_2\{A^2 \exp(2i\omega_0 T_0) \\ & + \Lambda_1^2 \exp(2i\Omega_1 T_0) + \Lambda_2^2 \exp(2i\Omega_2 T_0) + A\bar{A} + \Lambda_1\bar{\Lambda}_1 + \Lambda_2\bar{\Lambda}_2 \\ & + 2A\Lambda_1 \exp[i(\omega_0 + \Omega_1)T_0] + 2\bar{A}\Lambda_1 \exp[i(\Omega_1 - \omega_0)T_0] \\ & + 2A\Lambda_2 \exp[i(\omega_0 + \Omega_2)T_0] + 2\bar{A}\Lambda_2 \exp[i(\Omega_2 - \omega_0)T_0] \\ & + 2\Lambda_1\Lambda_2 \exp[i(\Omega_1 + \Omega_2)T_0] \\ & + 2\Lambda_1\bar{\Lambda}_2 \exp[i(\Omega_1 - \Omega_2)T_0]\} + cc \end{aligned} \quad (4.2.43)$$

With primary resonance excluded, (4.2.43) shows that the resonances which might exist to this order are

$\omega_0 \approx 2\Omega_1$ or $2\Omega_2$ — superharmonic resonance

$\omega_0 \approx \frac{1}{2}\Omega_1$ or $\frac{1}{2}\Omega_2$ — subharmonic resonance

$\omega_0 \approx \Omega_2 + \Omega_1$ or $\Omega_2 - \Omega_1$ — combination resonance

We note that for a multifrequency excitation more than one resonance can occur simultaneously, as in the case of cubic nonlinearity. With a two-frequency excitation at most two resonances can occur simultaneously. If these frequencies are denoted by Ω_1 and Ω_2, where Ω_2 is away from and larger than Ω_1, simultaneous resonances occur whenever

(a) $\Omega_2 \approx 4\Omega_1 \approx 2\omega_0$

(b) $\Omega_2 \approx 3\Omega_1 \approx \frac{3}{2}\omega_0$

(c) $\Omega_2 \approx \frac{3}{2}\Omega_1 \approx 3\omega_0$

As discussed at the beginning of this chapter, the case $\Omega_1 \approx \Omega_2$ can best be treated by considering $E(t)$ as a monofrequency, nonstationary excitation and then applying the analysis of Section 4.4. The case $\Omega_1 \approx 0$ is discussed in Exercises 4.37 and 4.38.

Since the individual primary, subharmonic, and superharmonic resonances were treated at length, we devote the rest of this section to the combination-resonance case $\omega_0 \approx \Omega_2 + \Omega_1$ and leave the simultaneous-resonance cases as exercises for the reader. Yamamoto and Nakao (1963) studied the case $\omega_0 \approx \Omega_2 + \Omega_1$, while Yamamoto and Hayashi (1964) studied the case $\omega_0 = \Omega_2 - \Omega_1$.

We let

$$\Omega_1 + \Omega_2 = \omega_0 + \epsilon\sigma \tag{4.2.44}$$

so that

$$(\Omega_1 + \Omega_2) T_0 = \omega_0 T_0 + \sigma T_1 \tag{4.2.45}$$

Then eliminating the terms in (4.2.43) that produce secular terms in u_1, we obtain

$$i\omega_0(A' + \mu A) + \alpha_2 \Lambda_1 \Lambda_2 \exp(i\sigma T_1) = 0 \tag{4.2.46}$$

whose solution is

$$A = a \exp(-\mu T_1) + \frac{i\alpha_2 \Lambda_1 \Lambda_2}{\omega_0(\mu + i\sigma)} \exp(i\sigma T_1) \tag{4.2.47}$$

where a is a constant. Therefore to the first approximation, the steady-state motion is given by

$$u = -\frac{\alpha_2 K_1 K_2}{2\omega_0(\omega_0^2 - \Omega_1^2)(\omega_0^2 - \Omega_2^2)\sqrt{\mu^2 + \sigma^2}} \sin[(\Omega_1 + \Omega_2)t + \theta_1 + \theta_2 - \gamma]$$
$$+ \frac{K_1}{\omega_0^2 - \Omega_1^2} \cos(\Omega_1 t + \theta_1) + \frac{K_2}{\omega_0^2 - \Omega_2^2} \cos(\Omega_2 t + \theta_2) + O(\epsilon) \tag{4.2.48}$$

where $\gamma = \tan^{-1}(\sigma/\mu)$. Thus steady-state motions exist in this case for all conditions. We note that the nonlinearity adjusts the frequency of the free-oscillation term so that it is exactly equal to the sum of the frequencies of the excitation. We note also that the response is periodic only when Ω_1 and Ω_2 are commensurable.

The results for the case in which $\omega_0 \approx \Omega_2 - \Omega_1$ can be obtained from the above results by simply changing the sign of Ω_1.

4.3. Systems with Self-Sustained Oscillations

In this section we consider systems that have free, self-sustained oscillations. Thus we consider systems governed by equations having the form

$$\ddot{u} + \omega_0^2 u = \epsilon(\dot{u} - \tfrac{1}{3}\dot{u}^3) + E(t) \tag{4.3.1}$$

where $E(t)$ is assumed to consist of a single harmonic in the next four subsections, and is assumed to consist of two harmonic terms with different frequencies in Section 4.3.5.

4.3.1 PRIMARY RESONANCES

To analyze the effect of primary-resonance excitations, we order the amplitude of the excitation so that it will appear in the same perturbation equation as the damping and nonlinear terms. Thus we assume the excitation to be soft and to have the form

$$E(t) = \epsilon k \cos \Omega t, \quad \Omega = \omega_0 + \epsilon\sigma \tag{4.3.2}$$

As in the previous section, we seek an approximate solution in the form

$$u(t;\epsilon) = u_0(T_0, T_1) + \epsilon u_1(T_0, T_1) + \cdots \tag{4.3.3}$$

Substituting this expansion into (4.3.1) and (4.3.2) and equating the coefficients of ϵ^0 and ϵ on both sides, we obtain

$$D_0^2 u_0 + \omega_0^2 u_0 = 0 \tag{4.3.4}$$

$$D_0^2 u_1 + \omega_0^2 u_1 = -2D_0 D_1 u_0 + D_0 u_0 - \tfrac{1}{3}(D_0 u_0)^3 + k \cos(\omega_0 T_0 + \sigma T_1) \tag{4.3.5}$$

The solution of (4.3.4) is given the form

$$u_0 = A(T_1) \exp(i\omega_0 T_0) + \bar{A}(T_1) \exp(-i\omega_0 T_0) \tag{4.3.6}$$

Hence (4.3.5) becomes

$$D_0^2 u_1 + \omega_0^2 u_1 = [-2i\omega_0 A' + i\omega_0 A - i\omega_0^3 A^2 \bar{A} + \tfrac{1}{2}k \exp(i\sigma T_1)] \cdot \exp(i\omega_0 T_0) + \tfrac{1}{3} i\omega_0^3 A^3 \exp(3i\omega_0 T_0) + cc \tag{4.3.7}$$

Secular terms will be eliminated from u_1 if

$$-2i\omega_0 A' + i\omega_0 A - i\omega_0^3 A^2 \bar{A} + \tfrac{1}{2}k \exp(i\sigma T_1) = 0 \tag{4.3.8}$$

Letting $A = \frac{1}{2}a \exp(i\beta)$ in (4.3.8), where an a and β are real, and separating real and imaginary parts, we obtain

$$a' = \tfrac{1}{2}(1 - \tfrac{1}{4}\omega_0^2 a^2)a + \frac{k}{2\omega_0} \sin\gamma \tag{4.3.9}$$

$$a\beta' = -\frac{k}{2\omega_0} \cos\gamma \tag{4.3.10}$$

where

$$\gamma = \sigma T_1 - \beta \tag{4.3.11}$$

Eliminating β from (4.3.10) and (4.3.11) gives

$$a\gamma' = a\sigma + \frac{k}{2\omega_0}\cos\gamma \tag{4.3.12}$$

Therefore for the first approximation

$$u = a\cos(\Omega t - \gamma) + O(\epsilon) \tag{4.3.13}$$

where a and γ are defined by (4.3.9) and (4.3.12).

For a steady-state motion, $a' = \gamma' = 0$. Then it follows from (4.3.9) and (4.3.12) that

$$\begin{aligned} \tfrac{1}{2}(1 - \tfrac{1}{4}\omega_0^2 a^2)a &= -\frac{k}{2\omega_0}\sin\gamma \\ a\sigma &= -\frac{k}{2\omega_0}\cos\gamma \end{aligned} \tag{4.3.14}$$

Moreover it follows from (4.3.13) that the steady-state motion is periodic, with a frequency equal to that of the excitation. Thus the response is synchronized at the frequency of the excitation. Squaring and adding equations (4.3.14) yields the frequency–response equation

$$\rho(1 - \rho)^2 + 4\sigma^2\rho = \tfrac{1}{4}k^2, \qquad \rho = \tfrac{1}{4}\omega_0^2 a^2 \tag{4.3.15}$$

In the $\rho\sigma$-plane, the frequency–response equation provides a one-parameter family of response curves that are symmetric with respect to the ρ-axis. For $k = 0$, the curves of the family degenerate into the line $\rho = 0$ and the point (0, 1), in agreement with the free-oscillation solution. As k increases, the curves first consist of two branches–a branch running near the σ-axis and a branch consisting of a closed curve (an oval) which can be approximated by an ellipse having its center at the point (0, 1). As k increases further, the ovals expand, and the branch near the σ-axis moves away from this axis. When k^2 reaches the critical value $\frac{16}{27}$, the two branches coalesce, and the resultant curve has a double point at $(0, \frac{1}{3})$ as shown in Figure 4-24. As k increases beyond this critical value, the response curves are open curves. However ρ is still not a single-valued function for all σ until k^2 exceeds the second critical value $k^2 = \frac{32}{27}$. Beyond this critical value the response curves are single-valued functions for all σ. The heavy broken line is the locus of vertical tangents and separates the stable from the unstable solutions. The stability of these solutions is discussed later.

The following discussion is based on Figure 4-24. For $0 < k^2 < \frac{16}{27}$, there are three periodic solutions for a given σ when $|\sigma|$ is small, and there is only one periodic solution when $|\sigma|$ is large. When three solutions exist, only the one

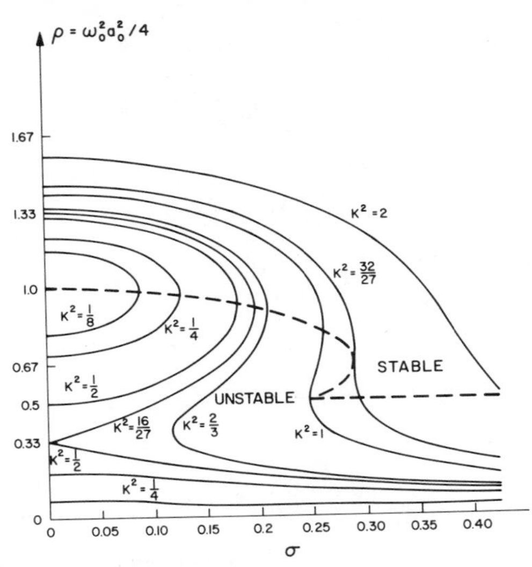

Figure 4-24. Frequency–response curves for primary resonances of the van der Pol oscillator.

having the largest amplitude is stable and hence corresponds to a realizable motion. When only one solution exists, it is unstable.

For $\frac{16}{27} < k^2 < 1$ there is a region of small $|\sigma|$ where only one periodic solution exists, followed by a region of larger $|\sigma|$ where three exist, and finally followed by a region of even larger $|\sigma|$ where only one exists. In the first region the only periodic solution is stable; in the second region only the solution having the largest amplitude is stable; and in the third region the only solution is unstable.

For $1 < k^2 < \frac{32}{27}$ one can again identify three regions as in the discussion directly above. However for this case, in the second, or middle, region there are two stable periodic solutions corresponding to the largest and smallest amplitudes. Here the initial conditions determine which of the solutions represents the actual motion.

For $k^2 > \frac{32}{27}$ there is only one periodic solution for a given σ. If $|\sigma|$ is small, the solution is stable; while if $|\sigma|$ is large, the solution is unstable.

To determine the stability of these periodic motions, we use the method of Andronov and Vitt (1930) and let

$$\begin{aligned} a &= a_0 + a_1 \\ \gamma &= \gamma_0 + \gamma_1 \end{aligned} \tag{4.3.16}$$

where a_0 and γ_0 are solutions of (4.3.14). Substituting (4.3.16) into (4.3.9) and (4.3.12), using (4.3.14), and keeping only the linear terms in a_1 and γ_1, we obtain

$$a_1' = \tfrac{1}{2}(1 - \tfrac{3}{4}\omega^2 a_0^2)a_1 + \frac{k}{2\omega_0}(\cos\gamma_0)\gamma_1$$
$$\gamma_1' = -\frac{k}{2\omega_0 a_0^2}(\cos\gamma_0)a_1 - \frac{k}{2\omega_0 a_0}(\sin\gamma_0)\gamma_1 \tag{4.3.17}$$

If we let $a_1 = a_{10}\exp(\lambda T_1)$ and $\gamma_1 = \gamma_{10}\exp(\lambda T_1)$, then

$$(\tfrac{1}{2} - \tfrac{3}{8}\omega_0^2 a_0^2 - \lambda)a_{10} + \frac{k}{2\omega_0}(\cos\gamma_0)\gamma_{10} = 0$$
$$\frac{k}{2\omega_0 a_0^2}(\cos\gamma_0)a_{10} + \left(\frac{k}{2\omega_0 a_0}\sin\gamma_0 + \lambda\right)\gamma_{10} = 0 \tag{4.3.18}$$

For a nontrivial solution the determinant of the coefficient matrix must vanish. Using (4.3.14), we write this condition as

$$\lambda^2 - (1 - 2\rho)\lambda + \Delta = 0 \tag{4.3.19}$$

where

$$\Delta = \tfrac{1}{4}(1 - 4\rho + 3\rho^2) + \sigma^2 \tag{4.3.20}$$

When $\Delta < 0$, the roots of (4.3.19) are real and have different signs; hence the predicted periodic motions correspond to saddle points and are unrealizable. These are shown as the interior points of the ellipse $\Delta = 0$ in Figure 4-25. It follows from differentiating (4.3.15) with respect to ρ that this ellipse is the locus of vertical tangents forming part of the boundary between stable and unstable solutions in Figure 4-24. Since the discriminant of (4.3.19) is $D = \rho^2 - 4\sigma^2$, periodic motions corresponding to $D < 0$ are focal points, while those corresponding to $D > 0$ are nodal points. These motions are stable or unstable depending on whether ρ is greater or less than $\frac{1}{2}$. A summary follows:

(a) $\rho^2 > 4\sigma^2$ $\begin{cases} \Delta > 0 & \begin{cases} \text{stable nodes if } \rho > \frac{1}{2} \\ \text{unstable nodes if } \rho < \frac{1}{2} \end{cases} \\ \Delta < 0 & \text{saddle points} \end{cases}$

(b) $\rho^2 = 4\sigma^2$ nodes $\begin{cases} \text{stable if } \rho > \frac{1}{2} \\ \text{unstable if } \rho < \frac{1}{2} \end{cases}$

(c) $\rho^2 < 4\sigma^2$ $\begin{cases} \text{stable focal points if } \rho > \frac{1}{2} \\ \text{unstable focal points if } \rho < \frac{1}{2} \\ \text{centers if } \rho = \frac{1}{2} \end{cases}$

This classification is shown in Figure 4-25 together with a hashed curve separating stable from unstable motions.

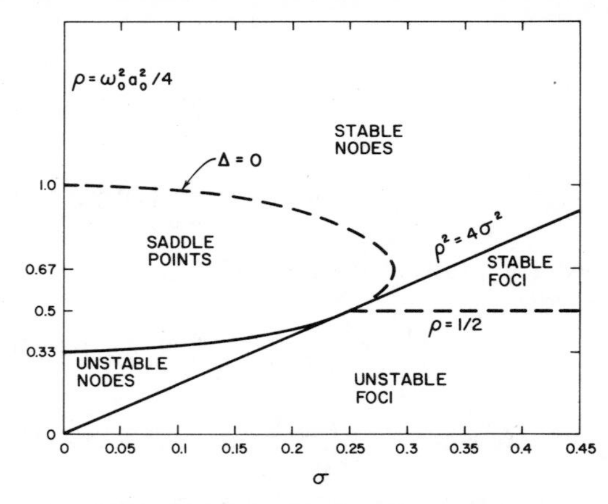

Figure 4-25. Stability of the steady-state responses of the van der Pol oscillator for primary resonances.

The results in Figure 4-24 indicate that a small change in the frequency of the excitation can cause a drastic change in the response. As an example we consider the case when $\epsilon = 0.1$, $\omega_0 = 1$, and $k^2 = 2$. According to Figure 4-24 the response is periodic when $|\Omega| < 1.0426$ and aperiodic when $|\Omega| > 1.0426$, approximately. To illustrate this result further, we integrated (4.3.1) and (4.3.2) numerically for $\Omega = 1.04$ and $\Omega = 1.05$. The results are shown in Figure 4-26. For the first approximation when $\Omega = 1.05$, we expect the response to consist of two terms

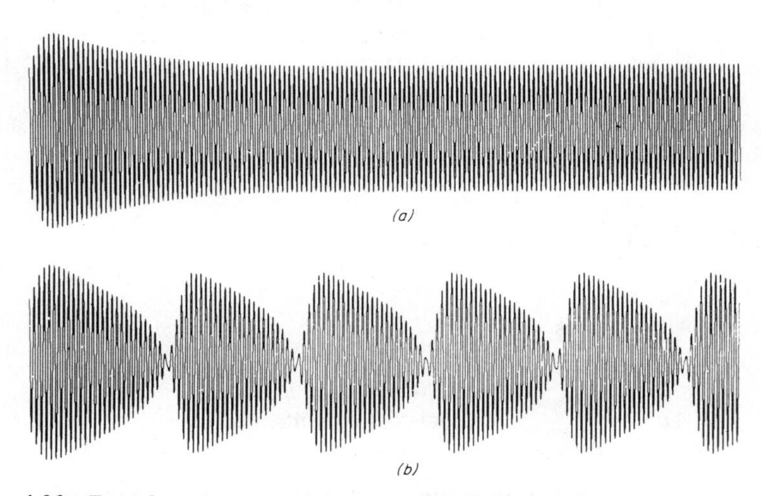

Figure 4-26. Forced response of the van der Pol oscillator when the frequency of the excitation is near the natural frequency: (*a*) response below the pull-out frequency; (*b*) response above the pull-out frequency.

having nearly the same frequencies. Thus we expect the response to exhibit a strong beating behavior, as the numerical integration shows. Finally we note that the amplitude of the response is very close to the value predicted by the perturbation solution.

When the two terms combine to yield a single-frequency response, the particular solution is said to *entrain* the homogeneous solution, or the homogeneous solution is said to *lock onto* the particular solution. When the two solutions separate to yield an aperiodic response, they are said to unlock. Associated with this separation is a *pull-out frequency*. There is a similar phenomenon associated with varying the amplitude of the excitation at a constant frequency. In this case one can identify a *pull-out amplitude*, below which the solution is aperiodic.

As the discussion above indicates, when the amplitude of the excitation is small [recall (4.3.2)], stable periodic solutions exist for only a narrow band of frequencies of the excitation around the natural frequency of the system. The width of this band increases as the amplitude of the excitation increases. In the next section we analyze the response of the system outside this region of resonance.

4.3.2. NONRESONANT EXCITATIONS

For a nonresonant excitation Ω must be away from ω_0, and to first order, as we shall show, Ω must be away from $\frac{1}{3}\omega_0$ and $3\omega_0$. Since Ω is away from ω_0, the excitation is assumed to be hard, and it is written as

$$E(t) = K \cos \Omega t \tag{4.3.21}$$

Substituting (4.3.3) into (4.3.1), using (4.3.21), and equating the coefficients of ϵ^0 and ϵ on both sides, we have

$$D_0^2 u_0 + \omega_0^2 u_0 = K \cos \Omega T_0 \tag{4.3.22}$$

$$D_0^2 u_1 + \omega_0^2 u_1 = -2D_0 D_1 u_0 + D_0 u_0 - \tfrac{1}{3}(D_0 u_0)^3 \tag{4.3.23}$$

The solution of (4.3.22) is written as

$$u_0 = A(T_1) \exp(i\omega_0 T_0) + \Lambda \exp(i\Omega T_0) + cc \tag{4.3.24}$$

where $\Lambda = \frac{1}{2}K(\omega_0^2 - \Omega^2)^{-1}$. Hence (4.3.23) becomes

$$\begin{aligned} D_0^2 u_1 + \omega_0^2 u_1 = {} & [-2i\omega_0 A' + i\omega_0(1 - 2\Omega^2\Lambda^2)A - i\omega_0^3 A^2\bar{A}] \exp(i\omega_0 T_0) \\ & + \tfrac{1}{3} i\{\omega_0^3 A^3 \exp(3i\omega_0 T_0) + \Omega^3\Lambda^3 \exp(3i\Omega T_0) \\ & + 3\Omega\Lambda(1 - \Omega^2\Lambda^2 - 2\omega_0^2 A\bar{A}) \exp(i\Omega T_0) + 3\omega_0^2\Omega\Lambda A^2 \\ & \cdot \exp[i(\Omega + 2\omega_0) T_0] + 3\omega_0^2\Omega\Lambda\bar{A}^2 \exp[i(\Omega - 2\omega_0) T_0] \\ & + 3\omega_0\Omega^2\Lambda^2 A \exp[i(2\Omega + \omega_0) T_0] - 3\omega_0\Omega^2\Lambda^2\bar{A} \\ & \cdot \exp[i(2\Omega - \omega_0) T_0]\} + cc \end{aligned} \tag{4.3.25}$$

Equation (4.3.25) shows that in addition to the primary resonances there might exist secondary resonances: superharmonic resonances (when $\Omega \approx \frac{1}{3}\omega_0$) and subharmonic resonances (when $\Omega \approx 3\omega_0$). In this section we treat the case of nonresonant excitations (i.e., Ω is away from ω_0, $\frac{1}{3}\omega_0$, and $3\omega_0$). Superharmonic resonances are treated in the next section, while subharmonic resonances are treated in Section 4.3.4.

For nonresonant excitations secular terms will be eliminated from u_1 if

$$-2A' + (1 - 2\Omega^2\Lambda^2)A - \omega_0^2 A^2\bar{A} = 0 \tag{4.3.26}$$

Letting $A = \frac{1}{2}a \exp(i\beta)$ and separating real and imaginary parts, we have

$$a' = \tfrac{1}{2}(\eta - \tfrac{1}{4}\omega_0^2 a^2)a, \qquad \beta' = 0 \tag{4.3.27}$$

where $\eta = 1 - \frac{1}{2}\Omega^2 K^2(\omega_0^2 - \Omega^2)^{-2}$. The solutions of (4.3.27) are $\beta = \text{constant}$ and

$$a^2 = \frac{4\eta}{\omega_0^2 + \left(\dfrac{4\eta}{a_0^2} - \omega_0^2\right)\exp(-\eta T_1)} \tag{4.3.28}$$

where a_0 is the initial amplitude. Therefore to the first approximation

$$u = a(t)\cos(\omega_0 t + \beta) + K(\omega_0^2 - \Omega^2)^{-1}\cos\Omega t + O(\epsilon) \tag{4.3.29}$$

where β is a constant and a is defined by (4.3.28).

Equation (4.3.28) shows that the steady-state motion depends on the sign of η. When $\eta > 0$ (i.e., $K < \sqrt{2}\,\Omega^{-1}|\omega_0^2 - \Omega^2|$), $a \to 2\omega_0^{-1}\sqrt{\eta}$ as $T_1 \to \infty$, and the steady-state motion consists of a combination of the forced (particular) and free (homogeneous) solutions according to (4.3.29). In general Ω and ω_0 are not commensurable, and hence the motion is not periodic. However when $\eta < 0$, $a \to 0$ as $T_1 \to \infty$ and the steady state is periodic, consisting of the forced solution only. These results for self-sustaining systems are at variance with those obtained for the response of a nonlinear spring in Section 4.1.2. Here a large force causes the free-oscillation term to decay, while a small force permits it to approach a nonzero value. An examination of (4.3.26) shows that the particular and homogeneous solutions of (4.3.22) interact in the cubic (positive) damping term, essentially causing the coefficient of the linear (negative) damping term to change. The process of increasing the amplitude of the excitation to cause the free-oscillation term to decay is called *quenching*. The problem of quenching was studied by Struble (1962), Pengilley and Milner (1967), Minorsky (1967), Fjeld (1968), Nayfeh (1968), Dewan and Lashinsky (1969), Mansour (1972), and Tondl (1975a, b, 1976b, d). When the amplitude of the excitation exceeds a certain critical value, the coefficient of the linear term changes sign and the free-oscillation term decays.

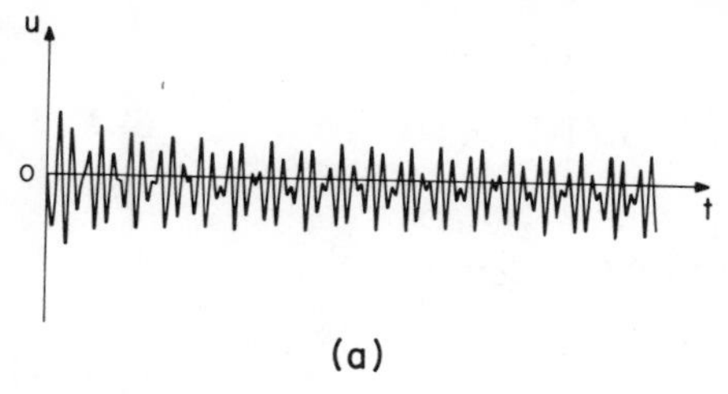

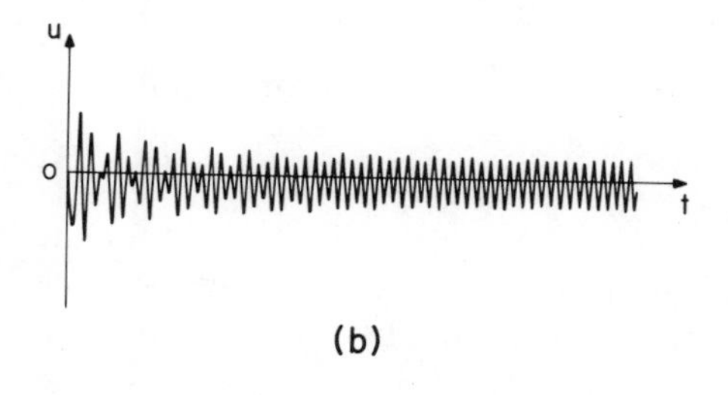

Figure 4-27. Forced response of the van der Pol oscillator illustrating the phenomenon of quenching: (*a*) unquenched response; (*b*) quenched response.

As an example we let $\omega_0 = 1$ and $\Omega = \sqrt{2}$. Then the critical value of K is unity. Using (4.3.21), we integrated (4.3.1) numerically for $K = 0.9$ and $K = 1.1$. According to the perturbation analysis, the free-oscillation term does not decay for the former and decays for the latter. The numerical results are shown in Figure 4-27; they agree with the analysis.

Finally we note how these results are consistent with the trend predicted in the previous section. Here the amplitude of the excitation is $O(1)$ [whereas in the previous section it was $O(\epsilon)$], and we find that it must exceed a certain critical value, which is proportional to $|\omega_0^2 - \Omega^2|$, in order to suppress the homogeneous solution and produce a periodic response.

4.3.3. SUPERHARMONIC RESONANCES

To analyze these resonances we let

$$3\Omega = \omega_0 + \epsilon\sigma \tag{4.3.30}$$

and express the resonant term involving $\exp(3i\Omega T_0)$ as $\exp[i(\omega_0 T_0 + \sigma T_1)]$. Then eliminating the terms that produce secular terms in (4.3.25) yields

$$-2\omega_0 A' + \omega_0 \eta A - \omega_0^3 A^2 \bar{A} + \tfrac{1}{3}\Omega^3 \Lambda^3 \exp(i\sigma T_1) = 0 \tag{4.3.31}$$

Letting $A = \frac{1}{2} a \exp[i(\sigma T_1 - \gamma)]$ and separating real and imaginary parts, we obtain

$$\begin{aligned} \omega_0 a' &= \tfrac{1}{2}(\eta - \tfrac{1}{4}\omega_0^2 a^2)\,\omega_0 a + \Gamma \cos\gamma \\ \omega_0 a \gamma' &= \sigma\omega_0 a - \Gamma \sin\gamma \end{aligned} \tag{4.3.32}$$

where $\Gamma = \frac{1}{3}\Omega^3\Lambda^3$. Therefore to the first approximation

$$u = a\cos(3\Omega t - \gamma) + K(\omega_0^2 - \Omega^2)^{-1}\cos\Omega t + O(\epsilon) \tag{4.3.33}$$

where a and γ are defined by (4.3.32).

The steady-state motions correspond to the solutions of

$$\begin{aligned} \tfrac{1}{2}(\eta - \tfrac{1}{4}\omega_0^2 a^2)\,\omega_0 a &= -\Gamma\cos\gamma \\ \sigma\omega_0 a &= \Gamma\sin\gamma \end{aligned} \tag{4.3.34}$$

Eliminating γ from these equations yields the frequency-response equation

$$\rho(\eta - \rho)^2 + 4\rho\sigma^2 = \Gamma^2, \qquad \rho = \tfrac{1}{4}\omega_0^2 a^2 \tag{4.3.35}$$

Equation (4.3.35) has the same form as (4.3.15), which was obtained for the case of primary resonance. However there is a significant difference; here the amplitude of the excitation appears in two places rather than one (both η and Γ are functions of K). Recalling that $\eta = 1 - \frac{1}{2}\Omega^2 K^2(\omega_0^2 - \Omega^2)^{-2}$, we note that, as K increases, η decreases and can become negative. Referring to (4.3.32), we see that the effect of decreasing η is one of decreasing the linear-, or negative-, damping coefficient. Thus in this case, as in the case of nonresonant excitations treated in the previous section, the nonlinear interaction between the particular and homogeneous solutions leads to an apparent change in the linear-damping coefficient. One should compare (4.3.32) and (4.3.27) with (4.3.9).

For a given value of Ω, ω_0, and K, one can calculate $\xi = \Omega\Lambda = \frac{1}{2}\Omega K(\omega_0^2 - \Omega^2)^{-1}$, $\eta = 1 - 2\Omega^2\Lambda^2$, and $\Gamma = \frac{1}{3}\Omega^3\Lambda^3$. Therefore for each harmonic excitation, (4.3.35) furnishes ρ and hence the amplitude of the response. In the $\rho\sigma$-plane the frequency-response equation provides a one-parameter family of curves with $\Omega\Lambda$ as the parameter, as shown in Figure 4-28. These curves are symmetric with respect to the ρ-axis. For $\xi = 0$, the curves of the family degenerate into the σ-axis and the point (0, 1). As ξ increases, the curves first consist of two branches—a branch running near the σ-axis and a branch consisting of a closed curve (an oval) which can be approximated by an ellipse having its center at the point $(0, \eta)$.

Because η is a function of the amplitude of the excitation (ξ essentially), all these ovals are not centered on the same point, or nested, as in the case of primary resonance. As ξ decreases, the ovals become smaller and are centered higher on the ρ-axis, and the branch near the σ-axis moves closer to this axis. Finally the branches collapse onto the point (0, 1) and the σ-axis. As ξ increases, the ovals expand, and the branch near the σ-axis moves away from this axis. When ξ reaches the critical value $[2 + (\frac{3}{4})^{1/3}]^{-1/2} \approx 0.586$, the branches coalesce and the curves have a double point at $(0, \frac{1}{3}\eta)$. Beyond this critical value, the

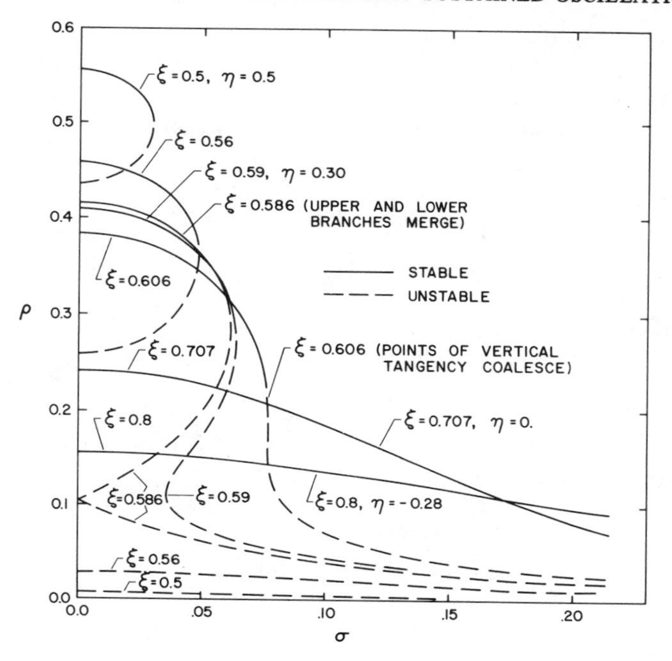

Figure 4-28. Response of the van der Pol oscillator to a superharmonic excitation.

response curves are open curves. The broken lines indicate the unstable portions of the frequency–response curves.

To study the stability of the steady-state solutions, we need to classify the singularities of the system (4.3.32). To do this, we let

$$a = a_0 + a_1$$
$$\gamma = \gamma_0 + \gamma_1 \tag{4.3.36}$$

Noting that a_0 and γ_0 are the solution of (4.3.34), we can write the following linear variational equations:

$$a_1' = \tfrac{1}{2}(\eta - \tfrac{3}{4}\omega_0^2 a^2)\, a_1 - \frac{\Gamma}{\omega_0}(\sin\gamma_0)\,\gamma_1$$
$$\gamma_1' = \frac{\Gamma}{\omega_0 a_0^2}(\sin\gamma_0)\, a_1 - \frac{\Gamma}{\omega_0 a_0}(\cos\gamma_0)\,\gamma_1 \tag{4.3.37}$$

According to Section 3.2.1, the character of the singular points depends on the determinant of the coefficient matrix of the right-hand sides of (4.3.37). Using (4.3.34) we can write this condition as

$$\lambda^2 - (\eta - 2\rho)\lambda + \Delta = 0 \tag{4.3.38}$$

where

$$\Delta = \tfrac{1}{4}(\eta^2 - 4\eta\rho + 3\rho^2) + \sigma^2$$

The discriminant of (4.3.38) is

$$\Delta = \rho^2 - 4\sigma^2$$

Using arguments similar to those used in Section 4.3.1, one can summarize the stability of the steady-state solutions as

$$\text{(a)}\quad \rho^2 > 4\sigma^2 \begin{cases} \Delta > 0 & \begin{cases} \text{stable nodes if } \rho > \frac{1}{2}\eta \\ \text{unstable nodes if } \rho < \frac{1}{2}\eta \end{cases} \\ \Delta < 0 & \quad\text{saddle points} \end{cases}$$

$$\text{(b)}\quad \rho^2 = 4\sigma^2 \quad \text{nodes} \quad \begin{cases} \text{stable if } \rho > \frac{1}{2}\eta \\ \text{unstable if } \rho < \frac{1}{2}\eta \end{cases}$$

$$\text{(c)}\quad \rho^2 < 4\sigma^2 \qquad \begin{cases} \text{stable focal points if } \rho > \frac{1}{2}\eta \\ \text{unstable focal points if } \rho < \frac{1}{2}\eta \\ \text{centers if } \rho = \frac{1}{2}\eta \end{cases}$$

We note that when $\eta < 0$ (i.e., $K > \sqrt{2}\,\Omega^{-1}|\omega_0^2 - \Omega^2|$), the steady-state solution is always stable.

4.3.4. SUBHARMONIC RESONANCES

Subharmonic resonances were studied by Cohen (1955); Matkowsky, Rogers, and Rubenfeld (1971); and Korolev and Postnikov (1973). To analyze these resonances we let

$$\Omega = 3\omega_0 + \epsilon\sigma \tag{4.3.39}$$

and express the resonant term involving $(\Omega - 2\omega_0)T_0$ as $(\omega_0 T_0 + \sigma T_1)$. Then eliminating the terms in (4.3.25) that produce secular terms in u_1 gives

$$-2A' + \eta A - \omega_0^2 A^2 \bar{A} + \omega_0 \xi \bar{A}^2 \exp(i\sigma T_1) = 0 \tag{4.3.40}$$

where $\xi = \Omega\Lambda$ and $\eta = 1 - 2\xi^2$. Letting $A = \frac{1}{2}a \exp[\frac{1}{3}i(\sigma T_1 - \gamma)]$ in (4.3.40) with real a and γ and separating real and imaginary parts, we obtain

$$a' = \tfrac{1}{2}(\eta - \tfrac{1}{4}\omega_0^2 a^2)a + \tfrac{1}{4}\omega_0 \xi a^2 \cos\gamma$$
$$a\gamma' = a\sigma - \tfrac{3}{4}\omega_0 \xi a^2 \sin\gamma \tag{4.3.41}$$

Thus to the first approximation

$$u = a\cos(\tfrac{1}{3}\Omega t - \tfrac{1}{3}\gamma) + K(\omega_0^2 - \Omega^2)^{-1}\cos\Omega t + O(\epsilon) \tag{4.3.42}$$

where a and γ are defined by (4.3.41).

Steady-state motions are periodic and correspond to the solutions of

$$\tfrac{1}{2}(\eta - \tfrac{1}{4}\omega_0^2 a^2)a = -\tfrac{1}{4}\omega_0 \xi a^2 \cos\gamma$$
$$a\sigma = \tfrac{3}{4}\omega_0 \xi a^2 \sin\gamma \tag{4.3.43}$$

Eliminating γ from these equations leads to the frequency-response equation

$$[\tfrac{4}{9}\sigma^2 + (\eta - \rho)^2]\rho = \rho^2\xi^2 \tag{4.3.44}$$

where $\rho = \frac{1}{4}\omega_0^2 a^2$. Equation (4.3.44) shows that there are two possibilities: either $\rho = 0$ or

$$\rho = 1 - \tfrac{3}{2}\xi^2 \pm (\xi^2 - \tfrac{7}{4}\xi^4 - \tfrac{4}{9}\sigma^2)^{1/2} \tag{4.3.45}$$

Equation (4.3.45) shows that nontrivial solutions exist only when certain restrictive conditions on σ and ξ are satisfied. Because ρ is real,

$$\xi^2 - \tfrac{7}{4}\xi^4 - \tfrac{4}{9}\sigma^2 \geqslant 0 \tag{4.3.46}$$

Thus in the $\xi\sigma$-plane, the boundary of the region where a subharmonic response can exist is defined by the equality in (4.3.46). This region is shown in Figure 4-29*a*.

It follows from (4.3.46) that

$$|\sigma| \leqslant \tfrac{3}{2}\xi(1 - \tfrac{7}{4}\xi^2)^{1/2}$$

and hence that

$$0 \leqslant \xi \leqslant \frac{2}{\sqrt{7}} \quad \text{and} \quad 0 \leqslant |\sigma| \leqslant \frac{3}{2\sqrt{7}} \tag{4.3.47}$$

The maximum value of σ occurs when $\xi = \sqrt{2/7}$. Some representative frequency-response curves are shown in Figure 4-29*b*. The curves are symmetric with

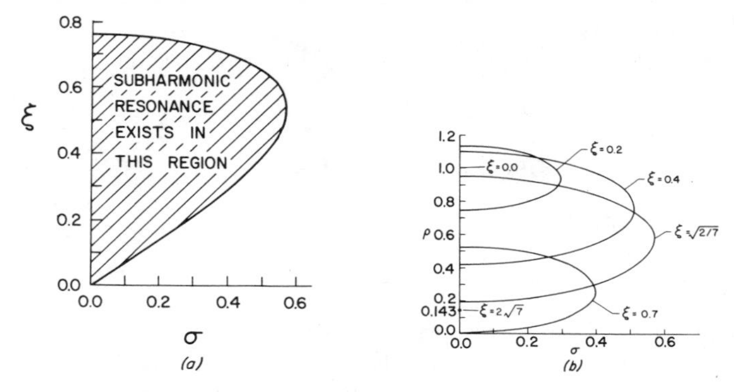

Figure 4-29. Response of the van der Pol oscillator to a subharmonic excitation: (*a*) region where subharmonic resonances exist; (*b*) frequency-response curves.

respect to the ρ-axis. As ξ increases from $\sqrt{2/7}$, the ovals become smaller and finally collapse on the point (0, 0.143) when $\xi = 2/\sqrt{7}$. For $\xi > 2/\sqrt{7}$, $\rho = 0$ according to (4.3.44). As ξ decreases from $\sqrt{2/7}$, the ovals become smaller and finally collapse on the point (0, 1) when $\xi = 0$.

4.3.5. COMBINATION RESONANCES

In this case the excitation has multifrequencies. For simplicity, we assume that it contains two frequencies only so that we can express it as

$$E(t) = K_1 \cos(\Omega_1 t + \theta_1) + K_2 \cos(\Omega_2 t + \theta_2) \tag{4.3.48}$$

Substituting the expansion (4.3.3) into (4.3.1), using (4.3.48), and equating the coefficients of ϵ^0 and ϵ on both sides, we obtain

$$D_0^2 u_0 + \omega_0^2 u_0 = K_1 \cos(\Omega_1 T_0 + \theta_1) + K_2 \cos(\Omega_2 T_0 + \theta_2) \tag{4.3.49}$$

$$D_0^2 u_1 + \omega_0^2 u_1 = -2D_0 D_1 u_0 + D_0 u_0 - \tfrac{1}{3}(D_0 u_0)^3 \tag{4.3.50}$$

The solution of (4.3.49) can be expressed in the form

$$u_0 = A(T_1)\exp(i\omega_0 T_0) + \Lambda_1 \Omega_1^{-1} \exp(i\Omega_1 T_0) + \Lambda_2 \Omega_2^{-1} \exp(i\Omega_2 T_0) + cc \tag{4.3.51}$$

where $\Lambda_m = \frac{1}{2} K_m \Omega_m (\omega_0^2 - \Omega_m^2)^{-1} \exp(i\theta_m)$. Then (4.3.50) becomes

$$\begin{aligned} D_0^2 u_1 + \omega_0^2 u_1 = {} & i\omega_0[-2A' + (1 - 2\Lambda_1\bar{\Lambda}_1 - 2\Lambda_2\bar{\Lambda}_2)A - \omega_0^2 A^2 \bar{A}] \exp(i\omega_0 T_0) \\ & + i\{\Lambda_1^2\Lambda_2 \exp[i(2\Omega_1 + \Omega_2) T_0 - \Lambda_1^2\bar{\Lambda}_2 \exp[i(2\Omega_1 - \Omega_2) T_0] \\ & + \Lambda_1\Lambda_2^2 \exp[i(\Omega_1 + 2\Omega_2) T_0] - \bar{\Lambda}_1\Lambda_2^2 \exp[i(-\Omega_1 + 2\Omega_2) T_0] \\ & + 2\omega_0\Lambda_1\Lambda_2 A \exp[i(\Omega_1 + \Omega_2 + \omega_0) T_0] - 2\omega_0\Lambda_1\Lambda_2\bar{A} \end{aligned}$$

$$\cdot \exp [i(\Omega_1 + \Omega_2 - \omega_0) T_0] + 2\omega_0 \bar{\Lambda}_1 \Lambda_2 \bar{A} \exp [i(-\Omega_1 + \Omega_2 - \omega_0) T_0] + 2\omega_0 \Lambda_1 \bar{\Lambda}_2 \bar{A} \exp (i(\Omega_1 - \Omega_2 - \omega_0) T_0]\} + cc$$
$$+ [\text{terms with frequencies } \pm\Omega_1, \pm\Omega_2, \pm 3\Omega_1, \pm 3\Omega_2, \pm 3\omega_0, \pm\Omega_1 \pm 2\omega_0, \pm\Omega_2 \pm 2\omega_0, \pm 2\Omega_1 \pm \omega_0, \pm 2\Omega_2 \pm \omega_0] \tag{4.3.52}$$

In addition to the primary, superharmonic, and subharmonic resonances, (4.3.52) shows the possibility of the existance to this order of the combination resonances $\omega_0 \approx 2\Omega_m \pm \Omega_n$ and $\omega_0 \approx \frac{1}{2}(\Omega_m \pm \Omega_n)$. We note that more than one resonance can occur simultaneously for any multifrequency excitation as discussed in detail in Sections 4.1.5 and 4.2.4.

The Case $\omega_0 \approx 2\Omega_1 + \Omega_2$. To analyze this case we let

$$\omega_0 = 2\Omega_1 + \Omega_2 - \epsilon\sigma \tag{4.3.53}$$

and express the term involving this combination resonance in (4.3.52) as

$$\exp [i(2\Omega_1 + \Omega_2) T_0] = \exp [i(\omega_0 T_0 + \sigma T_1)]$$

Then eliminating the terms that produce secular terms in (4.3.52) gives

$$-2A' + \eta A - \omega_0^2 A^2 \bar{A} + \omega_0^{-1} \Lambda_1^2 \Lambda_2 \exp (i\sigma T_1) = 0 \tag{4.3.54}$$

where

$$\eta = 1 - 2\Lambda_1 \bar{\Lambda}_1 - 2\Lambda_2 \bar{\Lambda}_2$$

Letting $A = \frac{1}{2} a \exp [i(\sigma T_1 - \gamma + 2\theta_1 + \theta_2)]$ in (4.3.54) with real a and γ and separating real and imaginary parts, we obtain

$$\begin{aligned} a' &= \tfrac{1}{2}(\eta - \tfrac{1}{4}\omega_0^2 a^2) a + \omega_0^{-1} \Gamma \cos \gamma \\ a\gamma' &= a\sigma - \omega_0^{-1} \Gamma \sin \gamma \end{aligned} \tag{4.3.55}$$

where

$$\Gamma = \tfrac{1}{8} K_1^2 K_2 \Omega_1^2 \Omega_2 (\omega_0^2 - \Omega_1^2)^{-2} (\omega_0^2 - \Omega_2^2)^{-1}$$

Therefore to the first approximation

$$u = a \cos [(2\Omega_1 + \Omega_2) t - \gamma + 2\theta_1 + \theta_2)] + K_1 (\omega_0^2 - \Omega_1^2)^{-1} \cos (\Omega_1 t + \theta_1) + K_2 (\omega_0^2 - \Omega_2^2)^{-1} \cos (\Omega_2 t + \theta_2) + O(\epsilon) \tag{4.3.56}$$

where a and γ are given by (4.3.55).

The steady-state motions correspond to the solutions of

$$\begin{aligned} \tfrac{1}{2}(\eta - \tfrac{1}{4}\omega_0^2 a^2) a &= -\Gamma \omega_0^{-1} \cos \gamma \\ a\sigma &= \Gamma \omega_0^{-1} \sin \gamma \end{aligned} \tag{4.3.57}$$

Eliminating γ from these equations leads to the frequency-response equation

$$4\sigma^2\rho + (\eta - \rho)^2\rho = \Gamma^2 \tag{4.3.58}$$

which is the same as (4.3.35) obtained for the superharmonic-resonant case except for the definitions of η and Γ; instead of depending on a single amplitude and a single frequency, they depend on two amplitudes and two frequencies. We note that the steady-state motions are not periodic unless the frequencies Ω_1 and Ω_2 are commensurable. However these combination resonances exist under all conditions.

The results for the case $\omega_0 \approx \Omega_1 + 2\Omega_2$ can be obtained from the above results by simply interchanging the subscripts 1 and 2, while the results for the combination resonance $\omega_0 \approx 2\Omega_1 - \Omega_2$ can be obtained from the above results by simply changing the sign of Ω_2.

The Case $\omega_0 \approx \frac{1}{2}(\Omega_1 + \Omega_2)$. To analyze this case we let

$$\Omega_1 + \Omega_2 = 2\omega_0 + \epsilon\sigma \tag{4.3.59}$$

and express the resonant term $\exp[i(\Omega_1 + \Omega_2 - \omega_0)T_0]$ as

$$\exp[i(\Omega_1 + \Omega_2 - \omega_0)T_0] = \exp[i(\omega_0 + \sigma T_1)] \tag{4.3.60}$$

Then eliminating the terms from (4.3.52) that produce secular terms in u_1 gives

$$-2A' + \eta A - \omega_0^2 A^2\bar{A} - 2\Lambda_1\Lambda_2\bar{A}\exp(i\sigma T_1) = 0 \tag{4.3.61}$$

Letting $A = \frac{1}{2}a\exp[\frac{1}{2}i(\sigma T_1 - \gamma + \theta_1 + \theta_2)]$ in (4.3.61), where a and γ are real, and separating real and imaginary parts, we obtain

$$\begin{aligned} a' &= \tfrac{1}{2}(\eta - \tfrac{1}{4}a^2\omega_0^2)a - \Gamma a\cos\gamma \\ a\gamma' &= \sigma a + 2\Gamma a\sin\gamma \end{aligned} \tag{4.3.62}$$

where $\Gamma = \frac{1}{4}K_1K_2\Omega_1\Omega_2(\omega_0^2 - \Omega_1^2)^{-1}(\omega_0^2 - \Omega_2^2)^{-1}$. Therefore to the first approximation

$$u = a\cos\{\tfrac{1}{2}[(\Omega_1 + \Omega_2)t - \gamma + \theta_1 + \theta_2]\} + K_1(\omega_0^2 - \Omega_1^2)^{-1}\cos(\Omega_1 t + \theta_1) + K_2(\omega_0^2 - \Omega_2^2)^{-1}\cos(\Omega_2 t + \theta_2) + O(\epsilon) \tag{4.3.63}$$

The steady-state motions correspond to $a' = \gamma' = 0$. Hence it follows from (4.3.62) that the frequency-response equation is

$$\sigma^2\rho + (\eta - \rho)^2\rho = 4\Gamma^2\rho \tag{4.3.64}$$

Equation (4.3.64) shows that there are two possibilities—either $\rho = 0$ or

$$\rho = \eta \pm (4\Gamma^2 - \sigma^2)^{1/2} \tag{4.3.65}$$

For the latter case to exist, $\sigma^2 \leqslant 4\Gamma^2$ and $\eta \geqslant -(4\Gamma^2 - \sigma^2)^{1/2}$. Therefore steady-state motions in which the free-oscillation term does not vanish exist for very special conditions as in all subharmonic resonances. Equation (4.3.63) shows that if these combination resonances exist, the response is not periodic unless Ω_1 and Ω_2 are commensurable.

4.4. Nonstationary Oscillations

So far we have considered excitations having constant frequencies and amplitudes. In this section we consider excitations that have time-dependent frequencies and amplitudes. Such excitations must lead to *nonstationary* (i.e., nonperiodic) resonances. Nonstationary studies are concerned with the deviation of the response of the system from the stationary response. It is expected that the most significant deviation occurs near the resonance conditions. When the frequency of the excitation is time dependent, the problem of the modification of the response near the resonance conditions is usually referred to as *passage through resonance*. Passage through resonance for single-degree-of-freedom systems was studied by Lewis (1932), Filippov (1956), Bogoliubov and Mitropolsky (1961, Section 8), Mitropolsky (1965), Hirano and Matsukura (1968), and Kevorkian (1971). Passage through resonance for multiple-degree-of-freedom systems was studied by Filippov (1956), Mitropolsky (1965), Agrawal and Evan-Iwanowski (1973), and Evan-Iwanowski (1976). Agrawal and Evan-Iwanowski (1973) demonstrated the existence of a "drag-out" phenomenon when the frequency of the external excitation passes through a combination resonance of the additive type. Passage of rotors, blades, and shafts through critical speeds was studied by Lewis (1932), Marples (1965), Mitropolsky (1965), Moseenkov (1957), Bodger (1967), and Quazi and McFarlane (1967). For more references we refer the reader to the review article of Evan-Iwanowski (1969).

For simplicity we restrict our attention to systems having constant parameters. However systems having slowly varying parameters can be treated as in Section 3.4. For a comprehensive treatment of nonstationary oscillations the reader is referred to the book of Mitropolsky (1965). Thus we consider

$$\ddot{u} + \omega_0^2 u = \epsilon f(u, \dot{u}) + 2\epsilon k(\epsilon t) \cos [\theta(t, \epsilon t)] \tag{4.4.1}$$

where $\dot{\theta} = \Omega$, the frequency of the excitation, is close to the natural frequency ω_0 during the time interval of interest.

Using the method of multiple scales, we seek an approximate solution in the form

$$u = u_0(T_0, T_1) + \epsilon u_1(T_0, T_1) + \cdots \tag{4.4.2}$$

Substituting (4.4.2) into (4.4.1), transforming the derivatives, and equating the coefficients of like powers of ϵ, we obtain

$$D_0^2 u_0 + \omega_0^2 u_0 = 0 \tag{4.4.3}$$

$$D_0^2 u_1 + \omega_0^2 u_1 = -2D_0 D_1 u_0 + f(u_0, Du_0) + 2k(T_1) \cos [\theta(T_0, T_1)] \tag{4.4.4}$$

We express the solution of (4.4.3) in the form

$$u_0 = A(T_1) \exp (i\omega_0 T_0) + cc \tag{4.4.5}$$

Hence (4.4.4) becomes

$$D_0^2 u_1 + \omega_0^2 u_1 = -2i\omega_0 A' \exp (i\omega_0 T_0) + k \exp (i\theta) + cc \\ + f[A \exp (i\omega_0 T_0) + cc, i\omega_0 A \exp (i\omega_0 T_0) + cc] \tag{4.4.6}$$

Because we are concerned with values of $\dot{\theta}$ near ω_0 and because $\dot{\theta}$ is slowly varying, we put

$$\theta = \omega_0 T_0 + \nu(T_1) \tag{4.4.7}$$

so that

$$\dot{\theta} = \omega_0 + \epsilon\nu'(T_1) = \omega_0 + \epsilon\sigma(T_1)$$

where the dot indicates the derivative with respect to t and the prime indicates the derivative with respect to T_1. Then eliminating the terms in (4.4.6) that produce secular terms in u_1 requires

$$2i\omega_0 A' - k \exp (i\nu) - f_1(A, \bar{A}) = 0 \tag{4.4.8}$$

Here we expanded $f(u_0, D_0 u_0)$ in a Fourier series according to

$$f(u_0, D_0 u_0) = \sum_{n=0}^{\infty} [f_n(A, \bar{A}) \exp (in\omega_0 T_0) + cc]$$

Thus

$$f_1(A, \bar{A}) = \frac{\omega_0}{2\pi} \int_0^{2\pi/\omega_0} f[A \exp (i\omega_0 T_0) + cc, i\omega_0 A \exp (i\omega_0 T_0) + cc] \\ \cdot \exp (-i\omega_0 T_0)\, dT_0 \tag{4.4.9}$$

Expressing A in the polar form $\frac{1}{2}a \exp (i\beta)$, where both a and β are real functions of T_1, we obtain

$$i\omega_0(a' + ia\beta') - k \exp [i\gamma(T_1)] - f_1(A, \bar{A}) \exp (-i\beta) = 0 \tag{4.4.10}$$

where

$$\gamma(T_1) = \nu(T_1) - \beta(T_1) \tag{4.4.11}$$

Letting $\psi = \omega_0 T_0 + \beta(T_1)$ so that $d\psi = [\omega_0 + O(\epsilon)]\ dT_0$, one finds from (4.4.9) that

$$f_1 \exp(-i\beta) = \frac{1}{2\pi} \int_{0+\beta}^{2\pi+\beta} f(a \cos \psi, -\omega_0 a \sin \psi)(\cos \psi - i \sin \psi)\, d\psi + O(\epsilon)$$

Separating (4.4.10) into real and imaginary parts leads to

$$a' = \frac{k \sin \gamma}{\omega_0} - \frac{1}{2\pi\omega_0} \int_0^{2\pi} f(a \cos \psi, -a\omega_0 \sin \psi) \sin \psi\, d\psi \tag{4.4.12}$$

and

$$\gamma' = \sigma + \frac{k \cos \gamma}{\omega_0 a} + \frac{1}{2\pi\omega_0 a} \int_0^{2\pi} f(a \cos \psi, -a\omega_0 \sin \psi) \cos \psi\, d\psi \tag{4.4.13}$$

Thus the response is given by

$$u_0 = a \cos(\theta - \gamma) + O(\epsilon) \tag{4.4.14}$$

where a and γ are given by (4.4.12) and (4.4.13). Next we specialize these results to the case of a system having a cubic nonlinearity.

We consider a system having a cubic nonlinearity and slight viscous damping. Thus

$$f(u, \dot{u}) = -2\mu\dot{u} - \alpha u^3 \tag{4.4.15}$$

Then (4.4.12) and (4.4.13) become

$$\begin{aligned} a' &= \frac{k \sin \gamma}{\omega_0} - \mu a \\ \gamma' &= \sigma - \frac{3\alpha a^2}{8\omega_0} + \frac{k \cos \gamma}{\omega_0 a} \end{aligned} \tag{4.4.16}$$

These equations have the same form as those obtained for the stationary case. However in the nonstationary case, steady-state solutions do not exist because k and ν are functions of T_1. Therefore the response of the system is aperiodic. In the remainder of this section, we discuss the passage of the system through resonance when the amplitude of the excitation is constant and the variation of the response for a fixed frequency when the amplitude of the excitation is varied.

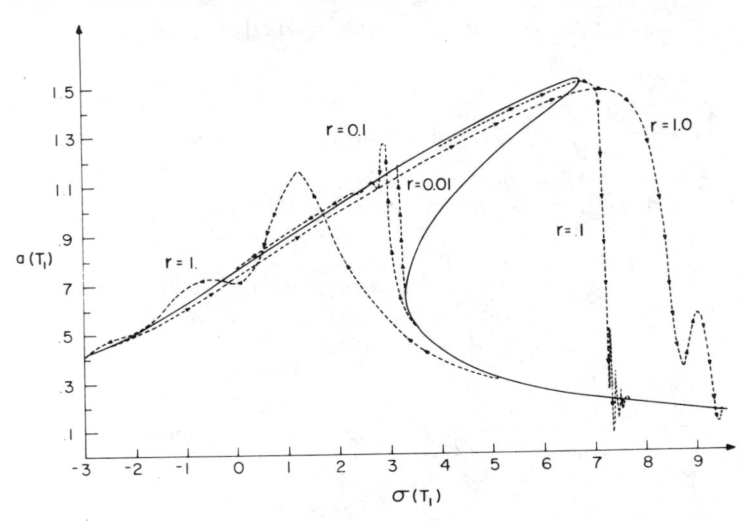

Figure 4-30. Comparison of nonstationary and stationary response curves: (————) stationary results; (- - - - - - - -) nonstationary results for several rates of changing σ.

As discussed in Section 4.1.1, the frequency–response curve for the stationary case is obtained by setting $a' = \gamma' = 0$ in (4.4.16) and eliminating γ from the resulting equations. The result is

$$\mu^2 a^2 + \left(\sigma - \frac{3\alpha a^2}{8\omega_0}\right)^2 a^2 = \frac{k^2}{\omega_0^2} \tag{4.4.17}$$

A representative frequency–response curve is represented by the solid curve in Figures 4-30. We refer to this as the *stationary curve*.

To determine the effects of varying the frequency of the excitation, we put, as an example,

$$\sigma(T_1) = \sigma_0 + rT_1 \tag{4.4.18}$$

where σ_0 and r are constants. Then (4.4.16) are integrated numerically. The detuning is varied from σ_0 to σ_1. These solutions of (4.4.16) are represented by the dotted curves in Figure 4-30. We note that these curves show the results of sweeping through resonance by both increasing and decreasing the detuning as well as the effect of varying the rate r.

First we consider increasing σ through resonance. The peak amplitude of the response decreases, and the frequency where this peak occurs increases as the rate of increasing the frequency of the excitation increases. The sharpness of the frequency–response curve decreases. After this peak is passed, a beat phe-

nomenon develops in the response and then decays. The faster the rate is, the easier it is to distinguish the subsequent lesser peaks.

Next we consider decreasing σ through resonance. The peak amplitude and the frequency where it occurs both decrease as the rate of decreasing σ increases. A beat phenomenon also develops after this peak is reached, and it also decays.

The smaller α becomes (the smaller the effect of the nonlinearity), the closer the frequency–response curves for increasing and decreasing σ come to being symmetric. The asymmetry is a nonlinear effect. Moreover the faster the rate of changing σ is, the closer the curves come to being symmetric. Thus the slower the passage through resonance is, the more apparent the nonlinear effects become.

To determine the effects of varying the amplitude of the excitation, we put, as an example,

$$k = k_0 + rT_1 \tag{4.4.19}$$

where both k_0 and r are constants. Then (4.4.16) are integrated numerically for a fixed value of σ. These solutions are represented by the dotted curves in Figure 4-31, while the solution of (4.4.17) is represented by the solid curve.

The faster the amplitude of the excitation k is increased, the greater is the peak amplitude of the response. Generally the sharpness of the curve is decreased as the rate of changing k is increased. We note that when $r = 1.0$, k

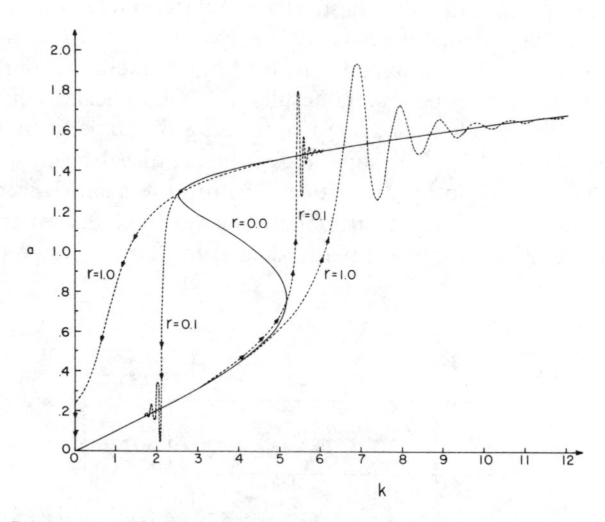

Figure 4-31. Comparison of nonstationary and stationary response curves: (————————) stationary results; (------------) nonstationary results for several rates of changing k.

reaches zero well ahead of the amplitude of the response. Here beat phenomena also develop.

4.5. Nonideal Systems

In the previous sections we considered the amplitude and the frequency of the excitation to be parameters of the systems (i.e., constants or prescribed functions of time), not state variables obtained as part of the solution. We did not consider the influence of the motion on the excitation. For many systems this is an acceptable simplification, but for many others it is not.

When the excitation is not influenced by the response, it is said to be an *ideal excitation*, or an *ideal source of energy*. On the other hand, when the excitation is influenced by the response, it is said to be *nonideal*. Depending on the excitation, one refers to systems as ideal or nonideal. Nonideal systems are classified as linear or nonlinear in the usual way, without regard to the excitation. Generally nonideal systems are those for which the power supply is limited. The behavior of the system departs farther from the ideal as the power supply becomes more limited. For nonideal systems one must add an equation that describes the motor to the equations that govern the corresponding ideal system. And the frequency of the response is unknown. Thus nonideal systems have one more degree of freedom than their ideal counterparts.

Examples of nonideal systems abound: Sommerfeld (1904) mounted an unbalanced electric motor on an elastically supported table and monitored the power input as well as the frequency and amplitude of the response. Here we briefly discuss a more recent experiment by Kononenko and Korablev (1959), who also compared their experimental results with theoretical results.

Kononenko and Korablev conducted tests using the experimental apparatus represented in Figure 4-32. The support for the cantilever beam was a massive slab, and the axis of the motor was vertical. The rotor was unbalanced by drilling the two holes indicated in the figure. There was no force due to gravity in the direction of the motion, which was perpendicular to the page.

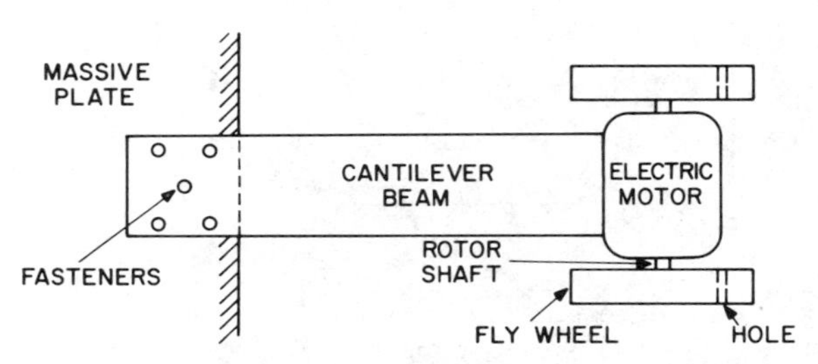

Figure 4-32. Experimental apparatus of Konoenko and Korablev for a nonideal system.

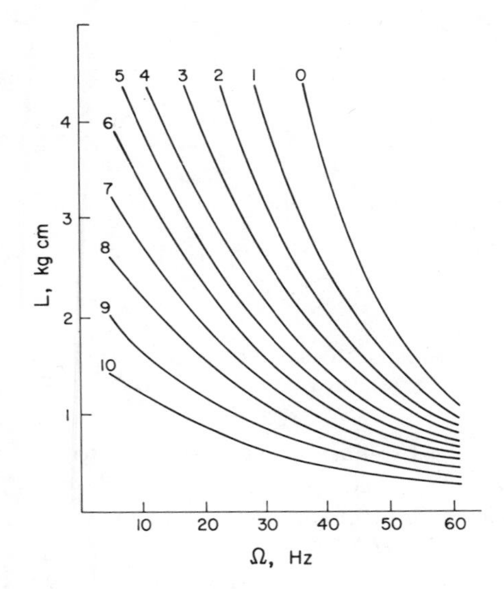

Figure 4-33. Data of Kononenko and Korablev: characteristics of the motor.

To account for the influence of the motion on the performance of the motor, one needs to know the *characteristics* of the motor, which are the net driving torques developed by the motor. Generally they are determined experimentally. In Figure 4-33 the data of Kononenko and Korablev are shown, giving the torque L as a function of the frequency or speed Ω of the rotor. Along each curve (labeled 0, 1, 2, . . . , 10) the control or regulator is constant. For each point on the characteristic curves the motor was running at a constant speed; thus these are the so-called static characteristics.

The net mechanical power available at the shaft of the motor is ΩL. The difference between the electrical power supplied to the motor and ΩL is the total loss in the motor—ohmic losses in the stator and the rotor, core losses due to eddy currents and hysteresis, mechanical losses due to friction in the bearings and windage on the rotor, and stray losses.

In Figure 4-34 the amplitude of the motion is plotted as a function of Ω. These curves were obtained by allowing the system to achieve a steady-state motion while the control was fixed. Then the amplitude of the steady-state response was measured. The control was then changed very slightly and held in the new position until a new steady state was achieved. These data are the small circles in the figures. The solid lines show the response predicted by the theory, which is discussed below. Figure 4-34*a* shows the results for increasing Ω, while Figure 4-34*b* shows the results for decreasing Ω. We note that there are gaps where no steady-state response exists. The gaps are not the same in the two figures, but there is some overlap.

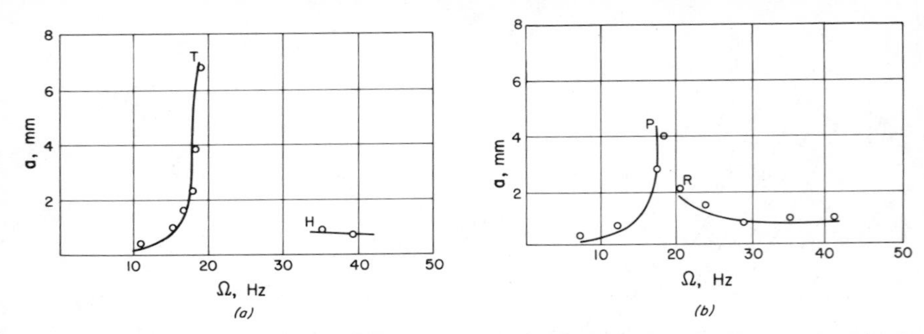

Figure 4-34. Comparison of experimental and theoretical frequency–response curves for a nonideal system from Kononenko and Korablev: (*a*) for increasing Ω; (*b*) for decreasing Ω.

In Figure 4-35 the solid line shows a typical frequency–response curve for an ideal linear system. The points labeled *P*, *R*, *T*, and *H* correspond to those in Figure 4-34. The arrows indicate the changes brought about by slowly increasing or decreasing the control setting in a nonideal system. We note that the nonideal system cannot be made to respond at a frequency between Ω_T and Ω_H by simply increasing the control setting from a low value. In contrast, an ideal system can respond at frequencies between Ω_T and Ω_H. When the control setting is continually decreased, the system cannot be made to respond between Ω_R and Ω_P. In other words, the right side of the resonance spike between Ω_T and Ω_R cannot be reached by either continually increasing or continually de-

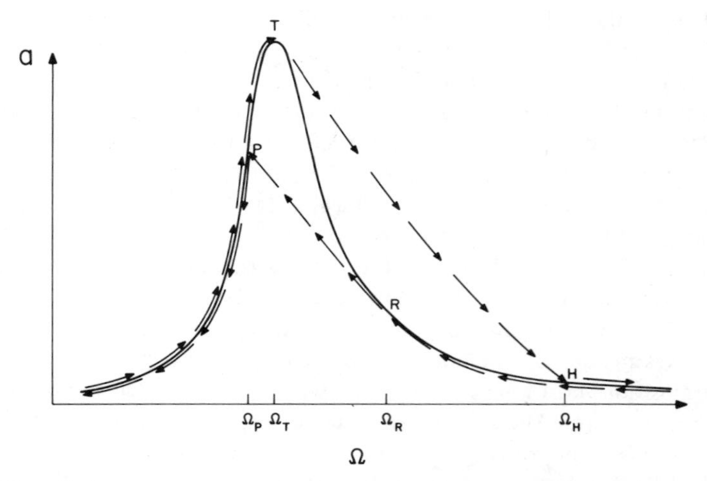

Figure 4-35. Frequency–response curves, comparing ideal and nonideal systems: (→→) nonideal; (————————) ideal.

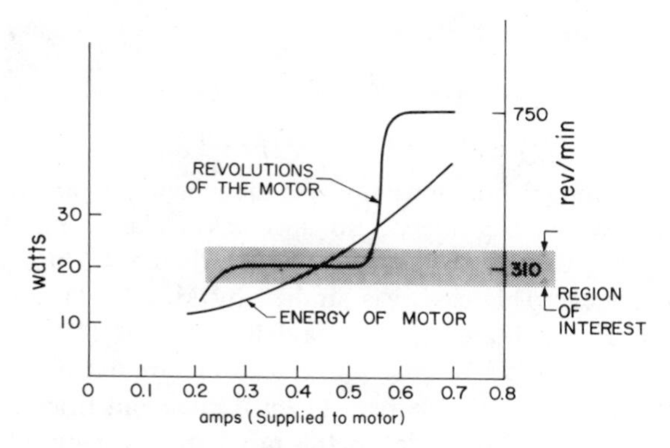

Figure 4-36. Sommerfeld's data.

creasing the control setting. Though the system is linear, the nonideal source causes a jump phenomenon to occur.

At the left-hand extremity of the frequency–response curve shown in Figure 4-35, the input power is relatively low. As the input power increases, the amplitude of the response increases noticeably while the frequency changes only slightly, especially along the portion of the curve between P and T. Here a relatively large increase in power causes a relatively large increase in the amplitude and practically no change in the frequency.

At T the character of the motion suddenly changes. An increase in the input power now causes the amplitude to decrease considerably and the frequency to increase considerably. This phenomenon, in all its manifestations, is called the *Sommerfeld effect*. Figure 4-36 shows Sommerfeld's data for a motor running on an elastically supported table. We shaded part of Sommerfeld's figure to show the region under consideration in the discussion of the theoretical model of such a system. The theoretical model is explained next by means of an example.

For a comprehensive treatment of nonideal systems we refer the reader to the book of Kononenko (1969), which is devoted to this topic and contains more references to his work as well as to that of other Russian scientists. Kononenko and Koval'chuk (1973a) studied the interaction of mechanisms generating oscillations in nonlinear systems. The passage through a critical speed of a motor with limited power was studied by Hübner (1965); Goeoskokov (1966); Kononenko (1969, Section 22); and Iwatsubo, Kanki, and Kawai (1972). Rajac and Evan-Iwanowski (1976) studied the interaction of a motor having a limited power supply with a dissipative hysteretic foundation.

The motion of a missile with slight asymmetries is another example of a system with a nonideal power source. The problem reduces mathematically to the analysis of equations of the form

$$\ddot{\xi} - ip\dot{\xi} + \omega_0^2 \xi = \epsilon K \exp(i\phi) + f(\xi, \bar{\xi}, \dot{\xi}, \dot{\bar{\xi}})$$

$$\dot{\phi} = p, \qquad \dot{p} = \epsilon g(p, \xi_i)$$

where f and g are known functions and ξ_i is the imaginary part of ξ. There exist conditions under which a primary resonance takes place, and the problem is usually referred to as the *roll-resonance problem*. When $g \equiv 0$, p is a constant and the excitation is ideal. This case was studied by Murphy (1957, 1963, 1965, 1968, 1971a, b, 1973), Clare (1971), Nayfeh and Saric (1971b, 1972a), and Murphy and Bradley (1975). When $g \neq 0$, p is a function of time. This latter case was studied by Platus (1969), Barbera (1969), Price and Ericsson (1970), and Nayfeh and Saric (1972a). A model of this problem was analyzed by Kevorkian (1974). For a body descending through the atmosphere, the parameters appearing in the equations are slowly varying functions of time. Nayfeh and Saric (1972a) used the method of multiple scales to study the nonideal case when the parameters are slowly varying with time.

An example of a nonideal linear system is shown in Figure 4-37, a dynamic system which is nearly equivalent to that of Kononenko and Korablev. The motion occurs in a horizontal plane and is constrained so that the motor executes a rectilinear motion along the x-axis.

The potential and kinetic energies are

$$V = \tfrac{1}{2} kx^2$$
$$T = \tfrac{1}{2} m_0 \dot{x}^2 + \tfrac{1}{2} I\dot{\phi}^2 + \tfrac{1}{2} m_1 (\dot{x} - r\dot{\phi}\cos\phi)^2 + \tfrac{1}{2} m_1 (r\dot{\phi}\sin\phi)^2 \tag{4.5.1}$$

where m_0 is the mass of the motor, m_1 is the unbalanced rotating mass, I is the moment of inertia of the rotor and other rotating parts except m_1, r is the distance between m_1 and the center of the motor, and k is the spring constant. Lagrange's equations have the form

$$\frac{d}{dt}\left(\frac{\partial \mathcal{L}}{\partial \dot{x}}\right) - \frac{\partial \mathcal{L}}{\partial x} = -c\dot{x}$$
$$\frac{d}{dt}\left(\frac{\partial \mathcal{L}}{\partial \dot{\phi}}\right) - \frac{\partial \mathcal{L}}{\partial \phi} = L(\dot{\phi}) - H(\dot{\phi}) \tag{4.5.2}$$

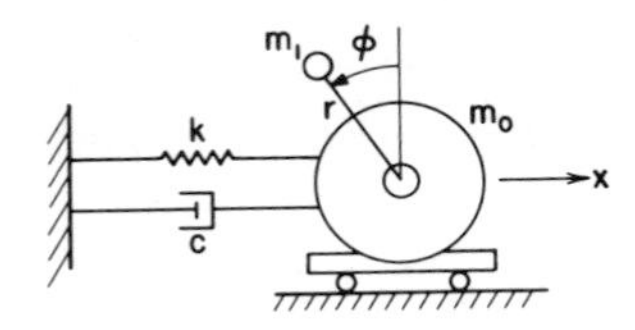

Figure 4-37. Example of a nonideal system.

where $\mathcal{L} = T - V$ is the Lagrangian, c is the coefficient for the dashpot, L is the characteristic of the motor, and H is the resisting moment due primarily to windage on the rotating parts outside the motor. Substituting (4.5.1) into (4.5.2) leads to

$$
\begin{aligned}
(m_0 + m_1)\ddot{x} - m_1 r(\ddot{\phi} \cos \phi - \dot{\phi}^2 \sin \phi) + kx &= -c\dot{x} \\
(I + m_1 r^2)\ddot{\phi} - m_1 r\ddot{x} \cos \phi &= L(\dot{\phi}) - H(\dot{\phi})
\end{aligned}
\tag{4.5.3}
$$

We note that equations (4.5.3) are autonomous and nonlinear.

It is convenient to rewrite (4.5.3) in terms of dimensionless variables as

$$
\begin{aligned}
x'' + x &= \epsilon(\phi'' \cos \phi - \phi'^2 \sin \phi - 2\mu x') \\
\phi'' &= \epsilon [bx'' \cos \phi + M(\phi')]
\end{aligned}
\tag{4.5.4}
$$

where

$$\epsilon = \frac{m_1 r}{X(m_0 + m_1)}$$

$$2\epsilon\mu = \frac{c}{[k(m_0 + m_1)]^{1/2}}$$

$$\epsilon b = \frac{m_1 r X}{I + m_1 r^2}$$

$$M(\phi') = \frac{L(\dot{\phi}) - H(\dot{\phi})}{I + m_1 r^2} \frac{m_0 + m_1}{k}$$

Here X is a length characteristic of the amplitude of the motion of the motor and the dimensionless independent variable is

$$\tau = \left(\frac{k}{m_0 + m_1}\right)^{1/2} t$$

The primes denote derivatives with respect to τ while the dots denote derivatives with respect to t.

We seek approximate solutions which are uniformly valid for small ϵ. Thus we are considering the case in which m_1 is small compared with m_0. As we have done before, we consider the damping force to be small compared with the restoring force and make the damping term appear at the same order as the nonlinearity and the excitation. And we are concerned with a relatively narrow band of frequencies which encloses the natural frequency of the system (unity in the dimensionless variables).

In this case the method of averaging (Section 2.3.5) is more convenient to use than the method of multiple scales. Accordingly we let

$$x = a \cos (\phi + \beta) \tag{4.5.5}$$

where a, ϕ, and β are functions of τ. Generally one cannot expect the frequency of the rectilinear motion $(\phi' + \beta')$ to be the same as the angular speed of the rotor (ϕ'); hence β is included in the argument.

We are considering motions near resonance, and it is convenient to introduce a detuning parameter Δ as follows:

$$\phi' = 1 + \Delta \tag{4.5.6}$$

Hence β is used to distinguish between the speed of the rotor and the actual frequency of the rectilinear motion, while Δ is used to distinguish between the speed of the rotor and the natural frequency of the rectilinear motion.

Using the method of variation of parameters, we put

$$a' \cos (\phi + \beta) - a(\Delta + \beta') \sin (\phi + \beta) = 0 \tag{4.5.7}$$

so that

$$x' = -a \sin (\phi + \beta) \tag{4.5.8}$$

and

$$x'' = -a' \sin (\phi + \beta) - a(1 + \Delta + \beta') \cos (\phi + \beta) \tag{4.5.9}$$

Substituting (4.5.6), (4.5.8), and (4.5.9) into (4.5.4) leads to

$$-a' \sin (\phi + \beta) - a(\Delta + \beta') \cos (\phi + \beta) = \epsilon\, [\Delta' \cos \phi - (1 + \Delta)^2 \sin \phi + 2\mu a \sin (\phi + \beta)] \tag{4.5.10}$$

and

$$\Delta' = \epsilon\, \{-b\, [a' \sin (\phi + \beta) + a(1 + \Delta + \beta') \cos (\phi + \beta)] \cos \phi + M\} \tag{4.5.11}$$

Solving (4.5.7) and (4.5.10) for a' and β' produces

$$a' = -\epsilon\, [\Delta' \cos \phi - (1 + \Delta)^2 \sin \phi + 2\mu a \sin (\phi + \beta)] \sin (\phi + \beta) \tag{4.5.12}$$

$$\beta' = -\Delta - \frac{\epsilon}{a}\, [\Delta' \cos \phi - (1 + \Delta)^2 \sin \phi + 2\mu a \sin (\phi + \beta)] \cos (\phi + \beta) \tag{4.5.13}$$

Equations (4.5.11) through (4.5.13) are equivalent to (4.5.4); no approximations have been made yet. These equations show that Δ' and a' are $O(\epsilon)$. If we restrict our attention to a narrow band of frequencies around the natural frequency (see Figure 4-35), then we may write

$$\Delta = \epsilon \sigma \tag{4.5.14}$$

Both Δ and Δ' are $O(\epsilon)$, and it follows from (4.5.13) that β' is also $O(\epsilon)$.

As a first simplification we neglect all terms $O(\epsilon^2)$ appearing in (4.5.11) through (4.5.13) and obtain

$$\Delta' = \epsilon\,[M - ab\cos(\phi + \beta)\cos\phi] \tag{4.5.15}$$

$$a' = \epsilon\,[\sin\phi - 2\mu a\sin(\phi + \beta)]\sin(\phi + \beta) \tag{4.5.16}$$

$$\beta' = -\epsilon\left\{\sigma + \frac{1}{a}\,[2\mu a\sin(\phi + \beta) - \sin\phi]\cos(\phi + \beta)\right\} \tag{4.5.17}$$

As a second simplification we can consider a, σ, and β to be constant over one cycle and integrate (average) the equations over one cycle. The result is

$$\Delta' = \epsilon\left(M - \frac{1}{2}ab\cos\beta\right) \tag{4.5.18}$$

$$a' = \epsilon\left(\frac{1}{2}\cos\beta - \mu a\right) \tag{4.5.19}$$

$$\beta' = -\epsilon\left(\sigma + \frac{1}{2a}\sin\beta\right) \tag{4.5.20}$$

We note the difference between ideal and nonideal systems. For the ideal system, (4.5.18) is not one of the governing equations, σ is specified, and (4.5.19) and (4.5.20) are solved for a and β. For the nonideal system, the control setting is specified (this is effected by means of M, as we discuss below), and (4.5.14) and (4.5.18) through (4.5.20) are solved for σ, a, Δ, and β.

For steady-state responses, Δ', a', and β' are zero. Then combining (4.5.19) and (4.5.20) yields

$$a = \tfrac{1}{2}(\mu^2 + \sigma^2)^{-1/2} \tag{4.5.21}$$

while combining (4.5.18) and (4.5.19) yields

$$M = \mu a^2 b \tag{4.5.22}$$

The phase is given by

$$\beta = \sin^{-1}(-2\sigma a) = \cos^{-1}(2\mu a) \tag{4.5.23}$$

Recalling the definitions of the dimensionless variables, we can rewrite these results in terms of the original physical variables as

$$(Xa) = \text{physical amplitude of the motion}$$

$$= \frac{\omega m_1 r}{[c^2 + 4(\omega - \dot{\phi})^2(m_0 + m_1)^2]^{1/2}} \tag{4.5.24}$$

$$\beta = \tan^{-1}\left[-\frac{2(m_0 + m_1)(\omega - \dot{\phi})}{c}\right] \tag{4.5.25}$$

where, from (4.5.23), $-\pi/2 < \beta < 0$ and

$$L(\dot{\phi}) - H(\dot{\phi}) = \frac{c\omega k m_1 r X}{2[c^2 + 4(\omega - \dot{\phi})^2(m_0 + m_1)^2]} \tag{4.5.26}$$

where ω is the natural frequency of the system:

$$\omega^2 = \frac{k}{m_0 + m_1} \tag{4.5.27}$$

Now the procedure is to solve (4.5.26) for $\dot{\phi}$ and then to determine the amplitude and phase from (4.5.24) and (4.5.25). This was the procedure followed by Kononenko and Korablev to determine the "theoretical" curve in Figure 4-34.

Generally $L(\dot{\phi})$ and $H(\dot{\phi})$ are only known in graphic form, having been determined experimentally. Consequently it is convenient to rewrite (4.5.26) as follows:

$$L(\dot{\phi}) = R(\dot{\phi}) \tag{4.5.28}$$

where

$$R(\dot{\phi}) = H(\dot{\phi}) + S(\dot{\phi}) \tag{4.5.29}$$

and

$$S(\dot{\phi}) = \frac{c\omega k m_1 r X}{2[c^2 + 4(\omega - \dot{\phi})^2(m_0 + m_1)^2]} \tag{4.5.30}$$

Then in the same graph one can construct the curve R versus $\dot{\phi}$ and the family of curves L versus $\dot{\phi}$. Each member of the latter corresponds to a constant setting of the control.

One can readily determine the nature of the curve R versus $\dot{\phi}$ as follows. Recalling that consideration is limited to a narrow band of frequencies around the natural frequency ω, we can represent H by a two-term expansion:

$$H(\dot{\phi}) = H(\omega) + [H'(\omega)](\dot{\phi} - \omega)$$

where both $H(\omega)$ and $H'(\omega)$ are positive. A graph of $S(\dot{\phi})$ given by (4.5.30) has the familiar shape associated with an ideal system. Both curves are shown in Figure 4-38, and the curve R versus $\dot{\phi}$ is shown in Figure 4-39. Also shown in Figure 4-39 is the family of characteristics of the motor. The corresponding curve of Xa versus $\dot{\phi}$ is shown in Figure 4-40 (see also Figures 4-34 and 4-35). The corresponding points in Figures 4-35 and 4-40 have the same label.

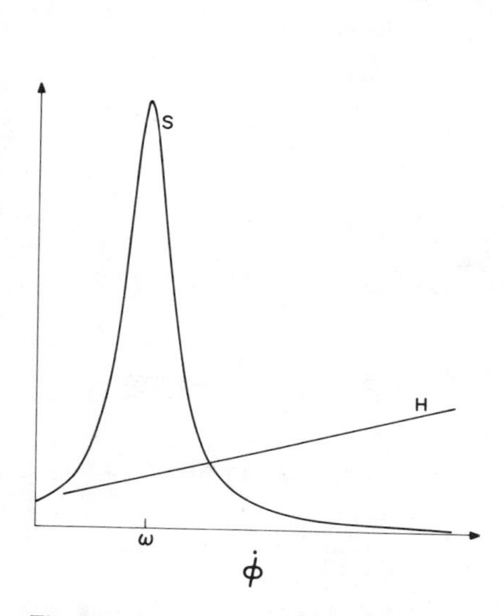

Figure 4-38. Functions H and S. See equation (4.5.30).

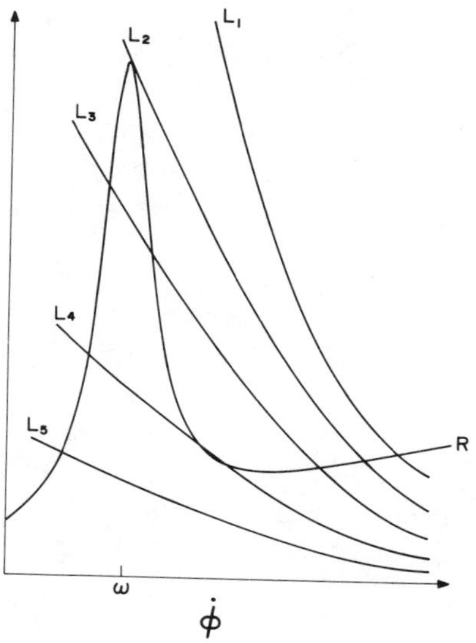

Figure 4-39. Characteristics of the motor and the function R. See equation (4.5.29).

For a given setting of the control there can be one, two, or three steady-state solutions. In the first approximation these solutions have the form

$$x = (Xa) \cos (\phi + \beta) \tag{4.5.31}$$

where $\dot{\phi}$ is determined from (4.5.28) and Xa and β are determined from (4.5.24) and (4.5.25), respectively. To determine which of these steady-state solutions actually corresponds to a realizable motion, we need to consider the stability of the solutions.

As we have done before, we determine the stability of the steady-state solutions by determining the nature of the singular points of (4.5.18) through (4.5.20) as in Section 3.2. To accomplish this, we let

$$a = a_0 + a_1, \qquad \beta = \beta_0 + \beta_1, \qquad \Delta = \Delta_0 + \Delta_1 \tag{4.5.32}$$

Substituting (4.5.32) into (4.5.18) through (4.5.20) and neglecting all but the linear terms in a_1, β_1, and Δ_1, we obtain

$$\begin{aligned} \Delta_1' &= \epsilon \frac{dM}{d\phi'} (\phi_0') \Delta_1 - (\tfrac{1}{2} b \cos \beta_0) a_1 + (\tfrac{1}{2} a_0 b \sin \beta_0) \beta_1 \\ a_1' &= \epsilon(-\tfrac{1}{2} \beta_1 \sin \beta_0 - \mu a_1) \\ \beta_1' &= -\epsilon \left(\frac{\cos \beta_0}{2a_0} \beta_1 - \frac{\sin \beta_0}{2a_0^2} a_1 + \sigma_1 \right) \end{aligned} \tag{4.5.33}$$

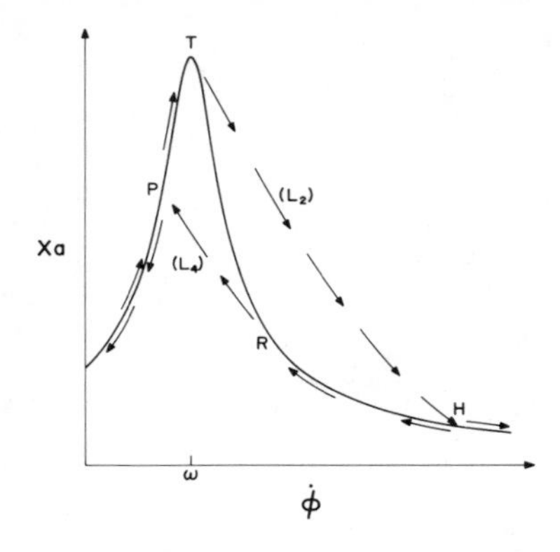

Figure 4-40. Frequency–response curve.

where $\epsilon\sigma_1 = \Delta_1$. Equations (4.5.33) are a system of linear equations having a solution in the form

$$(a_1, \beta_1, \Delta_1) = (a_{10}, \beta_{10}, \Delta_{10})\, e^{\lambda t}$$

where λ is an eigenvalue of the coefficient matrix and a_{10}, β_{10}, and Δ_{10} are constants. The solutions are stable, and hence the corresponding motions realizable, if the real part of each eigenvalue is negative or zero.

It turns out that the solutions between T and R are unstable while all those outside this region are stable (Figure 4-40). As the first equation of (4.5.33) indicates, the parameter that gives the influence of the motor on the stability is the slope of the characteristic.

Exercises

4.1. Consider the arrangement shown in Figure 3-1 with dry friction. When a harmonic excitation acts on the system, the equation of motion has the form

$$m\ddot{x} + k_1 x = d + F \cos \omega t - k_2 x^3$$

where

$$d = \begin{cases} -mg\mu & \text{if } \dot{x} > 0 \\ mg\mu & \text{if } \dot{x} < 0 \end{cases}$$

(a) Show that this equation can be put in the following convenient dimensionless form:

$$\ddot{u} + u = \epsilon(f - \alpha u^3) + 2k \cos \Omega t$$

where

$$f = \begin{cases} -1 & \text{if } \dot{u} > 0 \\ 1 & \text{if } \dot{u} < 0 \end{cases}$$

(b) For Ω near unity show that

$$u = a \cos [T_0 + \beta(T_1)] + O(\epsilon)$$

where

$$a' = -\frac{2}{\pi} + k \sin \gamma$$

$$a\gamma' = a\sigma - \frac{3\alpha a^3}{8} + k \cos \gamma$$

$$\gamma = \sigma T_1 - \beta$$

$$\epsilon\sigma = \Omega - 1 \quad \text{and} \quad \epsilon k = K$$

(c) Obtain the frequency-response equation. Plot the amplitude a and the phase γ as functions of σ for $k = 2/\pi$ and $k > 2/\pi$. What is the significance of $k = 2/\pi$? For $\sigma > 0$, plot the amplitude as a function of k. Is there a jump phenomenon associated with this motion?

4.2. For the arrangement shown in Figure 3-1 with square damping, the equation of motion can be put in the following convenient dimensionless form:

$$\ddot{u} + u = -\epsilon(|\dot{u}|\dot{u} + \alpha u^3) + 2K \cos \Omega t$$

(a) For Ω near unity, show that

$$u = a \cos [T_0 + \beta(T_1)] + O(\epsilon)$$

where

$$a' = -\frac{4a^2}{3\pi} + k \sin \gamma$$

$$a\gamma' = a\left(\sigma - \frac{3\alpha a^2}{8}\right) + k \cos \gamma$$

$$\gamma = \sigma T_1 - \beta \quad \text{and} \quad \epsilon\sigma = \Omega - 1$$

(b) Obtain the frequency-response equation. Plot a as a function of σ for constant k. Plot a as a function of k for $\sigma > 0$.

4.3. Consider the arrangement shown in Figure 3-2. The behavior of the bottom element is described by Figure 3-2b and (3.3.41). This is an example of bilinear hysteresis. When a harmonic excitation acts on this system (Caughey, 1960a; Iwan, 1965; Drew, 1974) the governing equation can be put in the following convenient dimensionless form:

$$\ddot{x} + x = \epsilon f + K \cos \Omega t$$

(a) For Ω near unity, show that

$$x = a(T_1) \cos [T_0 + \beta(T_1)] + O(\epsilon)$$

where

$$a' = \frac{2x_s}{a\pi}(x_s - a) + k \sin \gamma$$

$$\gamma' = \sigma - \frac{1}{\pi}\left\{\frac{1}{2} \cos^{-1} \frac{a - 2x_s}{a} - \left(1 - \frac{2x_s}{a}\right)\left[\frac{x_s}{a} - \left(\frac{x_s}{a}\right)^2\right]^{1/2}\right\} + \frac{k}{a} \cos \gamma$$

$$\gamma = \sigma T_1 - \beta, \quad \epsilon\sigma = \Omega - 1, \quad \text{and } 2\epsilon k = K$$

(b) Obtain the frequency-response equation. What is the significance of $k = 2x_s/\pi$? Explain why the present analysis only applies for $\frac{1}{2} + (k/x_s) > \sigma > \frac{1}{2} - (k/x_s)$.

(c) Plot a versus σ (for $k < 2x_s/\pi$ and $k > 2x_s/\pi$) and a versus k. Is there a jump phenomenon?

4.4. Consider the system governed by the following equation of motion:

$$\ddot{x} + x = \epsilon f + \mu_0 \operatorname{sgn} \dot{x} + K \cos \Omega t$$

where f is described by Figure 3-2b and equation (3.3.41).

(a) For Ω near unity, show that

$$x = a(T_1) \cos [T_0 + \beta(T_1)] + O(\epsilon)$$

where

$$a' = \frac{2x_s}{a\pi}(x_s - a) - \frac{2\mu_0}{\pi} + k \sin \gamma$$

$$\gamma' = \sigma - \frac{1}{\pi}\left\{\frac{1}{2} \cos^{-1}\left(\frac{a - 2x_s}{a}\right) - \left(1 - \frac{2x_s}{a}\right)\left[\frac{x_s}{a} - \left(\frac{x_s}{a}\right)^2\right]^{1/2}\right\} + \frac{k}{a} \cos \gamma$$

$$\gamma = \sigma T_1 - \beta, \quad \epsilon\sigma = \Omega - 1, \quad 2\epsilon k = K$$

(b) Obtain the frequency-response equation. What is the significance of $k = k_{crit} = 2(x_s + \mu_0)/\pi$. For what range of σ is the present analysis valid?

(c) Plot a versus σ for $k > k_{crit}$ and $k < k_{crit}$. Is there a jump phenomenon associated with this motion?

4.5. Consider the system governed by the following equation of motion:

$$\ddot{x} + x = \epsilon f - 2\epsilon\mu\dot{x} + K \cos \Omega t$$

where f is described by Figure 3-2b and (3.3.41).

(a) For Ω near unity, show that

$$x = a(T_1) \cos [T_0 + \beta(T_1)] + O(\epsilon)$$

where

$$a' = \frac{2x_s}{a\pi}(x_s - a) - \mu a + k \sin \gamma$$

$$\gamma' = \sigma - \frac{1}{\pi}\left\{\frac{1}{2} \cos^{-1}\left(\frac{a - 2x_s}{a}\right) - \left(1 - \frac{2x_s}{a}\right)\left[\frac{x_s}{a} - \left(\frac{x_s}{a}\right)^2\right]^{1/2}\right\} + \frac{k}{a} \cos \gamma$$

$$\gamma = \sigma T_1 - \beta, \quad \epsilon\sigma = \Omega - 1, \quad \text{and } 2\epsilon k = K$$

(b) Obtain the frequency-response equation. What is the minimum value of k for which a steady-state motion with $a \geqslant x_s$ is possible? Is it possible for a finite-amplitude force to produce an unbounded motion? What is the range of σ for which the present analysis is valid?

(c) Plot a versus σ and a versus k. Is there a jump phenomenon?

4.6. Consider the system governed by the following equation of motion:

$$\ddot{x} + x = \epsilon f - \epsilon\mu\dot{x}\,|\dot{x}\,| + K \cos \Omega t$$

where f is described by Figure 3-2b and (3.3.45).

(a) For Ω near unity, show that

$$x = a(T_1) \cos [T_0 + \beta(T_1)] + O(\epsilon)$$

where

$$a' = \frac{2x_s(x_s - a)}{\pi a} - \frac{4\mu}{3\pi}a^2 + k \sin \gamma$$

$$\gamma' = \sigma - \frac{1}{\pi}\left\{\frac{1}{2} \cos^{-1}\left(\frac{a - 2x_s}{a}\right) - \left(1 - \frac{2x_s}{a}\right)\left[\frac{x_s}{a} - \left(\frac{x_s}{a}\right)^2\right]^{1/2}\right\} + \frac{k}{a} \cos \gamma$$

$$\gamma = \sigma T_1 - \beta, \quad \epsilon\sigma = \Omega - 1, \quad \text{and } 2\epsilon k = K$$

(b) Obtain the frequency-response equation. For what range of σ is the present analysis valid? Explain why k must be $\geqslant 4\mu x_s^2/3\pi$ for the present analysis to be valid.

(c) Plot a versus σ and a versus k. Is there a jump phenomenon?

4.7. Consider a system for which the spring force is given by $f(x) = -k_1 x - k_2 x|x|$ and the damping force is given by $c(\dot{x}) = -c_1\dot{x} - c_2\dot{x}|\dot{x}|$. The equation of motion is

$$m\ddot{x} + c_1\dot{x} + c_2\dot{x}|\dot{x}| + k_1 x + k_2 x|x| = K\cos\Omega t$$

(a) Show that this equation can be written in the following convenient dimensionless form:

$$\ddot{x} + 2\epsilon\mu_1\dot{x} + \epsilon\mu_2\dot{x}|\dot{x}| + x + \epsilon\alpha x|x| = 2\epsilon k\cos\Omega t$$

where

$$2\epsilon\mu_1 = \frac{c_1}{m\omega_0}, \quad \epsilon\mu_2 = \frac{c_2 L}{m}, \quad \omega_0^2 = \frac{k_1}{m}$$

$$\epsilon\alpha = \frac{k_2 L}{m\omega_0^2}, \quad 2\epsilon k = \frac{K}{m\omega_0^2}, \quad \text{and } \Omega = \frac{\omega}{\omega_0}$$

(b) For Ω near unity, show that

$$x = a(T_1)\cos\phi + O(\epsilon)$$

where

$$\phi = T_0 + \beta(T_1)$$

$$a' = -\mu_1 a - \frac{4\mu_2}{3\pi}a^2 + k\sin\gamma$$

$$\gamma' = \sigma - \frac{4\alpha}{3\pi}a + \frac{k}{a}\cos\gamma$$

$$\gamma = \sigma T_1 - \beta, \quad \epsilon\sigma = \Omega - 1$$

(c) Using the approximate expression for x above, sketch $x|x|$ and $\dot{x}|\dot{x}|$ as functions of ϕ. Plot a versus σ and a versus k for stationary motion. Show that

$$a_{\max} = \frac{3\pi\mu_1}{8\mu_2}\left[-1 + \left(1 + \frac{16k\mu_2}{3\pi\mu_1^2}\right)^{1/2}\right]$$

(d) Assuming the expansion for x has the form

$$x = x_0 + \epsilon x_1 + \epsilon^2 x_2 + \cdots$$

explain why the effect of $x|x|$ can be determined from the expression for x_1, while the effect of x^2 can only be determined from x_2 (see Section 4.2).

4.8. Consider the system governed by

$$\ddot{\theta} + \sin\theta + 2\mu\theta^2\dot{\theta} = K\cos\Omega t$$

(a) When Ω is near unity, show that for small but finite amplitudes of the response

$$\theta = \epsilon a(T_2)\cos[T_0 + \beta(T_2)] + O(\epsilon^2)$$

where

$$a' = -\tfrac{1}{4}\mu a^3 + k\sin\gamma$$

$$\gamma' = \sigma + \tfrac{1}{16}a^2 + \frac{k}{a}\cos\gamma$$

$$\gamma = \sigma T_2 - \beta, \quad \epsilon^2\sigma = \Omega - 1, \quad \text{and } 2\epsilon^2 k = K$$

Here ϵ is a measure of the amplitude of the response. Obtain the frequency-response equation. Show that $a_{max} = (4k/\mu)^{1/3}$. How does this value of a_{max} compare with that for the case of linear viscous damping? Plot a versus σ and a versus k. Is there a jump phenomenon associated with this motion?

(b) When Ω is near one third (superharmonic resonance), show that

$$\theta = \epsilon a(T_2)\cos[T_0 + \beta(T_2)] + \tfrac{9}{8}K\cos(\Omega T_0) + O(\epsilon^2)$$

where

$$a' = -2\mu\left(\Lambda^2 + \frac{a^2}{8}\right)a - \Lambda^3(\tfrac{2}{3}\mu\cos\gamma - \tfrac{1}{6}\sin\gamma)$$

$$\gamma' = \sigma + \tfrac{1}{2}\left(\Lambda^2 + \frac{a^2}{8}\right) + \frac{\Lambda^3}{a}(\tfrac{2}{3}\mu\sin\gamma + \tfrac{1}{6}\cos\gamma)$$

where

$$\gamma = \sigma T_2 - \beta, \quad \epsilon^2\sigma = 3\Omega - 1, \quad \text{and } \Lambda = \tfrac{9}{16}K$$

Obtain the frequency-response equation. Plot a versus σ and a versus Λ. Is there a jump phenomenon associated with this motion?

(c) When Ω is near 3 (subharmonic resonance), show that

$$\theta = \epsilon a(T_2)\cos[T_0 + \beta(T_2)] - \tfrac{1}{8}K\cos(\Omega T_0) + O(\epsilon^2)$$

where

$$a' = -2\mu\left(\Lambda^2 + \frac{a^2}{8}\right)a + \Lambda a^2(\tfrac{1}{8}\sin\gamma - \tfrac{1}{2}\mu\cos\gamma)$$

$$\gamma' = \sigma + \tfrac{3}{2}\left(\Lambda^2 + \frac{a^2}{8}\right) + \Lambda a(\tfrac{3}{8}\cos\gamma + \tfrac{3}{2}\mu\sin\gamma)$$

where

$$\gamma = \sigma T_2 - 3\beta, \quad \epsilon^2\sigma = \Omega - 3, \quad \text{and } \Lambda = \frac{K}{16}$$

Obtain the frequency-response equation for the nontrivial solution. Plot a versus σ and a versus Λ. Is there a jump phenomenon associated with this motion?

4.9. Consider the system governed by

$$\ddot{u} + 2\epsilon\mu \sin \dot{u} + u = K \cos \Omega t$$

where $\mu = O(1)$ and $\epsilon \ll 1$. When Ω is near unity, show that

$$x = a \cos \phi$$

where

$$\phi = t + \beta(t)$$

$$\dot{a} = 2\epsilon\mu \sin(-a \sin\phi) \sin\phi - 2\epsilon k (\cos\gamma \cos\phi - \sin\gamma \sin\phi) \sin\phi$$

$$\dot{\gamma} = \epsilon\sigma - \frac{2\epsilon\mu}{a} \sin(-a \sin\phi) \cos\phi + \frac{2\epsilon k}{a}(\cos\gamma \cos\phi - \sin\gamma \sin\phi) \cos\phi$$

$$\gamma = \epsilon\sigma t - \beta, \quad \epsilon\sigma = \Omega - 1, \quad \text{and } 2\epsilon k = K$$

After averaging over one cycle, show that approximately

$$\dot{a} = -\epsilon[2\mu J_1(a) - k \sin\gamma]$$

$$\dot{\gamma} = \epsilon\left(\sigma + \frac{k}{a} \cos\gamma\right)$$

Hint:

$$\sin(a \sin\phi) = 2 \sum_{n=0}^{\infty} J_{2n+1}(a) \sin[(2n+1)\phi]$$

where J_m is the Bessel function of the first kind of order m. Obtain the frequency-response equation. Plot a versus σ. Is there a jump phénomenon associated with this motion? For a given k, is it possible to obtain any amplitude a by simply adjusting the frequency? For small a, expand $\sin(-a \sin\phi)$ and retain only the first two or three terms. Then after averaging over one cycle, show that approximately

$$\dot{a} = -\epsilon\left[\mu\left(a - \tfrac{1}{8}a^3 + \frac{a^5}{192}\right) + k \sin\gamma\right]$$

$$\dot{\gamma} = \epsilon\left(\sigma + \frac{k}{a} \cos\gamma\right)$$

Obtain the frequency-response equation. Plot a versus σ and a versus k.

4.10. Consider the system governed by the following equation of motion:

$$\ddot{u} - \tfrac{1}{2}(u - u^3) = K \cos \Omega t$$

Determine the frist term in a uniformly valid expansion of u describing the response near the center at $u = 1$ for

(a) Ω near unity
(b) Ω near 2
(c) Ω near $\frac{1}{2}$
(d) Ω near three
(e) Ω near one-third

4.11. Consider the system governed by the following equation of motion:

$$\ddot{u} + u + \alpha u^5 = K \cos \Omega t - 2\mu\dot{u}$$

(a) Determine the frequency-response equation when Ω is near unity. Make plots of the amplitude of the response as a function of the frequency and the amplitude of the excitation.

(b) Show that to first order, secondary resonances exist when $\Omega \approx 5, 3, 2, \frac{1}{2}, \frac{1}{3}$, and $\frac{1}{5}$.

(c) Determine the equations governing the phases and the amplitudes for the one-term expansions for each case of secondary resonance.

4.12. Consider the system governed by the following equation of motion:

$$\ddot{u} + f(u) = K \cos \Omega t$$

where $u(0) = u_0$ and $\dot{u}(0) = 0$.

(a) Show that, when $K = 0$, u is given implicitly by

$$t = \pm \int_{u_0}^{u} \frac{du}{[F(u_0) - F(u)]^{1/2}} = g_1(u)$$

where $\frac{1}{2}F'(u) = f(u)$.

(b) Use the above as the starting solution in an iteration scheme. Thus the second iteration is governed by

$$\ddot{u} + f_1(u) = 0$$

Determine $f_1(u)$ as a function of $f(u)$ and $g_1(u)$. Then show that for the second approximation the solution can be expressed as (Rauscher, 1938)

$$t = \pm \int_{u_0}^{u} \frac{du}{[F_1(u_0) - F_1(u)]^{1/2}}$$

4.13. Consider the system governed by the following equation of motion:

$$\ddot{u} + u + 2\epsilon\mu\dot{u} + \epsilon\alpha u^3 = \epsilon k \cos \Omega t$$

Using the method of harmonic balance, obtain the frequency response equation for Ω near unity. Compare your results with (4.1.17).

4.14. Consider the system governed by the following equation of motion:

$$\ddot{u} + u^5 = K \cos \Omega t$$

Using the method of harmonic balance, obtain the frequency-response equation.

4.15. Consider a system governed by an equation having the form

$$\ddot{u} + \omega_0^2 u + \epsilon(2\mu\dot{u} + \alpha u^3) = K_1 \cos(\Omega_1 t + \theta_1) + K_2 \cos(\Omega_2 t + \theta_2)$$

where

$$\Omega_1 + \Omega_2 \approx 2\omega_0$$

(a) Using the method of multiple scales to obtain a one-term, uniformly valid expansion for small ϵ, show that secular terms are eliminated from u_1 if

$$2i\omega_0(A' + \mu A) + \alpha(3A\bar{A} + 6\Lambda_1\bar{\Lambda}_1 + 6\Lambda_2\bar{\Lambda}_2)A + 6\alpha\Lambda_1\Lambda_2\bar{A} \exp(i\sigma T_1) = 0$$

where

$$\Lambda_n = \frac{K_n \exp(i\theta_n)}{2(\omega_0^2 - \Omega_n^2)}, \quad \epsilon\sigma = \Omega_1 + \Omega_2 - 2\omega_0$$

(b) Letting $A = \frac{1}{2}a \exp(i\beta)$, show that the solvability conditions are equivalent to

$$a' = -\mu a - \alpha\Gamma_1 a \sin\gamma$$

$$a\gamma' = (\sigma - 2\alpha\Gamma_2)a - \frac{3\alpha a^3}{4\omega_0} - 2\alpha\Gamma_1 a \cos\gamma$$

where

$$\Gamma_1 = \frac{3K_1K_2}{4\omega_0(\omega_0^2 - \Omega_1^2)(\omega_0^2 - \Omega^2)}$$

$$\Gamma_2 = \frac{3}{4\omega_0}\left[\frac{K_1^2}{(\omega_0^2 - \Omega_1^2)^2} + \frac{K_2^2}{(\omega_0^2 - \Omega_2^2)^2}\right]$$

$$\gamma = \sigma T_1 - 2\beta + \theta_1 + \theta_2$$

(c) Show that the steady-state amplitudes are given by

$$a = 0$$

or

$$a^2 = \frac{8}{3}\omega_0\left[\frac{\sigma}{2\alpha} - \Gamma_2 \pm \left(\Gamma_1^2 - \frac{\mu^2}{\alpha^2}\right)^{1/2}\right]$$

(d) Determine which of the possible steady-state amplitudes are stable. (Be sure to consider the trivial solution.)

(e) Use your results to explain Figure 4-41.

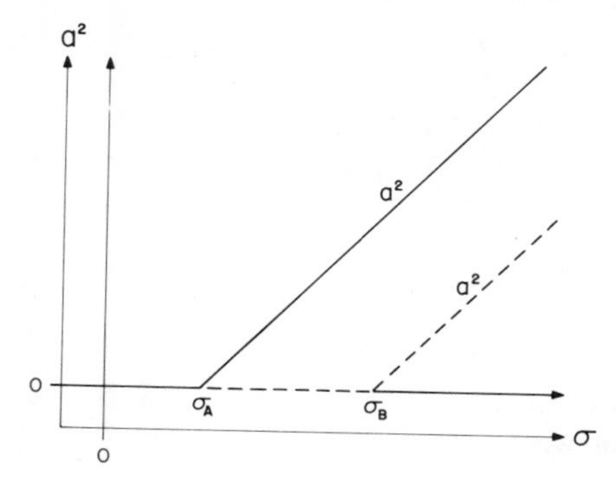

Figure 4-41. Response curves for the system described in Exercise 4.15.

4.16. Consider the system governed by the following equation of motion:

$$\ddot{u} + u + 2\epsilon\mu\dot{u} + \epsilon\alpha u^3 = K_1 \cos(\Omega_1 t + \theta_1) + K_2 \cos(\Omega_2 t + \theta_2)$$

where Ω_1 is near unity and Ω_2 is near 3.

(a) Using the method of multiple scales to obtain a one-term, uniformly valid expansion for small ϵ, show that secular terms are eliminated from u_1 if

$$a' = -\mu a + \tfrac{1}{2} k_1 \sin(\sigma_1 T_1 - \beta + \theta_1) - \tfrac{3}{4}\alpha\Lambda_2 a^2 \sin(\sigma_2 T_1 - 3\beta + \theta_2)$$

$$a\beta' = 3\alpha(\Lambda_2^2 + \tfrac{1}{8}a^2)a - \tfrac{1}{2}k_1 \cos(\sigma_1 T_1 - \beta + \theta_1) + \tfrac{3}{4}\alpha\Lambda_2 a^2 \cos(\sigma_1 T_1 - 3\beta + \theta_2)$$

where

$$\epsilon k_1 = K, \quad \epsilon\sigma_1 = \Omega_1 - 1, \quad \epsilon\sigma_2 = \Omega_2 - 3, \quad \text{and } \Lambda_2 = -\tfrac{1}{16}K_2$$

(b) Show that stationary solutions exist only when $\beta' = \sigma_1 = \frac{1}{3}\sigma_2$ and hence that they correspond to periodic responses.

(c) Plot a versus σ_1 and indicate the stable portions of the curves.

4.17. Consider the system governed by the following equation of motion:

$$\ddot{u} + u + 2\epsilon\mu\dot{u} + \epsilon\alpha u^3 = \sum_{n=1}^{3} K_n \cos(\Omega_n t + \theta_n)$$

where $\Omega_1 \approx 1$ and $2\Omega_2 + \Omega_3 \approx 1$.

(a) Using the method of multiple scales, show that

$$u = a(T_1)\cos[T_0 + \beta(T_1)] + 2\sum_{n=2}^{3} \Lambda_n \cos(\Omega_n T_0 + \theta_n) + O(\epsilon)$$

where

$$\Lambda_n = \frac{K_n}{2(1 - \Omega_n^2)}$$

$$a' = -\mu a + \tfrac{1}{2} k_1 \sin \gamma_1 - \alpha \Gamma_1 \sin \gamma_2$$
$$a\beta' = \alpha(\Gamma_2 + \tfrac{3}{8} a^2) a - \tfrac{1}{2} k_1 \cos \gamma_1 + \alpha \Gamma_1 \cos \gamma_2$$

where

$$\epsilon k_1 = K_1, \quad \epsilon \sigma_1 = \Omega_1 - 1, \quad \epsilon \sigma_2 = 2\Omega_2 + \Omega_3 - 1$$
$$\gamma_1 = \sigma_1 T_1 - \beta + \theta_1, \quad \gamma_2 = \sigma_2 T_1 - \beta + 2\theta_2 + \theta_3$$

and Γ_1 and Γ_2 are constants. Determine Γ_1 and Γ_2.

(b) Show that a stationary solution can exist only if $\beta' = \sigma_1 = \sigma_2$. Do stationary solutions correspond to periodic responses?

(c) Plot a versus σ_1 and indicate the stable portions of the curves.

4.18. Consider the system governed by the following equation of motion:

$$\ddot{u} + u + 2\epsilon\mu\dot{u} + \epsilon\alpha u^3 = \sum_{n=1}^{3} K_n \cos(\Omega_n t + \theta_n)$$

where $\Omega_1 \approx 1$, $\Omega_2 \approx \frac{1}{3}$, and $\Omega_3 \approx 3$.

(a) Using the method of multiple scales, show that

$$u = a(T_1) \cos[T_0 + \beta(T_1)] + 2 \sum_{n=2}^{3} \Lambda_n \cos(\Omega_n T_0 + \theta_n) + O(\epsilon)$$

where

$$\Lambda_n = \frac{K_n}{2(1 - \Omega_n^2)}$$
$$a' = \mu a + \tfrac{1}{2} k_1 \sin \gamma_1 - \alpha \Lambda_2^3 \sin \gamma_2 - \tfrac{3}{4} \alpha \Lambda_3 a^2 \sin \gamma_3$$
$$a\beta' = 3\alpha(\tfrac{1}{8} a^2 + \Lambda_2^2 + \Lambda_3^2) a - \tfrac{1}{2} k_1 \cos \gamma_1 + \alpha \Lambda_2^3 \cos \gamma_2 + \tfrac{3}{4} \alpha \Lambda_3 a^2 \cos \gamma_3$$
$$\epsilon k_1 = K_1, \quad \epsilon \sigma_1 = \Omega_1 - 1, \quad \epsilon \sigma_2 = 3\Omega_2 - 1$$
$$\epsilon \sigma_3 = \Omega_3 - 3, \gamma_1 = \sigma_1 T_1 - \beta + \theta_1$$
$$\gamma_2 = \sigma_2 T_1 - \beta + 3\theta_2, \quad \gamma_3 = \sigma_3 T_1 - 3\beta + \theta_3$$

(b) Show that a stationary solution can exist only if $\beta' = \sigma_1 = \sigma_2 = \frac{1}{3}\sigma_3$. Do stationary solutions correspond to periodic responses?

(c) Plot a versus σ_1 and indicate the stable portions of the curves.

4.19. Consider the system governed by the following equation of motion:

$$\ddot{u} + u + 2\,\epsilon\mu\dot{u} + \epsilon\alpha u^3 = \sum_{n=1}^{4} K_n \cos(\Omega_n t + \theta_n)$$

where $\Omega_1 \approx 1$ and $\Omega_2 + \Omega_3 + \Omega_4 \approx 1$.

(1) Using the method of multiple scales, show that

$$u = a(T_1)\cos[T_0 + \beta(T_1)] + 2\sum_{n=2}^{4} \Lambda_n \cos(\Omega_n t + \theta_n) + O(\epsilon)$$

where

$$\Lambda_n = \frac{K_n}{2(1 - \Omega_n^2)} \quad \text{for } n = 2, 3, \text{ and } 4$$

$$a' = -\mu a + \tfrac{1}{2}k_1 \sin\gamma_1 - \alpha\Gamma_1 \sin\gamma_2$$

$$a\beta' = \alpha(\tfrac{3}{8}a^2 + \Gamma_2)a - \tfrac{1}{2}k_1 \cos\gamma_1 + \alpha\Gamma_1 \cos\gamma_2$$

$$\epsilon\sigma_1 = \Omega_1 - 1, \quad \epsilon\sigma_2 = \Omega_2 + \Omega_3 + \Omega_4 - 1$$

$$\gamma_1 = \sigma_1 T_1 - \beta + \theta_1, \quad \gamma_2 = \sigma_2 T_1 - \beta + \theta_2 + \theta_3 + \theta_4$$

and Γ_1 and Γ_2 are constants. Determine Γ_1 and Γ_2.

(b) Show that stationary solutions can exist only if $\beta' = \sigma_1 = \sigma_2$. Do stationary solutions correspond to periodic responses?

(c) Plot a versus σ_1 for several values of Γ_1 and Γ_2 and indicate the stable portions of the curves.

4.20. Consider the system governed by the following equation of motion:

$$\ddot{u} + u + 2\epsilon\mu\dot{u} + \epsilon\alpha u^3 = \sum_{n=1}^{3} K_n \cos(\Omega_n t + \theta_n)$$

where $\Omega_1 \approx 1$ and $\Omega_2 + \Omega_3 \approx 2$.

(a) Using the method of multiple scales, show that

$$u = a(T_1)\cos[T_0 + \beta(T_1)] + 2\sum_{n=2}^{3} \Lambda_n \cos(\Omega_n T_0 + \theta_n) + O(\epsilon)$$

where

$$\Lambda_n = \frac{K_n}{2(1 - \Omega_n^2)}$$

$$a' = -\mu a + \tfrac{1}{2}k_1 \sin\gamma_1 - \alpha\Gamma_1 a \sin\gamma_2$$

$$a\beta' = \alpha(\tfrac{3}{8}a^2 + \Gamma_2)a - \tfrac{1}{2}k_1 \cos\gamma_1 + \alpha\Gamma_1 \cos\gamma_2$$

$$\epsilon k_1 = K_1, \quad \epsilon\sigma_1 = \Omega_1 - 1, \quad \epsilon\sigma_2 = \Omega_2 + \Omega_3 - 2$$

$$\gamma_1 = \sigma_1 T_1 - \beta + \theta_1, \quad \gamma_2 = \sigma_2 T_1 - 2\beta + \theta_2 + \theta_3$$

and Γ_1 and Γ_2 are constants. Determine Γ_1 and Γ_2.

(b) Show that stationary solutions can exist only if $\beta' = \sigma_1 = \frac{1}{2}\sigma_2$. Do stationary solutions correspond to periodic responses?

(c) Plot a versus σ_1 and indicate the unstable portions of the curves.

4.21. The response of a system to a harmonic excitation is governed by the following equation:

$$\ddot{u} + u + \hat{\mu} \operatorname{sgn} \dot{u} + \hat{\alpha} u^2 = K \cos \Omega t$$

where $\Omega \approx 1$. Assume all the coefficients are small. Introduce the parameter ϵ which is a measure of the smallness of these coefficients. Then introduce new coefficients, which are explicit functions of ϵ, recalling that the nonlinear, damping, and forcing terms must interact at the same order.

(a) Show that

$$u = a \cos(\Omega t - \gamma) + \tfrac{1}{2} \epsilon \alpha a^2 [\tfrac{1}{3} \cos(2\Omega t - 2\gamma) - 1] + O(\epsilon^2)$$

where

$$a' = -\frac{2\mu}{\pi} + \tfrac{1}{2} k \sin \gamma$$

$$a\gamma' = a\sigma + \tfrac{5}{12} \alpha^2 a^3 + \tfrac{1}{2} k \cos \gamma$$

$$\epsilon\sigma = \Omega - 1$$

(b) Obtain the frequency-response equation. Plot a versus σ for several values of k. Is there a jump phenomenon associated with this motion? Is this motion bounded?

4.22. The response of a system to a harmonic excitation is governed by the following equation:

$$\ddot{u} + u + \hat{\mu} \dot{u} |\dot{u}| + \hat{\alpha} u^2 = K \cos \Omega t$$

where $\Omega \approx 1$. Assume all the coefficients are small. Introduce the parameter ϵ which is a measure of the smallness of these coefficients. Then introduce new coefficients, which are explicit functions of ϵ, recalling that the nonlinear, damping, and forcing terms must interact at the same order.

(a) Show that

$$u = a \cos(\Omega t + \gamma) + \tfrac{1}{2} \epsilon \alpha a^2 [\tfrac{1}{3} \cos(2t - 2\gamma) - 1] + O(\epsilon^2)$$

where

$$a' = -\frac{4\mu a^2}{3\pi} + \tfrac{1}{2} k \sin \gamma$$

$$a\gamma' = a\sigma + \tfrac{5}{12} \alpha^2 a^3 + \tfrac{1}{2} k \cos \gamma$$

$$\epsilon\sigma = \Omega - 1$$

(b) Obtain the frequency-response equation. Plot a versus σ for several values of k. Is there a jump phenomenon associated with this motion? Is this motion bounded?

4.23. Consider the system governed by the following equation of motion:

$$\ddot{u} + u + 2\epsilon\mu\dot{u} + \epsilon\alpha u^2 = \sum_{n=1}^{3} K_n \cos(\Omega_n t + \theta_n)$$

where $\Omega_2 - \Omega_1 \approx 1$ and $\Omega_3 - \Omega_2 \approx 1$.

(a) Using the method of multiple scales, show that

$$u = A(T_1)\exp(iT_0) + cc + 2\sum_{n=1}^{3} \Lambda_n \cos(\Omega_n T_0 + \theta_n) + O(\epsilon)$$

where

$$\Lambda_n = \frac{K_n}{2(1 - \Omega_n^2)}$$

$$i(A' + \mu A) + \alpha\Lambda_1\Lambda_2 \exp(i\gamma_1) + \alpha\Lambda_2\Lambda_3 \exp(i\gamma_2) = 0$$

$$\epsilon\sigma_1 = \Omega_2 - \Omega_1 - 1, \quad \epsilon\sigma_2 = \Omega_3 - \Omega_2 - 1$$

$$\gamma_1 = \sigma_1 T_1 + \theta_2 - \theta_1, \quad \gamma_2 = \sigma_2 T_1 + \theta_3 - \theta_2$$

(b) Show that

$$A = C\exp(-\mu T_1) + \frac{i\alpha\Lambda_1\Lambda_2}{\mu + i\sigma_1}\exp(i\gamma_1) + \frac{i\alpha\Lambda_2\Lambda_3}{\mu + i\sigma_2}\exp(i\gamma_2)$$

where C is a constant of integration.

(c) Determine the steady-state response. Is it periodic?

4.24. Consider the system governed by the following equation of motion:

$$\ddot{u} + u + 2\epsilon\mu\dot{u} + \epsilon\alpha u^2 = \sum_{n=1}^{2} K_n \cos(\Omega_n t + \theta_n)$$

where $\Omega_1 \approx \frac{1}{2}$ and $\Omega_2 \approx 2$.

(a) Using the method of multiple scales, show that

$$u = A(T_1)\exp(iT_0) + cc + 2\sum_{n=1}^{2} \Lambda_n \cos(\Omega_n T_0 + \theta_n) + O(\epsilon)$$

where

$$\Lambda_n = \frac{K_n}{2(1 - \Omega_n^2)}$$

$$2i(A' + \mu A) + \alpha \Lambda_1^2 \exp(i\gamma_1) + 2\alpha \Lambda_2 \bar{A} \exp(i\gamma_2) = 0$$

$$\epsilon\sigma_1 = 2\Omega_1 - 1, \quad \epsilon\sigma_2 = \Omega_2 - 2$$

$$\gamma_1 = \sigma_1 T_1 + 2\theta_1, \quad \gamma_2 = \sigma_2 T_1 + \theta_2$$

(b) Let

$$A = (B_r + iB_i) \exp(\tfrac{1}{2} i\gamma_2)$$

where B_r and B_i are real functions of T_1, separate the solvability condition into real and imaginary parts, and obtain the equations governing B_r and B_i. Determine the conditions for which the solution becomes unbounded. Does the superharmonic resonance affect this condition?

4.25. The response of a system to a three-frequency excitation is governed by

$$\ddot{u} + u + 2\epsilon\mu\dot{u} + \epsilon\alpha u^2 = \sum_{n=1}^{3} K_n \cos(\Omega_n t + \theta_n)$$

(a) Show that

$$u = A(T_1) \exp(iT_0) + cc + 2\sum_{n=1}^{3} \Lambda_n \cos(\Omega_n T_0 + \theta_n) + O(\epsilon)$$

where

$$\Lambda_n = \frac{K_n}{2(1 - \Omega_n^2)}$$

$$i(A' + \mu A) + \alpha \bar{A} \Lambda_1 \exp(i\gamma_1) + \alpha \Lambda_2 \Lambda_3 \exp(i\gamma_2) = 0$$

$$\epsilon\sigma_1 = \Omega_1 - 2, \quad \epsilon\sigma_2 = \Omega_2 + \Omega_3 - 1$$

$$\gamma_1 = \sigma_1 T_1 + \theta_1, \quad \gamma_2 = \sigma_2 T_1 + \theta_2 + \theta_3$$

(b) Let

$$A = (B_r + iB_i) \exp(\tfrac{1}{2} i\gamma_1)$$

where B_r and B_i are real functions of T_1, separate the solvability condition into real and imaginary parts, and obtain the equations governing B_r and B_i. Determine the conditions for an unbounded solution.

4.26. The response of a system to a three-frequency excitation is governed by the following equation of motion:

$$\ddot{u} + u + 2\epsilon\mu\dot{u} + \epsilon\alpha u^2 = \sum_{n=1}^{3} K_n \cos(\Omega_n t + \theta_n)$$

where

$$\Omega_1 \approx \tfrac{1}{2} \quad \text{and} \quad \Omega_2 + \Omega_3 \approx 1.$$

(a) Show that

$$u = A(T_1) \exp(iT_0) + cc + 2\sum_{n=1}^{3} \Lambda_n \cos(\Omega_n T_0 + \theta_n) + O(\epsilon)$$

where

$$\Lambda_n = \frac{K_n}{2(1 - \Omega_n^2)}$$

$$2i(A' + \mu A) + \alpha\Lambda_1^2 \exp(i\gamma_1) + 2\alpha\Lambda_2\Lambda_3 \exp(i\gamma_2) = 0$$

$$\epsilon\sigma_1 = 2\Omega_1 - 1, \quad \epsilon\sigma_2 = \Omega_2 + \Omega_3 - 1$$

$$\gamma_1 = \sigma_1 T_1 + 2\theta_1, \quad \gamma_2 = \sigma_2 T_1 + \theta_2 + \theta_3$$

(b) Solve for A and determine the steady-state response. Is it periodic?

4.27. The response of a system to a five-frequency excitation is governed by

$$\ddot{u} + u + 2\epsilon\mu\dot{u} + \epsilon\alpha u^2 = \sum_{n=1}^{5} K_n \cos(\Omega_n t + \theta_n)$$

where

$$\Omega_1 \approx \tfrac{1}{2}, \quad \Omega_2 \approx 2, \quad \Omega_4 - \Omega_3 \approx 1, \quad \Omega_5 - \Omega_4 \approx 1$$

(a) Show that

$$u = A(T_1) \exp(iT_0) + cc + 2\sum_{n=1}^{5} \Lambda_n \cos(\Omega_n T_0 + \theta_n) + O(\epsilon)$$

where

$$\Lambda_n = \frac{K_n}{2(1 - \Omega_n^2)}$$

$$2i(A' + \mu A) + 2\alpha\bar{A}\Lambda_2 \exp(i\gamma_2) + \alpha\Lambda_1^2 \exp(i\gamma_1) + 2\alpha\Lambda_3\Lambda_4 \exp(i\gamma_3) + 2\alpha\Lambda_4\Lambda_5 \exp(i\gamma_4) = 0$$

$$\epsilon\sigma_1 = 2\Omega_1 - 1, \quad \epsilon\sigma_2 = \Omega_2 - 2, \quad \epsilon\sigma_3 = \Omega_4 - \Omega_3 - 1$$

$$\epsilon\sigma_4 = \Omega_5 - \Omega_4 - 1, \quad \gamma_1 = \sigma_1 T_1 + 2\theta_1$$

$$\gamma_2 = \sigma_2 T_1 + \theta_2, \quad \gamma_3 = \sigma_3 T_1 + \theta_4 - \theta_3$$

$$\gamma_4 = \sigma_4 T_1 + \theta_5 - \theta_4$$

(b) Solve for A and determine the condition for an unbounded solution. What is the effect of the superharmonic and combination resonances on this condition?

4.28. The response of a self-sustaining oscillator to a two-frequency excitation is governed by

$$\ddot{u} + u = \epsilon(\dot{u} - \tfrac{1}{3}\dot{u}^3) + \sum_{n=1}^{2} K_n \cos(\Omega_n t + \theta_n)$$

where Ω_1 is near unity and Ω_2 is near $\frac{1}{3}$.

(a) Show that

$$u = a(T_1)\cos[T_0 + \beta(T_1)] + 2\Lambda\cos(\Omega_2 T_0 + \theta_2) + O(\epsilon)$$

where

$$\Lambda = \tfrac{9}{16}K_2$$

$$a' = (\tfrac{1}{2} - \Omega_2^2\Lambda^2 - \tfrac{1}{8}a^2)a + k_1 \sin\gamma_1 + k_2\cos\gamma_2$$

$$a\beta' = -k_1\cos\gamma_1 + k_2\sin\gamma_2$$

$$2\epsilon k_1 = K_1, \quad \tfrac{1}{3}\Omega_2^3\Lambda^3 = k_2, \quad \gamma_1 = \sigma_1 T_1 - \beta + \theta_1$$

$$\gamma_2 = \sigma_2 T_1 - \beta + 3\theta_2, \quad \epsilon\sigma_1 = \Omega_1 - 1$$

$$\epsilon\sigma_2 = 3\Omega_2 - 1$$

(b) Show that stationary solutions exist only if $\beta' = \sigma_1 = \sigma_2$. Is the corresponding response periodic?

(c) Obtain the frequency-response equation:

$$\sigma = \pm\left[\frac{k_1^2 + 2k_1k_2\sin\psi + k_2^2}{a^2} - (\tfrac{1}{2} - \Omega_2^2\Lambda^2 - \tfrac{1}{8}a^2)^2\right]^{1/2}$$

where $\psi = \theta_1 - 3\theta_2$. Plot several frequency–response curves, indicating the stable and unstable portions.

4.29. The response of a self-sustaining system to a two-frequency excitation is governed by

$$\ddot{u} + u = \epsilon(\dot{u} - \tfrac{1}{3}\dot{u}^3) + \sum_{n=1}^{2} K_n \cos(\Omega_n t + \theta_n)$$

where Ω_1 is near unity and Ω_2 is near 3.

(a) Show that

$$u = a(T_1)\cos[T_0 + \beta(T_1)] + 2\Lambda\cos[\Omega_2 T_0 + \theta_2]$$

where

$$\Lambda = -\tfrac{1}{16}K_2$$

$$a' = (\tfrac{1}{2} - 16k_2^2 - \tfrac{1}{8}a^2)a + k_1\sin\gamma_1 + k_2 a^2\cos\gamma_2$$

$$a\beta' = -k_1\cos\gamma_1 + k_2 a^2\sin\gamma_2$$

$$2\epsilon k_1 = K_1, \quad k_2 = \tfrac{1}{4}\Omega_2 \Lambda, \quad \gamma_1 = \sigma_1 T_1 - \beta + \theta_1$$

$$\gamma_2 = \sigma_2 T_1 - 3\beta + \theta_2, \quad \epsilon\sigma_1 = \Omega_1 - 1, \quad \epsilon\sigma_2 = \Omega_2 - 3$$

(b) Show that stationary solutions exist only if $\beta' = \sigma_1 = \frac{1}{3}\sigma_2$. Is the corresponding response periodic?

4.30. The response of a self-sustaining system to a two-frequency excitation is governed by

$$\ddot{u} + u = \epsilon(\dot{u} - \tfrac{1}{3}\dot{u}^3) + \sum_{n=1}^{2} K_n \cos(\Omega_n t + \theta_n)$$

where Ω_1 is near 3 and Ω_2 is near $\frac{1}{3}$.

(a) Show that

$$u = a(T_1)\cos[T_0 + \beta(T_1)] + 2\sum_{n=1}^{2} \Lambda_n \cos(\Omega_n T_0 + \theta_n) + O(\epsilon)$$

where

$$\Lambda_1 = -\tfrac{1}{16}K_1, \quad \Lambda_2 = \tfrac{9}{16}K_2$$

$$a' = (\tfrac{1}{2} - \Omega_2^2\Lambda_2^2 - 16k_1^2 - \tfrac{1}{8}a^2)a + k_1 a^2 \cos\gamma_1 + \tfrac{1}{3}\Omega_2^3\Lambda_2^3 \cos\gamma_2$$

$$a\beta' = k_1 a^2 \sin\gamma_1 + \tfrac{1}{3}\Omega_2^3\Lambda_2^3 \sin\gamma_2$$

$$k_1 = \tfrac{1}{4}\Omega_1\Lambda_1, \quad \gamma_1 = \sigma_1 T_1 - 3\beta + \theta_1, \quad \gamma_2 = \sigma_2 T_1 - \beta + 3\theta_2$$

$$\epsilon\sigma_1 = \Omega_1 - 3, \quad \epsilon\sigma_2 = 3\Omega_2 - 1$$

(b) Show that stationary solutions exist only if $\beta' = \frac{1}{3}\sigma_1 = \sigma_2$. Is the corresponding response periodic?

4.31. The response of a self-sustaining system to a three-frequency excitation is governed by

$$\ddot{u} + u = \epsilon(\dot{u} - \tfrac{1}{3}\dot{u}^3) + \sum_{n=1}^{3} K_n \cos(\Omega_n t + \theta_n)$$

where Ω_1 and $2\Omega_2 + \Omega_3$ are near unity.

(a) Show that

$$u = a(T_1)\cos[T_0 + \beta(T_1)] + 2\sum_{n=2}^{3} \Lambda_n \cos(\Omega_n T_0 + \theta_n) + O(\epsilon)$$

where

$$\Lambda_n = \frac{K_n}{2(1 - \Omega_n^2)}$$

$$a' = \left(\tfrac{1}{2} - \sum_{n=2}^{3} \Omega_n^2 \Lambda_n^2 - \tfrac{1}{8}a^2\right) a + k_1 \sin \gamma_1 + k_2 \cos \gamma_2$$

$$a\beta' = -k_1 \cos \gamma_1 + k_2 \sin \gamma_2$$

$$2\epsilon k_1 = K_1, \quad k_2 = \Omega_2^2 \Omega_3 \Lambda_2^2 \Lambda_3$$

$$\gamma_1 = \sigma_1 T_1 - \beta + \theta_1, \quad \gamma_2 = \sigma_2 T_1 - \beta + 2\theta_2 + \theta_3$$

$$\epsilon\sigma_1 = \Omega_1 - 1, \quad \epsilon\sigma_2 = 2\Omega_2 + \Omega_3 - 1$$

(b) Show that stationary solutions exist only if $\sigma_1 = \sigma_2 = \beta'$. Is the corresponding response periodic?

(c) Obtain the frequency-response equation:

$$\sigma = \pm \left[\frac{k_1^2 + 2k_1 k_2 \sin \psi + k_2^2}{a^2} - \left(\tfrac{1}{2} - \sum_{n=2}^{3} \Omega_n^2 \Lambda_n^2 - \tfrac{1}{8}a^2\right)^2 \right]^{1/2}$$

where $\psi = \theta_1 - 2\theta_2 - \theta_3$. Plot several frequency–response curves, indicating the stable and unstable portions.

4.32. The response of a self-sustaining system to a three-frequency excitation is governed by

$$\ddot{u} + u = \epsilon(\dot{u} - \tfrac{1}{3}\mathring{u}^3) + \sum_{n=1}^{3} K_n \cos(\Omega_n T_0 + \theta_n) + O(\epsilon)$$

where Ω_1 is near unity and $\Omega_2 + \Omega_3$ is near 2.

(a) Show that

$$u = a(T_1) \cos [T_0 + \beta(T_1)] + 2\sum_{n=2}^{3} \Lambda_n \cos(\Omega_n T_0 + \theta_n) + O(\epsilon)$$

where

$$\Lambda_n = \frac{K_n}{2(1 - \Omega_n^2)}$$

$$a' = \left(\tfrac{1}{2} - \sum_{n=2}^{3} \Omega_n^2 \Lambda_n^2 - \tfrac{1}{8}a^2\right) a + k_1 \sin \gamma_1 - k_2 \cos \gamma_2$$

$$a\beta' = -k_1 \cos \gamma_1 - k_2 \sin \gamma_2$$

$$2\epsilon k_1 = K_1, \quad k_2 = \Omega_2 \Omega_3 \Lambda_2 \Lambda_3$$

$$\gamma_1 = \sigma_1 T_1 - \beta + \theta_1, \quad \gamma_2 = \sigma_2 T_1 - 2\beta + \theta_2 + \theta_3$$

$$\epsilon\sigma_1 = \Omega_1 - 1, \quad \epsilon\sigma_2 = \Omega_1 + \Omega_3 - 2$$

(b) Show that a stationary solution exists only if $\beta' = \sigma_1 = \frac{1}{2}\sigma_2$. Is the corresponding response periodic?

4.33. The response of a self-sustaining system to a three-frequency excitation is governed by

$$\ddot{u} + u = \epsilon(\dot{u} - \tfrac{1}{3}\dot{u}^3) + \sum_{n=1}^{3} K_n \cos(\Omega_n t + \theta_n)$$

where $\Omega_1 + \Omega_2 + \Omega_3$ is near unity.

(a) Show that

$$u = a(T_1)\cos[T_0 + \beta(T_1)] + 2\sum_{n=1}^{3} \Lambda_n \cos(\Omega_n T_0 + \theta_n) + O(\epsilon)$$

where

$$\Lambda_n = \frac{K_n}{2(1 - \Omega_n^2)}$$

$$a' = \left(\tfrac{1}{2} - \sum_{n=1}^{3} \Omega_n^2 \Lambda_n^2 - \tfrac{1}{8}a^2\right) a + k\cos\gamma$$

$$a\beta' = k\sin\gamma$$

$$k = 2\Omega_1\Omega_2\Omega_3\Lambda_1\Lambda_2\Lambda_3, \quad \gamma = \sigma_1 T_1 - \beta + \theta_1 + \theta_2 + \theta_3$$

$$\epsilon\sigma_1 = \Omega_1 + \Omega_2 + \Omega_3 - 1$$

(b) Obtain the frequency-response equation:

$$\sigma = \pm\left[\frac{k^2}{a^2} - \left(\tfrac{1}{2} - \sum_{n=1}^{3} \Omega_n^2\Lambda_n^2 - \tfrac{1}{8}a^2\right)^2\right]^{1/2}$$

(c) Plot several frequency-response curves, indicating the stable and unstable portions.

4.34. The response of a self-sustaining system to a three-frequency excitation is governed by

$$\ddot{u} + u = \epsilon(\dot{u} - \tfrac{1}{3}\dot{u}^3) + \sum_{n=1}^{3} K_n \cos(\Omega_n t + \theta_N)$$

where Ω_1 is near unity, Ω_2 is near $\frac{1}{3}$, and Ω_3 is near 3.

(a) Show that

$$u = a(T_1)\cos[T_0 + \beta(T_1)] + 2\sum_{n=2}^{3} \Lambda_n \cos(\Omega_n T_0 + \theta_n) + O(\epsilon)$$

where

$$\Lambda_n = \frac{K_n}{2(1 - \Omega_n^2)}$$

$$a' = \left(\tfrac{1}{2} - \sum_{n=2}^{3} \Omega_n^2 \Lambda_n^2 - \tfrac{1}{8}a^2\right) a + k_1 \sin \gamma_1 + k_2 \cos \gamma_2 + k_3 a^2 \cos \gamma_3$$

$$a\beta' = k_2 \sin \gamma_2 + k_3 a^2 \sin \gamma_3 - k_1 \cos \gamma_1$$

$$2\epsilon k_1 = K_1, \quad k_2 = \tfrac{1}{3}\Omega_2^3 \Lambda_2^3, \quad k_3 = \tfrac{1}{4}\Omega_3 \Lambda_3$$

$$\gamma_1 = \sigma_1 T_1 - \beta + \theta_1, \quad \gamma_2 = \sigma_2 T_1 - \beta + 3\theta_2$$

$$\gamma_3 = \sigma_3 T_1 - 3\beta + \theta_3, \quad \epsilon\sigma_1 + \Omega_1 - 1$$

$$\epsilon\sigma_2 = 3\Omega_2 - 1, \quad \epsilon\sigma_3 = \Omega_3 - 3$$

(b) Show that stationary solutions exist only if

$$\beta' = \sigma_1 = \sigma_2 = \tfrac{1}{3}\sigma_3$$

Is the corresponding response periodic?

4.35. The response of a self-sustaining system to a four-frequency excitation is governed by

$$\ddot{u} + u = \epsilon(\dot{u} - \tfrac{1}{3}\dot{u}^3) + \sum_{n=1}^{4} K_n \cos(\Omega_N T + \theta_n)$$

where Ω_1 and $\Omega_4 - \Omega_3 - \Omega_2$ are near unity.

(a) Show that

$$u = a(T_1)\cos[T_0 + \beta(T_1)] + 2\sum_{n=2}^{4} \Lambda_n \cos(\Omega_n T_0 + \theta_n) + O(\epsilon)$$

where

$$\Lambda_n = \frac{K_n}{2(1 - \Omega_n^2)}$$

$$a' = \left(\tfrac{1}{2} - \sum_{n=2}^{4} \Omega_n^2 \Lambda_n^2 - \tfrac{1}{8}a^2\right) a + k_1 \sin \gamma_1 + k_2 \cos \gamma_2$$

$$a\beta' = -k_1 \cos \gamma_1 + k_2 \sin \gamma_2$$

$$2\epsilon k_1 = K_1, \quad k_2 = 2\Omega_4 \Omega_3 \Omega_2 \Lambda_4 \Lambda_3 \Lambda_2$$

$$\gamma_1 = \sigma_1 T_1 - \beta + \theta_1, \quad \gamma_2 = \sigma_2 T_1 - \beta + \theta_4 - \theta_3 - \theta_2$$

$$\epsilon\sigma_1 = \Omega_1 - 1, \quad \epsilon\sigma_2 = \Omega_4 - \Omega_3 - \Omega_2 - 1$$

(b) Show that for a stationary solution to exist, $\beta' = \sigma_1 = \sigma_2$. Is the corresponding response periodic?

(c) Obtain the frequency-response equation. Plot several frequency–response curves, indicating the stable and unstable portions.

4.36. The response of a van der Pol oscillator with delayed amplitude limiting to a sinusoidal excitation is governed by (Golay, 1964)

$$\begin{aligned} \ddot{u} + \omega_0^2 u &= 2\epsilon[(1 - z)\dot{u} - \dot{z}u] - 2K\Omega \sin \Omega t \\ \tau\dot{z} + z &= u^2 \end{aligned} \tag{1}$$

where ω_0, ϵ, K, Ω, and τ are constants.

(a) For the case of primary resonance, let $\Omega = \omega_0 + \epsilon\sigma$ and $K = \epsilon k$. Use the method of multiple scales and obtain (Nayfeh, 1968)

$$u = a \cos (\omega_0 t + \beta) + O(\epsilon)$$

$$z = b \exp\left(\frac{-t}{\tau}\right) + \tfrac{1}{2}a^2 + \tfrac{1}{2}a^2(1 + 4\omega_0^2\tau^2)^{-1/2} \cos (2\omega_0 t + 2\beta - \tan^{-1} 2\omega_0\tau) + O(\epsilon) \tag{2}$$

where

$$\begin{aligned} \dot{a} &= \epsilon(1 - \tfrac{1}{4}\alpha_r a^2)a + \epsilon k \cos (\epsilon\sigma t - \beta) \\ \dot{\beta} &= -\tfrac{1}{4}\epsilon\alpha_i a^2 + \epsilon k a^{-1} \sin (\epsilon\sigma t - \beta). \end{aligned} \tag{3}$$

where α_r and α_i are known functions of $\omega_0 t$. Determine the stationary oscillations and their stability.

Forced oscillations of other third-order systems were studied by Srirangarajan and Srinivasan (1973, 1974) and Tondl (1974, 1976a).

(b) For the case of hard nonresonant excitations [i.e., $K = O(1)$, $\Omega - \omega_0 > O(\epsilon)$], show that

$$u = a \cos (\omega_0 t + \beta) - 2K\Omega(\omega_0^2 - \Omega^2)^{-1} \sin \Omega t + O(\epsilon) \tag{4}$$

where

$$\begin{aligned} \dot{a} &= (\eta - \tfrac{1}{4}\alpha_r a^2)a \\ \dot{\beta} &= -\tfrac{1}{4}\alpha_i a^2 \end{aligned} \tag{5}$$

and $\eta = 1 - 2K^2\Omega^2(\omega_0^2 - \Omega^2)^{-2}$. Solve for a and β. What are the stationary oscillations when $\eta > 0$, $\eta < 0$, and $\eta = 0$. What is the significance of $\eta = 0$?

(c) Determine a first-order solution for the subharmonic case (i.e., $\Omega \approx 3\omega_0$).

(d) Determine a first-order solution for the superharmonic case (i.e., $\Omega \approx \frac{1}{3}\omega_0$).

4.37. The forced response of a single-degree-of-freedom system is governed by (Arya, Bojadziev, and Farooqui, 1975)

$$\ddot{u} + \omega_0^2 u = -\epsilon\alpha u^3 - 2\epsilon\mu\dot{u} + \omega_0^2 f(T_1)$$

where $f(T_1) = O(1)$ and $T_1 = \epsilon t$.

(a) Show that

$$u = A(T_1) \exp(i\omega_0 T_0) + cc + f(T_1) + O(\epsilon)$$

where

$$2i\omega_0(A' + \mu A) + 3\alpha(A\bar{A} + f^2)A = 0$$

(b) Express A in the polar form $\frac{1}{2}a \exp(i\beta)$ and obtain

$$a' = -\mu a$$

$$\omega_0 \beta' = \tfrac{3}{2}\alpha(\tfrac{1}{4}a^2 + f^2)$$

4.38. The forced response of a self-excited system to a slowly varying external excitation is governed by

$$\ddot{u} + \omega_0^2 u = \epsilon(1 - u^2)\dot{u} + \omega_0^2 f(T_1)$$

where $f = O(1)$.

(a) Show that

$$u = A(T_1) \exp(i\omega_0 T_0) + cc + f(T_1) + O(\epsilon)$$

where

$$2A' = (1 - A\bar{A} - f^2)A$$

(b) Express A in polar form and determine the equations describing the amplitude and the phase.

4.39. The forced response of a self-excited system is governed by

$$\ddot{u} + \omega_0^2 u = \epsilon(1 - u^2)\dot{u} + K \cos \Omega t$$

where $K = O(1)$ and Ω is away from ω_0. Show that

$$u = A(T_1) \exp(i\omega_0 T_0) + \Lambda \exp(i\Omega T_0) + cc$$

where $\Lambda = \frac{1}{2}K(\omega_0^2 - \Omega^2)^{-1}$ and

(a) $2A' = A - 2\Lambda^2 A - A^2\bar{A}$ when Ω is away from 0, $3\omega_0$, and $\frac{1}{3}\omega_0$
(b) $2A' = A - 2\Lambda^2 A - A^2\bar{A} - \Omega\omega_0^{-1}\Lambda^3 \exp(-i\sigma T_1)$ when $\omega_0 = 3\Omega + \epsilon\sigma$
(c) $2A = A - 2\Lambda^2 - A^2\bar{A} + (2 - \Omega/\omega_0)\bar{A}^2\Lambda \exp(i\sigma T_1)$ when $\Omega = 3\omega_0 + \epsilon\sigma$
(d) $2A' = A - 2\Lambda^2 A - A^2\bar{A} - 2A\Lambda^2 \cos 2\sigma T_1$ when $\Omega = \epsilon\sigma$. Compare this result with that of Exercise 4.38.

4.40. The forced response of a single-degree-of-freedom system is governed by

$$\ddot{u} + \omega_0^2 u + 2\epsilon\mu\dot{u} + \epsilon u^4 = 2K \cos \Omega t$$

(a) Show that

$$u = A(T_1) \exp(i\omega_0 T_0) + \Lambda \exp(i\Omega T_0) + cc$$

where

(i) $2i\omega_0(A' + \mu A) + \Lambda^4 \exp(i\sigma T_1) = 0$ when $4\Omega = \omega_0 + \epsilon\sigma$
(ii) $2i\omega_0(A' + \mu A) + 4\Lambda\bar{A}^3 \exp(i\sigma T_1) = 0$ when $\Omega = 4\omega_0 + \epsilon\sigma$
(iii) $2i\omega_0(A' + \mu A) + 4\Lambda^3\bar{A} \exp(i\sigma T_1) = 0$ when $3\Omega = 2\omega_0 + \epsilon\sigma$
(iv) $2i\omega_0(A' + \mu A) + 6\Lambda^2\bar{A}^2 \exp(i\sigma T_1) = 0$ when $2\Omega = 3\omega_0 + \epsilon\sigma$
(v) $2i\omega_0(A' + \mu A) + 12(\Lambda^3\bar{A} + \Lambda\bar{A}^2 A) \exp(i\sigma T_1) = 0$ when $\Omega = 2\omega_0 + \epsilon\sigma$
(vi) $2i\omega_0(A' + \mu A) + (12\Lambda^2 A\bar{A} + 4\Lambda^4) \exp(i\sigma T_1) = 0$ when $2\Omega = \omega_0 + \epsilon\sigma$

(b) Determine the steady-state responses for each of the cases in (a).

4.41. The forced response of a single-degree-of-freedom system is given by (Proskuriakov, 1971)

$$\ddot{u} + \omega_0^2 u = -2\epsilon\mu\dot{u} + \epsilon\sum_{s=2}^{N} \alpha_s u^s + 2K \cos \Omega t$$

where Ω is away from ω_0.

(a) Determine all possible resonances for general N.
(b) When $N = 5$, determine the equations describing the amplitudes and the phases for all possible resonances.

4.42. The equation of motion of a gravity-stabilized, rigid satellite in an elliptic orbit around a spherical planet is

$$(1 + e \cos \theta)\psi'' - 2e\psi' \sin \theta + \tfrac{3}{2}K \sin 2\psi = 2e \sin \theta$$

Hablani and Shrivastava (1977) determined a third-order expansion for $\psi(\theta; e)$ for small e for ω_0 away from $\frac{1}{2}$ and 1, where $\omega_0^2 = 3K$. Determine all possible resonances to second order and determine uniform expansions for these cases (Alfriend, 1977).

CHAPTER 5

Parametrically Excited Systems

In this chapter, as in the preceding chapter, we consider motions that are the result of time-dependent excitations (actions) on the system. In contrast with the preceding and the following chapter, in which the excitations appear as inhomogeneities in the governing differential equations, in this chapter the excitations appear as coefficients in the governing differential equations. Thus mathematically one is led to differential equations with time-varying coefficients. In some branches of mechanics, one is led to the solution of partial differential equations with constant coefficients but spatially and/or temporally varying boundary conditions. Except for Exercises 5.27 and 5.28, we do not discuss problems with varying boundary conditions, and we refer the reader to the book of Brillouin (1956) and the detailed review article of Elachi (1976). Since the excitations when they are time independent appear as parameters in the governing equations, these excitations are called *parametric excitations*. Moreover in contrast with the case of external excitations in which a small excitation cannot produce a large response unless the frequency of the excitation is close to one of the natural frequencies of the system (primary resonance), a small parametric excitation can produce a large response when the frequency of the excitation is close to one half of one of the natural frequencies of the system (*principal parametric resonance*).

Faraday (1831) seems to be the first to observe the phenomenon of parametric resonance. He noted that surface waves in a fluid-filled cylinder under vertical excitation exhibited twice the period of the excitation itself. Melde (1859) tied a string between a rigid support and the extremity of the prong of a massive tuning fork of low pitch. He observed that the string could be made to oscillate laterally, although the exciting force is longitudinal, at one half the frequency of the fork under a number of critical conditions of string mass and tension and fork frequency and loudness. Strutt (1887) provided a theoretical basis for these observations and performed further experiments with a string attached to one end of the prong of a tuning fork. Stephenson (1906) amplified the results of Strutt (1887) and observed the possibility of exciting vibrations when the frequency of the applied axial excitation is a rational multiple of the fundamental frequency of the lateral vibration of the string. Raman (1912) presented a lengthy investigation which is beautifully and profusely illustrated with photographs of vibrating strings.

The problem of parametric resonance arises in many branches of physics and engineering. Examples are given in the next section. One of the important problems is that of dynamic instability which is the response of mechanical and elastic systems to time-varying loads, especially periodic loads. There are cases in which the introduction of a small vibrational loading can stabilize a system which is statically unstable or destabilize a system which is statically stable. Stephenson (1908) seems to be the first to point out that a column under the influence of a periodic load may be stable even though the steady value of the load is twice that of the Euler load. Beliaev (1924) analyzed the response of a straight elastic hinged-hinged column to an axial periodic load of the form $p(t) = p_0 + p_1 \cos \Omega t$. He obtained a Mathieu equation for the dynamic response of the column and determined the principal parametric resonance frequency of the column. The results show that a column can be made to oscillate with the frequency $\frac{1}{2}\Omega$ if it is close to one of the natural frequencies of the lateral motion even though the axial load may be below the static buckling load of the column. Beliaev's investigation was completed by Andronov and Leontovich (1927). Krylov and Bogoliubov (1935) used the Galérkin procedure to determine the dynamic response of a column with arbitrary boundary conditions to an axial periodic load, Chelomei (1939) studied the parametric resonance of a column, Kochin (1934) examined the mathematically related problem of the vibrations of a crankshaft, and Timoshenko (1955) and Bondarenko (1936) treated another mathematically related problem in connection with the vibrations of the driving system of an electric locomotive. Two basic references on the dynamic stability of elastic systems are the books of Bolotin (1963, 1964).

In spite of the relatively new history of the problem of parametric excitations, there are a number of books devoted to the analysis and applications of this problem. McLachlan (1947) discussed the theory and applications of the Mathieu functions, while Bondarenko (1936) and Magnus and Winkler (1966) discussed Hill's equation and its applications in engineering vibration problems. Bolotin (1964) discussed the influence of parametric resonances on the dynamic stability of elastic systems. Shtokalo (1961) discussed linear differential equations with variable coefficients; Arscott (1964), Erugin (1966), and Yakubovich and Starzhinskii (1975) discussed differential equations with periodic coefficients; and Schmidt (1974) discussed parametric resonances. In addition there are a number of books which deal with parametric excitations including those of Whittaker and Watson (1962); Den Hartog (1947); Minorsky (1947, 1962); Stoker (1950); Bellman (1953); Hayashi (1953a, 1964); Coddington and Levinson (1955); Malkin (1956); Cunningham (1958); Kauderer (1958); Bogoliubov and Mitropolsky (1961); Struble (1962); Hale (1963); McLachlan (1950);

Andronov, Vitt, and Khaikin (1966); Blaquière (1966); Kononenko (1969); Meirovitch (1970); Cesari (1971); Nayfeh (1973b); and Evan-Iwanowski (1976). The problem of parametric resonance and its associated problem of dynamic stability were reviewed by Beilin and Dzhanelidze (1952), Mettler (1962, 1967), and Evan-Iwanowski (1965).

First we consider some examples of parametrically excitated systems; then we consider the Floquet theory to obtain some characteristics that are common to all linear, parametrically excited systems, and we develop approximate solutions of linear systems having a single degree of freedom. In Section 5.4 we extend the analysis to linear systems having many degrees of freedom and distinct eigenfrequencies, while in Section 5.5 we consider systems having repeated eigenfrequencies. In Section 5.6 we consider linear, gyroscopic, parametrically excited systems. In the last section we consider nonlinear, parametrically excited systems.

5.1. Examples

5.1.1 A PENDULUM WITH A MOVING SUPPORT

As a first example we consider the motion of a particle of mass m attached to one end of a massless rod of length l, while the other end of the rod is attached to a point under the influence of a prescribed acceleration as shown in Figure 5-1. Applying Newton's second law of motion in the direction perpendicular to the rod leads to

$$ml\ddot{\theta} = -m[g - Y(t)] \sin\theta + mX(t)\cos\theta$$

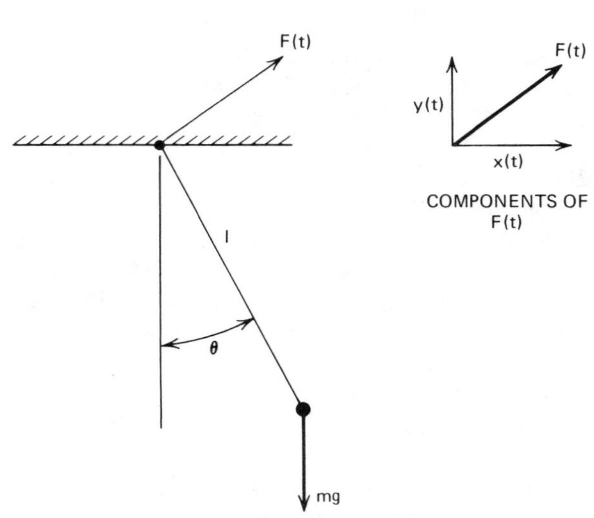

Figure 5-1. Pendulum with a moving support.

Hence

$$\ddot{\theta} + \left[\frac{g}{l} - \frac{Y(t)}{l}\right] \sin \theta - \frac{X(t)}{l} \cos \theta = 0 \tag{5.1.1}$$

which is an equation with variable coefficients. For small oscillations about $\theta = 0$, (5.1.1) can be linearized to yield

$$\ddot{\theta} + \left[\frac{g}{l} - \frac{Y(t)}{l}\right] \theta = \frac{X(t)}{l} \tag{5.1.2}$$

Stephenson (1908) seems to be the first to predict the possibility of converting the unstable equilibrium of a rigid rod standing on an end by applying a vertical periodic force at the bottom. Sethna (1973) showed that a pendulum can have stable motions in the neighborhood of the vertical up position for arbitrary vertical support motions provided that they are fast and the time average of the square of the velocity of the support motions is greater than the square of the time average of the velocity of these motions by a constant that depends on the system parameters. Consequently linear and nonlinear parametric excitations of a pendulum with a moving point of suspension were studied by many investigators including Hirsch (1930), Stoker (1950, pp. 189–213), Haacke (1951), Kapitza (1951), Malkin (1956, pp. 163–165), Kauderer (1958, pp. 524–536), Skalak and Yarymovych (1960), Bogoliubov and Mitropolsky (1961, pp. 404–408), Struble (1963), Phelps and Hunter (1965, 1966), Bogdanoff and Citron (1965), Ness (1967), Dugundji and Chhatpar (1970), Cheshankov (1971), Troger (1975), and Chester (1975). Tso and Asmis (1970) studied the parametric excitation of a pendulum with bilinear hysteresis, Ryland and Meirovitch (1977) studied the stability boundaries of a swinging spring with an oscillating support, Hemp and Sethna (1964) studied the effect of high-frequency support oscillations on the motion of a spherical pendulum, and Sethna and Hemp (1965) studied the nonlinear oscillations of a gyroscopic pendulum with an oscillating support.

5.1.2. A MECHANICAL-ELECTRICAL SYSTEM

As a second example we consider the mechanical-electrical system shown in Figure 5-2. It consists of an L-C circuit containing a constant inductor L connected in series with a capacitor whose plates can be moved mechanically in a

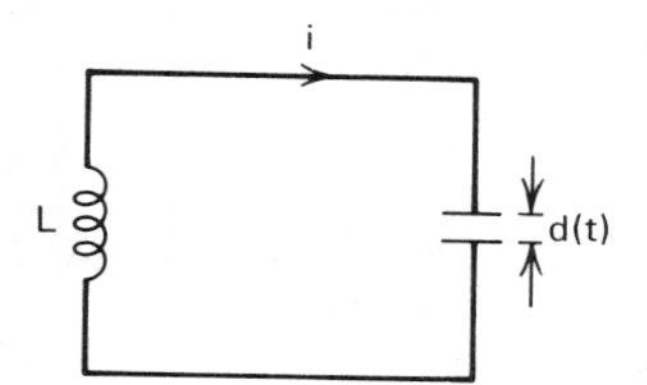

Figure 5-2. A mechanical-electrical system.

prescribed manner. If the charge on the capacitor is q, then the current i in the circuit is $\dot{q}$. The potential across the inductance is $L di/dt = L\dot{q}$, while the potential across the condenser is $q/C(t)$. The capacitance $C(t) = \epsilon S/d(t)$ in the MKS system, where ϵ is the dielectric permeativity of the material, S is the surface area of the plates, and $d(t)$ is the variable distance between the plates. Since the total potential across the two elements is zero, q is governed by

$$\ddot{q} + \frac{d(t)}{\epsilon S L} q = 0 \tag{5.1.3}$$

which is an equation with variable coefficients.

Brillouin (1897) studied parametric resonances in electric circuits. Similar experiments were performed by Mandelstam and Papalexi (1934) with a specially designed oscillating circuit which they called a parametric generator. They found out that if the circuit of the parametric generator is linear, the amplitude of the oscillation grows indefinitely until the insulation is destroyed by an excessive voltage. On the other hand, they found out that a stable stationary condition is reached if the circuit is nonlinear. For more references and applications to electric circuits, we refer the reader to the books of Minorsky (1962); Andronov, Vitt, and Khaikin (1966); and Blaquière (1966). Moreover for a history of parametric transducers, we refer the reader to Mumford (1960).

5.1.3. A DOUBLE PENDULUM

The two examples described above are systems having a single degree of freedom. In this section we describe a double pendulum, which has two degrees of freedom, and in the following section we consider a column experiencing transverse oscillations. The latter is an example of a system having infinite degrees of freedom.

We consider the motion of a double pendulum attached to a platform that has a prescribed vertical motion relative to an inertial frame as shown in Figure 5-3. Two particles of masses m_1 and m_2 are connected to massless rods of lengths l_1 and l_2 suspended from a platform that has a prescribed vertical motion $y(t)$ with respect to the inertial frame 0. The motion of the particles is constrained by springs that are initially horizontal and have the constants k_1 and k_2. The springs are unstretched when the particles lie vertically below the platform.

To derive the equations of motion, we form the Lagrangian and then write the Euler-Lagrange equations. To this end we observe that the velocities of m_1 and m_2 are

$$\mathbf{v}_1 = l_1\dot{\theta}_1 \cos\theta_1 \mathbf{i} + (l_1\dot{\theta}_1 \sin\theta_1 - \dot{y})\mathbf{j} \tag{5.1.4}$$

$$\mathbf{v}_2 = (l_1\dot{\theta}_1 \cos\theta_1 + l_2\dot{\theta}_2 \cos\theta_2)\mathbf{i} + (l_1\dot{\theta}_1 \sin\theta_1 + l_2\dot{\theta}_2 \sin\theta_2 - \dot{y})\mathbf{j} \tag{5.1.5}$$

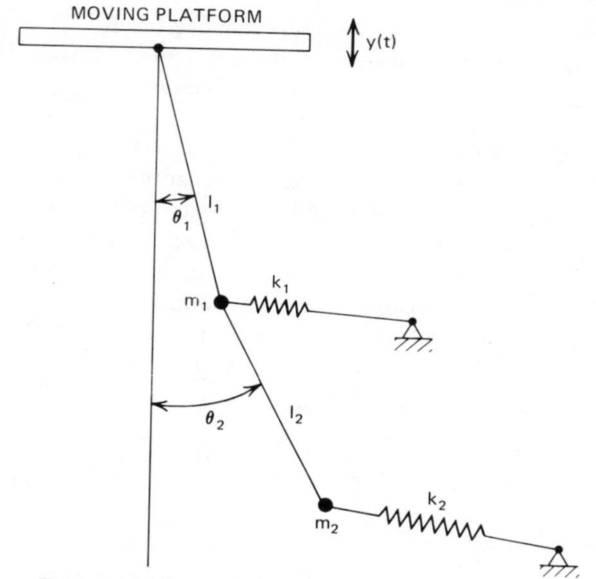

Figure 5-3. Double pendulum with a moving support.

The kinetic energy T and the potential energy V of the system are

$$T = \tfrac{1}{2}m_1 v_1^2 + \tfrac{1}{2}m_2 v_2^2 = \tfrac{1}{2}(m_1 + m_2) l_1^2 \dot{\theta}_1^2 - (m_1 + m_2)\, l_1 \dot{y} \dot{\theta}_1 \theta_1 + \tfrac{1}{2} m_2 l_2^2 \dot{\theta}_2^2$$
$$+ m_2 l_1 l_2 \dot{\theta}_1 \dot{\theta}_2 - m_2 l_2 \dot{y} \dot{\theta}_2 \theta_2 + \tfrac{1}{2}(m_1 + m_2)\, \dot{y}^2 + O(\theta_j^3) \quad (5.1.6)$$

$$V = -m_1 g(y + l_1 \cos \theta_1) - m_2 g(y + l_1 \cos \theta_1 + l_2 \cos \theta_2)$$
$$+ \tfrac{1}{2} k_1 l_1^2 \theta_1^2 + \tfrac{1}{2} k_2 (l_1 \theta_1 + l_2 \theta_2)^2 + O(\theta_j^3) \quad (5.1.7)$$

The cubic and higher-order terms in T and V are not needed in the linear problem. Expanding (5.1.7) for small θ and neglecting cubic and higher-order terms, we have

$$V = c(t) + \tfrac{1}{2} m_1 g l_1 \theta_1^2 + \tfrac{1}{2} m_2 g (l_1 \theta_1^2 + l_2 \theta_2^2) + \tfrac{1}{2} k_1 l_1^2 \theta_1^2$$
$$+ \tfrac{1}{2} k_2 (l_1 \theta_1 + l_2 \theta_2)^2 + O(\theta_j^3) \quad (5.1.8)$$

Then it follows that the equations of motion are

$$(m_1 + m_2)\, l_1 \ddot{\theta}_1 + m_2 l_2 \ddot{\theta}_2 - (m_1 + m_2)\, \ddot{y} \theta_1 + [(m_1 + m_2)\, g$$
$$+ (k_1 + k_2)\, l_1]\, \theta_1 + k_2 l_2 \theta_2 = 0 \quad (5.1.9)$$

$$m_2 l_2 \ddot{\theta}_2 + m_2 l_1 \ddot{\theta}_1 - m_2 \ddot{y} \theta_2 + k_2 l_1 \theta_1 + (m_2 g + k_2 l_2)\, \theta_2 = 0 \quad (5.1.10)$$

which are two linear equations having variable coefficients. Thus this is an example of a system having two degrees of freedom. Hsu and Cheng (1973) studied the effect of an axial impact load on a double pendulum.

5.1.4. DYNAMIC STABILITY OF ELASTIC SYSTEMS

As a fourth example we consider the transverse motion of a straight rod with a uniform cross section loaded by an axial time-varying force $P(t)$. We consider four typical boundary conditions as shown in Figure 5-4. We assume that plane sections remain plane, and we neglect transverse shear and rotary inertia. Assuming linear elasticity, one finds that the longitudinal inertia terms are negligible. Hence the axial force in the beam is uniform and equal to $P(t)$. Referring to Figure 5-5 one can write the pertinent equations of motion as

y-Momentum

$$-P\frac{\partial\theta}{\partial x}+\frac{\partial Q}{\partial x}=\rho A\frac{\partial^2 w}{\partial t^2} \tag{5.1.11}$$

Moment of momentum

$$-\frac{\partial M}{\partial x}+Q=0 \tag{5.1.12}$$

where ρ is the density per unit length and A is the cross-sectional area. Using

$$M=-EI\frac{\partial\theta}{\partial x} \quad \text{and} \quad \theta\approx\frac{\partial w}{\partial x}$$

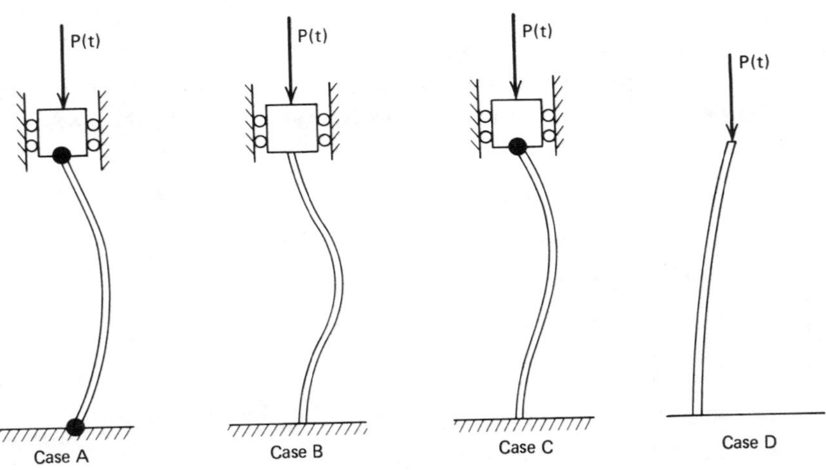

Figure 5-4. Dynamic stability of elastic columns.

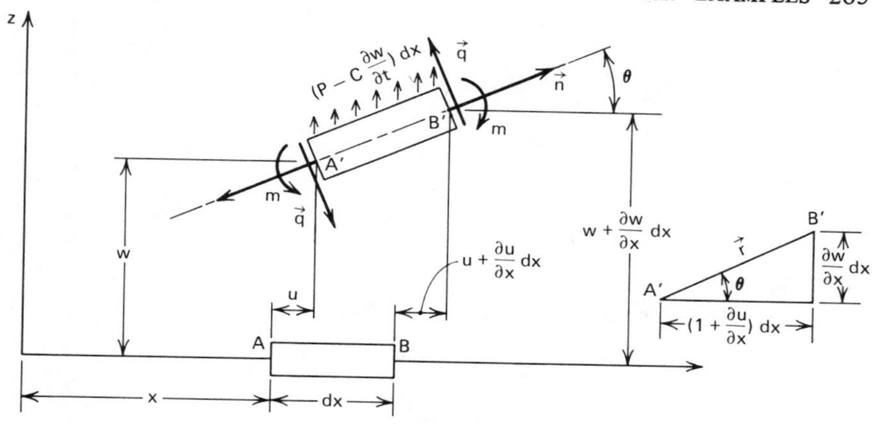

Figure 5-5. An element of a beam.

We can combine (5.1.11) and (5.1.12) to obtain

$$\rho A \frac{\partial^2 w}{\partial t^2} + P(t) \frac{\partial^2 w}{\partial x^2} + EI \frac{\partial^4 w}{\partial x^4} = 0 \tag{5.1.13}$$

In the following we assume that $P(t) = P_0 + P_1(t)$, where P_0 is constant. We follow Nayfeh and Mook (1977) and express the solution of (5.1.13) as an expansion in terms of the linear free-oscillation modes. That is,

$$w(x, t) = \sum_m u_m(t)\, \phi_m(x) \tag{5.1.14}$$

where the ϕ_m are the eigenfunctions of the problem

$$\phi^{iv} + p_0 \phi'' - \kappa^4 \phi = 0 \tag{5.1.15}$$

where $p_0 = P_0/EI$, $\kappa^4 = \rho A \omega^2/EI$, and ω is an eigenvalue (called a natural frequency), subject to one of the following sets of boundary conditions:

$$\phi = \phi'' = 0 \quad \text{at} \quad x = 0 \quad \text{and} \quad x = l \tag{5.1.16}$$

for case (a),

$$\phi = \phi' = 0 \quad \text{at} \quad x = 0 \quad \text{and} \quad x = l \tag{5.1.17}$$

for case (b),

$$\phi = \phi' = 0 \quad \text{at} \quad x = 0 \quad \text{and} \quad \phi = \phi'' = 0 \quad \text{at} \quad x = l \tag{5.1.18}$$

for case (c), and

$$\phi = \phi' = 0 \quad \text{at} \quad x = 0 \quad \text{and} \quad \phi'' = \phi''' = 0 \quad \text{at} \quad x = l \tag{5.1.19}$$

for case (d). One can easily show that the resulting eigenfunctions are orthogonal.

Substituting (5.1.14) into (5.1.13) yields

$$\sum_m [(\ddot{u}_m + \omega_m^2 u_m)\,\phi_m + p(t)\,u_m \phi_m''] = 0 \tag{5.1.20}$$

where

$$\omega_m^2 = \frac{EI\kappa_m^4}{\rho A}, \qquad p(t) = \frac{P_1(t)}{\rho A} \tag{5.1.21}$$

Multiplying (5.1.20) by ϕ_n, integrating the result from $x = 0$ to $x = l$, and using the orthogonality property of the ϕ_m, we obtain

$$\ddot{u}_n + \omega_n^2 u_n + p(t) \sum_m f_{nm} u_m = 0, \qquad n = 1, 2, 3, \ldots \tag{5.1.22}$$

where

$$f_{nm} = \left[\int_0^l \phi_n \phi_m''\, dx\right]\left[\int_0^l \phi_n^2\, dx\right]^{-1} \tag{5.1.23}$$

Equations (5.1.22) are an infinite set of coupled linear equations having variable coefficients. Thus this is an example of a parametrically excited system having infinite degrees of freedom.

We note that for a hinged-hinged column $\phi_m = \sin(m\pi x/l)$, and hence $f_{nm} = 0$ unless $n = m$. Consequently the system of equations (5.1.22) is uncoupled. However for the other boundary conditions, the system of equations is coupled. For general dynamic systems, Chelomei (1939) showed that the problem can be reduced to a system of coupled differential equations with variable coefficients. Brachkovskii (1942) and Bolotin (1953), using respectively the Galérkin procedure and the method of integral equations, discovered a class of problems that can be reduced exactly to a single second-order equation (i.e., the system of equations is uncoupled). This result was generalized by Dzhanelidze (1953, 1955) to the case of dissipative systems. In addition to the expansion in terms of the unperturbed natural modes mentioned above, a number of alternate approaches have been employed including the Galérkin procedure (Krylov and Bogoliubov, 1935; Iwatsubo, Sugiyama, and Ogino, 1974), analog and digital simulations (Moody, 1967; Sugiyama, Fujiwara, and Sekiya, 1970; Iwatsubo, Sugiyama, and Ishihara, 1972) and finite differences (Sugiyama, Katayama, and Sekiya, 1971; Iwatsubo, Saigo, and Sugiyama, 1973).

The problem of the transverse vibrations of a column with a time-dependent follower force is a part of the problem of *dynamic stability*. Mettler (1949, 1951) laid the foundation for analysing parametric responses of mechanical systems. Bernstein (1947) presented a formulation of the problem of dynamic stability, while Smirnov (1947) and Bolotin (1963, 1964) provided extensive studies of problems of elastic stability of various structures including columns, arches, rings, plates, and shells. Mettler (1962, 1967) and Evan-Iwanowski (1965, 1976) presented surveys of the state of the art.

As mentioned in the introduction, Stephenson (1908) pointed out that a column under the influence of an axial periodic load can be stable even though the steady value of the load is twice the Euler buckling load. Beliaev (1924) determined the principal parametric resonance for a hinged-hinged beam, Andronov and Leontovich (1927) completed Beliaev's analysis, and Lubkin and Stoker (1943) and Mettler (1940) presented detailed analyses of this problem. These results were verified experimentally by Gol'denblat (1947), Bolotin (1951), and Somerset and Evan-Iwanowski (1965). Krylov and Bogoliubov (1935) studied columns with various boundary conditions under the influence of multiharmonic axial forces. Mettler (1947) studied analytically while Burnashev (1954) and Sobolev (1954) studied experimentally the dynamic stability of the plane bending of a beam. Weidenhammer (1951) studied the stability of a clamped-clamped column, and Elmaraghy and Tabarrok (1975) studied the stability of an axially oscillating column. As mentioned earlier, the dynamic response of a column with boundary conditions other than hinged-hinged leads to a system of coupled equations with periodic coefficients. In addition to the usual resonances involving one degree of freedom, there exist combination and simultaneous resonances. These were studied analytically by Mettler (1949, 1967); Weidenhammer (1951); Sugiyama, Fugiwara, and Sekiya (1970); Sugiyama, Katayama, and Sekiya (1971); Iwatsubo, Sugiyama, and Ishihara (1972); Iwatsubo, Sugiyama, and Ogino (1974); and Nayfeh and Mook (1977) and were demonstrated experimentally by Iwatsubo, Saigo, and Sugiyama (1973) and Dugundji and Mukhopadhyay (1973).

Chelomei (1939) treated the case of time-varying loads distributed along the length of a column, Bondarenko (1936) and Schmidt (1964) examined the combined effect of longitudinal and lateral forces, Mettler and Weidenhammer (1956) studied the effect of an end mass, and Evensen and Evan-Iwanowski (1966) studied analytically and experimentally the effect of concentrated and distributed masses. Makushin (1947) treated the case of piecewise constant periodic loadings, Gastev (1949) treated the case of periodically repeated pulses, Finizio (1974) treated the case of periodic forces of the impulsive type, Infante and Plaut (1969) treated the case of a general time-dependent axial load, and Caughey and Gray (1964) treated the case of a random loading. Moody (1967) treated the case of imperfect columns, while Ahuja and Duffield (1975) treated the case of a column having a variable cross section and resting on an elastic foundation. Gol'denblat (1947), Bolotin (1964, pp. 291–304), Ghobarah and Tso (1972), Popelar (1972), and Ali Hasan and Barr (1974) treated columns of thin-walled sections.

The influence of damping on the boundaries of the instability was discussed by

Mettler (1941), Naumov (1946), Weidenhammer (1951), Grammel (1952), Schmidt and Weidenhammer (1961), Schmidt (1961a, b, 1974), Piszczek (1961), Bolotin (1964), Stevens (1966), Stevens and Evan-Iwanowski (1969), Mozer and Evan-Iwanowski (1972), and Evan-Iwanowski (1976). In most cases the damping forces are stabilizing. However Schmidt and Weidenhammer (1961), Piszczek (1961), and Valeev (1963) showed that for certain combination resonances the damping forces may alter a stable state into an unstable one. Moreover Stevens (1966) showed that some viscoelastic materials are destabilizing.

The linear theory is capable of determining the regions in which a small motion becomes dynamically unstable, and it predicts that the unstable motions grow without bound. However as the amplitude of the motion grows, the nonlinear effects come into play and limit the growth. Gol'denblat (1947) seems to be the first to point out the inadequacy of the linear theory for predicting the amplitudes in the unstable regions. Bolotin (1951, 1964), Weidenhammer (1952, 1956), Piszczek (1955), Tso and Caughey (1965), Sethna (1965), Mettler and Weidenhammer (1956), and Evan-Iwanowski, Sanford, and Kehagioglou (1970) treated the nonlinear dynamic problem of compressed columns. Tso (1968) studied the problem of longitudinal-torsional stability, while Mettler (1955) and Ghobarah and Tso (1972) studied the problem of bending-torsional stability of thin-walled beams. Evensen and Evan-Iwanowski (1966) studied analytically and experimentally the effect of midplane stretching, Evan-Iwanowski (1976) studied in detail columns as well as other elastic systems, and Tezak, Mook, and Nayfeh (1977) studied analytically the effect of midplane stretching, taking into account the effect of internal resonances. Hsu (1975b) analyzed the response of a parametrically excited string hanging in a fluid.

Schmidt (1961c) studied the lateral vibrations of a slightly curved bar under the influence of periodic eccentric loads. Malkina (1953) studied the dynamic stability of arches under the influence of longitudinal periodic loads, while Bolotin (1964, pp. 316–332) studied analytically and experimentally the linear and nonlinear dynamic stability of arches loaded by compression and bending. Salion (1956) studied arches under the influence of periodic moments, Bondar (1953) treated parabolic arches, and Dzhanelidze and Radstig (1940) and Woinowsky-Krieger (1942) studied the parametric resonance of rings. Schmidt (1963) and Bolotin (1964, pp. 358–381) studied the dynamic stability of trusses.

Einaudi (1936) seems to be the first to treat the response of a plate to periodic in-plane loads. Subsequently Chelomei (1939), Bodner (1938), Khalilov (1942), Kucharski (1950), Berezovskii and Shulezhko (1963), Bolotin (1964, Chapter 21), Somerset (1967), Somerset and Evan-Iwanowski (1967), Simons and Leissa (1971), and King and Lin (1974) studied the dynamic stability of isotropic plates under the influence of in-plane periodic loads. Ambartsumyan and Khachaturian (1960) treated anisotropic plates, while Duffield and Willems (1972) treated stiffened rectangular plates. Dzygadlo (1965), Dzygadlo and

Krzyanowski (1972), and Dzygadlo and Wielgus (1974) studied aeroelastic systems in supersonic flow.

Oniashvili (1951) presented a broad discussion of the dynamic stability of shells, Bolotin (1964, Chapter 22) gave an extensive treatment of the dynamic stability of shallow, cylindrical, and spherical shells, while Hsu (1974b) gave a review of the parametric excitation and snap-through instability phenomenon of shells. Ghobarah (1972) treated the nonlinear dynamic stability of monosymmetrical thin-walled structures. Bublik and Merkulov (1960) and Kana and Craig (1968) studied the dynamic stability of thin elastic shells filled with fluid, while Markov (1949) studied the dynamic stability of anisotropic shells. Federhofer (1954); Wenzke (1963); Yao (1963, 1965); Bieniek, Fan, and Lackman (1966); Vijayaraghavan and Evan-Iwanowski (1967); Adams and Evan-Iwanowski (1973); Popov, Antipov, and Krzhechkovskii (1973); and Vol'mir and Ponomarev (1973) studied cylindrical shells. Tani (1974, 1976) studied the dynamic stability of conical shells.

A number of physical systems contain pipes conveying fluid. The fluid velocity often has an unsteady component induced by the pumps. Thus parametric and combination instabilities might occur in such pipes. These were studied theoretically by Chen (1971), Ginsberg (1973), Paidoussis and Issid (1974), Bohn and Herrmann (1974), and Paidoussis and Sundararajan (1975) and experimentally by Paidoussis and Issid (1976). Beal (1965) studied the dynamic stability of a flexible missile under the influence of a pulsating thrust, while Ibrahim and Barr (1975a, b) studied autoparamteric resonances in structures containing liquids.

The coupled flap-lag and coupled flap-lag-torsional aeroelastic problems of rotary-wing systems were reviewed by Friedmann (1977) and studied by Horvay and Yuan (1947); Sissingh (1968); Sissingh and Kuczynski (1970); Peters and Hohenemser (1971); Hohenemser and Yin (1972); Friedmann and Tong (1973); Hammond (1974); Friedmann and Silverthorn (1974, 1975); Huber and Strehlow (1976); and Friedmann, Hammond, and Woo (1977). A detailed treatment of machinery and their parts is contained in the monograph of Tondl (1965). Wehrli (1963) studied parametric resonances in torsional and rotary motions. Ehrich (1971) observed combinational resonances in machinery. Naguleswaran and Williams (1968) and Rhodes (1971) treated belts, Mote (1968) treated an axially moving band, and Benedetti (1974) studied the dynamic stability of a beam loaded by a sequence of moving mass particles. Davydov (1970), Ho and Lai (1970), and Grybos (1972) treated gears, while Houben (1970) treated piston engines. Messal and Bonthron (1972) analytically and experimentally found combinational resonances in an asymmetric shaft. Rotating shafts were studied also by Smith (1933); Kellenberger (1955); Hull (1961); Dimentberg (1961); Bishop and Parkinson (1965); Black and McTernan (1968); and Iwatsubo, Tomito, and Kawai (1973).

5.1.5 STABILITY OF STEADY-STATE SOLUTIONS

The techniques available for determining the steady-state behavior of free and forced oscillations of dynamic and elastic systems can be divided into two groups. The first group includes the methods of averaging and multiple scales. With this group one determines first the equations describing the amplitudes and the phases. These equations are transformed into an autonomous system. Then the steady-state solutions correspond to the singular points of this autonomous system and the stability of these solutions corresponds to the stability of the singular points. The second group includes the Linstedt-Poincaré technique, the method of harmonic balance and the Galérkin procedure. With this group one determines directly the steady-state solutions and one investigates their stability by analyzing the solutions of the variational equations. Next we explain these points by using the forced Duffing equation as an example.

We consider the superharmonic response of a single-degree-of-freedom system that is governed by

$$\ddot{u} + \omega_0^2 u = -2\epsilon\mu\dot{u} - \epsilon\alpha u^3 + K \cos \Omega t \tag{5.1.24}$$

where $3\Omega = \omega_0 + \epsilon\sigma$. Using the method of multiple scales we find, as in Section 4.1.3, the following equations describing the response:

$$u = a \cos (\omega_0 t + \beta) + 2\Lambda \cos \Omega t + O(\epsilon) \tag{5.1.25}$$

where $\Lambda = \frac{1}{2}K(\omega_0^2 - \Omega^2)^{-1}$ and

$$\begin{aligned} a' &= -\mu a - \frac{\alpha\Lambda^3}{\omega_0} \sin (\sigma T_1 - \beta) \\ a\beta' &= \frac{3\alpha}{\omega_0}(\Lambda^2 + \tfrac{1}{8}a^2)a + \frac{\alpha\Lambda^3}{\omega_0} \cos (\sigma T_1 - \beta) \end{aligned} \tag{5.1.26}$$

To determine the steady-state response we first transform (5.1.26) from a non-autonomous system to an autonomous system by introducing the new variable

$$\gamma = \sigma T_1 - \beta \tag{5.1.27}$$

Eliminating β from (5.1.25) through (5.1.27) gives

$$u = a \cos (3\Omega t - \gamma) + 2\Lambda \cos \Omega t + O(\epsilon) \tag{5.1.28}$$

$$\begin{aligned} a' &= -\mu a - \frac{\alpha\Lambda^3}{\omega_0} \sin \gamma \\ a\gamma' &= \left(\sigma - \frac{3\alpha\Lambda^2}{\omega_0}\right)a - \frac{3\alpha}{8\omega_0}a^3 - \frac{\alpha\Lambda^3}{\omega_0} \cos \gamma \end{aligned} \tag{5.1.29}$$

Then the steady-state responses correspond to the singular (stationary) points of the system (5.1.29); that is, they correspond to the solutions of

$$-\mu a_0 - \frac{\alpha\Lambda^3}{\omega_0}\sin\gamma_0 = 0$$
$$\left(\sigma - \frac{3\alpha\Lambda^2}{\omega_0}\right)a_0 - \frac{3\alpha}{8\omega_0}\alpha_0^3 - \frac{\alpha\Lambda^3}{\omega_0}\cos\gamma_0 = 0 \tag{5.1.30}$$

Eliminating γ_0 from (5.1.30) leads to the frequency-response equation

$$\mu^2 a_0^2 + a_0^2\left[\sigma - \frac{3\alpha\Lambda^2}{\omega_0} - \frac{3\alpha}{8\omega_0}a_0^2\right]^2 = \frac{\alpha^2\Lambda^6}{\omega_0^2} \tag{5.1.31}$$

The stability of the steady-state solutions corresponds to the stability of the singular points. As in Chapter 3, the types of the singular points and hence their stability can be determined by superposing small perturbations on the singular-point solutions, that is, by letting

$$a = a_0 + a_1, \qquad \gamma = \gamma_0 + \gamma_1 \tag{5.1.32}$$

where a_1 and γ_1 are small compared with a_0 and γ_0. Substituting (5.1.32) into (5.1.29) and linearizing the resulting equations in a_1 and γ_1, we obtain

$$a_1' = -\mu a_1 - \frac{\alpha\Lambda^3}{\omega_0}\gamma_1\cos\gamma_0$$
$$\gamma_1' = -\frac{3\alpha}{8\omega_0}a_0 a_1 + \frac{\alpha\Lambda^3}{\omega_0 a_0^2}a_1\cos\gamma_0 + \frac{\alpha\Lambda^3}{\omega_0 a_0}\gamma_1\sin\gamma_0 \tag{5.1.33}$$

We seek a solution for (5.1.33) in the form

$$a_1 = a_{10}\exp(\lambda T_1) \quad \text{and} \quad \gamma_1 = \gamma_{10}\exp(\lambda T_1) \tag{5.1.34}$$

where a_{10}, γ_{10} and λ are constants. Hence

$$(\lambda + \mu)a_{10} + \left(\frac{\alpha\Lambda^3}{\omega_0}\cos\gamma_0\right)\gamma_{10} = 0$$
$$\left(\frac{3\alpha a_0}{8\omega_0} - \frac{\alpha\Lambda^3}{\omega_0 a_0^2}\cos\gamma_0\right)a_{10} + \left(\lambda - \frac{\alpha\Lambda^3}{\omega_0 a_0}\sin\gamma_0\right)\gamma_{10} = 0 \tag{5.1.35}$$

For a nontrivial solution the determinant of the coefficient matrix must vanish. Using the first of (5.1.30), we write this condition as

$$(\lambda + \mu)^2 = -\frac{\alpha\Lambda^3}{\omega_0}\cos\gamma_0\left(\frac{\alpha\Lambda^3}{\omega_0 a_0^2}\cos\gamma_0 - \frac{3\alpha a_0}{8\omega_0}\right) \tag{5.1.36}$$

Then the stability of the singular points and hence the steady-state solutions depends on the real parts of the roots of (5.1.36). If the real part of each root is negative or zero, the corresponding steady-state solution is stable. If the real part of at least one of the roots is positive definite, the corresponding steady-state solution is unstable.

The stability of the steady-state solutions in the large can also be studied using equations such as (5.1.29) in conjunction with a phase diagram in the case of a low dimensional problem or a Liapunov function if it can be found (Malkin, 1944; Sethna, 1973). These are not pursued further in this section.

As a representative of the second group, we discuss the method of harmonic balance. Thus we substitute a steady-state solution in the form

$$u = u_0 = A_1 \cos \Omega t + B_1 \sin \Omega t + A_3 \cos 3\Omega t + B_3 \sin 3\Omega t \tag{5.1.37}$$

into (5.1.24) and obtain

$$\begin{aligned}(\omega_0^2 - \Omega^2)A_1 \cos \Omega t + (\omega_0^2 - \Omega^2)B_1 \sin \Omega t + (\omega_0^2 - 9\Omega^2)A_3 \cos 3\Omega t \\ + (\omega_0^2 - 9\Omega^2)B_3 \sin 3\Omega t = 2\epsilon\mu\Omega A_1 \sin \Omega t - 2\epsilon\mu\Omega B_1 \cos \Omega t \\ + 6\epsilon\mu\Omega A_3 \sin 3\Omega t - 6\epsilon\mu\Omega B_3 \cos 3\Omega t - \epsilon\alpha[A_1 \cos \Omega t + B_1 \sin \Omega t \\ + A_3 \cos 3\Omega t + B_3 \sin 3\Omega t]^3 + K \cos \Omega t\end{aligned} \tag{5.1.38}$$

Expanding the term in the square brackets and equating the coefficients of each of $\cos \Omega t$, $\sin \Omega t$, $\cos 3\Omega t$, and $\sin 3\Omega t$ on both sides of (5.1.38), we obtain

$$\begin{aligned}(\omega_0^2 - \Omega^2)A_1 = K - 2\epsilon\mu\Omega B_1 - \tfrac{3}{4}\epsilon\alpha[A_1^3 + A_1B_1^2 \\ + (A_1^2 - B_1^2)A_3 + 2A_1B_1B_3 + 2(A_3^2 + B_3^2)A_1]\end{aligned} \tag{5.1.39}$$

$$\begin{aligned}(\omega_0^2 - \Omega^2)B_1 = 2\epsilon\mu\Omega A_1 - \tfrac{3}{4}\epsilon\alpha[B_1^3 + A_1^2B_1 + (A_1^2 - B_1^2)B_3 \\ - 2A_1B_1A_3 + 2(A_3^2 + B_3^2)B_1]\end{aligned} \tag{5.1.40}$$

$$(\omega_0^2 - 9\Omega^2)A_3 = -6\epsilon\mu\Omega B_3 - \tfrac{1}{4}\epsilon\alpha[A_1^3 - 3A_1B_1^2 + 6(A_1^2 + B_1^2)A_3 + 3A_3^3 + 3A_3B_3^2] \tag{5.1.41}$$

$$(\omega_0^2 - 9\Omega^2)B_3 = 6\epsilon\mu\Omega A_3 - \tfrac{1}{4}\epsilon\alpha[-B_1^3 + 3A_1^2B_1 + 6(A_1^2 + B_1^2)B_3 + 3B_3^3 + 3A_3^2B_3] \tag{5.1.42}$$

One usually solves (5.1.39) through (5.1.42) approximately for small ϵ or numerically to determine the A_m and B_m.

Once the steady-state solution (5.1.37) is calculated, its stability is usually investigated by superposing a small perturbation u_1 on u_0, that is, by letting

$$u = u_0 + u_1 \tag{5.1.43}$$

Substituting (5.1.43) into (5.1.24), using the fact that u_0 satisfies (5.1.24), and linearizing the resulting equation in u_1, we obtain

$$\ddot{u}_1 + \omega_0^2 u_1 = -2\epsilon\mu\dot{u}_1 - 3\epsilon\alpha[A_1 \cos \Omega t + B_1 \sin \Omega t + A_3 \cos 3\Omega t + B_3 \sin 3\Omega t]^2 u_1 \tag{5.1.44}$$

which is an equation with variable coefficients. Then the stability of the steady-state solutions corresponds to the stability of the solutions of (5.1.44). The stability in the large can be determined by keeping the nonlinear terms in (5.1.44) and using a Liapunov function if it can be found.

5.2. The Floquet Theory

Next we determine the behavior of systems governed by linear ordinary-differential equations with periodic coefficients. Single-degree-of-freedom systems are treated in Section 5.2.1, while multidegree-of-freedom systems are treated in Section 5.2.2. We describe the Floquet theory for characterizing the functional behavior of such systems (Floquet, 1883). Bloch (1928) generalized the results of Floquet to the case of partial-differential equations with periodic coefficients. The solutions of these equations are usually called *Bloch waves*, and they form the basis of the theory of electrons in crystals.

5.2.1. SINGLE-DEGREE-OF-FREEDOM SYSTEMS

In this section we consider systems governed by equations of the form

$$\ddot{u} + p_1(t)\dot{u} + p_2(t)u = 0 \tag{5.2.1}$$

where the p_n are periodic functions with a period T.

By introducing the transformation

$$u = x \exp\left[-\tfrac{1}{2}\int p_1(t)\, dt\right]$$

we rewrite (5.2.1) in the standard form

$$\ddot{x} + p(t)x = 0 \tag{5.2.2}$$

where

$$p(t) = p_2 - \tfrac{1}{4}p_1^2 - \tfrac{1}{2}\dot{p}_1$$

Thus for this transformation to be valid, p_1 must be differentiable. Equation (5.2.2) was discussed first by Hill (1886) in his determination of the motion of the lunar perigee, and it is called *Hill's equation*. When

$$p(t) = \delta + 2\epsilon \cos 2t$$

equation (5.2.2) reduces to

$$\ddot{x} + (\delta + 2\epsilon \cos 2t)x = 0 \tag{5.2.3}$$

which was discussed by Mathieu (1868) in connection with the problem of vibrations of an elliptic membrane and it is called *Mathieu's equation*.

Since (5.2.1) is a linear, second-order homogeneous differential equation, there exist two linear, nonvanishing independent solutions of this equation, $u_1(t)$ and $u_2(t)$. They are usually referred to as a *fundamental set* of solutions because every solution of (5.2.1) is a linear combination of these two solutions, that is,

$$u(t) = c_1 u_1(t) + c_2 u_2(t) \tag{5.2.4}$$

where c_1 and c_2 are constants. Since $p_j(t) = p_j(t + T)$,

$$\ddot{u}(t + T) = -p_1(t + T)\dot{u}(t + T) - p_2(t + T)u(t + T) = -p_1(t)\dot{u}(t + T) - p_2(t)u(t + T) \tag{5.2.5}$$

from which it follows that, if $u_1(t)$ and $u_2(t)$ are a fundamental set of solutions of (5.2.1), then $u_1(t + T)$ and $u_2(t + T)$ are also a fundamental set of solutions of the same equation. Hence

$$\begin{aligned} u_1(t + T) &= a_{11}u_1(t) + a_{12}u_2(t) \\ u_2(t + T) &= a_{21}u_1(t) + a_{22}u_2(t) \end{aligned} \tag{5.2.6}$$

where the a_{mn} are the elements of a constant nonsingular matrix $[A]$. This matrix is not unique; it depends on the fundamental set being used.

As shown below, there exist fundamental sets of solutions having the property

$$\begin{aligned} v_1(t + T) &= \lambda_1 v_1(t) \\ v_2(t + T) &= \lambda_2 v_2(t) \end{aligned} \tag{5.2.7}$$

where λ is a constant which may be complex. Such solutions are called *normal* or *Floquet solutions*. To show this, we note that any other fundamental set of solutions $v_1(t)$ and $v_2(t)$ is related to $u_1(t)$ and $u_2(t)$ by a nonsingular 2×2 constant matrix $[P]$ according to

$$\mathbf{u}(t) = [P]\,\mathbf{v}(t) \tag{5.2.8}$$

where $\mathbf{u}(t)$ and $\mathbf{v}(t)$ are column vectors whose elements are $u_1(t)$, $u_2(t)$ and $v_1(t)$, $v_2(t)$, respectively. Introducing (5.2.8) into (5.2.6) leads to

$$\mathbf{v}(t + T) = [P]^{-1}[A][P]\,\mathbf{v}(t) = [B]\,\mathbf{v}(t) \tag{5.2.9}$$

Since $[B] = [P]^{-1}[A][P]$, $[B]$ is similar to $[A]$, and they have the same eigenvalues (see Section 3.2). Moreover one can choose a matrix $[P]$ such that $[B]$ assumes its simplest possible form, the *Jordan canonical form*. The Jordan canonical form depends on the eigenvalues of $[A]$, that is, the solution of

$$|[A] - \lambda [I]| = 0 \tag{5.2.10}$$

If the roots of (5.2.10) are different, then $[B]$ has the form

$$[B] = \begin{bmatrix} \lambda_1 & 0 \\ 0 & \lambda_2 \end{bmatrix} \tag{5.2.11}$$

and (5.2.9) can be rewritten as

$$v_i(t + T) = \lambda_i v_i(t), \qquad i = 1 \text{ and } 2 \tag{5.2.12}$$

It follows from (5.2.12) that

$$v_i(t + nT) = \lambda_i^n \, v_i(t) \tag{5.2.13}$$

where n is an integer. Consequently, as $t \to \infty$ (i.e., $n \to \infty$),

$$v_i(t) \to \begin{cases} 0 & \text{if } |\lambda_i| < 1 \\ \infty & \text{if } |\lambda_i| > 1 \end{cases}$$

When $\lambda_i = 1$, v_i is periodic with the period T, and, when $\lambda_i = -1$, v_i is periodic with the period $2T$.

Multiplying (5.2.12) by $\exp[-\gamma_i(t + T)]$ yields

$$\exp[-\gamma_i(t + T)] \, v_i(t + T) = \lambda_i \exp(-\gamma_i T) \exp(-\gamma_i t) \, v_i(t) \tag{5.2.14}$$

Hence if we choose γ_i such that $\lambda_i = \exp(\gamma_i T)$, it follows from (5.2.14) that $\phi_i(t) = \exp(-\gamma_i t) \, v_i(t)$ is a periodic function with the period T. Thus the fundamental set of solutions $v_1(t)$ and $v_2(t)$ can be expressed in the normal form

$$\begin{aligned} v_1(t) &= \exp(\gamma_1 t) \, \phi_1(t) \\ v_2(t) &= \exp(\gamma_2 t) \, \phi_2(t) \end{aligned} \tag{5.2.15}$$

where $\phi_i(t + T) = \phi_i(t)$.

When $\lambda_1 = \lambda_2$, the Jordan canonical form is either the form

$$[B] = \begin{bmatrix} \lambda & 0 \\ 0 & \lambda \end{bmatrix} \tag{5.2.16}$$

or the form

$$[B] = \begin{bmatrix} \lambda & 0 \\ 1 & \lambda \end{bmatrix} \tag{5.2.17}$$

Corresponding to (5.2.16), the fundamental set of solutions can be expressed as in (5.2.15). When $[B]$ has the form (5.2.17),

$$v_1(t + T) = \lambda v_1(t) \tag{5.2.18a}$$

$$v_2(t+T) = \lambda v_2(t) + v_1(t) \tag{5.2.18b}$$

Using arguments similar to those above, one can show that $v_1(t)$ can be expressed in the normal form

$$v_1(t) = \exp(\gamma t)\phi_1(t) \tag{5.2.19}$$

where $\lambda = \exp(\gamma T)$ and $\phi_1(t+T) = \phi_1(t)$. Multiplying (5.2.18b) by exp $[-\gamma(t+T)]$ and using (5.2.19), we obtain

$$\exp[-\gamma(t+T)]\, v_2(t+T) = \exp(-\gamma t) v_2(t) + \frac{1}{\lambda}\phi_1(t) \tag{5.2.20}$$

Hence

$$v_2(t) = \exp(\gamma t)\left[\phi_2(t) + \frac{t}{\lambda T}\phi_1(t)\right] \tag{5.2.21}$$

where $\phi_2(t+T) = \phi_2(t)$.

The above results show that the motion is bounded when the real parts of γ_1 and γ_2 are not positive definite if $[B]$ has either of the forms (5.2.11) or (5.2.16), and the motion is also bounded when the real part of γ is negative if $[B]$ has the form (5.2.17). The parameter γ is usually referred to as the *characteristic exponent* and is related to λ by

$$\gamma = \frac{1}{T}\ln(\lambda)$$

Thus γ is unique to within a multiple of $2in\pi T^{-1}$, where n is an integer.

To show how the characteristic exponents can be determined, let us choose $u_1(t)$ and $u_2(t)$ to be the fundamental set of solutions of (5.2.1) that satisfy

$$\begin{aligned} u_1(0) &= 1, \quad \dot{u}_1(0) = 0 \\ u_2(0) &= 0, \quad \dot{u}_2(0) = 1 \end{aligned} \tag{5.2.22}$$

Setting $t = 0$ in (5.2.6) and using (5.2.22) yields

$$a_{11} = u_1(T) \quad \text{and} \quad a_{21} = u_2(T)$$

Differentiating (5.2.6) once with respect to t, setting $t = 0$ in the resulting equations, and using (5.2.22), we obtain

$$a_{12} = \dot{u}_1(T) \quad \text{and} \quad a_{22} = \dot{u}_2(T)$$

It follows from (5.2.10) that

$$\lambda^2 - 2\alpha\lambda + \Delta = 0 \tag{5.2.23}$$

where

$$\alpha = \tfrac{1}{2}[u_1(T) + \dot{u}_2(T)], \qquad \Delta = u_1(T)\dot{u}_2(T) - \dot{u}_1(T)u_2(T) \tag{5.2.24}$$

The parameter Δ is called the *Wronskian determinant* of $u_1(T)$ and $u_2(T)$. Solving (5.2.23) one can determine the λ's, and then

$$\gamma = \frac{1}{T} \ln(\lambda)$$

In the case of Hill's equation, the Wronskian is unity as shown below. Since u_1 and u_2 are solutions of (5.2.2), it follows that

$$\begin{aligned} \ddot{u}_1 + p(t)u_1 &= 0 \\ \ddot{u}_2 + p(t)u_2 &= 0 \end{aligned} \tag{5.2.25}$$

Substracting u_2 times the first of these equations from u_1 times the second yields

$$u_1\ddot{u}_2 - \ddot{u}_1 u_2 = 0$$

which can be integrated to yield

$$\Delta(t) = u_1(t)\dot{u}_2(t) - \dot{u}_1(t)u_2(t) = \text{constant} \tag{5.2.26}$$

Evaluating (5.2.26) at $t = 0$ leads to $\Delta(t) \equiv 1$.

With $\Delta = 1$, the roots of (5.2.23) are

$$\lambda_{1,2} = \alpha \pm \sqrt{\alpha^2 - 1} \tag{5.2.27}$$

and these roots are related by

$$\lambda_1 \lambda_2 = 1 \tag{5.2.28}$$

When $|\alpha| > 1$, the absolute value of one root is larger than unity while that of the other root is less than unity. Hence one of the normal solutions is unbounded and the other is bounded according to (5.2.28). When $|\alpha| < 1$, the roots are complex conjugates; and since $\lambda_1\lambda_2 = 1$, they have unit moduli. Consequently both normal solutions are bounded. It follows that the transition from stability to instability occurs for $|\alpha| = 1$, which corresponds to the repeated roots $\lambda_1 = \lambda_2 = \pm 1$. The case $\lambda_1 = \lambda_2 = 1$ corresponds to the existence of a periodic normal solution of period T, while the case $\lambda_1 = \lambda_2 = -1$ corresponds to the existence of a periodic normal solution of period $2T$.

In the case of the Mathieu equation, $\alpha = \alpha(\delta, \epsilon)$. The values of δ and ϵ for which $|\alpha| > 1$ are called unstable values, while those for which $|\alpha| = 1$ are called transition values. The locus of transition values separates the $\epsilon\delta$-plane into regions of stability and instability as shown in Figure 5-6. Along these curves at least one of the normal solutions is periodic, with the period π or 2π. Figure 5-6 is called the Strutt diagram, after Strutt (1928) and van der Pol and Strutt (1928).

The unbounded solutions can be divided qualitatively into two different types (Cunningham, 1958) as shown in Figure 5-7. The first type is oscillatory

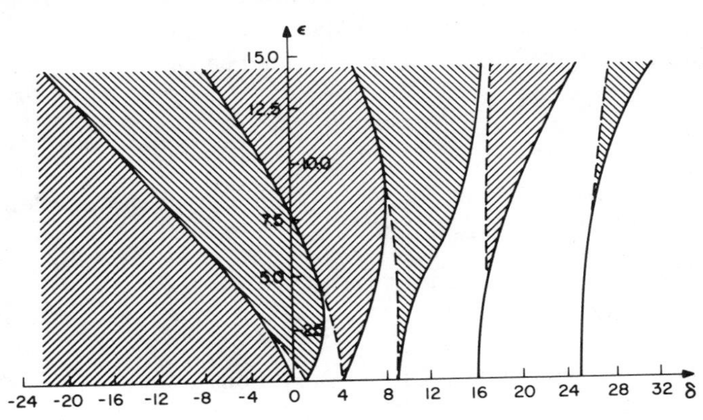

Figure 5-6. Stable and unstable (shaded) regions in the parameter plane for the Mathieu equation.

but with an amplitude that increases exponentially with time, while the second type is nonoscillatory and also increases exponentially with time. The bounded solutions are aperiodic varying with two frequencies—the imaginary part of γ and the frequency of the excitation 2. Depending on the ratio of these frequencies, the solution may exhibit many shapes besides the transition periodic shapes. Three of these shapes are shown in Figure 5-8. When the ratio is very small, the solution is almost periodic, with an amplitude and a phase having a high-frequency modulation. When the ratio is the same order, the shape of the solution is complicated.

The characteristic exponents for (5.2.1) can be obtained by numerically calculating two linear independent solutions of this equation having the initial conditions (5.2.22) during the first period of oscillation. Using the values and first derivatives of these solutions at $t = T$, one can calculate α and Δ from (5.2.24). Solving (5.2.23) one can then determine the λ's, which in turn yield the γ's since $\gamma = (1/T) \ln \lambda$. Using a Newton-Raphson procedure, one can determine the system parameters corresponding to $\lambda = \pm 1$, that is, the boundaries separating stability from instability. However this procedure may lead to serious computational difficulties, necessitating the use of approximate techniques to determine the characteristic exponents and hence the boundaries separating stability from instability. Some approximate techniques for determining the

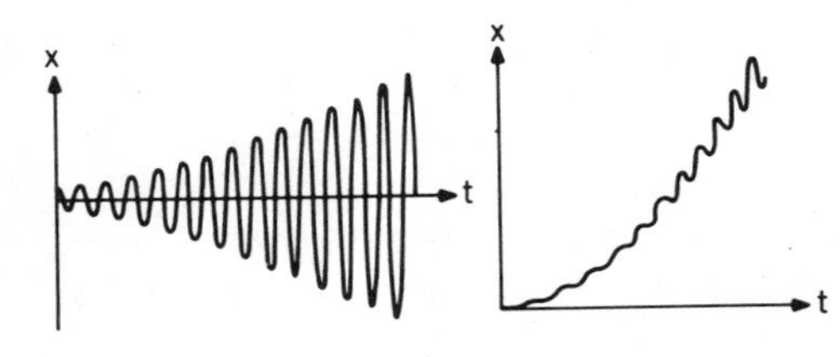

Figure 5-7. Unbounded solutions of the Mathieu equation.

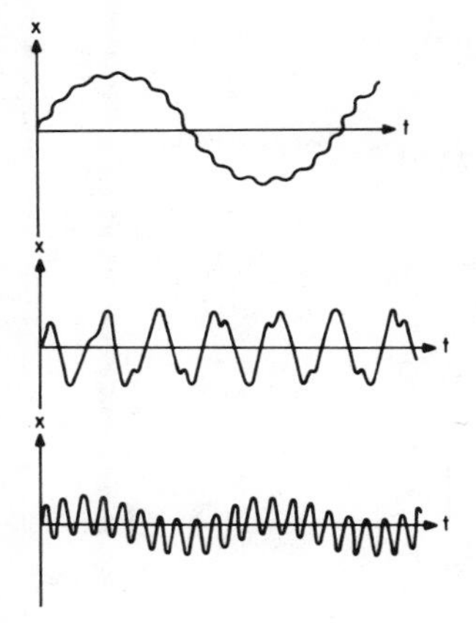

Figure 5-8. Bounded solutions of the Mathieu equation.

behavior of single-degree-of-freedom systems are discussed in Section 5.3, and those for determining the behavior of multidegree-of-freedom systems are discussed in Section 5.4.

5.2.2. MULTIDEGREE-OF-FREEDOM SYSTEMS

In this section we generalize the results of the previous section to multidegree-of-freedom systems described by equations of the form

$$\ddot{x}_n + \sum_{m=1}^{N} f_{nm}(t)x_m = 0 \tag{5.2.29}$$

where $f_{nm}(t + T) = f_{nm}(t)$. It is convenient to express (5.2.29) as a system of $2N$ first-order differential equations by defining

$$\begin{aligned} u_n &= x_n, \quad n = 1, 2, \ldots, N \\ u_n &= \dot{x}_n, \quad n = N + 1, N + 2, \ldots, 2N \end{aligned} \tag{5.2.30}$$

so that (5.2.29) becomes

$$\begin{aligned} \dot{u}_n &= u_{N+n}, \quad n = 1, 2, \ldots, N \\ \dot{u}_{N+n} &= -\sum_{m=1}^{N} f_{nm}(t)u_m, \quad n = 1, 2, \ldots, N \end{aligned} \tag{5.2.31}$$

These equations can be written in the compact form

$$\dot{\mathbf{u}} = [F(t)]\,\mathbf{u} \tag{5.2.32}$$

where $\mathbf{u}$ is a column vector with the components $u_1, u_2, \ldots, u_{2N}$ while $[F(t)]$ is an $2N \times 2N$ matrix such that $[F(t + T)] = [F(t)]$. In what follows, we discuss the behavior of solutions of general equations having the form (5.2.32) but with a general, periodic matrix $[F]$.

For the system (5.2.32), one can define a fundamental set of solutions

$$u_{1k}, u_{2k}, \ldots, u_{Mk}, \qquad k = 1, 2, \ldots, M \tag{5.2.33}$$

where $M = 2N$. This fundamental set can be expressed in the form of an $M \times M$ matrix $[U]$ called a *fundamental matrix solution* as

$$[U] = \begin{bmatrix} u_{11} & u_{21} & \ldots u_{M1} \\ u_{12} & u_{22} & \ldots u_{M2} \\ \cdot & \cdot & \cdots \cdot \\ \cdot & \cdot & \cdots \cdot \\ \cdot & \cdot & \cdots \cdot \\ u_{1M} & u_{2M} & \ldots u_{MM} \end{bmatrix} \tag{5.2.34}$$

Clearly $[U]$ satisfies the matrix equation

$$[\dot{U}] = [F(t)]\,[U] \tag{5.2.35}$$

Since $[F(t + T)] = [F(t)]$, $[U(t + T)]$ is also a fundamental matrix solution. Hence it is related to $[U(t)]$ by

$$[U(t + T)] = [A]\,[U(t)] \tag{5.2.36}$$

where $[A]$ is a nonsingular constant $M \times M$ matrix. Introducing the transformation $[U(t)] = [P]\,[V(t)]$, where $[P]$ is a nonsingular constant $M \times M$ matrix, we express (5.2.36) as

$$[V(t + T)] = [P]^{-1}\,[A]\,[P]\,[V(t)] = [B]\,[V(t)] \tag{5.2.37}$$

As in the single-degree-of-freedom case we choose $[P]$ so that $[B]$ has a Jordan canonical form. Again this form depends on the eigenvalues of $[A]$; they are the M roots of

$$\big|[A] - \lambda\,[I]\big| = 0 \tag{5.2.38}$$

When the roots λ_i of (5.2.38) are distinct, $[B]$ has the diagonal form

$$[B] = \begin{bmatrix} \lambda_1 & 0 & 0 & 0 & \ldots & 0 \\ 0 & \lambda_2 & 0 & 0 & \ldots & 0 \\ 0 & 0 & \lambda_3 & 0 & \ldots & 0 \\ 0 & 0 & 0 & \lambda_4 & \ldots & 0 \\ \cdot & \cdot & \cdot & \cdot & & \cdot \\ \cdot & \cdot & \cdot & \cdot & & \cdot \\ \cdot & \cdot & \cdot & \cdot & & \cdot \\ 0 & 0 & 0 & 0 & & \lambda_M \end{bmatrix} \tag{5.2.39}$$

Consequently (5.2.37) can be rewritten in component form as

$$\mathbf{v}_i(t+T) = \lambda_i \mathbf{v}_i(t) \qquad \text{for } i = 1, 2, \ldots, M \tag{5.2.40}$$

where $\mathbf{v}_i$ is the column solution whose elements are $v_{i1}, v_{i2}, \ldots, v_{iM}$. It follows from (5.2.40) that

$$\mathbf{v}_i(t+nT) = \lambda_i^n \mathbf{v}_i(t)$$

where n is an integer. Consequently as $t \to \infty$ (i.e., $n \to \infty$),

$$\mathbf{v}_i(t) \to \begin{cases} 0 & \text{if } |\lambda_i| < 1 \\ \infty & \text{if } |\lambda_i| > 1 \end{cases}$$

If $\lambda_i = 1$, $\mathbf{v}_i$ is periodic with the period T, while if $\lambda_i = -1$, $\mathbf{v}_i$ is periodic with the period $2T$.

Multiplying (5.2.40) with $\exp[-\gamma_i(t+T)]$ and letting $\lambda_i = \exp(\gamma_i T)$, we obtain

$$\exp[-\gamma_i(t+T)]\, \mathbf{v}_i(t+T) = \exp(-\gamma_i t)\mathbf{v}_i(t) \tag{5.2.41}$$

It follows from (5.2.41) that $\exp(-\gamma_i t)\mathbf{v}_i(t)$ is a periodic vector with the period T. Hence $\mathbf{v}_i$ can be expressed in the normal form

$$\mathbf{v}_i(t) = \exp(\gamma_i t)\phi_i(t) \tag{5.2.42}$$

where $\phi_i(t+T) = \phi_i(t)$

When the roots of (5.2.39) are not distinct, $[B]$ cannot be reduced in general to a diagonal form, but it can be reduced to the Jordan form

$$[B] = \begin{bmatrix} [B_0] & 0 & 0 \ldots\ldots\ldots\ldots & 0 \\ 0 & [B_1] & 0 \ldots\ldots\ldots\ldots & \\ \ldots\ldots & \ldots\ldots & \ldots\ldots\ldots\ldots & \\ \ldots\ldots & \ldots\ldots & \ldots\ldots\ldots\ldots 0 & [B_n] \end{bmatrix} \tag{5.2.43a}$$

where

$$[B_0] = \begin{bmatrix} \lambda_1 & 0 & 0 \ldots\ldots\ldots \\ 0 & \lambda_2 & 0 \ldots\ldots\ldots \\ \ldots & \ldots & \ldots\ldots\ldots \\ \ldots & \ldots & \ldots\ldots\ldots \lambda_q \end{bmatrix} \tag{5.2.43b}$$

and

$$[B_i] = \begin{bmatrix} \lambda_{q+i} & 0 & 0 & 0 \ldots\ldots\ldots & \\ 1 & \lambda_{q+i} & 0 & \ldots\ldots\ldots & \\ 0 & 1 & \lambda_{q+i} & \ldots\ldots\ldots & \\ \ldots & \ldots & \ldots & \ldots\ldots\ldots & \\ \ldots & \ldots & \ldots & \ldots\ldots\ldots 1 & \lambda_{q+i} \end{bmatrix} \tag{5.2.43c}$$

Thus $[B_0]$ contains the q distinct characteristic roots, while $[B_i]$ ($i = 1, 2, \ldots, n$) contains the repeated roots. The number of rows s_i in each of $[B_i]$ equals the number of times the characteristic root λ_{q+i} is repeated. Hence (5.2.37) can be decoupled into $n + 1$ groups. Associated with each group are s_k solutions, where s_0 is the number of distinct roots (q) and s_k for $k > 0$ is the number of times that the $(q + k)$th root is repeated. Thus $s_0 + s_1 + \cdots + s_n = M$.

It follows that

$$\mathbf{v}_1(t + T) = \lambda_1 \mathbf{v}_1(t) \tag{5.2.44a}$$

$$\cdots\cdots\cdots\cdots\cdots\cdots$$

$$\cdots\cdots\cdots\cdots\cdots\cdots$$

$$\mathbf{v}_q(t + T) = \lambda_q \mathbf{v}_q(t) \tag{5.2.44b}$$

$$\mathbf{v}_{q+1}(t + T) = \lambda_{q+1} \mathbf{v}_{q+1}(t) \tag{5.2.44c}$$

$$\cdots\cdots\cdots\cdots\cdots\cdots\cdots\cdots\cdots\cdots$$

$$\cdots\cdots\cdots\cdots\cdots\cdots\cdots\cdots\cdots\cdots$$

$$\mathbf{v}_{q+s_1}(t + T) = \lambda_{q+1} \mathbf{v}_{q+s_1}(t) + \mathbf{v}_{q+s_1-1}(t) \tag{5.2.44d}$$

$$\mathbf{v}_{q+s_1+1}(t + T) = \lambda_{q+2} \mathbf{v}_{q+s_1+1}(t) \tag{5.2.44e}$$

$$\mathbf{v}_{q+s_1+2}(t + T) = \lambda_{q+2} \mathbf{v}_{q+s_1+2}(t) + \mathbf{v}_{q+s_1+1}(t) \tag{5.2.44f}$$

$$\cdots\cdots\cdots\cdots\cdots\cdots\cdots\cdots\cdots\cdots$$

$$\cdots\cdots\cdots\cdots\cdots\cdots\cdots\cdots\cdots\cdots$$

$$\mathbf{v}_{q+s_2}(t + T) = \lambda_{q+2} \mathbf{v}_{q+s_2}(t) + \mathbf{v}_{q+s2-1}(t) \tag{5.2.44g}$$

$$\cdots\cdots\cdots\cdots\cdots\cdots\cdots\cdots\cdots\cdots$$

Hence

$$\mathbf{v}_k(t) = \exp(\gamma_k t)\phi_k(t) \tag{5.2.45a}$$

$$\mathbf{v}_{k+1}(t) = \exp(\gamma_k t)\left[\phi_{k+1}(t) + \frac{t}{T\lambda_k}\phi_k(t)\right] \tag{5.2.45b}$$

$$\mathbf{v}_{k+2}(t) = \exp(\gamma_k t)\left[\phi_{k+2}(t) + \frac{t}{T\lambda_k}\phi_{k+1}(t) + \frac{t(t-T)}{2T^2\lambda_k^2}\phi_k(t)\right] \tag{5.2.45c}$$

$$\cdots\cdots\cdots\cdots\cdots\cdots\cdots\cdots\cdots\cdots\cdots\cdots$$

$$\mathbf{v}_{k+n}(t) = \exp(\gamma_k t)\left[\phi_{k+n}(t) + \frac{t}{T\lambda_k}\phi_{k+n-1}(t) + \frac{t(t-T)}{2T^2\lambda_k^2}\phi_{k+n-2}(t)\right.$$
$$\left. + \cdots + \frac{t(t-T)\ldots(t+T-nT)}{n!\,T^n\lambda_k^n}\phi_k(t)\right] \tag{5.2.45d}$$

for $n = 1, 2, \ldots, s_k$, where

$$\phi_{k+j}(t+T) = \phi_{k+j}(t), \qquad j = 1, 2, \ldots, s_k$$

Since $\gamma_k = (1/T) \ln \lambda_k$, the solutions $\mathbf{v}_{k+1}, \mathbf{v}_{k+2}, \ldots, \mathbf{v}_{k+s_k}$ are bounded for all t when $|\lambda_k| < 1$ and unbounded as $t \to \infty$ if $|\lambda_k| > 1$.

To determine the eigenvalues and hence the characteristic exponents of (5.2.32), one can numerically calculate a fundamental set of solutions of (5.2.35) by using the initial conditions $[U(0)] = [I]$ during a period of oscillation. Then $[A] = [U(T)]$ according to (5.2.36). Then solving the characteristic equation (5.2.38) yields the λ's. With modern computers, numerical methods for the implementation of Floquet theory have been widely used (e.g., Kane and Sobala, 1963; Mingori, 1969b; Brockett, 1970; Peters and Hohenemser, 1971; Friedmann and Silverthorn, 1974). However the main deficiency of this method has been the computational effort required for evaluating the fundamental matrix for large systems. This computational effort necessitates n passes for the calculation of n linearly independent solutions over the period T of the system (5.2.35). To overcome this deficiency, Hsu (1972, 1974a) and Hsu and Cheng (1973) developed various methods for approximating the fundamental matrix during one period. The most efficient method seems to consist of approximating the periodic matrix $[F(t)]$ by a series of step functions. Friedmann, Hammond, and Woo (1977) developed a numerical scheme that yields the fundamental matrix in one pass rather than in n passes.

The results of the preceding numerical calculations yield information about the stability of the system for a given set of the system parameters, and the numerical calculations need to be repeated if any parameter in the set is changed. In practical problems one is not interested in the behavior of a given system with specified system parameters but in the behavior of a class of systems with parameters ranging over some domain. To determine the transition curves and surfaces separating the stable and unstable motions in the parameter space with the implementation of the Floquet theory, one is forced to use a Newton-Raphson procedure (e.g., Thurston, 1973) or establish a gridwork in the parameter space and separately assess the stability at each of the nodal points of the gridwork (e.g., Kane and Sobala, 1973; Mingori, 1969). As mentioned in the preceding section, such a procedure is expensive, time consuming, and may lead to serious computational difficulties. In Sections 5.3 and 5.4 we discuss alternate analytic approaches to the determination of the characteristic exponents and the boundaries separating stability from instability.

5.3. Single-Degree-of-Freedom Systems

Parametric excitations of single-degree-of-freedom systems have been studied extensively. As a result there exist a number of books devoted to such studies such as Bondarenko (1936), McLachlan (1947), and Magnus and Winkler (1966).

There exist also a number of books that treat such systems as discussed in the introduction of this chapter.

There are a number of analytic techniques available for the determination of the stability, characteristic exponents, and boundaries separating stability from instability. These techniques can be divided broadly into three classes. The first class uses Hill's method of infinite determinants (Hill, 1886). This technique was used extensively for single-degree-of-freedom systems, and recently it has been applied to multidegree-of-freedom systems (e.g., Proskuriakov, 1946; Valeev, 1960a, 1961; Bolotin, 1964; Meirovitch and Wallace, 1967; Lindh and Likins 1970; Brockett, 1970; Fu and Nemat-Nasser, 1972a, b; Yakubovich and Starzhinskii, 1975; Lee, 1976). The second class consists of perturbation methods that are based on the assumption that the variable-coefficient terms are small in some sense (e.g., Nayfeh, 1973b). The third class uses Liapunov's theory (LaSalle and Lefschetz, 1961; Hahn, 1963; Liapunov, 1966). The latter approach is limited by the ability to find a suitable Liapunov function. For canonical systems one might be able to use the Hamiltonian, but for other systems one might not be able to find such a function. With this approach one determines qualitatively the stability of the system in the large, but one cannot determine quantitatively the system response. This method was used by a number of investigators including Caughey and Gray (1964), Meirovitch and Wallace (1967), Dickerson and Gray (1969), Hsu and Lee (1971), and Lee and Hsu (1972). In this book we do not discuss the Liapunov method.

In Sections 5.3.1 through 5.3.3 we use the Mathieu equation to describe Hill's (1886) determinant, the Lindstedt-Poincaré technique, and the method of multiple scales. We treat Hill's equation in Section 5.3.4, effects of viscous damping in Section 5.3.5, and nonstationary excitations in Section 5.3.6.

5.3.1. HILL'S INFINITE DETERMINANT

In this section we use Hill's infinite determinant to obtain the stability boundaries of the Mathieu equation (Mathieu, 1868; Whittaker and Watson, 1962, pp. 413–416):

$$\ddot{u} + (\delta + 2\epsilon \cos 2t)u = 0 \tag{5.3.1}$$

According to the Floquet theory (see Section 5.2.1), (5.3.1) has normal solutions of the form

$$u = \exp(\gamma t)\phi(t) \tag{5.3.2}$$

where $\phi(t) = \phi(t + \pi)$. Expressing $\phi(t)$ in a Fourier series, we rewrite (5.3.2) as

$$u = \sum_{n=-\infty}^{\infty} \phi_n \exp[(\gamma + 2in)t] \tag{5.3.3}$$

where the ϕ_n are constants. Substituting (5.3.3) into (5.3.1) yields

$$\sum_{n=-\infty}^{\infty} \{[(\gamma + 2in)^2 + \delta]\, \phi_n \exp[(\gamma + 2in)t]\}$$

$$+ \epsilon \sum_{n=-\infty}^{\infty} \phi_n \{\exp[\gamma t + 2i(n+1)t] + \exp[\gamma t + 2i(n-1)t]\} = 0 \quad (5.3.4)$$

Equating each of the coefficients of the exponential functions to zero yields the following infinite set of linear, algebraic, homogeneous equations for the ϕ_m:

$$[(\gamma + 2im)^2 + \delta]\, \phi_m + \epsilon(\phi_{m-1} + \phi_{m+1}) = 0 \quad (5.3.5)$$

For a nontrivial solution the determinant of the coefficient matrix in (5.3.5) must vanish. Since the determinant is infinite, we divide the mth row by $\delta - 4m^2$ for convergence considerations and obtain the following Hill's determinant:

$$\Delta(\gamma) = \begin{vmatrix} \cdots & \cdot & \cdot & \cdot & \cdot & \cdot & \cdot & \cdot & \cdot & \cdot & \cdots \\ \cdots & 0 & \frac{\epsilon}{\delta - 4^2} & \frac{\delta + (\gamma - 4i)^2}{\delta - 4^2} & \frac{\epsilon}{\delta - 4^2} & 0 & 0 & 0 & 0 & 0 & \cdots \\ \cdots & 0 & 0 & \frac{\epsilon}{\delta - 2^2} & \frac{\delta + (\gamma - 2i)^2}{\delta - 2^2} & \frac{\epsilon}{\delta - 2^2} & 0 & 0 & 0 & 0 & \cdots \\ \cdots & 0 & 0 & 0 & \frac{\epsilon}{\delta} & \frac{\delta + \gamma^2}{\delta} & \frac{\epsilon}{\delta} & 0 & 0 & 0 & \cdots \\ \cdots & 0 & 0 & 0 & 0 & \frac{\epsilon}{\delta - 2^2} & \frac{\delta + (\gamma + 2i)^2}{\delta - 2^2} & \frac{\epsilon}{\delta - 2^2} & 0 & 0 & \cdots \\ \cdots & 0 & 0 & 0 & 0 & 0 & \frac{\epsilon}{\delta - 4^2} & \frac{\delta + (\gamma + 4i)^2}{\delta - 4^2} & \frac{\epsilon}{\delta - 4^2} & 0 & \cdots \\ \cdots & \cdot & \cdot & \cdot & \cdot & \cdot & \cdot & \cdot & \cdot & \cdot & \cdots \end{vmatrix} = 0$$

HILL'S INFINITE DETERMINANT

(5.3.6)

This determinant can be rewritten as (Whittaker and Watson, 1962, pp. 415-416)

$$\Delta(\gamma) = \Delta(0) - \frac{\sin^2(\frac{1}{2} i\pi\gamma)}{\sin^2(\frac{1}{2}\pi\sqrt{\delta})} \quad (5.3.7)$$

Since the characteristic exponents are the solutions of $\Delta(\gamma) = 0$, they are given by

$$\gamma = \pm \frac{2i}{\pi} \sin^{-1} [\Delta(0) \sin^2(\tfrac{1}{2}\pi\sqrt{\delta})]^{1/2} \quad (5.3.8)$$

Once γ is known, the ϕ_n for $n \neq 0$ can be related to ϕ_0 from (5.3.5). However the expression for γ involves the determination of an infinite determinant, which is not a trivial matter. When ϵ is small, approximate solutions can be obtained by considering only the central rows and columns (i.e., the rows and columns centered around the one corresponding to $m = 0$) of Hill's determinant (Bolotin, 1964).

Considering the central three rows and columns, we have the following approximate characteristic equation:

$$\begin{vmatrix} \delta + (\gamma - 2i)^2 & \epsilon & 0 \\ \epsilon & \delta + \gamma^2 & \epsilon \\ 0 & \epsilon & \delta + (\gamma + 2i)^2 \end{vmatrix} = 0$$

or

$$[\delta + (\gamma + 2i)^2](\delta + \gamma^2)[\delta + (\gamma - 2i)^2] - \epsilon^2 [\delta + (\gamma + 2i)^2] - \epsilon^2 [\delta + (\gamma - 2i)^2] = 0 \quad (5.3.9)$$

The transition curves separating stability from instability correspond to $\gamma = 0$ (i.e., periodic motions with the period π) or $\gamma = \pm i$ (i.e., periodic motions with the period 2π).

When $\gamma = 0$, (5.3.9) leads to the transition curves

$$\delta = -\tfrac{1}{2}\epsilon^2 \quad (5.3.10)$$

and

$$\delta = 4 + \tfrac{1}{2}\epsilon^2 \quad (5.3.11)$$

When $\gamma = i$, (5.3.9) leads to the curves

$$\delta = 1 \pm \epsilon \quad (5.3.12)$$

and

$$\delta = 9 + \tfrac{1}{8}\epsilon^2 \quad (5.3.13)$$

To determine better approximations to the above transition curves and to determine approximations to the other transition curves, one needs to consider higher-order determinants. It is clear that this approach is not systematic, and one does not know to what order the obtained expansions for the transition curves are valid. In fact, it is shown in the next section that the correct coefficient for ϵ^2 in (5.3.11) is $\frac{5}{12}$ and not $\frac{1}{2}$. Moreover the correct coefficient for ϵ^2 in (5.2.13) is $\frac{1}{16}$ and not $\frac{1}{8}$.

5.3.2. THE METHOD OF STRAINED PARAMETERS

In this section we consider an alternative of Hill's method, the method of strained parameters. As we shall see, this method is well suited for the determination of the transition curves when ϵ is small. Following this method one assumes, based on the Floquet theory, that the characteristic exponent is 0 or i (i.e., the solutions have periods of π or 2π) and then determines the values of the parameters for which the assumption is true. Thus this method does not yield a solution that is valid in a small neighborhood of a transition curve; rather it yields a solution that is valid right on a transition curve. In the next section the method of multiple scales is used to obtain solutions that are valid in small neighborhoods of transition curves.

We seek the solutions of (5.3.1) having periods of π and 2π and the equations for the transition curves $\delta = \delta(\epsilon)$ in the form of the following perturbation expansions:

$$u(t; \epsilon) = u_0(t) + \epsilon u_1(t) + \epsilon^2 u_2(t) + \cdots \tag{5.3.14}$$

$$\delta = \delta_0 + \epsilon\delta_1 + \epsilon^2\delta_2 + \cdots \tag{5.3.15}$$

Substituting (5.3.14) and (5.3.15) into (5.3.1) and equating coefficients of like powers of ϵ, we obtain

$$\ddot{u}_0 + \delta_0 u_0 = 0 \tag{5.3.16}$$

$$\ddot{u}_1 + \delta_0 u_1 = -\delta_1 u_0 - 2u_0 \cos 2t \tag{5.3.17}$$

$$\ddot{u}_2 + \delta_0 u_2 = -\delta_2 u_0 - \delta_1 u_1 - 2u_1 \cos 2t \tag{5.3.18}$$

The periodic solutions of (5.3.16) with period π are

$$u_0 = a \cos 2nt + b \sin 2nt, \qquad n = 0, 1, 2, \ldots \tag{5.3.19}$$

while the periodic solutions of (5.3.16) with period 2π are

$$u_0 = a \cos (2n - 1)t + b \sin (2n - 1)t, \qquad n = 1, 2, 3, \ldots \tag{5.3.20}$$

where a and b are constants. Here we treat the cases $\delta_0 = 0, 1$, and 4.

The Case $\delta_0 = 0$. In this case $u_0 = a$ and (5.3.17) becomes

$$\ddot{u}_1 = -a\delta_1 - 2a \cos 2t \tag{5.3.21}$$

In order that u_1 be periodic, $\delta_1 = 0$. Then the solution of (5.3.21) is

$$u_1 = \tfrac{1}{2} a \cos 2t \tag{5.3.22}$$

In this example we are able to determine a periodic solution without considering the complementary solutions of the u_n for $n \geqslant 1$. However this is not the case in general, as we shall see when we consider Hill's equation.

Substituting for u_0 and u_1 into (5.3.18) yields

$$\ddot{u}_2 = -\delta_2 a - \tfrac{1}{2} a(1 + \cos 4t) \tag{5.3.23}$$

To ensure that u_2 is periodic, we let $\delta_2 = -\frac{1}{2}$. Hence the transition curve emanating from the origin is given by

$$\delta = -\tfrac{1}{2}\epsilon^2 + O(\epsilon^3) \tag{5.3.24}$$

and along this curve,

$$u = a[1 + \tfrac{1}{2}\epsilon \cos 2t + O(\epsilon^2)] \tag{5.3.25}$$

This result is in agreement with (5.3.10).

The Case $\delta_0 = 1$. In this case

$$u_0 = a \cos t + b \sin t \tag{5.3.26}$$

Then (5.3.17) becomes

$$\ddot{u}_1 + u_1 = -a(\delta_1 + 1) \cos t - b(\delta_1 - 1) \sin t - a \cos 3t - b \sin 3t \tag{5.3.27}$$

In order that u_1 be periodic, the terms in (5.3.27) which lead to secular terms in u_1 must vanish. That is,

$$a(\delta_1 + 1) = 0 \tag{5.3.28}$$

and

$$b(\delta_1 - 1) = 0 \tag{5.3.29}$$

For a nontrivial solution it follows from (5.3.28) and (5.3.29) that either

$$\delta_1 = -1 \quad \text{and} \quad b = 0 \tag{5.3.30}$$

or

$$\delta_1 = 1 \quad \text{and} \quad a = 0 \tag{5.3.31}$$

When $\delta_1 = -1$ and $b = 0$, the particular solution of (5.3.27) is

$$u_1 = \tfrac{1}{8} a \cos 3t \tag{5.3.32}$$

Substituting for u_0, u_1, and δ_1 into (5.3.18) yields

$$\ddot{u}_2 + u_2 = -a(\delta_2 + \tfrac{1}{8}) \cos t + \tfrac{1}{8} a \cos 3t - \tfrac{1}{8} a \cos 5t \tag{5.3.33}$$

The periodicity of the solution demands that the terms in (5.3.33) which lead to secular terms in u_2 vanish. Thus

$$\delta_2 = -\tfrac{1}{8} \tag{5.3.34}$$

Hence one of the transition curves emanating from $\delta = 1$ is given by

$$\delta = 1 - \epsilon - \tfrac{1}{8}\epsilon^2 + O(\epsilon^3) \tag{5.3.35}$$

On this curve

$$u = a\left[\cos t + \tfrac{1}{8}\epsilon \cos 3t + O(\epsilon^2)\right] \tag{5.3.36}$$

When $\delta = 1$ and $a = 0$, the particular solution of (5.3.27) is

$$u_1 = \tfrac{1}{8} b \sin 3t \tag{5.3.37}$$

Substituting for u_0, u_1, and δ_1 into (5.3.18) yields

$$\ddot{u}_2 + u_2 = -b(\delta_2 + \tfrac{1}{8}) \sin t - \tfrac{1}{8} b \sin 3t - \tfrac{1}{8} b \sin 5t \tag{5.3.38}$$

The periodicity of the solution requires that

$$\delta_2 = -\tfrac{1}{8} \tag{5.3.39}$$

Hence the other transition curve emanating from $\delta = 1$ is given by

$$\delta = 1 + \epsilon - \tfrac{1}{8}\epsilon^2 + O(\epsilon^3) \tag{5.3.40}$$

On this curve

$$u = b\left[\sin t + \tfrac{1}{8}\epsilon \sin 3t + O(\epsilon^2)\right] \tag{5.3.41}$$

The Case $\delta_0 = 4$. In this case

$$u_0 = a \cos 2t + b \sin 2t \tag{5.3.42}$$

Then (5.3.17) becomes

$$\ddot{u}_1 + 4u_1 = -\delta_1(a \cos 2t + b \sin 2t) - a(1 + \cos 4t) - b \sin 4t \tag{5.3.43}$$

The periodicity condition demands that $\delta_1 = 0$. Then the particular solution of (5.3.43) becomes

$$u_1 = -\tfrac{1}{4}a + \tfrac{1}{12}a \cos 4t + \tfrac{1}{12} b \sin 4t \tag{5.3.44}$$

Substituting for u_0 and u_1 into (5.3.18) and recalling that $\delta_1 = 0$, we obtain

$$\ddot{u}_2 + 4u_2 = -a(\delta_2 - \tfrac{5}{12}) \cos 2t - b(\delta_2 + \tfrac{1}{12}) \sin 2t - \tfrac{1}{12} a \cos 6t - \tfrac{1}{12} b \sin 6t \tag{5.3.45}$$

The periodicity condition demands that

$$a(\delta_2 - \tfrac{5}{12}) = 0 \tag{5.3.46}$$

and

$$b(\delta_2 + \tfrac{1}{12}) = 0 \tag{5.3.47}$$

For a nontrivial solution it follows from (5.3.46) and (5.3.47) that either

$$\delta_2 = \tfrac{5}{12} \quad \text{and} \quad b = 0 \tag{5.3.48}$$

or

$$\delta_2 = -\tfrac{1}{12} \quad \text{and} \quad a = 0 \tag{5.3.49}$$

Hence the transition curves emanating from $\delta = 4$ are

$$\delta = 4 + \tfrac{5}{12}\epsilon^2 + O(\epsilon^3) \tag{5.3.50a}$$

and

$$\delta = 4 - \tfrac{1}{12}\epsilon^2 + O(\epsilon^3) \tag{5.3.50b}$$

on which

$$u = a\cos 2t - \tfrac{1}{4}\epsilon a(1 - \tfrac{1}{3}\cos 4t) + O(\epsilon^2) \tag{5.3.51a}$$

and

$$u = b\sin 2t + \tfrac{1}{12}\epsilon b\sin 4t + O(\epsilon^2) \tag{5.3.51b}$$

respectively. These results are not in agreement with (5.3.11).

5.3.3. THE METHOD OF MULTIPLE SCALES

In this section we consider another alternative of Hill's method, the method of multiple scales. Three cases are taken up: δ away from zero, unity, and four; δ near four; and δ near unity.

We seek a second-order uniform expansion for (5.3.1) in the form

$$u(t;\epsilon) = u_0(T_0, T_1, T_2) + \epsilon u_1(T_0, T_1, T_2) + \epsilon^2 u_2(T_0, T_1, T_2) + \cdots \tag{5.3.52}$$

where $T_n + \epsilon^n t$. Here we do not choose to expand δ as in (5.3.15); rather we effect the straining by introducing a detuning parameter. Substituting (5.3.52) into (5.3.1) and equating coefficients of like powers of ϵ, we obtain

$$D_0^2 u_0 + \delta u_0 = 0 \tag{5.3.53}$$

$$D_0^2 u_1 + \delta u_1 = -2D_0D_1u_0 - 2u_0\cos 2T_0 \tag{5.3.54}$$

$$D_0^2 u_2 + \delta u_2 = -2D_0D_2u_0 - D_1^2u_0 - 2D_0D_1u_1 - 2u_1\cos 2T_0 \tag{5.3.55}$$

where $D_n = \partial/\partial T_n$. In deriving (5.3.53) through (5.3.55), we assumed that δ is away from zero. The case $\delta \approx 0$ can be handled by letting $\delta = \epsilon^2\delta_2$ in (5.3.1) before equating the coefficients of like powers of ϵ.

The solution of (5.2.53) can be written as

$$u_0 = A(T_1, T_2)\exp(i\omega T_0) + \bar{A}(T_1, T_2)\exp(-i\omega T_0) \tag{5.3.56}$$

where $\delta = \omega^2$. Hence (5.3.54) becomes

$$D_0^2 u_1 + \omega^2 u_1 = -2i\omega D_1 A\exp(i\omega T_0) - A\exp[i(2+\omega)T_0] - \bar{A}\exp[i(2-\omega)T_0] + cc \tag{5.3.57}$$

In analyzing the particular solution of (5.3.57) we need to distinguish between two cases: ω away from 1 and $\omega \approx 1$.

The Case ω away from 1. Eliminating the terms that produce secular terms in (5.3.57) leads to $D_1 A = 0$ so that $A = A(T_2)$. Hence the particular solution of (5.3.57) is

$$u_1 = \frac{A}{4(\omega+1)} \exp\,[i(2+\omega)\,T_0] - \frac{\bar{A}}{4(\omega-1)} \exp\,[i(2-\omega)\,T_0] + cc \tag{5.3.58}$$

Substituting for u_0 and u_1 into (5.3.55) yields

$$D_0^2 u_2 + \omega^2 u_2 = \left[-2i\omega D_2 A + \frac{A}{2(\omega^2-1)}\right] \exp\,(i\omega T_0)$$
$$- \frac{A}{4(\omega+1)} \exp\,[i(4+\omega)\,T_0] + \frac{\bar{A}}{4(\omega-1)} \exp\,[i(4-\omega)]\,T_0 + cc \tag{5.3.59}$$

In determing the particular solution of (5.3.59) we need to distinguish between two cases: ω away from 2 and $\omega \approx 2$.

When ω is away from 2, elimination of terms that produce secular terms in (5.3.59) yields

$$2i\omega D_2 A - \frac{A}{2(\omega^2-1)} = 0 \tag{5.3.60}$$

Hence

$$A = \tfrac{1}{2} a \exp\left[-\frac{i}{4\omega(\omega^2-1)}\,T_2 + i\beta\right] \tag{5.3.61}$$

where a and β are constants. Substituting for u_0 and u_1 from (5.3.56) and (5.3.58) into (5.3.52), making use of (5.3.61), and replacing T_n by $\epsilon^n t$, we obtain the following approximate solution when ω is away from 1 and 2:

$$u = a \cos\phi + \tfrac{1}{4}\epsilon a\left[(\omega+1)^{-1}\cos\,(2t+\phi) - (\omega-1)^{-1}\cos\,(2t-\phi)\right] + O(\epsilon^2) \tag{5.3.62}$$

where

$$\phi = \left[\omega - \frac{\epsilon^2}{4\omega(\omega^2-1)}\right] t + \beta \tag{5.3.63}$$

When $\omega \approx 2$, we introduce a detuning parameter σ defined by

$$2 = \omega + \epsilon^2 \sigma \tag{5.3.64}$$

and express $(4-\omega)\,T_0$ in (5.3.59) as

$$(4-\omega)\,T_0 = \omega T_0 + 2\epsilon^2 \sigma T_0 = \omega T_0 + 2\sigma T_2 \tag{5.3.65}$$

Then elimination of the terms that produce secular terms in (5.3.59) gives

$$-2i\omega D_2 A + \frac{A}{2(\omega^2 - 1)} + \frac{\bar{A}}{4(\omega - 1)} \exp(2i\sigma T_2) = 0 \tag{5.3.66}$$

To obtain the solution of (5.3.66) we let

$$A(T_2) = B(T_2) \exp(i\sigma T_2)$$

and find

$$-2i\omega D_2 B + \left[2\omega\sigma + \frac{1}{2(\omega^2 - 1)}\right] B + \frac{\bar{B}}{4(\omega - 1)} = 0 \tag{5.3.67}$$

Putting $B = B_r + iB_i$ with real B_r and B_i in (5.3.67) and separating the real and imaginary parts, we have

$$\begin{aligned} -2\omega D_2 B_r + \left[2\omega\sigma - \frac{1}{4(\omega + 1)}\right] B_i &= 0 \\ 2\omega D_2 B_i + \left[2\omega\sigma + \frac{\omega + 3}{4(\omega^2 - 1)}\right] B_r &= 0 \end{aligned} \tag{5.3.68}$$

The solution of (5.3.68) can be expressed as

$$B_r = b_r \exp(\gamma T_2), \quad B_i = b_i \exp(\gamma T_2) \tag{5.3.69}$$

where b_r and b_i are constants and

$$\gamma^2 = -\frac{1}{4\omega^2}\left[2\omega\sigma - \frac{1}{4(\omega + 1)}\right]\left[2\omega\sigma + \frac{\omega + 3}{4(\omega^2 - 1)}\right] \tag{5.3.70}$$

The motion is unstable when γ^2 is positive definite. Thus the motion is unstable when

$$-\frac{\omega + 3}{8\omega(\omega^2 - 1)} < \sigma < \frac{1}{8\omega(\omega + 1)} \tag{5.3.71}$$

and otherwise is stable. Because $\omega \approx 2$, the transition curves correspond to

$$\sigma \approx \tfrac{1}{48} \quad \text{and} \quad \sigma \approx -\tfrac{5}{48}$$

Therefore the transition curves emanating from $\omega = 2$ are given by

$$\omega = 2 - \tfrac{1}{48}\epsilon^2 + \cdots \quad \text{and} \quad \omega = 2 + \tfrac{5}{48}\epsilon^2 + \cdots$$

Since $\delta = \omega^2$, the transition curves emanating from $\delta = 4$ are

$$\delta = 4 - \tfrac{1}{12}\epsilon^2 + \cdots \quad \text{and} \quad \delta = 4 + \tfrac{5}{12}\epsilon^2 + \cdots \tag{5.3.72}$$

in agreement with those obtained earlier by using the Lindstedt-Poincaré technique.

The Case $\omega \approx 1$. In this case we introduce a detuning parameter σ according to

$$1 = \omega + \epsilon\sigma \tag{5.3.73}$$

and we express $(2 - \omega)\, T_0$ as

$$(2 - \omega)\, T_0 = \omega T_0 + 2\epsilon\sigma T_0 = \omega T_0 + 2\sigma T_1 \tag{5.3.74}$$

Then in (5.3.57) secular terms in u_1 are eliminated if

$$2i\omega D_1 A + \bar{A} \exp(2i\sigma T_1) = 0 \tag{5.3.75}$$

Consequently the particular solution of (5.3.57) becomes

$$u_1 = \tfrac{1}{4} A(\omega + 1)^{-1} \exp[i(2 + \omega)\, T_0] + cc \tag{5.3.76}$$

Substituting for u_0 and u_1 from (5.3.56) and (5.3.76) into (5.3.55) yields

$$\begin{aligned} D_0^2 u_2 + \omega^2 u_2 = &- [2i\omega D_2 A + D_1^2 A + \tfrac{1}{4}(\omega + 1)^{-1} A] \exp(i\omega T_0) \\ &- \tfrac{1}{4} A(\omega + 1)^{-1} \exp[i(4 + \omega)\, T_0] \\ &- \tfrac{1}{2} i(2 + \omega)(\omega + 1)^{-1} D_1 A \exp[i(2 + \omega)\, T_0] + cc \end{aligned} \tag{5.3.77}$$

Eliminating the terms that produce secular terms from (5.3.77) and keeping in mind that $\omega \approx 1$, we obtain

$$2i\omega D_2 A + D_1^2 A + \tfrac{1}{4}(\omega + 1)^{-1} A = 0 \tag{5.3.78}$$

To determine A, we combine (5.3.75) and (5.3.78) as follows. From (5.3.75),

$$D_1^2 A = \tfrac{1}{2} i\omega^{-1} [2i\sigma\bar{A} + D_1\bar{A}] \exp(2i\sigma T_1) = -\sigma\omega^{-1}\bar{A} \exp(2i\sigma T_1) + \tfrac{1}{4}\omega^{-2} A \tag{5.3.79}$$

Eliminating $D_1^2 A$ from (5.3.78) and (5.3.79) gives

$$2i\omega D_2 A + \frac{\omega^2 + \omega + 1}{4\omega^2(\omega + 1)} A - \frac{\sigma}{\omega}\bar{A} \exp(2i\sigma T_1) = 0 \tag{5.3.80}$$

It can be easily shown that (5.3.75) and (5.3.80) result from a multiple-scales expansion of

$$2i\omega \frac{dA}{dt} + \epsilon\left(1 - \frac{\epsilon\sigma}{\omega}\right)\bar{A} \exp(2i\epsilon\sigma t) + \epsilon^2 \frac{\omega^2 + \omega + 1}{4\omega^2(\omega + 1)} A = 0 \tag{5.3.81}$$

We seek a solution for (5.3.81) in the form $A = (B_r + iB_i) \exp(i\epsilon\sigma t)$ with real B_r and B_i, separate real and imaginary parts, and obtain

$$2\omega \frac{dB_r}{dt} - \left[\epsilon + 2\epsilon\sigma\omega - \frac{\epsilon^2\sigma}{\omega} - \epsilon^2 \frac{\omega^2 + \omega + 1}{4\omega^2(\omega + 1)}\right] B_i = 0$$

$$2\omega \frac{dB_i}{dt} - \left[\epsilon - 2\epsilon\sigma\omega - \frac{\epsilon^2\sigma}{\omega} + \epsilon^2 \frac{\omega^2 + \omega + 1}{4\omega^2(\omega + 1)}\right] B_r = 0 \tag{5.3.82}$$

Equation (5.3.82) admit a solution in the form

$$(B_r, B_i) = (b_r, b_i) \exp(\gamma t) \tag{5.3.83}$$

with constant b_r and b_i, provided that

$$4\omega^2\gamma^2 = \epsilon^2 \left[1 + 2\sigma\omega - \frac{\epsilon\sigma}{\omega} - \frac{\epsilon(\omega + 1 + \omega^2)}{4\omega^2(\omega + 1)}\right]\left[1 - 2\sigma\omega - \frac{\epsilon\sigma}{\omega} + \frac{\epsilon(\omega + 1 + \omega^2)}{4\omega^2(\omega + 1)}\right] \tag{5.3.84}$$

The motion is unstable when γ^2 is positive definite. Therefore the motion is unstable when

$$\frac{1}{2\omega}\left[-1 + \frac{\epsilon(\omega^2 - \omega - 1)}{4\omega^2(\omega + 1)}\right] < \sigma < \frac{1}{2\omega}\left[1 + \frac{\epsilon(\omega^2 - \omega - 1)}{4\omega^2(\omega + 1)}\right] \tag{5.3.85}$$

and otherwise is stable. Because $\omega = 1 - \epsilon\sigma$, the transition curves emanating from $\delta = 1$ correspond to

$$\sigma = -\tfrac{1}{2} + \tfrac{3}{16}\epsilon + O(\epsilon^2) \tag{5.3.86}$$

and

$$\sigma = \tfrac{1}{2} + \tfrac{3}{16}\epsilon + O(\epsilon^2) \tag{5.3.87}$$

Hence the transition curves are given by

$$\delta = \omega^2 = 1 + \epsilon - \tfrac{1}{8}\epsilon^2 + O(\epsilon^3) \tag{5.3.88}$$

and

$$\delta = \omega^2 = 1 - \epsilon - \tfrac{1}{8}\epsilon^2 + O(\epsilon^3) \tag{5.3.89}$$

in agreement with those obtained earlier by the Lindstedt-Poincaré technique.

We note that one cannot expand (5.3.84) for small ϵ in the neighborhood of a transition curve where $1 + 2\sigma\omega$ is the same order as $\epsilon\{(\sigma/\omega) + (\omega^2 + \omega + 1)/[4\omega^2(\omega + 1)]\}$. This is the reason that equations (5.3.75) and (5.3.78) were combined into the single equation (5.3.81), which was used to determine the transition curves.

5.3.4. HILL'S EQUATION

In this section we consider a somewhat general form of the equation of motion for a parametrically excited system, namely Hill's equation (Hill, 1886)

$$\ddot{u} + \left[\delta + \sum_{m=1}^{\infty} \epsilon^m f_m(t)\right] u = 0 \tag{5.3.90}$$

where $f_m(t + \pi) = f_m(t)$.

Without loss of generality we may assume that

$$\int_0^{\pi} f_m(t)\, dt = 0 \tag{5.3.91}$$

and hence that each function f_m can be represented by a Fourier series having the form

$$f_m = \sum_{n=1}^{\infty} (\alpha_{mn} \cos 2nt + \beta_{mn} \sin 2nt) \tag{5.3.92}$$

Approximate solutions of this problem were obtained by Bondarenko (1936), Klotter and Kotowski (1943a, b), Struble and Fletcher (1962), Magnus and Winkler (1966), Yang and Rosenberg (1967), Rand (1969), Rand and Tseng (1969), Mostaghel and Sackman (1970), Hamer and Smith (1972), Nayfeh (1972), Rubenfeld (1973), and Karpasiuk (1973).

Here we limit our analysis to the determination of the transition curves for the system governed by (5.3.90) when ϵ is small. Thus it is most convenient to use the method of strained parameters. Accordingly we assume expansions of the form

$$u(t; \epsilon) = u_0(t) + \epsilon u_1(t) + \epsilon^2 u_2(t) + \cdots \tag{5.3.93}$$

$$\delta(\epsilon) = \delta_0 + \epsilon\delta_1 + \epsilon^2\delta_2 + \cdots \tag{5.3.94}$$

Substituting (5.3.92) through (5.3.94) into (5.3.90) and equating coefficients of like powers of ϵ, we obtain

$$\ddot{u}_0 + \delta_0 u_0 = 0 \tag{5.3.95}$$

$$\ddot{u}_1 + \delta_0 u_1 = -(\delta_1 + f_1) u_0 \tag{5.3.96}$$

$$\ddot{u}_2 + \delta_0 u_2 = -(\delta_1 + f_1) u_1 - (\delta_2 + f_2) u_0 \tag{5.3.97}$$

According to the Floquet theory, u has either the period π or 2π along a transition curve. Thus we write the solution of (5.3.95) in the form

$$u_0 = a_0 \cos nt + b_0 \sin nt \tag{5.3.98}$$

where n is a nonzero integer (the special case of n equals zero is considered later) related to δ_0 by

$$n^2 = \delta_0 \tag{5.3.99}$$

and a_0 and b_0 are arbitrary constants at this point.

Substituting for u_0 into (5.3.96) leads to

$$\begin{aligned}
\ddot{u}_1 + \delta_0 u_1 = &- [(\delta_1 + \tfrac{1}{2}\alpha_{1n})\, a_0 + \tfrac{1}{2}\beta_{1n} b_0] \cos nt \\
&- [(\delta_1 - \tfrac{1}{2}\alpha_{1n})\, b_0 + \tfrac{1}{2}\beta_{1n} a_0] \sin nt \\
&- \tfrac{1}{2} \sum_{m=1}^{\infty} [(a_0\alpha_{1m} - b_0\beta_{1m}) \cos (2m + n)\, t \\
&+ (a_0\beta_{1m} + b_0\alpha_{1m}) \sin (2m + n)\, t] \\
&- \tfrac{1}{2} \sum_{\substack{m=1 \\ m \neq n}}^{\infty} [(a_0\alpha_{1m} + b_0\beta_{1m}) \cos (2m - n)\, t \\
&+ (a_0\beta_{1m} - b_0\alpha_{1m}) \sin (2m - n)\, t]
\end{aligned} \tag{5.3.100}$$

To eliminate secular terms from u_1, we must put

$$(\delta_1 + \tfrac{1}{2}\alpha_{1n})\, a_0 + \tfrac{1}{2}\beta_{1n} b_0 = 0 \tag{5.3.101}$$

and

$$\tfrac{1}{2}\beta_{1n} a_0 + (\delta_1 - \tfrac{1}{2}\alpha_{1n})\, b_0 = 0 \tag{5.3.102}$$

For a nontrivial solution to exist,

$$\delta_1 = \pm\tfrac{1}{2}(\alpha_{1n}^2 + \beta_{1n}^2)^{1/2} \tag{5.3.103}$$

It follows from (5.3.101) and (5.3.102) that in general a_0 and b_0 are related by

$$a_0 = -\frac{\beta_{1n}}{2\delta_1 + \alpha_{1n}}\, b_0 \tag{5.3.104}$$

Only when δ_1 is zero (i.e., $\alpha_{1n} = \beta_{1n} = 0$) are a_0 and b_0 not related at this level of approximation.

The solution for u_1 has the form

$$u_1 = \sum_{m=1}^{\infty} \frac{1}{8m(m+n)} [(a_0\alpha_{1m} - b_0\beta_{1m}) \cos (2m+n)t + (a_0\beta_{1m} + b_0\alpha_{1m}) \sin (2m+n)t] + a_1 \cos nt + b_1 \sin nt + \sum_{\substack{m=1 \\ m \neq n}}^{\infty} \frac{1}{8m(m-n)} [(a_0\alpha_{1m} + b_0\beta_{1m}) \cos (2m-n)t + (a_0\beta_{1m} - b_0\alpha_{1m}) \sin (2m-n)t] \tag{5.3.105}$$

where a_1 and b_1 are arbitrary constants. We note that u_1 must contain the complementary solution if δ_1 is not zero, but not if δ_1 is zero. The reason will become obvious later.

Substituting (5.3.98) and (5.3.105) into (5.3.97) leads to

$$\ddot{u}_2 + \delta_0 u_2 = -[(\delta_1 + \tfrac{1}{2}\alpha_{1n}) a_1 + \tfrac{1}{2}\beta_{1n} b_1 - y_1 a_0 - y_2 b_0] \cos nt - [\tfrac{1}{2}\beta_{1n} a_1 + (\delta_1 - \tfrac{1}{2}\alpha_{1n}) b_1 - y_2 a_0 - y_3 b_0] \sin nt + NST \tag{5.3.106}$$

where NST denotes the terms which cannot lead to secular terms in u_2, and

$$y_1 = -\delta_2 - \tfrac{1}{2}\alpha_{2n} - \sum_{m=1}^{\infty} \frac{\alpha_{1m}^2 + \beta_{1m}^2 + 2(\alpha_{1m}\alpha_{1m+n} + \beta_{1m}\beta_{1m+n})}{16m(m+n)} - \sum_{\substack{m=1 \\ m \neq n}}^{\infty} \frac{\alpha_{1m}^2 + \beta_{1m}^2}{16m(m-n)} - \sum_{m=1}^{n-1} \frac{\alpha_{1n-m}\alpha_{1m} - \beta_{1n-m}\beta_{1m}}{16m(m-n)} \tag{5.3.107}$$

$$y_2 = -\tfrac{1}{2}\beta_{2n} - \sum_{m=1}^{\infty} \frac{\alpha_{1m}\beta_{1n+m} - \beta_{1m}\alpha_{1n+m}}{16m(m+n)} - \sum_{\substack{m=1 \\ m \neq n}}^{\infty} \frac{\alpha_{1m}\beta_{1m+n} - \beta_{1m}\alpha_{1m+n}}{16m(m+n)} - \sum_{m=1}^{n-1} \frac{\beta_{1m}\alpha_{1n-m} + \alpha_{1m}\beta_{1n-m}}{16m(m-n)} \tag{5.3.108}$$

$$y_3 = -\delta_2 + \tfrac{1}{2}\alpha_{2n} - \sum_{m=1}^{\infty} \frac{\alpha_{1m}^2 + \beta_{1m}^2 - 2(\beta_{1m}\beta_{1m+n} + \alpha_{1m}\alpha_{1m+n})}{16m(m+n)} - \sum_{\substack{m=1 \\ m \neq n}}^{\infty} \frac{\alpha_{1m}^2 + \beta_{1m}^2}{16m(m-n)} - \sum_{m=1}^{n-1} \frac{\beta_{1n-m}\beta_{1m} - \alpha_{1n-m}\alpha_{1m}}{16m(m-n)} \tag{5.3.109}$$

Thus to eliminate secular terms from u_2, we must put

$$(\delta_1 + \tfrac{1}{2}\alpha_{1n})a_1 + \tfrac{1}{2}\beta_{1n}b_1 = y_1a_0 + y_2b_0 \tag{5.3.110}$$

$$\tfrac{1}{2}\beta_{1n}a_1 + (\delta_1 - \tfrac{1}{2}\alpha_{1n})b_1 = y_2a_0 + y_3b_0 \tag{5.3.111}$$

In determining δ_2 from (5.3.110) and (5.3.111), we need to consider two cases: δ_1 is zero and δ_1 is nonzero.

The Case $\delta_1 = 0$. Referring to (5.3.103), we see that δ_1, α_{1n}, and β_{1n} are zero. Thus (5.3.110) and (5.3.111) reduce to

$$y_1a_0 + y_2b_0 = 0 \tag{5.3.112}$$

$$y_2a_0 + y_3b_0 = 0 \tag{5.3.113}$$

Recalling that the elimination of secular terms from u_1 does not impose a relationship between a_0 and b_0 when δ_1 is zero, we see that secular terms can be eliminated from u_2, and a nontrivial solution exists if, and only if,

$$y_1y_3 - y_2^2 = 0 \tag{5.3.114}$$

Equation (5.3.114) can be solved for δ_2.

The Case $\delta_1 \neq 0$. Considering a_1 and b_1 as the unknowns, we note that the determinant of the coefficient matrix is zero and that it is possible to obtain a solution only if

$$(\delta_1 + \tfrac{1}{2}\alpha_{1n})(y_2a_0 + y_3b_0) - \tfrac{1}{2}\beta_{1n}(y_1a_0 + y_2b_0) = 0 \tag{5.3.115}$$

In this case a_0 and b_0 are not independent, being related by (5.3.104). Thus (5.3.115) can be written as

$$\beta_{1n}[(2\delta_1 + \alpha_{1n})y_2 - \beta_{1n}y_1] - (2\delta_1 + \alpha_{1n})[(2\delta_1 + \alpha_{1n})y_3 - \beta_{1n}y_2] = 0 \tag{5.3.116}$$

Equation (5.3.116) can be solved for δ_2.

The Case $\delta_0 \approx 0$. In this case we do not know a priori the order of δ. Thus we write

$$\delta = \epsilon\delta_1 + \epsilon^2\delta_2 + \cdots \tag{5.3.117}$$

Then instead of (5.3.95) through (5.3.97) we obtain

$$\ddot{u}_0 = 0 \tag{5.3.118}$$

$$\ddot{u}_1 = -(\delta_1 + f_1)u_0 \tag{5.3.119}$$

$$\ddot{u}_2 = -(\delta_1 + f_2)u_1 - (\delta_2 + f_2)u_0 \tag{5.3.120}$$

The periodic solution of (5.3.118) is

$$u_0 = a_0 \tag{5.3.121}$$

where a_0 is an arbitrary constant. Then (5.3.119) becomes

$$\ddot{u}_1 = -\delta_1 a_0 - a_0 \sum_{m=1}^{\infty} (\alpha_{1m} \cos 2mt + \beta_{1m} \sin 2mt) \tag{5.3.122}$$

Thus to eliminate secular terms from u_1, we must put

$$\delta_1 = 0 \tag{5.3.123}$$

It follows that

$$u_1 = a_1 + \tfrac{1}{4} a_0 \sum_{m=1}^{\infty} \frac{1}{m^2} (\alpha_{1m} \cos 2mt + \beta_{1m} \sin 2mt) \tag{5.3.124}$$

where a_1 is an arbitrary constant.

Substituting (5.3.121), (5.3.123), and (5.3.124) into (5.3.120) lead to

$$\ddot{u}_2 = -a_0 \left[\tfrac{1}{8} \sum_{m=1}^{\infty} \frac{\alpha_{1m}^2 + \beta_{1m}^2}{m^2} + \delta_2 \right] + NST \tag{5.3.125}$$

Thus to eliminate secular terms we must put

$$\delta_2 = -\tfrac{1}{8} \sum_{m=1}^{\infty} \frac{\alpha_{1m}^2 + \beta_{1m}^2}{m^2} \tag{5.3.126}$$

Example. We consider

$$\ddot{u} + (\delta + \tfrac{8}{3} \epsilon \cos^3 2t) u = 0 \tag{5.3.127}$$

which can be written as

$$\ddot{u} + (\delta + 2\epsilon \cos 2t + \tfrac{2}{3} \epsilon \cos 6t) u = 0 \tag{5.3.128}$$

Thus

$$\alpha_{11} = 2, \quad \alpha_{13} = \tfrac{2}{3} \tag{5.3.129}$$

and all other α_{im} and β_{im} are zero. It follows that

$$\delta = -\tfrac{41}{81} \epsilon^2 + O(\epsilon^3) \tag{5.3.130}$$

$$\delta = 1 \pm \epsilon - \tfrac{19}{144} \epsilon^2 + O(\epsilon^3) \tag{5.3.131}$$

$$\delta = 4 + \tfrac{7}{20} \epsilon^2 + O(\epsilon^3) \tag{5.3.132}$$

$$\delta = 4 - \tfrac{7}{180} \epsilon^2 + O(\epsilon^3) \tag{5.3.133}$$

5.3.5. EFFECTS OF VISCOUS DAMPING

In this section we consider the effect of small viscous damping on the response of the parametrically excited system of the preceding section. Thus we modify (5.3.90) to

$$\ddot{u} + 2\mu\dot{u} + \left[\delta + \sum_{m=1}^{\infty} \epsilon^m f_m(t)\right] u = 0 \tag{5.3.134}$$

where $\mu > 0$. Gunderson, Rigas, and van Vleck (1974) proposed a technique for determining the stability regions of the damped Mathieu equation. Introducing the transformation $u = v \exp(-\mu t)$, we rewrite (5.3.134) as

$$\ddot{v} + \left[\delta - \mu^2 + \sum_{m=1}^{\infty} \epsilon^m f_m(t)\right] v = 0 \tag{5.3.135}$$

which has the same form as (5.3.90). Therefore the effects of the viscous damping are to decrease the growth rate by μ and to modify the natural frequency of the system from $\delta^{1/2}$ to $(\delta - \mu^2)^{1/2}$. Both of these effects are stabilizing.

To analyze the response of systems including the effects of viscous damping, one can either use the methods of the preceding sections directly on (5.3.134) or use the method of multiple scales or the method of averaging to determine a uniform expansion for v. In the remainder of this section let us use the Lindstedt-Poincaré technique directly on (5.3.134) and determine the transition curves for the principal-resonance case. To accomplish this, we set $\mu = \epsilon\hat{\mu}$.

Substituting (5.3.93) and (5.3.94) into (5.3.134) and equating the coefficients of ϵ^0 and ϵ to zero, we obtain

$$\ddot{u}_0 + \delta_0 u_0 = 0 \tag{5.3.136}$$

$$\ddot{u}_1 + \delta_0 u_1 = -(\delta_1 + f_1)\, u_0 - 2\hat{\mu}\dot{u}_0 \tag{5.3.137}$$

The solution of (5.3.136) is given by (5.3.98) where $\delta_0 = n^2$. Then (5.3.137) becomes

$$\ddot{u}_1 + n^2 u_1 = -\left[(\delta_1 + \tfrac{1}{2}\alpha_{1n})\, a_0 + (\tfrac{1}{2}\beta_{1n} + 2\hat{\mu}n)\, b_0\right] \cos nt$$
$$- \left[(\delta_1 - \tfrac{1}{2}\alpha_{1n})\, b_0 + (\tfrac{1}{2}\beta_{1n} - 2\hat{\mu}n)\, a_0\right] \sin nt + NST \tag{5.3.138}$$

Eliminating the terms that lead to secular terms yields

$$\begin{aligned} (\delta_1 + \tfrac{1}{2}\alpha_{1n}) a_0 + (\tfrac{1}{2}\beta_{1n} + 2\hat{\mu}n)\, b_0 &= 0 \\ (\tfrac{1}{2}\beta_{1n} - 2\hat{\mu}n) a_0 + (\delta_1 - \tfrac{1}{2}\alpha_{1n})\, b_0 &= 0 \end{aligned} \tag{5.3.139}$$

Hence

$$\delta_1^2 = \tfrac{1}{4}(\alpha_{1n}^2 + \beta_{1n}^2) - 4\hat{\mu}^2 n^2 \tag{5.3.140}$$

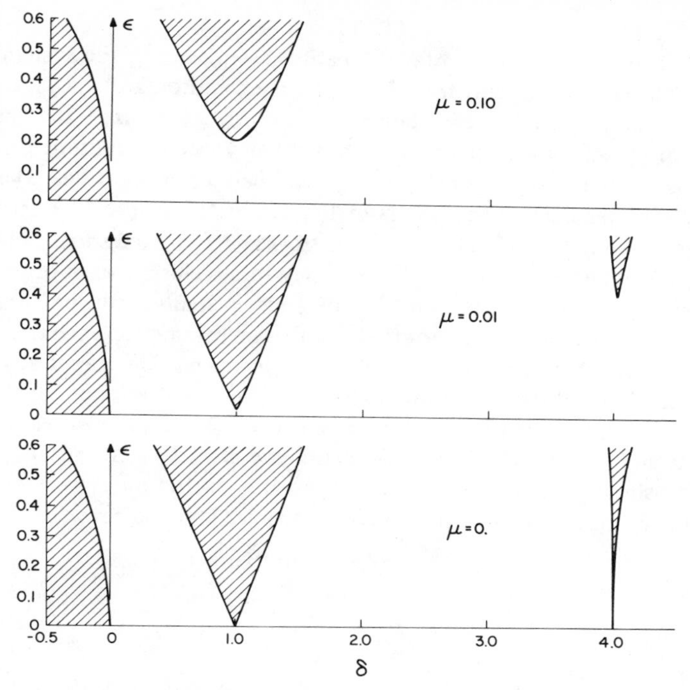

Figure 5-9. Effect of viscous damping on the stability of the solutions of the Mathieu equation. Shaded areas are unstable.

and the transition curves separating stability from instability are given by

$$\delta = n^2 \pm \tfrac{1}{2}\left[(\alpha_{1n}^2 + \beta_{1n}^2)\,\epsilon^2 - 16\mu^2 n^2\right]^{1/2} + \cdots \tag{5.3.141}$$

Therefore the motion is completely stabilized to first order by the viscous damping if $\mu \geqslant \mu_c = (4n)^{-1}\,(\alpha_{1n}^2 + \beta_{1n}^2)^{1/2}\,\epsilon$. Otherwise there is still a region of instability. However the viscous damping decreases this unstable region by lifting it from the δ-axis and narrowing its boundaries in the $\epsilon\delta$-plane as shown in Figure 5-9. This figure shows the transition curves to second order for the Mathieu equation. They are given by

$$\delta = -\tfrac{1}{2}\epsilon^2 + \cdots \tag{5.3.142}$$

$$\delta = 1 \pm (\epsilon^2 - 4\mu^2)^{1/2} - \tfrac{1}{8}\epsilon^2 + \cdots \tag{5.3.143}$$

$$\delta = 4 + \tfrac{1}{6}\epsilon^2 \pm (\tfrac{1}{16}\epsilon^4 - 16\mu^2)^{1/2} + \cdots \tag{5.3.144}$$

In deriving (5.3.144) we assumed that $\mu = O(\epsilon^2)$; if μ is bigger the motion near $\delta = 4$ will be completely stabilized. We note that the viscous damping does not affect the transition curves near $\delta = 0$.

5.3.6. NONSTATIONARY EXCITATIONS

In the examples discussed so far the natural frequency of the system is constant and the amplitude and frequency of the excitation are constant; that is, the oscillations are stationary. Bogdanoff (1962) generalized the results of Lowenstern (1932) on the effect of high-frequency, small-amplitude parametric excitations to the case of small, rapid, quasi-periodic excitations. Hemp and Sethna (1968) generalized the results of Bogdanoff to include the effects of external forces, simultaneous occurrence of slow and fast excitations, and occurrence of several fast excitations with frequency values close to each other. Moran (1970) studied transient motions in dynamic systems with high-frequency parametric excitations. Their results show that a parametric excitation having two frequencies that are close to each other in magnitude may destabilize an inverted pendulum that would otherwise be stable if it were not for the closeness of the two frequencies, in agreement with the experimental results of Bogdanoff and Citron (1965). In this section we consider nonstationary oscillations. For simplicity, we consider nonstationary excitations whose amplitude and frequency are slowly varying functions of time. Thus we consider

$$\ddot{u} + [\omega^2 + 2\epsilon k(\epsilon t)\cos\theta]\, u = 0 \tag{5.3.145}$$

where

$$\dot{\theta} = \Omega(\epsilon t) = 2\omega + \epsilon\sigma(\epsilon t) \tag{5.3.146}$$

We seek a first-order uniform solution to (5.3.145) in the form

$$u = u_0(T_0, T_1) + \epsilon u_1(T_0, T_1) + \cdots \tag{5.3.147}$$

where

$$T_n = \epsilon^n t$$

Substituting (5.3.147) into (5.3.145) and equating coefficients of like powers of ϵ yield

$$D_0^2 u_0 + \omega^2 u_0 = 0 \tag{5.3.148}$$

$$D_0^2 u_1 + \omega^2 u_1 = -2D_0 D_1 u_0 - 2u_0 k(T_1)\cos\theta \tag{5.3.149}$$

The solution of (5.3.148) can be expressed as

$$u_0 = A(T_1)\exp(i\omega T_0) + cc \tag{5.3.150}$$

Hence (5.3.149) becomes

$$D_0^2 u_1 + \omega^2 u_1 = -2i\omega A' \exp(i\omega T_0) - kA\exp[i(\omega T_0 + \theta)] - k\bar{A}\exp[i(\theta - \omega T_0)] + cc \tag{5.3.151}$$

Since $\dot{\theta} \approx 2\omega$, $\theta - 2\omega T_0$ is a slowly varying function of t, that is, it can be considered a function of T_1. Then eliminating the terms that produce secular terms in u_1 yields

$$2i\omega A' + k\bar{A} \exp[i(\theta - 2\omega T_0)] = 0 \tag{5.3.152}$$

Letting

$$A = \tfrac{1}{2}a \exp[\tfrac{1}{2}i(\gamma + \theta - 2\omega T_0)]$$

with real a and γ in (5.3.152) and separating real and imaginary parts, we obtain

$$a' = \frac{k(T_1)}{2\omega} \sin\gamma$$
$$\gamma' = -\sigma(T_1) + \frac{k(T_1)}{\omega} \cos\gamma \tag{5.3.153}$$

Alternatively we let $A = (x + iy) \exp(\tfrac{1}{2}i\int \sigma dT_1)$ in (5.3.152) with real x and y, separate real and imaginary parts, and obtain

$$x' - \tfrac{1}{2}\left(\sigma + \frac{k}{\omega}\right)y = 0$$
$$y' + \tfrac{1}{2}\left(\sigma - \frac{k}{\omega}\right)x = 0 \tag{5.3.155}$$

For constant σ and k, the general solution of (5.3.155) is

$$x = c_1 \exp(\gamma T_1) + c_2 \exp(-\gamma T_1)$$
$$y = 2\omega\gamma(k + \sigma\omega)^{-1} [c_1 \exp(\gamma T_1) - c_2 \exp(-\gamma T_1)] \tag{5.3.156}$$

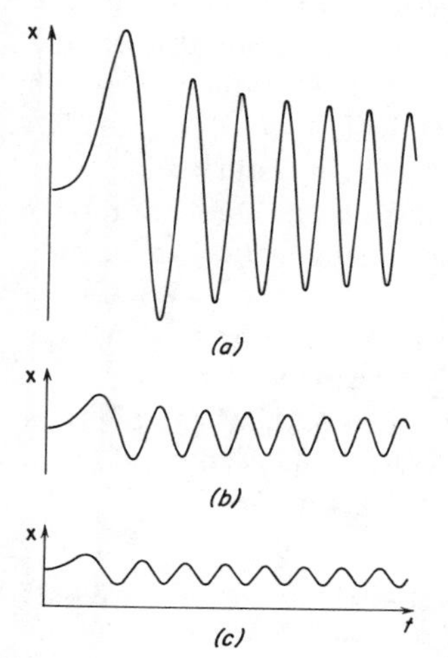

Figure 5-10. Nonstationary parametric excitation of the Mathieu equation: (*a*) $\mu = 0.01$; (*b*) $\mu = 0.015$; (*c*) $\mu = 0.2$.

where $\gamma = \frac{1}{2}(k^2\omega^{-2} - \sigma^2)^{1/2}$. When σ and k are time varying, the solution deviates from (5.3.156); the deviation increases with the rate of variation of σ and k.

Croll (1975) considered a system for which σ is a constant while $k = \exp(-\epsilon\mu t)$. For this case Figure 5-10 shows the variation of x with T_1 for $\omega = 1, \sigma = 0.8$, and three values of μ and for the initial conditions $x(0) = 1$ and $y(0) = 0$. We note that the three curves are drawn to the same scale. It is clear that x and y become periodic in all cases because the excitation causing the nonstationary behavior decays with time. An analysis of the stationary case shows that the motion is unbounded for $k > \sigma$ and bounded for $k < \sigma$. Since x and y grow initially and then decay, the growth depends on the rate at which k decays, that is on the value of μ. The smaller the value of μ is, the larger the growth is and hence the amplitude of the resulting periodic motion as shown in Figure 5-10.

5.4. Linear Systems Having Distinct Frequencies

In this section we determine approximate solutions to multidegree-of-freedom systems that are parametrically excited. Since combination as well as simple resonances might occur in these systems, the method of multiple scales is more suited for these problems than the method of strained parameters.

We consider nongyroscopic systems in this section and in Section 5.5; in Section 5.6 we consider a gyroscopic system. Thus we consider systems governed by

$$\ddot{\mathbf{x}} + [A]\,\mathbf{x} + 2\epsilon[B(t)]\,\mathbf{x} = 0 \tag{5.4.1}$$

where $\mathbf{x}$ is a column vector having N components, $[A]$ is an $N \times N$ constant matrix, and $[B(t)]$ is an $N \times N$ matrix whose elements are periodic functions of t. Theorems of boundedness and criteria for unboundedness of the solutions of (5.4.1) were given by Cesari (1940), Cesari and Hale (1954), Gambill (1954, 1955), and Hale (1954, 1957) when the eigenvalues of $[A]$ are all different from zero, distinct, and purely imaginary. Hale (1958) treated systems for which $[A]$ also has a number of zero eigenvalues, while Bailey and Cesari (1958) treated systems for which some of the eigenvalues of $[A]$ are distinct and purely imaginary and only one possibly zero and the remaining eigenvalues are real or complex with negative real parts. For comprehensive mathematical treatments of these systems and more references, we refer the reader to the books of Hale (1963) and Cesari (1971). In this section we consider only the case in which the eigenvalues of $[A]$ are different from zero and purely imaginary.

We introduce the linear transformation $\mathbf{x} = [P]\mathbf{u}$ in (5.4.1), where $[P]$ is a nonsingular constant matrix such that $[P]^{-1}[A][P]$ is a Jordan canonical form. Multiplying the result by $[P]^{-1}$ we obtain

$$\ddot{u}_n + \omega_n^2 u_n + 2\epsilon \sum_{m=1}^{N} g_{nm}(t) u_m = 0 \tag{5.4.2}$$

for $n = 1, 2, \ldots, N$ when the eigenvalues of $[A]$ are distinct and positive. The g_{nm} are the elements of the matrix $[G] = [P]^{-1}[B]P$. In general (5.4.2) represents a system of n coupled Hill's equations. However there is a restricted class of (5.4.1) for which $[G]$ is diagonal (i.e., $g_{nm} = 0$ if $n \neq m$) and (5.4.2) represents a system of n uncoupled Hill's equations. Hsu (1961) treated this restricted class and applied his results to two identical compound pendulums, double pendulums, masses on a noncircular shaft, and masses on a noncircular rotating shaft. We will not pursue the restricted class any further because its analysis involves the analysis of the individual Hill's equations as already done in Section 5.3.

In Section 5.5 we treat a case in which all the eigenvalues of $[A]$ are distinct except one, which has a multiplicity of 2. In what follows we consider the case

$$g_{nm}(t) = f_{nm} \cos \omega t \tag{5.4.3}$$

where the f_{nm} are constants. Thus we consider the system

$$\ddot{u}_n + \omega_n^2 u_n + 2\epsilon \cos \omega t \sum_{m=1}^{N} f_{nm} u_m = 0 \tag{5.4.4}$$

We exhibit the different resonant conditions by developing a straightforward expansion in the next section; we use the method of multiple scales to obtain first-and second-order uniform expansions in Sections 5.4.2 and 5.4.3, respectively. In Section 5.4.4 we apply the results to the analysis of the lateral deflections of columns subjected to periodic follower forces. This is the problem of dynamic buckling. The effects of viscous damping are discussed in Section 5.4.5.

5.4.1. THE STRAIGHTFORWARD EXPANSION

We seek an expansion of the form

$$u_n(t;\epsilon) = u_{n0}(t) + \epsilon u_{n1}(t) + \epsilon^2 u_{n2} + \cdots \tag{5.4.5}$$

Substituting (5.4.5) into (5.4.4) and equating coefficients of like powers of ϵ yields

$$\ddot{u}_{n0} + \omega_n^2 u_{n0} = 0 \tag{5.4.6}$$

$$\ddot{u}_{n1} + \omega_n^2 u_{n1} = -\sum_r f_{nr} u_{r0} [\exp(i\omega t) + cc] \tag{5.4.7}$$

$$\ddot{u}_{n2} + \omega_n^2 u_{n2} = -\sum_r f_{nr} u_{r1} [\exp(i\omega t) + cc] \tag{5.4.8}$$

where cc represents the complex conjugate of the preceding terms.

The general solution of (5.4.6) can be written in the form

$$u_{n0} = A_n \exp(i\omega_n t) + cc \tag{5.4.9}$$

where A_n is a complex constant. Substituting (5.4.9) into (5.4.7) yields

$$\ddot{u}_{n1} + \omega_n^2 u_n = -\sum_r f_{nr} A_r \{\exp[i(\omega_r + \omega)t] + \exp[i(\omega_r - \omega)t]\} + cc \quad (5.4.10)$$

A particular solution of (5.4.10) can be written in the form

$$u_{n1} = \sum_r f_{nr} A_r \left\{ \frac{\exp[i(\omega_r + \omega)t]}{(\omega_r + \omega)^2 - \omega_n^2} + \frac{\exp[i(\omega_r - \omega)t]}{(\omega_r - \omega)^2 - \omega_n^2} \right\} + cc \quad (5.4.11)$$

Substituting (5.4.11) into (5.4.8) yields

$$\ddot{u}_{n2} + \omega_n^2 u_{n2} = -\sum_r \sum_s f_{nr} f_{rs} A_s \cdot \left\{ \frac{\exp[i(\omega_s + 2\omega)t] + \exp(i\omega_s t)}{(\omega_s + \omega)^2 - \omega_r^2} + \frac{\exp[i(\omega_s - 2\omega)t] + \exp(i\omega_s t)}{(\omega_s - \omega)^2 - \omega_r^2} \right\} + cc \quad (5.4.12)$$

A particular solution of (5.4.12) can be written in the form

$$\begin{aligned} u_{n2} = {} & \sum_r \sum_s f_{nr} f_{rs} A_s \left\{ \frac{\exp[i(\omega_s + 2\omega)t]}{[(\omega_s + 2\omega)^2 - \omega_n^2][(\omega_s + \omega)^2 - \omega_r^2]} \right. \\ & \left. + \frac{\exp[i(\omega_s - 2\omega)t]}{[(\omega_s - 2\omega)^2 - \omega_n^2][(\omega_s - \omega)^2 - \omega_r^2]} \right\} + \sum_r \sum_{\substack{s \\ s \neq n}} f_{nr} f_{rs} A_s \\ & \cdot \left[\frac{1}{(\omega + \omega_s)^2 - \omega_r^2} + \frac{1}{(\omega - \omega_s)^2 - \omega_r^2} \right] \frac{A_s \exp(i\omega_s t)}{\omega_s^2 - \omega_n^2} + \tfrac{1}{2} i \sum_r f_{nr} f_{rn} A_n \\ & \cdot \left[\frac{1}{(\omega + \omega_n)^2 - \omega_r^2} + \frac{1}{(\omega - \omega_n)^2 - \omega_r^2} \right] \frac{t \exp(i\omega_n t)}{\omega_n} + cc \end{aligned} \quad (5.4.13)$$

It follows from (5.4.9) and (5.4.13) that the expansion (5.4.5) is only valid for short times because $\lim_{t \to \infty}(u_{n2}/u_{n0}) = \infty$; this is the result of the secular terms in u_{n2} which contain the factors $t \exp(\pm i\omega_n t)$. It follows from (5.4.11) and (5.4.13) that the expansion (5.4.5) is not valid if

$$p\omega \simeq \omega_n \pm \omega_m \quad (5.4.14)$$

where p, n, and m are integers because, when such resonant combinations of frequencies exist, some terms in u_{n1}, u_{n2}, etc. may contain small divisors. The existence of such combination resonances has been demonstrated both by experiment and by analog simulation (Yamamoto and Saito, 1970; Sugiyama, Fugiwara, and Sekiya, 1970; Iwatsubo, Saigo, and Sugiyama, 1973; Dugundji and Mukhopadhyay, 1973).

In order to eliminate the troublesome secular and small-divisor terms, one must modify the straightforward procedure. In the next section the modification is accomplished by using the method of multiple scales.

5.4.2. FIRST-ORDER EXPANSIONS

Following the method of multiple scales, one seeks a uniformly valid expansion having the form

$$u_n(t;\epsilon) = u_{n0}(T_0, T_1, T_2) + \epsilon u_{n1}(T_0, T_1, T_2) + \epsilon^2 u_{n2}(T_0, T_1, T_2) + \cdots \tag{5.4.15}$$

Substituting (5.4.15) into (5.4.4) and equating coefficients of like powers of ϵ yields

$$D_0^2 u_{n0} + \omega_n^2 u_{n0} = 0 \tag{5.4.16}$$

$$D_0^2 u_{n1} + \omega_n^2 u_{n1} = -2D_0 D_1 u_{n0} - \sum_r f_{nr} u_{r0} [\exp(i\omega T_0) + cc] \tag{5.4.17}$$

and

$$D_0^2 u_{n2} + \omega_n^2 u_{n2} = -2D_0 D_2 u_{n0} - D_1^2 u_{n0} - 2D_0 D_1 u_{n1} - \sum_r f_{nr} u_{r1} [\exp(i\omega T_0) + cc] \tag{5.4.18}$$

The general solution of (5.4.16) can be written in the form

$$u_{n0} = A_n(T_1, T_2) \exp(i\omega_n T_0) + cc \tag{5.4.19}$$

Substituting (5.4.19) into (5.4.17) yields

$$D_0^2 u_{n1} + \omega_n^2 u_{n1} = -2i\omega_n D_1 A_n \exp(i\omega_n T_0) - \sum_r f_{nr} A_r \cdot \{\exp[i(\omega_r + \omega) T_0] + \exp[i(\omega_r - \omega) T_0]\} + cc \tag{5.4.20}$$

Now A_n is to be chosen in such a way as to eliminate the troublesome terms from u_{n1}. This choice depends on the resonant combinations of frequencies; five different cases are considered.

The Case ω Away from $\omega_q \pm \omega_p$. When ω is away from $\omega_q \pm \omega_p$ for all possible values of q and p, small divisors cannot appear, and the troublesome terms will be eliminated from u_{n1} if

$$D_1 A_n = 0 \quad \text{or} \quad A_n = A_n(T_2) \text{ for all } n \tag{5.4.21}$$

Consequently a particular solution of (5.4.20) can be written in the form

$$u_{n1} = \sum_r f_{nr} A_r \left\{ \frac{\exp[i(\omega_r + \omega) T_0]}{(\omega_r + \omega)^2 - \omega_n^2} + \frac{\exp[i(\omega_r - \omega) T_0]}{(\omega_r - \omega)^2 - \omega_n^2} \right\} + cc \tag{5.4.22}$$

The Case ω *Near* $\omega_p + \omega_q$. When ω is near $\omega_p + \omega_q$, we speak of a combination resonance of the summed type. The possibility of the existence of such combination resonances was discussed by Lazarev (1937) and Simanov (1952). They were analyzed by using different techniques by a number of investigators including Mettler (1949, 1967), Yakubovich (1958), Iakubovich (1959), Valeev (1960a, b, 1961, 1963), Schmidt and Weidenhammer (1961), Piszczek (1961), Hsu (1963, 1965), Lion (1966), Stevens (1966), Yamamoto and Saito (1970), Grybos (1972), and van Dao (1973).

We express the nearness of ω to $\omega_p + \omega_q$ by introducing the detuning parameter σ, that is defined by

$$\omega = \omega_p + \omega_q + \epsilon\sigma \tag{5.4.23}$$

Then we can write

$$(\omega - \omega_q)T_0 = \omega_p T_0 + \sigma T_1 \tag{5.4.24}$$

and

$$(\omega - \omega_p)T_0 = \omega_q T_0 + \sigma T_1 \tag{5.4.25}$$

Then it follows from (5.4.20) that the troublesome terms are eliminated from u_{p1} if

$$2i\omega_p D_1 A_p + f_{pq}\bar{A}_q \exp(i\sigma T_1) = 0 \tag{5.4.26}$$

from u_{q1} if

$$2i\omega_q D_1 A_q + f_{qp}\bar{A}_p \exp(i\sigma T_1) = 0 \tag{5.4.27}$$

and from u_{n1} when $n \neq p$ and q if (5.4.21) is satisfied. In this case u_{p1}, u_{q1}, and u_{n1} for $n \neq p$ and q are given by (5.4.22), with the troublesome terms being deleted from the expressions for u_{p1} and u_{q1}.

Equations (5.4.26) and (5.4.27) admit nontrivial solutions having the form

$$A_p = a_p \exp(-i\lambda T_1) \quad \text{and} \quad A_q = a_q \exp[i(\bar{\lambda} + \sigma)T_1] \tag{5.4.28}$$

where a_p and a_q are complex functions of T_2, $\bar{\lambda}$ is the complex conjugate of λ,

$$\lambda = -\tfrac{1}{2}[\sigma \pm (\sigma^2 - \Lambda_{pq})^{1/2}] \tag{5.4.29}$$

and

$$\Lambda_{pq} = \frac{f_{pq} f_{pq}}{\omega_p \omega_q} \tag{5.4.30}$$

It follows from (5.4.28) and (5.4.29) that A_p and A_q are bounded if, and only if,

$$\sigma^2 \geqslant \Lambda_{pq} \tag{5.4.31}$$

It follows that the motion is always bounded when f_{pq} and f_{qp} have different signs. When the signs of f_{pq} and f_{qp} are the same, the transition curves in the $\epsilon\omega$-plane that separate stable from unstable solutions are defined by

$$\omega = \omega_p + \omega_q \pm \epsilon \left[\frac{f_{pq} f_{qp}}{\omega_p \omega_q}\right]^{1/2} + O(\epsilon^2) \tag{5.4.32}$$

When $p = q$, (5.4.32) reduces to

$$\omega = 2\omega_p \pm \epsilon \frac{f_{pp}}{\omega_p} + O(\epsilon^2) \tag{5.4.33}$$

which is the known result for the Mathieu equation (Section 5.3).

The Case ω Near $\omega_q - \omega_p$. When ω is near $\omega_q - \omega_p$, we speak of a combination resonance of the difference type. When there are no other resonances to this order, the results can be obtained from those above by simply changing the sign of ω_p. For this case unstable solutions occur only when f_{pq} and f_{qp} have different signs.

The Case ω Near $\omega_p + \omega_q$ and $\omega_s - \omega_q$. In this case ω is simultaneously near $\omega_p + \omega_q$ and $\omega_s - \omega_q$, there are no other resonances to this order. To express the nearness of ω to $\omega_p + \omega_q$ and $\omega_s - \omega_q$, one introduces the detuning parameters defined by

$$\omega = \omega_p + \omega_q + \epsilon\sigma_1 \quad \text{and} \quad \omega = \omega_s - \omega_q + \epsilon\sigma_2 \tag{5.4.34}$$

Then it follows that the troublesome terms are eliminated if

$$2i\omega_p D_1 A_p + f_{pq}\bar{A}_q \exp(i\sigma_1 T_1) = 0 \tag{5.4.35}$$

$$2i\omega_q D_1 A_q + f_{qp}\bar{A}_p \exp(i\sigma_1 T_1) + f_{qs}A_s \exp(-i\sigma_2 T_1) = 0 \tag{5.4.36}$$

$$2i\omega_s D_1 A_s + f_{sq}A_q \exp(i\sigma_2 T_1) = 0 \tag{5.4.37}$$

and for $n \neq p$, q, or s, the A_n satisfy (5.4.21). In this case u_{p1}, u_{q1}, u_{s1}, and u_{n1} for $n \neq p$, q, and s are given by (5.4.22), with the troublesome terms being deleted.

Equations (5.4.35) through (5.4.37) admit nontrivial solutions having the form

$$A_p = a_p \exp[-i(\bar{\lambda} - \sigma_1)T_1], \quad A_q = a_q \exp(i\lambda T_1), \quad A_s = a_s \exp[i(\lambda + \sigma_2)T_1] \tag{5.4.38}$$

where a_p, a_q, and a_s are complex functions of T_2 and

$$\lambda + \tfrac{1}{4}\Lambda_{pq}(\lambda - \sigma_1)^{-1} - \tfrac{1}{4}\Lambda_{qs}(\lambda + \sigma_2)^{-1} = 0 \tag{5.4.39}$$

Equation (5.4.39) is a cubic equation for λ and has closed-form solutions. It was obtained first by Hsu (1965) by using the method of averaging and later by

Nayfeh and Mook (1977) by using the method of multiple scales. The transition curves correspond to the value of ω for which λ has two real roots.

We note that as $\epsilon \to 0$ only one resonance can exist. When $\epsilon \to 0$ and σ_1 remains bounded, $\sigma_2 \to \infty$. In this case (5.4.39) reduces to (5.4.29). On the other hand, when $\epsilon \to 0$ and σ_2 remains bounded, (5.4.39) yields the expression for λ when ω is near $\omega_s - \omega_q$ and no other resonances exist to this order.

The Case ω Near $\omega_q - \omega_p$ and $\omega_s - \omega_q$. In this case ω is simultaneously near $\omega_q - \omega_p$ and $\omega_s - \omega_q$, and there are no other resonances to this order. The results can be obtained from those directly above by simply changing the sign of ω_p.

5.4.3. SECOND-ORDER EXPANSIONS

Substituting (5.4.19) and (5.4.22) into (5.4.18) yields

$$D_0^2 u_{n2} + \omega_n^2 u_{n2} = -(2i\omega_n D_2 A_n + D_1^2 A_n) \exp(i\omega_n T_0) - \sum_r \sum_s f_{nr} f_{rs} A_s \left\{ \frac{\exp[i(\omega_s + 2\omega) T_0] + \exp(i\omega_s T_0)}{(\omega_s + \omega)^2 - \omega_r^2} + \frac{\exp[i(\omega_s - 2\omega) T_0] + \exp(i\omega_s T_0)}{(\omega_s - \omega)^2 - \omega_r^2} \right\} + cc + NTT \tag{5.4.40}$$

where NTT stands for terms which do not produce troublesome terms in u_{n2} under any of the resonant conditions being considered. As in the previous section different cases need to be considered.

The Case ω Away From $\omega_q \pm \omega_p$. Here three subcases are considered: 2ω away from $\omega_l \pm \omega_k$, 2ω near $\omega_l + \omega_k$, and 2ω near $\omega_l - \omega_k$. Other resonances, such as 2ω simultaneously near $\omega_l - \omega_k$ and $\omega_m - \omega_l$, can be treated in the same manner as the cases being considered, but the results are not presented here. In this case A_n is a function of T_2 only [recall (5.4.21)].

When 2ω is away from $\omega_l \pm \omega_k$ for all possible values of l and k, the troublesome terms are eliminated if

$$i D_2 A_n + \chi_n A_n = 0 \tag{5.4.41}$$

where

$$2\omega_n \chi_n = \sum_r f_{nr} f_{rn} \left[\frac{1}{(\omega_n + \omega)^2 - \omega_r^2} + \frac{1}{(\omega_n - \omega)^2 - \omega_r^2} \right] \tag{5.4.42}$$

The solution of (5.4.41) can be written in the form

$$A_n = a_n \exp(i\chi_n T_2) \tag{5.4.43}$$

where a_n is a complex constant. Consequently to second order every mode is bounded for all times.

When 2ω is near $\omega_l + \omega_k$, we speak of a combination resonance of the summed type of second order. Such resonances were analyzed by Yamamoto and Saito (1970) using the method of averaging and by Nayfeh and Mook (1977) using the method of multiple scales. It is convenient to introduce a detuning parameter defined by

$$2\omega = \omega_l + \omega_k + \epsilon^2 \sigma \tag{5.4.44}$$

Then the troublesome terms are eliminated if

$$i D_2 A_k + \chi_k A_k + \mu_{kl} \bar{A}_l \exp(i\sigma T_2) = 0 \tag{5.4.45}$$

$$i D_2 A_l + \chi_l A_l + \mu_{lk} \bar{A}_k \exp(i\sigma T_2) = 0 \tag{5.4.46}$$

and for $n \neq l$ or k the A_n satisfy (5.4.41). In (5.4.45) and (5.4.46)

$$\mu_{kl} = \frac{1}{2\omega_k} \sum_r f_{kr} f_{rl} \left[\frac{1}{(\omega_l - \omega)^2 - \omega_r^2} \right] \tag{5.4.47}$$

Equations (5.4.45) and (5.4.46) admit nontrivial solutions having the form

$$A_k = a_k \exp(i\lambda T_2) \quad \text{and} \quad A_l = a_l \exp[-i(\bar{\lambda} - \sigma) T_2] \tag{5.4.48}$$

where a_k and a_l are complex constants and

$$\lambda^2 - \lambda(\sigma - \chi_l + \chi_k) + \chi_k(\sigma - \chi_l) + \mu_{kl}\mu_{lk} = 0 \tag{5.4.49}$$

Hence the transition curves correspond to the vanishing of the discriminant of (5.4.49); they correspond to

$$\sigma = \chi_k + \chi_l \pm 2(\mu_{kl}\mu_{lk})^{1/2} \tag{5.4.50}$$

Combining (5.4.44) and (5.4.50) leads to the following definition of the transition curves in the $\epsilon\omega$-plane:

$$\omega = \tfrac{1}{2}(\omega_l + \omega_k) + \epsilon^2 \left[\tfrac{1}{2}(\chi_k + \chi_l) \pm (\mu_{kl}\mu_{lk})^{1/2} \right] \tag{5.4.51}$$

Unstable solutions occur only when μ_{kl} and μ_{lk} have the same sign.

The case 2ω near $\omega_l - \omega_k$ is called a combination resonance of the difference type of second order. It can be obtained from the above results by simply changing the sign of ω_k.

The Case ω Near $\omega_p + \omega_q$. In this case ω is near $\omega_p + \omega_q$, and there are no other resonances. Moreover A_p and A_q are functions of T_1 and T_2; consequently the troublesome terms are eliminated from (5.4.40) if

$$2i\omega_p D_2 A_p + D_1^2 A_p + 2\omega_p \hat{\chi}_p A_p = 0 \tag{5.4.52}$$

$$2i\omega_q D_2 A_q + D_1^2 A_q + 2\omega_q \hat{\chi}_q A_q = 0 \tag{5.4.53}$$

and for $n \neq p$ or q the A_n satisfy (5.4.41). In (5.4.52)

$$2\omega_p \hat{\chi}_p = \sum_r f_{pr} f_{rp} [(\omega_p + \omega)^2 - \omega_r^2]^{-1} + \sum_{r \neq q} f_{pr} f_{rp} [(\omega_p - \omega)^2 - \omega_r^2]^{-1} \tag{5.4.54}$$

Comparing (5.4.42) with (5.4.54), one sees that $\hat{\chi}_p$ is formed by removing the terms containing small divisors from χ_p.

It is convenient, for reasons that are given below, to combine (5.4.26) and (5.4.52) as well as (5.4.27) and (5.4.53) into a single equation in terms of the original time scale. To accomplish this, one may use equations (5.4.26) and (5.4.27) to obtain

$$D_1^2 A_p = \tfrac{1}{4} \Lambda_{pq} A_p - \frac{\sigma f_{pq}}{2\omega_p} \bar{A}_q \exp(i\sigma T_1) \tag{5.4.55}$$

and

$$D_1^2 A_q = \tfrac{1}{4} \Lambda_{pq} A_q - \frac{\sigma f_{qp}}{2\omega_q} \bar{A}_p \exp(i\sigma T_1) \tag{5.4.56}$$

Substituting (5.4.55) and (5.4.56) into (5.4.52) and (5.4.53) yields

$$2i\omega_p D_2 A_p + (\tfrac{1}{4} \Lambda_{pq} + 2\omega_p \hat{\chi}_p) A_p - \frac{\sigma f_{pq}}{2\omega_p} \bar{A}_q \exp(i\sigma T_1) = 0 \tag{5.4.57}$$

and

$$2i\omega_q D_2 A_q + (\tfrac{1}{4} \Lambda_{pq} + 2\omega_q \hat{\chi}_q) A_q - \frac{\sigma f_{qp}}{2\omega_q} \bar{A}_p \exp(i\sigma T_1) = 0 \tag{5.4.58}$$

It can easily be verified that (5.4.26) and (5.4.57) are the first two terms in a multiple-scales expansion of

$$2i\omega_p \frac{dA_p}{dt} + \epsilon f_{pq} \left(1 - \frac{\epsilon\sigma}{2\omega_p}\right) \bar{A}_q \exp(i\epsilon\sigma t) + \epsilon^2 (\tfrac{1}{4} \Lambda_{pq} + 2\omega_p \hat{\chi}_p) A_p = 0 \tag{5.4.59}$$

Similarly (5.4.27) and (5.4.58) are the first two terms in a multiple-scales expansion of

$$2i\omega_q \frac{dA_q}{dt} + \epsilon f_{qp} \left(1 - \frac{\epsilon\sigma}{2\omega_q}\right) \bar{A}_p \exp(i\epsilon\sigma t) + \epsilon^2 (\tfrac{1}{4} \Lambda_{pq} + 2\omega_q \hat{\chi}_q) A_q = 0 \tag{5.4.60}$$

Equations (5.4.59) and (5.4.60) admit a nontrivial solution having the form

$$A_p = a_p \exp\,[i\epsilon(\lambda + \sigma)t] \qquad \text{and} \qquad A_q = a_q \exp\,(-i\epsilon\bar{\lambda}t) \tag{5.4.61}$$

where a_p and a_q are complex constants and

$$\lambda^2 + (\sigma + \epsilon\gamma_1) + \tfrac{1}{4}\Lambda_{pq} + \epsilon\sigma\gamma_2 = 0 \tag{5.4.62}$$

In (5.4.62)

$$\gamma_1 = \tfrac{1}{8}\Lambda_{pq}\left(\frac{1}{\omega_q} - \frac{1}{\omega_p}\right) + \hat{\chi}_q - \hat{\chi}_p \tag{5.4.63}$$

and

$$\gamma_2 = \hat{\chi}_q - \tfrac{1}{8}\frac{\Lambda_{pq}}{\omega_p} \tag{5.4.64}$$

Solving (5.4.62) gives

$$\lambda = -\tfrac{1}{2}\,\{\sigma + \epsilon\gamma_1 \pm [(\sigma + \epsilon\gamma_1)^2 - \Lambda_{pq} - 4\epsilon\sigma\gamma_2]^{1/2}\} \tag{5.4.65}$$

The transition curves correspond to the vanishing of the radical in (5.4.65). That is

$$\sigma + \epsilon\gamma_1 = \pm(\Lambda_{pq})^{1/2}\left(1 + \frac{2\epsilon\sigma\gamma_2}{\Lambda_{pq}}\right) + O(\epsilon^2) \tag{5.4.66}$$

Eliminating σ from (5.4.23) and (5.4.66) yields

$$\begin{aligned}\omega = \omega_p + \omega_q \pm \epsilon(\Lambda_{pq})^{1/2} - \tfrac{1}{2}\epsilon^2 \Bigg\{&\tfrac{1}{4}\Lambda_{pq}\left(\frac{1}{\omega_p} + \frac{1}{\omega_q}\right) \\ &- \sum_r \omega_r \left[\frac{\Lambda_{rq}}{(\omega_p + 2\omega_q)^2 - \omega_r^2} + \frac{\Lambda_{rp}}{(2\omega_p + \omega_q)^2 - \omega_r^2}\right] \\ &- \sum_{r \neq p} \frac{\omega_r\Lambda_{rq}}{\omega_p^2 - \omega_r^2} - \sum_{r \neq q} \frac{\omega_r\Lambda_{rp}}{\omega_q^2 - \omega_r^2}\Bigg\} + O(\epsilon^3)\end{aligned} \tag{5.4.67}$$

This result was obtained first by Valeev (1960, 1961) using the Floquet theory. When $p = q$, (5.4.67) reduces to

$$\omega = 2\omega_p \pm \epsilon\frac{f_{pp}}{\omega_p} - \epsilon^2\left[\frac{f_{pp}^2}{4\omega_p^3} - \sum_r \frac{\omega_r\Lambda_{rp}}{9\omega_p^2 - \omega_r^2} - \sum_{r \neq p} \frac{\omega_r\Lambda_{rp}}{\omega_p^2 - \omega_r^2}\right] + O(\epsilon^3) \tag{5.4.68}$$

If $f_{pr} = 0$ for $r \neq p$, then (5.4.68) reduces to

$$\omega = 2\omega_p \pm \epsilon \frac{f_{pp}}{\omega_p} - \epsilon^2 \frac{f_{pp}^2}{8\omega_p^3} + O(\epsilon^3) \tag{5.4.69}$$

which is the known result for the Mathieu equation (see Section 5.3.2).

Note that one cannot expand (5.4.65) for small ϵ in the neighborhood of the transition curves because $\sigma^2 - \Lambda_{pq}$ is the same order as $2\epsilon\sigma\gamma_1 - 4\epsilon\sigma\gamma_2$. This is the reason (5.4.26), (5.4.27), (5.4.52), and (5.4.53) were combined into (5.4.59) and (5.4.60), which were then used to determine an expansion valid on and near the transition curves. Thus one cannot determine a uniform expansion near the transition curves by expanding the characteristic exponents as well as the dependent variables.

The Case ω Near $\omega_q - \omega_p$. In this case ω is near $\omega_q - \omega_p$, and there are no other resonances. The results can be obtained from those above by simply changing the sign of ω_p.

The Case ω Near $\omega_p + \omega_q$ and 2ω Near $\omega_s - \omega_q$. It is convenient to define two detuning parameters, σ_1 and σ_2; σ_1 is defined according to (5.4.23) while σ_2 is defined according to the following equation:

$$2\omega = \omega_s - \omega_q + \epsilon^2 \sigma_2. \tag{5.4.70}$$

Then the troublesome terms are eliminated if

$$2i\omega_p D_2 A_p + D_1^2 A_p + 2\omega_p \hat{\chi}_p A_p = 0 \tag{5.4.71}$$

$$2i\omega_q D_2 A_q + D_1^2 A_q + 2\omega_q \hat{\chi}_q A_q + 2\omega_q \mu_{qs} A_s \exp(-i\sigma_2 T_2) = 0 \tag{5.4.72}$$

$$iD_2 A_s + \chi_s A_s + \hat{\mu}_{sq} A_q \exp(i\sigma_2 T_2) = 0 \tag{5.4.73}$$

where

$$\hat{\mu}_{sq} = \frac{1}{2\omega_s} \sum_r f_{sr} f_{rq} [(\omega_q + \omega)^2 - \omega_r^2]^{-1}$$

and for $n \neq p, q$, or s the A_n satisfy (5.4.41).

Substituting (5.4.55) and (5.4.56) into (5.4.71) and (5.4.72) yields

$$2i\omega_p D_2 A_p + (\tfrac{1}{4}\Lambda_{pq} + 2\omega_p \hat{\chi}_p) A_p - \frac{\sigma_1 f_{pq}}{2\omega_p} \bar{A}_q \exp(i\sigma_1 T_1) = 0 \tag{5.4.74}$$

and

$$2i\omega_q D_2 A_q + (\tfrac{1}{4}\Lambda_{pq} + 2\omega_q \hat{\chi}_q) A_q - \frac{\sigma_1 f_{qp}}{2\omega_q} \bar{A}_p \exp(i\sigma_1 T_1) + 2\omega_q \mu_{qs} A_s \exp(-i\sigma_2 T_2) = 0 \tag{5.4.75}$$

It can readily be verified that (5.4.26) and (5.4.74) result from a multiple-scale expansion of

$$2i\omega_p \frac{dA_p}{dt} + \epsilon f_{pq}\left(1 - \frac{\epsilon\sigma_1}{2\omega_p}\right)\bar{A}_q \exp(i\epsilon\sigma_1 t) + \epsilon^2(\tfrac{1}{4}\Lambda_{pq} + 2\omega_p\hat{\chi}_p)A_p = 0 \tag{5.4.76}$$

that equations (5.4.27) and (5.4.75) result from

$$2i\omega_q \frac{dA_q}{dt} + \epsilon f_{qp}\left(1 - \frac{\epsilon\sigma_1}{2\omega_q}\right)\bar{A}_p \exp(i\epsilon\sigma_1 t) + \epsilon^2(\tfrac{1}{4}\Lambda_{pq} + 2\omega_q\hat{\chi}_q)A_q + 2\epsilon^2\omega_q\mu_{qs}A_s \exp(-i\epsilon^2\sigma_2 t) = 0 \tag{5.4.77}$$

and that (5.4.21) and (5.4.73) result from

$$i\frac{dA_s}{dt} + \epsilon^2\chi_s A_s + \epsilon^2\hat{\mu}_{sq}A_q \exp(i\epsilon^2\sigma_2 t) = 0 \tag{5.4.78}$$

Equations (5.4.76) through (5.4.78) admit a solution having the form

$$A_p = a_p \exp[i\epsilon(\lambda + \sigma_1)t], \quad A_q = a_q \exp(-i\epsilon\bar{\lambda}t), \quad A_s = a_s \exp[-i\epsilon(\bar{\lambda} - \epsilon\sigma_2)t] \tag{5.4.79}$$

where a_p, a_q, and a_s are complex constants and

$$\lambda^3 + (\gamma_1 + \gamma_2 + \gamma_3)\lambda^2 + (\gamma_1\gamma_2 + \gamma_1\gamma_3 + \gamma_2\gamma_3 + \gamma_4 - \gamma_5)\lambda + \gamma_1\gamma_2\gamma_3 + \gamma_3\gamma_4 - \gamma_2\gamma_5 = 0 \tag{5.4.80}$$

where

$$\gamma_1 = \epsilon\left(\frac{\Lambda_{pq}}{8\omega_q} + \hat{\chi}_q\right), \quad \gamma_2 = \sigma_1 - \epsilon\left(\frac{\Lambda_{pq}}{8\omega_p} + \hat{\chi}_p\right), \quad \gamma_3 = \epsilon(\chi_s - \sigma_2)$$

$$\gamma_4 = \tfrac{1}{4}\Lambda_{pq}\left(1 - \frac{\epsilon\sigma_1}{2\omega_p}\right)\left(1 - \frac{\epsilon\sigma_1}{2\omega_q}\right), \quad \gamma_5 = \epsilon^2\hat{\mu}_{sq}\mu_{qs}$$

Equations (5.4.23) and (5.4.70) give σ_1 and σ_2 in terms of ϵ and ω. The transition curves are the loci of points in the $\epsilon\omega$-plane for which λ has two equal roots.

The case ω simultaneously near $\omega_p + \omega_q$ and 2ω near $\omega_q + \omega_s$ cannot be obtained from the results in this section by changing the sign of ω_q because there is an extra resonance (ω near $\omega_s - \omega_p$) at first order. A similar case is considered below.

The Case ω Near $\omega_p + \omega_q$ and $\omega_s - \omega_q$. In this case ω is simultaneously near $\omega_p + \omega_q$ and $\omega_s - \omega_q$ and thus 2ω is near $\omega_p + \omega_s$; there are no other resonances to this order. The troublesome terms are eliminated from (5.4.40) if

$$2i\omega_p D_2 A_p + D_1^2 A_p + 2\omega_p \hat{\chi}_p A_p + 2\omega_p \hat{\mu}_{ps} \bar{A}_s \exp\left[i(\sigma_1 + \sigma_2) T_1\right] = 0 \tag{5.4.81}$$

$$2i\omega_q D_2 A_q + D_1^2 A_q + 2\omega_q \hat{\chi}_q A_q = 0 \tag{5.4.82}$$

$$2i\omega_s D_2 A_s + D_1^2 A_s + 2\omega_s \hat{\chi}_s A_s + 2\omega_s \hat{\mu}_{sp} \bar{A}_p \exp\left[i(\sigma_1 + \sigma_2) T_1\right] = 0 \tag{5.4.83}$$

where σ_1 and σ_2 are defined by (5.4.34) and $\hat{\mu}_{ps}$ and $\hat{\mu}_{sq}$ are obtained from μ_{ps} and μ_{sq} by deleting the troublesome terms. Using (5.4.35) through (5.4.37), one finds that

$$D_1^2 A_p = -\frac{\sigma_1 f_{pq}}{2\omega_p} \bar{A}_q \exp(i\sigma_1 T_1) + \tfrac{1}{4} \Lambda_{pq} A_p + \frac{f_{pq} f_{qs}}{4\omega_p \omega_q} \bar{A}_s \exp\left[i(\sigma_1 + \sigma_2) T_1\right] \tag{5.4.84}$$

$$D_1^2 A_q = \tfrac{1}{4}(\Lambda_{pq} - \Lambda_{qs}) A_q - \frac{\sigma_1 f_{qp}}{2\omega_q} \bar{A}_p \exp(i\sigma_1 T_1) + \frac{\sigma_2 f_{qs}}{2\omega_q} A_s \exp(-i\sigma_2 T_1) \tag{5.4.85}$$

$$D_1^2 A_s = -\frac{\sigma_2 f_{sq}}{2\omega_s} A_q \exp(i\sigma_2 T_1) - \tfrac{1}{4} \Lambda_{qs} A_s - \frac{f_{sq} f_{qp}}{4\omega_q \omega_s} \bar{A}_p \exp\left[i(\sigma_1 + \sigma_2) T_1\right] \tag{5.4.86}$$

Using (5.4.84) through (5.4.86) in (5.4.81) through (5.4.83), one finds that the resulting equations together with (5.4.35) through (5.4.37) are the first two terms in a multiple-scales expansion of

$$2i\omega_p \frac{dA_p}{dt} + \epsilon f_{pq}\left(1 - \frac{\epsilon\sigma_1}{2\omega_p}\right) \bar{A}_q \exp(i\epsilon\sigma_1 t) + \epsilon^2(\tfrac{1}{4}\Lambda_{pq} + 2\omega_p \hat{\chi}_p) A_p + \epsilon^2\left(2\omega_p \hat{\mu}_{ps} + \frac{f_{pq} f_{qs}}{4\omega_p \omega_q}\right) \bar{A}_s \exp\left[i\epsilon(\sigma_1 + \sigma_2) t\right] = 0 \tag{5.4.87}$$

$$2i\omega_q \frac{dA_q}{dt} + \epsilon f_{qp}\left(1 - \frac{\epsilon\sigma_1}{2\omega_q}\right) \bar{A}_p \exp(i\epsilon\sigma_1 t) + \epsilon f_{qs}\left(1 + \frac{\epsilon\sigma_2}{2\omega_q}\right) A_s \exp(-i\epsilon\sigma_2 t) + \epsilon^2(\tfrac{1}{4}\Lambda_{pq} - \tfrac{1}{4}\Lambda_{qs} + 2\omega_q \hat{\chi}_q) A_q = 0 \tag{5.4.88}$$

$$2i\omega_s \frac{dA_s}{dt} + \epsilon f_{sq}\left(1 - \frac{\epsilon\sigma_2}{2\omega_s}\right) A_q \exp(i\epsilon\sigma_2 t) + \epsilon^2\left(2\omega_s \hat{\mu}_{sp} - \frac{f_{sq} f_{qp}}{4\omega_q \omega_s}\right) \bar{A}_p \times \exp\left[i\epsilon(\sigma_1 + \sigma_2) t\right] + \epsilon^2(2\omega_s \hat{\chi}_s - \tfrac{1}{4}\Lambda_{qs}) A_s = 0 \tag{5.4.89}$$

Equations (5.4.87) through (5.4.89) admit a solution of the form

$$A_p = a_p \exp\left[i\epsilon(\lambda + \sigma_1)t\right], \quad A_q = a_q \exp(-i\epsilon\bar{\lambda}t), \quad A_s = a_s \exp\left[-i\epsilon(\bar{\lambda} - \sigma_2)t\right] \tag{5.4.90}$$

where a_p, a_q, and a_s are complex constants and

$$\lambda^3 + (\sigma_1 - \sigma_2 + \epsilon\gamma_1)\lambda^2 + \left[\tfrac{1}{4}(\Lambda_{pq} - \Lambda_{qs}) - \sigma_1\sigma_2 + \epsilon(\sigma_1\gamma_2 + \sigma_2\gamma_3)\right]\lambda - \tfrac{1}{4}(\sigma_1\Lambda_{qs} + \sigma_2\Lambda_{pq}) + \epsilon(\sigma_1\sigma_2\gamma_4 + \gamma_5) = 0 \tag{5.4.91}$$

where

$$\gamma_1 = \tfrac{1}{8}\left[\Lambda_{pq}\left(\frac{1}{\omega_q} - \frac{1}{\omega_p}\right) - \Lambda_{qs}\left(\frac{1}{\omega_s} + \frac{1}{\omega_q}\right)\right] - \hat{\chi}_p + \hat{\chi}_q + \hat{\chi}_s$$

$$\gamma_2 = -\tfrac{1}{8}\left(\frac{\Lambda_{pq}}{\omega_p} + \frac{\Lambda_{qs}}{\omega_q} + \frac{\Lambda_{qs}}{\omega_s}\right) + \hat{\chi}_q + \hat{\chi}_s$$

$$\gamma_3 = \tfrac{1}{8}\left(\frac{\Lambda_{sq}}{\omega_s} + \frac{\Lambda_{pq}}{\omega_p} - \frac{\Lambda_{pq}}{\omega_q}\right) + \hat{\chi}_p - \hat{\chi}_q$$

$$\gamma_4 = \tfrac{1}{8}\left(\frac{\Lambda_{pq}}{\omega_p} + \frac{\Lambda_{qs}}{\omega_s}\right) - \hat{\chi}_q$$

and

$$\gamma_5 = \tfrac{1}{4}(\Lambda_{qs}\hat{\chi}_p + \Lambda_{pq}\hat{\chi}_s) - \frac{1}{4\omega_q}\left(\frac{\hat{\mu}_{sp} f_{pq} f_{qs}}{\omega_p} + \frac{\hat{\mu}_{ps} f_{qp} f_{sq}}{\omega_s}\right)$$

The transition curves correspond to the value of ω and ϵ for which λ has two equal roots.

Next the results of Sections 5.4.2 and 5.4.3 are applied to the dynamic buckling of a beam under the influence of a periodic follower force.

5.4.4. LATERAL VIBRATIONS OF A COLUMN PRODUCED BY A FOLLOWER FORCE

As a numerical example we consider the linearly elastic, uniform column shown in Figure 5-4. The partial-differential equation governing the small transverse motion without damping is

$$EI\frac{\partial^4 w}{\partial x^4} + m\frac{\partial^2 w}{\partial t^2} + P\cos\Omega t\,\frac{\partial^2 w}{\partial x^2} = 0 \tag{5.4.92}$$

where E is the elastic (Young's) modulus, I is the moment of inertia of the cross-sectional area about the centroidal axis, and m is the mass per unit of length.

It is convenient to introduce dimensionless variables (denoted with an asterisk). Here we put

$$x^* = \frac{x}{l} \quad \text{and} \quad t^* = \frac{t}{l^2}\sqrt{\frac{EI}{m}}$$

Then (5.4.92) becomes

$$\frac{\partial^4 w^*}{\partial x^{*4}} + \frac{\partial^2 w^*}{\partial t^{*2}} + 2\epsilon \cos \omega^* t^* \frac{\partial^2 w^*}{\partial x^{*2}} = 0 \tag{5.4.93}$$

where

$$2\epsilon = \frac{Pl^2}{EI} \quad \text{and} \quad \omega^* = \Omega l^2 \sqrt{\frac{m}{EI}}$$

To complete the statement of the problem, we need to specify the boundary conditions. At $x^* = 0$, there is a clamp; thus both the deflection and the slope are zero there, and hence

$$w^*(0, t) = 0 \quad \text{and} \quad \frac{\partial w^*}{\partial x^*}(0, t) = 0 \tag{5.4.94}$$

At $x^* = 1$, the end is free; thus both the moment and the shear force are zero there, and hence

$$\frac{\partial^2 w^*}{\partial x^{*2}}(1, t) = 0 \quad \text{and} \quad \frac{\partial^3 w^*}{\partial x^{*3}}(1, t) = 0 \tag{5.4.95}$$

We shall drop the asterisk in what follows.

We express the deflection as an expansion in terms of the free-oscillation modes. That is, we put

$$w(x, t) = \sum_{n=1}^{\infty} u_n(t)\, \phi_n(x) \tag{5.4.96}$$

where

$$\phi_n^{\text{iv}}(x) - \lambda_n^4\, \phi_n(x) = 0 \tag{5.4.97}$$

$$\phi_n(0) = \phi_n'(0) = 0 \tag{5.4.98}$$

and

$$\phi_n''(1) = 0, \quad \phi_n'''(1) = 0 \tag{5.4.99}$$

The solution of the eigenvalue problem defined by (5.4.97) through (5.4.99) can be written in the form

$$\phi_n = \cosh \lambda_n x - \cos \lambda_n x - \frac{\cosh \lambda_n + \cos \lambda_n}{\sinh \lambda_n + \sin \lambda_n}(\sinh \lambda_n x - \sin \lambda_n x) \tag{5.4.100}$$

where the λ_n are solutions of

$$\sinh^2 \lambda_n - \sin^2 \lambda_n - (\cosh \lambda_n + \cos \lambda_n)^2 = 0 \tag{5.4.101}$$

We note that the ϕ_n are orthogonal. The ϕ_n are called the free-oscillation modes, or simply modes, and the λ_n^2 are called the natural frequencies of the system. Next we shall obtain the equations governing the time-dependent coefficients u_n.

Substituting (5.4.96) into (5.4.93), multiplying by $\phi_m(x)$, and then integrating the result from $x = 0$ to $x = 1$, we obtain

$$\ddot{u}_m + \omega_m^2 u_m + 2\epsilon \cos \omega t \sum_{n=1}^{\infty} f_{mn} u_n = 0 \tag{5.4.102}$$

where

$$f_{mn} = \frac{\int_0^1 \phi_n'' \phi_m \, dx}{\int_0^1 \phi_m^2 \, dx} \tag{5.4.103}$$

and

$$\omega_m^2 = \lambda_m^4 \tag{5.4.104}$$

Figure 5-11, from Nayfeh and Mook (1977), shows the transition curves. The results for ω near $2\omega_1$ and $\omega_2 - \omega_1$ were computed by using (5.4.67), which

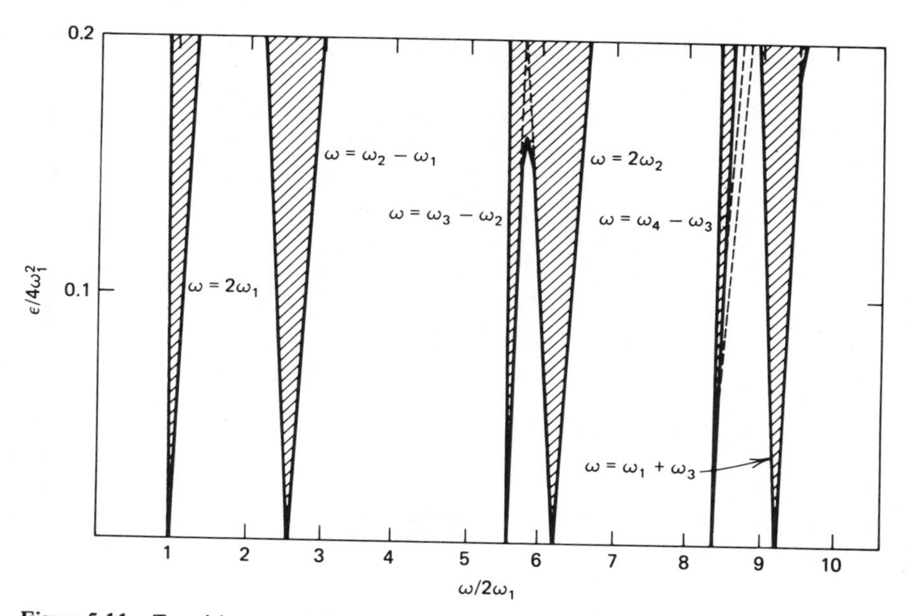

Figure 5-11. Transition curves for the dynamic buckling of a free-fixed column under the influence of a sinusoidal follower force.

applies when there is a single resonance. The results for ω simultaneously near $\omega_3 - \omega_2$ and $2\omega_2$ were computed by using both (5.4.67) (these are shown by dotted lines) and (5.4.80) (these are shown by solid lines); the latter applies when multiresonances occur simultaneously. Moreover the results for ω simultaneously near $\omega_4 - \omega_3$ and $\omega_1 + \omega_3$ were computed by using both (5.4.67) and (5.4.91); the solid lines account for the effect of simultaneous resonances. For the present results 20 terms were used to compute μ_{pq}, μ_{qs}, $\hat{\chi}_p$, $\hat{\chi}_q$, and $\hat{\chi}_s$.

For the case when ω is near $2\omega_2$ and $\omega_3 - \omega_2$ simultaneously, the results obtained from (5.4.80) show a rounded merger lower than the intersection obtained from (5.4.67). For the case when ω is near $\omega_4 - \omega_3$ and $\omega_1 + \omega_3$ simultaneously, the results obtained from (5.4.91) do not intersect, in contrast with the results obtained from (5.4.67).

We note that there is no discernible difference between the two sets of results when ϵ is less than 0.1.

5.4.5. EFFECTS OF VISCOUS DAMPING

It was shown in Section 5.3.5 that viscous damping is stabilizing. However Schmidt and Weidenhammer (1961), Piszczek (1961), and Valeev (1963) showed that viscous damping may have a destabilizing effect on combination resonances. Thus we restrict our attention in this section to exhibiting this destabilizing effect.

We modify (5.4.4) by the addition of viscous damping and obtain

$$\ddot{u}_n + 2\epsilon\mu_n \dot{u}_n + \omega_n^2 u_n + 2\epsilon \cos \omega t \sum_{m=1}^{N} f_{nm} u_m = 0 \qquad (5.4.105)$$

We only consider the resonant case $\omega \approx \omega_p + \omega_q$ and let $\omega = \omega_p + \omega_q + \epsilon\sigma$. Using the method of multiple scales as in Section 5.3.5, we find that

$$u_n = A_n(T_1) \exp(i\omega_n T_0) + cc + O(\epsilon) \qquad (5.5.106)$$

where $A_n' + \mu_n A_n = 0$ for $n \neq p$ and q and

$$\begin{aligned} 2i\omega_p(A_p' + \mu_p A_p) + f_{pq}\bar{A}_q \exp(i\sigma T_1) &= 0 \\ 2i\omega_q(A_q' + \mu_q A_q) + f_{qp}\bar{A}_p \exp(i\sigma T_1) &= 0 \end{aligned} \qquad (5.4.107)$$

We seek a solution for (5.4.107) in the form (5.4.28) and obtain

$$\lambda = -\tfrac{1}{2}\sigma - \tfrac{1}{2}i(\mu_p + \mu_q) \mp \tfrac{1}{2}[\sigma^2 - \Lambda_{pq} - (\mu_q - \mu_p)^2 + 2i\sigma(\mu_q - \mu_p)]^{1/2} \qquad (5.4.108)$$

It follows from (5.4.108) that the motion is stable if $\mu_p + \mu_q \geqslant y$ and unstable if $\mu_p + \mu_q < y$, where y is the imaginary part of the radical in (5.4.108). The transition curve separating stability from instability corresponds to $\mu_p + \mu_q = y$. In this case the radical in (5.4.108) is $x + i(\mu_p + \mu_q)$. Hence

$$[x + i(\mu_p + \mu_q)]^2 = \sigma^2 - \Lambda_{pq} - (\mu_q - \mu_p)^2 + 2i\sigma(\mu_q - \mu_p) \tag{5.4.109}$$

Equating real and imaginary parts in (5.4.109) leads to

$$\begin{aligned} x^2 - (\mu_p + \mu_q)^2 &= \sigma^2 - \Lambda_{pq} - (\mu_q - \mu_p)^2 \\ (\mu_p + \mu_q)x &= \sigma(\mu_q - \mu_p) \end{aligned} \tag{5.4.110}$$

Eliminating x from (5.4.110) and solving the resulting equation for σ^2, we obtain

$$\sigma^2 = (\mu_q + \mu_p)^2 \left(\frac{\Lambda_{pq}}{4\mu_p \mu_q} - 1 \right) \tag{5.4.111}$$

Hence the transition curves separating stability from instability emanating from $\omega = \omega_p + \omega_q$ are given by

$$\omega = \omega_p + \omega_q \pm \epsilon(\mu_q + \mu_p)\left[\Lambda_{pq}(4\mu_p\mu_q)^{-1} - 1\right]^{1/2} + \cdots \tag{5.4.112}$$

As in the case of no damping, the motion is completely stable if $\Lambda_{pq} < 0$. When $\Lambda_{pq} > 0$, there exists a region in the $\delta\omega$-plane in which the motion is unstable. Since $\sigma^2 = \Lambda_{pq}$ in the absence of damping, viscous damping increases the unstable region when $\sigma^2 > \Lambda_{pq}$ and decreases the unstable region when $\sigma^2 < \Lambda_{pq}$. When $\mu_p = \mu_q$, $\sigma^2 = \Lambda_{pq} - 4\mu_p^2$ and viscous damping is stabilizing. As $\mu_p \to 0$ while μ_q is fixed, $\sigma^2 \to \infty$ and viscous damping is destabilizing. For general values of μ_p and μ_q, viscous damping is destabilizing if

$$\Lambda_{pq} < (\mu_q + \mu_p)^2 \left(\frac{\Lambda_{pq}}{4\mu_p \mu_q} - 1 \right)$$

or

$$\Lambda_{pq} > 4\mu_p\mu_q(\mu_q + \mu_p)^2(\mu_q - \mu_p)^{-2} \tag{5.4.113}$$

Certainly (5.4.113) is satisfied if $\mu_p \to 0$ while μ_q is fixed, and it cannot be satisfied if $\mu_q = \mu_p$.

5.5. Linear Systems Having Repeated Frequencies

In contrast with the previous section, here we consider systems having repeated frequencies. We can expose the principal features of the analysis and limit the algebra to a minimum by considering a linear system having three degrees of freedom, with two of the frequencies being equal. Thus we consider

$$\ddot{x}_1 + \omega_1^2 x_1 + 2\epsilon \cos \omega t \sum_{n=1}^{3} f_{1n} x_n = 0 \tag{5.5.1}$$

$$\ddot{x}_2 + \omega_1^2 x_2 + x_1 + 2\epsilon \cos \omega t \sum_{n=1}^{3} f_{2n} x_n = 0 \tag{5.5.2}$$

$$\ddot{x}_3 + \omega_3^2 x_3 + 2\epsilon \cos \omega t \sum_{n=1}^{3} f_{3n} x_n = 0 \tag{5.5.3}$$

where ω_3 is away from ω_1 and ϵ is small. Fu and Nemat-Nasser (1972 a, b, 1975) were the first to analyze such systems.

In the absence of the parametric excitation, the system is unstable (the system is said to be in *flutter*) because x_2 contains a secular or resonant term of the form $t \sin(\omega_1 t + \beta)$, where β is a constant. We wish to determine if the parametric excitation can stabilize the system. To do this, we assume that all three modes are bounded and then, if possible, determine the values of the parameters which are consistent with this assumption, particularly those values at the boundaries of the region where the assumption is valid.

Although the parametric excitation might stabilize the motion, we still expect the amplitude of the x_2-*mode* to be much larger than that of the x_1-*mode.* We do not have any indication of the amplitude of the x_3-*mode.* To express our expectations systematically, we scale the dependent variables. Without loss of generality we put

$$x_1 = u_1, \quad x_2 = \epsilon^{-\lambda_2} u_2, \quad \text{and } x_3 = \epsilon^{-\lambda_3} u_3 \tag{5.5.4}$$

where the u_n are $O(1)$ and the λ_n are positive constants to be determined in the solution. Substituting (5.5.4) into (5.5.1) through (5.5.3) leads to

$$\ddot{u}_1 + \omega_1^2 u_1 + 2(\epsilon f_{11} u_1 + \epsilon^{1-\lambda_2} f_{12} u_2 + \epsilon^{1-\lambda_3} f_{13} u_3) \cos \omega t = 0 \tag{5.5.5}$$

$$\ddot{u}_2 + \omega_1^2 u_2 + \epsilon^{\lambda_2} u_1 + 2(\epsilon^{1+\lambda_2} f_{21} u_1 + \epsilon f_{22} u_2 + \epsilon^{1-\lambda_3+\lambda_2} f_{23} u_3) \cos \omega t = 0 \tag{5.5.6}$$

$$\ddot{u}_3 + \omega_3^2 u_3 + 2(\epsilon^{1+\lambda_3} f_{31} u_1 + \epsilon^{1-\lambda_2+\lambda_3} f_{32} u_2 + \epsilon f_{33} u_3) \cos \omega t = 0 \tag{5.5.7}$$

Here we are concerned with various combinations of the frequencies which can lead to a resonant response. Thus we use the method of multiple scales to determine a uniformly valid approximate solution which exhibits the effects of the repeated frequency and the resonant combinations of frequencies. We write the various time scales as

$$T_0 = t, \quad T_1 = \epsilon^{\lambda_4} t, \quad \text{and } T_2 = \epsilon^{2\lambda_4} t \tag{5.5.8}$$

where λ_4 is another constant to be determined. In terms of these scales,

$$\frac{d^2}{dt^2} = D_0^2 + 2\epsilon^{\lambda_4} D_0 D_1 + \epsilon^{2\lambda_4}(2D_0 D_2 + D_1^2) + \cdots \tag{5.5.9}$$

And we assume expansions in the form

$$u_n(t;\epsilon) = u_{n0}(T_0, T_1, T_2) + \epsilon^{\lambda_4} u_{n1}(T_0, T_1, T_2) + \epsilon^{2\lambda_4} u_{n2}(T_0, T_1, T_2) + \cdots \tag{5.5.10}$$

Substituting (5.5.8) through (5.5.10) into (5.5.5) through (5.5.7) leads to

$$\begin{aligned} &D_0^2 u_{10} + \omega_1^2 u_{10} + \epsilon^{\lambda_4}(D_0^2 u_{11} + \omega_1^2 u_{11} + 2D_0 D_1 u_{10}) \\ &\quad + \epsilon^{2\lambda_4}(D_0^2 u_{12} + \omega_1^2 u_{12} + 2D_0 D_2 u_{10} + D_1^2 u_{10} + 2D_0 D_1 u_{11}) \\ &\quad + 2(\epsilon f_{11} u_{10} + \epsilon^{1-\lambda_2} f_{12} u_{20} + \epsilon^{1-\lambda_2+\lambda_4} f_{12} u_{21} \\ &\quad + \epsilon^{1-\lambda_3} f_{13} u_{30} + \epsilon^{1-\lambda_3+\lambda_4} f_{13} u_{31}) \cos \omega T_0 + \cdots = 0 \end{aligned} \tag{5.5.11}$$

$$\begin{aligned} &D_0^2 u_{20} + \omega_1^2 u_{20} + \epsilon^{\lambda_4}(D_0^2 u_{21} + \omega_1^2 u_{21} + 2D_0 D_1 u_{20}) \\ &\quad + \epsilon^{2\lambda_4}(D_0^2 u_{22} + \omega_1^2 u_{22} + 2D_0 D_1 u_{21} + D_1^2 u_{20} + 2D_0 D_2 u_{20}) \\ &\quad + \epsilon^{\lambda_2} u_{10} + 2(\epsilon^{1+\lambda_2} f_{21} u_{10} + \epsilon f_{22} u_{20} + \epsilon^{1+\lambda_4} f_{22} u_{21} \\ &\quad + \epsilon^{1-\lambda_3+\lambda_2} f_{23} u_{30} + \epsilon^{1-\lambda_3+\lambda_2+\lambda_4} f_{23} u_{31}) \cos \omega T_0 + \cdots = 0 \end{aligned} \tag{5.5.12}$$

$$\begin{aligned} &D_0^2 u_{30} + \omega_3^2 u_{30} + \epsilon^{\lambda_4}(D_0^2 u_{31} + \omega_3^2 u_{31} + 2D_0 D_1 u_{30}) \\ &\quad + \epsilon^{2\lambda_4}(D_0^2 u_{32} + \omega_3^2 u_{32} + 2D_0 D_1 u_{31} + D_1^2 u_{30} + 2D_0 D_2 u_{30}) \\ &\quad + 2(\epsilon^{1+\lambda_3} f_{31} u_{10} + \epsilon^{1-\lambda_2+\lambda_3} f_{32} u_{20} + \epsilon^{1-\lambda_2+\lambda_3+\lambda_4} f_{32} u_{21} \\ &\quad + \epsilon f_{33} u_{30} + \epsilon^{1+\lambda_4} f_{33} u_{31}) \cos \omega T_0 + \cdots = 0 \end{aligned} \tag{5.5.13}$$

Next we consider the following resonant combinations: ω near $2\omega_1$, ω near $\omega_1 + \omega_3$, and ω near ω_1.

5.5.1. THE CASE OF ω NEAR $2\omega_1$

In this case the resonant combination is

$$\omega - \omega_1 = \omega_1 + 2\epsilon^{\lambda_5} \sigma \tag{5.5.14}$$

where λ_5 is another constant to be determined and σ is a familiar detuning parameter. To determine the effects of the resonant combination and the repeated frequency, we must at least include the effect of $f_{12}u_{20}$ in (5.5.11) and the effect of u_{10} in (5.5.12). Thus we put

$$\lambda_2 = \lambda_4 = \tfrac{1}{2} \tag{5.5.15}$$

And because the resonant term appears in the equation for u_{11}, we put

$$\lambda_5 = \tfrac{1}{2} \tag{5.5.16}$$

In this case λ_3 cannot be determined unless initial conditions are taken into account. It can be easily shown that the stability of the motion is independent of the value of λ_3. For simplicity we set $\lambda_3 = 0$.

Equating coefficients of like powers of $\epsilon^{1/2}$ in (5.5.11) through (5.5.13) leads to

Order ϵ^0

$$D_0^2 u_{10} + \omega_1^2 u_{10} = 0 \tag{5.5.17}$$

$$D_0^2 u_{20} + \omega_1^2 u_{20} = 0 \tag{5.5.18}$$

$$D_0^2 u_{30} + \omega_3^2 u_{30} = 0 \tag{5.5.19}$$

Order $\epsilon^{1/2}$

$$D_0^2 u_{11} + \omega_1^2 u_{11} + 2D_0 D_1 u_{10} + 2f_{12} u_{20} \cos \omega T_0 = 0 \tag{5.5.20}$$

$$D_0^2 u_{21} + \omega_1^2 u_{21} + 2D_0 D_1 u_{20} + u_{10} = 0 \tag{5.5.21}$$

$$D_0^2 u_{31} + \omega_3^2 u_{31} + 2D_0 D_1 u_{30} + 2f_{32} u_{20} \cos \omega T_0 = 0 \tag{5.5.22}$$

For a first approximation that includes the effects of the repeated frequency and the resonant combination, we do not need to determine the dependence of the result on the scale T_2. Thus we write the solutions of (5.5.17) through (5.5.19) as follows:

$$u_{10} = A_1(T_1) \exp(i\omega_1 T_0) + cc \tag{5.5.23}$$

$$u_{20} = A_2(T_1) \exp(i\omega_1 T_0) + cc \tag{5.5.24}$$

$$u_{30} = A_3(T_1) \exp(i\omega_3 T_0) + cc \tag{5.5.25}$$

Substituting (5.5.14) and (5.5.23) through (5.5.25) into (5.5.20) through (5.5.22) leads to the following conditions for the elimination of secular terms from u_{11}, u_{21}, and u_{31}:

$$2i\omega_1 A_1' + f_{12} \bar{A}_2 \exp(2i\sigma T_1) = 0 \tag{5.5.26}$$

$$2i\omega_1 A_2' + A_1 = 0 \tag{5.5.27}$$

$$2i\omega_3 A_3' = 0 \tag{5.5.28}$$

where the prime indicates the derivative with respect to T_1. In this case there is no strong interaction between all the modes. Rather the system responds as two independent smaller systems—one having two degrees of freedom and the other having a single degree of freedom.

Eliminating A_1 from (5.5.26) and (5.5.27) leads to

$$A_2'' + \frac{f_{12}}{4\omega_1^2} \bar{A}_2 \exp(2i\sigma T_1) = 0 \tag{5.5.29}$$

Letting

$$A_2 = (B_r + iB_i) \exp(i\sigma T_1) \tag{5.5.30}$$

where B_r and B_i are real functions of T_1 leads to

$$B_r'' - 2\sigma B_i' + \left(\frac{f_{12}}{4\omega_1^2} - \sigma^2\right) B_r = 0 \tag{5.5.31}$$

$$B_i'' + 2\sigma B_r' - \left(\frac{f_{12}}{4\omega_1^2} + \sigma^2\right) B_i = 0 \tag{5.5.32}$$

Equations (5.5.31) and (5.5.32) admit nontrivial solutions having the form

$$(B_r, B_i) = (b_r, b_i) \exp(\gamma T_1) \tag{5.5.33}$$

where b_r, b_i, and γ are constants provided that

$$\begin{vmatrix} \gamma^2 - \sigma^2 + \dfrac{f_{12}}{4\omega_1^2} & -2\sigma\gamma \\ 2\sigma\gamma & \gamma^2 - \sigma^2 - \dfrac{f_{12}}{4\omega_1^2} \end{vmatrix} = 0 \tag{5.5.34}$$

or

$$\gamma^2 = -\sigma^2 \pm \frac{f_{12}}{4\omega_1^2} \tag{5.5.35}$$

The motion is stable if γ is imaginary. Thus the motion is stable if

$$\sigma^2 > \frac{|f_{12}|}{4\omega_1^2} \tag{5.5.36}$$

and unstable if

$$\sigma^2 < \frac{|f_{12}|}{4\omega_1^2} \tag{5.5.37}$$

The transition curves correspond to

$$\sigma = \pm \frac{1}{2\omega_1} (|f_{12}|)^{1/2} \tag{5.5.38}$$

and hence the transition curves emanating from $\omega \approx 2\omega_1$ are

$$\omega = 2\omega_1 + 2\epsilon^{1/2}\sigma = 2\omega_1 \pm \frac{\epsilon^{1/2}}{\omega_1} (|f_{12}|)^{1/2} + O(\epsilon) \tag{5.5.39}$$

Thus the parametric excitation can be stabilizing when ω is near $2\omega_1$.

5.5.2. THE CASE OF ω NEAR $\omega_1 + \omega_3$

In this case the resonant combinations are written as

$$\omega - \omega_1 = \omega_3 + \epsilon^{\lambda_s}\sigma \quad \text{and} \quad \omega - \omega_3 = \omega_1 + \epsilon^{\lambda_s}\sigma \tag{5.5.40}$$

where λ_5 is to be determined. To determine the effects of the repeated frequency and the resonant combination, we must at least include $2f_{13}u_{30}\cos\omega T_0$ in (5.5.11), u_{10} in (5.5.12), and $2f_{32}u_{20}\cos\omega T_0$ in (5.5.13). Initially it may appear that we have two choices:

1. $1-\lambda_3=\lambda_4, \quad \lambda_2=\lambda_4, \quad \text{and } 1-\lambda_2+\lambda_3=\lambda_4$ (5.5.41)

2. $1-\lambda_3=2\lambda_4, \quad \lambda_2=2\lambda_4, \quad \text{and } 1-\lambda_2+\lambda_3=2\lambda_4$ (5.5.42)

The first choice leads to

$$\lambda_2=\lambda_4=\tfrac{2}{3} \quad \text{and} \quad \lambda_3=\tfrac{1}{3} \tag{5.5.43}$$

while the second leads to

$$\lambda_2=\tfrac{2}{3} \quad \text{and} \quad \lambda_3=\lambda_4=\tfrac{1}{3} \tag{5.5.44}$$

We note that the order of the second terms in expansion (5.5.10) is higher for the first choice than it is for the second choice. Thus it appears that one may exclude or miss a term with the first choice, and hence the resulting expansion may be inconsistent or incomplete. In this case we note that for the first choice the order of the term containing $2f_{12}u_{20}\cos(\omega T_0)$ is $\epsilon^{1/3}$. Thus equating coefficients of like powers of ϵ leads to the impossible condition

$$f_{12}u_{20}=0 \tag{5.5.45}$$

Here we proceed using the second choice (5.5.42). This is an example in which we must consider the first three terms in expansion (5.5.10) in order to obtain a first approximation which includes the effects of the repeated frequency and the resonant combination.

Equating coefficients of like powers of ϵ leads to

Order ϵ^0

$$D_0^2u_{10}+\omega_1^2u_{10}=0 \tag{5.5.46}$$

$$D_0^2u_{20}+\omega_1^2u_{20}=0 \tag{5.5.47}$$

$$D_0^2u_{30}+\omega_3^2u_{30}=0 \tag{5.5.48}$$

Order $\epsilon^{1/3}$

$$D_0^2u_{11}+\omega_1^2u_{11}+2D_0D_1u_{10}+2f_{12}u_{20}\cos\omega T_0=0 \tag{5.5.49}$$

$$D_0^2u_{21}+\omega_1^2u_{21}+2D_0D_1u_{20}=0 \tag{5.5.50}$$

$$D_0^2u_{31}+\omega_3^2u_{31}+2D_0D_1u_{30}=0 \tag{5.5.51}$$

Order $\epsilon^{2/3}$

$$D_0^2u_{12}+\omega_1^2u_{12}+2D_0D_1u_{11}+2D_0D_2u_{10}+D_1^2u_{10} + 2f_{12}u_{21}\cos\omega T_0+2f_{13}u_{30}\cos\omega T_0=0 \tag{5.5.52}$$

$$D_0^2 u_{22} + \omega_1^2 u_{22} + 2D_0 D_1 u_{21} + 2D_0 D_2 u_{20} + D_1^2 u_{20} + u_{10} = 0 \quad (5.5.53)$$

$$D_0^2 u_{32} + \omega_3^2 u_{32} + 2D_0 D_1 u_{31} + 2D_0 D_2 u_{30} + D_1^2 u_{30} + 2f_{32} u_{20} \cos \omega T_0 = 0 \quad (5.5.54)$$

The solutions of (5.5.46) to (5.5.48) can be written as

$$u_{10} = A_1(T_1, T_2) \exp(i\omega_1 T_0) + cc \quad (5.5.55)$$

$$u_{20} = A_2(T_1, T_2) \exp(i\omega_1 T_0) + cc \quad (5.5.56)$$

$$u_{30} = A_3(T_1, T_2) \exp(i\omega_3 T_0) + cc \quad (5.5.57)$$

Substituting into (5.5.49) through (5.5.51), one finds that secular terms are eliminated from u_{11}, u_{21}, and u_{31} if

$$D_1 A_1 = D_1 A_2 = D_1 A_3 = 0 \quad (5.5.58)$$

Thus the solution is independent of T_1, and hence we put

$$\lambda_5 = 2\lambda_4 = \tfrac{2}{3} \quad (5.5.59)$$

It follows that

$$u_{11} = f_{12} \left\{ \frac{A_2 \exp[i(\omega + \omega_1)T_0]}{\omega(\omega + 2\omega_1)} + \frac{\bar{A}_2 \exp[i(\omega - \omega_1)T_0]}{\omega(\omega - 2\omega_1)} \right\} + cc \quad (5.5.60)$$

$$u_{21} = u_{31} = 0 \quad (5.5.61)$$

Substituting all these results into (5.5.52) through (5.5.54), one finds that secular terms are eliminated from u_{12}, u_{22}, and u_{32} if

$$2i\omega_1 A_1' + f_{13}\bar{A}_3 \exp(i\sigma T_2) = 0 \quad (5.5.62)$$

$$2i\omega_1 A_2' + A_1 = 0 \quad (5.5.63)$$

$$2i\omega_3 A_3' + f_{32}\bar{A}_2 \exp(i\sigma T_2) = 0 \quad (5.5.64)$$

where the prime indicates the derivative with respect to T_2.

Eliminating A_2 and A_3 yields

$$A_1''' - i\sigma A_1'' - i\frac{f_{13} f_{32}}{8\omega_1^2 \omega_3} A_1 = 0 \quad (5.5.65)$$

Equation (5.5.65) admits solutions having the form

$$A_1 = a \exp(i\gamma T_2) \quad (5.5.66)$$

where a and γ are constants, provided that

$$\gamma^3 - \sigma\gamma^2 + \frac{f_{13} f_{32}}{8\omega_1^2 \omega_3} = 0 \quad (5.5.67)$$

Equation (5.5.67) has either three real roots or one real root and two complex roots that are conjugates. It follows from (5.5.66) that the motion is stable if, and only if, all the roots are real. Thus the transition from stable to unstable motion occurs when two of the roots are equal.

We denote the roots corresponding to transition as γ_1, γ_1, and γ_3. Then

$$2\gamma_1 + \gamma_3 = \sigma \tag{5.5.68}$$

$$\gamma_1(\gamma_1 + 2\gamma_3) = 0 \tag{5.5.69}$$

$$\gamma_1^2\gamma_3 = -\frac{f_{13}f_{32}}{8\omega_1^2\omega_3} \tag{5.5.70}$$

Equation (5.5.69) shows that either $\gamma_1 = 0$ or $\gamma_1 + 2\gamma_3 = 0$. We consider these possibilities next. When $\gamma_1 = 0$, it follows that $f_{13}f_{32} = 0$ and that $\gamma_3 = \sigma$. Thus there are three real roots, two of which are zero, for all values of σ. When $\gamma_1 = -2\gamma_3$, it follows from (5.5.63) and (5.5.70) that

$$\sigma = \tfrac{3}{2}\left(\frac{f_{13}f_{32}}{4\omega_1^2\omega_3}\right)^{1/3} \tag{5.5.71}$$

Combining (5.5.40) and (5.5.71) yields the following transition curve:

$$\omega = \omega_1 + \omega_3 + \tfrac{3}{2}\left(\frac{f_{13}f_{32}}{4\omega_1^2\omega_3}\right)^{1/3}\epsilon^{2/3} + \cdots \tag{5.5.72}$$

Thus the parametric excitation can be stabilizing when ω is near $\omega_1 + \omega_3$.

5.5.3. THE CASE OF ω NEAR ω_1

In this case the resonant combination is

$$2\omega - \omega_1 = \omega_1 + \epsilon^{\lambda_5}\sigma \tag{5.5.73}$$

Thus we must include the terms containing $f_{12}u_{21}\cos\omega T_0$ in (5.5.11) and u_{10} in (5.5.12). The resonant terms first appear in the equations governing u_{n1}. Consequently we put

$$\lambda_2 = \lambda_4 = \lambda_5 = 1 \tag{5.5.74}$$

It follows that λ_3 must be either zero or unity. It can be shown that the value of γ_3 does not affect the stability of the solution, so we set it equal to zero.

Equating coefficients of like powers of ϵ leads to

Order ϵ^0

$$D_0^2u_{10} + \omega_1^2u_{10} + 2f_{12}u_{20}\cos\omega T_0 = 0 \tag{5.5.75}$$

$$D_0^2 u_{20} + \omega_1^2 u_{20} = 0 \tag{5.5.76}$$

$$D_0^2 u_{30} + \omega_3^2 u_{30} + 2f_{32} u_{20} \cos \omega T_0 = 0 \tag{5.5.77}$$

Order ϵ^1

$$D_0^2 u_{11} + \omega_1^2 u_{11} + 2D_0 D_1 u_{10} + 2(f_{11} u_{10} + f_{12} u_{21} + f_{13} u_{30}) \cos \omega T_0 = 0 \tag{5.5.78}$$

$$D_0^2 u_{21} + \omega_1^2 u_{21} + 2D_0 D_1 u_{20} + u_{10} + 2f_{22} u_{20} \cos \omega T_0 = 0 \tag{5.5.79}$$

$$D_0^2 u_{31} + \omega_3^2 u_{31} + 2D_0 D_1 u_{30} + 2(f_{31} u_{10} + f_{32} u_{21} + f_{33} u_{30}) \cos \omega T_0 = 0 \tag{5.5.80}$$

It follows that u_{20} is given by (5.5.24). Then

$$u_{30} = A_3(T_1) \exp(i\omega_3 T_0) + f_{32} \left\{ \frac{A_2 \exp [i(\omega + \omega_1) T_0)}{(\omega + \omega_1)^2 - \omega_3^2} + \frac{\bar{A}_2 \exp [i(\omega - \omega_1) T_0]}{(\omega - \omega_1)^2 - \omega_3^2} \right\} + cc \tag{5.5.81}$$

$$u_{10} = A_1(T_1) \exp(i\omega_1 T_0) + f_{12} \left\{ \frac{A_2 \exp [i(\omega + \omega_1) T_0]}{\omega(\omega + 2\omega_1)} + \frac{\bar{A}_2 \exp [i(\omega - \omega_1) T_0]}{\omega(\omega - 2\omega_1)} \right\} + cc \tag{5.5.82}$$

Substituting these results into (5.5.79), one finds that secular terms are eliminated from u_{21} if

$$2i\omega_1 A_2' + A_1 = 0 \tag{5.5.83}$$

and that

$$u_{21} = \frac{A_2}{\omega(\omega + 2\omega_1)} \left[f_{22} + \frac{f_{12}}{\omega(\omega + 2\omega_1)} \right] \exp [i(\omega + \omega_1) T_0] + \frac{\bar{A}_2}{\omega(\omega - 2\omega_1)} \left[f_{22} + \frac{f_{12}}{\omega(\omega - 2\omega_1)} \right] \exp [i(\omega - \omega_1) T_0] + cc \tag{5.5.84}$$

Moreover it follows from (5.5.80) that secular terms are eliminated from u_{31} if

$$A_3' = 0 \tag{5.5.85}$$

Substituting (5.5.81), (5.5.82), and (5.5.84) into (5.5.78), we find that secular terms are eliminated from u_{11} if

$$iA_1' + 2\omega_1 \alpha_1 A_2 + 2\omega_1 \alpha_2 \bar{A}_2 \exp(i\sigma T_1) = 0 \tag{5.5.86}$$

where

$$4\omega_1^2\alpha_1 = \frac{2f_{12}(f_{11}+f_{22})}{\omega^2-4\omega_1^2} + \frac{f_{12}^2}{\omega^2}\left[\frac{1}{(\omega-2\omega_1)^2}+\frac{1}{(\omega+2\omega_1)^2}\right]$$
$$+ f_{13}f_{32}\left[\frac{1}{(\omega+\omega_1)^2-\omega_3^2}+\frac{1}{(\omega-\omega_1)^2-\omega_3^2}\right]$$
$$\approx -\frac{2f_{12}(f_{11}+f_{22})}{3\omega_1^2} + \frac{10f_{12}^2}{9\omega_1^4} - f_{13}f_{32}\left[\frac{1}{\omega_3^2}+\frac{1}{\omega_3^2-4\omega_1^2}\right] \tag{5.5.87}$$

$$4\omega_1^2\alpha_2 = \frac{f_{12}(f_{11}+f_{22})}{\omega(\omega-2\omega_1)} + \frac{f_{12}^2}{\omega^2(\omega-2\omega_1)^2} + \frac{f_{13}f_{32}}{(\omega-\omega_1)^2-\omega_3^2}$$
$$\approx \frac{f_{12}^2}{\omega_1^4} - \frac{f_{12}(f_{11}+f_{22})}{\omega_1^2} - \frac{f_{13}f_{32}}{\omega_3^2} \tag{5.5.88}$$

Eliminating A_1 from (5.5.83) and (5.5.86) yields

$$A_2'' + \alpha_1 A_2 + \alpha_2 \bar{A}_2 \exp(i\sigma T_1) = 0 \tag{5.5.89}$$

Letting

$$A_2 = (B_r + iB_i)\exp(\tfrac{1}{2} i\sigma T_1) \tag{5.5.90}$$

where B_r and B_i are real functions of T_1, leads to

$$B_r'' - \sigma B_i' + (\alpha_1 + \alpha_2 - \tfrac{1}{4}\sigma^2)B_r = 0$$
$$B_i'' + \sigma B_r' + (\alpha_1 - \alpha_2 - \tfrac{1}{4}\sigma^2)B_i = 0 \tag{5.5.91}$$

Equations (5.5.91) admit a solution having the form

$$(B_r, B_i) = (b_r b_i)\exp(\gamma T_1) \tag{5.5.92}$$

where b_r, b_i, and γ are constants, provided that

$$\begin{vmatrix} \gamma^2 + \alpha_1 - \frac{1}{4}\sigma^2 + \alpha_2 & -\sigma\gamma \\ \sigma\gamma & \gamma^2 + \alpha_1 - \frac{1}{4}\sigma^2 - \alpha_2 \end{vmatrix} = 0 \tag{5.5.93}$$

or

$$(\gamma^2 + \alpha_1 - \tfrac{1}{4}\sigma^2)^2 - \alpha_2^2 + \sigma^2\gamma^2 = 0 \tag{5.5.94}$$

Hence

$$\gamma^2 = -(\alpha_1 + \tfrac{1}{4}\sigma^2) \mp (\alpha_1\sigma^2 + \alpha_2^2)^{1/2} \tag{5.5.95}$$

Equations (5.5.90) and (5.5.92) show that A_2 is unbounded, and hence the motion is unstable if the real part of any of the γ's is positive definite. The transition curves separating stability from instability correspond to

$$(\alpha_1 - \tfrac{1}{4}\sigma^2)^2 = \alpha_2^2 \qquad \text{or} \qquad \sigma = \pm 2(\alpha_1 \pm \alpha_2)^{1/2} \tag{5.5.96}$$

Hence the transition curves emanating from $\omega = \omega_1$ are given by

$$\omega = \omega_1 \mp \epsilon(\alpha_1 \pm \alpha_2)^{1/2} + \cdots \tag{5.5.97}$$

Thus the parametric excitation can stabilize an otherwise unstable motion.

5.6. Gyroscopic Systems

In this section we consider linear systems governed by equations having the form

$$[M]\ddot{\mathbf{u}} + [G]\dot{\mathbf{u}} + [A]\mathbf{u} = 0 \tag{5.6.1}$$

where $\mathbf{u}$ is a column vector having n components, $[M]$ is an $n \times n$ symmetric matrix, $[G]$ is an $n \times n$ antisymmetric matrix, and $[A]$ is an $n \times n$ matrix. In this system $[M]\ddot{\mathbf{u}} + [G]\dot{\mathbf{u}}$ represents the inertia and $[G]\dot{\mathbf{u}}$ represents the portion that is due to gyroscopic effects. Systems governed by equations containing inertial terms such as $[G]\dot{\mathbf{u}}$ are called *gyroscopic systems* because their behavior is characteristic of the gyroscope.

Systems of this type were analyzed by a number of investigators. Smith (1933), Tondl (1965), Black and McTernan (1968), and Iwatsubo, Tomita, and Kawai (1973) studied the vibrations of asymmetric shafts and rotors supported by asymmetric bearings. Fedorchenko (1958, 1961) analyzed the motion of gyroscopes resting on vibrating supports. Danby (1964), Grebenikov (1964), Alfriend and Rand (1969), Luk'ianov (1969), Markeev (1970), Nayfeh and Kamel (1970a), and Nayfeh (1970a) analyzed the stability of the triangular points in the elliptic restricted problem of three bodies. Kane and Sobala (1963) and Nishikawa and Willems (1969) analyzed satellite attitude stability. Kane and Mingori (1965) investigated the effect of a rotor on the attitude of a satellite in a circular orbit, Mingori (1969) determined the effect of internal damping on the stability of dual-spin satellites, and Lindh and Likins (1970) used an infinite determinant method to analyze the problem studied by Mingori (1969). The stability of spinning asymmetric satellites in circular orbits was studied by Kane and Shippy (1963) and Meirovitch and Wallace (1967), while the stability of spinning symmetric satellites in elliptic orbits was studied by Kane and Barba (1966), Wallace and Meirovitch (1967), and Markeev (1967b).

For simplicity we consider a system having two degrees of freedom in order to illustrate the basic method of analysis and obtain the characteristics of the solution. Specifically we consider

$$\begin{aligned} \ddot{u}_1 + \lambda_1 \dot{u}_2 + \alpha_1 u_1 + 2\epsilon(f_{11}u_1 + f_{12}u_2)\cos\omega t = 0 \\ \ddot{u}_2 - \lambda_2 \dot{u}_1 + \alpha_2 u_2 + 2\epsilon(f_{21}u_1 + f_{22}u_2)\cos\omega t = 0 \end{aligned} \tag{5.6.2}$$

where ϵ, ω, λ_i, α_i, and f_{ij} are constants.

We seek a first-order uniform expansion of the solution of (5.6.2) for small ϵ in the form

$$u_m = u_{m0}(T_0, T_1) + \epsilon u_{m1}(T_0, T_1) + \cdots \tag{5.6.3}$$

where $T_n = \epsilon^n t$. Substituting (5.6.3) into (5.6.2), transforming the derivatives, and equating coefficients of like powers of ϵ, we obtain

Order ϵ^0

$$D_0^2 u_{10} + \lambda_1 D_0 u_{20} + \alpha_1 u_{10} = 0 \tag{5.6.4}$$

$$D_0^2 u_{20} - \lambda_2 D_0 u_{10} + \alpha_2 u_{20} = 0 \tag{5.6.5}$$

Order ϵ

$$D_0^2 u_{11} + \lambda_1 D_0 u_{21} + \alpha_1 u_{11} = -2D_0 D_1 u_{10} - \lambda_1 D_1 u_{20} - 2(f_{11}u_{10} + f_{12}u_{20})\cos\omega T_0 \tag{5.6.6}$$

$$D_0^2 u_{21} - \lambda_2 D_0 u_{11} + \alpha_2 u_{21} = -2D_0 D_1 u_{20} + \lambda_2 D_1 u_{10} - 2(f_{21}u_{10} + f_{22}u_{20})\cos\omega T_0 \tag{5.6.7}$$

We write the solutions of (5.6.4) and (5.6.5) in the form

$$u_{10} = A_1(T_1)\exp(i\omega_1 T_0) + A_2(T_1)\exp(i\omega_2 T_0) + cc \tag{5.6.8}$$

$$u_{20} = \frac{i(\alpha_1 - \omega_1^2)}{\lambda_1 \omega_1} A_1 \exp(i\omega_1 T_0) + \frac{i(\alpha_1 - \omega_2^2)}{\lambda_1 \omega_2} A_2 \exp(i\omega_2 T_0) + cc \tag{5.6.9}$$

where ω_1^2 and ω_2^2 are the solutions of

$$\omega^4 - (\alpha_1 + \alpha_2 + \lambda_1\lambda_2)\omega^2 + \alpha_1\alpha_2 = 0 \tag{5.6.10}$$

Here we assume that ω_1 and ω_2 are real and positive, and we make $\omega_2 > \omega_1$.

Substituting (5.6.8) and (5.6.9) into (5.5.6) and (5.6.7) leads to

$$\begin{aligned} D_0^2 u_{11} + \lambda_1 D_0 u_{21} + \alpha_1 u_{11} = & -\frac{i(\alpha_1 + \omega_1^2)}{\omega_1} A_1' \exp(i\omega_1 T_0) \\ & - \frac{i(\alpha_1 + \omega_2^2)}{\omega_2} A_2' \exp(i\omega_2 T_0) \\ & - \left[f_{11} + i \frac{(\alpha_1 - \omega_1^2) f_{12}}{\lambda_1 \omega_1} \right] A_1 \exp[i(\omega + \omega_1) T_0] \\ & - \left[f_{11} - \frac{i(\alpha_1 - \omega_1^2) f_{12}}{\lambda_1 \omega_1} \right] \bar{A}_1 \exp[i(\omega - \omega_1) T_0] \\ & - \left[f_{11} + \frac{i(\alpha_1 - \omega_2^2) f_{12}}{\lambda_1 \omega_2} \right] A_2 \exp[i(\omega + \omega_2) T_0] \\ & - \left[f_{11} - \frac{i(\alpha_1 - \omega_2^2) f_{12}}{\lambda_1 \omega_2} \right] \bar{A}_2 \exp[i(\omega - \omega_2) T_0] + cc \end{aligned} \tag{5.6.11}$$

$$\begin{aligned} D_0^2 u_{21} - \lambda_2 D_0 u_{11} + \alpha_2 u_{21} = & \frac{2\alpha_1 + \lambda_1 \lambda_2 - 2\omega_1^2}{\lambda_1} A_1' \exp(i\omega_1 T_0) \\ & + \frac{2\alpha_1 + \lambda_1 \lambda_2 - 2\omega_2^2}{\lambda_1} A_2' \exp(i\omega_2 T_0) \\ & - \left[f_{21} + \frac{i(\alpha_1 - \omega_1^2) f_{22}}{\lambda_1 \omega_1} \right] A_1 \exp[i(\omega + \omega_1) T_0] \\ & - \left[f_{21} - \frac{i(\alpha_1 - \omega_1^2) f_{22}}{\lambda_1 \omega_1} \right] \bar{A}_1 \exp[i(\omega - \omega_1) T_0] \\ & - \left[f_{21} + \frac{i(\alpha_1 - \omega_2^2) f_{22}}{\lambda_1 \omega_2} \right] A_2 \exp[i(\omega + \omega_2) T_0] \\ & - \left[f_{21} - \frac{i(\alpha_1 - \omega_2^2) f_{22}}{\lambda_1 \omega_2} \right] \bar{A}_2 \exp[i(\omega - \omega_2)] T_0 \end{aligned} \tag{5.6.12}$$

To this order there are several possible resonant combinations of ω, ω_1, and ω_2. These include (a) $\omega \approx 2\omega_1$, (b) $\omega \approx 2\omega_2$, (c) $\omega \approx \omega_1 + \omega_2$, and (d) $\omega \approx \omega_2 - \omega_1$. In the following subsections we consider the case when none of these resonant combinations exists as well as several resonant cases.

5.6.1. THE NONRESONANT CASE

Because (5.6.11) and (5.6.12) are linear equations, we can obtain particular solutions for each of the terms on the right-hand sides independently. To determine the solvability conditions, we seek particular solutions corresponding to the terms containing the factors $\exp(i\omega_n T_0)$ in the form

$$\begin{aligned} u_{11} &= P_1(T_1)\exp(i\omega_1 T_0) + Q_1(T_1)\exp(i\omega_2 T_0) \\ u_{21} &= P_2(T_1)\exp(i\omega_1 T_0) + Q_2(T_1)\exp(i\omega_2 T_0) \end{aligned} \tag{5.6.13}$$

Substituting (5.6.13) into (5.6.11) and (5.6.12) and equating the coefficients of $\exp(i\omega_n T_0)$ on both sides, we obtain

$$\begin{aligned} (\alpha_1 - \omega_1^2)P_1 + i\omega_1\lambda_1 P_2 &= R_1 \\ -i\omega_1\lambda_2 P_1 + (\alpha_2 - \omega_1^2)P_2 &= R_2 \end{aligned} \tag{5.6.14}$$

where

$$R_1 = -\frac{i(\alpha_1 + \omega_1^2)}{\omega_1}A_1', \qquad R_2 = \frac{2\alpha_1 + \lambda_1\lambda_2 - 2\omega_1^2}{\lambda_1}A_1' \tag{5.6.15}$$

and

$$\begin{aligned} (\alpha_1 - \omega_2^2)Q_1 + i\omega_2\lambda_1 Q_2 &= S_1 \\ -i\omega_2\lambda_2 Q_1 + (\alpha_2 - \omega_2^2)Q_2 &= S_2 \end{aligned} \tag{5.6.16}$$

where

$$S_1 = -\frac{i(\alpha_1 + \omega_2^2)}{\omega_2}A_2', \qquad S_2 = \frac{2\alpha_1 + \lambda_1\lambda_2 - 2\omega_2^2}{\lambda_1}A_2' \tag{5.6.17}$$

Because the coefficient matrices for (5.6.14) and (5.6.16) are singular according to (5.6.10), solutions of (5.6.14) and (5.6.16) do not exist unless

$$\begin{vmatrix} \alpha_1 - \omega_1^2 & R_1 \\ -i\omega_1\lambda_2 & R_2 \end{vmatrix} = 0 \tag{5.6.18a}$$

and

$$\begin{vmatrix} \alpha_1 - \omega_2^2 & S_1 \\ -i\omega_2\lambda_2 & S_2 \end{vmatrix} = 0 \tag{5.6.18b}$$

It follows that

$$(\alpha_1 - \omega_1^2)R_2 + i\omega_1\lambda_2 R_1 = 0 \tag{5.6.19a}$$

and

$$(\alpha_1 - \omega_2^2)S_2 + i\omega_2\lambda_2 S_1 = 0 \tag{5.6.19b}$$

Substituting (5.6.15) into (5.6.19a) leads to

$$A_1' = 0 \tag{5.6.20}$$

while substituting (5.6.17) into (5.6.19b) leads to

$$A_2' = 0 \tag{5.6.21}$$

Hence A_1 and A_2 are independent of T_1, and one has to continue the expansion to second order to determine the dependence of A_1 and A_2 on T_2.

5.6.2. THE CASE OF ω NEAR $2\omega_n$

Here we consider only the case of $\omega \approx 2\omega_1$. The case of $\omega \approx 2\omega_2$ can be treated in a similar fashion. Thus we set

$$\omega = 2\omega_1 + \epsilon\sigma \tag{5.6.22}$$

and then write

$$\exp\left[i(\omega - \omega_1)T_0\right] = \exp\left[i\omega_1 T_0 + i\sigma T_1\right] \tag{5.6.23}$$

In this case we also seek particular solutions which correspond to the terms containing the factors $\exp(i\omega_n T_0)$ and have the form of (5.6.13). But now, instead of (5.6.15), we obtain

$$\begin{aligned} R_1 &= -\frac{i(\alpha_1 + \omega_1^2)}{\omega_1} A_1' - \left[f_{11} - \frac{i(\alpha_1 - \omega_1^2)}{\lambda_1\omega_1} f_{12}\right]\bar{A}_1 \exp(i\sigma T_1) \\ R_2 &= \frac{2\alpha_1 + \lambda_1\lambda_2 - 2\omega_1^2}{\lambda_1} A_1' - \left[f_{21} - \frac{i(\alpha_1 - \omega_1^2)}{\lambda_1\omega_1} f_{22}\right]\bar{A}_1 \exp(i\sigma T_1) \end{aligned} \tag{5.6.24}$$

Then it follows from (5.6.19a) that

$$A_1' - \Gamma\bar{A}_1 \exp\left[i(\sigma T_1 + \tau)\right] = 0 \tag{5.6.25}$$

where Γ and τ are real constants such that

$$\begin{aligned} \Gamma\exp(i\tau) = \tfrac{1}{2}\left[\frac{(\alpha_1 - \omega_1^2)^2}{\lambda_1} + \alpha_1\lambda_2\right]^{-1} \Bigg[& i\omega_1\lambda_2 f_{11} - \frac{i(\alpha_1 - \omega_1^2)^2}{\lambda_1\omega_1} f_{22} \\ & + \frac{(\alpha_1 - \omega_1^2)\lambda_2}{\lambda_1} f_{12} + (\alpha_1 - \omega_1^2) f_{21}\Bigg] \end{aligned} \tag{5.6.26}$$

We substitute

$$A_1 = (B_r + iB_i)\exp\left[\tfrac{1}{2}i(\sigma T_1 + \tau)\right] \tag{5.6.27}$$

where B_r and B_i are real into (5.6.25), separate real and imaginary parts, and obtain

$$\begin{aligned} B_r' - \tfrac{1}{2}\sigma B_i - \Gamma B_r &= 0 \\ B_i' + \tfrac{1}{2}\sigma B_r + \Gamma B_i &= 0 \end{aligned} \tag{5.6.28}$$

The solution of (5.6.28) has the form

$$(B_r, B_i) = (b_r, b_i) \exp(\gamma T_1) \tag{5.6.29}$$

where b_r, b_i, and γ are constants. It follows that a solution exists only if

$$\gamma = \pm(\Gamma^2 - \tfrac{1}{4}\sigma^2)^{1/2} \tag{5.6.30}$$

The results for A_2 are the same as in the nonresonant case. Hence A_2 is independent of T_1.

Combining (5.6.30), (5.6.29), and (5.6.27) yields

$$A_1 = (b_r + ib_i) \exp[\gamma T_1 + \tfrac{1}{2}i(\sigma T_1 + \tau)] \tag{5.6.31}$$

It follows that A_1 is bounded, and hence the motion is bounded if, and only if, $\sigma \geqslant 2|\Gamma|$. Consequently the transition from stability to instability corresponds to $\sigma = \pm 2\Gamma$, and the transition curves are given by

$$\omega = 2\omega_1 \pm 2\epsilon\Gamma + O(\epsilon^2) \tag{5.6.32}$$

In contrast with the nongyroscopic case, the transition curves are in general functions of f_{12}, f_{21}, and f_{22} in addition to f_{11}.

5.6.3. THE CASE OF ω NEAR $\omega_2 - \omega_1$

We let

$$\omega = \omega_2 - \omega_1 + \epsilon\sigma \tag{5.6.33}$$

Hence we can write

$$\begin{aligned} \exp[i(\omega + \omega_1)T_0] &= \exp(i\omega_2 T_0 + i\sigma T_1) \\ \exp[i(\omega_2 - \omega)T_0] &= \exp(i\omega_1 - i\sigma T_1) \end{aligned} \tag{5.6.34}$$

We continue to seek particular solutions having the form of (5.6.13). This leads to equations having the form of (5.6.14) and (5.6.16). But now, instead of (5.6.15) and (5.6.17), we find that

$$\begin{aligned} R_1 &= -\frac{i(\alpha_1 + \omega_1^2)}{\omega_1} A_1' - \left[f_{11} + \frac{i(\alpha_1 - \omega_2^2) f_{12}}{\lambda_1 \omega_2} \right] A_2 \exp(-i\sigma T_1) \\ R_2 &= \frac{2\alpha_1 + \lambda_1\lambda_2 - 2\omega_1^2}{\lambda_1} A_1' - \left[f_{21} + \frac{i(\alpha_1 - \omega_2^2) f_{22}}{\lambda_1 \omega_2} \right] A_2 \exp(-i\sigma T_1) \end{aligned} \tag{5.6.35}$$

$$S_1 = -\frac{i(\alpha_1 + \omega_2^2)}{\omega_2} A_2' - \left[f_{11} + \frac{i(\alpha_1 - \omega_1^2)}{\lambda_1 \omega_1} f_{12}\right] A_1 \exp(i\sigma T_1)$$

$$S_2 = \frac{2\alpha_1 + \lambda_1 \lambda_2 - 2\omega_2^2}{\lambda_1} A_2' - \left[f_{21} + \frac{i(\alpha_1 - \omega_1^2)}{\lambda_1 \omega_1} f_{22}\right] A_1 \exp(i\sigma T_1) \tag{5.6.36}$$

Then it follows from (5.6.19a) that

$$A_1' = \Gamma_1 A_2 \exp(-i\sigma T_1) \tag{5.6.37}$$

and from (5.6.19b) that

$$A_2' = \Gamma_2 A_1 \exp(i\sigma T_1) \tag{5.6.38}$$

where

$$\Gamma_1 = \tfrac{1}{2}\left[\frac{(\alpha_1 - \omega_1^2)^2}{\lambda_1} + \alpha_1 \lambda_2\right]^{-1} \left\{(\alpha_1 - \omega_1^2) f_{21} + \frac{i(\alpha_1 - \omega_2^2)(\alpha_1 - \omega_1^2)}{\lambda_1 \omega_2} f_{22} + i\omega_1 \lambda_2 f_{11} - \frac{\omega_1 \lambda_2 (\alpha_1 - \omega_2^2)}{\omega_2 \lambda_1} f_{12}\right\} \tag{5.6.39}$$

$$\Gamma_2 = \tfrac{1}{2}\left[\frac{(\alpha_1 - \omega_2^2)^2}{\lambda_1} + \alpha_1 \lambda_2\right]^{-1} \left\{(\alpha_1 - \omega_2^2) f_{21} + \frac{i(\alpha_1 - \omega_1^2)(\alpha_1 - \omega_2^2)}{\omega_1 \lambda_1} f_{22} + i\omega_2 \lambda_2 f_{11} - \frac{\omega_2 \lambda_2 (\alpha_1 - \omega_1^2)}{\omega_1 \lambda_1} f_{12}\right\} \tag{5.6.40}$$

The solution of (5.6.37) and (5.6.38) has the form

$$A_1 = a_1 \exp(\gamma T_1), \qquad A_2 = a_2 \exp[(\gamma + i\sigma) T_1] \tag{5.6.41}$$

provided that

$$\gamma^2 + i\sigma\gamma - \Gamma_1 \Gamma_2 = 0 \tag{5.6.42}$$

and

$$a_2 = \frac{\gamma}{\Gamma_1} a_1 \tag{5.6.43}$$

The solution of (5.6.42) is

$$\gamma = -\tfrac{1}{2} i\sigma \mp \sqrt{\Gamma_1 \Gamma_2 - \tfrac{1}{4}\sigma^2} \tag{5.6.44}$$

Since $\Gamma_1 \Gamma_2$ is complex, in general, the real part of one of the roots of γ is always positive definite, and the motion is unstable. In special cases $\Gamma_1 \Gamma_2$ is real (such as $f_{11} = f_{22} = 0$, or $f_{12} = f_{21} = 0$), and the motion is stable only when $\sigma^2 \geqslant 4\Gamma_1 \Gamma_2$. The transition from stable to unstable motions corresponds to $\sigma = \pm 2\sqrt{\Gamma_1 \Gamma_2}$.

Hence the transition curve is given by

$$\omega = \omega_2 - \omega_1 \pm 2\epsilon\sqrt{\Gamma_1 \Gamma_2} + O(\epsilon^2) \tag{5.6.45}$$

Note that unstable motions occur only when Γ_1 and Γ_2 have the same sign. Moreover in contrast with the nongyroscopic case, the transition curves are functions of f_{11} and f_{22} in addition to f_{12} and f_{21}.

The case $\omega \approx \omega_2 + \omega_1$ can be obtained from the above results by simply changing the sign of ω_1.

5.7 Effects of Nonlinearities

In the previous sections we found that parametrically excited linear, undamped systems possess solutions that grow indefinitely with time. However actual systems possess some degree of damping which has a stabilizing effect, except that in some cases viscous damping may destabilize a system having a combination resonance (see Section 5.4.5). If the system is truly linear, the amplitude grows until the system is destroyed as happened to the specially designed linear oscillating circuit of Mandelstam and Papalexi (1934) whose amplitude of oscillation grew until the insulation was destroyed by an excessive voltage. However most systems possess some degree of nonlinearity which comes into play as soon as the amplitude of the motion becomes appreciable, and it modifies the response. In some instances, as the amplitude grows, the nonlinearity limits the growth, resulting in a limit cycle, as happened in the specially designed nonlinear oscillating circuit of Mandelstam and Papalexi (1934). Thus although the linear theory is useful in determining the initial growth or decay, it may be inadequate if the system possesses any nonlinearity. In this section and in Section 6.7 we discuss the effect of nonlinearities on systems having single- and multidegree-of-freedoms, respectively.

In addition to those mentioned in Section 5.1.4 that deal with the effects of nonlinearities on the parametric response of elastic systems, there are a number of investigations that deal with various dynamic systems. Pendulums with oscillating points of support were studied by Hirsch (1930), Kauderer (1958), Skalak and Yarymovych (1960), Struble (1963), Ness (1967), Dugundji and Chhatpar (1970), Troger (1975), and Chester (1975). The nonlinear motion of a gyroscopic pendulum with a moving point of support was studied by Sethna and Hemp (1965). Kauderer (1958), Bolotin (1964), Weidenhammer (1956), and Boston (1971) studied Mathieu's equation with cubic nonlinearities, while Tso and Caughey (1965) studied an equation of the form

$$\ddot{u} + \delta u + \epsilon\alpha_1 u^m + \epsilon\alpha_2 u^{n-1} - \epsilon\alpha_3 u^{n-1} \cos \omega t = 0$$

Hsu (1974c) determined exact solutions for a nonlinear Hill's equation when the parametric excitation is given in terms of a Jacobian elliptic function, while

Schwartz (1970) and Pun (1973) analyzed periodic solutions of a second-order nonlinear conservative differential equation with periodic coefficients, while Schneider (1972) analyzed periodic solutions of nonlinear differential equations with periodic coefficients. Tso and Asmis (1974) studied nonlinear parametric resonances in a system having two degrees of freedom. Effects of nonlinear damping were treated by Bogoliubov and Mitropolsky (1961), Bolotin (1964), Hagedorn (1968, 1969, 1970a, b), Hsu (1975a), Nguyen (1976a), and Tondl (1976c). Jong (1969) and Tso and Asmis (1970) studied parametric excitations of a circulatory system and a pendulum, respectively, with bilinear hysteretic damping. The combined influence of parametric and external excitations was investigated by Ness (1971), Hsu and Cheng (1974), Nguyen (1975b), and Troger and Hsu (1977).

To exhibit the influence of nonlinearities, we consider the behavior of solutions of

$$\ddot{u} + (\delta + 2\epsilon \cos 2t)\, u = \epsilon f(u, \dot{u}) \tag{5.7.1}$$

which is a modified Mathieu equation. In what follows we restrict our treatment to the case of principal resonance (i.e., $\delta \approx 1$) and obtain a first approximation only.

To determine the combined effect of nonlinearities and parametric excitations on the amplitude and phase, we use the method of multiple scales and let

$$u(t;\epsilon) = u_0(T_0, T_1) + \epsilon u_1(T_0, T_1) + \cdots \tag{5.7.2}$$

where $T_0 = t$ and $T_1 = \epsilon t$. Substituting (5.7.2) into (5.7.1) and equating the coefficients of ϵ^0 and ϵ on both sides, we obtain

$$D_0^2 u_0 + \delta u_0 = 0 \tag{5.7.3}$$

$$D_0^2 u_1 + \delta u_1 = -2D_0 D_1 u_0 - 2u_0 \cos 2T_0 + f(u_0, D_0 u_0) \tag{5.7.4}$$

The general solution of (5.7.3) is

$$u_0 = A(T_1) \exp(i\omega T_0) + \bar{A}(T_1) \exp(-i\omega T_0) \tag{5.7.5}$$

where $\delta = \omega^2$. Hence (5.7.4) becomes

$$\begin{aligned} D_0^2 u_1 + \omega^2 u_1 = {} & -2i\omega A' \exp(i\omega T_0) - \bar{A} \exp[i(-\omega + 2)\, T_0] \\ & - \bar{A} \exp[-i(\omega + 2)\, T_0] + cc \\ & + f\{A \exp(i\omega T_0) + \bar{A} \exp(-i\omega T_0), \\ & \quad i\omega [A \exp(i\omega T_0) - A \exp(-i\omega T_0)]\} \end{aligned} \tag{5.7.6}$$

where the primes indicate differentiation with respect to T_1. To express the nearness of δ to 1, we let

$$1 = \omega + \epsilon\sigma \tag{5.7.7}$$

so that we can express $(2 - \omega) T_0$ as

$$(2 - \omega) T_0 = \omega T_0 + 2\epsilon\sigma T_0 = \omega T_0 + 2\sigma T_1 \tag{5.7.8}$$

Using (5.7.8) in eliminating the terms from (5.7.6) that produce secular terms in u_1, we have

$$2i\omega A' = -\bar{A} \exp(2i\sigma T_1) + \frac{\omega}{2\pi} \int_0^{2\pi/\omega} f \exp(-i\omega T_0)\, dT_0 \tag{5.7.9}$$

Letting $A = \frac{1}{2} a \exp(i\beta)$ in (5.7.9) with real a and β and separating real and imaginary parts, we have

$$a' = -\frac{a}{2\omega} \sin(2\sigma T_1 - 2\beta) - \frac{1}{2\pi\omega} \int_0^{2\pi} \sin\phi\, f(a\cos\phi, -\omega a \sin\phi)\, d\phi \tag{5.7.10}$$

$$a\beta' = \frac{a}{2\omega} \cos(2\sigma T_1 - 2\beta) - \frac{1}{2\pi\omega} \int_0^{2\pi} \cos\phi\, f(a\cos\phi, -\omega a \sin\phi)\, d\phi \tag{5.7.11}$$

where $\phi = \omega T_0 + \beta$. Therefore to the first approximation

$$u = a\cos(\omega T_0 + \beta) + O(\epsilon) \tag{5.7.12}$$

where a and β are given by (5.7.10) and (5.7.11), which can be transformed into the autononous system

$$a' = -\frac{a}{2\omega} \sin\psi - \frac{1}{2\pi\omega} \int_0^{2\pi} \sin\phi\, f(a\cos\phi, -\omega a \sin\phi)\, d\phi \tag{5.7.13}$$

$$a\psi' = 2\sigma a - \frac{a}{\omega} \cos\psi + \frac{1}{\pi\omega} \int_0^{2\pi} \cos\phi\, f(a\cos\phi, -\omega a \sin\phi)\, d\phi \tag{5.7.14}$$

where

$$\psi = 2\sigma T_1 - 2\beta \tag{5.7.15}$$

Eliminating β from (5.7.12) and (5.7.15) yields

$$u = a\cos(t - \tfrac{1}{2}\psi) + O(\epsilon) \tag{5.7.16}$$

Next we use (5.7.13) and (5.7.14) to ascertain the influence of nonlinearities.

5.7.1. THE CASE OF QUADRATIC DAMPING

In this case $f(u, \dot{u}) = -\mu\dot{u}|\dot{u}|$, and (5.7.13) and (5.7.14) become

$$a' = -\frac{a}{2\omega}\sin\psi - \frac{4}{3\pi}\mu\omega a^2$$
$$a\psi' = 2\sigma a - \frac{a}{\omega}\cos\psi \tag{5.7.17}$$

In the absence of the nonlinearity, the system is stable (i.e., no energy is pumped into the system) if $|\sigma| > 1/2\omega \approx \frac{1}{2}$, and the system is unstable (i.e., energy is pumped into the system) if $|\sigma| < 1/2\omega \approx \frac{1}{2}$.

When the system is linearly stable, the phase ψ is such that no energy is being pumped into the system by the parametric excitation; that is, $\sin\psi > 0$. Since the damping term in the equation for a' is always negative, the motion dies out no matter how large the initial amplitude is. On the other hand, when the system is linearly unstable, energy is being pumped into the system by the parametric excitation (i.e., $\sin\psi < 0$) causing a to increase. But as a increases, the dissipation due to viscosity restricts the increase of a. Thus steady-state motions occur when the rate at which energy is being pumped into the system is exactly equal to the rate at which energy is being dissipated. When this occurs, $a' = \psi' = 0$, and the steady-state motions correspond to

$$a\left(\sin\psi + \frac{8}{3\pi}\mu\omega^2 a\right) = 0 \tag{5.7.18}$$

$$a(\cos\psi - 2\omega\sigma) = 0 \tag{5.7.19}$$

Equation (5.7.19) shows that a necessary condition for the existence of nontrivial steady-state amplitudes is $|\sigma| < 1/2\omega$ or approximately $|\sigma| < \frac{1}{2}$ because $\omega \approx 1$. Thus as anticipated above, steady-state motions exist only if the linear motion is unstable. That is, the phasing is such that energy is pumped into the system by the parametric excitation. When $a \neq 0$, the solution of (5.7.18) and (5.7.19) is

$$\psi = \cos^{-1}(2\omega\sigma)$$
$$a = \frac{3\pi}{8\mu\omega^2}(1 - 4\omega^2\sigma^2)^{1/2} \tag{5.7.20}$$

Since the steady-state motion corresponds to ψ = constant, (5.7.16) shows that the effect of the nonlinear damping is to limit the unstable linear motion to a finite-amplitude motion whose frequency is one half the frequency of the excitation. In other words, a subharmonic is generated by the system.

5.7.2 THE RAYLEIGH OSCILLATOR

For the Rayleigh oscillator, $f = \dot{u} - \dot{u}^3$, and (5.7.13) and (5.7.14) become

$$a' = -\frac{a}{2\omega}\sin\psi + \tfrac{1}{2}a(1 - \tfrac{3}{4}\omega^2 a^2) \tag{5.7.21}$$

$$a\psi' = 2\sigma a - \frac{a}{\omega}\cos\psi \tag{5.7.22}$$

Also in this case parametrically unstable linear motions do not grow indefinitely with time because the damping tends to limit the growth. Thus if the phasing is such that energy is being pumped into the system by the parametric excitation, a steady-state motion occurs whenever the rate at which energy is being pumped into the system is exactly equal to the rate at which energy is being dissipated. This occurs when $a' = \psi' = 0$. The steady-state motion corresponds to

$$a\sin\psi = a\omega(1 - \tfrac{3}{4}\omega^2 a^2) \tag{5.7.23}$$

$$a\cos\psi = 2\sigma\omega a \tag{5.7.24}$$

Equation (5.7.24) shows that finite-amplitude steady-state motions exist only if $|\sigma| \leqslant 1/2\omega \approx \frac{1}{2}$, that is, if the linear motion is unstable. In this case eliminating ψ from (5.7.23) and (5.7.24) leads to the frequency-response equation

$$4\sigma^2 + (1 - \tfrac{3}{4}\rho)^2 = \omega^{-2} \tag{5.7.25}$$

where $\rho = \omega^2 a^2 \approx a^2$. This response curve is shown in Figure 5-12. Note that finite-amplitude motions do not exist when $|\sigma| > \frac{1}{2}$, that is, when the linear motion is stable. Since for steady-state motions ψ is a constant, the nonlinearity limits the unstable linear motions to finite-amplitude motions whose frequency

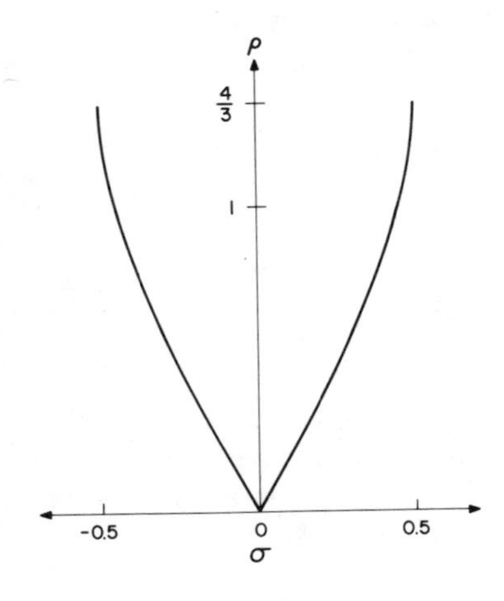

Figure 5-12. Frequency–response curves of a parametrically excited Rayleigh oscillator.

is exactly one half the frequency of the excitation; that is, a subharmonic is generated.

5.7.3. THE DUFFING EQUATION WITH SMALL DAMPING

Among other places, equations of the Duffing type arise in the study of the lateral vibrations of pin-ended columns subjected to periodic loads. In this context the second derivative represents the inertia, the linear term represents the restoring force due to bending, and the cubic term represents the restoring force due to stretching of the neutral axis.

For the Duffing equation with a small amount of viscous damping, $f = -\alpha u^3 - 2\mu\dot{u}$ and (5.7.13) and (5.7.14) become

$$a' = -\frac{a}{2\omega} \sin \psi - \mu a \tag{5.7.26}$$

$$a\psi' = 2\sigma a - \frac{a}{\omega} \cos \psi - \frac{3\alpha}{4\omega} a^3 \tag{5.7.27}$$

In this case the nonlinearity does not affect the amplitude directly, as in the previous two examples, but it affects the amplitude indirectly through changing the phase ψ.

In the absence of the nonlinearity, the motion is unstable if the parameters δ and ϵ correspond to a point above the curve in Figure 5-9; otherwise it is stable. If the parameters correspond to a point above the curve, $\sin \psi < -2\mu\omega$ initially; and no matter how small a is initially, provided it is different from zero, energy will be pumped into the system by the parametric excitation leading to an increase in a. However this increase will be accompanied by a change in the phase ψ according to (5.7.27) and hence a change in the rate of energy being pumped into the system. When the rate at which energy is being pumped into the system is exactly balanced by the rate at which energy is being dissipated by viscous effects, the system achieves a steady-state motion, thereby the amplitude will be limited by the nonlinearity to a finite value. The steady-state motions occur when $a' = \psi' = 0$, which for the nontrivial case corresponds to the solutions of

$$\begin{aligned} \sin \psi &= -2\omega\mu \\ \cos \psi &= 2\sigma\omega - \tfrac{3}{4}\alpha a^2 \end{aligned} \tag{5.7.28}$$

Hence, recalling that $\omega = 1 + \epsilon\sigma$, one finds that in the first approximation the steady-state amplitudes are given by

$$a = \left[\frac{8\sigma}{3\alpha} \pm \frac{4}{3\alpha}(1 - 4\mu^2)^{1/2}\right]^{1/2} \tag{5.7.29}$$

For a steady-state solution to exist, a^2 must be positive. Thus 2μ must be less than unity (i.e., the amplitude of the excitation must be greater than the damping coefficient to produce a sustained motion). Moreover we note that if the above conditions are satisfied, only one steady-state solution is possible if $|\sigma| < \frac{1}{2}(1 - 4\mu^2)^{1/2}$ while two are possible if $\sigma > \frac{1}{2}(1 - 4\mu^2)^{1/2}$ when $\alpha > 0$. If these steady-state solutions are also stable, then one should be able to observe them in an experiment. Next we consider the stability of these various possible solutions.

We note that (5.7.26) and (5.7.27) have the form of equation (3.2.3) and that what we are calling a steady-state solution here is a singluar point or an equilibrium state in the discussion of Chapter 3. Thus determining the stability of the steady-state solution is precisely the problem of determining the nature of the singular points, as described in Section 3.2.1.

In this case the coefficient matrix [see (3.2.16)] is

$$[A] = \begin{bmatrix} 0 & -\frac{1}{2}a_0 \cos \psi_0 \\ -\frac{3}{2}\alpha a_0 & -2\mu \end{bmatrix} \tag{5.7.30}$$

where the subscript 0 denotes steady-state values. The eigenvalues of $[A]$, as given by (3.2.23), are

$$\lambda_{1,2} = -\mu \pm (\mu^2 + \tfrac{3}{4}\alpha a_0^2 \cos \psi_0)^{1/2} \tag{5.7.31}$$

Thus if $\cos \psi_0$ is negative, the equilibrium point is a stable node or focal point; and if $\cos \psi_0$ is positive, the equilibrium point is a saddle point.

We note from equations (5.7.28) and (5.7.29) that, when there are two steady-state solutions, $\cos \psi_0$ is positive for the solution having the smaller amplitude, and thus it is unstable; while $\cos \psi_0$ is negative for the solution having the larger amplitude, and thus it is stable. When there is only one solution, $\cos \psi_0$ is negative, and thus it is stable.

For further discussion of these results, it is more direct to rewrite (5.7.29) as

$$a^2 = \frac{8\hat{\sigma}}{3\hat{\alpha}} \pm \frac{4}{3\hat{\alpha}}(\epsilon^2 - 4\hat{\mu}^2)^{1/2} \tag{5.7.32}$$

where we introduced the new notation $\hat{\sigma} = \epsilon\sigma$, $\hat{\alpha} = \epsilon\alpha$, and $\epsilon\mu = \hat{\mu}$. We recall that ω is the natural frequency of the linear system $(1 - \omega = \epsilon\sigma = \hat{\sigma})$, 2 is the frequency and ϵ is the amplitude of the excitation, $\hat{\alpha}$ is the coefficient of the nonlinear term, and $2\hat{\mu}$ is the coefficient of the damping term. Here we express the amplitude of the response as a function of the natural frequency and consider the frequency of the excitation to be unchangeable. By introducing a new independent variable, one can readily reorient the problem so that the response appears as a function of the frequency of the excitation.

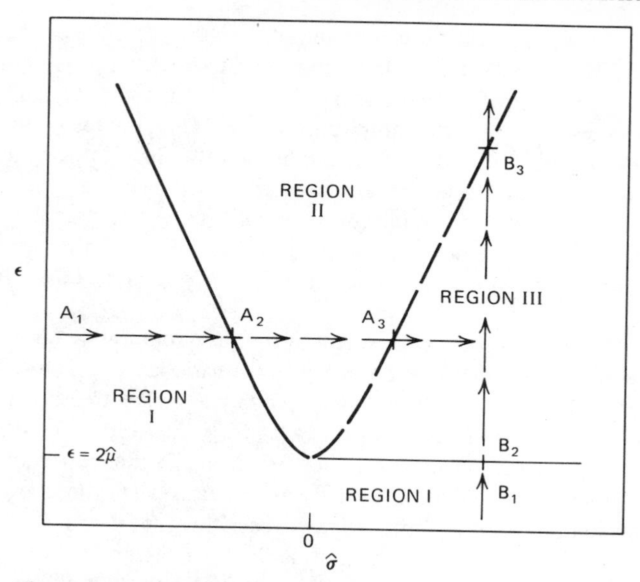

Figure 5-13. The various regions in parameter space for the classifications of the steady-state solutions of a parametrically excited Duffing equation.

In Figure 5-13, for a given $\hat{\mu}$ and $\hat{\alpha}$, the $\epsilon\sigma$-plane is divided into three regions by the three curves $\hat{\sigma} = \pm\frac{1}{2}(\epsilon^2 - 4\hat{\mu}^2)^{1/2}$ and $\epsilon = 2\hat{\mu}$. We note that the boundaries of these regions are independent of $\hat{\alpha}$, the coefficient of the nonlinear term. However the behavior of the solution in regions II and III is certainly dependent on $\hat{\alpha}$.

It appears that the responses to all initial disturbances, regardless of how large the amplitude, decay in region I. In region II, the response of the linear system to any initial disturbance grows without bound while the response of the nonlinear system is bounded. In region III, the response of the nonlinear system to an initial disturbance may either decay or achieve a sustained periodic motion, while the response of the linear system always decays. This behavior can be explained as follows: It appears that regardless of how large the initial amplitude is in region I, the phasing never becomes such that the rate at which work is being done is as large as the rate at which energy is being dissipated. Thus we conclude that, for the linear as well as the nonlinear system, the phasing is such that the force actually does negative work and thus contributes to the decay. In region II for a linear system the phasing is such that work is being done at a faster rate than energy is being dissipated, and thus the response to any initial disturbance grows without bound. For the nonlinear system the phasing for large amplitudes differs from the phasing in the linear system due to the presence of the nonlinear term in (5.7.27). The effect is to limit the rate at which work is being done to the rate at which energy is being dissipated and thereby to produce a bounded harmonic response. If the initial amplitude is very large, the response will decay until the steady-state solution is reached. On the other hand, if the initial disturbance is very small, the response will grow (the system being governed by the linear equations when the amplitude is small) until the nonlinear term in (5.7.27) becomes large enough to cause the phase shift. Thus in region II all initial disturbances produce the same steady-state response (i.e., a limit cycle exists).

In region III the response of the linear system to an initial disturbance always decays. The mechanism causing the decay is the same as in region I. The results for the nonlinear system show that only the larger of the two possible steady-state responses is stable. Thus it appears that for some initial disturbances, the nonlinear term in (5.2.27) does not have a strong influence on the resulting motion and the system behaves essentially as a linear system; the motion decays. On the other hand, for other initial disturbances the nonlinear term has a strong influence; phase changes such as those described for region II occur, and a nontrivial steady-state solution exists. Thus in region III there is the possibility of producing motions which have characteristics that are similar to those of the motions in region I as well as region II. The boundary separating the two regions is the unstable limit cycle corresponding to the smaller amplitude. The type of motion is determined by the amplitude of the initial disturbance. This is a rare example in which a nontrivial, steady-state response of the nonlinear system exists in a region where the response of the linear system decays.

Let us suppose that $\hat{\sigma}$ is increased while ϵ is held constant. This process is represented by the line through points A_1, A_2, and A_3. Between points A_1 and A_2 only the trivial solution exists, and it is stable. Between points A_2 and A_3 the trivial solution is unstable, and the only realizable solution is given by (5.7.32). Beyond point A_3 the trivial solution is again stable and so is the larger solution given by (5.7.32); hence two solutions are realizable.

Finally let us suppose ϵ is increased while $\hat{\sigma}$ is held constant. This process is represented by the line through points B_1, B_2, and B_3. Between points B_1 and B_2 only the trivial solution exists, and it is stable. Between points B_2 and B_3 two steady-state solutions are realizable–the trivial solution, which is still stable, and the larger one given by (5.7.32). Beyond point B_3 the trivial solution is unstable, and only the solution given by (5.7.32) is realizable.

For the first process, $\hat{a}$ is plotted as a function of $\hat{\sigma}$ in Figure 5-14. If the frequency decreases from a large value, we note that upon reaching point A_3, where the trivial solution becomes unstable, there can be a jump up to point A_3'. This process is indicated by the arrows.

For the second process, $\hat{a}$ is plotted as a function of ϵ in Figure 5-15. Using

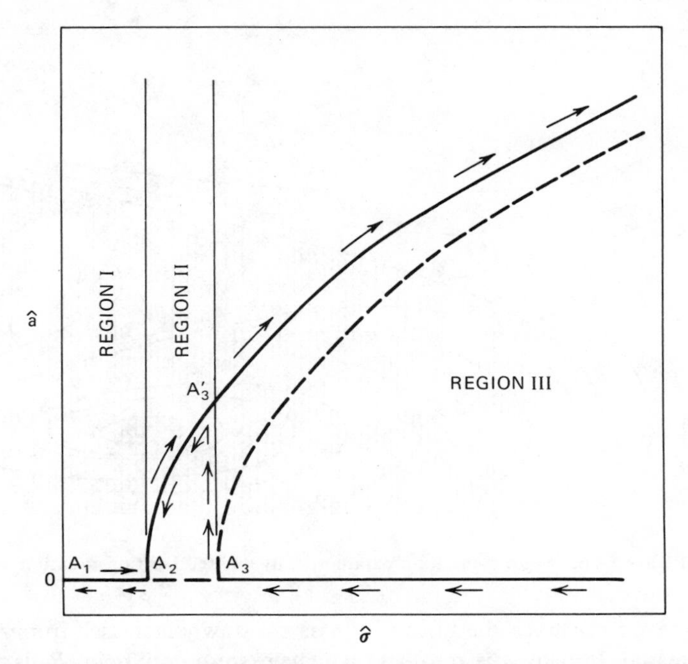

Figure 5-14. Frequency–response curves for a parametrically excited Duffing equation.

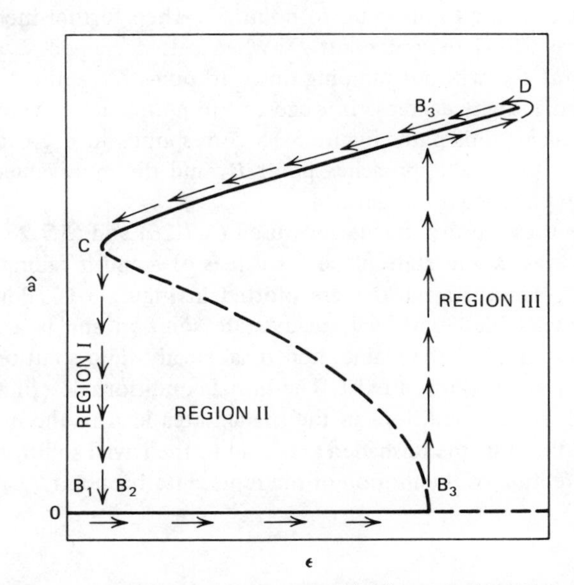

Figure 5-15. Response curves for a parametrically excited Duffing equation.

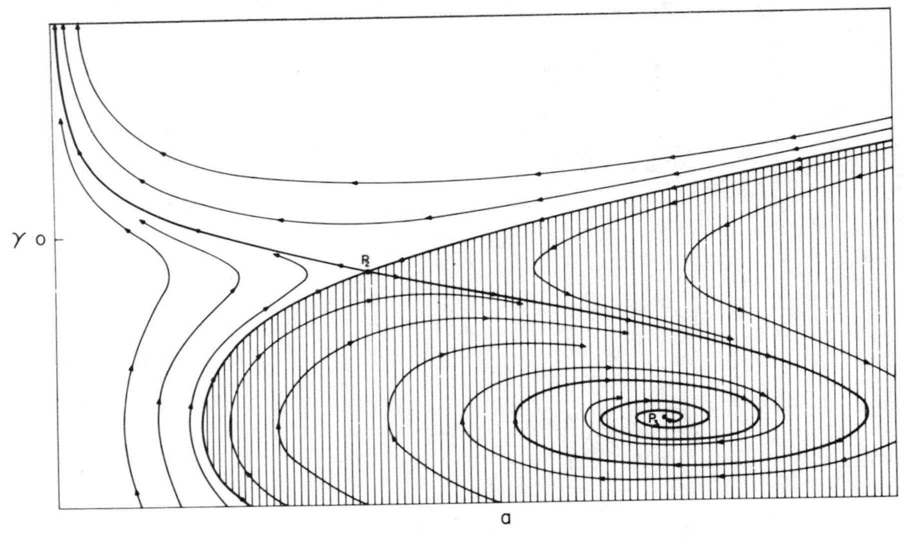

Figure 5-16. State plane for a parametrically excited Duffing equation.

Figure 5-15 we can trace the history of a as ϵ is slowly increased from zero and then decreased. Initially a is zero, and it remains zero until point B_3 is reached. At point B_3 the trivial solution becomes unstable, and hence a slight increase in ϵ at this point causes a to jump up to point B_3'. Then further increases cause a to follow curve $CB_3'D$ toward point D. When ϵ is decreased, a follows curve $DB_3'C$ past point B_3', without jumping down to point B_3, until it reaches point C. At this point a slight decrease in ϵ causes a to jump back to zero at point B_2. The arrows indicate this path. Figure 5-15 corresponds to $\hat{\sigma}$ greater than zero. As $\hat{\sigma}$ decreases, point C approaches point B_3 and the multivalued region, and hence the jump phenomenon, vanishes.

To illustrate these results further, we used (5.7.26) and (5.7.27) to calculate several trajectories in the state plane for values of ϵ and $\hat{\sigma}$ falling in region III of Figure 5-13. These trajectories are plotted in Figure 5-16. Point P_2 corresponds to the unstable, nontrivial steady-state solution and is a saddle point. Point P_3 corresponds to the stable, nontrivial steady-state solution when more than one steady-state solution exist. The initial conditions determine which one is reached. All initial conditions in the shaded area lead to the nontrivial solution, while all those in the unshaded area lead to the trivial solution. The arrows indicate the direction of the motion of the representative point.

Exercises

5.1. In analyzing the propagation of elastic waves in a harmonic inhomogeneous medium, Nayfeh and Nemat-Nasser (1972) encountered the following equation:

$$u'' + \tfrac{1}{4}(1 - \epsilon \cos x)^{-2} a^{-2} [2(1 - \epsilon \cos x)(2 - \epsilon a^2 \cos x) + \epsilon^2 a^2 \sin^2 x]\, u = 0$$

where a and ϵ are constants. Show that three of the transition curves are

$$a = 2 \pm \tfrac{1}{2}\epsilon + O(\epsilon^2)$$

$$a = 1 + \tfrac{1}{6}\epsilon^2 + O(\epsilon^4)$$

5.2. The response of an LRC circuit with sinusoidally varying resistance is governed by (Batchelor, 1976)

$$L\ddot{I} + R(t)\dot{I} + (\dot{R} + C^{-1})I = 0$$

where I is the current, L is the inductance, C is the capacitance, and $R = R_0(1 + \alpha \sin \Omega t)$ is the resistance.

(a) Let

$$I = w(t) \exp\left[-\tfrac{1}{2}L^{-1}\int_0^t R(\xi)\, d\xi\right]$$

and show that the governing equation can be put in the form

$$w'' + [\delta + \epsilon \cos 2\tau - \tfrac{1}{2}\epsilon^2\alpha^{-1} \sin 2\tau + \tfrac{1}{8}\epsilon^2 \cos 4\tau]\, w = 0$$

where primes denote differentiation with respect to $\tau = \frac{1}{2}\Omega t$ and

$$\epsilon = \frac{2\alpha R_0}{L\Omega'}, \qquad \delta = \frac{4}{\Omega^2}\left[\frac{1}{LC} - \frac{R_0^2}{4L^2}(1 + \tfrac{1}{2}\alpha^2)\right]$$

(b) Determine two terms for the transition curves separating stability from instability when

(i) $\delta \approx 1$
(ii) $\delta \approx 2$

5.3. Consider a pendulum with a vertically moving support (Figure 5-1). The governing equation is

$$\ddot{\theta} + \frac{g - \ddot{Y}}{l} \sin \theta = 0$$

(a) When $Y \equiv 0$, show that the equilibrium positions are $\theta = n\pi$, where n is an integer. Show that $n = 0$ corresponds to a center while $\theta = \pi$ corresponds to a saddle point.

(b) If $Y = \epsilon g \cos \Omega t$, determine the values of ϵ, g/l, and Ω for which the stable position $\theta = 0$ becomes unstable.

(c) Let $\theta = \pi + u$, $Y = \alpha g \cos \Omega t$, linearize the resulting equation, and obtain

$$\ddot{u} - \frac{g}{l}(1 - \alpha \cos \Omega t)\, u = 0$$

Determine the values of g/l, α, and Ω for which the unstable position $\theta = \pi$ becomes stable.

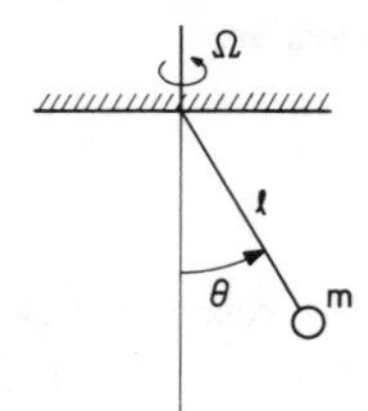

Figure 5-17. Simple pendulum attached to rotating base.

5.4. Consider the system shown in Figure 5-17.

(a) Show that the governing equation is

$$\ddot{\theta} + \frac{g}{l} \sin \theta - \tfrac{1}{2}\Omega^2 \sin 2\theta = 0$$

(b) Let $\Omega = \Omega_0 (1 + \epsilon \cos \omega t)$, where $\epsilon \ll 1$, linearize the governing equation, and obtain

$$\ddot{\theta} + \left(\frac{g}{l} - \Omega_0^2 - 2\Omega_0^2 \epsilon \cos \omega t - \Omega_0^2 \epsilon^2 \cos^2 \omega t\right) \theta = 0$$

(c) Determine second-order expansions for the transition curves separating stability from instability when

$$gl^{-1} - \Omega_0^2 \approx 0, \tfrac{1}{4}\omega^2, \omega^2$$

(d) If $\theta = O(\epsilon^{1/2})$, determine the influence of the nonlinear terms to first order when

$$gl^{-1} - \Omega_0^2 \approx \tfrac{1}{4}\omega^2$$

5.5. Consider the system shown in Figure 5-18 when the tension $T = T_0(1 + \epsilon \sin \omega t)$.

(a) Show that the governing equation is

$$m\ddot{x} + 2T_0 (1 + \epsilon \sin \omega t) x (l^2 + x^2)^{-1/2} = 0$$

(b) Linearize the governing equation to obtain

$$\ddot{x} + 2\omega_0^2 (1 + \epsilon \sin \omega t) x = 0, \qquad \omega_0^2 = \frac{T_0}{ml}$$

(c) Determine second-order expansions for the transition curves separating stability from instability when

$$\omega = 2\omega_0, \omega_0$$

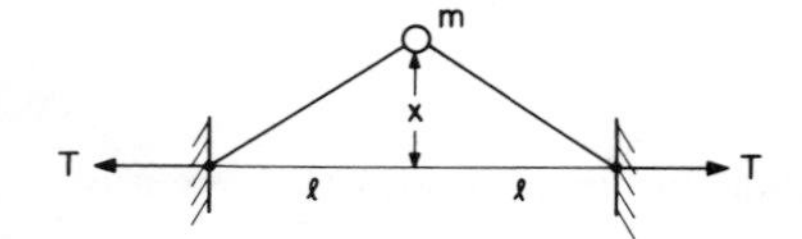

Figure 5-18. Particle attached to stretched string.

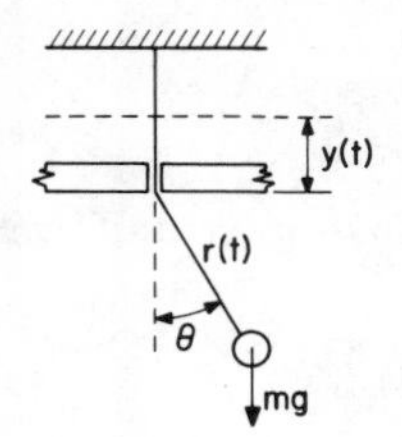

Figure 5-19. Pendulum with varying length.

(d) If $x = 0(\epsilon^{1/2})$, determine the influence of the nonlinearity to first order when $\omega \approx 2\omega_0$.

5.6. Consider the system shown in Figure 5-19 when $l = r(t) + y(t) =$ constant

(a) Show that the governing equation is

$$(l - y)\ddot{\theta} + g \sin \theta - 2\dot{y}\dot{\theta} - \ddot{y} \sin \theta = 0$$

(b) Linearize the governing equation to obtain

$$\ddot{\theta} + \frac{g - \ddot{y}}{l - y}\theta - \frac{2\dot{y}\dot{\theta}}{l - y} = 0$$

(c) If $y = \epsilon l \cos \Omega t$, determine second-order expansions for the transition curves separating stability from instability when $gl^{-1} \approx \frac{1}{2}\Omega, 2\Omega$.

(d) If $\theta = O(\epsilon^{1/2})$, determine the nonlinear influence to first order on the transition curves when $gl^{-1} \approx \frac{1}{2}\Omega$.

5.7. Consider the system shown in Figure 2-14 when

$$\Omega = \Omega_0(1 + \epsilon \cos \omega t)$$

(a) Show that the governing equation is

$$(1 + 4p^2x^2)\ddot{x} + 4p^2\dot{x}^2x + \omega_0^2 x - \Omega_0^2 (2\epsilon \cos \omega t + \epsilon^2 \cos^2 \omega t)x = 0$$

where $\omega_0^2 = 2gp - \Omega_0^2$.

(b) Determine the linearized transition curves separating stability from instability when $\omega_0 \approx 0, \frac{1}{2}\omega, \omega$.

(c) When $x = O(\epsilon^{1/2})$, determine the influence of the nonlinearity to first order on the transition curves for the case $\omega_0 \approx \frac{1}{2}\omega$.

5.8. The cylinder rolls without slip on the circular surface as in Figure 5-20. The displacement of the block is prescribed as $x(t)$ and $y(t)$; it does not rotate.

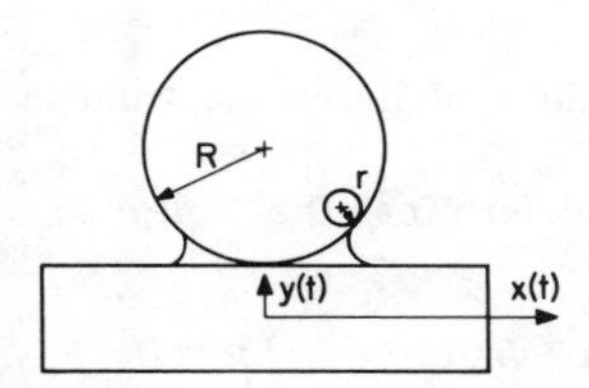

Figure 5-20. Cylinder rolling without slip on circular surface.

(a) Show that the governing equation is

$$\ddot{\theta} + \tfrac{2}{3}(R - r)^{-1}(g + \ddot{y}) \sin\theta + \tfrac{2}{3}(R - r)^{-1}\ddot{x} \cos\theta = 0$$

(b) When $\ddot{x} = \epsilon K \cos \Omega t$, $\ddot{y} = 0$, and $\theta = O(\epsilon^{1/2})$, determine first-order uniform expansions for the cases

(i) $\Omega \approx \omega_0$, $\omega_0^2 = \frac{2}{3}g(R - r)^{-1}$
(ii) $\Omega \approx 3\omega_0$
(iii) $\Omega \approx \frac{1}{3}\omega_0$

(c) When $\ddot{x} = \epsilon K \cos \Omega_1 t$, $\ddot{y} = \epsilon \cos \Omega_2 t$, and $\theta = O(\epsilon^{1/2})$, determine first-order uniform expansions for the cases

(i) $\Omega_1 \approx \omega_0$ and $\Omega_2 \approx 2\omega_0$
(ii) $\Omega_2 - \Omega_1 \approx \omega_0$
(iii) $\Omega_1 \approx 3\omega_0$ and $\Omega_2 \approx 2\omega_0$

5.9. Consider Duffing's equation

$$\ddot{u} + \omega_0^2 u + \alpha u^3 = K \cos \omega t$$

where ω_0, α, K, and ω are constants.

(a) Let $u_0(t) = a \cos \omega t$, use the method of harmonic balance, and show that the frequency-response equation is

$$(\omega_0^2 - \omega^2)a + \tfrac{3}{4}\alpha a^3 = K$$

(b) Let $u = u_0(t) + x(t)$ be a new solution to the same problem subject to slightly different initial conditions. Assume that $x \ll u_0$, substitute the new solution into the governing equation, use the fact that u_0 is a solution of the governing equation, and obtain the following so-called linearized variational equation for x:

$$\ddot{x} + (\omega_0^2 + 3\alpha u_0^2)x = 0$$

(c) Substitute for u_0 in the variational equation to show that the stability of the periodic solution is transformed into determining the stability of the solutions of the Mathieu equation

$$\ddot{x} + (\omega_0^2 + \tfrac{3}{2}\alpha a^2 + \tfrac{3}{2}\alpha a^2 \cos 2\omega t)x = 0$$

(d) Use the results of Section 5.2.1 to analyze the stability of the periodic solution in the $a\omega$-plane. Compare your results with those of Section 4.1.1.

5.10. Consider the stability of the periodic solutions of

$$\ddot{u} + u^3 = K \cos \omega t$$

(a) Let $u_0 = a \cos \omega t + b \cos 3\omega t$ and use the method of harmonic balance to determine two algebraic equations for a and b.

(b) Let $u = u_0(t) + x(t)$, where $x \ll u_0$, and determine the following variational equation for x:

$$\ddot{x} + \tfrac{3}{2}\,[a^2 + b^2 + (a^2 + 2ab) \cos 2\omega t + 2ab \cos 4\omega t + b^2 \cos 6\omega t]\, x = 0$$

(c) Analyze the stability of the solutions of the variational equation and relate the results to the stability of the periodic solutions of the original equation.

5.11. The nonlinear parametric excitation of a system is governed by

$$\ddot{u} + \omega_0^2 u + 2\epsilon\mu\dot{u} + 2\epsilon u^2 \cos 2t = 0, \qquad \epsilon << 1$$

(a) Show that

$$u = a \cos(\omega_0 t + \beta) + O(\epsilon)$$

where

(i) $a' = -\mu a, \beta' = 0$
when ω_0 is away from 2 and $\frac{2}{3}$,
(ii) $a' = -\mu a + \frac{1}{8}a^2 \sin\gamma$
$a\beta' = \frac{3}{8}a^2 \cos\gamma$
$\gamma = \sigma T_1 + \beta$
when $\omega_0 = 2 + \epsilon\sigma$, and
(iii) $a' = -\mu a + \frac{3}{8}a^2 \sin\gamma$
$a\beta' = \frac{3}{8}a^2 \cos\gamma$
$\gamma = \sigma T_1 + 3\beta$
when $3\omega_0 = 2 + \epsilon\sigma$.

(b) Determine the steady-state oscillations and their stability.

5.12. The nonlinear parametric excitation of a system is governed by

$$\ddot{u} + \omega_0^2 u + 2\epsilon\mu\dot{u} + 2\epsilon u^3 \cos 2t = 0, \qquad \epsilon << 1$$

(a) Show that

$$u = a \cos(\omega_0 t + \beta) + O(\epsilon)$$

where

(i) $a' = -\mu a, \beta' = 0$
when ω_0 is away from 1 and $\frac{1}{2}$,
(ii) $a' = -\mu a + \frac{1}{4}a^3 \sin\gamma$
$a\beta' = \frac{1}{2}a^3 \cos\gamma$
$\gamma = 2\sigma T_1 + 2\beta$
when $\omega_0 = 1 + \epsilon\sigma$, and
(iii) $a' = -\mu a + \frac{1}{4}a^3 \sin\gamma$
$a\beta' = \frac{1}{4}a^3 \cos\gamma$
$\gamma = 2\sigma T_1 + 4\beta$
when $2\omega_0 = 1 + \epsilon\sigma$.

(b) Determine the steady-state oscillations and their stability.

5.13. The nonlinear parametric excitation of a system is governed by

$$\ddot{u} + \omega_0^2 u + 2\epsilon\mu\dot{u} + 2\epsilon u^n \cos 2t = 0$$

where $\epsilon << 1$ and n is an integer.

(a) Show that parametric resonances occur to first order in ϵ when

$$\omega_0 \approx 1, \tfrac{2}{4}, \tfrac{2}{6}, \tfrac{2}{8}, \ldots, \frac{2}{n+1} \qquad \text{for odd } n$$

$$\omega_0 \approx 2, \tfrac{2}{3}, \tfrac{2}{5}, \tfrac{2}{7}, \ldots, \frac{2}{n+1} \qquad \text{for even } n$$

(b) When $(n+1)\omega_0 = 2 + \epsilon\sigma$, show that

$$u = a\cos(\omega_0 t + \beta) + O(\epsilon)$$

where

$$a' = -\mu a + (n+1)\, 2^{-(n+1)} a^n \sin\gamma$$
$$a\beta' = (n+1)\, 2^{-(n+1)} a^n \cos\gamma$$
$$\gamma = \sigma T_1 + (n+1)\beta$$

Determine the steady-state oscillations and their stability.

(c) When $(n-1)\omega_0 = 2 + \epsilon\sigma$, show that

$$u = a\cos(\omega_0 t + \beta) + O(\epsilon)$$

where

$$a' = -\mu a + (n-1)^2\, 2^{-(n+1)} a^n \sin\gamma$$
$$a\beta' = (n^2-1)\, 2^{-(n+1)} a^n \cos\gamma$$
$$\gamma = \sigma T_1 + (n-1)\beta$$

Determine the steady-state oscillations and their stability. Tso and Caughey (1965) treated this case by using the method of averaging.

5.14. The parametric excitation of a system under the influence of a nonlinear damping is governed by

$$\ddot{u} + \omega_0^2 u + \epsilon\mu\,|\dot{u}|^n \dot{u} + 2\epsilon u \cos 2t = 0$$

where $\epsilon \ll 1$.

(a) Show that, to first order, parametric excitations occur only when $\omega_0 \approx 1$.
(b) When ω_0 is away from 1, show that

$$u = a\cos(\omega_0 t + \beta) + O(\epsilon)$$

where

$$a' = -\mu\omega_0^n b a^{n+1}, \qquad \beta' = 0$$

$$b = \frac{1}{2\pi}\int_0^{2\pi} \sin^2\phi\,|\sin\phi|^n\, d\phi = \Gamma[\tfrac{1}{2}(n+3)]/\sqrt{\pi}\,\Gamma[\tfrac{1}{2}(n+4)]$$

with Γ being the gamma function. Solve for a and indicate its decay with time.

(c) When $\omega_0 = 1 + \epsilon\sigma$, show that

$$\begin{aligned} u &= a \cos(\omega_0 t + \beta) + O(\epsilon) \\ a' &= -\mu\omega_0^n b a^{n+1} + \tfrac{1}{2} a \sin \gamma \\ a\beta' &= \tfrac{1}{2} a \cos \gamma \\ \gamma &= 2\sigma T_1 + 2\beta \end{aligned}$$

Determine the steady-state oscillations and their stability.

5.15. The parametric excitation of a system is governed by

$$\ddot{u} + \omega_0^2 u + 2\epsilon\mu\dot{u} + \epsilon\alpha_1 u^n + 2\epsilon\alpha_2 u \cos 2t = 0$$

where $\epsilon \ll 1$ and n is an odd integer.

(a) Show that

$$u = a \cos(\omega_0 t + \beta) + O(\epsilon)$$

where

(i) $a' = -\mu a$

$$\beta' = \frac{b\alpha_1 a^{n-1}}{\omega_0 2^n}, \qquad b = \begin{Bmatrix} n \\ \tfrac{1}{2}(n+1) \end{Bmatrix} = \frac{n!}{\tfrac{1}{2}(n+1)!\,\tfrac{1}{2}(n-1)!}$$

when ω_0 is away from 1.

(ii) $a' = -\mu a + \tfrac{1}{2}\alpha_2 a \sin \gamma$

$\beta' = \alpha_1 b 2^{-n} a^{n-1} + \tfrac{1}{2}\alpha_2 \cos \gamma$

$\gamma = 2\sigma T_1 + 2\epsilon\sigma$

when $\omega_0 = 1 + \epsilon\sigma$.

(b) Determine the steady-state oscillations and their stability.

5.16. The response of a system to a parametric excitation is governed by

$$\ddot{u} + 2\epsilon\mu_1\dot{u} + \epsilon\mu_2 |\dot{u}|^n \dot{u} + (\omega_0^2 + \epsilon \cos 2t)(u + \epsilon\alpha u^3) = 0$$

where $\epsilon \ll 1$.

(a) When $\omega_0 = 1 + \epsilon\sigma$, show that

$$u = a \cos(\omega_0 t + \beta) + O(\epsilon)$$

where

$$\begin{aligned} a' &= -\mu_1 a - \mu_2 b a^{n+1} + \tfrac{1}{2} a \sin \gamma_1 \\ \beta' &= \tfrac{3}{8}\alpha a^2 + \tfrac{1}{4} \cos \gamma_1 \\ \gamma_1 &= 2\sigma T_1 + 2\beta \end{aligned}$$

Determine the constant b.

(b) Determine the steady-state oscillations and their stability.

5.17. The response of a system which is parametrically excited is governed by

$$\ddot{u} + \omega_0^2 u + \epsilon\mu |\dot{u}|^n \dot{u} + \epsilon\alpha_1 u^m + 2\epsilon\alpha_2 u^k \cos 2t = 0$$

where $\epsilon \ll 1$, k is an integer, and m is an odd integer.

(a) Determine all values of ω_0 for which parametric resonances exist to order ϵ.

(b) Show that

$$u = a \cos(\omega_0 t + \beta) + O(\epsilon)$$

where

(i) $a' = -\mu\omega_0^n b_1 a^{n+1} + \alpha_2(k-1)^2 2^{-(k+1)} a^k \sin\gamma$
$\beta' = b_2 \alpha_1 \omega_0^{-1} a^{m-1} + \alpha_2(k^2-1) 2^{-(k+1)} a^{k-1} \cos\gamma$
$\gamma = \sigma T_1 + (k-1)\beta$
when $(k-1)\omega_0 = 2 + \epsilon\sigma$

(ii) $a' = -\mu\omega_0^n b_1 a^{n+1} + \alpha_2(k+1) 2^{-(k+1)} a^k \sin\gamma$
$\beta' = b_2 \alpha_1 \omega_0^{-1} a^{m-1} + \alpha_2(k+1) 2^{-(k+1)} a^{k-1} \cos\gamma$
$\gamma = \sigma T_1 + (k+1)\beta$
when $(k+1)\omega_0 = 2 + \epsilon\sigma$

Determine the constants b_1 and b_2.

(c) Determine the steady-state solutions and their stability.

(d) Write down the equations describing a and β when ω_0 is away from those corresponding to parametric resonances.

5.18. The response of a system under the combined influence of parametric and external excitations is governed by

$$\ddot{u} + \omega_0^2 u + 2\epsilon\mu\dot{u} + \alpha_1 u^3 + 2\epsilon\alpha_2 u \cos 2t = K \cos \Omega t$$

where $\epsilon \ll 1$. Such problems were studied by Hsu and Cheng (1974), Nguyen (1975b), and Dimentberg (1976).

(a) When $\Omega = 1 + \epsilon\sigma_1$ and $\omega_0 = 1 + \epsilon\sigma_2$, show that

$$u = a \cos(\omega_0 t + \beta) + O(\epsilon)$$

where

$$a' = -\mu a + \tfrac{1}{2} k \sin\gamma_1 + \tfrac{1}{2}\alpha_2 a \sin\gamma_2$$

$$a\beta' = \tfrac{3}{8}\alpha_1 a^3 - \tfrac{1}{2} k \cos\gamma_1 + \tfrac{1}{2}\alpha_2 a \cos\gamma_2$$

$$K = \epsilon k, \qquad \gamma_1 = (\sigma_1 - \sigma_2) T - \beta, \qquad \gamma_2 = 2\sigma_2 T_1 + 2\beta$$

(b) When $\omega_0 = 1 + \epsilon\sigma_2$ and Ω is away from 1, show that

$$u = a \cos(\omega_0 t + \beta) + 2\Lambda \cos \Omega t + O(\epsilon)$$

where

(i) $a' = -\mu a + \tfrac{1}{2}\alpha_2 a \sin\gamma_2$
$a\beta' = 3\alpha_1(\Lambda^2 + \tfrac{1}{8}a^2) a + \tfrac{1}{2}\alpha_2 a \cos\gamma_2$
when Ω is away from $\tfrac{1}{3}$ and 3.

(ii) $a' = -\mu a - \alpha_1 \Lambda^3 \sin\gamma_1 + \tfrac{1}{2}\alpha_2 a \sin\gamma_2$
$a\beta' = 3\alpha_1(\Lambda^2 + \tfrac{1}{8}a^2) a + \alpha_1 \Lambda^3 \cos\gamma_1 + \tfrac{1}{2}\alpha_2 a \cos\gamma_2$
when $3\Omega = 1 + \epsilon\sigma_1$.

(iii) $a' = -\mu a - \frac{3}{4}\alpha_1 \Lambda a^2 \sin \gamma_1 + \frac{1}{2}\alpha_2 a \sin \gamma_2 - \alpha_2 \Lambda \sin \gamma_3$
$a\beta' = 3\alpha_1(\Lambda^2 + \frac{1}{8}a^2)a + \frac{3}{4}\alpha_1 \Lambda a^2 \cos \gamma_1 + \frac{1}{2}\alpha_2 a \cos \gamma_2 + \alpha_2 \Lambda \cos \gamma_3$
$\gamma_1 = (\sigma_1 - 3\sigma_2) T_1 - 3\beta, \qquad \gamma_3 = \sigma_1 T_1 - \frac{1}{2}\gamma_2$
when $\Omega = 3 + \epsilon\sigma_1$.

(c) Determine the steady-state responses and their stability.

(d) When ω_0 is away from 1, show that combination resonances occur if $\Omega \mp 2 \approx \omega_0$, $2 - \Omega \approx \omega_0$. Let $\Omega = 2 + \omega_0 + \epsilon\sigma_1$ and show that

$$u = a \cos(\omega_0 t + \beta) + 2\Lambda \cos \Omega t + O(\epsilon)$$

where

$$a' = -\mu a - \Lambda \alpha_2 \omega_0^{-1} \sin \gamma_1$$
$$\omega_0 a\beta' = 3\alpha_1(\Lambda^2 + \tfrac{1}{8}a^2)a + \Lambda a_2 \cos \gamma_1$$
$$\gamma_1 = \sigma_1 T_1 - \beta$$

5.19. The response of a system under the combined effect of parametric excitation and a multifrequency excitation is governed by

$$\ddot{u} + \omega_0^2 u + 2\epsilon\mu\dot{u} + \alpha_1 u^3 + 2\epsilon\alpha_2 u \cos 2t = \sum_{n=1}^{3} K_n \cos(\Omega_n t + \theta_n)$$

where $\epsilon \ll 1$.

(a) When $\omega_0 = 1 + \epsilon\sigma_2$, show that

$$u = a \cos(\omega_0 t + \beta) + \sum_{n=1}^{3} 2\Lambda_n \cos(\Omega_n t + \theta_n) + O(\epsilon)$$

where

$$a' = -\mu a - 3\alpha_1 \Lambda_1^2 \Lambda_2 \sin \gamma_1 + \tfrac{1}{2}\alpha_2 a \sin \gamma_2$$

$$a\beta' = 3\alpha_1 \left[\sum_{n=1}^{3} \Lambda_n^2 + \tfrac{1}{8}a^2 \right] a + 3\alpha_1 \Lambda_1^2 \Lambda_2 \cos \gamma_1 + \tfrac{1}{2}\alpha_2 a \cos \gamma_2$$

$$\gamma_1 = \sigma_1 T_1 - \beta + 2\theta_1 + \theta_2, \qquad \gamma_2 = 2\sigma_2 T_1 + 2\beta$$

when $2\Omega_1 + \Omega_2 = 1 + \epsilon\sigma_1$ and no other external resonances occur.

(b) Determine the steady-state oscillations and their stability.

(c) When $\omega_0 = 1 + \epsilon\sigma_2$, determine the equations describing the amplitude and the phase for the following resonant cases:

(i) $\Omega_1 + \Omega_2 + \Omega_3 = 1 + \epsilon\sigma_1$
(ii) $\Omega_1 + \Omega_2 = 2 + \epsilon\sigma_1$
(iii) $3\Omega_1 = 1 + \epsilon\sigma_1$, $\Omega_2 + \Omega_3 = 2 + \epsilon\sigma_3$
(iv) $3\Omega_1 = 1 + \epsilon\sigma_1$, $\Omega_2 = 3 + \epsilon\sigma_3$
(v) $\Omega_1 = 1 + \epsilon\sigma_1$, $3\Omega_2 = 1 + \epsilon\sigma_3$, $\Omega_3 = 3 + \epsilon\sigma_4$
(vi) $\Omega_1 = 1 + \epsilon\sigma_1$, $\Omega_2 + \Omega_3 = 2 + \epsilon\sigma_3$

5.20. The response of a system under the combined influence of parametric and external excitations is governed by

$$\ddot{u} + 2\epsilon\mu_1\dot{u} + \epsilon\mu_2 |\dot{u}|\dot{u} + (\omega_0^2 + \epsilon \cos 2t)(u + \epsilon\alpha u^3) = K \cos(\Omega t - \theta)$$

where $\epsilon \ll 1$ and $\omega_0 = 1 + \epsilon\sigma_2$.

(a) Show that

$$u = a \cos(\omega_0 t + \beta) + O(\epsilon)$$

where

$$a' = -\mu_1 a - \frac{4\mu_2}{3\pi} a^2 + \tfrac{1}{2}k \sin\gamma_1 + \tfrac{1}{4}a \sin\gamma_2$$

$$a\beta' = \tfrac{3}{8}\alpha a^3 - \tfrac{1}{2}k \cos\gamma_1 + \tfrac{1}{4}a \cos\gamma_2$$

$$K = \epsilon k, \qquad \gamma_1 = (\sigma_1 - \sigma_2) T_1 - \beta - \theta, \qquad \gamma_2 = 2\sigma_2 T_1 + 2\beta$$

when $\Omega = 1 + \epsilon\sigma_1$.

(b) When $\omega_0 = 1 + \epsilon\sigma_2$ and $\Omega = 2$, show that (Troger and Hsu, 1977)

$$u = a \cos(\omega_0 t + \beta) + 2\Lambda \cos(2t - \theta) + O(\epsilon)$$

where

$$a' = -\mu_1 a + \tfrac{1}{4}a \sin\gamma_2 + \frac{\mu_2}{2\pi}\int_0^{2\pi} |\dot{u}|^2 \operatorname{sgn}\dot{u} \sin t \, dt$$

$$a\beta' = \alpha_3(\Lambda^2 + \tfrac{3}{8}a^2)a + \tfrac{1}{4}a \cos\gamma_2 + \frac{\mu_2}{2\pi}\int_0^{2\pi} |\dot{u}|^2 \operatorname{sgn}\dot{u} \cos t \, dt$$

$$\dot{u} = -\omega_0 a \sin(t + \tfrac{1}{2}\gamma_2) - 4\Lambda \sin(2t - \theta)$$

Show that stationary oscillations correspond to $a' = \gamma' = 0$. In this case, the frequency-response equation is complicated because the integrals cannot be evaluated a priori.

5.21. Consider the double pendulum shown in Figure 5-3.

(a) Show that (5.1.9) and (5.1.10) reduce to the following equations when $l_1 = l_2 = l$, $m_1 = m_2 = m$, and $k_1 = k_2 = 0$:

$$\ddot{\theta}_1 + \frac{g - \ddot{y}}{l}(2\theta_1 - \theta_2) = 0$$

$$\ddot{\theta}_2 + 2\frac{g - \ddot{y}}{l}(\theta_2 - \theta_1) = 0$$

(b) When $\ddot{y} \equiv 0$, determine the linear natural frequencies ω_1 and ω_2.

(c) When $y = \epsilon l \cos\Omega t$, where $\epsilon \ll 1$, determine first-order uniform expansions for θ_1 and θ_2 and hence determine the transition curves separating stability from instability for the cases

(i) $\Omega \approx 2\omega_1$
(ii) $\Omega \approx \omega_1 + \omega_2$
(iii) $\Omega \approx \omega_2 - \omega_1$

(d) When $y = \epsilon l[y_1 \cos(\Omega_1 t + \theta_1) + y_2 \cos(\Omega_2 t + \theta_2)]$, where the y_n and θ_n are constants, determine first-order uniform expansions for θ_1 and θ_2, and hence determine the transition curves separating stability from instability for the cases

(i) $\Omega_1 \approx \omega_2 - \omega_1$ and $\Omega_2 \approx 2\omega_1$
(ii) $\Omega_1 \approx \omega_2 - \omega_1$ and $\Omega_2 \approx \omega_2 + \omega_1$

5.22. A two-frequency parametric excitation of a multidegree-of-freedom system is governed by

$$\ddot{u}_n + \omega_n^2 u_n + 2\epsilon \sum_{m=1}^{N} [f_{nm} \cos \Omega_1 t + g_{nm} \cos(\Omega_2 t + \theta)] u_m = 0$$

where the ω_n are distinct. Determine first-order uniform expansions and hence the transition curves separating stability from instability when

(a) $\Omega_1 \approx \omega_2 - \omega_1$ and $\Omega_2 \approx \omega_3 - \omega_2$
(b) $\Omega_1 \approx 2\omega_1, \Omega_2 \approx \omega_1 + \omega_2, \Omega_2 \approx \omega_3 - \omega_1$

5.23. Consider the system governed by (5.5.1) to (5.5.3). Determine first-order uniform expansions for the cases

(a) $\omega \approx 2\omega_1$ and $\omega \approx \omega_3 - \omega_1$ (i.e., $\omega_3 \approx 3\omega_1$)
(b) $\omega \approx \omega_1$ and $\omega \approx \omega_3 - \omega_1$ (i.e., $\omega_3 \approx 2\omega_1$)

5.24. A two-frequency parametric excitation of a three-degrees-of-freedom system with repeated frequencies is governed by

$$\ddot{x}_1 + \omega_1^2 x_1 + 2 \sum_{n=1}^{3} [f_{1n} \cos \Omega_1 t + g_{1n} \cos(\Omega_2 t + \theta)] x_n = 0$$

$$\ddot{x}_2 + \omega_1^2 x_2 + x_1 + 2 \sum_{n=1}^{3} [f_{2n} \cos \Omega_1 t + g_{2n} \cos(\Omega_2 t + \theta)] x_n = 0$$

$$\ddot{x}_3 + \omega_3^2 x_3 + 2 \sum_{n=1}^{3} [f_{3n} \cos \Omega_1 t + g_{3n} \cos(\Omega_2 t + \theta)] x_n = 0$$

where ω_3 is different from ω_1.

(a) Determine first-order uniform expansions for the cases

(i) $\omega_1 \approx \frac{1}{2}\Omega_1$ and $\Omega_3 - \omega_1 \approx \Omega_2$
(ii) $\omega_1 \approx \Omega_1$ and $\omega_3 - \omega_1 \approx \Omega_2$

(b) What other resonances exist to first order?

5.25. Show that the equations describing the motion of a swinging spring with a moving support, Figure 6-2, are

$$\ddot{u} + \omega_1^2 u - (1+u)\dot{\theta}^2 + \delta(1 - \cos\theta) - \ddot{y}\cos\theta = 0$$

$$(1+u)\ddot{\theta} + 2\dot{u}\dot{\theta} + (\delta + \ddot{y})\sin\theta = 0$$

where $u = x/l$, $\omega_1^2 = k/m$, and $\delta = g/l$.

(a) To determine the stability boundaries for infinitesimal θ motions, neglect the nonlinear θ terms and obtain

$$\ddot{u} + \omega_1^2 u = \ddot{y}$$

$$(1+u)\ddot{\theta} + 2\dot{u}\dot{\theta} + (\delta + \ddot{y})\theta = 0$$

(b) When $\ddot{y} = 2\epsilon\cos 2t$, show that

$$u_e = a\cos(\omega_1 t + \beta) + 2\epsilon(\omega_1^2 - 4)^{-1}\cos 2t, \qquad \omega_1 \neq 2$$

where a and β are determined from the initial conditions. Then the equation for θ becomes

$$(1+u_e)\ddot{\theta} + 2\dot{u}_e\dot{\theta} + (\delta + 2\epsilon\cos 2t)\theta = 0$$

Letting $a = 0$, analyze the transition curves separating stability from instability (Ryland and Meirovitch, 1977). What are the limitations imposed by letting $a = 0$? When $\omega_1^2 \approx 4\delta$ there is an internal resonance condition, and hence show that the nonlinear terms cannot be neglected (see Exercise 6.15).

(c) Let $\theta = (1+u)^{-1}\psi$ in part (a) and obtain

$$\ddot{\psi} + (\delta + \ddot{y} - u)(1+u)^{-1}\psi = 0$$

With u known from part (a), the linearized stability problem reduces to that of analyzing the solutions of this equation for ψ.

(d) When

$$\ddot{y} = \epsilon \sum_{n=1}^{N} y_n \cos(\Omega_n t + \theta_n)$$

solve for u from part (a), substitute the result into the equation for ψ, and then analyze the stability of the swinging spring.

5.26. Consider the buckling of the column shown in Figure 5-21 under the influence of a nonideal energy source.

(a) Neglecting the longitudinal and rotary inertia and the transverse shear, show that the governing equations for the fundamental mode are (Kononenko, 1969, Section 14)

$$\ddot{\psi} + 2\mu\dot{\psi} + (\omega^2 - \alpha_1\sin\phi)\psi + \alpha_2\psi^3 = 0$$

$$I\ddot{\phi} = M(\dot{\phi}) - k_1\left(u_0 + r\sin\phi - \frac{\pi^2}{4l}\psi^2\right) r\cos\phi = 0$$

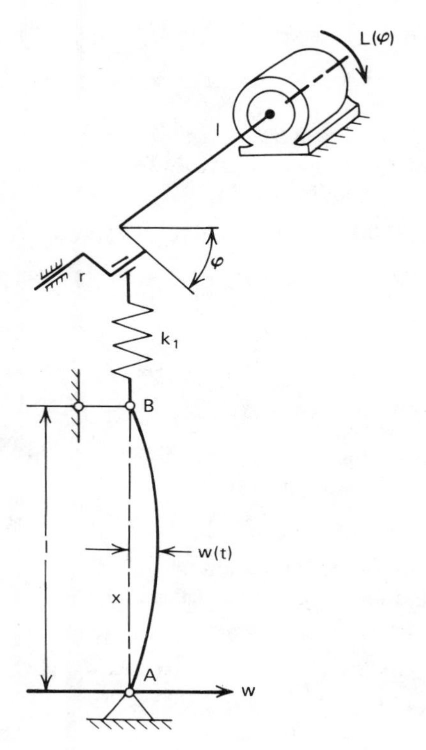

Figure 5-21. Buckling of beam under influence of nonideal energy source.

where

$$w(x, t) = \psi(t) \sin \frac{\pi x}{l}, \qquad u = \frac{\pi^2}{4l} \psi^2$$

$$\omega^2 = \frac{\pi^4}{l^4} \frac{EI_x}{m_1} \left(1 - \frac{k_1 u_0 l^2}{\pi^2 EI_x}\right), \qquad \alpha_1 = \frac{\pi^2 r k_1}{l^2 m_1}$$

$$\alpha_2 = \frac{\pi^4 k_1}{4l^3 m_1}$$

Here m_1 is the mass per unit length of the beam, E is the modulus of elasticity, l is the length, k_1 is the spring constant, I_x is the moment of inertia of the cross section, and μ is a damping coefficient.

(b) Determine a uniform first-order expansion and use it to determine the steady-state periodic motions and their stability.

5.27. Consider the harmonic time variation of sound waves in two-dimensional ducts with sinusoidal walls. The mathematical statement of the problem is

$$\nabla^2 \phi + \omega^2 \phi = 0 \tag{1}$$

$$\phi_y = \epsilon \phi_x k_w \cos k_w x \qquad \text{at} \qquad y = \epsilon \sin k_w x \tag{2}$$

$$\phi_y = 0 \qquad \text{at} \qquad y = 1 \tag{3}$$

where $\epsilon \ll 1$.

(a) Show that (1) through (3) possess the following straightforward expansion (Isakovitch, 1957; Samuels, 1959; Salant, 1973; Nayfeh, 1974):

$$\phi = A \cos(n\pi y) \exp(ik_n x) + \tfrac{1}{2} i\epsilon A \{(k_n k_w - n^2\pi^2)\Phi_1(y) \exp[i(k_n + k_w)x] + (k_n k_w + n^2\pi^2)\Phi_2(y) \exp[i(k_n - k_w)x]\} + O(\epsilon^2) \tag{4}$$

where $k_n^2 = \omega^2 - n^2\pi^2$,

$$\Phi_m = (\kappa_m \sin \kappa_m)^{-1} [\sin \kappa_m \sin \kappa_m y + \cos \kappa_m \cos \kappa_m y] \tag{5}$$

$$\kappa_1^2 = \omega^2 - (k_n + k_w)^2, \qquad \kappa_2^2 = \omega^2 - (k_n - k_w)^2 \tag{6}$$

Show that this expansion breaks down when $\kappa_m - m\pi = O(\epsilon)$. Show that this condition is equivalent to

$$k_w = k_n \pm k_m + O(\epsilon) \tag{7}$$

where k_n and k_m are the wavenumbers of two modes in the duct.

(b) When $k_w = k_n - k_m + \epsilon\sigma$, use the method of multiple scales and seek an expansion in the form (Nayfeh, 1974)

$$\phi(x, y) = \phi_0(x_0, x_1, y) + \epsilon\phi_1(x_0, x_1, y) + \cdots, \qquad x_0 = x, \qquad x_1 = \epsilon x \tag{8}$$

Take ϕ_0 to contain the resonant modes, that is,

$$\phi_0 = A_m(x_1) \cos(m\pi y) \exp(ik_m x_0) + A_n(x_1) \cos(n\pi y) \exp(ik_n x_0) \tag{9}$$

Then show that ϕ_1 is governed by

$$\frac{\partial^2 \phi_1}{\partial y^2} + \frac{\partial^2 \phi_1}{\partial x_0} + \omega^2 \phi_1 = -2ik_m A_m' \cos(m\pi y) \exp(ik_m x_0) - 2ik_n A_n' \cos(n\pi y) \exp(ik_n x_0) \tag{10}$$

$$\phi_{1y} = \tfrac{1}{2} i \sum_{j=m,n} A_j (k_j k_w - j^2\pi^2) \exp[i(k_j + k_w)x_0] + \tfrac{1}{2} i \sum_{j=m,n} A_j (k_j k_w + j^2\pi^2) \exp[i(k_j - k_w)x_0] \qquad \text{at } y = 0 \tag{11}$$

$$\phi_{1y} = 0 \qquad \text{at } y = 1 \tag{12}$$

Show that the solvability conditions of (10) through (12) are

$$\begin{aligned} A_m' &= \tfrac{1}{2} k_m^{-1} (k_n k_w + n^2\pi^2) A_n \exp(-i\sigma x_1) \\ A_n' &= \tfrac{1}{2} k_n^{-1} (k_m k_w - m^2\pi^2) A_m \exp(i\sigma x_1) \end{aligned} \tag{13}$$

(c) Show that equations (13) possess a solution in the form

$$A_m = a_m \exp(sx_1), \qquad A_n = a_n \exp[(s + i\sigma)x_1] \tag{14}$$

Determine s. Show that s is pure imaginary when k_n and k_m have the same sign (i.e., the modes propagate in the same direction). When k_n and k_m have opposite signs (i.e., the modes propagate in opposite directions), s may be complex and the modes are cut off.

5.28. Consider the time harmonic variation of electromagnetic waves propagating in a two-dimensional waveguide. The mathematical statement of the problem is

$$\nabla^2 \psi + \omega^2 \psi = 0 \tag{1}$$

$$\psi_{xx} + \omega^2 \psi = -\epsilon k_w \psi_{xy} \cos k_w x \quad \text{at} \quad y = \epsilon \sin k_w x \tag{2}$$

$$\psi_{xx} + \omega^2 \psi = 0 \quad \text{at} \quad y = 1 \tag{3}$$

where $\epsilon \ll 1$.

(a) Show that (1) through (3) possess the straightforward expansion (Nayfeh and Asfar, 1974)

$$\psi = A \sin n\pi y \exp(ik_n x) + \tfrac{1}{2}\epsilon i n\pi A \{(n^2\pi^2 - k_n k_w)\Psi_1(y) \cdot \exp[i(k_n + k_w)x] - (n^2\pi^2 + k_n k_w)\Psi_2(y)\exp[i(k_n - k_w)x]\} + O(\epsilon^2) \tag{4}$$

where $k_n^2 = \omega^2 - n^2\pi^2$,

and

$$\Psi_m = (\kappa_m^2 \sin \kappa_m)^{-1}[\sin \kappa_m \cos \kappa_m y - \cos \kappa_m \sin \kappa_m y] \tag{5}$$

$$\kappa_1^2 = \omega^2 - (k_n + k_w)^2, \; \kappa_2^2 = \omega^2 - (k_n - k_w)^2 \tag{6}$$

Show that this expansion breaks down when

$$\kappa_m - m\pi = O(\epsilon) \quad \text{or} \quad k_w = k_n \pm k_m + O(\epsilon) \tag{7}$$

where k_n and k_m are the wavenumbers of two modes propagating in the waveguide.

(b) When $k_w = k_n - k_m + \epsilon\sigma$, use the method of multiple scales and seek an expansion in the form (Nayfeh and Asfar, 1974)

$$\psi = \psi_0(x_0, x_1, y) + \epsilon\psi_1(x_0, x_1, y) + \cdots, \quad x_0 = x, \quad x_1 = \epsilon x \tag{8}$$

Take ψ_0 to contain the two resonating modes; that is,

$$\psi_0 = A_m(x_1)\sin(m\pi y)\exp(ik_m x_0) + A_n(x_1)\sin(n\pi y)\exp(ik_n x_0) \tag{9}$$

Then show that

$$\frac{\partial^2 \psi_1}{\partial x_0^2} + \frac{\partial^2 \psi_1}{\partial y^2} + \omega^2\psi_1 = -2ik_m A_m' \sin(m\pi y)\exp(ik_m x_0) - 2ik_n A_n' \sin(n\pi y)\exp(ik_n x_0) \tag{10}$$

$$\frac{\partial^2 \psi_1}{\partial x_0^2} + \omega^2\psi_1 = \tfrac{1}{2}i\pi \sum_{j=m,n} j(j^2\pi^2 - k_j k_w)A_j \exp[i(k_j + k_w)x_0] - \tfrac{1}{2}i\pi \sum_{j=m,n} j(j^2\pi^2 + k_j k_w)A_j \exp[i(k_j - k_w)x_0] \quad \text{at } y = 0 \tag{11}$$

$$\frac{\partial^2 \psi_1}{\partial x_0^2} + \omega^2 \psi_1 = 0 \qquad \text{at } y = 1 \tag{12}$$

Show that the solvability conditions of (10) through (12) are

$$A_m' = \frac{1}{2} \frac{n}{m k_m} (k_n k_w + n^2 \pi^2) A_n \exp(-i\sigma x_1)$$
$$A_n' = \frac{1}{2} \frac{m}{n k_n} (k_m k_w - m^2 \pi^2) A_m \exp(i\sigma x_1) \tag{13}$$

(c) Show that equations (13) possess a solution in the form

$$A_m = a_m \exp(s x_1), \qquad A_n = a_n \exp[(s + i\sigma) x_1] \tag{14}$$

Determine s. Show that s is pure imaginary when $k_n k_m > 0$ and s may be complex depending on σ when $k_n k_m < 0$.

CHAPTER 6

Systems Having Finite Degrees of Freedom

In this chapter we discuss discrete nonlinear systems having finite degrees of freedom. The discussion is limited to weakly nonlinear systems, and solutions are obtained by using a perturbation technique, the method of multiple scales. In the case of strongly nonlinear systems perturbation methods can be used in cases for which a basic exact nonlinear solution exists. For the other cases recourse is often made to geometrical methods to obtain a qualitative description of the behavior of the system including its stability and/or to numerical methods. Rosenberg (1966) presented a survey of the geometrical methods and results concerning the vibration of a strongly nonlinear mass-spring system governed by the following equations in the case of two degrees of freedom:

$$m_1\ddot{u}_1 + \sum_{k=1}^{n} \alpha_{1k} u_1^k + \sum_{k=1}^{n} \Gamma_k (u_1 - u_2)^k = 0$$

$$m_2\ddot{u}_2 + \sum_{k=1}^{n} \alpha_{2k} u_2^k - \sum_{k=1}^{n} \Gamma_k (u_1 - u_2)^k = 0$$

with k being an odd integer. In particular, he surveyed the "normal mode concept" for such systems. Rosenberg and Atkinson (1959), Rosenberg (1960, 1961), and Atkinson (1962) found "linear" modal solutions (normal modes) which are related by $u_2 = cu_1$, where c is a constant. These modal solutions are not discussed further in this book except in Exercise 6.7. For more depth the reader is referred to the studies of Rosenberg and Atkinson (1959), Rosenberg (1960, 1961, 1962, 1964, 1966, 1968), Atkinson (1962), Rosenberg and Kuo (1964), Anand (1972), Vito (1973), Rand (1974), Mishra and Singh (1974), and Yen (1974).

In contrast with a single-degree-of-freedom system, which has only a single linear natural frequency and a single mode of motion, an n-degree-of-freedom system has n linear natural frequencies and n corresponding modes. Let us denote these frequencies by $\omega_1, \omega_2, \ldots, \omega_n$ and assume that all of them are real and different from zero. An important case occurs whenever two or more are commensurable or nearly commensurable. Examples of near-commensurability are

$$\omega_2 \approx 2\omega_1, \quad \omega_2 \approx 3\omega_1, \quad \omega_3 \approx \omega_2 \pm \omega_1,$$
$$\omega_3 \approx 2\omega_2 \pm \omega_1, \quad \omega_4 \approx \omega_3 \pm \omega_2 \pm \omega_1$$

Depending on the order of the nonlinearity in the system, these commensurable relationships of frequencies can cause the corresponding modes to be strongly coupled, and an *internal resonance* is said to exist. For example, if the system has quadratic nonlinearities, then to first order an internal resonance can exist if $\omega_m \approx 2\omega_k$ or $\omega_q \approx \omega_p \pm \omega_m$. For a system with cubic nonlinearities, to first order an internal resonance can exist if $\omega_m \approx 3\omega_k$ or $\omega_q \approx 2\omega_p \pm \omega_m$ or $\omega_q \approx \omega_p \mp \omega_m \mp \omega_k$. When an internal resonance exists in a free system, energy imparted initially to one of the modes involved in the internal resonance will be continuously exchanged among all the modes involved in that internal resonance. If damping is present in the system, then the energy will be continuously reduced as it is being exchanged.

In a conservative nongyroscopic single-degree-of-freedom system, if the linear motion is oscillatory, then the nonlinear motion is bounded and hence stable. For a conservative gyroscopic multidegree-of-freedom system, the nonlinear motion may be unbounded and hence unstable if an internal resonance exists.

If a harmonic external excitation of frequency Ω acts on a multidegree-of-freedom system, then in addition to all the primary and secondary resonances ($p\Omega \approx q\omega_m$, with p and q being integers) of a single-degree-of-freedom system, there might exist other *resonant combinations* of the frequencies in the form

$$p\Omega = a_1\omega_1 + a_2\omega_2 + \cdots + a_N\omega_N$$

where p and the a_n are integers such that

$$p + \sum_{n=1}^{N} |a_n| = M$$

where M is the order of the nonlinearity plus one and N is the number of degrees of freedom. The type of combination resonance which might exist in a system depends on the order of the nonlinearity in the system. For a system having quadratic nonlinearities, to first order the combination resonances that might exist involve two frequencies in addition to Ω. That is, $\Omega \approx \omega_m + \omega_k$ or $\Omega \approx \omega_m - \omega_k$. The first of these is called a summed combination resonance, while the second is called a difference combination resonance. These types of combination resonances were predicted theoretically by Malkin (1956) and found experimentally by Yamamoto (1957, 1960). For a system having cubic nonlinearities, to first order the combination resonances that might exist involve either two or three of the natural frequencies in addition to Ω. That is,

$$\Omega \approx \omega_p \pm \omega_m \pm \omega_k, \quad \Omega \approx 2\omega_p \pm \omega_m, \quad \Omega \approx \omega_p \pm 2\omega_m, \quad 2\Omega \approx \omega_m \pm \omega_k$$

Under some conditions an external resonance, which might involve one or more modes, exists in a system that has an internal resonance. For a system having quadratic nonlinearities in which $\omega_2 \approx 2\omega_1$, Nayfeh, Mook, and Marshall (1973) showed that a saturation phenomenon exists when $\Omega \approx \omega_2$. When the amplitude of the excitation k is small, only the second mode with frequency ω_2 is excited. As k reaches a critical value k_c, which depends on the damping coefficients of the two modes and the detunings, this mode saturates and the first mode begins to grow. As k increases further, all the additional energy goes into the first mode (although $\Omega \approx \omega_2$) due to the internal resonance. For a system having cubic nonlinearities, the saturation phenomenon does not exist although there is a tendency for the energy to flow from the higher to the lower modes involved in the internal resonance.

For a system having quadratic nonlinearities, if $\Omega \approx \omega_1$ and $\omega_2 \approx 2\omega_1$, Nayfeh, Mook, and Marshall (1973) found that under some conditions there exists no steady-state motion in spite of the presence of damping. In this case the energy is continuously exchanged between these two modes without being attenuated. This behavior should be contrasted with that of systems having one degree of freedom, which always possess steady-state periodic motions when acted on by periodic excitations in the presence of positive damping.

When the models involved in an internal resonance are also involved in a combination resonance with an external excitation, two or more fractional harmonics might exist in the response, depending on the order of nonlinearity. For a system having quadratic nonlinearities, the internal and combination resonances $\omega_2 \approx 2\omega_1$ and $\Omega \approx \omega_2 + \omega_1$ might exist. Thus the fractional-harmonic pair $(\frac{1}{3}\Omega, \frac{2}{3}\Omega)$ might occur in the response. For a system with cubic nonlinearities, $\omega_2 \approx 3\omega_1$ and $\Omega \approx \omega_2 + 2\omega_1$ or $\Omega \approx \frac{1}{2}(\omega_2 + \omega_1)$ might exist. Then one of the fractional-harmonic pairs $(\frac{1}{5}\Omega, \frac{3}{5}\Omega)$ or $(\frac{1}{2}\Omega, \frac{3}{2}\Omega)$ might exist. Such fractional-harmonic pairs were observed in a variety of physical systems. Dallos and Linnell (1966a, b) and Dallos (1966) observed fractional-harmonic pairs in the cochlear microphonics of chinchilla ears that were excited by sound pressure levels above 110 dB (re 0.0002 dyne/cm^2). Luukkala (1967) observed fractional-harmonic pairs in standing waves in quartz transducers. Adler and Breazeale (1970) observed fractional-harmonic pairs in underwater standing waves. Eller (1973) observed fractional-harmonic pairs in the response of two coupled nonlinear electronic oscillators.

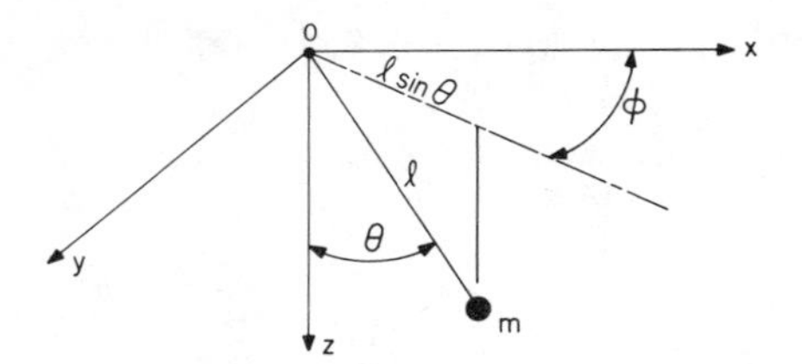

Figure 6-1. Spherical pendulum.

6.1. Examples

6.1.1. THE SPHERICAL PENDULUM

As a first example we consider the motion under gravity of a particle of mass m attached to a fixed point O by an inextensible, massless rod of length l. The particle is free to move on a sphere of radius l as shown in Figure 6-1.

Using the angles θ and ϕ as the generalized coordinates, we find that the kinetic and potential energies are

$$\begin{aligned} T &= \tfrac{1}{2} m(l^2 \dot{\theta}^2 + l^2 \dot{\phi}^2 \sin^2 \theta) \\ V &= mgl(1 - \cos \theta) \end{aligned} \tag{6.1.1}$$

so that the Lagrangian is given by

$$L = T - V = \tfrac{1}{2} ml^2 (\dot{\theta}^2 + \dot{\phi}^2 \sin^2 \theta) - mgl(1 - \cos \theta) \tag{6.1.2}$$

Hence Lagrange's equations have the form

$$\begin{aligned} \frac{d}{dt}\left(\frac{\partial L}{\partial \dot{\theta}}\right) - \frac{\partial L}{\partial \theta} &= 0 \\ \frac{d}{dt}\left(\frac{\partial L}{\partial \dot{\phi}}\right) - \frac{\partial L}{\partial \phi} &= 0 \end{aligned} \tag{6.1.3}$$

Substituting (6.1.2) into (6.1.3) and simplifying the result, we obtain the following set of two coupled equations:

$$\ddot{\theta} + \frac{g}{l} \sin \theta - \tfrac{1}{2} \dot{\phi}^2 \sin 2\theta = 0 \tag{6.1.4}$$

$$\ddot{\phi} \sin \theta + 2\dot{\phi}\dot{\theta} \cos \theta = 0 \tag{6.1.5}$$

We note that (6.1.5) has the integral

$$\dot{\phi} \sin^2 \theta = p, \quad \text{a constant} \tag{6.1.6}$$

so that (6.1.4) becomes

$$\ddot{\theta} + \frac{g}{l} \sin \theta - \frac{p^2 \cos \theta}{\sin^3 \theta} = 0 \tag{6.1.7}$$

Thus the equations governing the motion of this system can be uncoupled; one can solve (6.1.7) for θ and then use the result to solve (6.1.6) for ϕ. The original two-degree-of-freedom system is reduced to two single-degree-of-freedom systems.

Tissot (1852) obtained a closed-form solution to this problem in terms of the theta and eta functions of Jacobi. Later Whittaker (1961, Section 55) also obtained a closed-form solution to this problem in terms of the elliptic, sigma, and zeta functions of Weirstrass. Johansen and Kane (1969) determined a first-order expansion of the solution of this problem by using the method of averaging with canonical variables. Miles (1962) analyzed the response of a spherical pendulum to a harmonic excitation in a plane. He found that the planar motion is unstable over a major portion of the resonant peak, the nonplanar motion is stable in a spectral neighborhood above resonance, and no stable harmonic motions are possible in a finite neighborhood of the natural frequency. Hemp and Sethna (1964) studied the response of a spherical pendulum whose support moves vertically.

6.1.2. THE SPRING PENDULUM

As a second example we consider the motion of a mass m attached to a spring that is swinging in a vertical plane as shown in Figure 6-2.

The equations of motion can be conveniently derived by writing the Lagrangian and then writing the Lagrange equations. The kinetic and potential energies of the mass m are

$$T = \tfrac{1}{2} m [\dot{x}^2 + (l + x)^2 \dot{\theta}^2] \tag{6.1.8}$$

$$V = \tfrac{1}{2} kx^2 + mg(l + x)(1 - \cos\theta) - mgx \tag{6.1.9}$$

where x is the stretch in the spring beyond its equilibrium. Therefore

$$L = T - V = \tfrac{1}{2} m[\dot{x}^2 + (l + x)^2 \dot{\theta}^2] - \tfrac{1}{2} kx^2 - mg(l + x)(1 - \cos\theta) + mgx \tag{6.1.10}$$

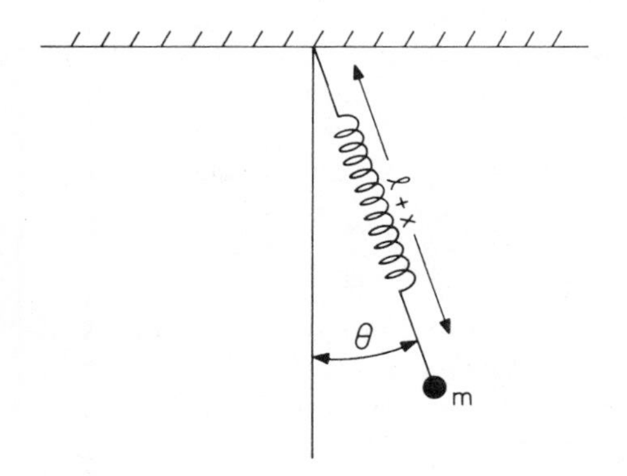

Figure 6-2. Spring pendulum.

Hence Lagrange's equations have the form

$$\frac{d}{dt}\left(\frac{\partial L}{\partial \dot{x}}\right) - \frac{\partial L}{\partial x} = 0$$
$$\frac{d}{dt}\left(\frac{\partial L}{\partial \dot{\theta}}\right) - \frac{\partial L}{\partial \theta} = 0 \tag{6.1.11}$$

Substituting (6.1.10) into (6.1.11) and simplifying the results, we obtain the following two equations:

$$\ddot{x} + \omega_2^2 x - (l + x)\dot{\theta}^2 - g\cos\theta = 0$$
$$\ddot{\theta} + \frac{g\sin\theta + 2\dot{x}\dot{\theta}}{l + x} = 0 \tag{6.1.12}$$

where $\omega_2^2 = k/m$.

If the nonlinear terms are neglected in (6.1.12), there result two uncoupled modes of oscillation—a spring mode with frequency ω_2 and a pendulum mode with frequency $\omega_1 = (g/l)^{1/2}$. Thus based on the linear theory, if an experiment is conducted, one expects to find that the two modes are uncoupled. However Gorelik and Witt (1933) found experimentally that when $\omega_2 \approx 2\omega_1$ the two modes are coupled. If one starts the motion when $\theta = \theta_0$, where $\theta_0 \neq 0$ but very small, by pulling the mass m down, one finds that the mass oscillates up and down first, and that then a pendulum-type component of motion develops and grows at the expense of the spring-type motion. After a while the pendulum-type motion starts to decrease and the spring-type motion starts to grow. Thus the energy is transferred continuously back and forth between the two modes of oscillation. Besides Gorelik and Witt (1933), this problem has been studied by many investigators including Minorsky (1962, Section 7), Heinbockel and Struble (1963), Mettler (1968, 1975), Kane and Kahn (1968), van der Burgh (1968, 1975, 1976), Nayfeh (1973b, Sections 5.53, 5.75, 6.27), Broucke and Baxa (1973), Srinivasan and Sankar (1974), and Olsson (1976). Crespo da Silva (1977) examined the motion of a swinging spring with a spinning support, while Novikov and Kharlamov (1973) and Ryland and Meirovitch (1977) studied a swinging spring with an oscillatory support. Sethna (1965) treated a pendulum consisting of two masses connected by a spring, and Sevin (1961) and Struble and Heinbockel (1963) treated the related problem of a pendulum-type vibration absorber.

6.1.3. A RESTRICTED SHIP MOTION

As a third problem we consider the motion of a ship restricted to pitch and roll only. Consistent equations of motion can be conveniently determined by using a Lagrangian formulation (Lamb, 1932, Chapter 6; Nayfeh, Mook, and

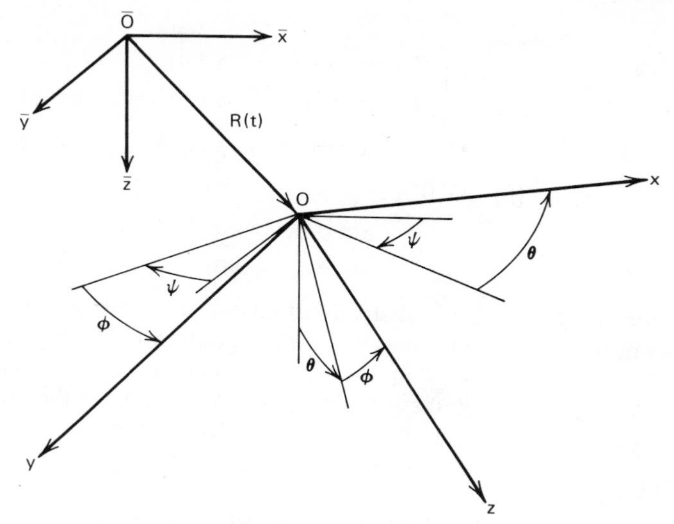

Figure 6-3. Coordinate systems for ship motion.

Marshall, 1974) in which the ship and the sea are regarded as a single dynamic system. We suppose that the motion of the fluid is entirely due to the motion of the ship and neglect the effects of viscosity. Then the whole effect of the fluid might be represented by added inertia and radiation damping. The ship is assumed to possess lateral symmetry. To avoid lengthy algebra, we derive the equations neglecting cubic and higher-order terms.

We employ two Cartesian coordinate systems–one fixed in an inertial space and the other fixed in the ship with its origin at the mass center as shown in Figure 6-3. Initially the two coordinate systems coincide. The orientation of the ship is described by the Euler angles associated with the following sequence: (a) a yawlike rotation about the initial position of the z-axis through the angle ψ, (b) a pitchlike rotation about the new position of the y-axis through the angle θ, and (c) a roll-like rotation about the final position of the x-axis through the angle ϕ. If p, q, and r denote the angular velocities about the final positions of the x-, y-, and z-axes, then

$$\{\Pi\} = [\alpha]\ \{\dot{\gamma}\}, \quad \{\dot{\gamma}\} = [\beta]\ \{\Pi\} \tag{6.1.13}$$

where

$$\{\Pi\} = \begin{Bmatrix} p \\ q \\ r \end{Bmatrix}, \quad \{\gamma\} = \begin{Bmatrix} \phi \\ \theta \\ \psi \end{Bmatrix} \tag{6.1.14}$$

$$[\alpha] = \begin{bmatrix} 1 & 0 & -\sin\theta \\ 0 & \cos\phi & \sin\phi\cos\theta \\ 0 & -\sin\phi & \cos\phi\cos\theta \end{bmatrix} \tag{6.1.15}$$

$$[\beta] = \begin{bmatrix} 1 & \sin\phi\tan\theta & \cos\phi\tan\theta \\ 0 & \cos\phi & -\sin\phi \\ 0 & \dfrac{\sin\phi}{\cos\theta} & \dfrac{\cos\phi}{\cos\theta} \end{bmatrix} \tag{6.1.16}$$

The kinetic energy T and the dissipation function μ must be positive definite for every motion. And because the undisturbed position is a stable equilibrium position, the potential energy V must increase with every displacement from the undisturbed position. All of these functions must account for the lateral symmetry of the ship. Consequently they have the following form:

$$T = \tfrac{1}{2}(I_{xx} + I_1 + I_2\theta)\,p^2 + \tfrac{1}{2}(I_{yy} + I_3 + I_4\theta)\,q^2 + \tfrac{1}{2}(I_{zz} + I_5 + I_6\theta)\,r^2 + I_7\phi pq - (I_{xz} + I_8 + I_9\theta)\,pr + I_{10}\phi qr + \text{h.o.t.} \tag{6.1.17}$$

$$V = \tfrac{1}{2}(V_1 + V_2\theta)\,\phi^2 + \tfrac{1}{2}(V_3 + V_4\theta)\,\theta^2 + \text{h.o.t.} \tag{6.1.18}$$

$$\mu = \tfrac{1}{2}(\mu_1 p^2 + \mu_2 q^2) + \text{h.o.t.} \tag{6.1.19}$$

where h.o.t. stands for higher-order terms, I_{xx}, I_{yy}, I_{zz}, and I_{xz} are the moments and the product of inertia, and the I_n are coefficients for the added inertia. We note that we kept the terms involving r in T at this stage, although r will be set identically equal zero if the ship is restricted to pitch and roll only. As will be evident from the following development, there is a difference between setting $r = 0$ in (6.1.17) and setting $r = 0$ in the equation of motion.

We note that the energy is a function of true coordinates ϕ, θ, and ψ and quasi-coordinates p, q, and r. The first are called *true coordinates* because if $\dot{\phi}$, $\dot{\theta}$, and $\dot{\psi}$ are known, an integration with respect to time yields the coordinates. On the other hand, an integration with respect to time of $\dot{p}, \dot{q}$, or $\dot{r}$ does not yield the coordinates. Since T is a function of true and *quasi-coordinates*, there are two approaches for writing Lagrange's equations. In the first approach one first substitutes for the quasi-coordinates in terms of the true coordinates from (6.1.13) through (6.1.16) and then writes Lagrange's equations. Thus one expresses T and μ as

$$\begin{aligned} T = {} & \tfrac{1}{2}(I_{xx} + I_1 + I_2\theta)(\dot{\phi} - \theta\dot{\psi})^2 + \tfrac{1}{2}(I_{yy} + I_3 + I_4\theta)(\dot{\theta} + \dot{\psi}\phi)^2 \\ & + \tfrac{1}{2}(I_{zz} + I_5 + I_6\theta)(\dot{\psi} - \dot{\theta}\phi)^2 + I_7\phi(\dot{\phi} - \theta\dot{\psi})(\dot{\theta} + \dot{\psi}\phi) \\ & - (I_{xz} + I_8 + I_9\theta)(\dot{\phi} - \theta\dot{\psi})(\dot{\psi} - \dot{\theta}\phi) \\ & + I_{10}\phi(\dot{\theta} + \dot{\psi}\phi)(\dot{\psi} - \dot{\theta}\phi) + \text{h.o.t.} \end{aligned} \tag{6.1.20}$$

$$\mu = \tfrac{1}{2}\mu_1(\dot{\phi} - \theta\dot{\psi})^2 + \tfrac{1}{2}\mu_2(\dot{\theta} + \dot{\psi}\phi)^2 + \text{h.o.t.} \tag{6.1.21}$$

Substituting for T, V, and μ from (6.1.20), (6.1.18), and (6.1.21) into Lagrange's equations, we obtain

$$\begin{aligned} \frac{d}{dt}\left(\frac{\partial T}{\partial \dot{\phi}}\right) - \frac{\partial T}{\partial \phi} + \frac{\partial V}{\partial \phi} + \frac{\partial \mu}{\partial \dot{\phi}} &= 0 \\ \frac{d}{dt}\left(\frac{\partial T}{\partial \dot{\theta}}\right) - \frac{\partial T}{\partial \theta} + \frac{\partial V}{\partial \theta} + \frac{\partial \mu}{\partial \dot{\theta}} &= 0 \end{aligned} \tag{6.1.22}$$

Setting $\dot{\psi} = \dot{\theta}\phi$ (i.e., $r = 0$) and neglecting cubic and higher-order terms, we obtain

$$(I_{xx} + I_1)\ddot{\phi} + V_1\phi + \mu_1(\dot{\phi} - \theta\dot{\psi}) + V_2\phi\theta - (I_{xz} + I_2 + I_8)\dot{\phi}\dot{\theta} + I_7\phi\ddot{\theta} + I_2\ddot{\phi}\theta = 0 \tag{6.1.23}$$

$$\begin{aligned} (I_{yy} + I_3)\ddot{\theta} &+ V_3\theta + \mu_2(\dot{\theta} + \phi\dot{\psi}) + \tfrac{1}{2}V_2\phi^2 + \tfrac{3}{2}V_4\theta^2 \\ &+ (I_{xz} + \tfrac{1}{2}I_2 + I_7 + I_8)\dot{\phi}^2 + \tfrac{1}{2}I_4\dot{\theta}^2 + I_7\phi\ddot{\phi} + I_4\theta\ddot{\theta} = 0 \end{aligned} \tag{6.1.24}$$

We note the presence of the terms involving $I_{xz} + I_8$ in (6.1.23) and (6.1.24). These terms would have been absent had we set $r = 0$ (i.e., no yaw angular velocity) in (6.1.17) rather than after the derivation of the equations.

In the second approach one keeps the kinetic energy and dissipation function expressed in mixed true and quasi-coordinates but uses the following modified form of Lagrange's equations (Whittaker 1961, Section 30; Meirovitch 1970, Section 4.12):

$$\frac{d}{dt}\left(\frac{\partial T}{\partial \Pi}\right) + [\Gamma]\frac{\partial T}{\partial \Pi} - [\beta]^T\left(\frac{\partial T}{\partial q} - \frac{\partial V}{\partial q}\right) + \frac{\partial \mu}{\partial \Pi} = 0 \tag{6.1.25}$$

where

$$[\Gamma] = \begin{bmatrix} 0 & -r & q \\ r & 0 & -p \\ -q & p & 0 \end{bmatrix} \tag{6.1.26}$$

Substituting for T, V, and μ from (6.1.17) through (6.1.19), letting $r = 0$, using (6.1.13) through (6.1.16) in the resulting equations, and neglecting cubic and higher-order terms, we obtain exactly equations (6.1.23) and (6.1.24).

If the nonlinear terms in (6.1.23) and (6.1.24) are neglected, the pitch and roll modes of the oscillations are uncoupled. However the nonlinearity couples the two modes, especially when the pitch frequency is approximately twice the roll frequency. In the latter case Froude (1863) observed that such ships have undesirable roll characteristics. This phenomenon was explained by Nayfeh,

Mook, and Marshall (1973), who showed that there exists a *saturation phenomenon* in the response of such ships. They found that as the amplitude of an excitation having a frequency equal, or nearly equal, to that of the pitch mode increases, the pitch amplitude increases until a critical value is reached. At that point further increases in the amplitude of the excitation cause the roll mode to develop and do not further affect the pitch mode. The pitch mode is *saturated*, and all the additional energy fed into the system by increasing the amplitude of the excitation goes into rolling motion. Paulling and Rosenberg (1959) solved the linearized form of (6.1.24), substituded the pitch solution into (6.1.23), and analyzed the linearized form of the resulting equation. Such an approach recognizes the similarity to parametric resonance; but because it basically uncouples the equations, it cannot reveal the essential feature of the motion–the saturation phenomenon. Mook, Marshall, and Nayfeh (1974) analyzed the subharmonic, superharmonic, combination, and ultraharmonic responses of ships that are restrained to pitch and roll only. Marshall and Morrow (1975) treated the related problem of wave-induced instabilities of semi-submersible oil-drilling platforms.

6.1.4 SELF-SUSTAINING OSCILLATORS

As a fourth example we follow Theodorchik (1948) and consider a circuit consisting of two RLC oscillators coupled to a vacuum tube as shown in Figure 6-4. The vacuum tube consists of a cathode (filament) F heated by a battery so that it will emit electrons, a plate P charged positively (anode) so that it will attract the electrons emitted by the filament, and a grid G that consists of a coarse mesh for controlling the flow of electrons from the cathode to the anode. This control is accomplished by maintaining the grid voltage the same as that across the capacitor C_1. The current in the grid is maintained negligibly small by connecting it with a large resistor R_G.

The equations describing the motion of the oscillators are

$$L_1 \frac{di_1}{dt} + R_1 i_1 + \frac{1}{C_1} \int i_1 \, dt + \frac{1}{C} \int (i_1 - i_2) \, dt = M_1 \frac{di_p}{dt} \tag{6.1.27}$$

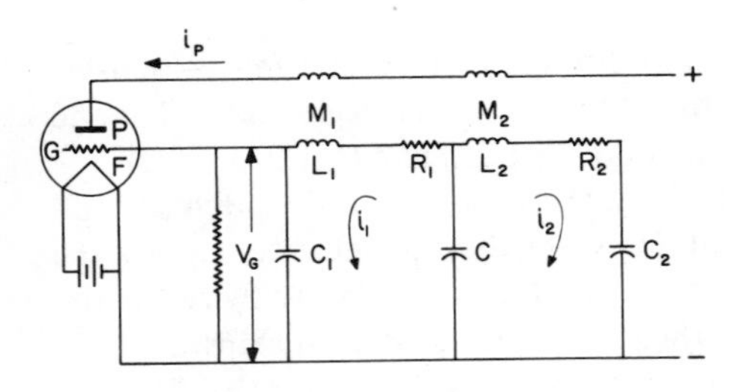

Figure 6-4. Coupled self-sustaining oscillators.

$$L_2 \frac{di_2}{dt} + R_2 i_2 + \frac{1}{C_2}\int i_2\, dt + \frac{1}{C}\int (i_2 - i_1)\, dt = M_2 \frac{di_p}{dt} \tag{6.1.28}$$

$$V_G = \frac{1}{C_1}\int i_1\, dt \tag{6.1.29}$$

where L is the inductance, R is the resistance, and C is the capacitance. The mutual inductances M_1 and M_2 are positive. We let

$$x_n = \frac{1}{C_n}\int i_n\, dt \tag{6.1.30}$$

so that $V_G = x_1$. Moreover we assume that the plate current i_p is related to the grid voltage $V_G = x_1$ by

$$i_p = \alpha_1 x_1 - \tfrac{1}{3}\alpha_3 x_1^3 \tag{6.1.31}$$

where α_1 and α_2 are positive. Substituting for x_1, x_2, and i_p from (6.1.30) and (6.1.31) into (6.1.27) and (6.1.28) yields the following two equations:

$$\ddot{x}_1 + \frac{R_1}{L_1}\dot{x}_1 + \frac{1}{L_1 C_1}\left(1 + \frac{C_1}{C}\right)x_1 - \frac{C_2}{L_1 C C_1}x_2 = \frac{M_1}{L_1 C_1}(\alpha_1 - \alpha_3 x_1^2)\,\dot{x}_1 \tag{6.1.32}$$

$$\ddot{x}_2 + \frac{R_2}{L_2}\dot{x}_2 + \frac{1}{L_2 C_2}\left(1 + \frac{C_2}{C}\right)x_2 - \frac{C_1}{L_2 C C_2}x_1 = \frac{M_2}{L_2 C_2}(\alpha_1 - \alpha_3 x_1^2)\,\dot{x}_1 \tag{6.1.33}$$

A number of investigators analyzed the so-called mutual synchronized solution of (6.1.32) and (6.1.33); see Minorsky (1962, Section 6), Butenin (1965, Section 4), and Tondl (1970b) for a discussion and additional references. The problem of mutual synchronization is addressed in Exercise 6-20. Interactions of self-excited-oscillations were studied by Kononenko and Koval'chuk (1973b) and Hayashi and Kuramitsu (1974), while interactions between forced and self-excited oscillations in multidegree-of-freedom systems were studied by Nguyen (1975a). Quenching of limit cycles was analyzed by Mansour (1972) and Tondl (1975a, b).

6.1.5. THE STABILITY OF THE TRIANGULAR POINTS IN THE RESTRICTED PROBLEM OF THREE BODIES

As a last example we consider the planar motion of three bodies of masses m_1, m_2, and m_3 when $m_3 << m_1$ and $m_3 << m_2$ so that the motion of m_3 does not affect the motion of the other two masses. The masses m_1 and m_2 are as-

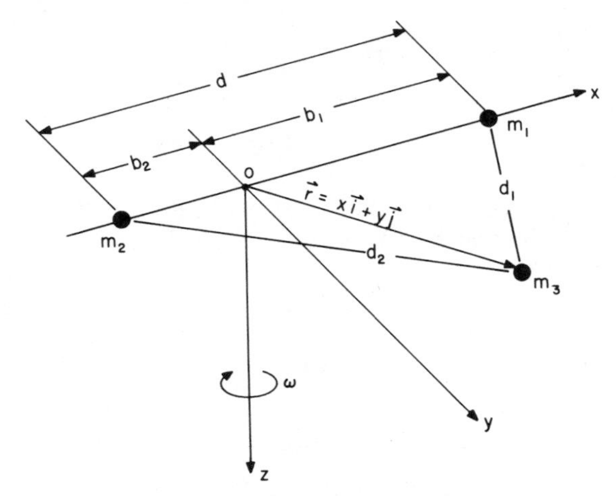

Figure 6-5. Restricted three-body problem.

sumed to move in coplanar circular orbits about their center of mass O with a constant angular velocity ω as shown in Figure 6-5. We introduce a rotating Cartesian coordinate system xyz centered at O such that the masses m_1 and m_2 are located on the x-axis and the z-axis is normal to the plane of motion.

We first relate the angular velocity ω to d, m_1, m_2, and the gravitational constant G. The distances of m_1 and m_2 from O are

$$b_1 = \frac{m_2 d}{m_1 + m_2}, \quad b_2 = \frac{m_1 d}{m_1 + m_2}$$

Since the mass m_3 is negligible, the only forces acting on the masses m_1 and m_2 are the mutual attraction forces f_{12} and f_{21} along the x-axis, and they are given by

$$f_{12} = -f_{21} = -\frac{Gm_1 m_2}{d^2} \tag{6.1.34}$$

Since ω and d are constants, the equation of motion of the mass m_2 is

$$-\frac{m_1 m_2}{m_1 + m_2} d\omega^2 = -\frac{Gm_1 m_2}{d^2} \tag{6.1.35}$$

Hence

$$\omega^2 = \frac{G(m_1 + m_2)}{d^3} \tag{6.1.36}$$

To determine the equations describing the motion of m_3, we write the Lagrangian and then Lagrange's equations. To do this, we note that the velocity of m_3 in an inertial frame is given by

$$\mathbf{v} = (\dot{x} - \omega y)\,\mathbf{i} + (\dot{y} + \omega x)\,\mathbf{j} \tag{6.1.37}$$

where **i** and **j** are unit vectors along the x- and y-axes. Hence

$$T = \tfrac{1}{2} m_3 \left[(\dot{x} - \omega y)^2 + (\dot{y} + \omega x)^2\right] \tag{6.1.38}$$

The potential energy is given by

$$V = -\frac{Gm_1 m_3}{d_1} - \frac{Gm_2 m_3}{d_2} \tag{6.1.39}$$

where

$$\begin{aligned} d_1^2 &= (b_1 - x)^2 + y^2 \\ d_2^2 &= (b_2 + x)^2 + y^2 \end{aligned} \tag{6.1.40}$$

Hence the Lagrangian is

$$L = \tfrac{1}{2} m_3 \left[(\dot{x} - \omega y)^2 + (\dot{y} + \omega x)^2\right] + \frac{Gm_1 m_3}{d_1} + \frac{Gm_2 m_3}{d_2} \tag{6.1.41}$$

Therefore Lagrange's equations have the form

$$\begin{aligned} \frac{d}{dt}\left(\frac{\partial L}{\partial \dot{x}}\right) - \frac{\partial L}{\partial x} &= 0 \\ \frac{d}{dt}\left(\frac{\partial L}{\partial \dot{y}}\right) - \frac{\partial L}{\partial y} &= 0 \end{aligned} \tag{6.1.42}$$

Substituting (6.1.41) into (6.1.42) and making lengths and time dimensionless by using d and ω^{-1}, respectively, we obtain the following two dimensionless equations:

$$\begin{aligned} \ddot{x} - 2\dot{y} - x &= -\frac{\mu(x - 1 + \mu)}{d_1^3} - \frac{(1 - \mu)(x + \mu)}{d_2^3} \\ \ddot{y} + 2\dot{x} - y &= -\frac{\mu y}{d_1^3} - \frac{(1 - \mu)\, y}{d_2^3} \end{aligned} \tag{6.1.43}$$

where x, y, and t are dimensionless and $\mu = m_1/(m_1 + m_2)$.

Equations (6.1.43) have five equilibrium-point solutions (e.g., Szebehely 1967, pp. 231–318); they are usually called Lagrange's points and are denoted by $L_n, n = 1, 2, 3, 4, 5$. They correspond to the solutions of

$$\begin{aligned} x - \frac{\mu(x - 1 + \mu)}{d_1^3} - \frac{(1 - \mu)(x + \mu)}{d_2^3} &= 0 \\ y - \frac{\mu y}{d_1^3} - \frac{(1 - \mu)\, y}{d_2^3} &= 0 \end{aligned} \tag{6.1.44}$$

It can be easily verified that $d_1 = d_2 = 1$ satisfies (6.1.44) so that the coordinates of two of these points (L_4 and L_5) are

$$x_0 = \tfrac{1}{2} - \mu, \quad y_0 = \frac{\sqrt{3}}{2}$$
$$x_0 = \tfrac{1}{2} - \mu, \quad y_0 = -\frac{\sqrt{3}}{2} \tag{6.1.45}$$

The other three equilibrium points lie along the x-axis.

Next we consider the motion and stability of these equilibrium points. As an example we consider L_4 and we let

$$x = \tfrac{1}{2} - \mu + u_1, \quad y = \frac{\sqrt{3}}{2} + u_2 \tag{6.1.46}$$

Substituting (6.1.46) into (6.1.43) and expanding for small but finite values of u_1 and u_2, we obtain

$$\ddot{u}_1 - 2\dot{u}_2 - \tfrac{3}{4}u_1 - \eta u_2 = -\frac{\partial U}{\partial u_1}$$
$$\ddot{u}_2 + 2\dot{u}_1 - \eta u_1 - \tfrac{9}{4}u_2 = -\frac{\partial U}{\partial u_2} \tag{6.1.47}$$

where $\eta = (3\sqrt{3}/4)(1 - 2\mu)$ and

$$U = \frac{3\sqrt{3}}{16} u_2(u_1^2 + u_2^2) + \frac{\sqrt{3}}{36}\eta u_1(33u_2^2 - 7u_1^2) + \text{h.o.t.} \tag{6.1.48}$$

The stability of the points L_4 and L_5 (usually called the triangular points) received considerable attention. The solution of the linearized equations (6.1.47) indicates that L_4 and L_5 are stable for all $\mu < \mu_c = \frac{1}{2}(1 - \frac{1}{9}\sqrt{69})$. However using the nonlinear equations, Leontovich (1962) proved that L_4 and L_5 are stable for all $\mu < \mu_c$ except on a set of measure zero. Deprit and Deprit-Bartholomé (1967) proved that the exceptional set contains at most four values of μ, including the values μ_2 and μ_3 corresponding to a two-to-one and a three-to-one internal resonance, respectively. Using a Hamiltonian formulation, Markeev (1968, 1969a, 1972) proved the instability of L_4 and L_5 for $\mu \approx \mu_2$ and μ_3. The nonlinear motion near L_4 and L_5 for the case $\mu \approx \mu_3$ was studied by Breakwell and Pringle (1966b), Deprit (1969), Kamel (1969), Nayfeh and Kamel (1970b), and Alfriend (1971a). The nonlinear motion near L_4 and L_5 for the case $\mu \approx \mu_2$ was studied by Alfriend (1970), Henrard (1970), and Nayfeh (1971b). These studies show that although a gyroscopic system is linearly stable it may be nonlinearly unstable if there are internal resonances. The linear stability of the motion near L_4 and L_5 in the elliptic restricted problem of three bodies was studied by Danby (1964), Grebenikov (1964), Bennett (1966). Alfriend and Rand (1969), Luk'ianov (1969), Nayfeh and Kamel (1970a), Markeev (1970), and Nayfeh (1970a).

6.2. Free Oscillations of Systems Having Quadratic Nonlinearities

Governing equations with quadratic nonlinearities are associated with many physical systems such as betatron oscillations (Blaquière, 1966, p. 140), the motion of a swinging spring, the motion of a ship, the motion of a fluid interface, the motion of a rotating shaft, the vibrations of shells and composite plates, the vibration of a structure about a loaded static equilibrium configuration, and the coupled longitudinal and transverse oscillations of a column. Free oscillations of nongyroscopic systems having quadratic nonlinearities were treated by using a variety of methods by Beth (1913); Paulling and Rosenberg (1959); Sevin (1961); Mettler (1963, 1968, 1975); Struble and Heinbockel (1963); Heinbockel and Struble (1963); Kane and Kahn (1968); van der Burgh (1968, 1975, 1976); Markeev (1969a, b); Khazin and Tsel'man (1970); Tsel'man (1970, 1971); Alfriend (1970, 1971c); Nayfeh (1971b); Habakow (1972); Nayfeh (1973b, Sections 5.5.3, 5.7.5, and 6.2.7); Broucke and Baxa (1973); Nayfeh, Mook, and Marshall (1973); Kononenko and Koval'chuk (1973b); Srinivasan and Sankar (1974); Cheshankov (1974a); Marshall and Morrow (1975); and Crespo da Silva (1974).

It is shown in this section that the behavior of systems having quadratic nonlinearities is the same as the linear behavior to second order (i.e., the frequencies are independent of the amplitudes and the modes of oscillation are uncoupled) unless the frequencies ω_n are commensurable or nearly commensurable, that is, unless $\omega_n \approx 2\omega_m$ or $\omega_n \approx \omega_m \mp \omega_k$. When one or more of these conditions are satisfied, an *internal resonance* is said to exist.

To describe the main physical features of such systems without involving a great deal of algebra, we consider a two-degree-of freedom system governed by

$$\begin{aligned} \ddot{u}_1 + \omega_1^2 u_1 &= -2\hat{\mu}_1 \dot{u}_1 + \alpha_1 u_1 u_2 \\ \ddot{u}_2 + \omega_2^2 u_2 &= -2\hat{\mu}_2 \dot{u}_2 + \alpha_2 u_1^2 \end{aligned} \tag{6.2.1}$$

In this book we use the method of multiple scales and seek a first-order approximate solution of (6.2.1) for small but finite amplitudes in the form

$$\begin{aligned} u_1 &= \epsilon u_{11}(T_0, T_1) + \epsilon^2 u_{12}(T_0, T_1) + \cdots \\ u_2 &= \epsilon u_{21}(T_0, T_1) + \epsilon^2 u_{22}(T_0, T_1) + \cdots \end{aligned} \tag{6.2.2}$$

where ϵ is a small dimensionless parameter the order of the amplitudes of oscillation and $T_n = \epsilon^n t$. In order to have the damping and nonlinear terms appear in the same perturbation equations, we scale the damping coefficients by letting $\hat{\mu}_j = \epsilon\mu_j$. Substituting (6.2.2) into (6.2.1) and equating coefficients of like powers of ϵ, we obtain

Order ϵ

$$\begin{aligned} D_0^2 u_{11} + \omega_1^2 u_{11} &= 0 \\ D_0^2 u_{21} + \omega_2^2 u_{21} &= 0 \end{aligned} \tag{6.2.3}$$

Order ϵ^2

$$\begin{aligned} D_0^2 u_{12} + \omega_1^2 u_{12} &= -2D_0(D_1 u_{11} + \mu_1 u_{11}) + \alpha_1 u_{11} u_{21} \\ D_0^2 u_{22} + \omega_2^2 u_{22} &= -2D_0(D_1 u_{21} + \mu_2 u_{21}) + \alpha_2 u_{11}^2 \end{aligned} \tag{6.2.4}$$

where $D_n = \partial/\partial T_n$.

The solutions of (6.2.3) can be written in the form

$$\begin{aligned} u_{11} &= A_1(T_1) \exp(i\omega_1 T_0) + cc \\ u_{21} &= A_2(T_1) \exp(i\omega_2 T_0) + cc \end{aligned} \tag{6.2.5}$$

Substituting (6.2.5) into (6.2.4) leads to

$$D_0^2 u_{12} + \omega_1^2 u_{12} = -2i\omega_1(A_1' + \mu_1 A_1) \exp(i\omega_1 T_0) + \alpha_1 \{A_1 A_2 \exp[i(\omega_1 + \omega_2)T_0] + A_2 \bar{A}_1 \exp[i(\omega_2 - \omega_1)T_0]\} + cc \tag{6.2.6}$$

$$D_0^2 u_{22} + \omega_2^2 u_{22} = -2i\omega_2(A_2' + \mu_2 A_2) \exp(i\omega_2 T_0) + \alpha_2 [A_1^2 \exp(2i\omega_1 T_0) + A_1 \bar{A}_1] + cc \tag{6.2.7}$$

When $2\omega_1 \approx \omega_2$ there is an extra link, or term, connecting u_1 and u_2. This is referred to as an internal resonance. In analyzing the particular solutions of (6.2.6) and (6.2.7) we need to distinguish between the resonant case in which $\omega_2 \approx 2\omega_1$ and the nonresonant case in which ω_2 is away from $2\omega_1$.

6.2.1. THE NONRESONANT CASE

In this case the solvability conditions (the conditions for the elimination of secular terms) become

$$A_1' + \mu_1 A_1 = 0 \quad \text{and} \quad A_2' + \mu_2 A_2 = 0 \tag{6.2.8}$$

where the prime denotes the derivative with respect to T_1. It follows that

$$A_1 = a_1 \exp(-\mu_1 T_1) \quad \text{and} \quad A_2 = a_2 \exp(-\mu_2 T_1) \tag{6.2.9}$$

where a_1 and a_2 are complex constants and

$$\begin{aligned} u_1 &= \epsilon \exp(-\epsilon\mu_1 t)[a_1 \exp(i\omega_1 t) + cc] + O(\epsilon^2) \\ u_2 &= \epsilon \exp(-\epsilon\mu_2 t)[a_2 \exp(i\omega_2 t) + cc] + O(\epsilon^2) \end{aligned} \tag{6.2.10}$$

Thus both modes decay, and the steady-state solutions are

$$u_1 = u_2 = 0 \tag{6.2.11}$$

6.2.2. THE RESONANT CASE (INTERNAL RESONANCE)

In this case we introduce a detuning parameter σ according to

$$\omega_2 = 2\omega_1 + \epsilon\sigma \tag{6.2.12}$$

and set

$$\begin{aligned} 2\omega_1 T_0 &= \omega_2 T_0 - \epsilon\sigma T_0 = \omega_2 T_0 - \sigma T_1 \\ (\omega_2 - \omega_1)T_0 &= \omega_1 T_0 + \epsilon\sigma T_0 = \omega_1 T_0 + \sigma T_1 \end{aligned} \tag{6.2.13}$$

In this case it follows from (6.2.6), (6.2.7), and (6.2.13) that the solvability conditions are

$$\begin{aligned} -2i\omega_1(A_1' + \mu_1 A_1) + \alpha_1 A_2 \bar{A}_1 \exp(i\sigma T_1) &= 0 \\ -2i\omega_2(A_2' + \mu_2 A_2) + \alpha_2 A_1^2 \exp(-i\sigma T_1) &= 0 \end{aligned} \tag{6.2.14}$$

It is convenient to introduce polar notation. Thus we put

$$A_m = \tfrac{1}{2} a_m \exp(i\theta_m) \qquad \text{for } m = 1 \text{ and } 2 \tag{6.2.15}$$

where a_m and θ_m are real functions of T_1. Substituting (6.2.15) into (6.2.14) and separating the result into real and imaginary parts, we obtain

$$a_1' = -\mu_1 a_1 + \frac{\alpha_1}{4\omega_1} a_1 a_2 \sin\gamma \tag{6.2.16}$$

$$a_2' = -\mu_2 a_2 - \frac{\alpha_2}{4\omega_2} a_1^2 \sin\gamma \tag{6.2.17}$$

$$a_1\theta_1' = -\frac{\alpha_1}{4\omega_1} a_1 a_2 \cos\gamma \tag{6.2.18}$$

$$a_2\theta_2' = -\frac{\alpha_2}{4\omega_2} a_1^2 \cos\gamma \tag{6.2.19}$$

where

$$\gamma = \theta_2 - 2\theta_1 + \sigma T_1 \tag{6.2.20}$$

Eliminating θ_1 and θ_2 from (6.2.18) through (6.2.20) yields

$$a_2\gamma' = \sigma a_2 + \left(\frac{\alpha_1 a_2^2}{2\omega_1} - \frac{\alpha_2 a_1^2}{4\omega_2}\right)\cos\gamma \tag{6.2.21}$$

Thus the problem is reduced to the solution of (6.2.16), (6.2.17), and (6.2.21).

The steady-state response corresponds to

$$a_1' = a_2' = \gamma' = 0 \tag{6.2.22}$$

Thus it corresponds to the solutions of

$$-\mu_1 a_1 + \frac{\alpha_1}{4\omega_1} a_1 a_2 \sin\gamma = 0 \tag{6.2.23}$$

$$-\mu_2 a_2 - \frac{\alpha_2}{4\omega_2} a_1^2 \sin\gamma = 0 \tag{6.2.24}$$

$$\left(\frac{\alpha_1}{2\omega_1} a_2^2 - \frac{\alpha_2}{4\omega_2} a_1^2\right) \cos\gamma + \sigma a_2 = 0 \tag{6.2.25}$$

Eliminating γ from (6.2.23) and (6.2.24) leads to

$$a_1^2 + \frac{\mu_2\omega_2\alpha_1}{\mu_1\omega_1\alpha_2} a_2^2 = 0 \tag{6.2.26}$$

Thus, if α_1 and α_2 have different signs, a_1 and a_2 can differ from zero. This can be seen by manipulating (6.2.23) through (6.2.26) to obtain

$$\cot\gamma = -\frac{\sigma}{\mu_2 + 2\mu_1}, \quad a_2 \sin\gamma = \frac{4\mu_1\omega_1}{\alpha_1} \tag{6.2.27}$$

Equations (6.2.27) can be solved for γ and a_2; a_1 can then be found from (6.2.26). This occurs only if a *regenerative element* (see Sections 3.1.7 and 6.1.4) exists in the system.

When there is no internal resonance, the first approximation of the solution is essentially the solution of the linear problem, regardless of the signs of α_1 and α_2. On the other hand, when there is an internal resonance and α_1 and α_2 have opposite signs, the equations admit self-sustained oscillations in spite of the presence of damping and in contrast with the solution of the linear problem.

In the absence of damping (i.e., $\mu_1 = \mu_2 = 0$) the exact solution of (6.2.16), (6.2.17), and (6.2.21) can be expressed in terms of elliptic functions as follows. Multiplying (6.2.16) by a_1 and (6.2.17) by νa_2, where $\nu = \alpha_1\omega_2/\alpha_2\omega_1$, adding the resulting equations, and integrating, we obtain

$$a_1^2 + \nu a_2^2 = E \tag{6.2.28}$$

where E is a constant of integration proportional to the initial energy in the system. If α_1 and α_2 have the same sign, (6.2.28) shows that a_1 and a_2 are always bounded. However when α_1 and α_2 have opposite signs, a_1 and a_2 may grow with time even though $a_1^2 - |\nu| a_2^2$ is bounded, as shown in Section 6.4. In what follows we assume that the system does not contain regenerative elements so that α_1 and α_2 have the same sign.

Changing the independent variable from T_1 to a_2 in (6.2.21) by using (6.2.17) yields

$$-a_2 a_1^2 \sin\gamma \frac{d\gamma}{da_2} = \frac{4\omega_2\sigma}{\alpha_2} a_2 + (2\nu a_2^2 - a_1^2)\cos\gamma \tag{6.2.29}$$

or

$$-4\omega_2\sigma\alpha_2^{-1} a_2\, da_2 + a_2 a_1^2\, d(\cos\gamma) - 2\nu a_2^2 \cos\gamma\, da_2 + a_1^2 \cos\gamma\, da_2 = 0 \tag{6.2.30}$$

But from (6.2.28)

$$a_1\, da_1 = -\nu a_2\, da_2 \tag{6.2.31}$$

Hence (6.2.30) can be rewritten as

$$a_2 a_1^2\, d(\cos\gamma) + a_1^2 \cos\gamma\, da_2 + 2a_1 a_2 \cos\gamma\, da_1 - 4\omega_2\sigma\alpha_2^{-1} a_2\, da_2 = 0$$

or

$$d(a_1^2 a_2 \cos\gamma) - 2\omega_2\sigma\alpha_2^{-1}\, d(a_2^2) = 0 \tag{6.2.32}$$

which can be integrated to yield

$$a_2 a_1^2 \cos\gamma - 2\omega_2\sigma\alpha_2^{-1} a_2^2 = L \tag{6.2.33}$$

where L is a constant of integration.

To determine a single equation for a_1, we let

$$a_1^2 = E\xi \tag{6.2.34}$$

Hence it follows from (6.2.28) that

$$\nu a_2^2 = E(1-\xi) \tag{6.2.35}$$

Using (6.2.33) to eliminate γ from (6.2.16) and expressing a_1^2 and a_2^2 in terms of ξ, we obtain

$$\frac{4\nu\omega_1^2}{\alpha_1^2 E}\left(\frac{d\xi}{dT_1}\right)^2 = \xi^2(1-\xi) - \frac{\nu}{E^3}\left[L + \frac{2\omega_2\sigma E}{\alpha_2\nu}(1-\xi)\right]^2 = F^2(\xi) - G^2(\xi) \tag{6.2.36}$$

where

$$F = \pm\xi\sqrt{1-\xi}, \qquad G = \pm\left(\frac{\nu}{E^3}\right)^{1/2}\left[L + \frac{2\omega_2\sigma E}{\alpha_2\nu}(1-\xi)\right] \tag{6.2.37}$$

The functions F and G are shown schematically in Figure 6-6. For real motions, $F^2 \geqslant G^2$. The points where G meets F correspond to the vanishing of ξ'. In general, the curve G meets the branches of F at three points corresponding to

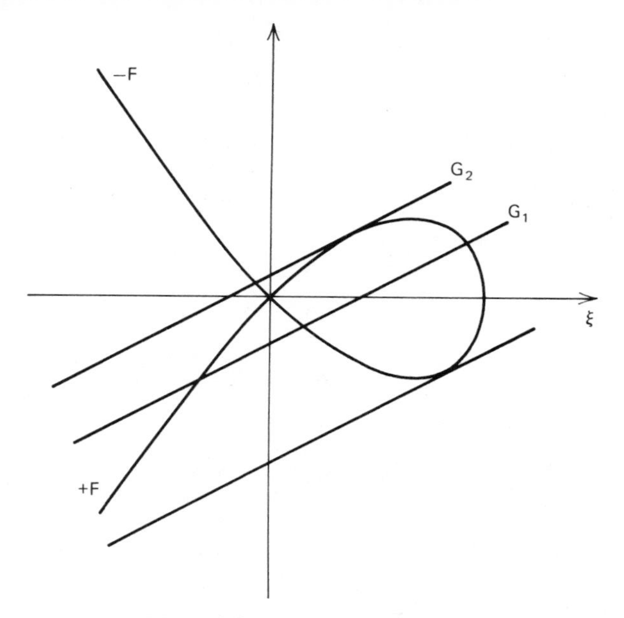

Figure 6-6. Schematic of F and G.

the three roots ξ_1, ξ_2, and ξ_3 of the right-hand side of (6.2.36). Let $\xi_1 \leqslant \xi_2 \leqslant \xi_3$. Since $\xi = a_1^2/E$, the motion is confined between ξ_2 and ξ_3.

When the three roots are distinct corresponding to a curve such as G_1, ξ is periodic and oscillates between ξ_2 and ξ_3, and the motion is aperiodic. In this case the solution for ξ can be expressed in terms of Jacobi elliptic functions as follows. First, in terms of the ξ_n, we express (6.2.36) as

$$\frac{4\nu\omega_1^2}{\alpha_1^2 E}\left(\frac{d\xi}{dT_1}\right)^2 = (\xi_3 - \xi)(\xi - \xi_2)(\xi - \xi_1) \tag{6.2.38}$$

Introducing the transformation

$$\xi_3 - \xi = (\xi_3 - \xi_2)\sin^2\chi \tag{6.2.39}$$

into (6.2.38) we obtain

$$\frac{4\omega_1\sqrt{\nu}}{\alpha_1\sqrt{E}}\frac{d\chi}{dT_1} = \pm\sqrt{\xi_3 - \xi_1}\,(1 - \eta^2\sin^2\chi)^{1/2} \tag{6.2.40}$$

where

$$\eta = \sqrt{\frac{\xi_3 - \xi_2}{\xi_3 - \xi_1}} \tag{6.2.41}$$

Separating the variables in (6.2.40), putting $T_1 = \epsilon t$, and integrating the resulting equation, we obtain

$$\kappa(t - t_0) = \int_0^\chi \frac{d\chi}{\sqrt{1 - \eta^2 \sin^2 \chi}} \tag{6.2.42}$$

or

$$\sin \chi = \text{sn}\, [\kappa(t - t_0); \eta] \tag{6.2.43}$$

where t_0 corresponds to $\chi = 0$, sn is a Jacobi elliptic function, and

$$\kappa = \frac{\epsilon \alpha_1}{4\omega_1} \left[\frac{E(\xi_3 - \xi_1)}{\nu} \right]^{1/2} \tag{6.2.44}$$

Combining (6.2.34), (6.2.39), and (6.2.43) yields

$$\xi = \frac{a_1^2}{E} = \xi_3 - (\xi_3 - \xi_2)\, \text{sn}^2\, [\kappa(t - t_0); \eta] \tag{6.2.45}$$

Thus (6.2.45) and (6.2.35) show that, in the absence of damping and for the conditions such that the ξ_n are distinct, the energy in the system continues to be exchanged undamped between the two modes of oscillation as shown in Figure 6-7 for two different initial conditions.

In the presence of damping, Figure 6-8 shows numerical integrations of (6.2.16), (6.2.17), and (6.2.21). In this case the energy continues to be exchanged between the two modes but it is continuously dissipated.

The internal resonance can be used to adjust the rate at which a given mode

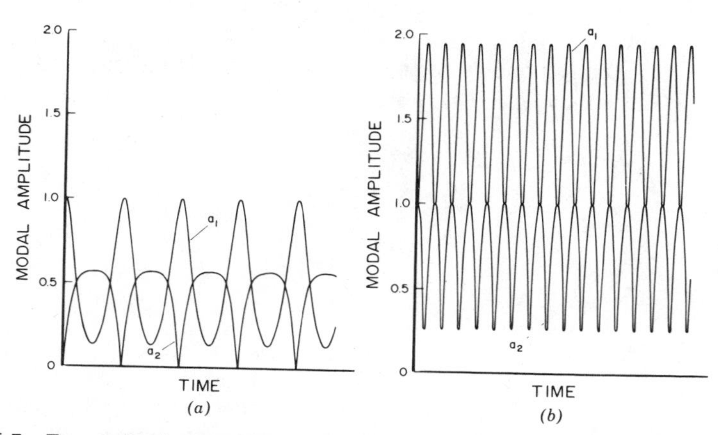

Figure 6-7. Free-oscillation amplitudes of a conservative two-degree-of-freedom system with quadratic nonlinearities; $\omega_2 \simeq 2\omega_1$: *(a)* $a_1(0) = 1, a_2(0) = 0$; *(b)* $a_1(0) = a_2(0) = 1$.

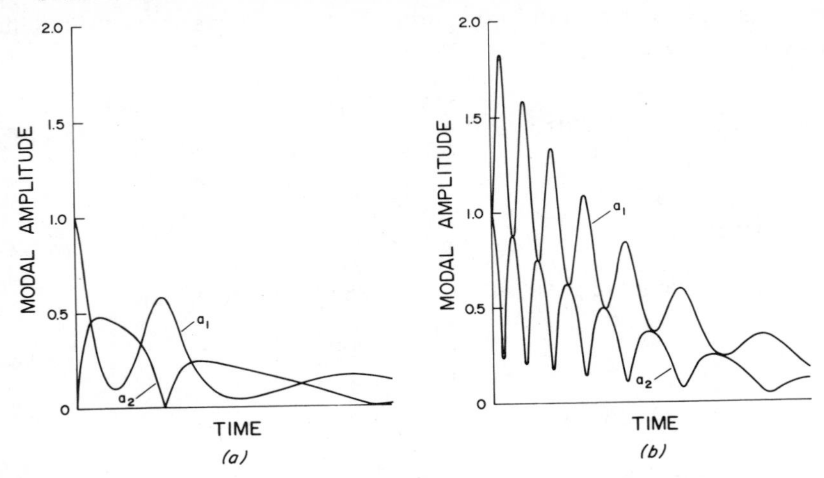

Figure 6-8. Free-oscillation amplitudes of a damped two-degree-of-freedom system with quadratic nonlinearities; $\omega_2 \simeq 2\omega_1$: (*a*) $a_1(0) = 1, a_2(0) = 0$; (*b*) $a_1(0) = a_2(0) = 1$.

decays. To see this, we first suppose that there is no internal resonance. In the first approximation a_1 decays exponentially with time as in the linear case. This curve is shown in Figure 6-9. Also shown in Figure 6-9 are two curves giving a_1 as a function of time in the presence of an internal resonance. One curve corresponds to $\mu_2 < \mu_1$, while the other corresponds to $\mu_2 > \mu_1$. The corresponding curves for a_2 are not shown. We note that in all cases only the first mode is excited initially. These results suggest that one can adjust the rate at which a given mode decays by coupling it with another mode through an internal resonance.

When $\xi_2 = \xi_3$ corresponding to the curve G_2 which is tangent to one of the branches of F, $\xi = \xi_3$ is a constant according to (6.2.45), and hence $a_1 = \sqrt{E\xi_2}$

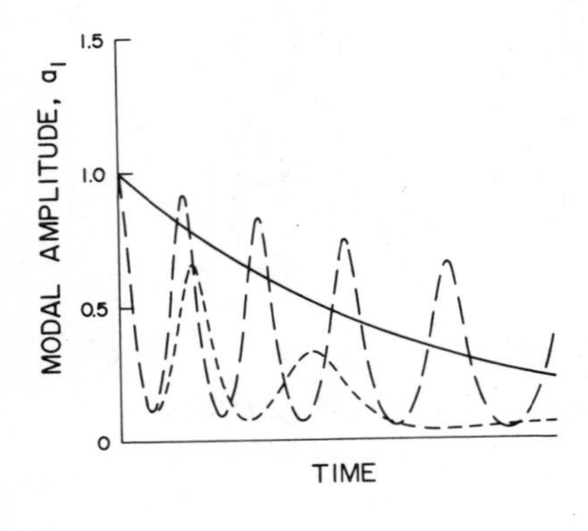

Figure 6-9. Effect of internal resonance and damping coefficients on the rate of decay: (——) no internal resonance; (— —) $\mu_2 < \mu_1$; (- - - -) $\mu_2 > \mu_1$.

and $a_2 = [E(1 - \xi_2)/\nu]^{1/2}$. In this case the motion is periodic. However any small disturbance would lead to a curve such as G_1 where the roots are distinct, and hence the motion is aperiodic as discussed above.

When $\xi_2 = \xi_1$, G coincides with the ξ-axis, and hence $L = \sigma = 0$ according to (6.2.37). Consequently it follows from (6.2.36) that $\xi_1 = \xi_2 = 0$ and $\xi_3 = 1$. The solution in this case can be obtained by introducing the transformation

$$\xi = \operatorname{sech}^2 \phi \tag{6.2.46}$$

into (6.2.36) with $L = \sigma = 0$. The result is

$$\frac{d\phi}{dt} = \kappa \tag{6.2.47}$$

whose solution is

$$\phi = \kappa(t - t_0) \tag{6.2.48}$$

Therefore

$$a_1 = \sqrt{E\xi} = \sqrt{E} \operatorname{sech} [\kappa(t - t_0)] \tag{6.2.49}$$

and it follows from (6.2.35) that

$$a_2 = \sqrt{\frac{E}{\nu}} \tanh [\kappa(t - t_0)] \tag{6.2.50}$$

We note that $L = \sigma = 0$ demands that $\cos \gamma = 0$ according to (6.2.33). Hence it follows from (6.2.18) and (6.2.19) that $\theta_1' = \theta_2' = 0$; that is, the phases are constant. Consequently the motion consists of only amplitude-modulated motions. As $t \to \infty$, $a_1 \to 0$ while $a_2 \to (E/\nu)^{1/2}$, leading to a motion that is independent of the lower mode. Thus the coupling that results from an internal resonance leads to a complete transfer of energy from the lower mode to the higher mode. However Figure 6-6 shows that such a motion is unstable because any small disturbance applied to it would lead to a motion corresponding to a curve such as G_1 for which the amplitude and the phase are modulated.

6.3. Free Oscillations of Systems Having Cubic Nonlinearities

Governing equations with cubic nonlinearities are associated with many physical systems such as the vibration of strings, beams, membranes, and plates for which stretching is significant (Chapter 7), dynamic vibration-isolation systems, dynamic vibration absorbers (Section 6.6), the motion of spherical, centripetal, and double pendulums, and the motion of masses connected by nonlinear springs. Tobias (1959) discussed the design of nonlinear vibration-isolation units. The general motion of these units is governed by a system of six coupled nonlinear differential equations with cubic nonlinearities. Henry and Tobias (1959, 1961), Gilchrist (1961), and Henry (1962) determined the modes of oscillation and their stability for a two-degree-of-freedom model. Grossley (1952) and Newland (1965) analyzed the motion of a centripetal pendulum. Other systems with two degrees of freedom were studied by Proskuriakov (1960b, 1965), Butenin (1965, Section 9), Blaquière (1966, Chapter 4), Yang and Rosenberg (1967), Hori (1967), Rand and Vito (1972), Varga and Aks (1974), Kuroda (1974), and Month and Rand (1977). Systems with several degrees of freedom were studied by Proskuriakov (1960a, b, 1962, 1965, 1968); Bogoliubov and Mitropolsky (1961, Sections 20 and 21); Sethna (1963b); Walker and Ford (1969); Lansdowne and Soudack (1971); Habakow (1972); Nustrov (1974); Rangacharyulu, Srinivasan, and Dasarathy (1974); Postnikov (1974); Helleman and Montroll (1974); Ford (1975); Hoogstraten and Kaper (1975); Montroll and Helleman (1975); and Eminhizer, Helleman, and Montroll (1976). For a comprehensive review and a discussion of the convergence of available techniques of determining periodic solutions of nonintegrable systems, we refer the reader to the book of Moser (1973) and the Proceedings of the AIP Conference edited by Jorna (1978) and in particular to the articles of Moser (1978), Helleman (1978), and Berry (1978).

As in the preceding section, we exhibit the main features of the behavior of systems having cubic nonlinearities by discussing a simple system, namely

$$\begin{aligned} \ddot{u}_1 + \omega_1^2 u_1 &= -2\hat{\mu}_1 \dot{u}_1 + \alpha_1 u_1^3 + \alpha_2 u_1^2 u_2 + \alpha_3 u_1 u_2^2 + \alpha_4 u_2^3 \\ \ddot{u}_2 + \omega_2^2 u_2 &= -2\hat{\mu}_2 \dot{u}_2 + \alpha_5 u_1^3 + \alpha_6 u_1^2 u_2 + \alpha_7 u_1 u_2^2 + \alpha_8 u_2^3 \end{aligned} \tag{6.3.1}$$

We seek an approximate solution of (6.3.1) for small but finite amplitudes in the form

$$\begin{aligned} u_1 &= \epsilon u_{11}(T_0, T_2) + \epsilon^3 u_{13}(T_0, T_2) + \cdots \\ u_2 &= \epsilon u_{21}(T_0, T_2) + \epsilon^3 u_{23}(T_0, T_2) + \cdots \end{aligned} \tag{6.3.2}$$

where ϵ is a small, dimensionless parameter the order of the amplitudes and $T_n = \epsilon^n t$. Note that the slow scale $T_1 = \epsilon t$ as well as the terms $\epsilon^2 u_{12}$ and $\epsilon^2 u_{22}$ are absent from (6.3.2) because the nonlinearity is cubic. Had we kept T_1, u_{12}, and u_{22}, we would have found that the solution is independent of T_1 and that u_{12} and u_{22} satisfy exactly the same equations as u_{11} and u_{21}. Hence u_{12} and u_{22} can be omitted without loss of generality. Moreover in order for the effect of the damping to balance the effect of the nonlinearity, the damping coefficients must be ordered so that the damping terms appear in the same perturbation equations as the nonlinear terms. Hence we set $\hat{\mu}_m = \epsilon^2 \mu_m$. Then substituting

(6.3.2) into (6.3.1) and equating coefficients of like powers of ϵ, we obtain

Order ϵ

$$D_0^2 u_{11} + \omega_1^2 u_{11} = 0$$
$$D_0^2 u_{21} + \omega_2^2 u_{21} = 0 \tag{6.3.3}$$

Order ϵ^3

$$D_0^2 u_{13} + \omega_1^2 u_{13} = -2D_0(D_2 u_{11} + \mu_1 u_{11}) + \alpha_1 u_{11}^3 + \alpha_2 u_{11}^2 u_{21} + \alpha_3 u_{11} u_{21}^2 + \alpha_4 u_{21}^3 \tag{6.3.4}$$

$$D_0^2 u_{23} + \omega_2^2 u_{23} = -2D_0(D_2 u_{21} + \mu_2 u_{21}) + \alpha_5 u_{11}^3 + \alpha_6 u_{11}^2 u_{21} + \alpha_7 u_{11} u_{21}^2 + \alpha_8 u_{21}^3 \tag{6.3.5}$$

where $D_n = \partial/\partial T_n$.

The solutions of (6.3.3) can be written in the form

$$u_{11} = A_1(T_2) \exp(i\omega_1 T_0) + cc$$
$$u_{21} = A_2(T_2) \exp(i\omega_2 T_0) + cc \tag{6.3.6}$$

Substituting (6.3.6) into (6.3.4) and (6.3.5) yields

$$\begin{aligned} D_0^2 u_{13} + \omega_1^2 u_{13} = {} & [-2i\omega_1(A_1' + \mu_1 A_1) + 3\alpha_1 A_1^2 \bar{A}_1 + 2\alpha_3 A_2 \bar{A}_2 A_1] \\ & \cdot \exp(i\omega_1 T_0) + (2\alpha_2 A_1 \bar{A}_1 + 3\alpha_4 A_2 \bar{A}_2) A_2 \exp(i\omega_2 T_0) \\ & + \alpha_1 A_1^3 \exp(3i\omega_1 T_0) + \alpha_4 A_2^3 \exp(3i\omega_2 T_0) \\ & + \alpha_2 A_1^2 A_2 \exp[i(2\omega_1 + \omega_2)T_0] + \alpha_2 \bar{A}_1^2 A_2 \\ & \cdot \exp[i(\omega_2 - 2\omega_1)T_0] + \alpha_3 A_1 A_2^2 \exp[i(\omega_1 + 2\omega_2)T_0] \\ & + \alpha_3 A_1 \bar{A}_2^2 \exp[i(\omega_1 - 2\omega_2)T_0] + cc \end{aligned} \tag{6.3.7}$$

$$\begin{aligned} D_0^2 u_{23} + \omega_2^2 u_{23} = {} & [-2i\omega_2(A_2' + \mu_2 A_2) + 3\alpha_8 A_2^2 \bar{A}_2 + 2\alpha_6 A_1 \bar{A}_1 A_2] \\ & \cdot \exp(i\omega_2 T_0) + (2\alpha_7 A_2 \bar{A}_2 + 3\alpha_5 A_1 \bar{A}_1) A_1 \exp(i\omega_1 T_0) \\ & + \alpha_5 A_1^3 \exp(3i\omega_1 T_0) + \alpha_8 A_2^3 \exp(3i\omega_2 T_0) \\ & + \alpha_6 A_1^2 A_2 \exp[i(2\omega_1 + \omega_2)T_0] + \alpha_6 \bar{A}_1^2 A_2 \\ & \cdot \exp[i(\omega_2 - 2\omega_1)T_0] + \alpha_7 A_1 A_2^2 \exp[i(\omega_1 + 2\omega_2)T_0] \\ & + \alpha_7 A_1 \bar{A}_2^2 \exp[i(\omega_1 - 2\omega_2)T_0] + cc \end{aligned} \tag{6.3.8}$$

where primes denote differentiation with respect to T_2. In analyzing the particular solutions of (6.3.7) and (6.3.8), we need to distinguish among three cases:

$\omega_2 \approx 3\omega_1$, $\omega_2 \approx \frac{1}{3}\omega_1$, and ω_2 away from $3\omega_1$ and $\frac{1}{3}\omega_1$ (i.e., nonresonance). If one of the first two cases occurs, an internal resonance is said to exist. In what follows we treat the internal-resonance case and the nonresonant case, starting with the latter.

6.3.1. THE NONRESONANT CASE

In this case the only terms that produce secular terms are the terms proportional to $\exp(\pm i\omega_1 T_0)$ in (6.3.7) and the terms proportional to $\exp(\pm i\omega_2 T_0)$ in (6.3.8). Thus eliminating the terms that produce secular terms in (6.3.7) and (6.3.8) yields

$$-2i\omega_1(A_1' + \mu_1 A_1) + 3\alpha_1 A_1^2 \bar{A}_1 + 2\alpha_3 A_2 \bar{A}_2 A_1 = 0$$
$$-2i\omega_2(A_2' + \mu_2 A_2) + 3\alpha_8 A_2^2 \bar{A}_2 + 2\alpha_6 A_1 \bar{A}_1 A_2 = 0 \tag{6.3.9}$$

Putting $A_m = \frac{1}{2}a_m \exp(i\theta_m)$ in (6.3.9) and separating real and imaginary parts we obtain

$$a_1' + \mu_1 a_1 = 0 \tag{6.3.10}$$

$$a_2' + \mu_2 a_2 = 0 \tag{6.3.11}$$

$$\theta_1' = -\left(\frac{3\alpha_1}{8\omega_1} a_1^2 + \frac{\alpha_3}{4\omega_1} a_2^2\right) \tag{6.3.12}$$

$$\theta_2' = -\left(\frac{3\alpha_8}{8\omega_2} a_2^2 + \frac{\alpha_6}{4\omega_2} a_1^2\right) \tag{6.3.13}$$

The solutions of (6.3.10) through (6.3.13) are

$$a_1 = a_{10} \exp(-\epsilon^2 \mu_1 t) \tag{6.3.14}$$

$$a_2 = a_{20} \exp(-\epsilon^2 \mu_2 t) \tag{6.3.15}$$

$$\theta_1 = \frac{3\alpha_1}{16\omega_1\mu_1} a_{10}^2 \exp(-2\epsilon^2\mu_1 t) + \frac{\alpha_3}{8\omega_1\mu_2} a_{20}^2 \exp(-2\epsilon^2\mu_2 t) + \theta_{10} \tag{6.3.16}$$

$$\theta_2 = \frac{3\alpha_8}{16\omega_2\mu_2} a_{20}^2 \exp(-2\epsilon^2\mu_2 t) + \frac{\alpha_6}{8\omega_2\mu_1} a_{10}^2 \exp(-2\epsilon^2\mu_1 t) + \theta_{20} \tag{6.3.17}$$

where a_{10}, a_{20}, θ_{10}, and θ_{20} are constants of integration. Consequently the amplitudes and hence the motion decay with time. However while the motion is decaying, the phases θ_1 and θ_2 and hence the frequencies of both modes are dependent on both amplitudes of oscillation.

In the absence of damping (i.e., $\mu_1 = \mu_2 = 0$), (6.3.14) through (6.3.17) reduce to $a_1 = a_{10}, a_2 = a_{20}$, and

$$\theta_1 = -\left(\frac{3\alpha_1}{8\omega_1} a_{10}^2 + \frac{\alpha_3}{4\omega_1} a_{20}^2\right) \epsilon^2 t + \theta_{10}$$
$$\theta_2 = -\left(\frac{3\alpha_8}{8\omega_2} a_{20}^2 + \frac{\alpha_6}{4\omega_2} a_{10}^2\right) \epsilon^2 t + \theta_{20} \tag{6.3.18}$$

which show that the phases and hence the frequencies are functions of the amplitudes. The frequencies increase or decrease with the amplitudes depending on the signs and relative magnitudes of α_1, α_3, α_6, and α_8 as well as on the ratio of a_{10} to a_{20}.

6.3.2. THE RESONANT CASE (INTERNAL RESONANCE)

We consider only one case, namely $\omega_2 \approx 3\omega_1$. The case $\omega_1 \approx 3\omega_2$ can be treated by following the present procedure.

To express quantitatively the nearness of ω_2 to $3\omega_1$, we introduce a detuning parameter σ according to

$$\omega_2 = 3\omega_1 + \epsilon^2 \sigma \tag{6.3.19}$$

and write

$$\omega_2 T_0 = 3\omega_1 T_0 + \epsilon^2 \sigma T_0 = 3\omega_1 T_0 + \sigma T_2 \tag{6.3.20}$$

Inspection of the right-hand sides of (6.3.7) and (6.3.8) reveals that in addition to the terms proportional to $\exp(\pm i\omega_m T_0)$ secular terms are produced by the terms proportional to $\exp[\pm i(\omega_2 - 2\omega_1)T_0]$ in (6.3.7) and the terms proportional to $\exp(\pm 3i\omega_1 T_0)$ in (6.3.8). To exhibit this secular behavior we express these factors as

$$\exp[\pm i(\omega_2 - 2\omega_1)T_0] = \exp(\pm i\omega_1 T_0 \pm i\sigma T_2)$$
$$\exp(\pm 3i\omega_1 T_0) = \exp(\pm i\omega_2 T_0 \mp i\sigma T_2) \tag{6.3.21}$$

Then the secular terms are eliminated from u_{13} and u_{23} if

$$-2i\omega_1(A_1' + \mu_1 A_1) + 3\alpha_1 A_1^2 \bar{A}_1 + 2\alpha_3 A_2 \bar{A}_2 A_1 + \alpha_2 \bar{A}_1^2 A_2 \exp(i\sigma T_2) = 0$$
$$-2i\omega_2(A_2' + \mu_2 A_2) + 3\alpha_8 A_2^2 \bar{A}_2 + 2\alpha_6 A_1 \bar{A}_1 A_2 + \alpha_5 A_1^3 \exp(-i\sigma T_2) = 0 \tag{6.3.22}$$

Putting $A_m = \frac{1}{2} a_m \exp(i\theta_m)$ in (6.3.22) and separating real and imaginary parts we have

$$8\omega_1(a_1' + \mu_1 a_1) = \alpha_2 a_1^2 a_2 \sin\gamma \tag{6.3.23}$$

$$8\omega_2(a_2' + \mu_2 a_2) = -\alpha_5 a_1^3 \sin\gamma \tag{6.3.24}$$

$$8\omega_1 a_1 \theta_1' = -(3\alpha_1 a_1^2 + 2\alpha_3 a_2^2) a_1 - \alpha_2 a_1^2 a_2 \cos\gamma \tag{6.3.25}$$

$$8\omega_2 a_2 \theta_2' = -(3\alpha_8 a_2^2 + 2\alpha_6 a_1^2) a_2 - \alpha_5 a_1^3 \cos\gamma \tag{6.3.26}$$

where

$$\gamma = \theta_2 - 3\theta_1 + \sigma T_2 \tag{6.3.27}$$

Eliminating θ_1 and θ_2 from (6.3.25) through (6.3.27) yields

$$a_2\gamma' = a_2\sigma + \left(\frac{3\alpha_3}{4\omega_1} - \frac{3\alpha_8}{8\omega_2}\right) a_2^3 + \left(\frac{9\alpha_1}{8\omega_1} - \frac{\alpha_6}{4\omega_2}\right) a_1^2 a_2 + \left(\frac{3\alpha_2}{8\omega_1} a_1 a_2^2 - \frac{\alpha_5}{8\omega_2} a_1^3\right)\cos\gamma \tag{6.3.28}$$

Thus the problem is reduced to one of finding the solutions of (6.3.23), (6.3.24), and (6.3.28).

Multiplying (6.3.23) by $\omega_1^{-1} a_1$ and (6.3.24) by $\omega_2^{-1} \nu a_2$, where $\nu = (\alpha_2\omega_2/\alpha_5\omega_1)$, and adding the results, we obtain

$$a_1 a_1' + \nu a_2 a_2' = -\mu_1 a_1^2 - \mu_2 \nu a_2^2 \tag{6.3.29}$$

For steady-state motions, $a_1' = a_2' = 0$, and it follows from (6.3.29) that

$$\mu_1 a_1^2 + \mu_2 \nu a_2^2 = 0 \tag{6.3.30}$$

Hence, unless $\nu > 0$, nontrivial steady-state free oscillations can exist, which is physically unrealistic unless there is a regenerative element (see Sections 3.1.7 and 6.1.4) in the system. In what follows we assume that such elements do not exist in the system and hence that $\nu > 0$ or α_2 and α_5 have the same sign.

Equation (6.3.29) can be integrated if $\mu_1 = \mu_2 = \mu$. The result is

$$a_1^2 + \nu a_2^2 = E \exp(-2\epsilon^2 \mu t) \tag{6.3.31}$$

where E is a constant of integration proportional to the initial energy of the system. Thus, as $t \to \infty$, $a_1^2 + \nu a_2^2 \to 0$. That is, the energy in the system decays exponentially with time.

If $\mu_1 = \mu_2 = 0$ (i.e., in the absence of damping), a second integral of the motion can be found as follows. Changing the independent variable in (6.3.28) from T_2 to a_2 and using (6.3.24), we obtain

$$-a_1^3 a_2 \sin\gamma \frac{d\gamma}{da_2} = \frac{8\omega_2\sigma}{\alpha_5} a_2 + \Gamma_1 a_2^3 + \Gamma_2 a_1^2 a_2 + (3\nu a_1 a_2^2 - a_1^3)\cos\gamma \tag{6.3.32}$$

where

$$\Gamma_1 = \frac{6\alpha_3\omega_2}{\alpha_5\omega_1} - \frac{3\alpha_8}{\alpha_5}, \qquad \Gamma_2 = \frac{9\alpha_1\omega_2}{\alpha_5\omega_1} - \frac{2\alpha_6}{\alpha_5}$$

Using (6.3.31) with $\mu = 0$, we can integrate (6.3.32) to obtain

$$a_1^3 a_2 \cos\gamma - \frac{1}{2}\left(\frac{8\omega_2\sigma}{\alpha_5} + \Gamma_2 E\right) a_2^2 + \frac{1}{4}(\nu\Gamma_2 - \Gamma_1) a_2^4 = L \tag{6.3.33}$$

where L is a constant of integration.

With (6.3.31) and (6.3.33) one can reduce the problem to a single first-order equation. To accomplish this we let $a_1^2 = E\xi$. Then it follows from (6.3.31) that $a_2^2 = E\nu^{-1}(1 - \xi)$. Using these expressions for a_1^2 and a_2^2 and eliminating γ from (6.3.23) and (6.3.33), we obtain

$$\frac{16\omega_1^2 \nu}{E^2 \alpha_2^2} \xi'^2 = F^2(\xi) - G^2(\xi) \tag{6.3.34}$$

where

$$F = \pm\sqrt{\xi^3(1 - \xi)} \tag{6.3.35}$$

$$G = \frac{\sqrt{\nu}}{E^2}\left\{L + \frac{E}{2\nu}\left(\frac{8\omega_2\sigma}{\alpha_5} + \Gamma_2 E\right)(1 - \xi) - \frac{E^2}{4\nu^2}(\nu\Gamma_2 - \Gamma_1)(1 - \xi)^2\right\} \tag{6.3.36}$$

In contrast with the case of quadratic nonlinearities, the exact solution of (6.3.34) through (6.3.36) is not available yet, and so (6.3.34) is solved numerically.

The functions $F(\xi)$ and $G(\xi)$ are shown schematically in Figure 6-10. Since a_1 and hence ξ must be real, $F^2(\xi) \geqslant G^2(\xi)$. The points where G meets F corre-

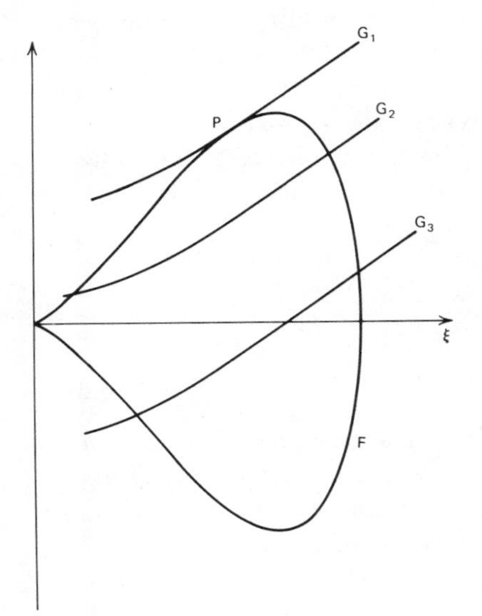

Figure 6-10. Schematic of F and G.

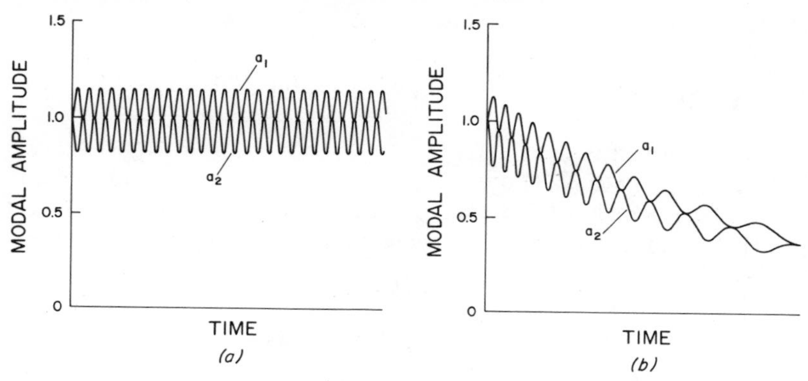

Figure 6-11. Free-oscillation amplitudes of a two-degree-of-freedom system with cubic nonlinearities; $\omega_2 \simeq 3\omega_1$: (*a*) no damping; (*b*) $\mu_1, \mu_2 > 0$.

spond to the vanishing of ξ' and hence to the vanishing of a_1' and a_2'. A curve such as G_2 which meets one branch of F at two different points or a curve G_3 which meets both branches corresponds to a periodic solution for ξ and hence a_1 and a_2. Consequently the motion is aperiodic. Figure 6-11 shows that in the absence of damping the energy is continuously exchanged between the two modes. Numerical integration of (6.3.23), (6.3.24), and (6.3.28) shows the effect of the damping. The energy in a given mode can be more effectively damped if that mode is coupled via an internal resonance to another mode having a large coefficient of damping.

On the other hand, a point such as P where G_1 touches F represents a stationary solution for ξ and hence a_1 and a_2. Consequently the motion corresponding to such a point is periodic, and the effect of the nonlinearity is to modulate the phase in such a way that the nonlinear frequencies are commensurable. To analyze the stationary solutions of a_1 and a_2, we set $a_1' = a_2' = \gamma' = 0$ in (6.3.23), (6.3.24), and (6.3.28). In the absence of damping, the stationary solutions are given by

$$\sin\gamma = 0 \quad \text{or} \quad \gamma = n\pi \tag{6.3.37}$$

$$a_2\sigma + \left(\frac{3\alpha_3}{4\omega_1} - \frac{3\alpha_8}{8\omega_2}\right)a_2^3 + \left(\frac{9\alpha_1}{8\omega_1} - \frac{\alpha_6}{4\omega_2}\right)a_1^2a_2 + \left(\frac{3\alpha_2}{8\omega_1}a_1a_2^2 - \frac{\alpha_5}{8\omega_2}a_1^3\right)\cos n\pi = 0 \tag{6.3.38}$$

where n is an integer. Equation (6.3.38) is a cubic equation for a_2 in terms of $a_1 \cos n\pi$. Thus for a given σ and $a_1 \cos n\pi$, (6.3.38) has either one real root or three real roots. Thus the periodic motion may consist of a unique motion, or it may be one of three possible motions. Figure 6-10 shows that the periodic motion is unstable because any small disturbance would lead to a curve G similar to G_2 where it intersects one branch of F in two different points, and hence it would lead to an aperiodic motion.

To show that the nonlinear motions in the case of stationary solutions for a_1 and a_2 are periodic, we note that the frequencies are given by

$$\hat{\omega}_1 = \omega_1 + \epsilon^2 \theta_1', \qquad \hat{\omega}_2 = \omega_2 + \epsilon^2 \theta_2' \tag{6.3.39}$$

Then

$$\hat{\omega}_2 - 3\hat{\omega}_1 = \omega_2 - 3\omega_1 + \epsilon^2(\theta_2' - 3\theta_1') = \epsilon^2 \sigma + \epsilon^2(\theta_2' - 3\theta_1') = \epsilon^2 \gamma' = 0 \tag{6.3.40}$$

Thus the nonlinearity adjusts the phases such that the frequencies are exactly in the ratio of 3 to 1, and hence the motion is periodic.

So far we have discussed phase-modulated motions (periodic motions) and both amplitude- and phase-modulated motions. The question arises whether pure amplitude-modulated motions can exist as in the case of quadratic nonlinearities discussed in the preceding section. If θ_1 and θ_2 are constants, then it follows from (6.3.25) and (6.3.26) that

$$\begin{aligned} (3\alpha_1 a_1^2 + 2\alpha_3 a_2^2)a_1 + \alpha_2 a_1^2 a_2 \cos\gamma &= 0 \\ (3\alpha_8 a_2^2 + 2\alpha_6 a_1^2)a_2 + \alpha_5 a_1^3 \cos\gamma &= 0 \end{aligned} \tag{6.3.41}$$

Eliminating $\cos\gamma$ from (6.3.41) yields

$$\alpha_5(3\alpha_1 a_1^2 + 2\alpha_3 a_2^2)a_1^2 - \alpha_2(3\alpha_8 a_2^2 + 2\alpha_6 a_1^2)a_2^2 = 0 \tag{6.3.42}$$

But in the absence of damping

$$a_1^2 + \nu a_2^2 = E \tag{6.3.43}$$

Equations (6.3.42) and (6.3.43) can be solved to determine constant values for a_1 and a_2. Therefore pure amplitude-modulated motions do not exist in the case of cubic nonlinearities, in contrast with the case of quadratic nonlinearities.

6.4. Free Oscillations of Gyroscopic Systems

A discussion of the motion of coupled rigid bodies is given in the book of Leimanis (1965), while a discussion of problems of rotor dynamics is given in the book of Tondl (1965). Kyner (1969) discusses the occurrence of nonlinear resonances in physical systems such as time-varying torques, coupled pendulums, and artificial satellites. Sethna and Balachandra (1976) presented a survey of nonlinear gyroscopic systems. The motion of general gyroscopic systems was studied by Cherry (1924), Moser (1958), Bert (1961), Arnold (1963), Butenin (1965, Section 10), Hori (1966), Gustavson (1966), Garfinkel (1966), Markeev

(1968, 1969a), Deprit (1969), Kamel (1969, 1970, 1971), Henrard (1970), Alfriend (1971b, c), Tsel'man (1971), Khazin (1971), Alfriend and Richardson (1973), Balachandra (1973), Sethna and Balachandra (1974b). Rangacharyulu and Srinivasan (1973), and Bhansali and Thiruvenkatachar (1975). The problem of the stability of the triangular points and references dealing with it are given in Section 6.1.5.

Junkins, Jacobson, and Blanton (1973) reduced the motion of any torque-free rigid body to the solution of three uncoupled Duffing's equations. Goodstein (1959) studied the free and forced vibrations of a gyroscope, Thorne (1961) solved numerically the equations governing the motion of a gyroscope under constant acceleration and a correcting torque, and Poli and Budynas (1971) studied the stability of a symmetric gyroscope.

The motion of a satellite about an oblate earth was studied by Liu (1974). The problem of a symmetric satellite in a nearly circular orbit was studied by Pringle (1964), Likins (1965), Breakwell and Pringle (1966a), Markeev (1967a), and Hitzl (1969, 1971). The effect of a rotor on the attitude stability of a satellite was examined by Kane and Mingori (1965), Kane (1966), and Crespo da Silva (1972b). Dual-spin satellites were studied by Likins (1967); Mingori (1969); Pringle (1969, 1973); Likins, Tseng, and Mingori (1971); Mingori, Tseng, and Likins (1972); Scher and Farrenkopf (1974); Gebman and Mingori (1976); Fujii (1976); and Cochran (1977). Pringle (1968) suggested the exploitation of nonlinear resonances in damping the librations of a dumbbell satellite, Likins and Wrout (1969) and Schneider and Likins (1973) suggested the use of internal resonances as an internal kinetic-energy exchange mechanism to accelerate the dissipation of the libration energy of a satellite, and Fujii (1976) investigated the effect of resonances on the attenuation of the librations of a satellite having a finite mass and connected to an energy damper. The effect of gravity-gradient perturbations on the attitude motion of satellites was analyzed by Beletskii (1960, 1968), Kane (1966), Breakwell and Pringle (1966a), White and Likins (1969), Modi and Brereton (1969a, b), Crespo da Silva (1970, 1972a), Hitzl and Breakwell (1971), Cochran (1972), and Nishinaga and Likins (1974).

The motion of a rotating shaft with gyroscopic moments and nonlinear restoring springs leads to a system of nonlinear gyroscopic equations. Genin and Maybee (1969, 1970) and Mingori (1973) analyzed the stability of whirling shafts with internal and external damping. Using a forced rotating shaft, Yamamoto (1957, 1960) demonstrated the occurrence of combination resonances of the summed and difference type. These resonances were studied by Yamamoto (1961a); Yamamoto and Ishida (1974, 1977); and Yamamoto, Ishida, and Kawasumi (1975, 1977).

In this section we consider the free oscillations of a conservative system having gyroscopic forces and simple quadratic nonlinearities. Namely we consider the system

$$\begin{aligned} \ddot{u}_1 - \lambda\dot{u}_2 + \alpha_1 u_1 + \alpha_3 u_2 &= 2u_1 u_2 \\ \ddot{u}_2 + \lambda\dot{u}_1 + \alpha_3 u_1 + \alpha_2 u_2 &= u_1^2 \end{aligned} \tag{6.4.1}$$

We seek a first-order solution for small but finite amplitudes in the form

$$\begin{aligned} u_1 &= \epsilon u_{11}(T_0, T_1) + \epsilon^2 u_{12}(T_0, T_1) + \cdots \\ u_2 &= \epsilon u_{21}(T_0, T_1) + \epsilon^2 u_{22}(T_0, T_1) + \cdots \end{aligned} \tag{6.4.2}$$

where ϵ is a small, dimensionless parameter the order of the amplitudes and $T_n = \epsilon^n t$. Substituting (6.4.2) into (6.4.1) and equating coefficients of like powers of ϵ we obtain

Order ϵ

$$\begin{aligned} D_0^2 u_{11} - \lambda D_0 u_{21} + \alpha_1 u_{11} + \alpha_3 u_{21} &= 0 \\ D_0^2 u_{21} + \lambda D_0 u_{11} + \alpha_3 u_{11} + \alpha_2 u_{21} &= 0 \end{aligned} \tag{6.4.3}$$

Order ϵ^2

$$\begin{aligned} D_0^2 u_{12} - \lambda D_0 u_{22} + \alpha_1 u_{12} + \alpha_3 u_{22} &= -2D_0 D_1 u_{11} + \lambda D_1 u_{21} + 2u_{11} u_{21} \\ D_0^2 u_{22} + \lambda D_0 u_{12} + \alpha_3 u_{12} + \alpha_2 u_{22} &= -2D_0 D_1 u_{21} - \lambda D_1 u_{11} + u_{11}^2 \end{aligned} \tag{6.4.4}$$

The solution of (6.4.3) can be expressed in the form

$$\begin{aligned} u_{11} &= A_1(T_1) \exp(i\omega_1 T_0) + A_2(T_1) \exp(i\omega_2 T_0) + cc \\ u_{21} &= \Lambda_1 A_1(T_1) \exp(i\omega_1 T_0) + \Lambda_2 A_2(T_1) \exp(i\omega_2 T_0) + cc \end{aligned} \tag{6.4.5}$$

where the ω_n^2 are the roots of

$$\omega^4 - (\alpha_1 + \alpha_2 + \lambda^2)\omega^2 + \alpha_1\alpha_2 - \alpha_3^2 = 0 \tag{6.4.6}$$

$$\Lambda_n = -\frac{\alpha_3 + i\omega_n\lambda}{\alpha_2 - \omega_n^2} \tag{6.4.7}$$

and the ω_n are assumed to be distinct.

Substituting u_{11} and u_{21} into (6.4.4) yields

$$\begin{aligned} D_0^2 u_{12} - \lambda D_0 u_{22} + \alpha_1 u_{12} + \alpha_3 u_{22} &= -(2i\omega_1 - \lambda\Lambda_1)A_1' \exp(i\omega_1 T_0) \\ &- (2i\omega_2 - \lambda\Lambda_2)A_2' \exp(i\omega_2 T_0) + 2\Lambda_1 A_1^2 \exp(2i\omega_1 T_0) \\ &+ 2\Lambda_2 A_2^2 \exp(2i\omega_2 T_0) + 2(\Lambda_1 + \Lambda_2)A_1 A_2 \exp[i(\omega_1 + \omega_2)T_0] \\ &+ 2(\bar{\Lambda}_1 + \Lambda_2)\bar{A}_1 A_2 \exp[i(\omega_2 - \omega_1)T_0 + 2\bar{\Lambda}_1 A_1 \bar{A}_1 + 2\bar{\Lambda}_2 A_2 \bar{A}_2 + cc \end{aligned} \tag{6.4.8}$$

$$D_0^2 u_{22} + \lambda D_0 u_{12} + \alpha_3 u_{12} + \alpha_2 u_{22} = -(2i\omega_1 \Lambda_1 + \lambda) A_1' \exp(i\omega_1 T_0) - (2i\omega_2 \Lambda_2 + \lambda) A_2' \exp(i\omega_2 T_0) + A_1^2 \exp(2i\omega_1 T_0) + A_2^2 \exp(2i\omega_2 T_0) + 2A_1 A_2 \exp[i(\omega_1 + \omega_2) T_0] + 2\bar{A}_1 A_2 \exp[i(\omega_2 - \omega_1) T_0] + A_1 \bar{A}_1 + A_2 \bar{A}_2 + cc \tag{6.4.9}$$

In determining the solvability conditions of (6.4.8) and (6.4.9) and hence the equations that describe the A_n, we need to distinguish between resonant (i.e., $\omega_2 \approx 2\omega_1$ or $\frac{1}{2}\omega_1$) and nonresonant (ω_2 is away from $2\omega_1$ or $\frac{1}{2}\omega_1$) situations.

In the nonresonant case the solvability conditions of (6.4.8) and (6.4.9) yield $A_n' = 0$ or $A_n = \frac{1}{2} a_n \exp(i\theta_n)$, where a_n and θ_n are real constants.

We analyze only the resonant case $\omega_2 \approx 2\omega_1$ because the other resonant case can be treated in a similar fashion and because the physical features can be obtained by the analysis of one of the two cases. Thus we introduce a detuning parameter σ defined by

$$\omega_2 = 2\omega_1 + \epsilon\sigma \tag{6.4.10}$$

and express $2\omega_1 T_0$ and $(\omega_2 - \omega_1) T_0$ as

$$\begin{aligned} 2\omega_1 T_0 &= \omega_2 T_0 - \sigma T_1 \\ (\omega_2 - \omega_1) T &= \omega_1 T_0 + \sigma T_1 \end{aligned} \tag{6.4.11}$$

To determine the solvability conditions of (6.4.8) and (6.4.9), we seek a particular solution in the form

$$\begin{aligned} u_{12} &= P_{11} \exp(i\omega_1 T_0) + P_{12} \exp(i\omega_2 T_0) \\ u_{22} &= P_{21} \exp(i\omega_1 T_0) + P_{22} \exp(i\omega_2 T_0) \end{aligned} \tag{6.4.12}$$

Substituting (6.4.12) into (6.4.8) and (6.4.9), using (6.4.11), and equating the coefficients of $\exp(i\omega_1 T_0)$ and $\exp(i\omega_2 T_0)$ on both sides, we obtain

$$\begin{aligned} (\alpha_1 - \omega_n^2) P_{1n} + (\alpha_3 - i\omega_n \lambda) P_{2n} &= R_{1n} \\ (\alpha_3 + i\omega_n \lambda) P_{1n} + (\alpha_2 - \omega_n^2) P_{2n} &= R_{2n} \end{aligned} \tag{6.4.13}$$

where

$$\begin{aligned} R_{11} &= -(2i\omega_1 - \lambda\Lambda_1) A_1' + 2(\bar{\Lambda}_1 + \Lambda_2) A_2 \bar{A}_1 \exp(i\sigma T_1) \\ R_{21} &= -(2i\omega_1 \Lambda_1 + \lambda) A_1' + 2A_2 \bar{A}_1 \exp(i\sigma T_1) \end{aligned} \tag{6.4.14}$$

$$\begin{aligned} R_{12} &= -(2i\omega_2 - \lambda\Lambda_2) A_2' + 2\Lambda_1 A_1^2 \exp(-i\sigma T_1) \\ R_{22} &= -(2i\omega_2 \Lambda_2 + \lambda) A_2' + A_1^2 \exp(-i\sigma T_1) \end{aligned} \tag{6.4.15}$$

Thus the problem of determining the solvability conditions of (6.4.8) and (6.4.9) is reduced to that of determining the solvability conditions of (6.4.13).

Since the determinant of the coefficient matrix of (6.4.13) is zero according to (6.4.6), the solvability conditions are

$$\begin{vmatrix} R_{1n} & \alpha_3 - i\omega_n\lambda \\ R_{2n} & \alpha_2 - \omega_n^2 \end{vmatrix} = 0 \tag{6.4.16}$$

or

$$R_{1n} = -\overline{\Lambda}_n R_{2n} \tag{6.4.17}$$

on account of (6.4.7).

Substituting (6.4.14) and (6.4.15) into (6.4.17) and rearranging, we obtain

$$\begin{aligned} A_1' &= -i\Gamma_1 A_2 \overline{A}_1 \exp\,[i(\sigma T_1 + \tau)] \\ A_2' &= -\tfrac{1}{2} i\Gamma_2 A_1^2 \exp\,[-i(\sigma T_1 + \tau)] \end{aligned} \tag{6.4.18}$$

where

$$\begin{aligned} \Gamma_n &= |\Lambda_2 + 2\overline{\Lambda}_1|\,(\alpha_2 - \omega_n^2)\,\omega_n^{-1}(\alpha_1 + \alpha_2 + \lambda^2 - 2\omega_n^2)^{-1} \\ \tau &= \text{imaginary part of } \log\,(\Lambda_2 + 2\overline{\Lambda}_1) \end{aligned} \tag{6.4.19}$$

Putting $A_n = \frac{1}{2} a_n \exp\,(i\theta_n)$ with real a_n and θ_n in (6.4.18) and separating the real and imaginary parts yields

$$a_1' = \tfrac{1}{2}\Gamma_1 a_1 a_2 \sin\gamma \tag{6.4.20}$$

$$a_2' = -\tfrac{1}{4}\Gamma_2 a_1^2 \sin\gamma \tag{6.4.21}$$

$$\theta_1' = -\tfrac{1}{2}\Gamma_1 a_2 \cos\gamma \tag{6.4.22}$$

$$a_2\theta_2' = -\tfrac{1}{4}\Gamma_2 a_1^2 \cos\gamma \tag{6.4.23}$$

where

$$\gamma = \theta_2 - 2\theta_1 + \sigma T_1 + \tau \tag{6.4.24}$$

Eliminating θ_1 and θ_2 from (6.4.22) through (6.4.24) gives

$$a_2\gamma' = a_2\sigma - \tfrac{1}{4}\Gamma_2 a_1^2 \cos\gamma + \Gamma_1 a_2^2 \cos\gamma \tag{6.4.25}$$

Equations (6.4.20) through (6.4.25) have the same form as (6.2.16) through (6.2.21) derived in Section 6.2 for nongyroscopic systems when $\mu_1 = \mu_2 = 0$.

Eliminating γ from (6.4.20) and (6.4.21) yields $\nu a_2 a_2' + a_1 a_1' = 0$ where $\nu = (2\Gamma_1/\Gamma_2)$. Hence

$$\nu a_2^2 + a_1^2 = E \tag{6.4.26}$$

where E is a constant of integration. As discussed in Section 6.2, the response will be bounded if $\nu > 0$, but it may be unbounded, depending on the detuning σ, as shown below if $\nu < 0$. When $\nu > 0$, the analysis and behavior of the system are exactly as in Section 6.2. Hence we restrict the discussion in this section to the case $\nu < 0$.

It follows from (6.4.19) that

$$\nu = \frac{2\Gamma_1}{\Gamma_2} = \frac{2\omega_2(\alpha_2 - \omega_1^2)(\alpha_1 + \alpha_2 + \lambda^2 - 2\omega_2^2)}{\omega_1(\alpha_2 - \omega_2^2)(\alpha_1 + \alpha_2 + \lambda^2 - 2\omega_1^2)} \tag{6.4.27}$$

Since $\alpha_1 + \alpha_2 + \lambda^2 = \omega_1^2 + \omega_2^2$ from (6.4.6),

$$\nu = -\frac{2\omega_2(\alpha_2 - \omega_1^2)}{\omega_1(\alpha_2 - \omega_2^2)} \tag{6.4.28}$$

Certainly $\nu < 0$ when $\alpha_2 < 0$. But if $\alpha_2 < 0$, it follows from (6.4.6) that $\alpha_1 < 0$ in order that ω_1 and ω_2 be real. This is exactly the condition that the Hamiltonian is not positive definite. To see this, we note that the Hamiltonian corresponding to the linear parts of equations (6.4.1) is

$$H = \tfrac{1}{2}(p_1 + \tfrac{1}{2}\lambda u_2)^2 + \tfrac{1}{2}(p_2 - \tfrac{1}{2}\lambda u_1)^2 + \tfrac{1}{2}\alpha_1 u_1^2 + \alpha_3 u_1 u_2 + \tfrac{1}{2}\alpha_2 u_2^2 \tag{6.4.29}$$

where p_1 and p_2 are the generalized momenta. Substituting H into Hamilton's equations

$$\dot{u}_n = \frac{\partial H}{\partial p_n}, \quad \dot{p}_n = -\frac{\partial H}{\partial u_n} \tag{6.4.30}$$

yields the linearized parts of (6.4.1). Equation (6.4.29) can be rewritten as

$$H = \tfrac{1}{2}(p_1 + \tfrac{1}{2}\lambda u_2)^2 + \tfrac{1}{2}(p_2 - \tfrac{1}{2}\lambda u_1)^2 + \tfrac{1}{2}\alpha_1\left(u_1 + \frac{\alpha_3}{\alpha_1}u_2\right)^2 + \frac{\alpha_1\alpha_2 - \alpha_3^2}{2\alpha_1}u_2^2 \tag{6.4.31}$$

Since $\alpha_1\alpha_2 - \alpha_3^2 > 0$ in order that ω_1 and ω_2 be real, H is positive definite unless $\alpha_1 < 0$ and hence $\alpha_2 < 0$.

Following an analysis similar to that of Section 6.2, we can write the following second integral for (6.4.20), (6.4.21), and (6.4.25):

$$a_2 a_1^2 \cos\gamma - \frac{2\sigma}{\Gamma_2}a_2^2 = L \tag{6.4.32}$$

where L is a constant of integration. To determine a single equation for a_2 we let

$$a_2^2 = |E|\xi \tag{6.4.33}$$

It follows from (6.4.26) that

$$a_1^2 = |E|(\pm 1 + \hat{\nu}\xi), \quad \hat{\nu} = -\nu > 0 \tag{6.4.34}$$

where the plus and minus signs inside the parentheses correspond to positive and negative values of E, respectively. Using (6.4.32) to eliminate γ from (6.4.21)

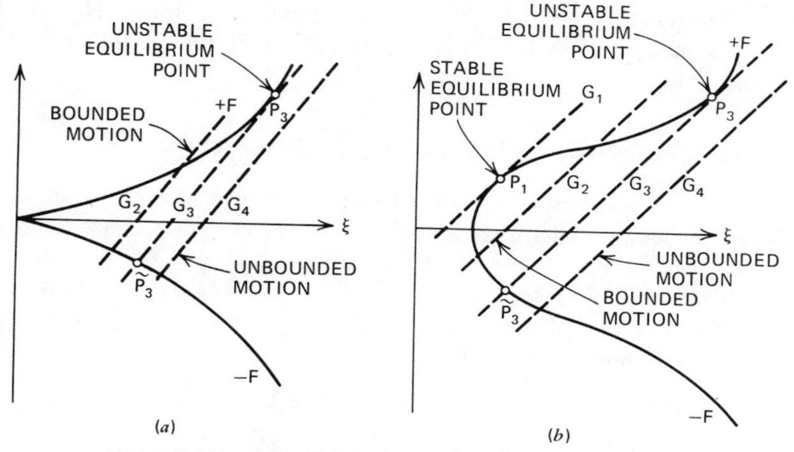

Figure 6-12. Schematic of the motion: (*a*) $E > 0$; (*b*) $E < 0$.

and expressing a_1^2 and a_2^2 in terms of ξ we obtain

$$\left(\frac{d\hat{\xi}}{d\tau}\right)^2 = F^2(\hat{\xi}) - G^2(\hat{\xi}) \tag{6.4.35}$$

where $\hat{\xi} = \hat{\nu}\xi$,

$$F = \pm\hat{\xi}(\hat{\xi} \pm 1), \quad G = \hat{\sigma}(\hat{\xi} + \hat{L}) \tag{6.4.36}$$

$$\tau = \tfrac{1}{2}\Gamma_2\sqrt{\hat{\nu}|E|}\,T_1, \quad \hat{\sigma} = \frac{2\sigma}{\Gamma_2\sqrt{\hat{\nu}|E|}}, \quad \hat{L} = \frac{L\Gamma_2\hat{\nu}}{2\sigma|E|} \tag{6.4.37}$$

The functions F and G are shown schematically in Figure 6-12. Since a_2, and hence $\hat{\xi}$, must be real, $F^2 \geqslant G^2$ must hold. The points where G meets F correspond to the vanishing of both a_1' and a_2'. A curve such as G_2 which meets both branches of F, or meets one branch at two different points, corresponds to a bounded aperiodic motion. In this case $\hat{\xi}$ and hence a_2^2 can be expressed in terms of Jacobi elliptic functions as in Section 6.2.

On the other hand, a curve such as G_4, which meets F at one point only, represents an unbounded motion. However the points P_1 and P_3 where G_1 and G_3 touch F represent equilibrium (periodic motions). A point such as P_1 corresponds to a stable periodic motion, whereas a point such as P_3 corresponds to an unstable periodic motion.

When $\sigma = 0$ (i.e., the case of perfect resonance), G = constant and the curve G intersects the curve F in one point only. Consequently the nonlinearity causes the motion to be unbounded though it is bounded according to the linear theory.

Since $\nu < 0$ when α_1 and α_2 are negative, the system under consideration is unstable linearly if $\alpha_3 = 0$ in the absence of the gyroscopic forces. However the nonlinearity causes the system to be unstable if the two linear frequencies are commensurable.

6.5. Forced Oscillations of Systems Having Quadratic Nonlinearities

In this section we consider the forced response of a system having quadratic nonlinearities. For simplicity we consider only the case of a single-frequency excitation. van Dooren (1971a, b, 1973b) studied the case of a two-frequency excitation and obtained results for the cases $\omega_n = \Omega_2 \pm \Omega_1$, where ω_n is one of the natural frequencies of the system and Ω_1 and Ω_2 are the frequencies of the excitation. Thus we consider

$$\begin{aligned} \ddot{u}_1 + \omega_1^2 u_1 &= -2\hat{\mu}_1 \dot{u}_1 + u_1 u_2 + F_1 \cos(\Omega t + \tau_1) \\ \ddot{u}_2 + \omega_2^2 u_2 &= -2\hat{\mu}_2 \dot{u}_2 + u_1^2 + F_2 \cos(\Omega t + \tau_2) \end{aligned} \tag{6.5.1}$$

with ω_2 being larger than ω_1. We follow Nayfeh, Mook, and Marshall (1973) and seek a first-order uniform expansion by using the method of multiple scales in the form

$$\begin{aligned} u_1 &= \epsilon u_{11}(T_0, T_1) + \epsilon^2 u_{12}(T_0, T_1) + \cdots \\ u_2 &= \epsilon u_{21}(T_0, T_1) + \epsilon^2 u_{22}(T_0, T_1) + \cdots \end{aligned} \tag{6.5.2}$$

where ϵ is a small, dimensionless parameter related to the amplitudes and $T_n = \epsilon^n t$. We order the damping coefficients so that the effects of the damping and the nonlinearity appear in the same perturbation equations. Thus we let $\hat{\mu}_n = \epsilon \mu_n$. As before, we consider two major categories—primary ($\Omega \approx \omega_n$) and secondary resonances ($\Omega \approx 2\omega_n$, $\Omega = \omega_1 \pm \omega_2$, or $\Omega \approx \frac{1}{2}\omega_n$). Moreover we consider a number of cases within each category. Primary resonances were analyzed by Mettler and Weidenhammer (1962); Sethna (1965); Tondl (1966); Nayfeh, Mook, and Marshall (1973); and Marshall and Morrow (1975). Secondary resonances were analyzed by Kruschul (1960); Yamamoto (1961a, b); Yamamoto and Nakao (1963); Yamamoto and Hayashi (1963, 1964); van Dooren (1971a, 1973b); Eller (1973); Mook, Marshall, and Nayfeh (1974); Agrawal (1975); and Evan-Iwanowski (1976).

6.5.1. THE CASE OF Ω NEAR ω_2

To analyze primary resonances, we order the forcing term so that it appears in the same perturbation equation as the nonlinear terms and the damping. First we consider the case in which $\Omega \approx \omega_2$. Thus we let $F_1 = \epsilon f_1$ and $F_2 = \epsilon^2 f_2$. Substituting (6.5.2) into (6.5.1), recalling that $\hat{\mu} = \epsilon \mu_n$, and equating coefficients of like powers of ϵ, we obtain

Order ϵ

$$\begin{aligned} D_0^2 u_{11} + \omega_1^2 u_{11} &= f_1 \cos(\Omega T_0 + \tau_1) \\ D_0^2 u_{21} + \omega_2^2 u_{21} &= 0 \end{aligned} \tag{6.5.3}$$

Order ϵ^2

$$\begin{aligned} D_0^2 u_{12} + \omega_1^2 u_{12} &= -2D_0(D_1 u_{11} + \mu_1 u_{11}) + u_{11} u_{21} \\ D_0^2 u_{22} + \omega_2^2 u_{22} &= -2D_0(D_1 u_{21} + \mu_2 u_{21}) + u_{11}^2 + f_2 \cos(\Omega T_0 + \tau_2) \end{aligned} \tag{6.5.4}$$

The solutions of (6.5.3) can be expressed in the form

$$\begin{aligned} u_{11} &= A_1(T_1) \exp(i\omega_1 T_0) + \Lambda \exp[i(\Omega T_0 + \tau_1)] + cc \\ u_{21} &= A_2(T_1) \exp(i\omega_2 T_0) + cc \end{aligned} \tag{6.5.5}$$

where A_1 and A_2 are arbitrary functions at this level of approximation and $\Lambda = f_1/2(\omega_1^2 - \Omega^2)$. Substituting (6.5.5) into (6.5.4) yields

$$\begin{aligned} D_0^2 u_{12} + \omega_1^2 u_{12} = {} & -2i\omega_1(A_1' + \mu_1 A_1) \exp(i\omega_1 T_0) + A_2 A_1 \exp[i(\omega_2 + \omega_1)T_0] \\ & + A_2 \bar{A}_1 \exp[i(\omega_2 - \omega_1)T_0] + \Lambda A_2 \exp[i(\Omega + \omega_2)T_0 + i\tau_1] \\ & + \Lambda \bar{A}_2 \exp[i(\Omega - \omega_2)T_0 + i\tau_1] \\ & - 2i\mu_1 \Omega \Lambda \exp[i(\Omega T_0 + \tau_1)] + cc \end{aligned} \tag{6.5.6}$$

$$\begin{aligned} D_0^2 u_{22} + \omega_2^2 u_{22} = {} & -2i\omega_2(A_2' + \mu_2 A_2) \exp(i\omega_2 T_0) + A_1^2 \exp(2i\omega_1 T_0) \\ & + A_1 \bar{A}_1 + \Lambda^2 + 2A_1 \Lambda \exp[i(\omega_1 + \Omega)T_0 + i\tau_1] \\ & + 2\bar{A}_1 \Lambda \exp[i(\Omega - \omega_1)T_0 + i\tau_1] + \Lambda^2 \exp[2i(\Omega T_0 + \tau_1)] \\ & + \tfrac{1}{2} f_2 \exp[i(\omega_2 T_0 + \sigma_1 T_1 + \tau_2)] + cc \end{aligned} \tag{6.5.7}$$

where

$$\Omega = \omega_2 + \epsilon\sigma_1 \tag{6.5.8}$$

As in Sections 6.2 and 6.4 we need to distinguish between the case of internal resonance $\omega_2 \approx 2\omega_1$ and the case of no internal resonance (i.e., ω_2 is away from ω_1). In the latter case none of the nonlinear terms produces a secular term, and the solvability conditions are

$$A_1' + \mu_1 A_1 = 0 \tag{6.5.9}$$

$$2i\omega_2(A_2' + \mu_2 A_2) = \tfrac{1}{2} f_2 \exp[i(\sigma_1 T_1 + \tau_2)] \tag{6.5.10}$$

whose solutions are

$$A_1 = \tfrac{1}{2}a_1 \exp(-\mu_1 T_1 + i\theta_1) \tag{6.5.11}$$

$$A_2 = \tfrac{1}{2}a_2 \exp(-\mu_2 T_1 + i\theta_2) - \tfrac{1}{4} i f_2 \omega_2^{-1}(\mu_2 + i\sigma_1)^{-1} \exp[i(\sigma_1 T_1 + \tau_2)] \tag{6.5.12}$$

where the a_n and θ_n are constants.

As $t \to \infty$, $T_1 \to \infty$ and

$$A_1 \to 0, A_2 \to -\tfrac{1}{4} i f_2 \omega_2^{-1}(\mu_2 + i\sigma_1)^{-1} \exp[i(\sigma_1 T_1 + \tau_2)] \tag{6.5.13}$$

Substituting (6.5.13) into (6.5.5) and (6.5.2) and expressing the result in terms of the original variables, we obtain the following steady-state response:

$$\begin{aligned} u_1 &= F_1(\omega_1^2 - \Omega^2)^{-1} \cos(\Omega t + \tau_1) + O(\epsilon^2) \\ u_2 &= \tfrac{1}{2}\epsilon^{-1} F_2 \omega_2^{-1}(\mu_2^2 + \sigma_1^2)^{-1/2} \sin(\Omega t + \tau_2 - \gamma_0) + O(\epsilon^2) \end{aligned} \tag{6.5.14}$$

where $\gamma_0 = \arctan(\sigma_1/\mu_2)$. Thus when there is no internal resonance, the first approximation is not influenced by the nonlinear terms; it is essentially the solution of the corresponding linear problem. As we shall see next, when there is an internal resonance the solution can differ drastically from (6.5.14).

When $\omega_2 \approx 2\omega_1$, the solvability conditions of (6.5.6) and (6.5.7) are

$$\begin{aligned} &-2i\omega_1(A_1' + \mu_1 A_1) + A_2 \bar{A}_1 \exp(-i\sigma_2 T_1) = 0 \\ &-2i\omega_2(A_2' + \mu_2 A_2) + A_1^2 \exp(i\sigma_2 T_1) + \tfrac{1}{2} f_2 \exp[i(\sigma_1 T_1 + \tau_2)] = 0 \end{aligned} \tag{6.5.15}$$

where

$$\omega_2 = 2\omega_1 - \epsilon\sigma_2 \tag{6.5.16}$$

As before, we introduce polar notation $A_n = \tfrac{1}{2}a_n \exp(i\theta_n)$ and obtain

$$a_1' = -\mu_1 a_1 + \tfrac{1}{4}\omega_1^{-1} a_1 a_2 \sin\gamma_2 \tag{6.5.17}$$

$$a_2' = -\mu_2 a_2 - \tfrac{1}{4}\omega_2^{-1} a_1^2 \sin\gamma_2 + \tfrac{1}{2}\omega_2^{-1} f_2 \sin\gamma_1 \tag{6.5.18}$$

$$a_1\theta_1' = -\tfrac{1}{4}\omega_1^{-1} a_2 a_1 \cos\gamma_2 \tag{6.5.19}$$

$$a_2\theta_2' = -\tfrac{1}{4}\omega_2^{-1} a_1^2 \cos\gamma_2 - \tfrac{1}{2}\omega_2^{-1} f_2 \cos\gamma_1 \tag{6.5.20}$$

where

$$\begin{aligned} \gamma_1 &= \sigma_1 T_1 - \theta_2 + \tau_2 \\ \gamma_2 &= \theta_2 - 2\theta_1 - \sigma_2 T_1 \end{aligned} \tag{6.5.21}$$

For the steady-state response, $a_n' = \gamma_n' = 0$. We find two possibilities. The first is given by (6.5.13), and it is essentially the solution of the linear problem. The second is

$$a_1 = 2[-\Gamma_1 \pm (\tfrac{1}{4}f_2^2 - \Gamma_2^2)^{1/2}]^{1/2} \tag{6.5.22}$$

$$a_2 = a_2^* = 2\omega_1 [4\mu_1^2 + (\sigma_1 - \sigma_2)^2]^{1/2} \tag{6.5.23}$$

where

$$\begin{aligned} \Gamma_1 &= 2\omega_1\omega_2 [\sigma_1(\sigma_2 - \sigma_1) + 2\mu_1\mu_2] \\ \Gamma_2 &= 2\omega_1\omega_2 [2\sigma_1\mu_1 - \mu_2(\sigma_2 - \sigma_1)] \end{aligned} \tag{6.5.24}$$

We note here, and discuss later, a very interesting feature of the response: a_2, the amplitude of the only mode that is directly excited by the external excitation, is independent of f_2, the amplitude of the excitation.

It follows from (6.5.5), (6.5.21), (6.5.22), and (6.5.23) that

$$\begin{aligned} u_1 &= F_1(\omega_1^2 - \Omega^2)^{-1} \cos(\Omega t + \tau_1) \\ &\quad + 2\epsilon[-\Gamma_1 \pm (\tfrac{1}{4}f_2^2 - \Gamma_2^2)^{1/2}]^{1/2} \cos[\tfrac{1}{2}(\Omega t + \tau_2 - \gamma_1 - \gamma_2)] + O(\epsilon^2) \end{aligned} \tag{6.5.25}$$

$$u_2 = 2\epsilon\omega_1 [4\mu_1^2 + (\sigma_1 - \sigma_2)^2]^{1/2} \cos(\Omega t + \tau_2 - \gamma_1) + O(\epsilon^2) \tag{6.5.26}$$

Thus the nonlinearity produces perfect tuning for the primary (external) resonance as well as the internal resonance.

Next we determine when the roots of (6.5.22) and (6.5.23) are real. We begin by defining two critical values of f_2, namely

$$f_2 = \zeta_1 = 2|\Gamma_2| \quad \text{and} \quad f_2 = \zeta_2 = 2(\Gamma_1^2 + \Gamma_2^2)^{1/2} \tag{6.5.27}$$

Clearly ζ_2 must be greater than ζ_1. Then there are two possibilities: $\Gamma_1 \geqslant 0$ and $\Gamma_1 < 0$. For the former, one real solution exists if

$$f_2 > \zeta_2 \tag{6.5.28}$$

For the latter, two solutions exist if

$$f_2 < \zeta_2 \tag{6.5.29}$$

and one solution exists if

$$f_2 > \zeta_2 \tag{6.5.30}$$

Consequently when $f_2 < \zeta_1$, the response must be given by (6.5.14). When $\Gamma_1 < 0$ and $\zeta_1 < f_2 < \zeta_2$, the response is one of the three possibilities predicted by (6.5.14) and (6.5.25) and (6.5.26). And when $f_2 > \zeta_2$, the response is one of the two possibilities predicted by (6.5.14) and (6.5.25) and (6.5.26).

Next we consider the stability of the various steady-state solutions. The governing equations for the amplitudes and phases have the form of (3.2.5). Thus we want to determine the nature of the various singular (or steady-state) points in the state space. Proceeding as in Chapter 3, we expand the right sides of (6.5.17) through (6.5.21) about the singular point, obtaining a set of linear equations having constant coefficients, which govern the components of the disturbance. If the real part of each eigenvalue of the coefficient matrix is not positive definite, then the point is stable; otherwise it is unstable.

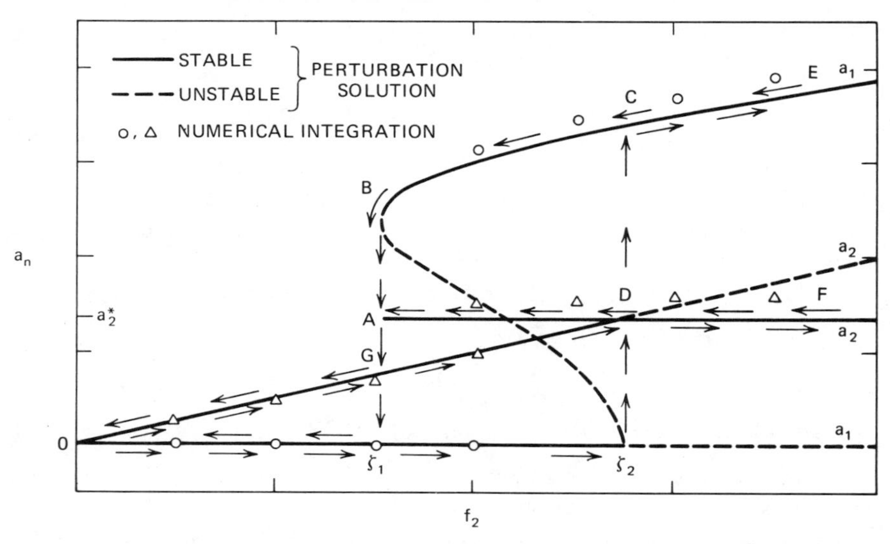

Figure 6-13. Amplitudes of the response as functions of the amplitude of the excitation; $\Gamma_1 < 0; \Omega \simeq \omega_2$.

In order to illustrate the basic character of the possible responses, Nayfeh, Mook, and Marshall (1973) arbitrarily chose values for the parameters and computed the solutions. Moreover to promote confidence in the method of multiple scales, they integrated (6.5.1) numerically and compared these results with the approximate, analytic solution.

In Figures 6-13 and 6-14 a_1 and a_2 are plotted as functions of f_2. In Figure 6-13 there is a small detuning of the external resonance, while the internal resonance is perfectly tuned; this combination renders Γ_1 negative. The values of ζ_1 and ζ_2 defined in (6.5.27) and a_2^* defined in (6.5.23) are indicated. One can clearly see the different solutions in the regions defined by (6.5.28) through (6.5.30). For $\zeta_1 < f_2 < \zeta_2$, two of the three solutions are stable according to the approximate analysis. The initial conditions determine which of these solutions gives the response. In the other regions there is only one stable solution. These conclusions were verified by the numerical results. In Figure 6-14, both resonances are perfectly tuned. This combination renders $\Gamma_1 > 0$.

Returning to (6.5.1) we see that u_2 is essentially a parametric excitation for u_1. The internal resonance (i.e., ω_2 being nearly $2\omega_1$) is also a parametric resonance. Such internal resonances are sometimes referred to as *autoparametric resonances*. Typical of parametrically excited linear systems, when the amplitude of the excitation (u_2 in this case) exceeds a critical value, the trivial homogeneous solution becomes unstable. However atypical of linear systems, the phasing between

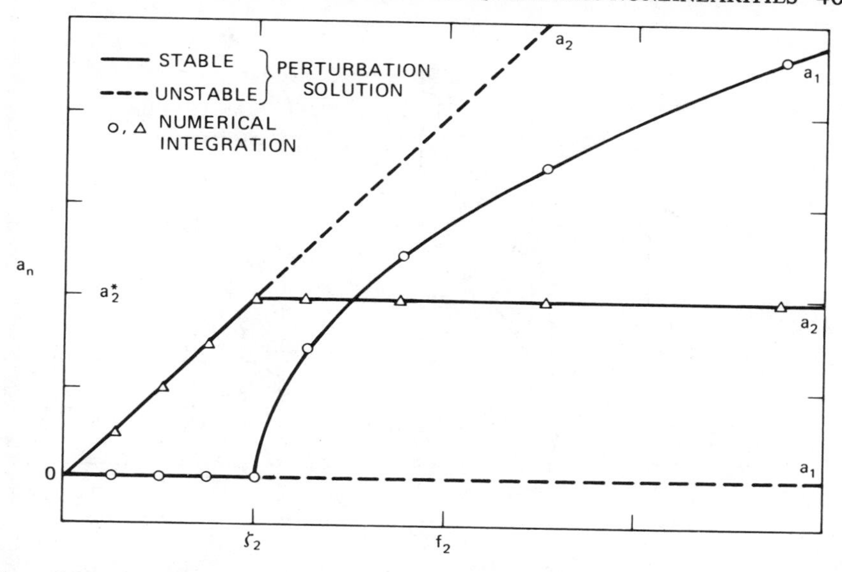

Figure 6-14. Amplitudes of the response as functions of the amplitude of the excitation; $\sigma_1 = \sigma_2 = 0$; $\Omega \simeq \omega_2$.

the response and excitation begins to change, as shown by (6.5.20). This change is significant because it limits the amplitude of the response to a finite value.

In Figures 6-13 and 6-14 one can clearly see a *saturation phenomenon*. As f_2 increases from zero, so does a_2 until it reaches the value a_2^*, while a_1 is zero. This agrees with the solution of the corresponding linear problem. At this point, however, a_2 has its maximum value, and further increases in f_2 will not produce further increases in a_2 because the solution given by (6.5.14) is unstable and the solution given by (6.5.25) and (6.5.26) is stable. The u_2 mode is saturated. Further increases in f_2 cause a_1 to increase, as clearly shown in Figure 6-14 and indicated in (6.5.22).

When there are multiple stable solutions such as the situation illustrated in Figure 6-13, there is a jump phenomenon associated with varying the amplitude of the excitation. Referring to Figure 6-13, we note that when f_2 increases slowly from zero, a_2 follows along the line through O and D and a_1 is zero. As f_2 increases beyond ζ_2, a_2 continues to have the value a_2^* (saturation) and a_1 jumps from ζ_2 to C. For further increases in f_2, a_1 follows to the right along the curve through B, C, and E. When f_2 decreases slowly from a large value well beyond ζ_2, a_2 has a constant value, following along the line from F through D to A, and a_1 follows along the curve from E through C to B. When f_2 decreases below ζ_1, a_2 jumps down from A to G and a_1 jumps down from B to ζ_1. Then

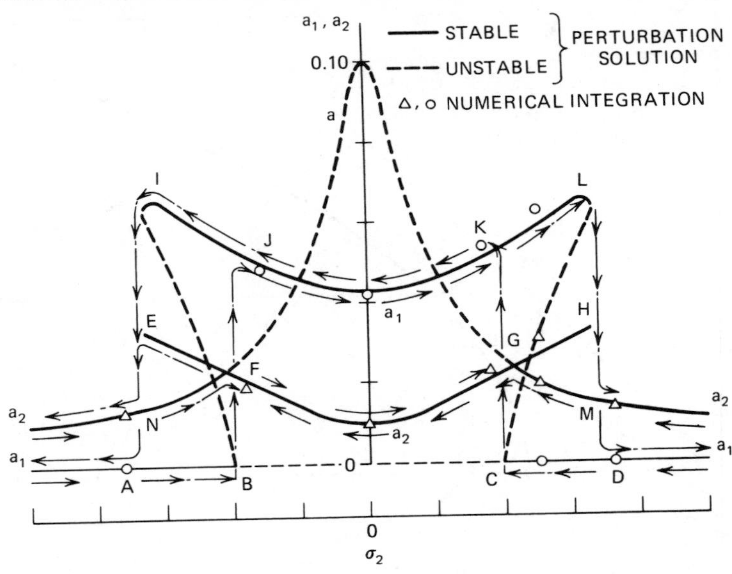

Figure 6-15. Frequency–response curves; $\sigma_1 = 0$, $\Omega \simeq \omega_2$.

both a_1 and a_2 follow the linear solution back to the origin as f_2 continues to decrease.

In Figures 6-15 and 6-16, a_1 and a_2 are plotted as a function of σ_2 for $\sigma_1 = 0$ and $\sigma_1 > 0$, respectively. The dotted curves having peaks at $\sigma_2 = 0$ correspond

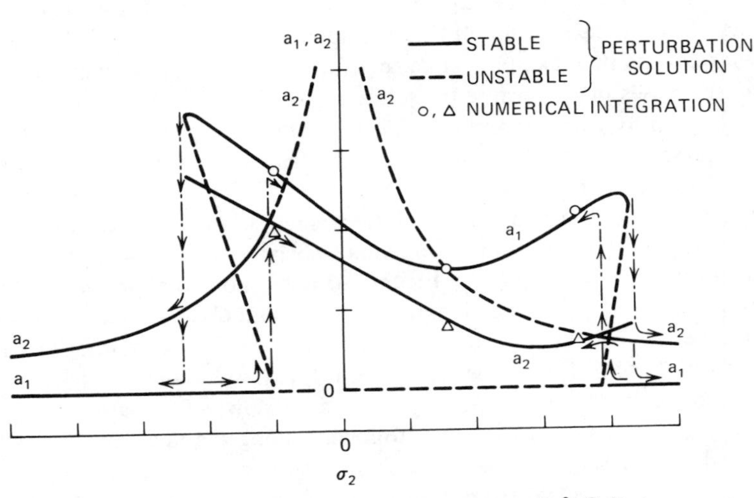

Figure 6-16. Frequency–response curves; $\sigma_1 > 0$, $\Omega \simeq \omega_2$.

to $a_1 = 0$, and they are the solution of the corresponding linear problem. The unstable regions for the "linear" solution depend on the values of σ_1 and σ_2. Hence at a given value of σ_2, the linear solution may be either a stable or an unstable solution, depending on the value of σ_1. The regions in which there are two stable solutions correspond to Figure 6-13, while the center regions correspond to Figure 6-14.

In Figures 6-15 and 6-16 the jump phenomenon associated with varying the frequency Ω of the excitation is indicated by the arrows. When Ω is such that σ_2 is to the left of A, the response is always given by the "linear" solution (6.5.14). As σ_2 increases slowly, the response follows the linear curves–a_2 along the solid line through N to F and a_1 along the solid line through A to B. As σ_2 increases to the right of B, a_2 follows along the curve FGH and a_1 jumps up from B to J and then follows along the solid curve through K to L, as given by (6.5.25) and (6.5.26). When σ_2 increases to the right of D, a_2 jumps down from H to M and a_1 jumps down from L to D. Then both a_1 and a_2 follow the linear curve as σ_2 continues to increase.

6.5.2. THE CASE OF Ω NEAR ω_1

In this case we put

$$F_1 = \epsilon^2 f_1, \quad F_2 = \epsilon f_2 \tag{6.5.31}$$

and let

$$\Omega = \omega_1 + \epsilon\sigma_1 \tag{6.5.32}$$

Instead of the first of (6.5.21) we put

$$\gamma_1 = \sigma_1 T_1 - \theta_1 + \tau_1 \tag{6.5.33}$$

Using polar notation, we now obtain the following solvability conditions:

$$a_1' = -\mu_1 a_1 + \tfrac{1}{4}\omega_1^{-1} a_1 a_2 \sin\gamma_2 + \tfrac{1}{2}\omega_1^{-1} f_1 \sin\gamma_1 \tag{6.5.34}$$

$$a_2' = -\mu_2 a_2 - \tfrac{1}{4}\omega_2^{-1} a_1^2 \sin\gamma_2 \tag{6.5.35}$$

$$a_1\theta_1' = -\tfrac{1}{4}\omega_1^{-1} a_1 a_2 \cos\gamma_2 - \tfrac{1}{2}\omega_1^{-1} f_1 \cos\gamma_1 \tag{6.5.36}$$

$$a_2\theta_2' = -\tfrac{1}{4}\omega_2^{-1} a_1^2 \cos\gamma_2 \tag{6.5.37}$$

where γ_2, μ_1, μ_2, and σ_2 are the same as in Section 6.5.1.

For the steady-state solution, $a_n' = \gamma_n' = 0$ and (6.5.34) through (6.5.37) can be combined into

$$a_2^3 + 8\omega_1(\sigma_1\cos\gamma_2 - \mu_1\sin\gamma_2)a_2^2 + 16\omega_1^2(\mu_1^2 + \sigma_1^2)a_2 - \Gamma\omega_2^{-1} f_1^2 = 0 \tag{6.5.38}$$

$$a_1 = 2\sqrt{\frac{\omega_2 a_2}{\Gamma}} \tag{6.5.39}$$

$$\sin \gamma_2 = -\mu_2 \Gamma, \quad \cos \gamma_2 = -(\sigma_2 + 2\sigma_1)\Gamma \tag{6.5.40}$$

where

$$\Gamma = [\mu_2^2 + (\sigma_2 + 2\sigma_1)^2]^{-1/2} \tag{6.5.41}$$

Thus to the first approximation

$$u_1 = 2\epsilon \sqrt{\frac{\omega_2 a_2}{\Gamma}} \cos (\Omega t - \gamma_1 + \tau_1) + O(\epsilon^2) \tag{6.5.42}$$

$$u_2 = F_1(\omega_2^2 - \Omega^2)^{-1} \cos (\Omega t + \tau_2) + \epsilon a_2 \cos (2\Omega t + 2\tau_1 + \gamma_2 - 2\gamma_1) + O(\epsilon^2) \tag{6.5.43}$$

If there is no internal resonance, the solution is essentially that of the corresponding linear problem.

In Figures 6-17 and 6-18, a_1 and a_2 are plotted as functions of σ_2 for $\sigma_1 = 0$ and $\sigma_1 > 0$, respectively. For some combinations of the parameters, there is a region near the center dip of these curves where no stable, steady-state solution exists. In these cases u_1 and u_2 are plotted for very large values of t in Figure 6-19. This figure shows a continuous exchange of energy back and forth between the two modes.

In Figures 6-17 and 6-18 the jump phenomenon, which is similar to that in Figures 6-15 and 6-16, is indicated by the arrows. There is no saturation phenomenon in this case. However from (6.5.39) it follows that as f_1 tends to

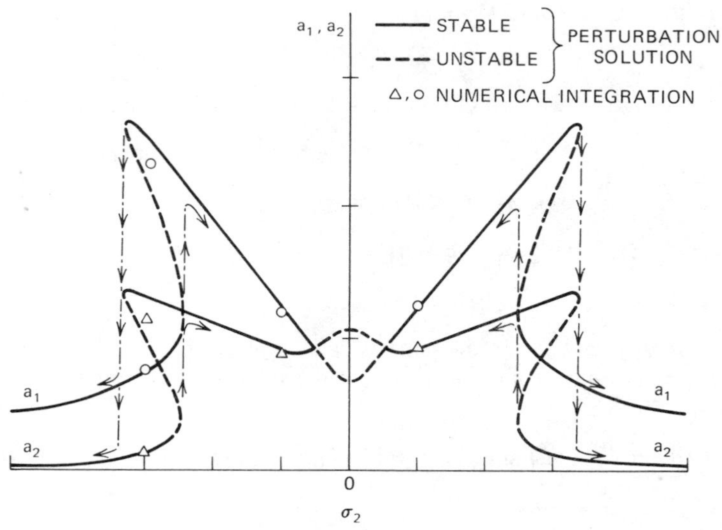

Figure 6-17. Frequency-response curves; $\sigma_1 = 0$, $\Omega \simeq \omega_1$.

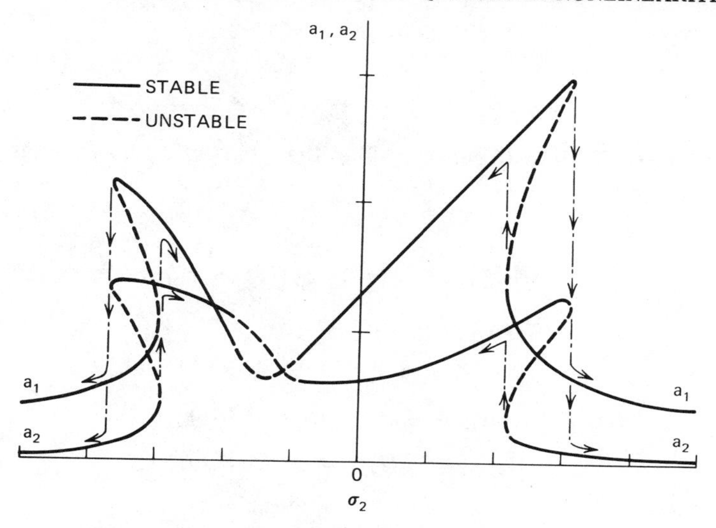

Figure 6-18. Frequency–response curves; $\sigma_1 > 0, \Omega \simeq \omega_1$.

zero, a_2 tends to zero faster than a_1, and hence this solution also agrees with the solution of the linear problem for very small amplitudes of the response.

The value of an analytic solution is made apparent here. One can easily imagine the difficulty in obtaining the dominant characteristics of the solution by numerical methods alone.

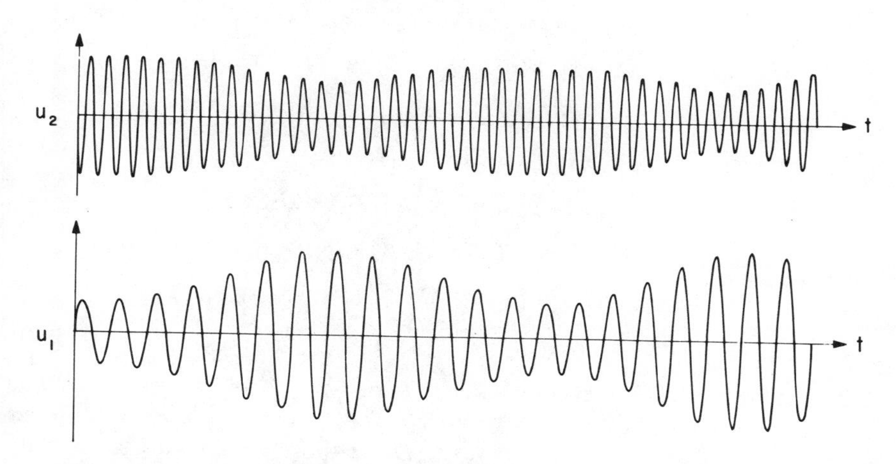

Figure 6-19. Nonexistence of periodic motions in a system with quadratic nonlinearities.

6.5.3. THE CASE OF NONRESONANT EXCITATIONS

In this case we put

$$F_n = \epsilon f_n \tag{6.5.44}$$

Substituting (6.5.2) into (6.5.1), recalling that $\hat{\mu}_n = \epsilon \mu_n$, and equating coefficients of like powers of ϵ we obtain

Order ϵ

$$\begin{aligned} D_0^2 u_{11} + \omega_1^2 u_{11} &= f_1 \cos(\Omega T_0 + \tau_1) \\ D_0^2 u_{21} + \omega_2^2 u_{21} &= f_2 \cos(\Omega T_0 + \tau_2) \end{aligned} \tag{6.5.45}$$

Order ϵ^2

$$\begin{aligned} D_0^2 u_{12} + \omega_1^2 u_{12} &= -2D_0(D_1 u_{11} + \mu_1 u_{11}) + u_{11} u_{21} \\ D_0^2 u_{22} + \omega_2^2 u_{22} &= -2D_0(D_1 u_{21} + \mu_2 u_{21}) + u_{11}^2 \end{aligned} \tag{6.5.46}$$

The solutions of (6.5.45) can be expressed in the form

$$\begin{aligned} u_{11} &= A_1(T_1) \exp(i\omega_1 T_0) + \Lambda_1 \exp(i\Omega T_0) + cc \\ u_{21} &= A_2(T_1) \exp(i\omega_2 T_0) + \Lambda_2 \exp(i\Omega T_0) + cc \end{aligned} \tag{6.5.47}$$

where

$$\Lambda_n = \tfrac{1}{2} f_n (\omega_n^2 - \Omega^2)^{-1} \exp(i\tau_n) \tag{6.5.48}$$

Hence (6.5.46) become

$$\begin{aligned} D_0^2 u_{12} + \omega_1^2 u_{12} = {} & -2i\omega_1(A_1' + \mu_1 A_1) \exp(i\omega_1 T_0) - 2i\Omega\mu_1 \Lambda_1 \exp(i\Omega T_0) \\ & + A_2 A_1 \exp[i(\omega_1 + \omega_2) T_0] + A_2 \Lambda_1 \exp[i(\omega_2 + \Omega) T_0] \\ & + A_2 \bar{A}_1 \exp[i(\omega_2 - \omega_1) T_0] + A_2 \bar{\Lambda}_1 \exp[i(\omega_2 - \Omega) T_0] \\ & + A_1 \Lambda_2 \exp[i(\omega_1 + \Omega) T_0] + \Lambda_1 \Lambda_2 \exp(2i\Omega T_0) \\ & + \Lambda_2 \bar{A}_1 \exp[i(\Omega - \omega_1) T_0] + \Lambda_2 \bar{\Lambda}_1 + cc \end{aligned} \tag{6.5.49}$$

$$\begin{aligned} D_0^2 u_{22} + \omega_2^2 u_{22} = {} & -2i\omega_2(A_2' + \mu_2 A_2) \exp(i\omega_2 T_0) - 2i\Omega\mu_2 \Lambda_2 \exp(i\Omega T_0) \\ & + A_1^2 \exp(2i\omega_1 T_0) + 2A_1 \Lambda_1 \exp[i(\omega_1 + \Omega) T_0] \\ & + 2A_1 \bar{\Lambda}_1 \exp[i(\omega_1 - \Omega) T_0] + \Lambda_1^2 \exp(2i\Omega T_0) \\ & + A_1 \bar{A}_1 + \Lambda_1 \bar{\Lambda}_1 + cc \end{aligned} \tag{6.5.50}$$

Inspecting the inhomogeneous terms in (6.5.49) and (6.5.50), one finds that in addition to the terms proportional to $\exp(\pm i\omega_n T_0)$, secular terms result when $\Omega \approx 2\omega_1$, $\Omega \approx \frac{1}{2}\omega_1$, $\Omega \approx \frac{1}{2}\omega_2$, $\Omega \approx \omega_2 - \omega_1$, $\Omega = \omega_2 + \omega_1$, and $\omega_2 \approx 2\omega_1$. The last case is the internal resonance case, while the first six cases correspond to secondary resonances (external resonances).

In the absence of secondary and internal resonances the solvability conditions are

$$A_1' + \mu_1 A_1 = 0$$
$$A_2' + \mu_2 A_2 = 0 \tag{6.5.51}$$

whose solutions decay with T_1. Hence to the first approximation, u_1 and u_2 are approximately the solutions of the linear problem.

In the absence of secondary resonances and in the presence of internal resonance ($\omega_2 = 2\omega_1 - \epsilon\sigma_2$), the solvability conditions are the same as (6.2.14) when $\alpha_1 = \alpha_2 = 1$. The analysis of the motion in this case is the same as in Section 6.2.

6.5.4. THE CASE OF 2Ω NEAR ω_1

When the system does not possess an internal resonance the two modes are uncoupled, and there is a superharmonic resonance for the u_1 mode alone. The analysis, which is carried out in detail in Section 4.2.2, is not repeated here.

When the system possesses an internal resonance, we introduce two detuning parameters as follows:

$$\omega_2 = 2\omega_1 - \epsilon\sigma_2 \quad \text{and} \quad 2\Omega = \omega_1 + \epsilon\sigma_1 \tag{6.5.52}$$

It follows from (6.5.49) and (6.5.50) that secular terms are eliminated from u_{12} and u_{22} if

$$-2i\omega_1(A_1' + \mu_1 A_1) + A_2\bar{A}_1 \exp(-i\sigma_2 T_1) + \Lambda_1\Lambda_2 \exp(i\sigma_1 T_1) = 0$$
$$-2i\omega_2(A_2' + \mu_2 A_2) + A_1^2 \exp(i\sigma_2 T_1) = 0 \tag{6.5.53}$$

As before, we introduce polar notation into (6.5.53) and obtain

$$a_1' = -\mu_1 a_1 + \tfrac{1}{4}\omega_1^{-1} a_1 a_2 \sin\gamma_2 + \tfrac{1}{2}\omega_1^{-1}\Gamma \sin\gamma_1 \tag{6.5.54}$$

$$a_2' = -\mu_2 a_2 - \tfrac{1}{4}\omega_2^{-1} a_1^2 \sin\gamma_2 \tag{6.5.55}$$

$$a_1\theta_1' = -\tfrac{1}{4}\omega_1^{-1} a_1 a_2 \cos\gamma_2 - \tfrac{1}{2}\omega_1^{-1}\Gamma\cos\gamma_1 \tag{6.5.56}$$

$$a_2\theta_2' = -\tfrac{1}{4}\omega_2^{-1} a_1^2 \cos\gamma_2 \tag{6.5.57}$$

where γ_2 is defined in (6.5.21) and

$$\Gamma = \tfrac{1}{2} f_1 f_2 (\omega_1^2 - \Omega^2)^{-1} (\omega_2^2 - \Omega^2)^{-1}$$
$$\gamma_1 = \sigma_1 T_1 - \theta_1 + \tau_1 + \tau_2 \tag{6.5.58}$$

Since (6.5.54) through (6.5.58) have the same form as (6.5.33) through (6.5.37), the steady-state solution in this case is given by (6.5.38) through (6.5.43) if f_1 is replaced by Γ and τ_1 is replaced by $\tau_1 + \tau_2$. Therefore the response curves are qualitatively the same as those appearing in Figures 6-18 and

6-17. In particular, for some combinations of the parameters there exists a region where steady-state solutions do not exist.

6.5.5. THE CASE OF Ω NEAR $\omega_1 + \omega_2$

In the case of combination resonance the two modes are coupled irrespective of the existence of an internal resonance. Hence we consider both cases and let

$$\Omega = \omega_2 + \omega_1 + \epsilon\sigma_1 \tag{6.5.59}$$

In the absence of an internal resonance the solvability conditions for (6.5.49) and (6.5.50) are

$$\begin{aligned} -2i\omega_1(A_1' + \mu_1 A_1) + \bar{A}_2 \Lambda_1 \exp(i\sigma_1 T_1) &= 0 \\ -i\omega_2(A_2' + \mu_2 A_2) + \bar{A}_1 \Lambda_1 \exp(i\sigma_1 T_1) &= 0 \end{aligned} \tag{6.5.60}$$

Equations (6.5.60) admit a solution of the form

$$A_1 = b_1 \exp(\lambda T_1 + i\sigma_1 T_1 + i\tau_1) \quad \text{and} \quad A_2 = b_2 \exp(\bar{\lambda} T_1) \tag{6.5.61}$$

provided that

$$\lambda^2 + (\mu_1 + \mu_2 + i\sigma_1)\lambda + \mu_2(\mu_1 + i\sigma_1) - \tfrac{1}{2}\omega_1^{-1}\omega_2^{-1}|\Lambda_1|^2 = 0 \tag{6.5.62}$$

Hence,

$$\lambda = -\tfrac{1}{2}(\mu_1 + \mu_2 + i\sigma_1) \pm \tfrac{1}{2}[(\mu_1 - \mu_2 + i\sigma_1)^2 + 2\omega_1^{-1}\omega_2^{-1}|\Lambda_1|^2]^{1/2} \tag{6.5.63}$$

The real parts of all values of λ must be negative for the homogeneous solutions to decay. If the real part of any of the λ's is positive definite, the homogeneous solution grows and is part of the first approximation of the response. This situation is similar to the subharmonic resonance for a single degree of freedom discussed in Section 4.2.3. The case of $\Omega \approx \omega_2 - \omega_1$ can be obtained from the above results by simply changing the sign of ω_1. Combination resonances of the summed type $\Omega \approx \omega_1 + \omega_2$ were found experimentally by Yamamoto (1960).

In the presence of internal resonance, $\omega_2 = 2\omega_1 - \epsilon\sigma_2$, the solvability conditions of (6.5.49) and (6.5.50) are

$$\begin{aligned} -2i\omega_1(A_1' + \mu_1 A_1) + A_2\bar{A}_1 \exp(-i\sigma_2 T_1) + \bar{A}_2\Lambda_1 \exp(i\sigma_1 T_1) &= 0 \\ -2i\omega_2(A_2' + \mu_2 A_2) + A_1^2 \exp(i\sigma_2 T_1) + 2\bar{A}_1\Lambda_1 \exp(i\sigma_1 T_1) &= 0 \end{aligned} \tag{6.5.64}$$

Introducing polar notation we rewrite (6.5.64) as

$$a_1' = -\mu_1 a_1 + \tfrac{1}{4}\omega_1^{-1} a_1 a_2 \sin\gamma_2 + \tfrac{1}{2}\omega_1^{-1} a_2 |\Lambda_1| \sin\gamma_1 \tag{6.6.65}$$

$$a_2' = -\mu_2 a_2 - \tfrac{1}{4}\omega_2^{-1} a_1^2 \sin\gamma_2 + \omega_2^{-1} a_1 |\Lambda_1| \sin\gamma_1 \tag{6.5.66}$$

$$a_2\theta_1' = -\tfrac{1}{4}\omega_1^{-1}a_1a_2\cos\gamma_2 - \tfrac{1}{2}\omega_1^{-1}a_2|\Lambda_1|\cos\gamma_1 \tag{6.5.67}$$

$$a_2\theta_2' = -\tfrac{1}{4}\omega_2^{-1}a_1^2\cos\gamma_2 - \omega_2^{-1}a_1|\Lambda_1|\cos\gamma_1 \tag{6.5.68}$$

where

$$\gamma_1 = \sigma_1 T_1 - \theta_2 - \theta_1 + \tau_1 \quad \text{and} \quad \gamma_2 = -\sigma_2 T_1 - 2\theta_1 + \theta_2 \tag{6.5.69}$$

For the steady-state response $a_n' = 0$ and $\gamma_n' = 0$. It follows from (6.5.69) that $\theta_1' = \frac{1}{3}(\sigma_1 - \sigma_2)$ and $\theta_2' = \frac{1}{3}(2\sigma_1 + \sigma_2)$. There are two possibilities: either $a_1 = a_2 = 0$ or $a_1 \neq 0$ and $a_2 \neq 0$. In the latter case solving for the circular functions of γ_1 and γ_2 from the steady form of (6.5.65) through (6.5.68) yields

$$\tfrac{3}{2}a_1a_2|\Lambda_1|\sin\gamma_1 = \omega_1\mu_1a_1^2 + \omega_2\mu_2a_2^2 \tag{6.5.70}$$

$$\tfrac{3}{4}a_1^2a_2\sin\gamma_2 = 2\omega_1\mu_1a_1^2 - \omega_2\mu_2a_2^2 \tag{6.5.71}$$

$$\tfrac{1}{2}a_1a_2|\Lambda_1|\cos\gamma_1 = \tfrac{1}{3}\omega_1(\sigma_1 - \sigma_2)a_1^2 - \tfrac{1}{3}\omega_2(2\sigma_1 + \sigma_2)a_2^2 \tag{6.5.72}$$

$$\tfrac{1}{4}a_1^2a_2\cos\gamma_2 = -\tfrac{2}{3}\omega_1(\sigma_1 - \sigma_2)a_1^2 + \tfrac{1}{3}\omega_2(2\sigma_1 + \sigma_2)a_2^2 \tag{6.5.73}$$

Eliminating the γ_n from (6.5.70) through (6.5.73) yields

$$\omega_1^2[\mu_1^2 + (\sigma_1 - \sigma_2)^2]a_1^4 + \omega_2^2[\mu_2^2 + (2\sigma_1 + \sigma_2)^2]a_2^4 + 2\omega_1\omega_2[\mu_1\mu_2 - (\sigma_1 - \sigma_2)(2\sigma_1 + \sigma_2)]a_1^2a_2^2 = \tfrac{9}{4}a_1^2a_2^2|\Lambda_1|^2 \tag{6.5.74}$$

$$4\omega_1^2[\mu_1^2 + (\sigma_1 - \sigma_2)^2]a_1^4 + \omega_2^2[\mu_2^2 + (2\sigma_1 + \sigma_2)^2]a_2^4 - 4\omega_1\omega_2[\mu_1\mu_2 + (\sigma_1 - \sigma_2)(2\sigma_1 + \sigma_2)]a_1^2a_2^2 = \tfrac{9}{16}a_2^2a_1^4 \tag{6.5.75}$$

Eliminating a_2^4 from (6.5.74) and (6.5.75) yields

$$3\omega_1^2[\mu_1^2 + (\sigma_1 - \sigma_2)^2]a_1^2 - 2\omega_1\omega_2[3\mu_1\mu_2 + (\sigma_1 - \sigma_2)(2\sigma_1 + \sigma_2)]a_2^2 = \tfrac{9}{16}a_2^2a_1^2 - \tfrac{9}{4}|\Lambda_1|^2a_2^2 \tag{6.5.76}$$

while eliminating a_1^4 from the the left-hand sides of (6.5.74) and (6.5.75) yields

$$3\omega_2^2[\mu_2^2 + (2\sigma_1 + \sigma_2)^2]a_2^2 + 4\omega_1\omega_2[\mu_1\mu_2 - (\sigma_1 - \sigma_2)(2\sigma_1 + \sigma_2)]a_1^2 = 9|\Lambda_1|^2a_1^2 - \tfrac{9}{16}a_1^4 \tag{6.5.77}$$

Eliminating a_2^2 from (6.5.76) and (6.5.77), one obtains a quadratic equation in a_1^2 whose solution is

$$a_1^2 = 5|\Lambda_1|^2 - 16\omega_1\omega_2[\mu_1\mu_2 - \tfrac{2}{9}(\sigma_1 + \tfrac{1}{2}\sigma_2)(\sigma_1 - \sigma_2)] \pm 2|\Lambda_1|\{9|\Lambda_1|^2 - 16\omega_1\omega_2[\mu_1\mu_2 - 2(\sigma_1 + \tfrac{1}{2}\sigma_2)(\sigma_1 - \sigma_2)] - \tfrac{256}{9}\omega_1^2\omega_2^2[\mu_2^2(\sigma_1 - \sigma_2)^2 + 4\mu_1^2(\sigma_1 + \tfrac{1}{2}\sigma_2)^2 + 4\mu_1\mu_2(\sigma_1 + \tfrac{1}{2}\sigma_2)(\sigma_1 - \sigma_2)]\}^{1/2} \tag{6.5.78}$$

When a_1 and a_2 differ from zero, the steady-state response is

$$u_1 = \frac{\epsilon f_1}{\omega_1^2 - \Omega^2} \cos(\Omega t + \tau_1) + \epsilon a_1 \cos\left[\tfrac{1}{3}(\Omega t + \tau_1 - \gamma_1 - \gamma_2)\right] + O(\epsilon^2) \tag{6.5.79}$$

$$u_2 = \frac{\epsilon f_2}{\omega_2^2 - \Omega^2} \cos(\Omega t + \tau_2) + \epsilon a_2 \cos\left[\tfrac{2}{3}(\Omega t + \tau_1 + \tfrac{1}{2}\gamma_2 - \gamma_1)\right] + O(\epsilon^2) \tag{6.5.80}$$

Consequently a *fractional-harmonic pair* is produced by the excitation. Such fractional-harmonic pairs were observed in a vareity of physical systems by Dallos and Linnell (1966a, b). Dallos (1966), Luukkala (1967), Adler and Breazeale (1970), and Eller (1973).

In Figure 6-20, a_1 and a_2 are plotted as functions of f_1. This figure reveals that

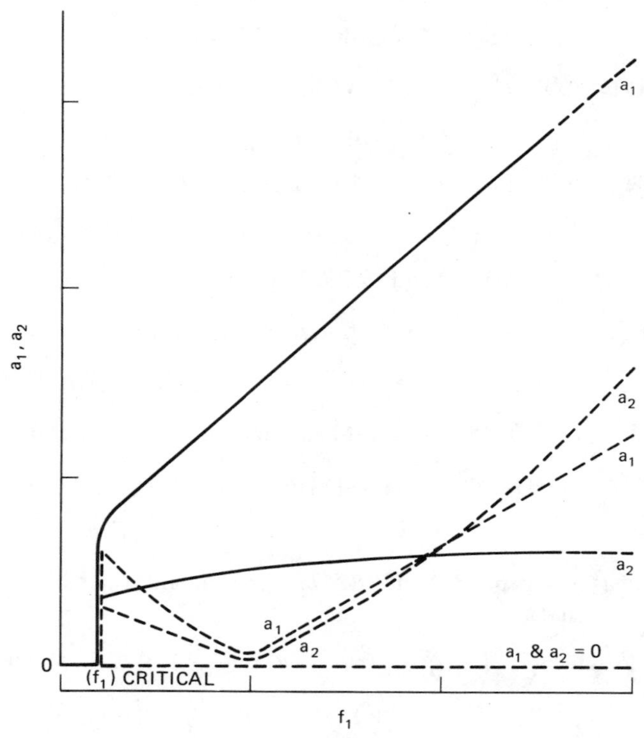

Figure 6-20. Amplitudes of the response as functions of the amplitude of the excitation; $\omega_2 \simeq 2\omega_1$, $\Omega \simeq \omega_1 + \omega_2$.

f_1 must increase beyond a minimum value before a_1 and a_2 can differ from zero. Moreover beyond this critical value the solution for which $a_1 = a_2 = 0$ is unstable, and thus the response is given by (6.5.79) and (6.5.80). We note a saturationlike phenomenon here. After f_1 exceeds the critical value, a_2 varies only slightly, while a_1 increases rapidly as f_1 increases. Finally we note that for large values of f_1 all the steady-state solutions are unstable.

6.6. Forced Oscillations of Systems Having Cubic Nonlinearities

We consider the forced oscillations of the system (6.3.1) considered in Section 6.3. Namely we consider

$$\ddot{u}_1 + \omega_1^2 u_1 = -2\hat{\mu}_1 \dot{u}_1 + \alpha_1 u_1^3 + \alpha_2 u_1^2 u_2 + \alpha_3 u_1 u_2^2 + \alpha_4 u_2^3 + F_1 \cos(\Omega t + \tau_1) \tag{6.6.1}$$

$$\ddot{u}_2 + \omega_2^2 u_2 = -2\hat{\mu}_2 \dot{u}_2 + \alpha_5 u_1^3 + \alpha_6 u_1^2 u_2 + \alpha_7 u_1 u_2^2 + \alpha_8 u_2^3 + F_2 \cos(\Omega t + \tau_2) \tag{6.6.2}$$

Primary resonances of systems having two degrees of freedom were studied by Roberson (1952); Pipes (1953); Sethna (1954); Arnold (1955); Huang (1955); Rosenberg (1955); Atkinson (1956); Carter and Liu (1961); Plotnikova (1963a); Kinney and Rosenberg (1965); Williams (1966); Efstathiades and Williams (1967); Janssens, van Dooren, and Melchambre (1969); van Dooren (1972a, b, 1973c); and Srirangarajan and Srinivasan (1973). Primary resonances of systems having many degrees of freedom were studied by Klotter (1954), Hovanessian (1959), Sethna (1960, 1963a), Szulkin (1960), Bogoliubov and Mitropolsky (1961, Section 22), Butenin (1965, Section 17), Loud and Sethna (1966), Bainov (1966), Mettler (1967), Sethna and Moran (1968), Akulenko (1968a), Szemplinska-Stupnicka (1970, 1972), Tondl (1972), Ginsberg (1972a, b), Agrawal and Evan-Iwanowski (1973), Mikhlin (1974), Eckhaus (1975), and Evan-Iwanowski (1976).

The methods used to determine the periodic response of a system can be divided into two classes (Section 5.1.5). The first class includes the method of multiple scales and the method of averaging. With these methods one determines a set of time-dependent equations that govern the time variation of the amplitudes and the phases. These equations are usually transformed into an autonomous system. Then the stationary points of this set correspond to the periodic motions and the stability of these stationary points correspond to the stability of the periodic motions. The second class includes the method of harmonic balance, the Galérkin–Ritz procedure, and the Lindstedt–Poincaré technique. With these methods one determines directly the periodic motions. To investigate the stability of these periodic motions, one usually perturbs them and determines variational equations having periodic coefficients. Thus determination of the stability of the periodic motions is transformed into the stability of the solutions of a set of equations with periodic coefficients. The latter approach is used by Plotnikova (1965) to determine the stability of periodic solutions having two degrees of freedom and by Akulenko (1968b), Szemplinska-Stupnicka (1973), and Ponzo and Wax (1974) to determine the stability of systems having many degrees of freedom.

Secondary resonances of systems with cubic nonlinearities include subharmonic, superharmonic, and combinational resonances. One or more of these resonances might be excited in the presence or absence of internal resonances. When a combinational resonance is excited in the presence of an internal resonance, fractional-harmonic pairs might be excited. Secondary resonances were studied by Huang (1954); Kruschul (1960); Yamamoto (1961b); Tondl (1963a, b, 1964, 1965); Yamamoto and Hayashi (1963); Efstathiades (1968, 1969); Szemplinska-Stupnicka (1969, 1972, 1975); Bauer (1970); Asmis and Tso (1972); Agrawal and Evan-Iwanowski (1973); Yamamoto and Yasuda (1974, 1977); Cheshankov (1974b); Stevanovich and Rashkovich (1974); Szemplinska-Stupnicka (1974); van Dooren (1975); Sridhar, Mook, and Nayfeh (1975); van Dooren and Bouc (1975); Evan-Iwanowski (1976); and Yamamoto, Yasuda, and Nagasaka (1977).

We restrict our attention to the cases of primary resonances, that is $\omega_1 \approx \Omega$ and $\omega_2 \approx \Omega$. Moreover we assume that $\omega_2 > \omega_1$. The cases of secondary resonances are left for exercises. As in Section 6.3, we seek an asymptotic expansion in the form

$$\begin{aligned} u_1 &= \epsilon u_{11}(T_0, T_2) + \epsilon^3 u_{13}(T_0, T_2) + \cdots \\ u_2 &= \epsilon u_{21}(T_0, T_2) + \epsilon^3 u_{23}(T_0, T_2) + \cdots \end{aligned} \tag{6.6.3}$$

As before, we note that the terms $O(\epsilon^2)$ and the scale T_1 are missing from (6.6.3) because the effect of the nonlinearity appears at $O(\epsilon^3)$. Moreover we set $\hat{\mu}_n = \epsilon^2 \mu_n$ and $F_n = \epsilon^2 f_n$ so that in the case of primary resonances the effect of the damping, nonlinearity, and excitation appear in the same perturbation equations. Then substituting (6.6.3) into (6.6.1) and (6.6.2) and equating coefficients of like powers of ϵ we obtain

Order ϵ

$$\begin{aligned} D_0^2 u_{11} + \omega_1^2 u_{11} &= 0 \\ D_0^2 u_{21} + \omega_2^2 u_{21} &= 0 \end{aligned} \tag{6.6.4}$$

Order ϵ^3

$$\begin{aligned} D_0^2 u_{13} + \omega_1^2 u_{13} = -2D_0(D_1 u_{11} + \mu_1 u_{11}) + \alpha_1 u_{11}^3 + \alpha_2 u_{11}^2 u_{21} + \alpha_3 u_{11} u_{21}^2 \\ + \alpha_4 u_{21}^3 + f_1 \cos(\Omega T_0 + \tau_1) \end{aligned} \tag{6.6.5}$$

$$D_0^2 u_{23} + \omega_2 u_{23} = -2D_0(D_1 u_{21} + \mu_2 u_{21}) + \alpha_5 u_{11}^3 + \alpha_6 u_{11}^2 u_{21} + \alpha_7 u_{11} u_{21}^2 + \alpha_8 u_{21}^3 + f_2 \cos(\Omega T_0 + \tau_2) \quad (6.6.6)$$

The solutions of (6.6.4) can be expressed in the form

$$u_{11} = A_1(T_2) \exp(i\omega_1 T_0) + cc$$
$$u_{21} = A_2(T_2) \exp(i\omega_2 T_0) + cc \quad (6.6.7)$$

Substituting for u_{11} and u_{21} in (6.6.5) and (6.6.6) we obtain equations (6.3.7) and (6.3.8) except for the addition of the two terms, $\frac{1}{2} f_1 \exp[i(\Omega T_0 + \tau_1)]$ and $\frac{1}{2} f_2 \exp[i(\Omega T_0 + \tau_2)]$, respectively.

In the absence of internal resonances the steady-state motions are uncoupled, and they are treated in detail in Section 4.1. Therefore we restrict our attention in this section to the case of internal resonances. Since the nonlinearity is cubic, an internal resonance to first order occurs when $\omega_2 \approx 3\omega_1$. Hence we introduce a detuning parameter σ_1 according to

$$\omega_2 = 3\omega_1 + \epsilon^2 \sigma_1 \quad (6.6.8)$$

Next we consider the cases $\omega_1 \approx \Omega$ and $\omega_2 \approx \Omega$, in order.

6.6.1. THE CASE OF Ω NEAR ω_1

We introduce a second detuning parameter σ_2 according to

$$\Omega = \omega_1 + \epsilon^2 \sigma_2 \quad (6.6.9)$$

With (6.6.8) and (6.6.9) the solvability conditions of the modified equations (6.3.7) and (6.3.8) are

$$-2i\omega_1(A_1' + \mu_1 A_1) + 3\alpha_1 A_1^2 \bar{A}_1 + 2\alpha_3 A_2 \bar{A}_2 A_1 + \alpha_2 A_2 \bar{A}_1^2 \exp(i\sigma_1 T_2) + \tfrac{1}{2} f_1 \exp[i(\sigma_2 T_2 + \tau_1)]$$
$$-2i\omega_2(A_2' + \mu_2 A_2) + \alpha_5 A_1^3 \exp(-i\sigma_1 T_2) + 3\alpha_8 A_2^2 \bar{A}_2 + 2\alpha_6 A_1 \bar{A}_1 A_2 = 0 \quad (6.6.10)$$

Introducing polar notation we rewrite (6.6.10) as

$$8\omega_1(a_1' + \mu_1 a_1) = \alpha_2 a_1^2 a_2 \sin\gamma_1 + 4f_1 \sin\gamma_2 \quad (6.6.11)$$

$$8\omega_2(a_2' + \mu_2 a_2) = -\alpha_5 a_1^3 \sin\gamma_1 \quad (6.6.12)$$

$$8\omega_1 a_1 \theta_1' = -(3\alpha_1 a_1^2 + 2\alpha_3 a_2^2) a_1 - \alpha_2 a_1^2 a_2 \cos\gamma_1 - 4f_1 \cos\gamma_2 \quad (6.6.13)$$

$$8\omega_2 a_2 \theta_2' = -(3\alpha_8 a_2^2 + 2\alpha_6 a_1^2) a_2 - \alpha_5 a_1^3 \cos\gamma_1 \quad (6.6.14)$$

where

$$\gamma_1 = \sigma_1 T_2 + \theta_2 - 3\theta_1, \quad \gamma_2 = \sigma_2 T_2 - \theta_1 + \tau_1 \quad (6.6.15)$$

For the steady-state response, $a_n' = \gamma_n' = 0$. Hence it corresponds to the solution of

$$8\omega_1 \mu_1 a_1 - \alpha_2 a_1^2 a_2 \sin \gamma_1 - 4f_1 \sin \gamma_2 = 0 \tag{6.6.16}$$

$$8\omega_2 \mu_2 a_2 + \alpha_5 a_1^3 \sin \gamma_1 = 0 \tag{6.6.17}$$

$$8\omega_1 a_1 \sigma_2 + (3\alpha_1 a_1^2 + 2\alpha_3 a_2^2) a_1 + \alpha_2 a_1^2 a_2 \cos \gamma_1 + 4f_1 \cos \gamma_2 = 0 \tag{6.6.18}$$

$$8\omega_2 a_2 (3\sigma_2 - \sigma_1) + (3\alpha_8 a_2^2 + 2\alpha_6 a_1^2) a_2 + \alpha_5 a_1^3 \cos \gamma_1 = 0 \tag{6.6.19}$$

Nayfeh, Mook, and Sridhar (1974) derived equations similar to these for the problem of primary resonances of a clamped-hinged beam. They solved these equations numerically.

Figures 6-21 and 6-22 show the variation of a_1 and a_2 with σ_2 for several values of the amplitude of the excitation when $\Omega \approx \omega_1$. The results for a_1 are similar to the single-degree-of-freedom case, see Figure 4-1. Equations (6.6.1) and (6.6.2) were integrated numerically, and these results are shown by the small circles. Although a_2 cannot be zero, it is small compared with a_1, and the response is practically described by the first mode. These graphs were obtained using coefficients that are typical of beam vibrations. This behavior is typical

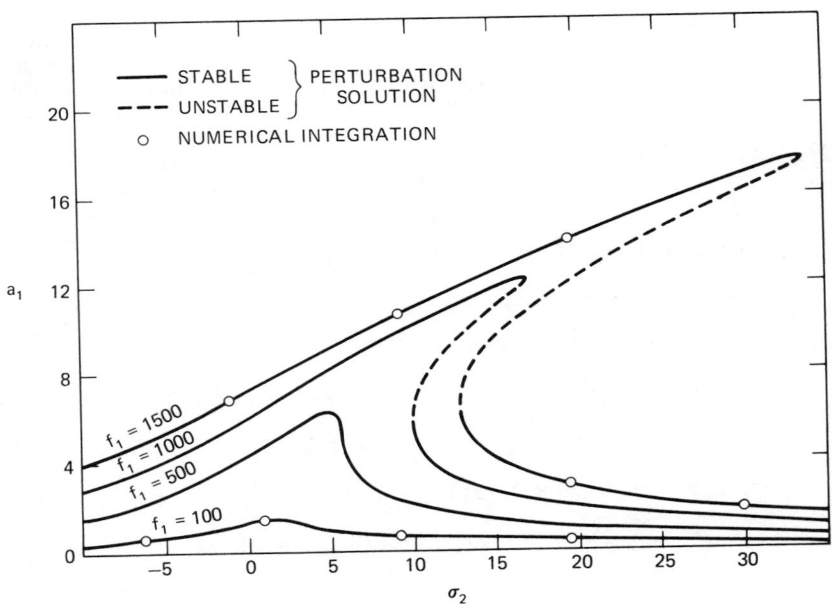

Figure 6-21. Frequency–response curves; $\omega_2 \simeq 3\omega_1$, $\Omega \simeq \omega_1$.

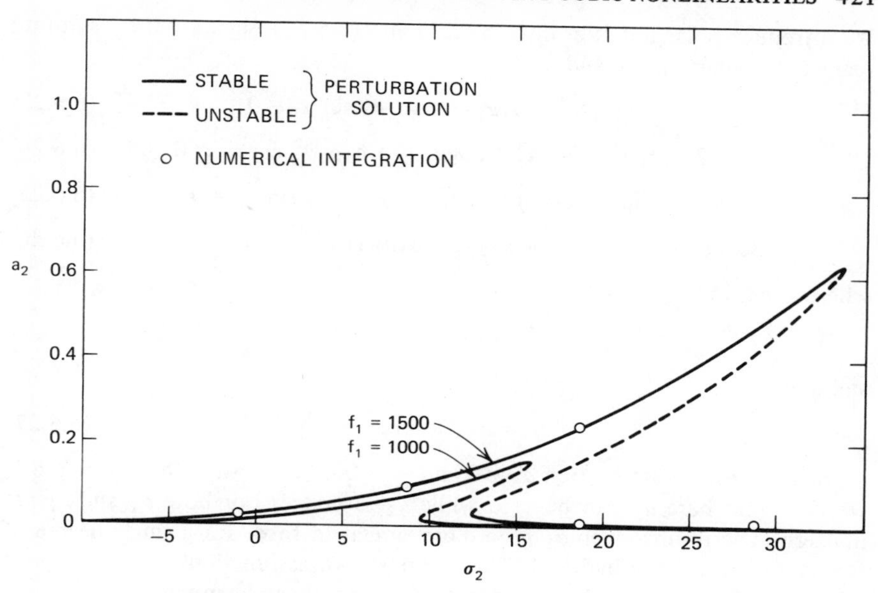

Figure 6-22. Frequency–response curves; $\omega_2 \simeq 3\omega_1$, $\Omega \simeq \omega_1$.

of the response of systems having many degrees of freedom; that is, an excitation of the fundamental mode does not produce significant responses in the higher modes even though they are coupled to the fundamental mode through an internal resonance. On the other hand, it is shown in the next section that the excitation of a high mode may produce significant responses in the low modes, especially the fundamental mode, if they are coupled with the excited mode through an internal resonance.

6.6.2. THE CASE OF Ω NEAR ω_2

Here we introduce a second detuning parameter according to

$$\Omega = \omega_2 + \epsilon^2 \sigma_2 \tag{6.6.20}$$

Equation (6.6.8) is still used to define σ_1. It follows from (6.6.5) and (6.6.6) that secular terms are eliminated from u_{12} and u_{22} if

$$-2i\omega_1(A_1' + \mu_1 A_1) + 3\alpha_1 A_1^2 \overline{A}_1 + 2\alpha_3 A_1 A_2 \overline{A}_2 + \alpha_2 A_2 \overline{A}_1^2 \exp(i\sigma_1 T_2) = 0 \tag{6.6.21}$$

$$-2i\omega_2(A_2' + \mu_2 A_2) + 3\alpha_8 A_2^2 \overline{A}_2 + 2\alpha_6 A_2 A_1 \overline{A}_1 + \alpha_5 A_1^3 \exp(-i\sigma_1 T_2) + \tfrac{1}{2} f_2 \exp(i\sigma_2 T_2 + \tau_2) = 0 \tag{6.6.22}$$

We introduce polar notation into (6.6.21) and (6.6.22), separate the result into real and imaginary parts, and obtain

$$8\omega_1(a_1' + \mu_1 a_1) - \alpha_2 a_2 a_1^2 \sin \gamma_1 = 0 \tag{6.6.23}$$

$$8\omega_1 a_1 \theta_1' + 3\alpha_1 a_1^3 + 2\alpha_3 a_1 a_2^2 + \alpha_2 a_2 a_1^2 \cos \gamma_1 = 0 \tag{6.6.24}$$

$$8\omega_2(a_2' + \mu_2 a_2) + \alpha_5 a_1^3 \sin \gamma_1 - 4f_2 \sin \gamma_2 = 0 \tag{6.6.25}$$

$$8\omega_2 a_2 \theta_2' + 3\alpha_8 a_2^3 + 2\alpha_6 a_2 a_1^2 + \alpha_5 a_1^3 \cos \gamma_1 + 4f_2 \cos \gamma_2 = 0 \tag{6.6.26}$$

where as before

$$\gamma_1 = \sigma_1 T_2 - 3\theta_1 + \theta_2$$

and now

$$\gamma_2 = \sigma_2 T_2 - \theta_2 + \tau_2 \tag{6.6.27}$$

For the steady-state motion $a_n' = 0$ and $\gamma_n' = 0$. In contrast with Section 6.6.1, we note that here a_1 can be zero while a_2 is nonzero or both a_1 and a_2 are nonzero. The results when a_1 is zero are similar to those for a single degree of freedom obtained in Chapter 4. The a_n are shown as functions of σ_2 in Figures 6-23 and 6-24. We note that when a_1 is not zero, it can be much larger than a_2. This means that the internal resonance provides the mechanism for transferring energy down from a high to a low mode.

Figure 6-24 shows how complicated the solution can be when the high mode is excited, that is, $\Omega \approx \omega_2$. However many portions of these curves (dashed parts)

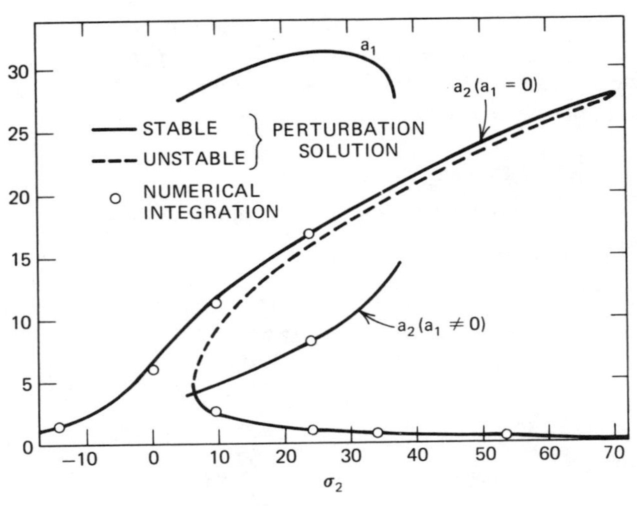

Figure 6-23. Frequency-response curves; $\omega_2 \simeq 3\omega_1$, $\Omega \simeq \omega_2$.

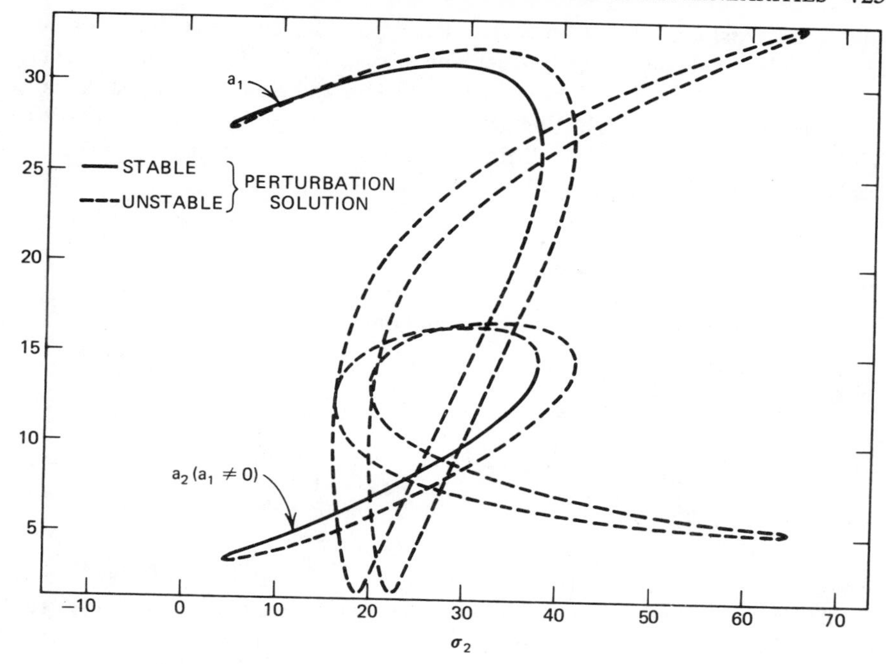

Figure 6-24. Frequency–response curves; $\omega_2 \simeq 3\omega_1$, $\Omega \simeq \omega_2$.

correspond to unstable solutions, and hence they cannot be realized in practice. Figure 6-23 shows only the stable portions (i.e., the possible responses) as a function of σ_2. We note that although the second mode is excited (i.e., $\Omega \approx \omega_2$) the amplitude of the fundamental mode can be as much as five times the amplitude of the excited mode, depending on the detuning. However in contrast with the case of quadratic nonlinearity, no saturation phenomenon exists in a system with a cubic nonlinearity.

There arises the question of what effect changes in the system parameters (e.g., damping, excitation amplitude and frequency, strength of the nonlinearity) have on the response. Specifically, can the response of the system be forced from one steady-state solution to another by sufficiently large disturbances? Thus the problem is one of establishing the region of stability of a solution relative to disturbances, that is, the degree of stability of a given solution (Holzer, 1974). Although the convenience of the phase-plane analysis is lost in the case of more than two equations, one can still study the behavior of the integral curves by using projections onto suitable two-coordinate planes (e.g., Subramanian and Kronauer, 1972, 1973; Blaquière, 1966). In conservative mechanical systems, where the original governing equations are derivable from an integral formulation, there exist convention functions such as potential energies (Thompson and Hunt, 1973), Lagrangians (Kronauer and Musa, 1966a; Musa and Kronauer, 1968) and Hamiltonians (Markeev, 1968; Nayfeh and Kamel 1970b; Tsel'man, 1970, 1971; Nayfeh, 1971b; Dysthe and Gudmestad, 1975; Novosiolov, 1976). The inverse problem is a more difficult one, that is, given a system of equations, find an integral of the motion.

6.7. Parametrically Excited Systems

We consider the following system of equations:

$$\ddot{u}_1 + \omega_1^2 u_1 + \epsilon[2\cos(\Omega t)(f_{11}u_1 + f_{12}u_2) - (\alpha_1 u_1^3 + \alpha_2 u_2 u_1^2 + \alpha_3 u_1 u_2^2 + \alpha_4 u_2^3 + 2\mu_1 \dot{u}_1] = 0 \quad (6.7.1)$$

$$\ddot{u}_2 + \omega_2^2 u_2 + \epsilon[2\cos(\Omega t)(f_{21}u_1 + f_{22}u_2) - (\alpha_5 u_1^3 + \alpha_6 u_1^2 u_2 + \alpha_7 u_1 u_2^2 + \alpha_8 u_2^3) + 2\mu_2 \dot{u}_2] \quad (6.7.2)$$

Among other places, equations of this type arise in the study of finite-amplitude oscillations of hinged-clamped columns subjected to a harmonic load. The presence of the terms proportional to α_2 and α_5 produce an internal resonance when $\omega_2 \approx 3\omega_1$.

Tso and Asmis (1974) considered a similar system. However they did not include the effects of an internal resonance. Their results, which were obtained using the method of averaging, do not apply to this situation.

Following the familiar procedure, we seek an approximate solution by using the method of multiple scales. We assume

$$u_1(t;\epsilon) = u_{10}(T_0, T_1) + \epsilon u_{11}(T_0, T_1) + \cdots \quad (6.7.3)$$

$$u_2(t;\epsilon) = u_{20}(T_0, T_1) + \epsilon u_{21}(T_0, T_1) + \cdots \quad (6.7.4)$$

Substituting (6.7.3) and (6.7.4) into (6.7.1) and (6.7.2) leads to

$$u_{10} = A_1(T_1)\exp(i\omega_1 T_0) + cc \quad (6.7.5)$$

$$u_{20} = A_2(T_1)\exp(i\omega_2 T_0) + cc \quad (6.7.6)$$

$$D_0^2 u_{11} + \omega_1^2 u_{11} = [-2i\omega_1(A_1' + \mu_1 A_1) + 3\alpha_1 A_1^2 \bar{A}_1 + 2\alpha_3 A_2 \bar{A}_2 A_1]\exp(i\omega_1 T_0) + \alpha_2 A_2 \bar{A}_1^2 \exp[i(\omega_2 - 2\omega_1)T_0] - f_{11}A_1\exp[i(\Omega + \omega_1)T_0] - f_{11}\bar{A}_1\exp[i(\Omega - \omega_1)T_0]$$

$$- f_{12}A_2 \exp [i(\Omega + \omega_2) T_0] - f_{12}\bar{A}_2 \exp [i(\Omega - \omega_2) T_0] + cc + NST \tag{6.7.7}$$

$$\begin{aligned} D_0^2 u_{21} + \omega_2^2 u_{21} = {} & [-2i\omega_2(A_2' + \mu_2 A_2) + 3\alpha_8 A_2^2 \bar{A}_2 + 2\alpha_6 A_1 \bar{A}_1 A_2] \exp(i\omega_2 T_0) \\ & + \alpha_5 A_1^3 \exp(3i\omega_1 T_0)] - f_{21}A_1 \exp [i(\Omega + \omega_1) T_0] \\ & - f_{21}\bar{A}_1 \exp [i(\Omega - \omega_1) T_0] - f_{22}A_2 \exp [i(\Omega + \omega_2) T_0] \\ & - f_{22}\bar{A}_2 \exp [i(\Omega - \omega_2) T_0] + cc \end{aligned} \tag{6.7.8}$$

We introduce a detuning parameter for the internal resonance according to

$$\omega_2 = 3\omega_1 + \epsilon\sigma \tag{6.7.9}$$

For the parametric resonance we note that there are three possibilities: (i) Ω near $2\omega_1 \approx \omega_2 - \omega_1$, (ii) Ω near $2\omega_2$, and (iii) Ω near $\omega_2 + \omega_1$. Each case is considered separately next.

6.7.1. THE CASE Ω NEAR $2\omega_1$

In this case we introduce a detuning parameter ρ for the parametric resonance according to

$$\Omega = 2\omega_1 + \epsilon\rho \tag{6.7.10}$$

Then it follows that secular terms are eliminated from u_{11} if

$$8\omega_1(a_1' + \mu_1 a_1) - \alpha_2 a_1^2 a_2 \sin\gamma_1 + 4f_{11}a_1 \sin\gamma_2 = 0 \tag{6.7.11}$$

$$8\omega_1 a_1 \beta_1' + 3\alpha_1 a_1^3 + 2\alpha_3 a_1 a_2^2 + \alpha_2 a_1^2 a_2 \cos\gamma_1 - 4f_{11}a_1 \cos\gamma_2 = 0 \tag{6.7.12}$$

where

$$A_n = \tfrac{1}{2} a_n(T_1) \exp [i\beta_n(T_1)] \tag{6.7.13}$$

$$\gamma_1 = \sigma T_1 + \beta_2 - 3\beta_1 \quad \text{and} \quad \gamma_2 = \rho T_1 - 2\beta_1 \tag{6.7.14}$$

Secular terms are eliminated from u_{21} if

$$8\omega_2(a_2' + \mu_2 a_2) + \alpha_5 a_1^3 \sin\gamma_1 + 4f_{21}a_1 \sin(\gamma_2 - \gamma_1) = 0 \tag{6.7.15}$$

$$8\omega_2 a_2 \beta_2' + 2\alpha_6 a_1^2 a_2 + 3\alpha_8 a_2^3 + \alpha_5 a_1^3 \sin\gamma_1 - 4f_{21}a_1 \cos(\gamma_2 - \gamma_1) = 0 \tag{6.7.16}$$

For a steady-state solution, a_1, a_2, γ_1, and γ_2 are constants. Equations (6.7.14) can be used to eliminate γ_1' and γ_2'. The result is

$$a_1 [8\omega_1\mu_1 - \alpha_2 a_1 a_2 \sin\gamma_1 + 4f_{11} \sin\gamma_2] = 0 \tag{6.7.17}$$

$$a_1 [4\omega_1\rho + 3\alpha_1 a_1^2 + 2\alpha_3 a_2^2 + \alpha_2 a_1 a_2 \cos\gamma_1 - 4f_{11} \cos\gamma_2] = 0 \tag{6.7.18}$$

$$8\omega_2\mu_2 a_2 + \alpha_5 a_1^3 \sin\gamma_1 + 4f_{21} a_1 \sin(\gamma_2 - \gamma_1) = 0 \tag{6.7.19}$$

$$8\omega_2(\tfrac{3}{2}\rho - \sigma) a_2 + 2\alpha_6 a_1^2 a_2 + 3\alpha_8 a_2^3 + \alpha_5 a_1^3 \cos\gamma_1 - 4f_{21} a_1 \cos(\gamma_2 - \gamma_1) = 0 \tag{6.7.20}$$

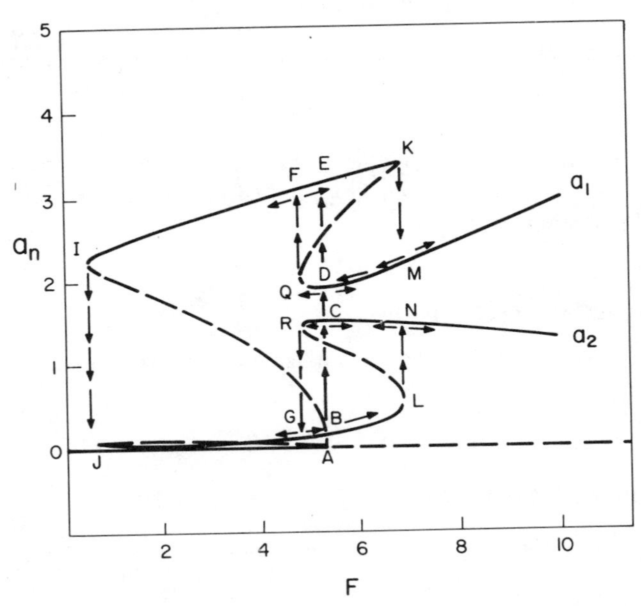

Figure 6-25. Amplitudes of the modes versus amplitude of the excitation; $\omega_2 \simeq 3\omega_1$, $\Omega \simeq 2\omega_1$.

We note that there are two possibilities: either a_1 and a_2 are zero, or neither one is zero. In the case of the latter, these equations can be solved for a_1, a_2, γ_1, and γ_2 using a Newton-Raphson technique. The characteristics of the various singular points (i.e., stable or unstable) can be determined in the usual manner.

In Figure 6-25 some typical results show a_1 and a_2 as functions of the amplitude of the excitation F. For columns, the f_{nm} are proportional to a single constant, F. For more details the reader is referred to the paper by Tezak, Mook, and Nayfeh (1978).

We note the multiplicity of possible jumps; jumps are indicated by the arrows. We can trace the histories of a_1 and a_2 as F slowly increases from zero to a large value and then slowly returns to zero. Initially both a_1 and a_2 are zero, and they remain zero until point A is reached. At this point the trivial solution becomes unstable. Then either a_1 jumps up to point E and a_2 jumps up to point B, or a_1 jumps up to point D and a_2 jumps up to point C. In the first instance a_1 follows curve $IFEK$ and a_2 follows curve GBL. Then a_1 jumps down from point K to point M, and a_2 jumps up from point L to point N. In the second instance a_1 follows the branch through points D and M, and a_2 follows the branch through points C and N. Upon returning, either a_1 jumps up from point Q to point F and a_2 jumps down from point R to point G, or both a_1 and a_2 jump back to zero. If a_1 jumps up, then a_1 moves along curve $KEFI$ while a_2 moves along the curve LBG until both jump back to zero at point J.

6.7.2. THE CASE Ω NEAR $2\omega_2$

In this case we introduce a detuning parameter ρ for the parametric resonance:

$$\Omega = 2\omega_2 + \epsilon\rho \tag{6.7.21}$$

For the steady-state motion we find

$$a_1(8\omega_1\mu_1 - \alpha_2 a_1 a_2 \sin\gamma_1) = 0 \tag{6.7.22}$$

$$a_1\left[\tfrac{8}{3}\omega_1(\sigma + \tfrac{1}{2}\rho) + 3\alpha_1 a_1^2 + 2\alpha_3 a_2^2 + \alpha_2 a_1 a_2 \cos\gamma_1\right] = 0 \tag{6.7.23}$$

$$a_2(8\omega_2\mu_2 + 4f_{22}\sin\gamma_2) + \alpha_5 a_1^3 \sin\gamma_1 = 0 \tag{6.7.24}$$

$$a_2(4\omega_2\rho + 2\alpha_6 a_1^2 + 3\alpha_8 a_2^2 - 4f_{22}\cos\gamma_2) + \alpha_5 a_1^3 \cos\gamma_1 = 0 \tag{6.7.25}$$

where

$$\gamma_2 = \rho T_1 - 2\beta_2 \tag{6.7.26}$$

and as before

$$\gamma_1 = \sigma T_1 - 3\beta_1 + \beta_2 \tag{6.7.27}$$

In contrast with the case when Ω is near $2\omega_1$, it appears that there are two possibilities: either a_1 is zero and a_2 is nonzero, or both a_1 and a_2 are nonzero. For the first instance the problem reduces to one having a single degree of freedom, which was discussed in Section 5.7.3. For the second instance some typical results are given in Figure 6-26. We note that over a fairly wide range, a_1 is much larger than a_2, in spite of the fact that only u_2 is directly excited by the parametric excitation. The internal resonance alone is responsible for this phenomenon; it follows from (6.7.22) that if α_2 is zero (i.e., no internal resonance), then so must a_1 be zero.

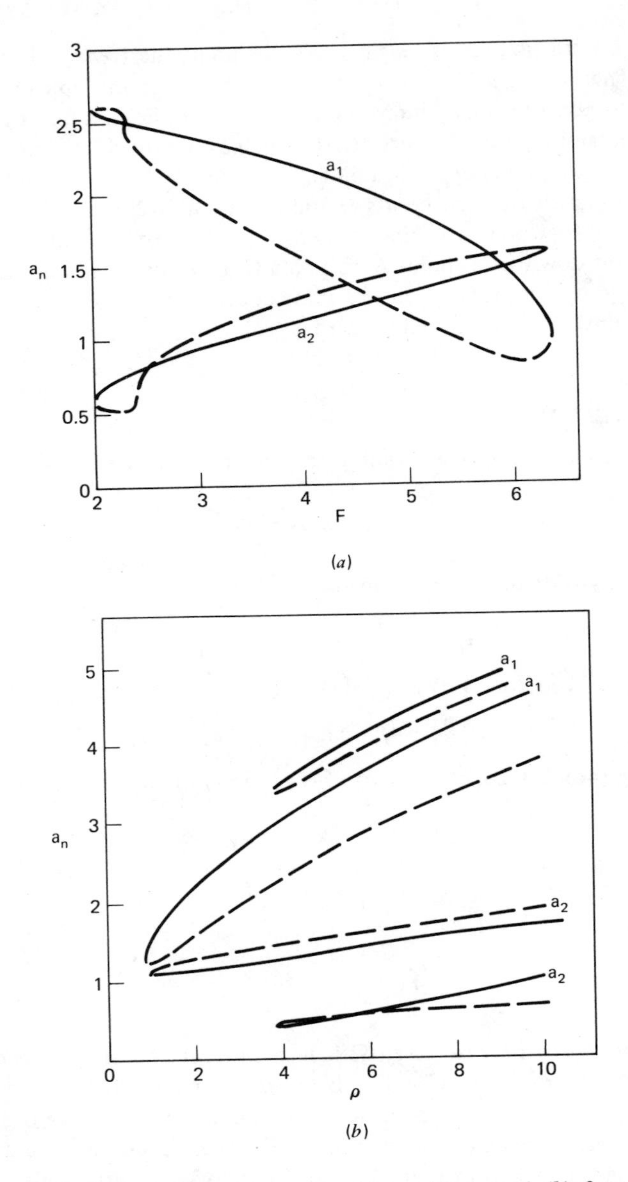

Figure 6-26. Amplitudes of the modes versus (*a*) amplitude and (*b*) frequency of the excitation; $\omega_2 \simeq 3\omega_1$, $\Omega \simeq 2\omega_2$.

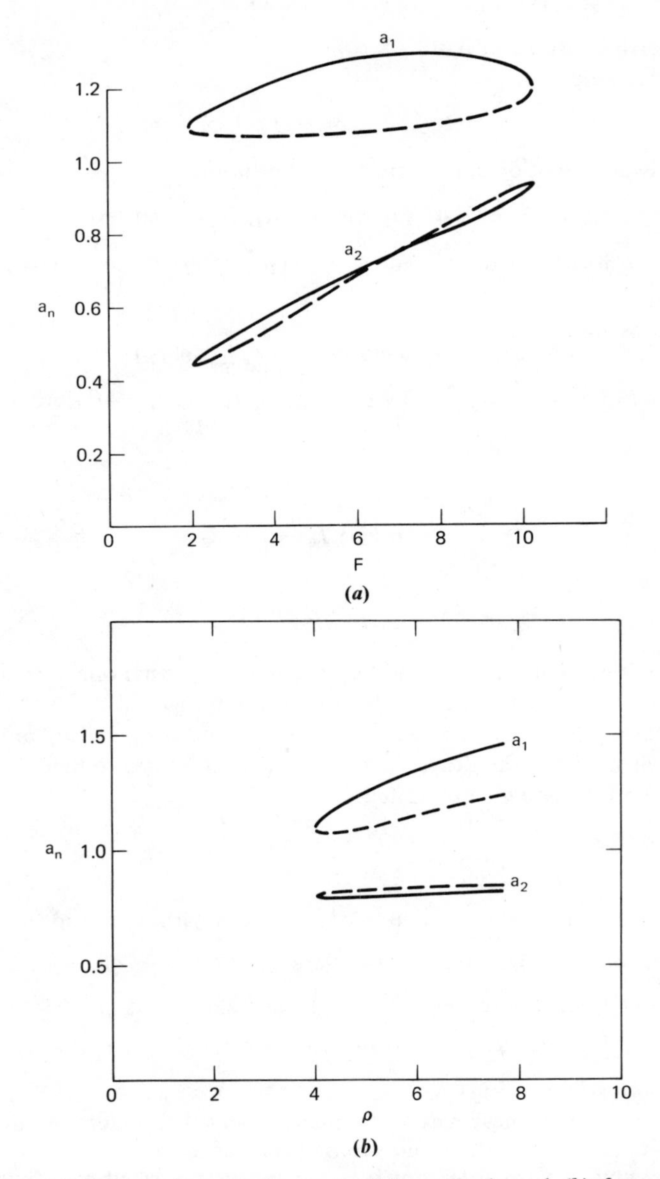

Figure 6-27. Amplitudes of the modes versus (*a*) amplitude and (*b*) frequency of the excitation; $\omega_2 \simeq 3\omega_1$, $\Omega \simeq \omega_1 + \omega_2$.

6.7.3. THE CASE Ω NEAR $\omega_1 + \omega_2$

In this case we put

$$\Omega = \omega_2 + \omega_1 + \epsilon(\rho + \sigma) \tag{6.7.28}$$

Then steady-state motion corresponds to the solution of

$$8\omega_1\mu_1 a_1 + 4f_{12}a_2 \sin\gamma_2 - \alpha_2 a_1^2 a_2 \sin\gamma_1 = 0 \tag{6.7.29}$$

$$2\omega_1 a_1(\rho + 2\sigma) + 3\alpha_1 a_1^3 + 2\alpha_3 a_1 a_2^2 - 4f_{12}a_2 \cos\gamma_2 + \alpha_2 a_1^2 a_2 \cos\gamma_1 = 0 \tag{6.7.30}$$

$$8\omega_2\mu_2 a_2 + a_1(\alpha_5 a_1^2 \sin\gamma_1 + 4f_{21}\sin\gamma_2) = 0 \tag{6.7.31}$$

$$a_2[2\omega_2(3\rho + 2\sigma) + 2\alpha_6 a_1^2 + 3\alpha_8 a_2^2] + a_1(\alpha_5 a_1^2 \cos\gamma_1 - 4f_{21}\cos\gamma_2) = 0 \tag{6.7.32}$$

where

$$\gamma_2 = (\rho + \sigma)T_1 - \beta_1 - \beta_2 \tag{6.7.33}$$

and as before

$$\gamma_1 = \sigma T_1 + \beta_2 - 3\beta_1 \tag{6.7.34}$$

Two possibilities exist: either a_1 and a_2 are zero, or neither one is zero. Typical results for the case when neither is zero are shown in Figure 6-27.

In summary, when Ω is near $2\omega_1$ or $\omega_1 + \omega_2$, both u_1 and u_2 are excited. However when Ω is near $2\omega_2$, there are two possibilities: either u_2 alone is excited, or both u_1 and u_2 are excited.

Exercises

6.1. Consider the double pendulum of Figure 5-3 without the springs.

(a) Show that the governing equations are

$$\ddot{\theta}_1 + gl_1^{-1}\sin\theta_1 + \alpha\ddot{\theta}_2\cos(\theta_2 - \theta_1) - \alpha\dot{\theta}_2^2\sin(\theta_2 - \theta_1) = 0$$

$$\ddot{\theta}_2 + gl_2^{-1}\sin\theta_2 + l_1 l_2^{-1}\ddot{\theta}_1\cos(\theta_2 - \theta_1) + l_1 l_2^{-1}\dot{\theta}_1^2\sin(\theta_2 - \theta_1) = 0$$

where $\alpha = m_2 l_2 l_1^{-1}(m_1 + m_2)^{-1}$.

(b) Determine the linear natural frequencies and then determine the system parameters that yield a three-to-one internal resonance.

(c) Determine a first-order uniform expansion for small but finite amplitudes including the cases of internal resonances.

6.2. A uniform rod of length l_2 and mass m is hanging from a massless chord as shown in Figure 6-28.

Figure 6-28. Exercise 6.2.

Figure 6-29. Exercise 6.3.

(a) Show that the governing equations are

$$l_1\ddot{\theta}_1 + \tfrac{1}{2}l_2\ddot{\theta}_2 \cos(\theta_2 - \theta_1) - \tfrac{1}{2}l_2\dot{\theta}_2^2 \sin(\theta_2 - \theta_1) + g\sin\theta_1 = 0$$

$$\tfrac{1}{3}l_2\ddot{\theta}_2 + \tfrac{1}{2}l_1\ddot{\theta}_1 \cos(\theta_2 - \theta_1) + \tfrac{1}{2}l_1\dot{\theta}_1^2 \sin(\theta_2 - \theta_1) + \tfrac{1}{2}g\sin\theta_2 = 0$$

(b) Determine the linear natural frequencies and then determine all possible resonances to first order.

(c) Choose l_2/l_1 to produce a three-to-one internal resonance and determine for this case a uniform first-order expansion for small but finite amplitudes.

6.3. A mass m is attached to a massless rod which is attached in turn to the rim of a wheel which is free to rotate about its center as shown in Figure 6-29.

(a) Show that the equations of motion are

$$(R^2 + \rho^2)\ddot{\theta}_1 + Rr\ddot{\theta}_2 \cos(\theta_2 - \theta_1) - Rr\dot{\theta}_2^2 \sin(\theta_2 - \theta_1) + gR\sin\theta_1 = 0$$

$$r\ddot{\theta}_2 + R\ddot{\theta}_1 \cos(\theta_2 - \theta_1) + R\dot{\theta}_1^2 \sin(\theta_2 - \theta_1) + g\sin\theta_2 = 0$$

where $I = m\rho^2$ is the moment of inertia of the wheel.

(b) Determine the linear natural frequencies. What are the conditions for modal coupling to first order?

(c) Choose r, R, and ρ so that the system possesses a three-to-one internal resonance and determine for this case a first-order uniform expansion for small but finite-amplitude motions.

6.4. A particle of mass m_1 is attached to a light rigid rod of length l which is free to rotate in the vertical plane as shown in Figure 6-30. A bead of mass m_2 is free to slide along the smooth rod under the action of the spring.

(a) Show that the governing equations are

$$\ddot{u} + \omega_1^2 u - u\dot{\theta}^2 + \omega_2^2(1 - \cos\theta) = \omega_2^2 u_e$$

$$(1 + mu^2)\ddot{\theta} + (1 + mu)\omega_2^2 \sin\theta + 2\,mu\dot{u}\dot{\theta} = 0$$

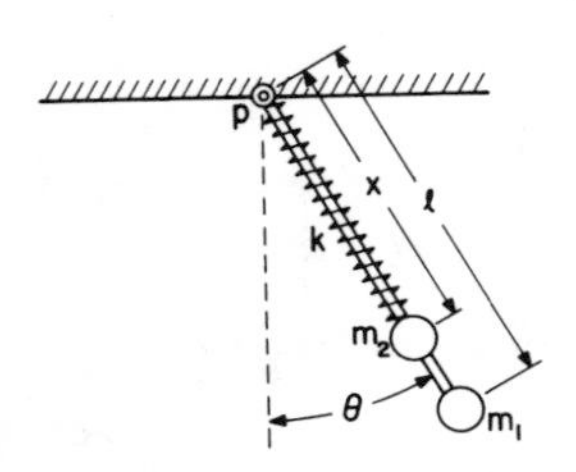

Figure 6-30. Exercise 6.4.

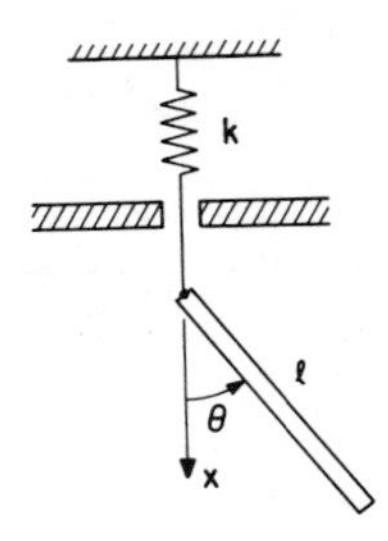

Figure 6-31. Exercise 6.5.

where $\omega_1^2 = k/m_2$, $\omega_2^2 = g/l$, $m = m_2/m_1$, $u = x/l$, and u_e is the equilibrium position.

(b) Determine a first-order uniform expansion when $\omega_1 \approx 2\omega_2$.

(c) What other resonances exist to first order?

6.5. A uniform rod of length l and mass m is hanging from a spring that is constrained to move vertically as shown in Figure 6-31.

(a) Show that the governing equations of motion are

$$\ddot{u} + \omega_1^2 u = \tfrac{1}{2}\ddot{\theta}\sin\theta + \tfrac{1}{2}\dot{\theta}^2\cos\theta$$

$$\ddot{\theta} + \omega_2^2\theta = \tfrac{3}{2}\ddot{u}\sin\theta$$

where $u = (x - x_e)/l$, $\omega_1^2 = k/m$, $\omega_2^2 = 3g/2l$, and x_e is the equilibrium position.

(b) Determine a first-order uniform expansion for small but finite amplitudes when $\omega_1 \approx 2\omega_2$.

(c) If one carries out the solution to second order, would the internal resonance $\omega_1 \approx 3\omega_2$ appear?

6.6. A rigid beam is supported by springs having the constants k_1 and k_2, the system is constrained to remain in the vertical plane of the figure, and the center of gravity G can move only vertically as shown in Figure 6-32.

(a) Show that the governing equations of motion are

$$m\ddot{x} + (k_1 + k_2)x + (k_1 l_1 - k_2 l_2)\sin\theta = -mg$$

$$I\ddot{\theta} + (k_1 l_1 - k_2 l_2)x\cos\theta + \tfrac{1}{2}(k_1 l_1^2 + k_2 l_2^2)\sin 2\theta = 0$$

where I is the moment of inertia of the beam and m is its mass.

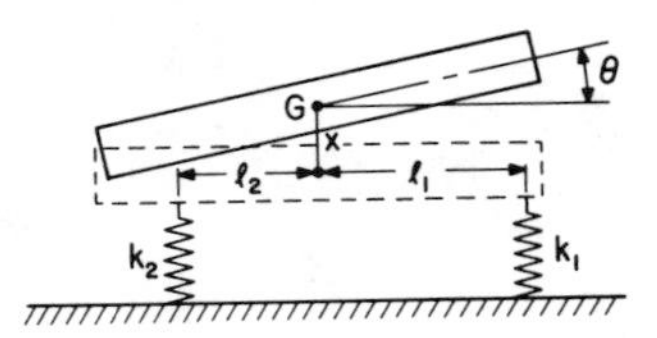

Figure 6-32. Exercise 6.6.

(b) Determine the linear natural frequencies and then determine all possible resonances.

(c) Determine a first-order uniform expansion for small but finite amplitudes including the case of modal coupling.

6.7. Consider the system of equations

$$m_1\ddot{u}_1 + k_1 u_1^n + \alpha_1(u_1 - u_2)^n = 0$$

$$m_2\ddot{u}_2 + k_2 u_2^n - \alpha_1(u_1 - u_2)^n = 0$$

where n is an odd integer.

(a) Show that this system possesses solutions related by $u_1 = cu_2$, where c is a constant defined by

$$m_2[k_1 c^n + \alpha_1(c-1)^n] = m_1 c[k_2 - \alpha_1(c-1)^n]$$

(b) Hence show that u_2 is governed by

$$\ddot{u}_2 + [k_2 - \alpha_1(c-1)^n]\, m_2^{-1} u_2^n = 0$$

Find the exact solution for this equation in quadratures and then find u_1.

6.8. In analyzing the motion of a rotating shaft with a rotor, Yamamoto (1957) encountered the following equations:

$$\ddot{u}_1 + I\Omega\dot{u}_2 + u_1 = -2\epsilon\mu_1\dot{u}_1 - \Omega^2 \cos\Omega t$$

$$\ddot{u}_2 - I\Omega\dot{u}_1 + u_2 = -2\epsilon\mu_2\dot{u}_2 - \epsilon\Omega^2 u_2^3$$

(a) Determine the linear, undamped natural frequencies ω_1 and ω_2.
(b) Determine the equations describing the amplitudes and the phases when

(i) $\omega_1 + \omega_2 = 2\Omega + \epsilon\sigma$
(ii) $3\omega_1 = \Omega + \epsilon\sigma_1$ and $\omega_2 = 3\Omega + \epsilon\sigma_2$

(c) Determine the steady-state oscillations and their stability.

6.9. In analyzing the nonlinear vibrations of a buckled beam under harmonic excitation, Mettler and Weidenhammer (1962) and Tseng and Dugundji (1971) encountered a system of equations having the form

$$\ddot{u}_1 + 2\mu_1\dot{u}_1 + \omega_1^2 u_1 + \tfrac{3}{2}\omega_1^2 u_1^2 + \tfrac{1}{2}\omega_1^2 u_1^3 + \alpha_1\omega_1^2(u_1 + 1)u_2^2 = K\cos\Omega t$$

$$\ddot{u}_2 + 2\mu_2\dot{u}_2 + \omega_2^2 u_2 + \alpha_2 u_2^3 + \alpha_3(u_1^2 + 2u_1)u_2 = 0$$

Determine first-order uniform solutions for small but finite amplitudes when

(a) $\Omega \approx \omega_1$ and ω_2 is away from $\frac{1}{2}\omega_1$
(b) $\Omega \approx \omega_1$ and $\omega_2 \approx 2\omega_1$
(c) $\Omega \approx \frac{1}{2}\omega_1$ and $\omega_2 \approx 2\omega_1$
(d) $\Omega \approx 2\omega_1$ and $\omega_2 \approx 2\omega_1$

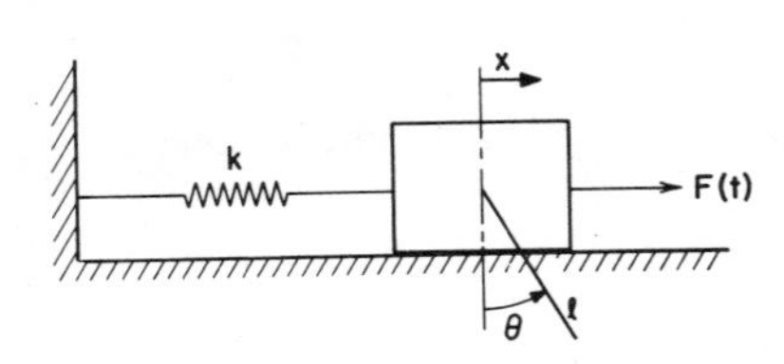

Figure 6-33. Exercise 6.10.

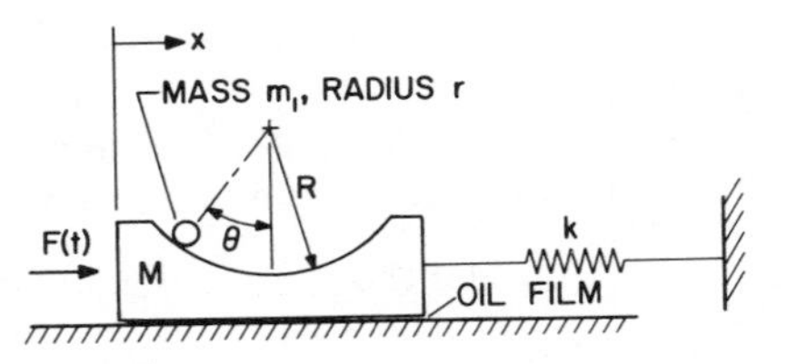

Figure 6-34. Exercise 6.11.

6.10. A uniform rod of length l and mass m is hanging from a cart of negligible mass whose motion is constrained by the spring as shown in Figure 6-33.

(a) Show that the governing equations are

$$\ddot{u} + \omega_1^2 u + \tfrac{1}{2}\ddot{\theta}\cos\theta - \tfrac{1}{2}\dot{\theta}^2 \sin\theta = F(t)$$

$$\ddot{\theta} + \omega_2^2 \sin\theta + \tfrac{3}{2}\ddot{u}\cos\theta = 0$$

where $u = x/l$, $\omega_1^2 = k/m$, and $\omega_2^2 = 3g/2l$.

(b) When $F(t) = 0$, determine a first-order uniform expansion for small but finite amplitudes when $\omega_1 \approx 3\omega_2$.

(c) When $F(t) = K \cos \Omega t$, determine first-order uniform expansions for small-but finite-amplitude motions for the following cases in the presence and absence of internal resonances:

(i) $\Omega \approx \omega_1$
(ii) $\Omega \approx \tfrac{1}{3}\omega_1$
(iii) $\Omega \approx 3\omega_1$

(d) Are there any other resonances beside those in (c)?

6.11. The cylinder rolls without slip on the circular surface as shown in Figure 6-34.

(a) Neglecting the friction between the sliding block and the ground, show that the equations of motion are

$$\ddot{u} + m\ddot{\theta}\cos\theta - m\dot{\theta}^2 \sin\theta + \omega_1^2 u = f(t)$$

$$\ddot{\theta} + \omega_2^2 \sin\theta + \tfrac{2}{3}\ddot{u}\cos\theta = 0$$

where $u = x/(R-r)$, $\omega_1^2 = k/(m_1 + M)$, $\omega_2^2 = \tfrac{2}{3}g/(R-r)$, $m = m_1/(m_1 + M)$, and $f(t) = F(t)/[(R-r)(m_1 + M)]$.

(b) Include the viscous effects (assume the friction force has the form $-c\dot{x}$) and obtain the equations of motion.

(c) Show that internal resonances occur to first order when $\omega_1 \approx 3\omega_2$.

(d) If $f(t) = \Sigma_{n=1}^{3} K_n \cos(\Omega_n t + \theta_n)$, determine the steady-state oscillations and their stability in the presence of viscous effects if the following resonances exist only:

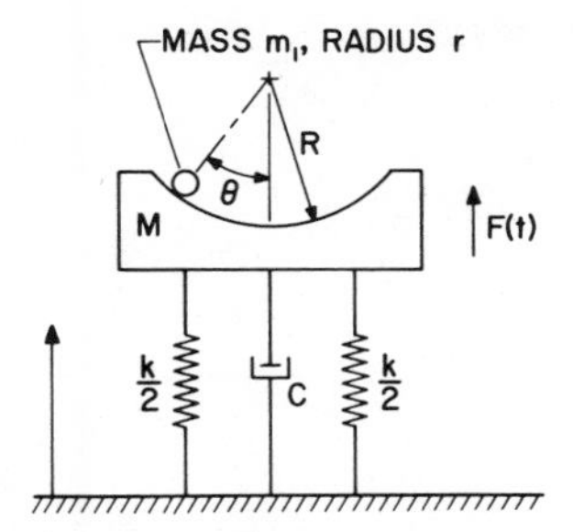

Figure 6-35. Exercise 6.12.

(i) $\Omega_1 \approx \omega_1$
(ii) $\Omega_1 \approx 3\omega_1$
(iii) $\Omega_1 \approx \frac{1}{3}\omega_1$
(iv) $\Omega_1 \approx \omega_1$ and $\omega_1 \approx 3\omega_2$
(v) $\Omega_1 \approx 3\omega_2$ and $\omega_1 \approx 3\omega_2$
(vi) $\Omega_1 \approx \frac{1}{3}\omega_2$ and $\omega_1 \approx 3\omega_2$
(vii) $\Omega_1 + \Omega_2 + \Omega_3 \approx \omega_1$
(viii) $\Omega_1 + \Omega_2 + \Omega_3 \approx \omega_1$ and $\omega_1 \approx 3\omega_2$
(ix) $\Omega_1 + \Omega_2 \approx \omega_1 + \omega_2$
(x) $\Omega_1 + \Omega_2 \approx \omega_1 + \omega_2$ and $\omega_1 \approx 3\omega_2$
(xi) $\Omega_1 + \Omega_2 \approx 2\omega_1$ and $\omega_1 \approx 3\omega_2$
(xii) $\Omega_1 + \Omega_2 \approx 2\omega_1$, $\omega_2 \approx \frac{1}{3}\Omega_2$, and $\omega_1 \approx 3\omega_2$

6.12 The cylinder rolls without slip on the circular surface as shown in Figure 6-35.

(a) Show that the equations of motion are

$$\ddot{u} + \omega_1^2 u + m(\ddot{\theta} \sin\theta + \dot{\theta}^2 \cos\theta) + 2\mu\dot{u} = f(t)$$

$$\ddot{\theta} + (\omega_2^2 + \tfrac{2}{3}\ddot{u}) \sin\theta = 0$$

where μ is a dimensionless coefficient, $u = (y - y_c)/(R - r)$, $m = m_1/(m_1 + M)$, $\omega_1^2 = k/(m_1 + M)$, $\omega_2^2 = \frac{2}{3}g/(R - r)$, and $f(t) = F(t)/[(R - r)(m_1 + M)]$.

(b) Show that internal resonances exist to first-order only when $\omega_1 \approx 3\omega_2$.

(c) If $f(t) = \Sigma_{n=1}^{3} K_n \cos(\Omega_n t + \theta_n)$, determine the steady-state oscillations and their stability in the presence of viscous effects if the following resonances exist only:

(i) $\Omega_1 \approx \omega_1$
(ii) $\Omega_1 \approx 3\omega_1$
(iii) $\Omega_1 \approx \frac{1}{3}\omega_1$
(iv) $\Omega_1 \approx \omega_1$ and $\omega_1 \approx 3\omega_2$
(v) $\Omega_1 \approx 3\omega_2$ and $\omega_1 \approx 3\omega_2$

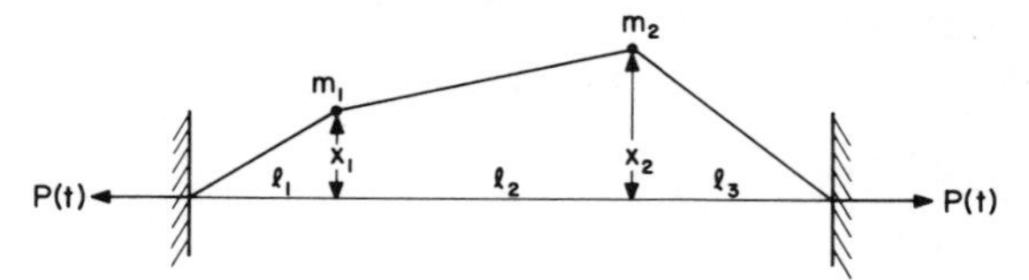

Figure 6-36. Exercise 6.13.

(vi) $\Omega_1 \approx \frac{1}{3}\omega_2$ and $\omega_1 \approx 3\omega_2$
(vii) $\Omega_1 + \Omega_2 + \Omega_3 \approx \omega_1$
(viii) $\Omega_1 + \Omega_2 + \Omega_3 \approx \omega_1$ and $\omega_1 \approx 3\omega_2$
(ix) $\Omega_1 + \Omega_2 \approx \omega_1 + \omega_2$
(x) $\Omega_1 + \Omega_2 \approx \omega_1 + \omega_2$ and $\omega_1 \approx 3\omega_2$
(xi) $\Omega_1 + \Omega_2 \approx 2\omega_1$ and $\omega_1 \approx 3\omega_2$
(xii) $\Omega_1 + \Omega_2 \approx 2\omega_1$, $\omega_2 \approx \frac{1}{3}\Omega_2$, and $\omega_1 \approx 3\omega_2$

6.13. The system shown in Figure 6-36 consists of a stretched wire carrying two particles of masses m_1 and m_2 that are constrained to move normal to the unstretched position.

(a) Show that the governing equations are

$$m_1\ddot{x}_1 + Px_1(l_1^2 + x_1^2)^{-1/2} + P(x_1 - x_2)[l_2^2 + (x_2 - x_1)^2]^{-1/2} = 0$$

$$m_2\ddot{x}_2 + Px_2(l_3^2 + x_2^2)^{-1/2} + P(x_2 - x_1)[l_2^2 + (x_2 - x_1)^2]^{-1/2} = 0$$

(b) When $P = P_0$ = constant, determine the linear natural frequencies ω_1 and ω_2 and determine the system parameters that may yield an internal resonance.

(c) Determine a first-order uniform expansion for small but finite amplitudes when $P = P_0$ = constant including internal resonances.

(d) When $P = P_0(1 + \epsilon \cos \Omega t)$ and $x_n = O(\epsilon^{1/2})$, determine first-order uniform expansions for the cases

(i) $\Omega \approx 2\omega_1$ in the absence of internal resonances
(ii) $\Omega \approx \omega_2 - \omega_1$ in the absence of internal resonances
(iii) $\Omega \approx \omega_2 - \omega_1$ and $\omega_2 \approx 3\omega_1$

6.14. Consider the system shown in Figure 6-2 when the support has the motion $ly(t)$ in the vertical direction.

(a) Show that the equations of motion (6.1.12) become

$$\ddot{u} + \omega_1^2 u - (1 + u)\dot{\theta}^2 + \omega_2^2(1 - \cos\theta) - \ddot{y}\cos\theta = 0$$

$$(1 + u)\ddot{\theta} + 2\dot{u}\dot{\theta} + (\omega_2^2 + \ddot{y})\sin\theta = 0$$

where $u = x/l$, $\omega_1^2 = k/m$, and $\omega_2^2 = g/l$.

(b) When $\ddot{y} \equiv 0$, show that small- but finite-amplitude motions are given by

$$u = a_1 \sin(\omega_1 t + \beta_1) + \cdots$$

$$\theta = a_2 \sin(\omega_2 t + \beta_2)$$

where $\omega_1 \approx 2\omega_2$ and

$$\dot{a}_1 = \tfrac{3}{8}\omega_2^2\omega_1^{-1}a_2^2 \cos\gamma$$
$$\dot{a}_2 = -\tfrac{3}{4}\omega_2 a_1 a_2 \cos\gamma$$
$$a_1\dot{\beta}_1 = -\tfrac{3}{8}\omega_2^2\omega_1^{-2}a_2^2 \sin\gamma$$
$$a_2\dot{\beta}_2 = -\tfrac{3}{4}\omega_2 a_1 a_2 \sin\gamma$$
$$\gamma = \beta_1 - 2\beta_2 + (\omega_1 - 2\omega_2)t$$

(c) When $\ddot{y} = 2\epsilon K \cos \Omega t$, where $\Omega = 2\omega_2 + \epsilon\sigma$ and Ω is away from ω_1, show that

$$u = \epsilon A_1 \exp(i\omega_1 t) + cc + 2K(\omega_1^2 - \Omega^2)^{-1} \cos\Omega t + O(\epsilon^2)$$
$$\theta = \epsilon A_2 \exp(i\omega_2 t) + cc + O(\epsilon^2)$$

where $\dot{A}_1 = 0$ and

$$2i\omega_2\dot{A}_2 + \epsilon K(2\omega_2\Omega - \omega_2^2 + \omega_1^2 - \Omega^2)(\omega_1^2 - \Omega^2)^{-1}\bar{A}_2 \exp(i\epsilon\sigma t) = 0$$

Solve for A_2 and show that the transition curves separating stability from instability of the $\theta = 0$ position are given by

$$K(2\omega_2\Omega - \omega_2^2 + \omega_1^2 - \Omega^2) = \pm\omega_2(\omega_1^2 - \Omega^2)\sigma$$

(d) When $\Omega = \omega_1 + \epsilon\sigma_1$ and $\omega_1 = 2\omega_2 + \epsilon\sigma_2$, let $K = \epsilon k$ and obtain

$$u = \epsilon a_1 \cos(\omega_1 t + \beta_1) + O(\epsilon^2)$$
$$\theta = \epsilon a_2 \cos(\omega_2 t + \beta_2) + O(\epsilon^2)$$

where

$$a_1' = \tfrac{3}{16}\omega_2 a_2^2 \sin\gamma_1 + \frac{k}{\omega_1}\sin\gamma_2$$
$$a_2' = -\tfrac{3}{4}\omega_2 a_1 a_2 \sin\gamma_1$$
$$a_1\beta_1' = \tfrac{3}{16}\omega_2 a_2^2 \cos\gamma_1 - \frac{k}{\omega_1}\cos\gamma_2$$
$$a_2\beta_2' = \tfrac{3}{4}\omega_2 a_1 a_2 \cos\gamma_1$$
$$\gamma_1 = \sigma_2 T_1 + \beta_1 - 2\beta_2, \quad \gamma_2 = \sigma_1 T_1 - \beta_1$$

Determine the steady-state solutions and their stability. Does the saturation phenomenon exist in this case?

6.15. Consider the following equations for the response of a ship that is constrained to pitch and roll only to a multifrequency excitation:

$$\ddot{\phi} + \omega_1^2\phi = 2\phi\theta - 2\mu_1\dot{\phi} + \sum_{n=1}^{N} K_n \cos(\Omega_n t + \theta_n)$$

$$\ddot{\theta} + \omega_2^2 \theta = \phi^2 - 2\mu_2 \dot{\theta} + \sum_{n=1}^{N} M_n \cos (\Omega_n t + \tau_n)$$

(a) Demonstrate the existence of the saturation phenomenon when $\Omega_1 \approx \omega_2$ and $\omega_2 \approx 2\omega_1$ and no other resonances exist.

(b) Demonstrate the possibility of the nonexistence of steady-state periodic motions when $\Omega_1 \approx \omega_1$ and $\omega_2 \approx 2\omega_1$ and no other resonances exist.

(c) Determine the steady-state response and its stability when the following resonances exist only:

(i) $\Omega_1 + \Omega_2 \approx \omega_1$
(ii) $\Omega_1 + \Omega_2 \approx \omega_1$ and $\omega_2 \approx 2\omega_1$
(iii) $\Omega_1 \approx \omega_2 - \omega_1$ and $\Omega_2 \approx 2\omega_1$
(iv) $\Omega_1 \approx \omega_2 - \omega_1$ and $\omega_2 \approx 2\omega_1$
(v) $\Omega_1 \approx 2\omega_1$ and $\omega_2 \approx 2\omega_1$
(vi) $\Omega_1 + \Omega_2 \approx \omega_1$ and $\Omega_3 - \Omega_2 \approx \omega_2$
(vii) $\Omega_1 + \Omega_2 \approx \omega_1$, $\omega_2 \approx 2\omega_1$, and $\Omega_4 - \omega_3 \approx \omega_2$

Van Dooren (1971a) analyzed combination tones of the summed type in a nonlinear damped vibratory system with two degrees of freedom.

6.16. Consider the response of the following system (Rubenfeld, 1977):

$$\ddot{u}_1 + \omega_1^2(\tau) u_1 = \epsilon \alpha_1 u_1 u_2$$

$$\ddot{u}_2 + \omega_2^2(\tau) u_2 = \epsilon \alpha_2 u_1^2$$

where $\tau = \epsilon t$ and $\epsilon << 1$.

(a) Use the generalized version of the method of multiple scales (Nayfeh 1973b, Section 6.4.6) and seek a solution in the form

$$u_1 = u_{10}(\eta_1, \tau) + \epsilon u_{11}(\eta_1, \eta_2, \tau) + \cdots$$

$$u_2 = u_{20}(\eta_2, \tau) + \epsilon u_{21}(\eta_1, \eta_2, \tau) + \cdots$$

where $\dot{\eta}_n = \omega_n$. Show that

$$u_{10} = A_1(\tau) \exp (i\eta_1) + cc$$

$$u_{20} = A_2(\tau) \exp (i\eta_2) + cc$$

(b) Determine the equations describing u_{11} and u_{21}. When $\omega_2 \approx 2\omega_1$, show that

$$2i\omega_1 A_1' + i\omega_1' A_1 - \alpha_1 A_2 \bar{A}_1 \exp [i\sigma(\tau)] = 0$$

$$2i\omega_2 A_2' + i\omega_2' A_2 - \alpha_2 A_1^2 \exp [-i\sigma(\tau)] = 0$$

where $\dot{\sigma} = \omega_2 - 2\omega_1$.

(c) Express A_n in the polar form $\frac{1}{2} a_n \exp (i\beta_n)$ and separate real and imaginary parts. Use the initial conditions $u_1(0) = 1$, $\dot{u}_1(0) = u_2(0) = \dot{u}_2(0) = 0$ and integrate numerically the equations describing the a_n and β_n.

6.17. The forced response of a two-degrees-of-freedom system is governed by (Plotnikova, 1963a)

$$\ddot{x} + x = -4 \cos 2t + \epsilon[\tfrac{4}{3}\dot{y} - (1 - x^2)\dot{x}]$$

$$\ddot{y} + \tfrac{1}{4} y = 5 \cos 2t + \tfrac{4}{3}\epsilon[(1 - x^2)\dot{x} - \dot{y}]$$

Determine a first-order uniform expansion for $\epsilon \ll 1$. Determine the steady-state motion.

6.18. Consider the spherical pendulum of Figure 6-1 when its point of support has the vertical motion $\epsilon l \sin \Omega t$.

(a) Show that the equations of motion become (Hemp and Sethna, 1964)

$$\ddot{\theta} + \left(\frac{g}{l} - \epsilon\Omega^2 \sin \Omega t - \dot{\phi}^2 \cos \theta\right) \sin \theta = 0$$

$$\ddot{\phi} \sin \theta + 2\dot{\phi}\dot{\theta} \cos \theta = 0$$

(b) Show that these equations can be combined to give

$$\ddot{\theta} + \left(\frac{g}{l} + \epsilon\Omega^2 \sin \Omega t\right) \sin \theta - \frac{p^2 \cos \theta}{\sin^3 \theta} = 0$$

(c) When $\epsilon = 0$, determine the equilibrium points.

(d) Determine the effect of the motion of the support on the stability of these equilibrium points.

6.19. The forced response of a two-degree-of-freedom system is governed by (Plotnikova, 1963a)

$$\ddot{x} + x = \epsilon[\lambda_0 \sin t + \alpha_1(1 - x^2)\dot{x} + \alpha_2\dot{y}]$$

$$\ddot{y} + \tfrac{1}{4} y = \epsilon[-\tfrac{1}{4}\lambda_0 \sin t + \alpha_3(1 - x^2)\dot{x} + \alpha_4\dot{y}]$$

Determine a first-order uniform expansion for $\epsilon \ll 1$. Determine the steady-state motion.

6.20. The free oscillations of a self-excited system with two degrees of freedom are governed by

$$\ddot{u}_1 + \omega_1^2 u_1 = \epsilon(\alpha_1 - \alpha_2 u_1^2)\dot{u}_1 + \epsilon\alpha_3 u_2$$

$$\ddot{u}_2 + \omega_2^2 u_2 = \epsilon(\alpha_1 - \alpha_2 u_2^2)\dot{u}_2 + \epsilon\alpha_4 u_1$$

where $\omega_2 = \omega_1 + \epsilon\sigma$.

(a) Show that to the first approximation

$$u_1 = A_1(T_1) \exp(i\omega_1 T_0) + cc, \quad u_2 = A_2(T_1) \exp(i\omega_2 T_0) + cc$$

where

$$2i\omega_1 A_1' = i\omega_1(\alpha_1 - \alpha_2 A_1 \bar{A}_1)A_1 + \alpha_3 A_2 \exp(i\sigma T_1)$$

$$2i\omega_2 A_2' = i\omega_2(\alpha_1 - \alpha_2 A_2 \bar{A}_2)A_2 + \alpha_4 A_1 \exp(-i\sigma T_1)$$

(b) Express the A_n in polar from and determine the steady-state motion.

6.21. Consider the forced response of a three-degree-of-freedom system governed by

$$\ddot{u}_1 + \omega_1^2 u_1 = -2\epsilon\mu_1 \dot{u}_1 - \alpha_1 u_1 u_3 - \tfrac{1}{2}\alpha_2 u_1 \ddot{u}_3 + 2K_1 \cos \Omega t$$
$$\ddot{u}_2 + \omega_2^2 u_2 = -2\epsilon\mu_2 \dot{u}_2 - \alpha_1 u_2 u_3 - \tfrac{1}{2}\alpha_2 u_2 \ddot{u}_3$$
$$\ddot{u}_3 + \omega_3^2 u_3 = -2\epsilon\mu_3 \dot{u}_3 - \tfrac{1}{2}\alpha_1 (u_1^2 + u_2^2) - \tfrac{1}{2}\alpha_2 \frac{d}{dt}(u_1\dot{u}_1 + u_2\dot{u}_2) + 2K_3 \cos \Omega t$$

where $\omega_2 = \omega_1$, $\omega_3 = 2\omega_1 + \epsilon\sigma$, and $\Omega = \omega_s + \epsilon\sigma_s$ for some s.

(a) Show that

$$u_n = \epsilon A_n(T_1) \exp(i\omega_n T_0) + cc + O(\epsilon^2)$$

where

$$2i\omega_1(A_1' + \mu_1 A_1) + (\alpha_1 - \tfrac{1}{2}\omega_3^2\alpha_2)A_3\bar{A}_1 \exp(i\sigma T_1) - k_1\delta_{1s} \exp(i\sigma_1 T_1) = 0$$
$$2i\omega_2(A_2' + \mu_2 A_2) + (\alpha_1 - \tfrac{1}{2}\omega_3^2\alpha_2)A_3\bar{A}_2 \exp(i\sigma T_1) = 0$$
$$2i\omega_3(A_3' + \mu_3 A_3) + (\tfrac{1}{2}\alpha_1 - \omega_1^2\alpha_2)(A_1^2 + A_2^2) \exp(-i\sigma T_1) - k_3\delta_{3s} \cdot \exp(i\sigma_3 T_1) = 0$$

and $K_n = \epsilon k_n$.

(b) When $s = 3$, express the A_n in polar form and determine the steady-state motion. Does saturation take place?

(c) When $s = 1$, express the A_n in polar form and determine the equations describing the long-time behavior. Do steady-state stable periodic solutions exist for all parameters?

6.22. The forced response of a system with three degrees of freedom is governed by

$$\ddot{u}_1 + \omega_1^2 u_1 = -2\epsilon\mu_1 \dot{u}_1 + \epsilon u_2 u_3 + 2\epsilon k_1 \cos \Omega t$$
$$\ddot{u}_2 + \omega_2^2 u_2 = -2\epsilon\mu_2 \dot{u}_2 + \epsilon u_1 u_3 + 2\epsilon k_2 \cos \Omega t$$
$$\ddot{u}_3 + \omega_3^2 u_3 = -2\epsilon\mu_3 \dot{u}_3 + \epsilon u_1 u_2 + 2\epsilon k_3 \cos \Omega t$$

where $\omega_3 = \omega_1 + \omega_2 + \epsilon\sigma$ and $\Omega = \omega_n + \epsilon\sigma_n$ for some n.

(a) Show that

$$u_n = \epsilon A_n \exp(i\omega_n T_0) + cc + O(\epsilon^2)$$

where

$$2i\omega_1(A_1' + \mu_1 A_1) - A_3\bar{A}_2 \exp(i\sigma T_1) - k_1\delta_{1n} \exp(i\sigma_1 T_1) = 0$$
$$2i\omega_2(A_2' + \mu_2 A_2) - A_3\bar{A}_1 \exp(i\sigma T_1) - k_2\delta_{2n} \exp(i\sigma_2 T_1) = 0$$
$$2i\omega_3(A_3' + \mu_3 A_3) - A_2 A_1 \exp(-i\sigma T_1) - k_3\delta_{3n} \exp(i\sigma_3 T_1) = 0$$

(b) When $n = 3$, express the A_n in polar form and determine the long-time behavior of the motion. Does a_3 saturate?

(c) When $n = 1$, express the A_n in polar form and determine the equations describing the long-time behavior. Are there stationary solutions for a_n that are unstable?

6.23. The problem of the stability of the nodal patterns in disk vibration can be reduced to (Williams, 1966)

$$\ddot{u} + u + 8\epsilon(u^3 + uv^2) = 2\epsilon k \cos \omega t$$

$$\ddot{v} + v + 8\epsilon(v^3 + u^2 v) = 0$$

(a) Show that

$$u = a \cos (t + \alpha) + O(\epsilon)$$

$$v = b \cos (t + \beta) + O(\epsilon)$$

where $\epsilon\sigma = \omega - 1$ and

$$a' = -ab^2 \sin \gamma + k \sin \nu$$

$$a\alpha' = 3a^3 + 2ab^2 + ab^2 \cos \gamma - k \cos \nu$$

$$b' = a^2 b \sin \gamma$$

$$b\beta' = 3b^3 + 2a^2 b + a^2 b \cos \gamma$$

$$\gamma = 2\beta - 2\alpha, \quad \nu = \sigma T_1 - \alpha$$

(b) For the steady-state motion, show that either $b = 0$ and $a \neq 0$ or $b \neq 0$ and $a \neq 0$.

(c) When $b = 0$, show that the frequency-response equation is

$$\sigma = 3a^2 - \frac{k}{a} \cos \nu, \quad \nu = 0 \text{ or } \pi$$

Perturb this solution and show that it is unstable for all $a > a_A = (\frac{1}{2} k)^{1/3}$ when $\nu = 0$ and for all $a > a_C = (\frac{1}{6} k)^{1/3}$ when $\nu = \pi$.

(d) When $b \neq 0$, show that the steady-state solution is given by $\gamma = \pi$ and

$$\sigma = 4a^2 - \frac{3k}{2a} \cos \nu$$

$$b^2 = a^2 - \frac{k}{2a} \cos \nu$$

where $\nu = 0$ or π. Perturb this solution and show that it is stable for all $a > a_A$ when $\nu = 0$ and for $a_E < a \leqslant a_D = (\frac{3}{16} k)^{1/3}$ when $\nu = \pi$ where $a_E \approx 0.564 k^{1/3}$.

6.24. Consider the spherical pendulum of Figure 6-1 when it is subjected to the horizontal force $mf(t)$.

(a) Introduce Cartesian coordinates and show that the governing equations can be put in the form

$$\ddot{u} + \omega_0^2 u + u(\dot{u}^2 + \dot{v}^2 + u\ddot{u} + v\ddot{v}) + \tfrac{1}{2}\omega_0^2 u(u^2 + v^2) = f(t)$$

$$\ddot{v} + \omega_0^2 v + v(\dot{u}^2 + \dot{v}^2 + u\ddot{u} + v\ddot{v}) + \tfrac{1}{2}\omega_0^2 v(u^2 + v^2) = 0$$

(b) When $f(t) = 2K \cos \Omega t$, where $\Omega \approx \omega_0$, use the method of multiple scales to determine a first-order uniform expansion for small but finite amplitudes.

(c) Determine the steady-state motions and follow the procedure in the preceding exercise to determine the conditions under which the planar motion becomes unstable and gives way to a nonplanar motion.

6.25. The nonlinear motion of a rolling reentry body with a constant spin rate p is govened by (Murphy, 1968, 1971a, b; Nayfeh and Saric, 1971b)

$$\ddot{\xi} - ip\dot{\xi} + \omega_0^2 \xi = \epsilon K \exp\,[i(pt + \phi_0)] + \gamma\,|\xi^2|\,\xi + \epsilon^2 \mu_1 \dot{\xi} \tag{1}$$

where $\xi = \beta + i\alpha$ is the complex angle of attack. Here ω_0, ϕ_0, γ, μ_n, and χ_n are constants and $\epsilon \ll 1$. The term proportional to K is a result of asymmetries in the body. Similar mathematical problems arise in the analysis of induced nutational motions of spinning spacecrafts (Tsuchiya and Saito, 1974; Tsuchiya, 1975, 1977).

(a) When $\epsilon = 0$, show that the linear natural frequencies are

$$\omega_{1,2} = \tfrac{1}{2}p \pm (\tfrac{1}{4}p^2 + \omega_0^2)^{1/2}$$

(b) When $K \equiv 0$, show that (Nayfeh and Saric, 1971b)

$$\xi = \epsilon a_1(T_2) \exp\,[i\omega_1 T_0 + i\theta_1(T_2)] + \epsilon a_2(T_2) \exp\,[i\omega_2 T_0 + i\theta_2(T_2)] + O(\epsilon^3) \tag{2}$$

where

$$a_1' = (\omega_1 - \omega_2)^{-1}\omega_1 \mu_1 a_1 \tag{3}$$

$$a_2' = -(\omega_1 - \omega_2)^{-1}\omega_2 \mu_1 a_2 \tag{4}$$

$$(\omega_1 - \omega_2)(\theta_1', -\theta_2') = -\gamma(a_1^2 + 2a_2^2, 2a_1^2 + a_2^2) = (\Lambda_3, \Lambda_4) \tag{5}$$

Discuss the long-time behavior of the motion.

(c) When $K = \epsilon^2 k$ and $p = \omega_1 + \epsilon^2 \sigma$, show that (Nayfeh and Saric, 1971b), to the first approximation, ξ is given by (2) where

$$a_1' = \Lambda_1 + k(\omega_1 - \omega_2)^{-1} \sin(\sigma T_2 - \theta_1 + \phi_0) \tag{6}$$

$$(\omega_1 - \omega_2)\theta_1' = \Lambda_3 - k a_1^{-1} \cos(\sigma T_2 - \theta_1 + \phi_0) \tag{7}$$

and a_2 and θ_2 are given by (4) and (5). Determine the steady-state oscillations and their stability.

(d) When $K = O(1)$, determine first-order uniform expansions (Nayfeh and Saric, 1971b) for the cases

(i) $p \approx 0$
(ii) $p \approx 2\omega_1 - \omega_2$. This is the case of subharmonic resonance; it was studied also by Murphy (1973) using the method of substitution.
(iii) nonresonant case

6.26. The motion of a rolling reentry body descending through the atmosphere is governed by

$$\ddot{\xi} - iI_x I^{-1} p\dot{\xi} + \omega_0^2 \xi = \epsilon K \exp [i(\phi + \phi_0)] + \gamma |\xi|^2 \xi + \epsilon^2 \mu_1 \dot{\xi}$$

$$\dot{\phi} = p$$

$$\dot{p} = \epsilon^2 \nu_0 + \epsilon^2 \nu_2 p + \epsilon \nu_1 \operatorname{Imag}[\xi \exp(-i\phi)]$$

where I_x and I are the longitudinal and transverse moments of inertia, $\xi = \beta + i\alpha$ is the complex angle of attack, and ϕ_0 is a constant. The parameters K, γ, μ_1, and ν_n are functions of the slow time scale $T_2 = \epsilon^2 t$. This system can be considered a system with a nonideal source of energy because the excitation depends on the response, and vice versa.

(a) When $K = \epsilon^2 k$ and $p \approx \omega_1$, seek a uniform expansion in the form (Nayfeh and Saric, 1972a)

$$\xi(t; \epsilon) = \epsilon \xi_1(\eta_1, \eta_2, T_2) + \epsilon^3 \xi_3(\eta_1, \eta_2, T_2) + \cdots$$

$$p(t; \epsilon) = p_0(T_2) + \epsilon^2 p_2(\eta_1, \eta_2, \phi, T_2) + \cdots$$

where $\dot{\phi} = p$ and

$$(\dot{\eta}_1, \dot{\eta}_2) = (\omega_1, \omega_2) = \frac{p_0 I_x}{2I} \pm \left[\left(\frac{p_0 I_x}{2I}\right)^2 + \omega_0^2\right]^{1/2}$$

Then show that

$$\xi_1 = a_1(T_2) \exp [i\eta_1 + i\theta_1(T_2)] + a_2(T_2) \exp [i\eta_2 + i\theta_2(T_2)]$$

where

$$a_1' = (\omega_1 - \omega_2)^{-1} [\omega_1 \mu_1 a_1 + k \sin (\phi - \eta_1 - \theta_1 + \phi_0)]$$

$$a_2' = -(\omega_1 - \omega_2)^{-1} \omega_2 \mu_1 a_2$$

$$\theta_1' = -\gamma(\omega_1 - \omega_2)^{-1}(a_1^2 + 2a_2^2) - k(\omega_1 - \omega_2)^{-1} a_1^{-1} \cos (\phi - \eta_1 - \theta_1 + \phi_0)$$

$$\theta_2' = \gamma(\omega_1 - \omega_2)^{-1}(2a_1^2 + a_2^2)$$

$$p_0' = \nu_0 + \nu_2 p_0 + \nu_1 a_1 \sin (\eta_1 - \phi + \theta_1)$$

Investigate the conditions under which the roll rate p_0 locks onto ω_1. This condition is referred to as roll resonance.

(b) When $K = O(1)$, determine first-order uniform expansions for the cases

(i) $p_0 \approx 0$
(ii) $p_0 \approx 2\omega_1 - \omega_2$
(iii) nonresonant case

CHAPTER 7

Continuous Systems

This chapter deals with nonlinear vibrations of continuous systems such as beams, strings, plates, membranes, discs, and shells, in contrast with the preceding five chapters, which deal with discrete systems. Eisley (1966) surveyed the nonlinear deformation of elastic beams, rings, and strings; Sathyamoorthy and Pandalai surveyed the large-amplitude vibrations of discs, membranes, and rings (1972a) and of plates and shells (1973b); and Crandall (1974) discussed the role of nonlinearities in structural dynamics.

For small oscillations the response of a deformable body can be adequately described by linear equations and boundary conditions. However as the amplitude of oscillation increases, nonlinear effects come into play. The source of the nonlinearities may be geometric, inertial, or material in nature. The geometric nonlinearity may be caused by nonlinear stretching or large curvatures. Nonlinear stretching of the midplane of a deformable body accompanies its transverse vibrations if it is supported in such a way as to restrict movement of its ends and/or edges. This stretching leads to a nonlinear relationship between the strain and the displacement. If the large-amplitude vibrations are accompanied by large changes in the curvature, it is necessary to employ a nonlinear relationship between the curvature and the displacement. Nonlinear inertial effects are caused by the presence of concentrated or distributed masses. Material nonlinearity occurs whenever the stresses are nonlinear functions of the strains.

Most of the existing studies deal with nonlinear stretching. However there are a number of studies that treat other nonlinearities and other stress–strain laws. Hu and Kirmser (1971) analyzed the transverse vibrations of a free-free beam using a nonlinear curvature–displacement relationship, Grammel (1953) considered the torsional vibrations of nonlinearly elastic circular shafts and the transverse vibrations of nonlinearly elastic beams, while Bolotin (1964, Section 13) discussed the planar motion of beams with nonlinear elasticity. McQueary and Clark (1967) analyzed periodic oscillations of a class of nonautonomous nonlinear elastic continua. Sinitsyn (1976) analyzed the oscillations of a cylindrical shell of nonlinear elastic materials, while Sun and Shafey (1976) assessed the influence of material nonlinearity on the dynamic behavior of composite plates. Steele (1967) and Tang and Yen (1970) studied the response of a beam on an elastic foundation to a moving load and included the effects of geometric and material nonlinearities. McIvor (1964) investigated the effect of longitudinal inertia on the transverse vibrations of beams, Atluri (1973) included nonlinear inertia and curvature-displacement relationships in his analysis of the vibrations of a hinged beam, and Arya and Lardner (1974) investigated the effect of longitudinal inertia on the transverse vibrations of a string. Porter and Billet (1965) analyzed the forced response of a continuous system having nonlinear constraints, while Dokainish and Elmadany (1978) analyzed the response of a helical spring having a nonlinear constraint. Watanabe (1978) treated linear beams with nonlinear boundary conditions, while Pasley and Gurtin (1960) studied systems resting on nonlinear springs. Viscoelastic structures were treated by Eringen (1955); Sethna (1962); Kravchuk, Morgunov, and Troyanovskii (1974); and Nambudiripad and Neis (1976); while beams with nonlinear internal friction and relaxation were studied by Osinski (1962).

Since exact solutions are in general not available, recourse has been to approximate analyses by using purely numerical techniques, purely analytic techniques and numerical-analytic techniques.

The principal numerical methods available deal with dynamic deformations (Tillerson, 1975). They may be divided into (a) finite difference in space and time (Witmer, Balmer, Leech, and Pian, 1963; Morino, Leech, and Witmer, 1971; Bayles, Lowery, and Boyd, 1973), (b) finite-element representation in space and finite difference in time (Stricklin, Martinez, Tillerson, Hong, and Haisler, 1971; Wu and Witmer, 1971, 1973; Oden, Key, and Fost, 1974), and (c) finite-element representation in both space and time (Fried, 1969; Argyris and Scharpf, 1969; Zienkiewicz and Parekh, 1970). Application of these purely numerical techniques to vibration problems may be costly in terms of computation time. In addition, there is the difficulty of obtaining the complicated responses, such as those described in Sections 7.4 and 7.6, which may occur for even simple structural elements.

In the purely analytic methods, the spatial as well as the temporal problems are solved analytically. The temporal problem is usually solved by using a perturbation method such as the method of harmonic balance, the Lindstedt-Poincaré technique, the method of averaging, or the method of multiple scales. The spatial problem is usually solved by using the method of weighted residuals (the Ritz and the Galérkin methods) or the expansion of the solution in terms of the normal modes of the linearized problem. These techniques are usually suited to simple structures with simple boundaries. Except for slowly varying material and geometric properties (e.g., Nayfeh, 1973a), the application of the normal-mode expansion has been limited to structures with uniform material and geometric properties. The method of weighted residuals has been used to determine the nonlinear frequencies of a number of structures. However the results depend somewhat on the assumed spatial variations.

The numerical-analytic approaches can be divided into three groups. In the first group one postulates the dependance of the solution on time in the form $w(\mathbf{r}, t) = \psi(t)\phi(\mathbf{r})$. Usually ψ is assumed to be harmonic. Then one uses the method of harmonic balance to obtain a nonlinear boundary-value problem for the determination of the spatial variation, $\phi(\mathbf{r})$. The solution of the spatial problem is usually obtained by using a numerical method in conjunction with an iterative procedure. Mei (1972, 1973a, b) used finite elements and iterated on the in-plane forces developed in the structure as a result of its large vibrations, while Venkateswara Rao, Raju, and Kanaka Raju (1976a), Kanaka Raju and Venkateswara Rao (1976a), and Kanaka Raju (1977) used finite elements but iterated on a linearized strain–displacement relationship. So far these techniques have been applied to the determination of the free-oscillation frequencies. One can use finite differences instead of finite elements. In the case of a single spatial variable (beams, symmetric vibrations of plates), Huang and Sandman (1971), Huang and Meng (1972), Huang (1973), and Huang, Woo, and Walker (1976) solved the spatial two-point boundary-value problem, using a forward-integration technique and iterating on the initial conditions to satisfy the boundary conditions (a shooting technique). In this group one needs to know, or assume, the temporal functional form $\psi(t)$.

In the second group (Hwang and Pi, 1972) one uses a finite-element technique to obtain the following nonlinear matrix equation:

$$M\ddot{\mathbf{q}} + C\dot{\mathbf{q}} + [K_0 + K_1(\mathbf{q}) + K_2(\mathbf{q}^2) + \cdots + K_n(\mathbf{q}^n)]\,\mathbf{q} = \mathbf{f}$$

where M, C, and K_n are $N \times N$ matrices and $\mathbf{q}$ and $\mathbf{f}$ are column vectors with N elements. Then one solves for $\mathbf{q}$, using the method of equivalent linearization.

In the third group (Busby and Weingarten, 1972; Nayfeh, Mook, and Lobitz, 1974; Nayfeh, 1976b; Lobitz, Nayfeh, and Mook, 1977), one expresses the displacement in the form

$$w(\mathbf{r}, t) = \sum_{n=1}^{N} \psi_n(t)\phi_n(\mathbf{r})$$

where the ϕ_n are the linear undamped natural modes of the system. Then one uses a finite-element technique to determine the ϕ_n, substitutes for w in the governing equations, uses the orthogonality property of the ϕ_n, and obtains a system of N coupled second-order ordinary-differential equations for the ψ_n. These equations are solved by using a perturbation technique such as the method of averaging or the method of multiple scales. This numerical-perturbation approach has been used to determine the nonlinear response, including the effect of internal resonance, of beams with varying properties (Nayfeh, Mook, and Lobitz, 1974) and of elliptic plates (Lobitz, Nayfeh, and Mook, 1977).

To concentrate on the physical mechanisms and effects, we restrict our attention here to uniform systems with simple boundary conditions whose linear natural modes can be obtained analytically. Then we use the method of multiple scales to solve the equations describing the functions $\psi_n(t)$. Beams are discussed in Sections 7.1 through 7.4. Strings are discussed in Section 7.5; their responses may be more complicated than those of beams owing to the many possibilities of modal interactions (internal resonances) and the likelihood of out-of-plane responses. Plates are discussed in Section 7.6.

7.1. Beams

7.1.1. GENERAL INTRODUCTION

As a first example we consider the nonlinear transverse vibrations of uniform beams supported in such a way as to restrict the movement at the ends and hence produce midplane stretching. Nayfeh (1973a) used the generalized version of the method of multiple scales to analyze the nonlinear free vibrations of a beam with slowly varying properties along its length; Nayfeh, Mook, and Lobitz (1974) used a numerical-perturbation technique to determine the forced response of a nonuniform beam; Verma and Murthy (1974) analyzed the nonlinear vibrations of nonuniform beams; and Raju, Venkateswara Rao, and Kanaka Raju (1976b) used a combination of the Galérkin procedure and the method of harmonic balance to determine the nonlinear frequencies of tapered beams. Saito, Sato, and Yutani (1976) studied the nonlinear forced response of a beam carrying a concentrated mass.

We consider beams that are straight initially. Eisley (1964a), Tseng and Dugundji (1971), and Min and Eisley (1972) analyzed the nonlinear vibrations of buckled beams; Mettler (1967) examined the stability and the vibrations of a sine arch under harmonic excitation; while Rehfield (1974b) and Singh and Ali (1975) analyzed the nonlinear flexural vibrations of shallow arches. Nonlinear vibrations of circular rings were studied by Federhofer (1959), Shkenev (1960), Evensen (1966), and Dowell (1967b), while the nonlinear vibrations of oval rings were studied by Sathyamoorthy and Pandalai (1971). Donaldson (1973a) treated flat skin-stringer-frame structures.

Here we simply state the equations of motion. The nonlinear coupling terms for beams are essentially the same as those for strings, and later in Section 7.5 we provide the details of the derivation of the string equations. Assuming that plane sections remain plane and assuming a linear stress–strain law, one can

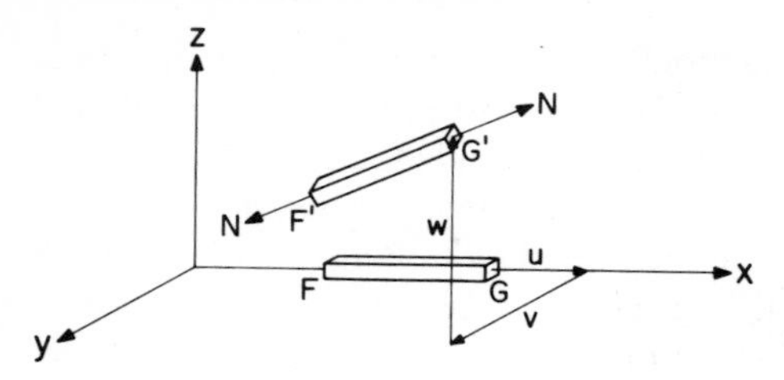

Figure 7-1. Element of a beam in the deflected and undeflected positions.

derive the following equations of motion for beams (see Figure 7-1 for the notation):

$$\rho A u_{tt} - EAu_{xx} = \tfrac{1}{2}(EA - N)\frac{\partial}{\partial x}\left[v_x^2 + w_x^2 - 2u_x(v_x^2 + w_x^2)\right] + G \tag{7.1.1}$$

$$\rho A w_{tt} - Nw_{xx} + EI_1 w_{xxxx} = (EA - N)\frac{\partial}{\partial x}(ew_x) + F_1 \tag{7.1.2}$$

$$\rho A v_{tt} - Nv_{xx} + EI_2 v_{xxxx} = (EA - N)\frac{\partial}{\partial x}(ev_x) + F_2 \tag{7.1.3}$$

$$e = u_x - u_x^2 + \tfrac{1}{2}v_x^2 + \tfrac{1}{2}w_x^2 \tag{7.1.4}$$

where ρ is the beam density; A, I_1, and I_2 are, respectively, the area and moments of inertia of the beam cross section; E is Young's modulus; N is the prescribed axial load, which may be time-dependent; G is the prescribed axial load; and the F_n are prescribed transverse loads. The subscripts indicate partial derivatives. Equations (7.1.1) through (7.1.4) describe the forced response of a beam allowing for nonplanar motions, longitudinal inertia, and stretching. For these equations to apply, the wavelength of the transverse vibrations must be long compared with the radius of gyration of the cross section; otherwise the effects of transverse shear and rotary inertia cannot be neglected. Eringen (1952), Wrenn and Mayers (1970), Nayfeh (1973a), and Venkateswara Rao, Raju, and Kanaka Raju (1976b) included the effects of rotary inertia and transverse shear in their analysis of the nonlinear free vibrations of beams. Chu (1961b) examined the influence of transverse shear on the nonlinear free vibrations of sandwich beams with honeycomb cores.

In discussing beams we restrict our attention to nonrotating and to planar motions (i.e., $v \equiv 0$). Nonlinear vibrations of rotating bars were studied by Ansari (1974, 1975) and Anderson (1975b), while nonlinear torsional vibrations of thin-walled beams of open sections were studied by Rao (1975). Nonplanar motions are discussed in connection with strings in Section 7.5. Haight and King (1972) investigated the planar harmonic forced excitation of a cantilever beam and found both analytically and experimentally that in many instances the response was nonplanar. Haight and King (1971) investigated experimentally and analytically the stability of planar motions under parametric excitations. Rodgers and Warner (1973) examined the dynamic stability of out-of-plane motion of curved elastic rods excited by harmonic end forces applied normal to the cross sections. Ho, Scott, and Eisely (1975, 1976) analyzed the nonplanar, nonlinear forced and free oscillations of beams. Their results confirm the existence of stable nonplanar (ballooning or whirling) motions.

In (7.1.1) through (7.1.4) damping is ignored. However in any physical situation there is always some damping. There are a number of models for damping (Lazan, 1968). In this chapter we consider only linear viscous damping. Sethna (1962), Davy and Ames (1973), Ivanova, Morgunov, and Troyanovskii (1974), Kadyrbekov (1975), and Hyer, Anderson, and Scott (1976) studied nonlinear vibrations of viscoelastic beams. Kovac, Anderson, and Scott (1971) studied the case of damped sandwich beams.

In order to obtain (7.1.1) through (7.1.4) one must retain through the cubic terms when accounting for the motion-induced stretching. If we want only a first correction which accounts for stretching, then $2u_x(v_x^2 + w_x^2)$ should be neglected in (7.1.1) and u_x^2 should be neglected in (7.1.4). Under some circumstances it is appropriate to consider u to be $O(w)$; then e is given by

$$e = u_x \tag{7.1.5}$$

Under other circumstances it is appropriate to neglect the out-of-plane motion and consider u to be $O(w^2)$; then e is given by

$$e = u_x + \tfrac{1}{2} w_x^2 \tag{7.1.6}$$

Thus instead of (7.1.4) one may use either (7.1.5) or (7.1.6). The appropriate choice, depending on the circumstances, is discussed below.

At this point it is convenient to introduce dimensionless variables (denoted by asterisks):

$$x^* = \frac{x}{L}, \quad t^* = \frac{r}{L^2}\sqrt{\frac{E}{\rho}}\, t, \quad N^* = \frac{NL^2}{r^2 EA}$$

$$w^* = \frac{w}{L}, \quad u^* = \frac{u}{L}, \quad l^* = \frac{l}{L}, \quad r^* = \frac{r}{L}$$

where r is the radius of gyration of the cross section, l is the actual length of the beam, and L is a characteristic length. An appropriate choice for L is the wavelength of the highest mode of the transverse oscillations being considered. Substituting these dimensionless quantities into (7.1.1) through (7.1.6), making the simplifications discussed above, adding damping and forcing terms, and dropping the asterisks, we obtain

$$r^2 u_{tt} - u_{xx} = (1 - r^2 N) w_x w_{xx} - 2\nu u_t + G(x, t) \tag{7.1.7}$$

$$r^2 (w_{tt} - N w_{xx} + w_{xxxx}) = (1 - r^2 N) \frac{\partial}{\partial x} (e w_x) - 2r^2 \mu w_t + F(x, t) \tag{7.1.8}$$

Equations (7.1.5) and (7.1.6) remain unchanged.

An exact solution of the problem defined by (7.1.6) through (7.1.8) subject to initial and boundary conditions does not yet exist, and recourse has been made to approximate methods that can be applied to weakly nonlinear equations. The parameter that influences the way the approximation must be made is the slenderness ratio r. If we consider r to be small, but not extremely small, then in (7.1.7) the longitudinal inertia and the restoring force u_{xx} are the same order, and it follows that $u = O(w)$. If we consider r to be extremely small, then in (7.1.7) the longitudinal inertia is small compared with the restoring force, and it follows that $u = O(w^2)$. By far, most of the studies of nonlinear beam vibrations are based on this simplification. Each case is considered separately below.

7.1.2. THE GOVERNING EQUATIONS FOR THE CASE $u = O(w)$

In this case we consider only hinged-hinged beams. Other boundary conditions are considered for the case $u = O(w^2)$. For a hinged-hinged beam with immovable ends

$$\begin{aligned} u &= 0 \quad \text{at } x = 0 \quad \text{and } x = l \\ w &= w_{xx} = 0 \quad \text{at } x = 0 \quad \text{and } x = l \end{aligned} \tag{7.1.9}$$

Thus we assume expansions for u and w in the form

$$\begin{aligned} w &= \sum_{n=1}^{\infty} \eta_n(t) \sin \frac{n\pi x}{l} \\ u &= \sum_{n=1}^{\infty} \xi_n(t) \sin \frac{n\pi x}{l} \end{aligned} \tag{7.1.10}$$

In general it is convenient to assume expansions for the displacements in terms of the linear free-oscillation modes. These functions and the natural frequencies are intrinsic properties. Moreover infinitesimal motions are well described by them. An analysis such as the one used here shows how the linear modes combine as the amplitude of the motion grows.

Substituting (7.1.10) into (7.1.5), (7.1.7), and (7.1.8) leads to

$$\sum_{m=1}^{\infty} \left(r^2 \ddot{\xi}_m + 2\nu \dot{\xi}_m + \frac{m^2 \pi^2}{l^2} \xi_m \right) \sin \frac{m\pi x}{l} = -\pi^3 \kappa \sum_{m=1}^{\infty} \sum_{k=1}^{\infty} k^2 m \eta_m \eta_k \cdot \cos \frac{m\pi x}{l} \sin \frac{k\pi x}{l} + G(x, t) \tag{7.1.11}$$

$$r^2 \sum_{m=1}^{\infty} (\ddot{\eta}_m + 2\mu\dot{\eta}_m + \omega_m^2 \eta_m) \sin \frac{m\pi x}{l} = -\pi^3 \kappa \sum_{m=1}^{\infty} \sum_{k=1}^{\infty} mk\eta_k \xi_m$$

$$\cdot \left(k \cos \frac{m\pi x}{l} \sin \frac{k\pi x}{l} + m \sin \frac{m\pi x}{l} \cos \frac{k\pi x}{l} \right) + F(x, t) \quad (7.1.12)$$

where

$$\omega_m = \frac{m\pi}{l} \left(\frac{m^2\pi^2}{l^2} + N \right)^{1/2} \quad \text{and} \quad \kappa = \frac{1 - r^2 N}{l^3} \quad (7.1.13)$$

Multiplying (7.1.11) by $\sin (n\pi x/l)$ leads to

$$\sum_{m=1}^{\infty} \left(r^2 \ddot{\xi}_m + 2\nu\dot{\xi}_m + \frac{m^2\pi^2}{l^2} \xi_m \right) \left[\cos \frac{(m-n)\pi x}{l} - \cos \frac{(m+n)\pi x}{l} \right]$$

$$= -\frac{\pi^3 \kappa}{2} \sum_{m=1}^{\infty} \sum_{k=1}^{\infty} mk^2 \eta_m \eta_k \left[\cos \frac{(m+k-n)\pi x}{l} - \cos \frac{(m+k+n)\pi x}{l} \right.$$

$$\left. - \cos \frac{(k-m+n)\pi x}{l} + \cos \frac{(k-m-n)\pi x}{l} \right] + 2G(x, t) \frac{\sin n\pi x}{l} \quad (7.1.14)$$

Integrating over the interval $[0, l]$ leads to

$$r^2 \ddot{\xi}_n + 2\nu_n \dot{\xi}_n + \frac{n^2\pi^2}{l^2} \xi_n = G_n(t) - \frac{\pi^3 \kappa}{2} \sum_{m=1}^{\infty} m^2 \eta_m$$

$$\cdot \left[|m-n| \eta_{|m-n|} - (m+n) \eta_{m+n} \right] \quad (7.1.15)$$

where

$$lG_n = 2 \int_0^l G(x, t) \sin \frac{n\pi x}{l} \, dx \quad (7.1.16)$$

and we assumed modal damping. Equation (7.1.15) can be put in a more convenient form in the following way. The last term in (7.1.15) can also be written as follows:

$$-\frac{\pi^3 \kappa}{2} \left[\sum_{m=1}^{\infty} m(m+n)^2 \eta_m \eta_{m+n} + \sum_{m=1}^{n-1} m(n-m)^2 \eta_n \eta_{n-m} \right.$$

$$\left. - \sum_{m=n+1}^{\infty} m(m-n)^2 \eta_n \eta_{m-n} \right] \quad (7.1.17)$$

If we use one half the last term in (7.1.15) and one half of (7.1.17), (7.1.15) reduces to the following convenient form:

$$r^2 \ddot{\xi}_n + 2\nu_n \dot{\xi}_n + \frac{n^2 \pi^2}{l^2} \xi_n = G_n(t) - \frac{n\pi^3}{4} \kappa \sum_{m=1}^{\infty} m\eta_m \cdot [|n - m| \eta_{|n-m|} + (m+n) \eta_{m+n}] \quad (7.1.18)$$

In a similar way one can obtain the following equation from (7.1.12):

$$r^2 (\ddot{\eta}_n + 2\mu_n \dot{\eta}_n + \omega_n^2 \eta_n) = F_n(t) - \frac{n\pi^3 \kappa}{4} \sum_{m=1}^{\infty} m\xi_m \cdot [|m - n| \eta_{|m-n|} + (m+n) \eta_{m+n}] \quad (7.1.19)$$

where

$$lF_n(t) = 2 \int_0^l F(x, t) \sin \frac{n\pi x}{l} \, dx \quad (7.1.20)$$

and again we assumed modal damping.

We let ϵ be a measure of the amplitude of the motion, which is small but finite. Then we put

$$\begin{aligned} \xi_n &= \epsilon \hat{\xi}_n \quad &\text{and} \quad \eta_n &= \epsilon \hat{\eta}_n \\ r^{-2} \nu_n &= \epsilon \hat{\nu}_n \quad &\text{and} \quad \mu_n &= \epsilon \hat{\mu}_n \\ r^{-2} G_n &= \epsilon \hat{G}_n \quad &\text{and} \quad r^{-2} F_n &= \epsilon \hat{F}_n \end{aligned} \quad (7.1.21)$$

and $\pi^3 \kappa (4r^2)^{-1} = \hat{\kappa}$. Here, following the usual practice, we ordered the damping coefficients so that the nonlinear terms and the damping terms appear at the same order. Substituting (7.1.21) into (7.1.18) and (7.1.19) and dropping the hats in the results, we obtain

$$\ddot{\xi}_n + \lambda_n^2 \xi_n = \epsilon \left[-2\nu_n \dot{\xi}_n - n\kappa \sum_{m=1}^{\infty} m\eta_n (p\eta_p + q\eta_q) \right] + G_n(t) \quad (7.1.22)$$

$$\ddot{\eta}_n + \omega_n^2 \eta_n = \epsilon \left[-2\mu_n \dot{\eta}_n - n\kappa \sum_{m=1}^{\infty} m\xi_m (p\eta_p + q\eta_q) \right] + F_n(t) \quad (7.1.23)$$

where $\lambda_n = n\pi/rl$, $p = |n - m|$, and $q = n + m$.

Equations (7.1.22) and (7.1.23) are the starting point for the discussion in Section 7.2, where solutions valid for small but finite values of ϵ are constructed.

7.1.3. GOVERNING EQUATION FOR THE CASE $u = O(w^2)$

In this case r is taken to be very small, and consequently the longitudinal inertia is neglected. The longitudinal damping is also neglected. Moreover we consider the case for which the longitudinal excitation and the axial force N are zero. As a consequence one can integrate (7.1.7) to obtain

$$u_x = e(t) - \tfrac{1}{2} w_x^2 \tag{7.1.24}$$

where $e(t)$ is the same function defined by (7.1.6). A second integration yields

$$u = c(t) + xe(t) - \tfrac{1}{2}\int_0^x w_x^2 \, dx \tag{7.1.25}$$

For the boundary conditions on u we take

$$u(0, t) = 0 \quad \text{and} \quad u(l, t) = lP(t) \tag{7.1.26}$$

where $P(t)$ is a prescribed function. Ray and Bert (1969), Wrenn and Mayers (1970), and Nayfeh (1973a) found that a spring support reduces the effect of the nonlinearity. Substituting (7.1.25) into (7.1.26) leads to

$$e = P(t) + \frac{1}{2l}\int_0^l w_x^2 \, dx \tag{7.1.27}$$

Then (7.1.8) becomes

$$r^2(w_{tt} + 2\mu w_t + w_{xxxx}) = \left[P(t) + \frac{1}{2l}\int_0^l w_x^2 \, dx\right] w_{xx} + F(x, t) \tag{7.1.28}$$

Equation (7.1.28) is the basis for most of the studies of nonlinear beam vibrations.

Because r is a small, dimensionless parameter, the weakly nonlinear assumption demands that $w = O(r^k)$, where $k > 1$. Moreover we order the damping and the parametric excitation so that they appear at the same order as the nonlinear terms.

Again we express the solution as an expansion in terms of the linear free-oscillation modes $\phi_m(x)$. That is,

$$w(x, t) = r^k \sum_{m=1}^{\infty} u_m(t)\, \phi_m(x) \tag{7.1.29}$$

The $\phi_m(x)$ are solutions of the following eigenvalue problem:

$$\phi_m^{iv} - \omega_m^2 \phi_m = 0 \tag{7.1.30}$$

where

$$\phi_m = \phi_m'' = 0 \tag{7.1.31}$$

at a hinged end and

$$\phi_m = \phi_m' = 0 \tag{7.1.32}$$

at a clamped end. Aravamudan and Murthy (1973) investigated the case of time-dependent boundary conditions. The eigenvalues ω_m are the natural frequencies.

The eigenfunctions are orthogonal. To show this, we consider (7.1.30) for two different values of m:

$$\phi_i^{iv} - \omega_i^2 \phi_i = 0 \quad \text{and} \quad \phi_j^{iv} - \omega_j^2 \phi_j = 0 \tag{7.1.33}$$

It follows that

$$(\omega_j^2 - \omega_i^2) \int_0^l \phi_i \phi_j \, dx = \int_0^l (\phi_i \phi_j^{iv} - \phi_j \phi_i^{iv}) \, dx \tag{7.1.34}$$

Integrating by parts leads to

$$(\omega_j^2 - \omega_i^2) \int_0^l \phi_i \phi_j \, dx = [\phi_i \phi_j''' - \phi_i' \phi_j'' + \phi_i'' \phi_j' - \phi_i''' \phi_j]_0^l \tag{7.1.35}$$

It follows from (7.1.31) and 7.1.32) that

$$\int_0^l \phi_i \phi_j \, dx = 0 \quad \text{if } i \neq j \tag{7.1.36}$$

For convenience, we choose the amplitudes of the ϕ_n such that

$$\int_0^l \phi_i \phi_j \, dx = \delta_{ij} = \begin{cases} 1 & i = j \\ 0 & i \neq j \end{cases} \tag{7.1.37}$$

Substituting (7.1.29) into (7.1.28), multiplying by ϕ_n, and integrating over the length we obtain

$$\ddot{u}_n + \omega_n^2 u_n = \epsilon \left(\sum_{m=1}^{\infty} p_{mn} u_m + \sum_{m,p,q=1}^{\infty} \Gamma_{nmpq} u_m u_p u_q - 2\mu_n \dot{u}_n \right) + f_n(t) \quad \text{for } n = 1, 2, \ldots \tag{7.1.38}$$

where $\epsilon = r^{2(k-1)}$, the damping is assumed to be modal, and

$$r^{2+k} f_n(t) = \int_0^l F(x,t) \phi_n \, dx \quad \text{and} \quad r^{2k} p_{nm}(t) = P \int_0^l \phi_m'' \phi_n \, dx \tag{7.1.39}$$

$$\Gamma_{nmpq} = \tfrac{1}{2} \left(\int_0^l \phi_n \phi_m'' \, dx \right) \left(\int_0^l \phi_p' \phi_q' \, dx \right) \quad \text{and} \quad \epsilon \mu_n = \int_0^l \mu \phi_n^2 \, dx \tag{7.1.40}$$

As mentioned above, μ and P are ordered so that they appear with the non-linearity. For a given r, one can choose the value of k that makes ϵ small. For example, if $r = 0.05$, then $k = \frac{3}{2}$ yields $\epsilon = 0.05$. If $r = 0.2$, then $k = 2$ yields $\epsilon = 0.04$. Integrating by parts we obtain the following from (7.1.40):

$$\Gamma_{nmpq} = -\tfrac{1}{2}\left[\int_0^l \phi_n' \phi_m' \, dx\right]\left[\int_0^l \phi_p' \phi_q' \, dx\right] \tag{7.1.41}$$

We note that

$$\Gamma_{nmpq} = \Gamma_{mnpq} = \Gamma_{mnqp} = \Gamma_{qpmn} \tag{7.1.42}$$

Equation (7.1.38) is the starting point for the discussion in Sections 7.3 and 7.4, where solutions valid for small values of ϵ are constructed.

7.2. Coupled Longitudinal and Transverse Oscillations

In this section we consider primary resonances for the case $u = O(w)$.

Hoff (1951) considered a column having an end displaced at a constant rate, neglecting the longitudinal inertia. Using a numerical procedure, Sevin (1960) obtained the response of a column when the end is displaced at a constant rate, including the longitudinal inertia. McIvor (1964) obtained the response of a column to pulsating loads, and McIvor and Bernard (1973) obtained the response to short-duration axial loads. In this section we formulate the problem for both axial and transverse loads. Then we provide some of the details for a harmonic axial load acting alone.

Here we consider the excitation to have the form

$$G_n(t) = 2\epsilon g_n \cos(\Omega_1 t - \tau) \text{ and } F_n(t) = 2\sum_{m=1}^{M} f_{nm} \cos(\Omega_{2m} t - \theta_m) \tag{7.2.1}$$

where τ, θ_m, g_n, and f_{nm} are constants. As discussed in Chapters 4 through 6, we need to order the damping coefficients and the amplitudes of the excitation in such a way that the effects of damping, nonlinearity, and excitation all appear at the same order. This was the reason for ordering the damping as in Section 7.1.2.

We seek a solution of (7.1.22) and (7.1.23) in the form

$$\begin{aligned} \xi_n &= \xi_{n0}(T_0, T_1) + \epsilon\xi_{n1}(T_0, T_1) + \cdots \\ \eta_n &= \eta_{n0}(T_0, T_1) + \epsilon\eta_{n1}(T_0, T_1) + \cdots \end{aligned} \tag{7.2.2}$$

Substituting (7.2.1) and (7.2.2) into (7.1.22) and (7.1.23) and equating coefficients of like powers of ϵ we obtain

Order ϵ^0

$$\begin{aligned} D_0^2 \xi_{n0} + \lambda_n^2 \xi_{n0} &= 0 \\ D_0^2 \eta_{n0} + \omega_n^2 \eta_{n0} &= 0 \end{aligned} \tag{7.2.3}$$

Order ϵ

$$D_0^2 \xi_{n1} + \lambda_n^2 \xi_{n1} = -2D_0 D_1 \xi_{n0} - 2\nu_n D_0 \xi_{n0} - n\kappa \sum_{m=1}^{\infty} m\eta_{m0}(p\eta_{p0} + q\eta_{q0}) + 2g_n \cos(\Omega_1 T_0 - \tau) \tag{7.2.4}$$

$$D_0^2 \eta_{n1} + \omega_n^2 \eta_{n1} = -2D_0 D_1 \eta_{n0} - 2\mu_n D_0 \eta_{n0} - n\kappa \sum_{m=1}^{\infty} m\xi_{m0}(p\eta_{p0} + q\eta_{q0}) + 2\sum_{k=1}^{M} f_{nk} \cos(\Omega_{2k} T_0 - \theta_k) \tag{7.2.5}$$

where

$$\lambda_n = \frac{n\pi}{rl}, \quad \omega_n = \frac{n\pi}{l}\left(\frac{n^2\pi^2}{l^2} + N\right)^{1/2}, \quad p = |n - m|, \quad \text{and } q = n + m \tag{7.2.6}$$

The solutions of (7.2.3) can be written in the form

$$\begin{aligned} \xi_{n0} &= A_n(T_1) \exp(i\lambda_n T_0) + cc \\ \eta_{n0} &= B_n(T_1) \exp(i\omega_n T_0) + cc \end{aligned} \tag{7.2.7}$$

Substituting (7.2.7) into (7.2.4) and (7.2.5) leads to

$$\begin{aligned} D_0^2 \xi_{n1} + \lambda_n^2 \xi_{n1} &= -2i\lambda_n(D_1 A_n + \nu_n A_n) \exp(i\lambda_n T_0) \\ &\quad - n\kappa \sum_{m=1}^{\infty} mB_m \{pB_p \exp[i(\omega_m + \omega_p)T_0] \\ &\quad + p\bar{B}_p \exp[i(\omega_m - \omega_p)T_0] + qB_q \exp[i(\omega_q + \omega_m)T_0] \\ &\quad + q\bar{B}_q \exp[i(\omega_m - \omega_q)T_0]\} + g_n \exp[i(\Omega_1 T_0 - \tau)] + cc \end{aligned} \tag{7.2.8}$$

$$\begin{aligned} D_0^2 \eta_{n1} + \omega_n^2 \eta_{n1} &= -2i\omega_n(D_1 B_n + \mu_n B_n) \exp(i\omega_n T_0) \\ &\quad - n\kappa \sum_{m=1}^{\infty} mA_m \{pB_p \exp[i(\lambda_m + \omega_p)T_0] \\ &\quad + p\bar{B}_p \exp[i(\lambda_m - \omega_p)T_0] + qB_q \exp[i(\lambda_m + \omega_q)]\, T_0] \\ &\quad + q\bar{B}_q \exp[i(\lambda_m - \omega_q)T_0]\} + \sum_{k=1}^{M} f_{nk} \exp[i(\Omega_{2k} T_0 - \theta_k)] \end{aligned} \tag{7.2.9}$$

Inspection of (7.2.8) and (7.2.9) shows that there is an internal resonance if

$$\lambda_n \simeq \omega_m \pm \omega_p \quad \text{or} \quad \lambda_n \simeq \omega_m \pm \omega_q \tag{7.2.10}$$

Recalling the definitions, we can rewrite (7.2.10) as follows:

$$\frac{n}{r} \simeq m\left(\frac{m^2\pi^2}{l^2} + N\right)^{1/2} \pm |n - m|\left[\frac{(n-m)^2\pi^2}{l^2} + N\right]^{1/2}$$

or (7.2.11)

$$\frac{n}{r} \simeq m\left(\frac{m^2\pi^2}{l^2} + N\right)^{1/2} \pm (n + m)\left[\frac{(n+m)^2\pi^2}{l^2} + N\right]^{1/2}$$

Equations (7.2.11) show that r is the parameter that tunes the internal resonance.

Because the slenderness ratio must be small for beam theory to apply, we restrict our attention to the fundamental longitudinal mode (i.e., $n = 1$). Moreover conditions (7.2.11) are first satisfied for the positive sign. Finally, to gain an estimate of the values of l/r and m which lead to resonance, we put $N \equiv 0$ and obtain

$$\frac{l}{r} \simeq (2m^2 + 2m + 1)\,\pi \tag{7.2.12}$$

In Table 7-1 the approximate values of l/r are given for three values of m. The column labeled MSR gives the modified slenderness ratio obtained by choosing the characteristic length to be the actual length divided by the number of the highest transverse mode being considered. The value of 14 for MSR is probably at the limit of the region where beam theory applies. When $N \neq 0$, the values of MSR are higher, though N has more influence on the frequency when r is very small. And when one or both ends are clamped, the values of MSR are slightly higher.

As an example we consider the case when

$$\begin{aligned} \lambda_1 &= \omega_2 + \omega_3 + \epsilon\sigma_I, & \Omega_1 &= \lambda_1 + \epsilon\sigma_1 \\ \Omega_{22} &= \omega_2 + \epsilon\sigma_2, & \Omega_{23} &= \omega_3 + \epsilon\sigma_3 \end{aligned} \tag{7.2.13}$$

Then secular terms are eliminated from the ξ_{n1} and η_{n1} if

$$2i\lambda_1(D_1A_1 + \nu_1A_1) + 6\kappa B_2B_3 \exp(-i\sigma_I T_1) - g_1 \exp(i\sigma_1 T_1) = 0 \tag{7.2.14}$$

TABLE 7-1. Variation of Slenderness Ratios with Mode Number

m	l/r	MSR
1	16	8
2	41	14
3	79	20

$$2i\lambda_n(D_1 A_n + \nu_n A_n) = 0 \qquad \text{for } n \geqslant 2 \tag{7.2.15}$$

$$2i\omega_2(D_1 B_2 + \mu_2 B_2) + 6\kappa A_1 \overline{B}_3 \exp(i\sigma_I T_1) - f_{22} \exp(i\sigma_2 T_1) = 0 \tag{7.2.16}$$

$$2i\omega_3(D_1 B_3 + \mu_3 B_3) + 6\kappa A_1 \overline{B}_2 \exp(i\sigma_I T_1) - f_{33} \exp(i\sigma_3 T_1) = 0 \tag{7.2.17}$$

$$2i\omega_n(D_1 B_n + \mu_n B_n) = 0 \qquad \text{for } n \neq 2 \text{ or } 3 \tag{7.2.18}$$

It follows immediately from (7.2.15) and (7.2.18) that

$$A_n \to 0 \quad \text{if } n > 1 \quad \text{and} \quad B_n \to 0 \quad \text{if } n \neq 2 \text{ or } 3 \tag{7.2.19}$$

as T_1 (and hence t) $\to \infty$.

Again we introduce the polar notation

$$A_n = \tfrac{1}{2} a_n \exp(i\alpha_n) \quad \text{and} \quad B_n = \tfrac{1}{2} b_n \exp(i\beta_n) \tag{7.2.20}$$

where the a_n, b_n, α_n, and β_n are real functions of T_1. Then for the steady-state response we need to solve the following:

$$\lambda_1 \nu_1 a_1 - k b_2 b_3 \sin \gamma_I - g_1 \sin \gamma_1 = 0 \tag{7.2.21}$$

$$\lambda_1 a_1 \alpha_1' - k b_2 b_3 \cos \gamma_I + g_1 \cos \gamma_1 = 0 \tag{7.2.22}$$

$$\omega_2 \mu_2 b_2 + k a_1 b_3 \sin \gamma_I - f_{22} \sin \gamma_2 = 0 \tag{7.2.23}$$

$$\omega_2 b_2 \beta_2' - k a_1 b_3 \cos \gamma_I + f_{22} \cos \gamma_2 = 0 \tag{7.2.24}$$

$$\omega_3 \mu_3 b_3 + k a_1 b_2 \sin \gamma_I - f_{33} \sin \gamma_3 = 0 \tag{7.2.25}$$

$$\omega_3 b_3 \beta_3' - k a_1 b_2 \cos \gamma_I + f_{33} \cos \gamma_3 = 0 \tag{7.2.26}$$

where

$$\begin{aligned} k &= \tfrac{3}{2}\kappa \\ \gamma_I &= \sigma_I T_1 + \alpha_1 - \beta_2 - \beta_3 \\ \gamma_1 &= \sigma_1 T_1 - \alpha_1 \\ \gamma_2 &= \sigma_2 T_1 - \beta_2 \\ \gamma_3 &= \sigma_3 T_1 - \beta_3 \end{aligned} \tag{7.2.27}$$

Also for the steady-state response γ_I, γ_1, γ_2, and γ_3 are constants. Solutions of these equations are discussed next.

We begin by considering f_{22} and f_{33} to be zero and leave the other cases as exercises. Then it follows that there are two possible solutions: either $a_1 \neq 0$ and $b_2 = b_3 = 0$, or a_1, b_2, and b_3 are all nonzero. In both cases

$$\alpha_1' = \sigma_1 \tag{7.2.28}$$

When $b_2 = b_3 = 0$, the solution is given by

$$a_1 = \frac{g_1}{\lambda_1(\nu_1^2 + \sigma_1^2)^{1/2}} \tag{7.2.29}$$

and

$$\gamma_1 = \sin^{-1}\left(\frac{\lambda_1 \nu_1 a_1}{g_1}\right) = \cos^{-1}\left(-\frac{\lambda_1 \sigma_1 a_1}{g_1}\right) \tag{7.2.30}$$

which is essentially the solution of the linearized problem.

When a_1, b_2, and b_3 are nonzero, we find that in addition to (7.2.28)

$$\beta_2' + \beta_3' = \sigma_I + \sigma_1 \tag{7.2.31}$$

Then (7.2.21) through (7.2.26) can be reduced to

$$\lambda_1 \nu_1 a_1 - k b_2 b_3 \sin \gamma_I - g_1 \sin \gamma_1 = 0 \tag{7.2.32}$$

$$\lambda_1 \sigma_1 a_1 - k b_2 b_3 \cos \gamma_I + g_1 \cos \gamma_1 = 0 \tag{7.2.33}$$

$$\omega_2 \mu_2 b_2 + k a_1 b_3 \sin \gamma_I = 0 \tag{7.2.34}$$

$$\omega_3 \mu_3 b_3 + k a_1 b_2 \sin \gamma_I = 0 \tag{7.2.35}$$

$$\sigma_I + \sigma_1 - k a_1 \left(\frac{b_3}{\omega_2 b_2} + \frac{b_2}{\omega_3 b_3}\right) \cos \gamma_I = 0 \tag{7.2.36}$$

It follows from (7.2.34) and (7.2.35) that

$$\mu_2 \omega_2 b_2^2 = \mu_3 \omega_3 b_3^2 \tag{7.2.37}$$

From (7.2.34) and (7.2.36) one can obtain

$$\left(\frac{\mu_2 \omega_2 b_2}{b_3}\right)^2 + \left[\frac{(\sigma_I + \sigma_1)\omega_2 \omega_3 b_2 b_3}{\omega_3 b_3^2 + \omega_2 b_2^2}\right]^2 = k^2 a_1^2 \tag{7.2.38}$$

Substituting (7.2.37) into (7.2.38) leads to

$$a_1 = \frac{1}{k}\left\{\mu_2 \mu_3 \omega_2 \omega_3 \left[1 + \left(\frac{\sigma_I + \sigma_1}{\mu_2 + \mu_3}\right)^2\right]\right\}^{1/2} \tag{7.2.39}$$

We note that a_1 is not a function of g_1, the amplitude of the excitation. There is a saturation phenomenon associated with this motion. The saturation phenomenon was discussed in Section 6.5. Here we may expect the beam to exhibit behavior which is similar to that of the system discussed in Section 6.5.1. In addition to saturation, we may expect that when the excitation is transverse rather than longitudinal, no steady-state motion exists for certain values of the parameters in spite of the presence of damping. Instead there is a continual exchange of energy back and forth between the longitudinal and transverse motions.

7-3. Hinged-Hinged Beams

Woinowsky-Krieger (1950) and Burgreen (1951) considered the free oscillations of a beam having hinged ends a fixed distance apart. Expressing the solution as a product of a function of time and a linear free-oscillation mode, they solved the nonlinear equation for the temporal function exactly in terms of Jacobi elliptic functions. Burgreen also considered vibrations about the buckled configuration and conducted an experimental study which basically supported the conclusions of the analysis. These results were also confirmed by the experiment of Ray and Bert (1969). The same problem was treated using different techniques by Wagner (1965), Srinivasan (1965, 1967), Woodall (1966), Evensen (1968), Rehfield (1973, 1975), and Lou and Sikarskie (1975). Wrenn and Mayers (1970) determined the influence of an axial restraint on the frequency of a single mode. Reiss and Matkowsky (1971) and Rubenfeld (1974) examined the stability of a compressed simply-supported column.

McDonald (1955) worked with the same governing equations but did not consider axial prestressing, as Burgreen did. He improved the analysis by letting the deflection curve be represented at any instant by a Fourier expansion in terms of the eigenfunctions of the linear problem, that is, the linear, free-oscillation modes. McDonald was able to solve the nonlinear equations for the coefficients in terms of elliptic functions. He commented that the problem is inherently nonlinear even for small-amplitude vibrations and there is always dynamic coupling of the modes.

Single-mode forced responses of simply-supported beams were studied by Eisley (1964b), Morris (1965), Srinivasan (1966a), Bennett and Rinkel (1972), Bert and Fisher (1972), and Rehfield (1974a). Multimodal forced responses of simply-supported beams were studied by Eisley and Bennett (1970), Bennett and Eisley (1970), and Busby and Weingarten (1972). Mojaddidy, Mook, and Nayfeh (1977) analyzed the multimode response of a simply-supported beam to multifrequency excitations.

Here we choose the characteristic length to be the actual length of the beam and let $p_{nm} = 0$. Then it follows from (7.1.30) and (7.1.31) that the natural frequencies $\omega_n = n^2\omega^2$ and

$$\phi_n = \sqrt{2}\sin n\pi x \qquad \text{for } n = 1, 2, 3, \ldots \tag{7.3.1}$$

Hence (7.1.38) becomes

$$\ddot{u}_n + n^4\pi^4 u_n = -\epsilon\left(2\mu_n\dot{u}_n + \tfrac{1}{2}\pi^4 n^2 u_n \sum_{m=1}^{\infty} m^2 u_m^2\right) + f_n(t) \qquad \text{for } n = 1, 2, 3, \ldots \tag{7.3.2}$$

We note that although there are commensurable linear natural frequencies such as $\omega_3 = 2\omega_2 + \omega_1$, where $\omega_n = n^2\pi^2$, there are no internal resonances in this case because of the vanishing of the coupling coefficients leading to internal resonance. This is not the case in the hinged-clamped and clamped-clamped cases. Next we consider both primary and secondary resonances. In the case of secondary resonances, we discuss superharmonic, subharmonic, and combination resonances.

7.3.1. PRIMARY RESONANCES

In this section we consider the primary resonance of the lowest mode. Thus we put

$$f_n(t) = 2\epsilon\pi^2 k_n \cos(\Omega t) \tag{7.3.3}$$

and

$$\Omega = \pi^2 + \epsilon\sigma_1 \tag{7.3.4}$$

Following the method of multiple scales we assume that

$$u_n(t;\epsilon) = u_{n0}(T_0, T_1) + \epsilon u_{n1}(T_0, T_1) + \cdots \tag{7.3.5}$$

Substituting (7.3.3) and (7.3.5) into (7.3.2), expanding the derivatives, and equating coefficients of like powers of ϵ, we obtain

$$D_0^2 u_{n0} + n^4\pi^4 u_{n0} = 0 \tag{7.3.6}$$

$$D_0^2 u_{n1} + n^4\pi^4 u_{n1} = -2D_0(D_1 u_{n0} + \mu_n u_{n0}) - \tfrac{1}{2}\pi^4 n^2 u_{n0} \sum_{m=1}^{\infty} m^2 u_{m0}^2 + 2\pi^2 k_n \cos(\Omega T_0) \quad \text{for } n = 1, 2, 3, \ldots \tag{7.3.7}$$

Writing the solution of (7.3.6) as

$$u_{n0} = A_n(T_1)\exp(in^2\pi^2 T_0) + cc \tag{7.3.8}$$

we find that

$$\begin{aligned} D_0^2 u_{n1} + n^4\pi^4 u_{n1} = {} & -2in^2\pi^2(A_n' + \mu_n A_n)\exp(in^2\pi^2 T_0) - \tfrac{1}{2}\pi^4 n^2 A_n \sum_{m=1}^{\infty} m^2 \\ & \cdot \{A_m^2 \exp[i(n^2 + 2m^2)\pi^2 T_0] + 2A_m\bar{A}_m \exp(in^2\pi^2 T_0) \\ & + \bar{A}_m^2 \exp[i(n^2 - 2m^2)\pi^2 T_0]\} + \pi^2 k_n \exp(i\Omega T_0) + cc \end{aligned} \tag{7.3.9}$$

Secular terms are eliminated from the u_{n1} if

$$2i(A_1' + \mu_1 A_1) + \tfrac{3}{2}\pi^2 A_1^2\bar{A}_1 + \pi^2 A_1 \sum_{m=2}^{\infty} m^2 A_m\bar{A}_m - k_1\exp(i\sigma T_1) = 0 \tag{7.3.10}$$

and for $n \geqslant 2$,

$$2i(A_n' + \mu A_n) + \tfrac{3}{2}n^2\pi^2 A_n^2\bar{A}_n + \pi^2 A_n \sum_{m\neq n}^{\infty} m^2 A_m\bar{A}_m = 0 \tag{7.3.11}$$

Letting

$$A_n = \tfrac{1}{2} a_n(T_1) \exp [i\beta_n(T_1)] \tag{7.3.12}$$

and

$$\gamma = \sigma T_1 - \beta_1 \tag{7.3.13}$$

where both a_n and β_n are real, substituting into (7.3.10) and (7.3.11), and separating the result into real and imaginary parts, we obtain

$$\begin{aligned} a_1' + \mu_1 a_1 - k_1 \sin \gamma &= 0 \\ a_1 \gamma' - a_1 \sigma + \tfrac{3}{16} \pi^2 a_1^3 + \tfrac{1}{8} \pi^2 a_1 \sum_{m=1}^{\infty} m^2 a_m^2 - k_1 \cos \gamma &= 0 \end{aligned} \tag{7.3.14}$$

and for $n \geqslant 2$,

$$\begin{aligned} a_n' + \mu_n a_n &= 0 \\ a_n \beta_n' - \tfrac{3}{16} n^2 \pi^2 a_n^3 - \tfrac{1}{8} \pi^2 a_n \sum_{m \neq n}^{\infty} m^2 a_m^2 &= 0 \end{aligned} \tag{7.3.15}$$

It follows from (7.3.15) that

$$a_n \propto \exp(-\mu_n T_1) \qquad \text{for } n \geqslant 2 \tag{7.3.16}$$

and hence that all a_n, except a_1, decay.

The steady-state motion ($a_1' = 0$ and $\gamma' = 0$) is described by

$$\begin{aligned} \mu_1 a_1 - k_1 \sin \gamma &= 0 \\ a_1(\sigma - \tfrac{3}{16} \pi^2 a_1^2) + k_1 \cos \gamma &= 0 \end{aligned} \tag{7.3.17}$$

and for $n \geqslant 2$,

$$a_n = 0 \tag{7.3.18}$$

Combining (7.3.17) leads to the following frequency-response equation:

$$\sigma = \tfrac{3}{16} \pi^2 a_1^2 \pm \left(\frac{k_1^2}{a_1^2} - \mu_1^2 \right)^{1/2} \tag{7.3.19}$$

This equation has the same form as (4.1.19). Thus this system is equivalent to a nonlinear hardening spring. The frequency-response curves are similar to those shown in Figure 4-1.

We note that

$$\omega_1 T_0 + \beta_1 = \omega_1 T_0 + \sigma T_1 - \gamma = (\omega_1 + \epsilon\sigma) T_0 - \gamma \tag{7.3.20}$$

and hence that the first approximation of the steady-state deflection is given by

$$w(x, t) = \sqrt{2}\, a_1 \cos(\Omega t - \gamma) \sin \pi x + O(\epsilon) \tag{7.3.21}$$

where a_1 and γ are given by (7.3.17).

When Ω is near ω_n, for $n \geqslant 2$, the results can be obtained in a similar fashion.

7.3.2. SUPERHARMONIC RESONANCES

In this section we consider the superharmonic resonance of the lowest mode. Bennett (1973) studied ultraharmonic motions of viscously damped beams. Thus we put

$$f_n(t) = 2K_n \cos(\Omega t) \tag{7.3.22}$$

where

$$3\Omega = \pi^2 + \epsilon\sigma \tag{7.3.23}$$

Substituting (7.3.5) and (7.3.22) into (7.3.2) and equating coefficients of like powers of ϵ we obtain

$$D_0^2 u_{n0} + n^4\pi^4 u_{n0} = 2K_n \cos \Omega T_0 \tag{7.3.24}$$

$$D_0^2 u_{n1} + n^4\pi^4 u_{n1} = -2D_0(D_1 u_{n0} + \mu_n u_{n0}) - \tfrac{1}{2}\pi^4 n^2 u_{n0} \sum_{m=1}^{\infty} m^2 u_{m0}^2 \tag{7.3.25}$$

The solution of (7.3.24) can be expressed as

$$u_{n0} = A_n(T_1)\exp(in^2\pi^2 T_0) + \Lambda_n \exp(i\Omega T_0) + cc \tag{7.3.26}$$

where

$$\Lambda_n = \frac{K_n}{\omega_n^2 - \Omega^2} = \frac{9K_n}{(9n^4 - 1)\pi^4} + O(\epsilon) \tag{7.3.27}$$

Then we find that secular terms are eliminated from the u_{n1} if

$$2i(A_1' + \mu_1 A_1) + \tfrac{3}{2}\pi^2 A_1(A_1\bar{A}_1 + 2\Lambda_1^2) + \pi^2 A_1 \sum_{m=2}^{\infty} m^2(A_m\bar{A}_m + \Lambda_m^2)$$
$$+ \tfrac{1}{2}\pi^2\Lambda_1\left(\sum_{m=1}^{\infty} m^2\Lambda_m^2\right)\exp(i\sigma T_1) = 0 \tag{7.3.28}$$

and for $n \geqslant 2$,

$$2i(A_n' + \mu_n A_n) + \tfrac{3}{2}n^2\pi^2 A_n(A_n\bar{A}_n + 2\Lambda_n^2) + \pi^2 A_n \sum_{m \neq n}^{\infty} m^2(A_m\bar{A}_m + \Lambda_m^2) = 0 \tag{7.3.29}$$

Substituting (7.3.12) and (7.3.13) into (7.3.28) and (7.3.29) leads to

$$a_1' + \mu_1 a_1 + F\sin\gamma = 0 \tag{7.3.30}$$

$$a_1(\gamma' - \sigma) + \tfrac{3}{2}\pi^2 a_1(\tfrac{1}{8}a_1^2 + \Lambda_1^2) + \tfrac{1}{2}\pi^2 a_1 \sum_{m=2}^{\infty} m^2(\tfrac{1}{4}a_m^2 + \Lambda_m^2) + F\cos\gamma = 0 \tag{7.3.30}$$

and for $n \geqslant 2$,

$$
\begin{gathered}
a_n' + \mu_n a_n = 0 \\
a_n \beta_n' - \tfrac{3}{2}\pi^2 n^2 a_n(\tfrac{1}{8}a_n^2 + \Lambda_n^2) - \tfrac{1}{2}\pi^2 a_n \sum_{m \neq n}^{\infty} m^2(\tfrac{1}{4}a_m^2 + \Lambda_m^2) = 0
\end{gathered}
\tag{7.3.31}
$$

where

$$
F = \tfrac{1}{2}\pi^2 \Lambda_1 \sum_{m=1}^{\infty} m^2 \Lambda_m^2 \tag{7.3.32}
$$

Again we find that for $n \geqslant 2$

$$
a_n \propto \exp(-\mu_n T_1) \tag{7.3.33}
$$

and hence that all a_n, except a_1, decay.

The steady-state motion ($a_1' = 0$ and $\gamma' = 0$) is described by

$$
\begin{gathered}
\mu_1 a_1 + F \sin \gamma = 0 \\
a_1(\sigma - H) - \tfrac{3}{16}\pi^2 a_1^3 - F \cos \gamma = 0
\end{gathered}
\tag{7.3.34}
$$

where

$$
H = \tfrac{1}{2}\pi^2 \left(3\Lambda_1^2 + \sum_{m=2}^{\infty} m^2 \Lambda_m^2 \right)
$$

Eliminating γ from (7.3.34) leads to the following frequency-response equation:

$$
\sigma = H + \tfrac{3}{16}\pi^2 a_1^2 \pm \left(\frac{F^2}{a_1^2} - \mu_1^2 \right)^{1/2} \tag{7.3.35}
$$

This equation has the same form as (4.1.43). The frequency-response curves are similar to those in Figure 4-10. The amplitude and frequency of the excitation appear in the coefficient of a_1 in (7.3.34). Otherwise these equations have the same form as (7.3.17). Thus in the case of a superharmonic resonance, increasing the amplitude of the excitation while the frequency is held constant can have the effect of detuning the resonance and may cause the amplitude of the free-oscillation term to decrease. This is illustrated in Figure 4-11.

The first approximation of the steady-state deflection is given by

$$
\begin{aligned}
w(x, t) = \sqrt{2}\, a_1 \cos(3\Omega t - \gamma) \sin \pi x + \frac{\sqrt{2}\, 18}{\pi^4} (\cos \Omega t) \\
\cdot \left[\sum_{n=1}^{\infty} K_n (9n^4 - 1)^{-1} \sin n\pi x \right] + O(\epsilon)
\end{aligned}
\tag{7.3.36}
$$

where a_1 and γ are given by (7.3.34).

7.3.3. SUBHARMONIC RESONANCES

In this section we consider the subharmonic resonance of the lowest mode. Thus we put

$$f_n(t) = 2K_n \cos(\Omega t) \tag{7.3.37}$$

where

$$\Omega = 3\pi^2 + \epsilon\sigma \tag{7.3.38}$$

Also we note that

$$3\Omega = 9\pi^2 + 3\epsilon\sigma = \omega_3 + 3\epsilon\sigma \tag{7.3.39}$$

Thus there is also a superharmonic resonance involving the third mode. Consequently in this case subharmonic and superharmonic resonances exist simultaneously. Again we assume the form given by (7.3.5). Hence u_{no} and u_{n1} are governed by (7.3.24) and (7.3.25). The solution of (7.3.24) has the form (7.3.26) but now

$$\Lambda_n = \frac{K_n}{\omega_n^2 - \Omega^2} = \frac{K_n}{(n^4 - 9)\pi^4} + O(\epsilon) \tag{7.3.40}$$

We find that secular terms are eliminated from the u_{n1} if

$$2i(A_1' + \mu_1 A_1) + \tfrac{3}{2}\pi^2 A_1(A_1\bar{A}_1 + 2\Lambda_1^2) + \pi^2 A_1 \sum_{m=2}^{\infty} m^2(A_m\bar{A}_m + \Lambda_m^2) + \tfrac{3}{2}\pi^2 \Lambda_1 \bar{A}_1^2 \exp(i\sigma T_1) = 0 \tag{7.3.41}$$

$$2i(A_3' + \mu_3 A_3) + \tfrac{27}{2}\pi^2 A_3(A_3\bar{A}_3 + 2\Lambda_3^2) + \pi^2 A_3 \sum_{m\neq 3}^{\infty} m^2(A_m\bar{A}_m + \Lambda_m^2) + \tfrac{1}{2}\pi^2 \Lambda_3 \sum_{m=1}^{\infty} m^2 \Lambda_m^2 \exp(3i\sigma T_1) = 0 \tag{7.3.42}$$

For all n except $n = 1$ and 3, the A_n are governed by (7.3.29).

Introducing the polar notation (7.3.12), one finds that

$$a_1' + \mu_1 a_1 + \tfrac{3}{8}\pi^2 \Lambda_1 a_1^2 \sin\gamma_1 = 0 \tag{7.3.43}$$

$$a_1\gamma_1' - a_1\left[\sigma - \tfrac{9}{2}\pi^2(\tfrac{1}{8}a_1^2 + \Lambda_1^2)\right] + \tfrac{3}{2}\pi^2 a_1 \sum_{m=2}^{\infty} m^2(\tfrac{1}{4}a_m^2 + \Lambda_m^2) + \tfrac{9}{8}\pi^2 \Lambda_1 a_1^2 \cos\gamma_1 = 0 \tag{7.3.44}$$

and

$$a_3' + \mu_3 a_3 + F\sin\gamma_3 = 0 \tag{7.3.45}$$

$$a_3\gamma_3' - 3a_3[\sigma - \tfrac{9}{2}\pi^2(\tfrac{1}{8}a_3^2 + \Lambda_3^2)] + \tfrac{1}{2}\pi^2 a_3 \sum_{m\neq 3}^{\infty} m^2(\tfrac{1}{4}a_m^2 + \Lambda_m^2) + F\cos\gamma_3 = 0 \tag{7.3.46}$$

where

$$\gamma_1 = \sigma T_1 - 3\beta_1, \qquad \gamma_3 = 3\sigma T_1 - \beta_3 \tag{7.3.47}$$

and

$$F = \tfrac{1}{2}\pi^2\Lambda_3 \sum_{m=1}^{\infty} m^2\Lambda_m^2 \tag{7.3.48}$$

It follows from (7.3.31) that for the steady-state response, all a_n except a_1 and a_3 must be zero. Thus the steady-state response is governed by

$$a_1(\mu_1 + \tfrac{3}{8}\pi^2\Lambda_1 a_1 \sin\gamma_1) = 0 \tag{7.3.49}$$

$$a_1[-\sigma + \tfrac{9}{16}\pi^2 a_1^2 + \tfrac{27}{8}\pi^2 a_3^2 + H_1 + \tfrac{9}{8}\pi^2\Lambda_1 a_1 \cos\gamma_1] = 0 \tag{7.3.50}$$

and

$$\mu_3 a_3 + F\sin\gamma_3 = 0 \tag{7.3.51}$$

$$a_3[-3\sigma + \tfrac{27}{16}\pi^2 a_3^2 + H_3 + \tfrac{1}{8}\pi^2 a_1^2] + F\cos\gamma_3 = 0 \tag{7.3.52}$$

where

$$H_1 = \tfrac{9}{2}\pi^2\Lambda_1^2 + \tfrac{3}{2}\pi^2 \sum_{m=2}^{\infty} m^2\Lambda_m^2 \quad \text{and} \quad H_3 = \tfrac{27}{2}\pi^2\Lambda_3^2 + \tfrac{1}{2}\pi^2 \sum_{m\neq 3}^{\infty} m^2\Lambda_m^2 \tag{7.3.53}$$

Two solutions are possible: either $a_1 = 0$ and $a_3 \neq 0$, or $a_1 \neq 0$ and $a_3 \neq 0$. For the former the solution is essentially given in Section 7.3.2. The remainder of the discussion applies to the latter.

Because γ_1 and γ_3 are constants,

$$\beta_3' = 9\beta_1' = 3\sigma \tag{7.3.54}$$

Thus the deflection has the following form:

$$w(x,t) = \sqrt{2}\,[a_1\cos(\tfrac{1}{3}\Omega t - \tfrac{1}{3}\gamma_1)\sin\pi x + a_3\cos(3\Omega t - \gamma_3)\sin 3\pi x] + \frac{2\sqrt{2}}{\pi^4}(\cos\Omega t)\left[\sum_{n=1}^{\infty} K_n(n^4 - 9)^{-1}\sin n\pi x\right] + O(\epsilon) \tag{7.3.55}$$

where a_1, a_3, γ_1, and γ_3 are obtained from (7.3.49) through (7.3.52).

As an example let us suppose that the force having magnitude B is applied at the midpoint of the beam. Then

$$f(x, t) = B\delta(x - \tfrac{1}{2}) \cos \Omega t \tag{7.3.56}$$

where δ is the delta function, and it follows that

$$f_n(t) = \int_0^1 f\phi_n \, dx = \sqrt{2}\, B \sin(\tfrac{1}{2}n\pi) \cos \Omega t \quad \text{and} \quad 2K_n = \sqrt{2}\, B \sin(\tfrac{1}{2}n\pi)$$

$$\Lambda_n = \frac{\sqrt{2}\, B \sin(\tfrac{1}{2}n\pi)}{2(n^4 - 9)\pi^4} \tag{7.3.57}$$

7.3.4. COMBINATION RESONANCES

In this section we put

$$f_n(t) = 2K_n \cos \Omega t \tag{7.3.58}$$

where

$$2\Omega = \omega_1 + \omega_2 + \epsilon\sigma = 5\pi^2 + \epsilon\sigma \tag{7.3.59}$$

Again we assume the form given by (7.3.5). Hence u_{no} and u_{n1} are governed by (7.3.24) and (7.3.25), respectively. The solution of (7.3.24) has the form given by (7.3.26); but now

$$\Lambda_n = \frac{4K_n}{(4n^4 - 25)\pi^4} \tag{7.3.60}$$

We find that secular terms are eliminated from the u_{n1} if

$$2i(A_1' + \mu_1 A_1) + \tfrac{3}{2}\pi^2 A_1(A_1\bar{A}_1 + 2\Lambda_1^2) + \pi^2 A_1 \sum_{m=2}^{\infty} m^2(A_m\bar{A}_m + \Lambda_m^2) + 4\pi^2 \Lambda_1 \Lambda_2 \bar{A}_2 \exp(i\sigma T_1) = 0 \tag{7.3.61}$$

$$2i(A_2' + \mu_2 A_2) + 6\pi^2 A_2(A_2\bar{A}_2 + 2\Lambda_2^2) + \pi^2 A_2 \sum_{m \neq 2}^{\infty} m^2(A_m\bar{A}_m + \Lambda_m^2) + \pi^2 \Lambda_1 \Lambda_2 \bar{A}_1 \exp(i\sigma T_1) = 0 \tag{7.3.62}$$

The remaining A_n are governed by (7.3.29).

Introducing the polar notation (7.3.12) we find that

$$a_1' + \mu_1 a_1 + 4Fa_2 \sin \gamma = 0 \tag{7.3.63}$$

$$a_2' + \mu_2 a_2 + Fa_1 \sin \gamma = 0 \tag{7.3.64}$$

$$-a_1\beta_1' + \tfrac{3}{16}\pi^2 a_1^3 + \frac{\pi^2}{8} a_1 \sum_{m=2}^{\infty} m^2 a_m^2 + H_1 a_1 + 4Fa_2 \cos \gamma = 0 \tag{7.3.65}$$

$$-a_2\beta_2' + \tfrac{3}{4}\pi^2 a_2^3 + \tfrac{1}{8}\pi^2 a_2 \sum_{m \neq 2} m^2 a_m^2 + H_2 a_2 + F a_1 \cos\gamma = 0 \quad (7.3.66)$$

$$\gamma = \sigma T_1 - \beta_1 - \beta_2 \quad (7.3.67)$$

where

$$F = \tfrac{1}{2}\pi^2 \Lambda_1 \Lambda_2, \qquad H_1 = \tfrac{1}{2}\pi^2 \left(3\Lambda_1^2 + \sum_{m=2}^{\infty} m^2 \Lambda_m^2\right),$$

$$H_2 = \pi^2 \left(6\Lambda_2^2 + \tfrac{1}{2} \sum_{m \neq 2} m^2 \Lambda_m^2\right) \quad (7.3.68)$$

The remaining a_n decay as T_1 (and hence t) increases. Thus it follows that there are two possibilities for the steady-state response: either all the a_n are zero, or all the a_n except a_1 and a_2 are zero.

When a_1 and a_2 are nonzero, (7.3.65) through (7.3.67) can be combined, and we find that the steady-state response is governed by

$$\mu_1 a_1 + 4F a_2 \sin\gamma = 0 \quad (7.3.69)$$

$$\mu_2 a_2 + F a_1 \sin\gamma = 0 \quad (7.3.70)$$

$$\sigma - \frac{5\pi^2}{16} a_1^2 - \frac{5\pi^2}{4} a_2^2 - F\left(\frac{4a_2}{a_1} + \frac{a_1}{a_2}\right) \cos\gamma - H_1 - H_2 = 0 \quad (7.3.71)$$

One can solve (7.3.69) through (7.3.71) for a_1, a_2, and γ and then calculate the constants β_1' and β_2' from (7.3.65) and (7.3.66). The deflection is given by

$$w(x, t) = \sqrt{2}\, a_1 \cos\left[(\omega_1 + \epsilon\beta_1')t - \delta_1\right] \sin \pi x + \sqrt{2}\, a_2 \cos\left[\omega_2 + \epsilon\beta_2')t - \delta_2\right]$$

$$\cdot \sin 2\pi x + 8\sqrt{2}\pi^{-4} \cos(\Omega t) \sum_{n=1}^{\infty} K_n \left[(4n^2 - 25)^{-1} \sin n\pi x\right] + O(\epsilon) \quad (7.3.72)$$

where δ_1 and δ_2 are constants to be determined from the initial conditions. We note that

$$(\omega_1 + \epsilon\beta_1')t - \delta_1 + (\omega_2 + \epsilon\beta_2')t - \delta_2 = 2\Omega t - \gamma \quad (7.3.73)$$

A mode is said to be directly excited if its frequency is involved with the frequency of the excitation in a resonant combination. As we have just seen, for hinged-hinged beams only directly excited modes appear in the first approximation of the deflection because of the vanishing of the coupling due to internal resonance. Next we consider an example in which a mode that is not directly excited can appear in the first approximation due to internal resonance. It frequently turns out that such a mode dominates the deflection.

7.4. Hinged-Clamped Beams

As in the case of hinged-hinged beams, most of the earlier studies on the nonlinear vibrations of hinged-clamped beams were limited to the study of undamped, free vibrations. Special consideration was given to the determination of the nonlinear frequency-amplitude relationship of single modes (Evensen, 1968; Mei, 1972, 1973a; Nayfeh, 1973a). For hinged-clamped uniform beams, the second natural frequency is approximately three times the first natural frequency. Since the coupling coefficients do not vanish, an internal resonance exists. The influence of this internal resonance on the nonlinear forced response to a single-frequency excitation was examined by Nayfeh, Mook, and Sridhar (1974); Nayfeh, Mook, and Lobitz (1974); and Sridhar, Nayfeh, and Mook (1975). Although the frequency of the excitation may be near the frequency of the second mode, the response most likely will be dominated by the first mode as a result of the internal resonance. In the next two sections we discuss primary resonances, in Section 7.4.3 we discuss superharmonic resonances, and in Section 7.4.4 we discuss combination resonances.

Here we choose the characteristic length to be one half the actual length of the beam. Then it follows from (7.1.30) through (7.1.32) that for a beam hinged at $x = 0$ and clamped at $x = 2$,

$$\phi_n = C_n [\sin(\eta_n x) \sinh(\eta_n l) - \sin(\eta_n l) \sinh(\eta_n x)] \tag{7.4.1}$$

where the C_n are chosen to make the ϕ_n orthonormal, $\omega_n = \eta_n^2$, and the η_n are the roots of

$$\tan(2\eta_n) - \tanh(2\eta_n) = 0 \tag{7.4.2}$$

The five lowest natural frequencies ω_n are

$$\omega_1 = 3.8545, \quad \omega_2 = 12.491, \quad \omega_3 = 26.062, \quad \omega_4 = 44.568, \quad \omega_5 = 68.007 \tag{7.4.3}$$

We note that ω_1 and ω_2 are nearly in the ratio of 1 to 3. Thus there is an internal resonance because the coefficients of the coupling terms do not vanish in this case. The significance of this will become apparent as the analysis proceeds. To express the nearness of $3\omega_1$ to ω_2 quantitatively, we introduce a detuning parameter σ according to

$$\omega_2 = 3\omega_1 + \epsilon\sigma_1 \tag{7.4.4}$$

where $\epsilon\sigma_1 = 0.9275$.

Next we consider primary and secondary resonances.

7.4.1. PRIMARY RESONANCES, THE CASE OF Ω NEAR ω_1

In this case we put

$$f_n = 2\epsilon k_n \cos \Omega t \tag{7.4.5}$$

where

$$\Omega = \omega_1 + \epsilon\sigma_2 \tag{7.4.6}$$

Again we assume

$$u_n(t;\epsilon) = u_{n0}(T_0, T_1) + \epsilon u_{n1}(T_0, T_1) + \cdots \tag{7.4.7}$$

Substituting (7.4.7) into (7.1.38) and equating coefficients of like powers of ϵ, we obtain

$$D_0^2 u_{n0} + \omega_n^2 u_{n0} = 0 \tag{7.4.8}$$

$$D_0^2 u_{n1} + \omega_n^2 u_{n1} = -2D_0 D_1 u_{n0} - 2\mu_n D_0 u_{n0} + \sum_{m,p,q} \Gamma_{nmpq} u_{m0} u_{p0} u_{q0} + 2k_n \cos \Omega T_0 \tag{7.4.9}$$

We note that (7.4.9) does not have the simple form of (7.3.7).

We write the solution of (7.4.8) as follows:

$$u_{n0} = A_n(T_1) \exp(i\omega_n T_0) + cc \tag{7.4.10}$$

Then (7.4.9) becomes

$$\begin{aligned} D_0^2 u_{n1} + \omega_n^2 u_{n1} = {} & -2i\omega_n(A_n' + \mu_n A_n) \exp(i\omega_n T_0) + \sum_{m,p,q} \Gamma_{nmpq} \\ & \cdot \{A_m A_p A_q \exp[i(\omega_m + \omega_p + \omega_q) T_0] \\ & + \bar{A}_m A_p A_q \exp[i(\omega_p + \omega_q - \omega_m) T_0] \\ & + A_m \bar{A}_p A_q \exp[i(\omega_m + \omega_q - \omega_p) T_0] + A_m A_p \bar{A}_q \\ & \cdot \exp[i(\omega_m + \omega_p - \omega_q) T_0]\} + k_n \exp(i\Omega T_0) + cc \end{aligned} \tag{7.4.11}$$

It follows that secular terms are eliminated from the u_{n1} if

$$\begin{aligned} -2i\omega_1(A_1' + \mu_1 A_1) + A_1 \sum_{j=1}^{\infty} \alpha_{1j} A_j \bar{A}_j + Q_1 \bar{A}_1^2 A_2 \exp(i\sigma_1 T_1) \\ + k_1 \exp(i\sigma_2 T_1) = 0 \end{aligned} \tag{7.4.12}$$

$$-2i\omega_2(A_2' + \mu_2 A_2) + A_2 \sum_{j=1}^{\infty} \alpha_{2j} A_j \bar{A}_j + Q_2 A_1^3 \exp(-i\sigma_1 T_1) = 0 \tag{7.4.13}$$

and for $n \geqslant 3$

$$-2i\omega_n(A'_n + \mu_n A_n) + A_n \sum_{j=1}^{\infty} \alpha_{nj} A_j \bar{A}_j = 0 \tag{7.4.14}$$

where

$$Q_1 = 3\Gamma_{1112} = -2.311, \quad Q_2 = \Gamma_{2111} = \tfrac{1}{3} Q_1 = -0.7703 \tag{7.4.15}$$

and

$$\alpha_{jn} = \alpha_{nj} = \begin{cases} 2(2\Gamma_{njnj} + \Gamma_{nnjj}) & \text{for } n \neq j \\ 3\Gamma_{nnnn} \end{cases} \tag{7.4.16}$$

It follows from (7.4.14) that all the A_n, except A_1 and A_2, decay. Though the second mode is not directly excited (i.e., ω_2 is not involved in a resonant combination with Ω), it does not follow that A_2 decays. The equation governing A_2 contains the term $Q_2 A_1^3 \exp(-i\sigma_1 T_1)$, which is a consequence of the internal resonance. As a rule all modes except those which are directly excited and those which are involved with a directly excited mode through an internal resonance decay.

For the steady-state motions only the following α_{ij} are needed:

$$\alpha_{11} = 3\Gamma_{1111} = -18.64, \quad \alpha_{22} = 3\Gamma_{2222} = -86.26,$$

$$\alpha_{33} = 3\Gamma_{3333} = -414.45, \quad \alpha_{12} = 2(\Gamma_{1122} + 2\Gamma_{1212}) = \alpha_{21} = -16.58,$$

$$\alpha_{13} = 2(\Gamma_{1133} + 2\Gamma_{1313}) = \alpha_{31} = -34.73,$$

$$\alpha_{23} = 2(\Gamma_{2233} + 2\Gamma_{2323}) = \alpha_{32} = -129.9$$

We let

$$A_n = \tfrac{1}{2} a_n(T_1) \exp[i\beta_n(T_1)] \tag{7.4.17}$$

$$\gamma_1 = \sigma_1 T_1 - 3\beta_1 + \beta_2, \quad \gamma_2 = \sigma_2 T_1 - \beta_1 \tag{7.4.18}$$

Then for steady-state motions ($a'_1 = a'_2 = 0$ and $\gamma'_1 = \gamma'_2 = 0$), (7.4.12) and (7.4.13) become

$$-\omega_1 \mu_1 a_1 + \tfrac{1}{8} Q_1 a_1^2 a_2 \sin\gamma_1 + k_1 \sin\gamma_2 = 0 \tag{7.4.19}$$

$$\omega_1 \sigma_2 a_1 + \tfrac{1}{8}(\alpha_{11} a_1^2 + \alpha_{12} a_2^2) a_1 + \tfrac{1}{8} Q_1 a_1^2 a_2 \cos\gamma_1 + k_1 \cos\gamma_2 = 0 \tag{7.4.20}$$

$$\omega_2 \mu_2 a_2 + \tfrac{1}{8} Q_2 a_1^3 \sin\gamma_1 = 0 \tag{7.4.21}$$

$$\omega_2(3\sigma_2 - \sigma_1) a_2 + \tfrac{1}{8}(\alpha_{21} a_1^2 + \alpha_{22} a_2^2) a_2 + \tfrac{1}{8} Q_2 a_1^3 \cos\gamma_1 = 0 \tag{7.4.22}$$

We note that the terms containing a_1^3 in (7.4.21) and (7.4.22), which are a result of the internal resonance, are similar to the forcing terms in a superharmonic resonance. Hence neither a_1 nor a_2 can be zero. Moreover it follows from (7.4.18) that the nonlinearity adjusts the frequencies of the first and second modes so that they are precisely in the ratio of 1 to 3 and the frequency of the first mode equals the frequency of the excitation. This can be shown as follows:

$$\omega_1 T_0 + \beta_1 = \omega_1 T_0 + \sigma_2 T_1 - \gamma_2 = (\omega_1 + \epsilon\sigma_2)t - \gamma_2 = \Omega t - \gamma_2 \tag{7.4.23}$$

$$\omega_2 T_0 + \beta_2 = \omega_2 T_0 + (3\sigma_2 - \sigma_1)T_1 + \gamma_1 - 3\gamma_2 = [(\omega_2 - \epsilon\sigma_1) + 3\epsilon\sigma_2]\,t$$
$$+ \gamma_1 - 3\gamma_2 = 3(\omega_1 + \epsilon\sigma_2)t + \gamma_1 - 3\gamma_2 = 3\Omega t + \gamma_1 - 3\gamma_2 \tag{7.4.24}$$

In Figure 7-2 the amplitudes are plotted as functions of the frequency of the excitation for several values of ϵk_1. Though a_2 cannot be zero, it is small compared with a_1 and the response is dominated by the first mode. Equations (7.4.12) and (7.4.13) were integrated numerically when $A_n \equiv 0$ for $n \geqslant 3$. These results are shown by the small circles.

In Figure 7-3 the variation of a_1 with ϵk_1 is shown. The log-log plot in Figure 7-3*b* compares qualitatively with the plot of the experimental data of Jacobson and van der Heyde (1972), though they experimented with honeycomb panels.

Finally we note that in the first approximation the deflection is given by

$$w(x, t) = a_1 \cos(\Omega t - \gamma_2)\,\phi_1(x) + a_2 \cos(3\Omega t + \gamma_1 - 3\gamma_2)\,\phi_2(x) + O(\epsilon) \tag{7.4.25}$$

where a_1, a_2, γ_1, and γ_2 are obtained from (7.4.19) through (7.4.22).

7.4.2. PRIMARY RESONANCES, THE CASE OF Ω NEAR ω_2

In this case we put

$$\Omega = \omega_2 + \epsilon\sigma_2 \tag{7.4.26}$$

Instead of (7.4.18) through (7.4.22), we obtain

$$-\omega_1\mu_1 a_1 + \tfrac{1}{8}Q_1 a_1^2 a_2 \sin\gamma_1 = 0 \tag{7.4.27}$$

$$\omega_1(\sigma_1 + \sigma_2)a_1 + \tfrac{3}{8}(\alpha_{11}a_1^2 + \alpha_{12}a_2^2)a_1 + \tfrac{3}{8}Q_1 a_1^2 a_2 \cos\gamma_1 = 0 \tag{7.4.28}$$

$$-\omega_2\mu_2 a_2 - \tfrac{1}{8}Q_2 a_1^3 \sin\gamma_1 + k_2 \sin\gamma_2 = 0 \tag{7.4.29}$$

$$\omega_2\sigma_2 a_2 + \tfrac{1}{8}(\alpha_{21}a_1^2 + \alpha_{22}a_2^2)a_2 + \tfrac{1}{8}Q_2 a_1^3 \cos\gamma_1 + k_2 \cos\gamma_2 = 0 \tag{7.4.30}$$

where

$$\gamma_1 = \sigma_1 T_1 - 3\beta_1 + \beta_2 \quad \text{and} \quad \gamma_2 = \sigma_2 T_1 - \beta_2 \tag{7.4.31}$$

We note that a_2 cannot be zero, as in Section 7.4.1. But in contrast with Section 7.4.1, $a_1 = 0$ is a possible solution. The terms proportional to $a_1^2 a_2$ in (7.4.27) and (7.4.28), which are a result of the internal resonance, are similar to the forcing terms in a subharmonic resonance. Thus there are two possibilities:

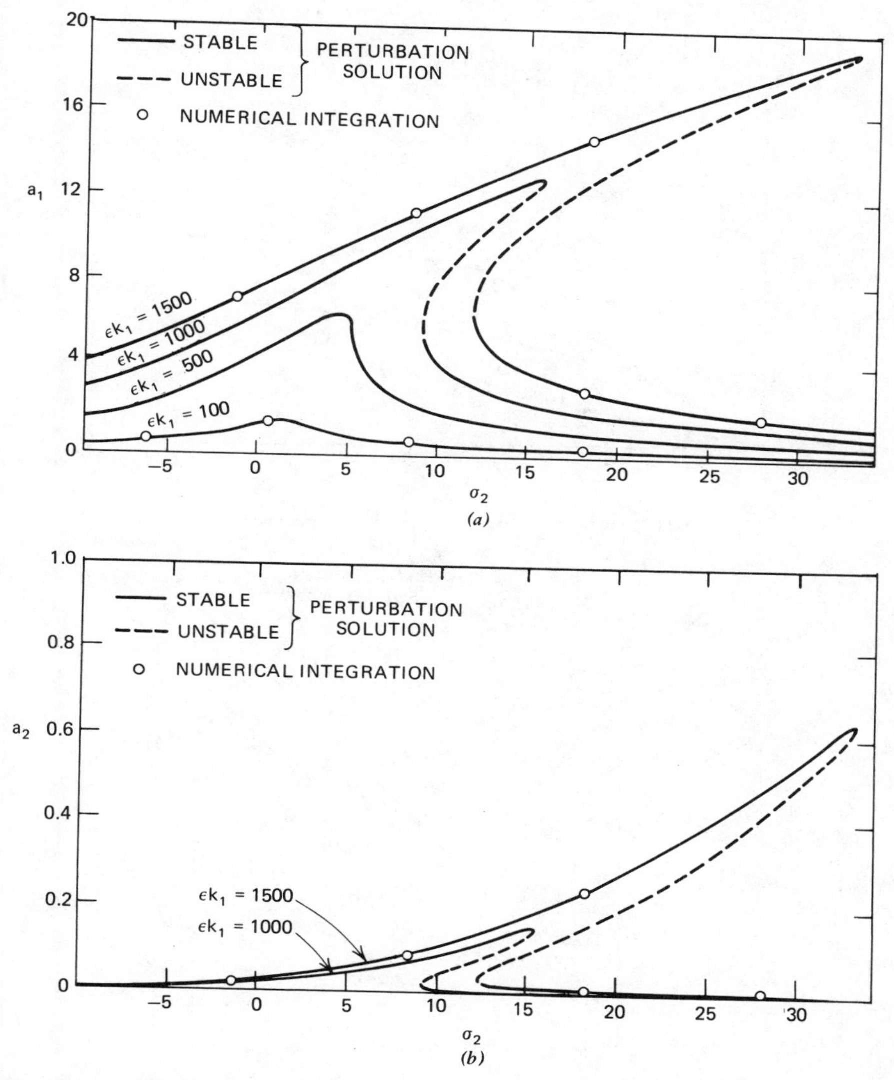

Figure 7-2. Frequency–response curves for a hinged-clamped beam for the case of a primary resonance of the fundamental mode: (*a*) first mode; (*b*) second mode.

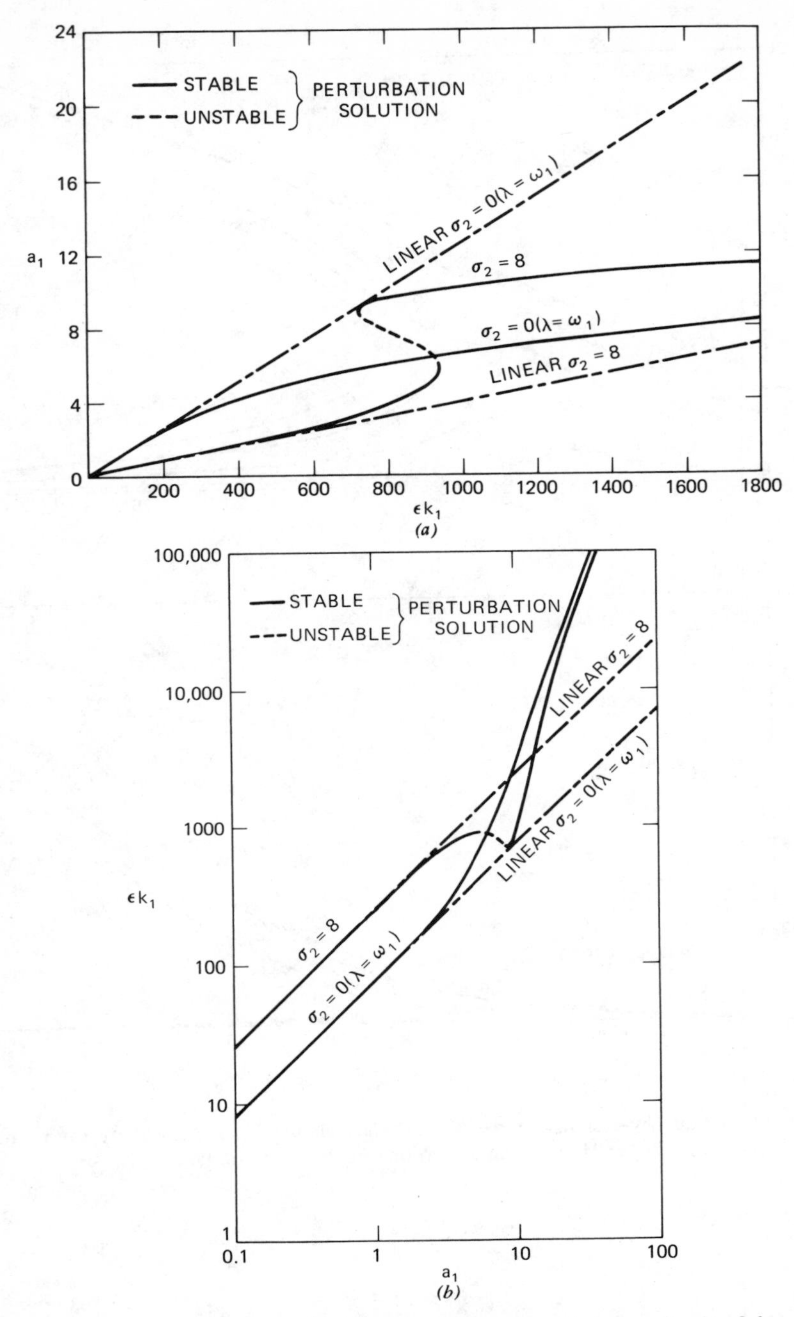

Figure 7-3. Variation of the amplitude of the first mode with the amplitude of the excitation: (*a*) linear scale; (*b*) logarithmic scale.

(1) $a_1 = 0$ and $a_2 \neq 0$, and (2) $a_1 \neq 0$ and $a_2 \neq 0$. When $a_1 = 0$, the solution closely resembles the single-mode solutions discussed in Section 7.3. However, when $a_1 \neq 0$, the results can differ surprisingly from any expectations based on a linear analysis or a single-mode solution.

The variation of a_1 and a_2 with the frequency of the excitation is shown in Figure 7-4*a* for the case when $a_1 \neq 0$ and in Figure 7-4*b* for the case when $a_1 = 0$. For comparison the stable portions of Figure 7-4*a* are repeated in Figure 7-4*b*. The variation of a_1 and a_2 with the amplitude of the excitation is shown in Figure 7-5. For the case when $a_1 \neq 0$, only the stable portions are shown. Also shown in Figure 7-5 are results obtained by numerical integration. We note that the amplitude of the excitation must exceed a critical value before the first mode can be excited. Moreover it is interesting to note that when $a_1 \neq 0$, it can be much larger than a_2. Thus the response can be dominated by the first mode in spite of the fact that the frequency of the excitation is near the second natural frequency.

It follows from (7.4.31) that

$$\omega_1 T_0 + \beta_1 = [\omega_1 + \tfrac{1}{3}\epsilon(\sigma_1 + \sigma_2)]\, t - \tfrac{1}{3}(\gamma_1 + \gamma_2)] = \tfrac{1}{3}(\Omega t - \gamma_1 - \gamma_2) \qquad (7.4.32)$$

and

$$\omega_2 T_0 + \beta_2 = (\omega_2 + \epsilon\sigma_2)\, t - \gamma_2 = \Omega t - \gamma_2 \qquad (7.4.33)$$

Thus the nonlinearity adjusts the frequencies of the first and second modes so that the frequency of the first is exactly one third that of the second and the frequency of the second is equal to that of the excitation.

In the first approximation the deflection is given by

$$w(x, t) = a_2 \cos(\Omega t - \gamma_2)\, \phi_2(x) + O(\epsilon) \qquad (7.4.34)$$

when $a_1 = 0$ and by

$$w(x, t) = a_1 \cos[\tfrac{1}{3}(\Omega t - \gamma_1 - \gamma_2)]\, \phi_1(x) + a_2 \cos(\Omega t - \gamma_2)\, \phi_2(x) + O(\epsilon) \qquad (7.4.35)$$

where a_1, a_2, γ_1, and γ_2 are obtained from (7.4.27) through (7.4.30).

For values of the amplitude and frequency of the excitation considered in Figures 7-4 and 7-5, we used (7.4.34) and (7.4.35) to construct deflection curves at various times. These are shown in Figure 7-6. In Figure 7-6*a*, $a_1 = 0$; while in Figure 7-6*b*, $a_1 \neq 0$. Figure 7-6*b* shows how closely the actual response resembles the first mode, further illustrating the far-reaching effects of the internal resonance. In Figure 7-6*a* nearly one cycle is completed when $t = 0.45$, while in Figure 7-6*b* approximately one half of a cycle is completed when $t = 0.75$. The deflection and the length are not plotted to the same scale.

The initial conditions determine whether (7.4.34) or (7.4.35) describes the actual deflections.

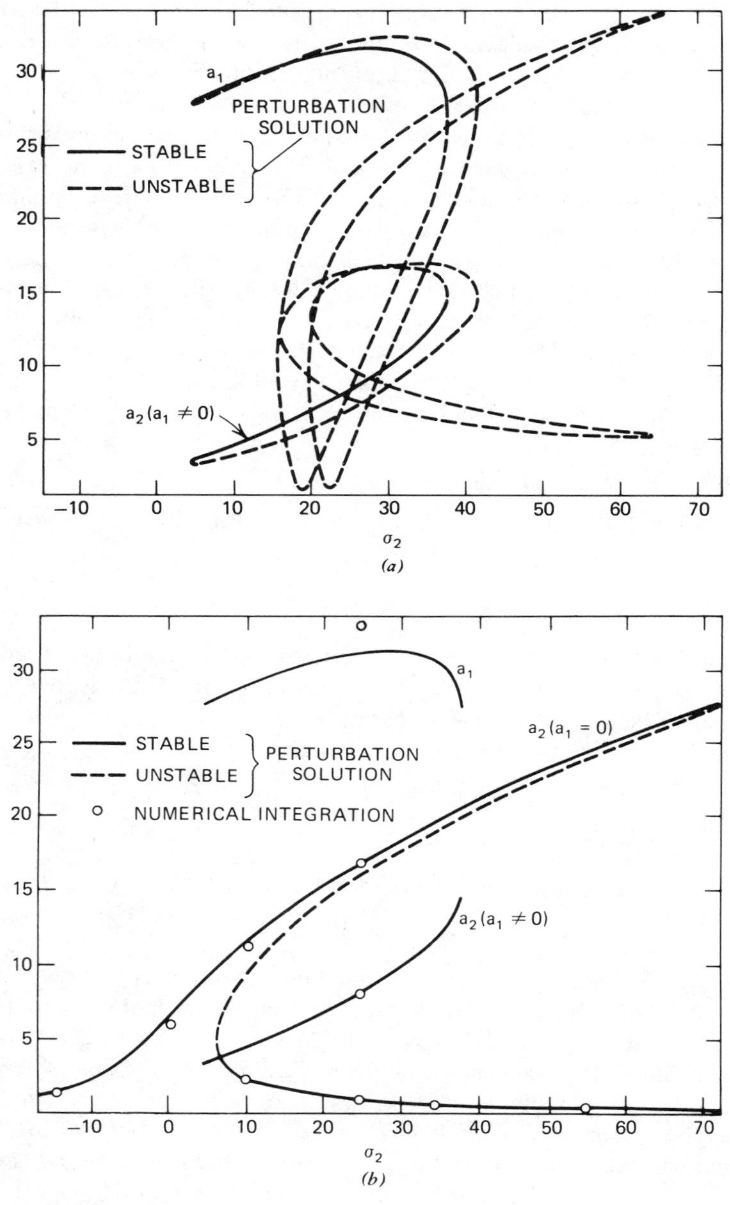

Figure 7-4. Frequency–response curves for a hinged-clamped beam for the case of a primary resonance of the second mode: (*a*) entire solution; (*b*) stable portion of the solution only.

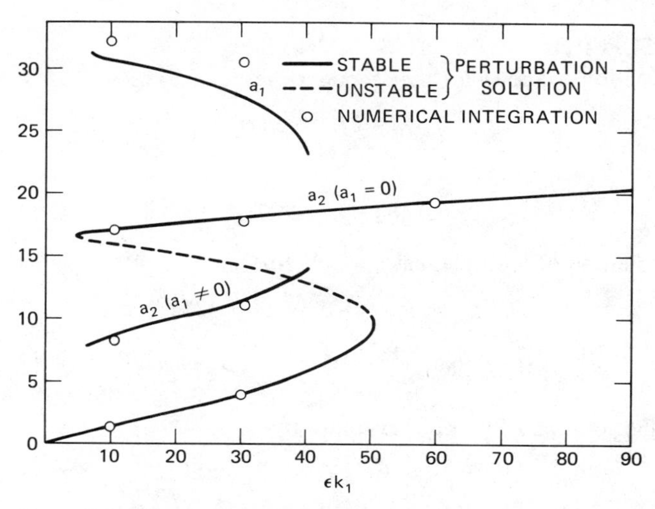

Figure 7-5. Response curves for a hinged-clamped beam; $\Omega \simeq \omega_1$.

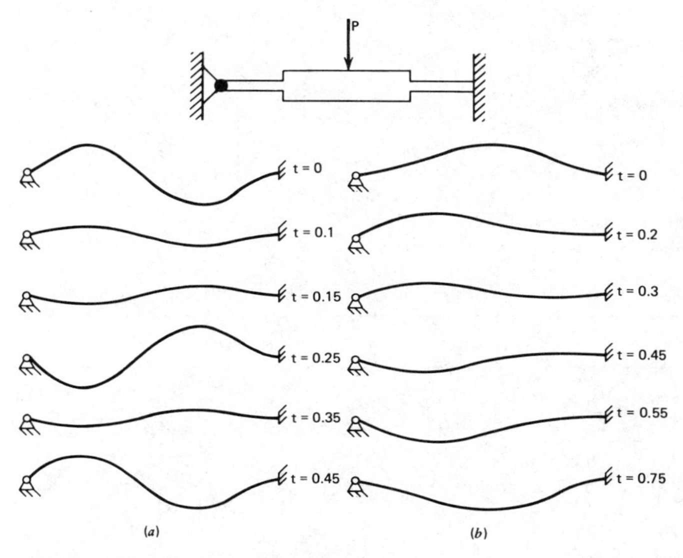

Figure 7-6. Deflection curves: $\omega_2 \simeq 3\omega_1$, $\Omega \simeq \omega_2$: (*a*) $a_1 = 0$ and (*b*) $a_1 \neq 0$.

7.4.3. SUPERHARMONIC RESONANCES

In this section we consider the secondary resonance case of $3\Omega \approx \omega_1$ and put

$$f_n = 2K_n \cos \Omega t \tag{7.4.36}$$

where

$$3\Omega = \omega_1 + \epsilon\sigma_2 \tag{7.4.37}$$

Using the method of multiple scales we obtain

$$D_0^2 u_{n0} + \omega_n^2 u_{n0} = K_n \exp(i\Omega T_0) + cc \tag{7.4.38}$$

$$D_0^2 u_{n1} + \omega_n^2 u_{n1} = -2D_0 D_1 u_{n0} - 2\mu_n D_0 u_{n0} + \sum_{m,p,q} \Gamma_{nmpq} u_{m0} u_{p0} u_{q0} \tag{7.4.39}$$

We write the solution of (7.4.38) as follows:

$$u_{n0} = A_n(T_1) \exp(i\omega_n T_0) + \Lambda_n \exp(i\Omega T_0) + cc \tag{7.4.40}$$

where

$$\Lambda_n = K_n(\omega_n^2 - \Omega^2)^{-1} \tag{7.4.41}$$

Substituting (7.4.40) into (7.4.39) leads to

$$D_0^2 u_{n1} + \omega_n^2 u_{n1} = -2i\omega_n(A_n' + \mu_n A_n) \exp(i\omega_n T_0) - 2i\Omega\mu_n \Lambda_n \exp(i\Omega T_0) + \sum_{m,p,q} \Gamma_{nmpq} \left[\sum_{j=1}^{27} S_j \exp(i\lambda_j T_0)\right] \tag{7.4.42}$$

where the S_j and λ_j are listed in Table 7-2.

TABLE 7-2. Coefficients for Equation (7.4.42)

j	S_j	λ_j	j	S_j	λ_j
1	$A_m A_p A_q$	$\omega_m + \omega_p + \omega_q$	15	$A_q \Lambda_m \Lambda_p$	$-2\Omega + \omega_q$
2	$A_m A_p \bar{A}_q$	$\omega_m + \omega_p - \omega_q$	16	$A_m A_p \Lambda_q$	$\Omega + \omega_m + \omega_p$
3	$A_m \bar{A}_p A_q$	$\omega_m - \omega_p + \omega_q$	17	$A_m A_p \Lambda_q$	$-\Omega + \omega_m + \omega_p$
4	$\bar{A}_m A_p A_q$	$-\omega_m + \omega_p + \omega_q$	18	$A_p A_q \Lambda_m$	$\Omega + \omega_p + \omega_q$
5	$2A_m \Lambda_p \Lambda_q$	ω_m	19	$A_p A_q \Lambda_m$	$-\Omega + \omega_p + \omega_q$
6	$2A_p \Lambda_q \Lambda_m$	ω_p	20	$A_q A_m \Lambda_p$	$\Omega + \omega_q + \omega_m$
7	$2A_q \Lambda_m \Lambda_p$	ω_q	21	$A_q A_m \Lambda_p$	$-\Omega + \omega_q + \omega_m$
8	$3\Lambda_m \Lambda_p \Lambda_q$	Ω	22	$A_m \bar{A}_p \Lambda_q$	$\Omega + \omega_m - \omega_p$
9	$\Lambda_m \Lambda_p \Lambda_q$	3Ω	23	$A_m \bar{A}_p \Lambda_q$	$-\Omega + \omega_m - \omega_p$
10	$A_m \Lambda_p \Lambda_q$	$2\Omega + \omega_m$	24	$A_p \bar{A}_q \Lambda_m$	$\Omega + \omega_p - \omega_q$
11	$A_m \Lambda_p \Lambda_q$	$-2\Omega + \omega_m$	25	$A_p \bar{A}_q \Lambda_m$	$-\Omega + \omega_p - \omega_q$
12	$A_p \Lambda_q \Lambda_m$	$2\Omega + \omega_p$	26	$A_q \bar{A}_m \Lambda_p$	$\Omega + \omega_q - \omega_m$
13	$A_p \Lambda_q \Lambda_m$	$-2\Omega + \omega_p$	27	$A_q \bar{A}_m \Lambda_p$	$-\Omega + \omega_q - \omega_m$
14	$A_q \Lambda_m \Lambda_p$	$2\Omega + \omega_q$			

It follows from (7.4.42) and Table 7-2 that secular terms will be eliminated from the u_{n1} if

$$-2i\omega_1(A_1' + \mu_1 A_1) + 2H_{11}A_1 + A_1 \sum_{j=1}^{\infty} \alpha_{1j}A_j\bar{A}_j + Q_1\bar{A}_1^2 A_2 \exp(i\sigma_1 T_1) + F_1 \exp(i\sigma_2 T_1) = 0 \quad (7.4.43)$$

$$-2i\omega_2(A_2' + \mu_2 A_2) + 2H_{22}A_2 + A_2 \sum_{j=1}^{\infty} \alpha_{2j}A_j\bar{A}_j + Q_2 A_1^3 \exp(-i\sigma_1 T_1) = 0 \quad (7.4.44)$$

and for $n \geqslant 3$

$$-2i\omega_n(A_n' + \mu_n A_n) + 2H_{nn}A_n + A_n \sum_{j=1}^{\infty} \alpha_{nj}A_j\bar{A}_j = 0 \quad (7.4.45)$$

where Q_1, Q_2, and the α_{nj} are defined as before in (7.4.15) and (7.4.16), and

$$F_n = \sum_{m,p,q} \Gamma_{nmpq}\Lambda_m\Lambda_p\Lambda_q \quad (7.4.46)$$

$$H_{nk} = \sum_{m,j} (\Gamma_{nkmj} + 2\Gamma_{nmkj})\Lambda_m\Lambda_j \quad (7.4.47)$$

As before all A_n except A_1 and A_2 decay. Writing A_1 and A_2 in the polar form (7.4.17), we find the following equations which govern the steady-state motion:

$$-\omega_1\mu_1 a_1 + \tfrac{1}{8}Q_1 a_1^2 a_2 \sin\gamma_1 + F_1 \sin\gamma_2 = 0 \quad (7.4.48)$$

$$(\omega_1\sigma_2 + H_{11})a_1 + \tfrac{1}{8}(\alpha_{11}a_1^2 + \alpha_{12}a_2^2)a_1 + \tfrac{1}{8}Q_1 a_1^2 a_2 \cos\gamma_1 + F_1 \cos\gamma_2 = 0 \quad (7.4.49)$$

$$\omega_2\mu_2 a_2 + \tfrac{1}{8}Q_2 a_1^3 \sin\gamma_1 = 0 \quad (7.4.50)$$

$$[\omega_2(3\sigma_2 - \sigma_1) + H_{22}]\, a_2 + \tfrac{1}{8}(\alpha_{21}a_1^2 + \alpha_{22}a_2^2)a_2 + \tfrac{1}{8}Q_2 a_1^3 \cos\gamma_1 = 0 \quad (7.4.51)$$

where $\gamma_1 = \sigma_1 T_1 - 3\beta_1 + \beta_2$ and $\gamma_2 = \sigma_2 T_1 - \beta_1$.

We note that the nonlinearity adjusts the frequencies of the first and second modes so that they are exactly in the ratio of 1 to 3 and the frequency of the first mode equals three times that of the excitation.

As an example we consider a uniformly distributed load acting along the entire length of the beam. Thus

$$2K_n = B\int_0^2 \phi_n(x)\, dx \quad (7.4.52)$$

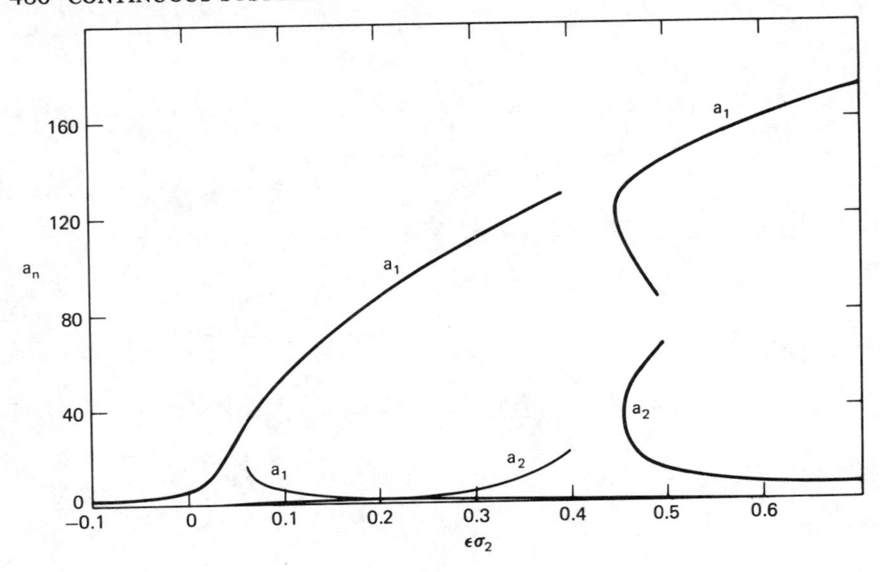

Figure 7-7. Frequency–response curves: $\omega_2 \simeq 3\omega_1$; $3\Omega \simeq \omega_1$.

Then for $B = 300$, it follows from (7.4.41), (7.4.46), and (7.4.47) that

$$F_1 = -9459, \qquad H_{11} = -1710, \qquad H_{22} = -2264 \tag{7.4.53}$$

For some arbitrary values of the amplitude of the excitation B and the damping coefficients, the variation of a_1 and a_2 with the frequency of the excitation is shown in Figure 7-7. Only the stable portions of the solution are shown. We note that a_1 is always greater than a_2.

In the first approximation the deflection is given by

$$w(x, t) = a_1 \cos(3\Omega t - \gamma_2)\,\phi_1(x) + a_2 \cos(9\Omega t - 3\gamma_2 + \gamma_1)\,\phi_2(x) + 2\cos\Omega t \sum_{n=1}^{\infty} K_n(\omega_n^2 - \Omega^2)^{-1}\phi_n(x) + O(\epsilon) \tag{7.4.54}$$

7.4.4. COMBINATION RESONANCES

In this section we consider the combination resonance case of $2\Omega \approx \omega_2 + \omega_3$ and put

$$f_n = 2K_n \cos\Omega t \tag{7.4.55}$$

where

$$2\Omega = \omega_2 + \omega_3 + \epsilon\sigma_2 \tag{7.4.56}$$

Referring to Table 7-2, we find that secular terms will be eliminated from the u_{n1} if

$$-2i\omega_1(A_1' + \mu_1 A_1) + 2H_{11}A_1 + A_1 \sum_{j=1}^{\infty} \alpha_{1j}A_j\bar{A}_j + Q_1\bar{A}_1^2 A_2 \exp(i\sigma_1 T_1) = 0 \tag{7.4.57}$$

$$-2i\omega_2(A_2' + \mu_2 A_2) + 2H_{22}A_2 + A_2 \sum_{j=1}^{\infty} \alpha_{2j}A_j\bar{A}_j + Q_2 A_1^3 \exp(-i\sigma_1 T_1) + H_{23}\bar{A}_3 \exp(i\sigma_2 T_1) = 0 \tag{7.4.58}$$

$$-2i\omega_3(A_3' + \mu_3 A_3) + 2H_{33}A_3 + A_3 \sum_{j=1}^{\infty} \alpha_{3j}A_j\bar{A}_j + H_{32}\bar{A}_2 \exp(i\sigma_2 T_1) = 0 \tag{7.4.59}$$

and for $n \geqslant 4$

$$-2i\omega_n(A_n' + \mu_n A_n) + 2H_{nn}A_n + A_n \sum_{j=1}^{\infty} \alpha_{nj}A_j\bar{A}_j = 0 \tag{7.4.60}$$

where the H_{nk} are defined by (7.4.47).

Here we find that all the A_n, except A_1, A_2, and A_3, decay. Ignoring all A_n for $n \geqslant 4$ and using the polar form (7.4.17), we find the following equations which govern the amplitudes and phases of the first three modes:

$$-\omega_1(a_1' + \mu_1 a_1) + \tfrac{1}{8}Q_1 a_1^2 a_2 \sin\gamma_1 = 0 \tag{7.4.61}$$

$$\omega_1 a_1 \beta_1' + H_{11}a_1 + \tfrac{1}{8}Q_1 a_1^2 a_2 \cos\gamma_1 + \tfrac{1}{8}a_1(\alpha_{11}a_1^2 + \alpha_{12}a_2^2 + \alpha_{13}a_3^2) = 0 \tag{7.4.62}$$

$$-\omega_2(a_2' + \mu_2 a_2) - \tfrac{1}{8}Q_2 a_1^3 \sin\gamma_1 + \tfrac{1}{2}H_{23}a_3 \sin\gamma_2 = 0 \tag{7.4.63}$$

$$\omega_2 a_2 \beta_2' + H_{22}a_2 + \tfrac{1}{8}Q_2 a_1^3 \cos\gamma_1 + \tfrac{1}{8}a_2(\alpha_{21}a_1^2 + \alpha_{22}a_2^2 + \alpha_{23}a_3^2) + \tfrac{1}{2}H_{23}a_3 \cos\gamma_2 = 0 \tag{7.4.64}$$

$$-\omega_3(a_3' + \mu_3 a_3) + \tfrac{1}{2}H_{32}a_2 \sin\gamma_2 = 0 \tag{7.4.65}$$

$$\omega_3 a_3 \beta_3' + H_{33}a_3 + \tfrac{1}{8}a_3(\alpha_{31}a_1^2 + \alpha_{32}a_2^2 + \alpha_{33}a_3^2) + \tfrac{1}{2}H_{32}a_2 \cos\gamma_2 = 0 \tag{7.4.66}$$

where

$$\gamma_1 = \sigma_1 T_1 - 3\beta_1 + \beta_2, \gamma_2 = \sigma_2 T_1 - \beta_2 - \beta_3 \tag{7.4.67}$$

For the steady-state motion, a_1, a_2 and, a_3 as well as γ_1 and γ_2 are constants. Thus (7.4.61) through (7.4.67) reduce to

$$a_1(\omega_1\mu_1 - \tfrac{1}{8}Q_1 a_1 a_2 \sin\gamma_1) = 0 \tag{7.4.68}$$

$$a_1 [\omega_1 \beta_1' + H_{11} + \tfrac{1}{8}(Q_1 a_1 a_2 \cos\gamma_1 + \alpha_{11} a_1^2 + \alpha_{12} a_2^2 + \alpha_{13} a_3^2)] = 0 \tag{7.4.69}$$

$$\omega_2 \mu_2 a_2 + \tfrac{1}{8} Q_2 a_1^3 \sin\gamma_1 - \tfrac{1}{2} H_{23} a_3 \sin\gamma_2 = 0 \tag{7.4.70}$$

$$\omega_2 a_2 \beta_2' + H_{22} a_2 + \tfrac{1}{8} Q_2 a_1^3 \cos\gamma_1 + \tfrac{1}{8} a_2 (\alpha_{21} a_1^2 + \alpha_{22} a_2^2 + \alpha_{23} a_3^2) + \tfrac{1}{2} H_{23} a_3 \cos\gamma_2 = 0 \tag{7.4.71}$$

$$\omega_3 \mu_3 a_3 - \tfrac{1}{2} H_{32} a_2 \sin\gamma_2 = 0 \tag{7.4.72}$$

$$\omega_3 a_3 \beta_3' + H_{33} a_3 + \tfrac{1}{8} a_3 (\alpha_{31} a_1^2 + \alpha_{32} a_2^2 + \alpha_{33} a_3^2) + \tfrac{1}{2} H_{32} a_2 \cos\gamma_2 = 0 \tag{7.4.73}$$

$$\sigma_1 - 3\beta_1' + \beta_2' = 0, \qquad \sigma_2 - \beta_2' - \beta_3' = 0 \tag{7.4.74}$$

We note that there are three possibilities: (1) $a_1 \neq 0$, $a_2 \neq 0$, and $a_3 \neq 0$; (2) $a_1 = 0$, $a_2 \neq 0$, and $a_3 \neq 0$; and (3) $a_1 = 0$, $a_2 = 0$, and $a_3 = 0$.

When $a_1 \neq 0$, $a_2 \neq 0$ and $a_3 \neq 0$, one can combine (7.4.69), (7.4.71), (7.4.73), and (7.4.74) to obtain

$$\sigma_1 + \tfrac{1}{8}\left(\frac{3\alpha_{11}}{\omega_1} - \frac{\alpha_{21}}{\omega_2}\right) a_1^2 + \tfrac{1}{8}\left(\frac{3\alpha_{12}}{\omega_1} - \frac{\alpha_{22}}{\omega_2}\right) a_2^2 + \tfrac{1}{8}\left(\frac{3\alpha_{13}}{\omega_1} - \frac{\alpha_{23}}{\omega_2}\right) a_3^2 + \frac{3H_{11}}{\omega_1} - \frac{H_{22}}{\omega_2} - \frac{H_{23} a_3}{2\omega_2 a_2} \cos\gamma_2 + \tfrac{1}{8}\left(\frac{3Q_1}{\omega_1} a_1 a_2 - \frac{Q_2 a_1^3}{\omega_2 a_2}\right) \cos\gamma_1 = 0 \tag{7.4.75}$$

and

$$\sigma_2 + \tfrac{1}{8}\left(\frac{\alpha_{21}}{\omega_2} + \frac{\alpha_{31}}{\omega_3}\right) a_1^2 + \tfrac{1}{8}\left(\frac{\alpha_{22}}{\omega_2} + \frac{\alpha_{32}}{\omega_3}\right) a_2^2 + \tfrac{1}{8}\left(\frac{\alpha_{23}}{\omega_2} + \frac{\alpha_{33}}{\omega_3}\right) a_3^2 + \frac{H_{22}}{\omega_2} + \frac{H_{33}}{\omega_3} + \frac{Q_2 a_1^3}{8\omega_2 a_2} \cos\gamma_1 + \tfrac{1}{2}\left(\frac{H_{23} a_3}{\omega_2 a_2} + \frac{H_{32} a_2}{\omega_3 a_3}\right) \cos\gamma_2 = 0 \tag{7.4.76}$$

One can solve (7.4.68), (7.4.70), (7.4.72), (7.4.75), and (7.4.76) for a_1, a_2, a_3, γ_1, and γ_2. Then one can obtain β_1', β_2', and β_3' from (7.4.69), (7.4.71), and (7.4.73). However one can only determine β_1, β_2, and β_3 to within constants that depend on the initial conditions.

When $a_1 = 0$, $a_2 \neq 0$, and $a_3 \neq 0$, one can solve (7.4.70), (7.4.72), (7.4.75), and (7.4.76) for a_2, a_3, γ_1, and γ_2. Then β_2' and β_3' can be obtained from (7.4.71) and (7.4.73). Again one can only determine β_2 and β_3 to within constants that depend on the initial conditions.

It follows from (7.4.74) that the nonlinearity adjusts the frequencies of the second and third modes so that the resonant frequency combination is satisfied exactly:

$$\omega_2 + \epsilon\beta_2' + \omega_3 + \epsilon\beta_3' = \omega_2 + \omega_3 + \epsilon\sigma_2 = 2\Omega$$

Also the nonlinearity adjusts the frequencies of the first and second modes so that they are exactly in the ratio of 1 to 3.

For the first approximation, when $a_1 = a_2 = a_3 = 0$,

$$w(x, t) = 2 \cos \Omega t \sum_{n=1}^{\infty} K_n(\omega_n^2 - \Omega^2)^{-1} \phi_n(x) \tag{7.4.77}$$

Otherwise

$$w(x, t) = \sum_{n=N}^{3} a_n \cos [(\omega_n + \epsilon\beta_n')t + \nu_n] \phi_n(x) + 2 \cos \Omega t \sum_{n=1}^{\infty} K_n(\omega_n^2 - \Omega^2)^{-1} \phi_n(x)$$

where the ν_n are arbitrary constants depending on the initial conditions and $N = 1$ or 2.

As an example we use the same loading as in Section 7.4.3 and select some arbitrary values for B. We find that for $B = 5000$,

$$H_{11} = -819.1, \quad H_{22} = -1914, \quad H_{33} = -9921, \quad H_{23} = H_{32} = -1054 \tag{7.4.78}$$

In Figure 7-8, a_2 and a_3 are plotted as functions of the frequency of the excitation ($\epsilon\sigma_2$) when $a_1 = 0$. And in Figure 7-9, a_1, a_2, and a_3 are plotted as functions of $\epsilon\sigma_2$. We note that $\epsilon\sigma_2$ must be greater than 0.3 approximately if only the second and third modes are to be strongly excited. Otherwise either all three modes are strongly excited or none is (recall that a trivial solution is possible).

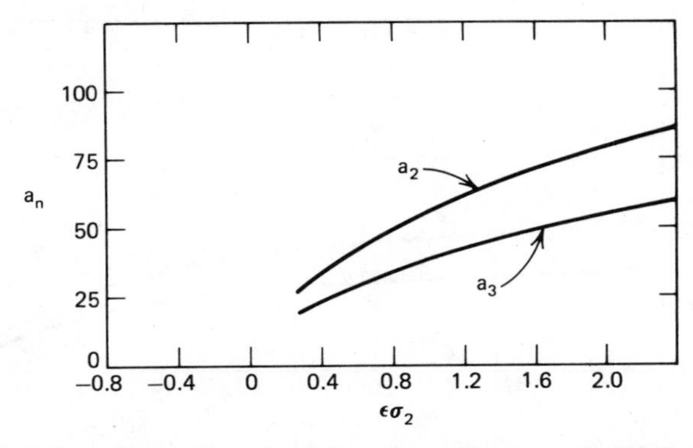

Figure 7-8. Variation of second- and third-mode amplitudes as functions of the detuning of the excitation when $a_1 = 0$; $\omega_2 \simeq 3\omega_1$, $\Omega \simeq \omega_2 + \omega_3$.

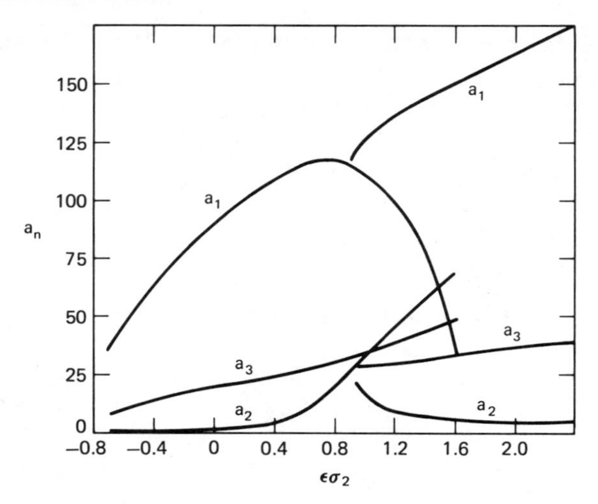

Figure 7-9. Variation of first-, second- and third-amplitudes as functions of the detuning of the excitation; $\omega_2 \simeq 3\omega_1$, $\Omega \simeq \omega_2 + \omega_3$.

Only the stable portions of these frequency-response curves are plotted. When $a_1 \neq 0$, there is a region where two stable solutions exist. In this region the initial conditions determine which solution represents the response. In Figures 7-10 and 7-11 the corresponding plots of the modal amplitudes as functions of the amplitudes of the excitation are shown.

We note that when only the second and third modes are strongly excited, a_2 is larger than a_3. And when all three modes are strongly excited, a_1 is generally much larger than either a_2 or a_3. Thus the internal resonance provides the means

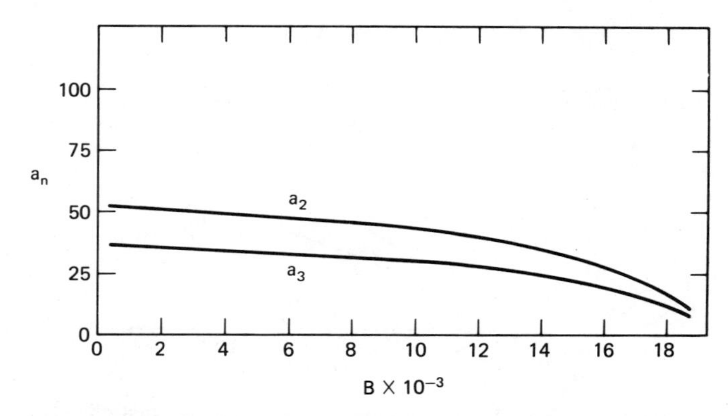

Figure 7-10. Variation of second- and third-mode amplitudes as functions of the amplitude of the excitation when $a_1 = 0$; $\omega_2 \simeq 3\omega_1$, $\Omega \simeq \omega_2 + \omega_3$.

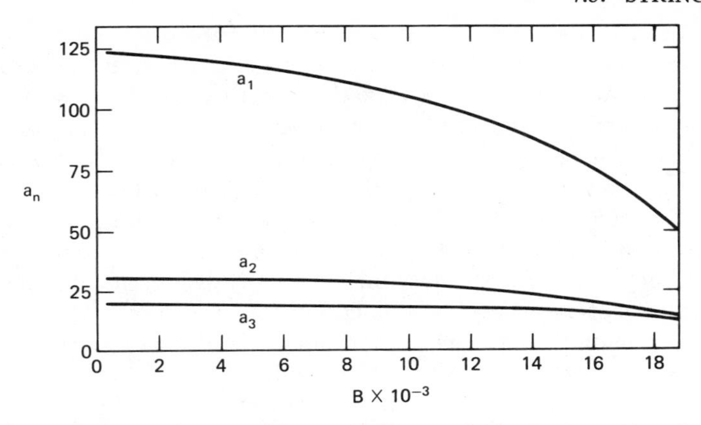

Figure 7-11. Variation of first-, second-, and third-mode amplitudes as functions of the amplitude of the excitation, $\omega_2 \simeq 3\omega_1$, $\Omega \simeq \omega_2 + \omega_3$.

for passing energy from the excitation through the directly excited modes down to the fundamental mode, which is not directly excited.

For all the graphs above, we took $\mu_1 = \mu_2 = \mu_3 = \mu$.

7.5. Strings

In an experiment intended for classroom demonstration, J. K. Hunton, as reported by Harrison (1948), excited a metallic wire (string) by placing it in a magnetic field and passing an alternating current through it. He observed that though the forcing function is planar, conditions exist under which the planar motion becomes unstable and the wire begins to whirl like a jump rope. This nonplanar (out-of-plane) motion is also referred to as *ballooning*, or *tubular motion*. Lee (1957) ran an experiment using essentially the apparatus described by Harrison. No mention was made of the nonplanar motion, but all the expected spontaneous jump phenomena were observed. He attempted a comparison with theory and found some disagreement which he attributed to subharmonic resonances; it is most likely due to the presence of a nonplanar motion. Murthy and Ramakrishna (1965) conducted an experiment in which they excited a string in a plane and observed the occurrence of the nonplanar motion and the expected jump phenomena. Kohut and Mathews (1971) experimentally studied the motion of a bowed violin string. Eller (1972) also conducted an experiment to determine the conditions under which transverse oscillations of a horizontal string driven in the vertical plane become unstable. He also observed the expected jump phenomena.

A summary of the experimental observations is as follows. The response of a string to a plane excitation is planar provided the response amplitude (which is a function of the amplitude and frequency of the excitation and the phase difference between the response and the excitation) is smaller than a certain critical value. Above this critical value the planar motion becomes unstable and gives way to a nonplanar, whirling motion. As the excitation frequency is varied from a value far below to a value far above the linear natural frequency, the response of the string is at first confined to the plane of the driving force; but as soon as the amplitude of the response exceeds a critical value, the planar motion becomes unstable and the string acquires an out-of-plane motion. As the frequency is increased further, the amplitudes of both the in-plane and the out-of-plane motions increase until a critical frequency is reached at which the out-of-plane motion disappears and the amplitude of the in-plane motion spontaneously becomes much smaller. On the other hand, if the frequency of the excitation is decreased from a value far above resonance, the response is first confined to the plane of the excitation until a critical frequency is reached at which the response amplitude spontaneously becomes much larger and the string acquires an out-of-plane motion. As the frequency of the excitation is decreased further, the response amplitude decreases, and when it goes below the critical value, the out-of-plane component disappears. The frequency at which the upward jump occurs is below that at which the downward jump occurs, indicating a hardening of the string due to the nonlinearity. These experiments show that it is not trivial to drive a string so that it will vibrate in a plane, and perhaps this is the reason why Oplinger (1960) constrained his string to vibrate in a slot.

Carrier (1945, 1949) examined free, undamped planar and nonplanar motions. The planar motion was studied also by Lee (1957) and Oplinger (1960). Zabusky (1962) and Kruskal and Zabusky (1964) solved model equations describing the nonlinear planar motion of strings. Rosen (1965) and Mackie (1968) showed the equivalence of the planar nonlinear string problem and that of one-dimensional gas flow.

Murthy and Ramakrishna (1965) studied the forced undamped response of a string to interpret their experimental data, but they did not study the stability of the various solutions. Miles (1965) used the method of harmonic balance to examine the stability of the various in-plane and out-of-plane motions for undamped forced motions. Since the undamped solutions cannot predict the amplitude and frequency for the downward jump, Anand (1966) and Shih (1971) reexamined the nonplanar problem by including the effect of linear viscous damping. However it may be more appropriate to model the damping by a nonlinear (quadratic) term as Hsu (1975b) did in his analysis of the parametric excitation of a string hanging in fluid. The analysis of nonplanar motions was further extended by Narashimha (1968) and Anand (1969a, b, 1973). Lee and Ames (1973) obtained a class of general solutions to the nonlinear dynamic equations of elastic strings. Roderman, Longcope, and Shampine (1976) determined the response of a string to an accelerating mass.

In this section we concentrate on the nonplanar motion in order to explain the experimentally observed whirling motions. The existence of the whirling phenomenon was also observed in the nonlinear motion of beams as discussed in Section 7.1. As mentioned in Section 7.1, here we provide the details of the derivation of the governing equations, including longitudinal inertia, in Section 7.5.1. In Section 7.5.2 we use the method of multiple scales to determine a uniform first-order nonplanar solution, neglecting longitudinal inertia; and in Section 7.5.3 we analyze the stability of the various possible motions.

7.5.1. GOVERNING EQUATIONS

Referring to Figure 7-1, we label the ends of an infinitesimal length of string by F and G in the undeformed position and F' and G' in the deformed position. The displacement of F is given by

$$\mathbf{\Delta}_F = u(x, t)\mathbf{i} + v(x, t)\mathbf{j} + w(x, t)\mathbf{k} \tag{7.5.1}$$

and the displacement of G is given by

$$\mathbf{\Delta}_G = (u + u_x\, dx)\mathbf{i} + (v + v_x\, dx)\mathbf{j} + (w + w_x\, dx)\mathbf{k} \tag{7.5.2}$$

It follows from Figure 7-1 that

$$\mathbf{\Delta}_F + \mathbf{\Delta}_{F'G'} = \mathbf{\Delta}_G + dx\mathbf{i} \tag{7.5.3}$$

where $\mathbf{\Delta}_{F'G'}$ is the vector giving the position of G' relative to F'. The length of the deformed segment is given by

$$|\mathbf{\Delta}_{F'G'}| = ds = [(1 + u_x)^2 + v_x^2 + w_x^2]^{1/2}\, dx \tag{7.5.4}$$

and the unit vector parallel to the deformed segment is given by

$$\frac{\mathbf{\Delta}_{F'G'}}{|\mathbf{\Delta}_{F'G'}|} = \boldsymbol{\delta} = [(1 + u_x)\mathbf{i} + v_x\mathbf{j} + w_x\mathbf{k}]\, \frac{dx}{ds} \tag{7.5.5}$$

The string has the tension N_0 initially. During the motion the length of the string changes and hence the tension in the string changes. The instantaneous value of the tension is given by

$$N = N_0 + \frac{EA(ds - dx)}{dx} \tag{7.5.6}$$

where E is Young's modulus and A is the cross-sectional area in the rest position.

The momentum equation for the string is given by

$$\frac{\partial}{\partial x}(N\boldsymbol{\delta}) = \rho A(u_{tt}\mathbf{i} + v_{tt}\mathbf{j} + w_{tt}\mathbf{k}) \tag{7.5.7}$$

where ρ is the density and A is the cross-sectional area of the string in the rest position. (Conservation of mass requires $\rho A\, ds$ to be constant during the motion;

here ρ and A are the instantaneous values of density and area, respectively.) Using (7.5.5) and (7.5.6) to eliminate $\boldsymbol{\delta}$ from (7.5.7) gives

$$\frac{\partial}{\partial x}\left\{\left[(N_0 - EA)\frac{dx}{ds} + EA\right][(1 + u_x)\mathbf{i} + v_x\mathbf{j} + w_x\mathbf{k}]\right\} = \rho A(u_{tt}\mathbf{i} + v_{tt}\mathbf{j} + w_{tt}\mathbf{k}) \tag{7.5.8}$$

Using (7.5.4) to eliminate dx/ds from (7.5.8), expanding for small u, v, and w, and keeping up to cubic terms, we obtain

$$u_{tt} - c_1^2 u_{xx} = (c_1^2 - c_2^2)\frac{\partial}{\partial x}\left[(\tfrac{1}{2} - u_x)(v_x^2 + w_x^2)\right] \tag{7.5.9}$$

$$v_{tt} - c_2^2 v_{xx} = (c_1^2 - c_2^2)\frac{\partial}{\partial x}\left[v_x(u_x - u_x^2 + \tfrac{1}{2}v_x^2 + \tfrac{1}{2}w_x^2)\right] \tag{7.5.10}$$

$$w_{tt} - c_2^2 w_{xx} = (c_1^2 - c_2^2)\frac{\partial}{\partial x}\left[w_x(u_x - u_x^2 + \tfrac{1}{2}v_x^2 + \tfrac{1}{2}w_x^2)\right] \tag{7.5.11}$$

where c_1 and c_2 are the longitudinal and transverse speeds of sound defined by

$$c_1^2 = \frac{E}{\rho} \quad \text{and} \quad c_2^2 = \frac{N_0}{\rho A} \tag{7.5.12}$$

For a string with fixed ends the boundary conditions are

$$u = v = w = 0 \qquad \text{at } x = 0 \text{ and } l \tag{7.5.13}$$

Hence the solutions of the linearized equations (7.5.9) through (7.5.11) subject to the boundary conditions (7.5.13) are

$$u = \sum_{n=1}^{\infty} a_{1n}\cos\left(\frac{n\pi c_1 t}{l} + \beta_{1n}\right)\sin\frac{n\pi x}{l} \tag{7.5.14}$$

$$v = \sum_{n=1}^{\infty} a_{2n}\cos\left(\frac{n\pi c_2 t}{l} + \beta_{2n}\right)\sin\frac{n\pi x}{l} \tag{7.5.15}$$

$$w = \sum_{n=1}^{\infty} a_{3n}\cos\left(\frac{n\pi c_2 t}{l} + \beta_{3n}\right)\sin\frac{n\pi x}{l} \tag{7.5.16}$$

Carrying out a multiple-scale analysis for small- but finite-amplitude motions shows that an mth longitudinal mode of oscillation interacts with an nth transverse mode of oscillation to first order if the linear frequency of the longitudinal mode is approximately twice the linear frequency of the transverse mode (internal resonance), that is, if $mc_1 \approx 2nc_2$. Using (7.5.12) we rewrite this condition as $n = \frac{1}{2}m(EA/N_0)^{1/2}$. For metal strings typical values of $(EA/N_0)^{1/2}$ lie in the range of 20 to 32. Hence there can be an interaction between the 10th through 16th transverse modes and the fundamental longitudinal mode. For non-metallic strings a lower-order transverse mode can interact with the fundamental longitudinal mode. Arya and Lardner (1974) used the method of averaging to analyze the interaction between a longitudinal mode and a planar transverse mode.

If we restrict our attention to metallic strings and consider the lower-order transverse modes, no interaction will occur between these transverse modes and the longitudinal modes, and the longitudinal inertia u_{tt} can be neglected as a consequence. Moreover since $c_1^2/c_2^2 = O(400\text{–}1000)$, c_2^2 is negligible compared with c_1^2 in (7.5.9) through (7.5.11). In this case (7.5.9) can be integrated once to give

$$u_x = e(t) - \tfrac{1}{2}\left[v_x^2 + w_x^2 - 2u_x(v_x^2 + w_x^2)\right] \tag{7.5.17}$$

where $e(t)$ is an arbitrary function. Equation (7.5.17) shows that $u_x = O(e) = O(v_x^2) = O(w_x^2)$. Hence, to $O(v^3, w^3)$,

$$u_x = e(t) - \tfrac{1}{2}(v_x^2 + w_x^2) \tag{7.5.18}$$

To allow for parametric excitations we now take the boundary conditions on u in the form

$$u = 0 \quad \text{at } x = 0 \qquad \text{and} \qquad u = lP(t) \quad \text{at } x = l \tag{7.5.19}$$

Integrating (7.5.18) and using these boundary conditions, we have

$$u = xe(t) - \tfrac{1}{2}\int_0^x (v_x^2 + w_x^2)\, dx \tag{7.5.20}$$

where

$$e(t) = P(t) + \frac{1}{2l}\int_0^l (v_x^2 + w_x^2)\, dx \tag{7.5.21}$$

Consequently

$$e(t) = u_x + \tfrac{1}{2}(v_x^2 + w_x^2) = P(t) + \frac{1}{2l}\int_0^l (v_x^2 + w_x^2)\, dx \tag{7.5.22}$$

Using (7.5.22) in (7.5.10) and (7.5.11), neglecting c_2^2 compared with c_1^2, and keeping terms through cubic in v and w, we obtain

$$v_{tt} - c_2^2 v_{xx} = c_1^2 P v_{xx} + \frac{c_1^2}{2l} v_{xx} \int_0^l (v_x^2 + w_x^2)\, dx \tag{7.5.23}$$

$$w_{tt} - c_2^2 w_{xx} = c_1^2 P w_{xx} + \frac{c_1^2}{2l} w_{xx} \int_0^l (v_x^2 + w_x^2)\, dx \tag{7.5.24}$$

Adding linear damping forces and external excitations we rewrite (7.5.23) and (7.5.24) in the form

$$v_{tt} - c_2^2 v_{xx} + 2\mu v_t = c_1^2 P v_{xx} + \frac{c_1^2}{2l} v_{xx} \int_0^l (v_x^2 + w_x^2)\, dx + g(x, t) \tag{7.5.25}$$

$$w_{tt} - c_2^2 w_{xx} + 2\mu w_t = c_1^2 P w_{xx} + \frac{c_1^2}{2l} w_{xx} \int_0^l (v_x^2 + w_x^2)\, dx + f(x, t) \tag{7.5.26}$$

We seek a solution of (7.5.25) and (7.5.26), subject to the boundary conditions (7.5.13), in the form of an expansion in terms of the linear modes. That is, we let

$$v = l \sum_{n=1}^{\infty} \epsilon^{1/2} \zeta_n(t) \sin \frac{n\pi x}{l}, \qquad w = l \sum_{n=1}^{\infty} \epsilon^{1/2} \eta_n(t) \sin \frac{n\pi x}{l} \tag{7.5.27}$$

where $\epsilon^{1/2}$ is a small dimensionless quantity the order of the amplitudes of the motion. Substituting (7.5.27) into (7.5.25) and (7.5.26) and using the orthogonality of the linear modes, we obtain

$$\ddot{\zeta}_n + \omega_n^2 \zeta_n = -\epsilon \left[2\mu_n \dot{\zeta}_n + n^2 p \zeta_n + \Gamma n^2 \zeta_n \sum_{m=1}^{\infty} m^2 (\zeta_m^2 + \eta_m^2) \right] + g_n(t) \tag{7.5.28}$$

$$\ddot{\eta}_n + \omega_n^2 \eta_n = -\epsilon \left[2\mu_n \dot{\eta}_n + n^2 p \eta_m + \Gamma n^2 \eta_n \sum_{m=1}^{\infty} m^2 (\zeta_m^2 + \eta_m^2) \right] + f_n(t) \tag{7.5.29}$$

where

$$\omega_n = \frac{n\pi c_2}{l}, \qquad \Gamma = \frac{\pi^4 c_1^2}{4l^2}, \qquad \epsilon p = \frac{\pi^2 P}{l^2} \tag{7.5.30}$$

$$\sqrt{\epsilon} f_n = \frac{2}{l^2} \int_0^1 f(x, t) \sin \frac{n\pi x}{l}\, dx, \sqrt{\epsilon} g_n = \frac{2}{l^2} \int_0^1 g(x, t) \sin \frac{n\pi x}{l}\, dx \tag{7.5.31}$$

In (7.5.28) and (7.5.29) the damping is assumed to be modal. Moreover the damping and the parametric excitation were ordered so that they are the same order as the nonlinearity. Next we obtain a first-order nonplanar approximation of the response of the string to a planar primary resonant excitation.

7.5.2. PRIMARY RESONANCES

In this section we consider the case of a primary resonant planar excitation. That is, we let

$$g_n(t) = 2\epsilon k_n \cos \Omega t, \qquad f_n = 0, \qquad p = 0 \tag{7.5.32}$$

where

$$\Omega = \omega_s + \epsilon\sigma \tag{7.5.33}$$

for fixed s. Using the method of multiple scales we assume that

$$\begin{aligned} \zeta_n(t;\epsilon) &= \zeta_{n0}(T_0, T_1) + \epsilon\zeta_{n1}(T_0, T_1) + \cdots \\ \eta_n(t;\epsilon) &= \eta_{n0}(T_0, T_1) + \epsilon\eta_{n1}(T_0, T_1) + \cdots \end{aligned} \tag{7.5.34}$$

Substituting (7.5.32) and (7.5.34) into (7.5.28) and (7.5.29) and equating coefficients of like powers of ϵ, we obtain

Order ϵ^0

$$\begin{aligned} D_0^2\zeta_{n0} + \omega_n^2\zeta_{n0} &= 0 \\ D_0^2\eta_{n0} + \omega_n^2\eta_{n0} &= 0 \end{aligned} \tag{7.5.35}$$

Order ϵ

$$D_0^2\zeta_{n1} + \omega_n^2\zeta_{n1} = -2D_0D_1\zeta_{n0} - 2\mu_n D_0\zeta_{n0} - \Gamma n^2\zeta_{n0} \sum_{m=1}^{\infty} m^2(\zeta_{m0}^2 + \eta_{m0}^2) + 2k_n \cos \Omega T_0 \tag{7.5.36}$$

$$D_0^2\eta_{n1} + \omega_n^2\eta_{n1} = -2D_0D_1\eta_{n0} - 2\mu_n D_0\eta_{n0} - \Gamma n^2\eta_{n0} \sum_{m=1}^{\infty} m^2(\zeta_{m0}^2 + \eta_{m0}^2) \tag{7.5.37}$$

The solutions of (7.5.35) can be written in the following convenient forms:

$$\begin{aligned} \zeta_{n0} &= A_n(T_1) \exp(i\omega_n T_0) + cc \\ \eta_{n0} &= B_n(T_1) \exp(i\omega_n T_0) + cc \end{aligned} \tag{7.5.38}$$

Then (7.5.36) and (7.5.37) become

$$D_0^2\zeta_{n1} + \omega_n^2\zeta_{n1} = -2i\omega_n(A_n' + \mu_n A_n)\exp(i\omega_n T_0) - \Gamma n^2 A_n \exp(i\omega_n T_0)\sum_{m=1}^{\infty} m^2[(A_m^2 + B_m^2)\exp(2i\omega_m T_0) + 2A_m\bar{A}_m + 2B_m\bar{B}_m + (\bar{A}_m^2 + \bar{B}_m^2)\exp(-2i\omega_m T_0)] + k_n\exp(i\Omega T_0) + cc \tag{7.5.39}$$

$$D_0^2\eta_{n1} + \omega_n^2\eta_{n1} = -2i\omega_n(B_n' + \mu_n B_n)\exp(i\omega_n T_0) - \Gamma n^2 B_n \exp(i\omega_n T_0)\sum_{m=1}^{\infty} m^2[(A_m^2 + B_m^2)\exp(2i\omega_m T_0) + 2A_m\bar{A}_m + 2B_m\bar{B}_m + (\bar{A}_m^2 + \bar{B}_m^2)\exp(-2i\omega_m T_0)] + cc \tag{7.5.40}$$

Using (7.5.33) one finds that secular terms are eliminated from the ζ_{n1} and η_{n1} if

$$-2i\omega_n(A_n' + \mu_n A_n) - 2\Gamma n^2 A_n \sum_{m=1}^{\infty} m^2(A_m\bar{A}_m + B_m\bar{B}_m) - \Gamma n^4 \bar{A}_n(A_n^2 + B_n^2) + \delta_{ns}k_s\exp(i\sigma T_1) = 0 \tag{7.5.41}$$

$$-2i\omega_n(B_n' + \mu_n B_n) - 2\Gamma n^2 B_n \sum_{m=1}^{\infty} m^2(A_m\bar{A}_m + B_m\bar{B}_m) - \Gamma n^4 \bar{B}_n(A_n^2 + B_n^2) = 0 \tag{7.5.42}$$

It is convenient to introduce the polar form

$$(A_n, B_n) = \tfrac{1}{2}(a_n, b_n)\exp[i(\alpha_n, \beta_n)] \tag{7.5.43}$$

Substituting (7.5.43) into (7.5.41) and (7.5.42) and separating real and imaginary parts, we obtain

$$\omega_n(a_n' + \mu_n a_n) + \tfrac{1}{8}\Gamma n^4 a_n b_n^2 \sin\gamma_n - \delta_{ns}k_s \sin\nu = 0 \tag{7.5.44}$$

$$\omega_n a_n \alpha_n' - \tfrac{1}{8}\Gamma n^4 a_n^3 - \tfrac{1}{4}\Gamma n^2 a_n \sum_{m=1}^{\infty} m^2(a_m^2 + b_m^2) - \tfrac{1}{8}\Gamma n^4 a_n b_n^2 \cos\gamma_n + \delta_{ns}k_s\cos\nu = 0 \tag{7.5.45}$$

$$\omega_n(b_n' + \mu_n b_n) - \tfrac{1}{8}\Gamma n^4 a_n^2 b_n \sin\gamma_n = 0 \tag{7.5.46}$$

$$\omega_n b_n \beta_n' - \tfrac{1}{8}\Gamma n^4 b_n^3 - \tfrac{1}{4}\Gamma n^2 b_n \sum_{m=1}^{\infty} m^2(a_m^2 + b_m^2) - \tfrac{1}{8}\Gamma n^4 a_n^2 b_n \cos\gamma_n = 0 \tag{7.5.47}$$

where

$$\gamma_n = 2(\beta_n - \alpha_n), \qquad \nu = \sigma T_1 - \alpha_s \tag{7.5.48}$$

For steady-state solutions a_n, b_n, γ_n, and ν are constants. Thus (7.5.44) and (7.5.46) reduce to

$$a_n[\omega_n \mu_n + \tfrac{1}{8}\Gamma n^4 b_n^2 \sin\gamma_n] - \delta_{ns} k_s \sin\nu = 0 \tag{7.5.49}$$

$$b_n[\omega_n \mu_n - \tfrac{1}{8}\Gamma n^4 a_n^2 \sin\gamma_n] = 0 \tag{7.5.50}$$

When $n \neq s$, it follows from (7.5.49) and (7.5.50) that $a_n = b_n = 0$. To show this, we note that if a_n and b_n differ from zero, (7.5.49) and (7.5.50) demand that

$$\begin{aligned} \omega_n \mu_n + \tfrac{1}{8}\Gamma n^4 b_n^2 \sin\gamma_n &= 0 \\ \omega_n \mu_n - \tfrac{1}{8}\Gamma n^4 a_n^2 \sin\gamma_n &= 0 \end{aligned} \tag{7.5.51}$$

Eliminating $\sin\gamma_n$ from (7.5.51) yields $a_n^2 + b_n^2 = 0$, which demands that $a_n = b_n = 0$. This conclusion is also valid if the damping coefficients in two perpendicular planes are different. If the damping coefficient in the first of equations (7.5.51) is $\hat{\mu}_n$, then $\hat{\mu}_n a_n^2 + \mu_n b_n^2 = 0$, which also demands that $a_n = b_n = 0$.

When $n = s$, (7.5.44) through (7.5.48) can be rewritten as

$$a' + \mu a + ab^2 \sin\gamma - k \sin\nu = 0 \tag{7.5.52}$$

$$\sigma - \nu' - 3a^2 - 2b^2 - b^2 \cos\gamma + ka^{-1}\cos\nu = 0 \tag{7.5.53}$$

$$b' + \mu b - a^2 b \sin\gamma = 0 \tag{7.5.54}$$

$$b[\sigma + \tfrac{1}{2}\gamma' - \nu' - 3b^2 - 2a^2 - a^2\cos\gamma] = 0 \tag{7.5.55}$$

where $(a, b) = (s^4\Gamma/8\omega_s)^{1/2}(a_s, b_s)$, $k = (\Gamma s^4/8\omega_s^3)^{1/2} k_s$, $\mu = \mu_s$, and $\gamma = \gamma_s$. There are two possible steady-state solutions in this case: (1) $b = 0$ and $a \neq 0$, and the motion is planar; and (2) $a \neq 0$ and $b \neq 0$, and the motion is nonplanar.

For steady-state planar motions (7.5.52) and (7.5.53) can be combined into

$$\sigma = 3a^2 - \frac{k}{a}\cos\nu, \qquad \sin\nu = \frac{\mu a}{k} \tag{7.5.56}$$

or

$$\sigma = 3a^2 \pm \left(\frac{k^2}{a^2} - \mu^2\right)^{1/2} \tag{7.5.57}$$

which has the same form as (4.1.19). Thus this system is equivalent to a hardening spring–mass system. The frequency-response curves are similar to those shown in Figure 4-1. However the planar response becomes unstable when the response amplitude exceeds a critical value, as shown in the next section, and the response becomes nonplanar (i.e., a and b differ from zero).

For steady-state nonplanar motions (7.5.53) and (7.5.55) can be combined into

$$\sigma = (5 + \cos\gamma)a^2 - \frac{3k}{a(1 - \cos\gamma)}\cos\nu \tag{7.5.58}$$

$$b^2 = a^2 - \frac{k}{a(1 - \cos\gamma)}\cos\nu \tag{7.5.59}$$

Moreover it follows from (7.5.52) and (7.5.54) that

$$\sin\gamma = \frac{\mu}{a^2}, \qquad \sin\nu = \frac{\mu(a^2 + b^2)}{ka} \tag{7.5.60}$$

For a given μ and k, one can solve (7.5.58) through (7.5.60) numerically to determine a and b as functions of σ.

If the damping is assumed to be so small that it can be completely neglected, the general character of the motion, which is still predicted by the analysis, can be readily obtained without an undue amount of algebra. Thus when $\mu = 0$, it follows from (7.5.60) that $\gamma = 0$ or π and $\nu = 0$ or π. The case $\gamma = 0$ must be ruled out because it corresponds to $k = 0$ according to (7.5.59). With $\gamma = \pi$, (7.5.58) and (7.5.59) simplify to

$$\sigma = 4a^2 - \frac{3k}{2a}\cos\nu \tag{7.5.61}$$

$$b^2 = a^2 - \frac{k}{2a}\cos\nu \tag{7.5.62}$$

Moreover the planar response in the absence of damping is given by (7.5.56) where $\nu = 0$ or π. There are two branches for both the planar and nonplanar motions corresponding to the two values 0 and π of ν. These are discussed next.

When $\nu = 0$ it follows from (7.5.56) that $-\infty < \sigma < \infty$ for $0 < a < \infty$. But from (7.5.62) we find that in order for b to be real, $a \geqslant a_A = (k/2)^{1/3}$. Then it follows from (7.5.61) that $\sigma \geqslant \sigma_A = (k/2)^{2/3}$. We note that (σ_A, a_A) is also a solution of (7.5.56); thus the point A is a bifurcation point.

When $\nu = \pi$ it follows from (7.5.56) that in the case of planar motion, $\sigma \geqslant \sigma_C = 9(k/6)^{2/3}$ corresponding to $a = a_C = (k/6)^{1/3}$. For nonplanar motion it follows from (7.5.61) that $\sigma \geqslant \sigma_D = 12(3k/16)^{2/3}$ corresponding to $a = a_D = (3k/16)^{1/3}$ and $b_D = (11/3)^{1/2}a_D$. The frequency-response curves are shown in

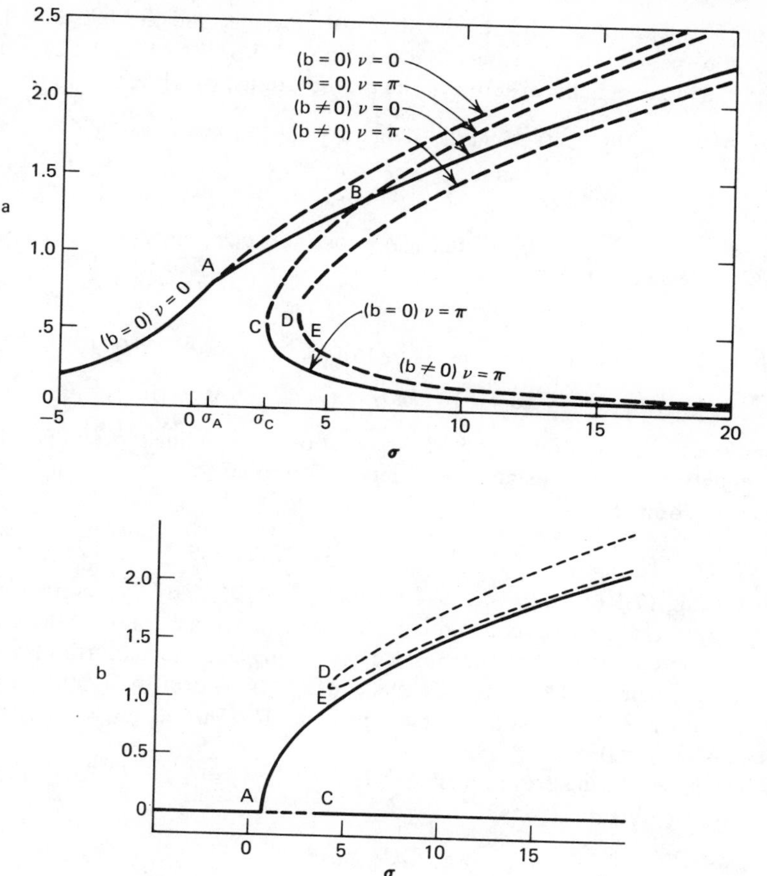

Figure 7-12. Frequency–response curves for the string.

Figure 7-12. The broken lines represent the unstable solutions, while the solid lines represent the stable solutions. (The stability of the various solutions is discussed below.)

7.5.3. STABILITY OF POSSIBLE SOLUTIONS

To determine which of the various possible solutions are stable, we slightly perturb the steady-state solutions. That is, we let

$$a = a_0 + a_1(T_1), \qquad b = b_0 + b_1(T_1)$$
$$\gamma = \gamma_0 + \gamma_1(T_1), \qquad \nu = \nu_0 + \nu_1(T_1) \tag{7.5.63}$$

where the subscript 0 indicates the steady-state values and the subscript 1 indicates the perturbation values. Substituting (7.5.63) into (7.5.52) through (7.5.55) and keeping linear terms in the perturbations, we obtain

$$a_1' + (\mu + b_0^2 \sin \gamma_0)a_1 + 2a_0 b_0 b_1 \sin \gamma_0 + a_0 b_0^2 \gamma_1 \cos \gamma_0 - k\nu_1 \cos \nu_0 = 0 \tag{7.5.64}$$

$$\nu_1' + \left(6a_0 + \frac{k \cos \nu_0}{a_0^2}\right)a_1 + 2b_0 b_1(2 + \cos \gamma_0) - b_0^2 \gamma_1 \sin \gamma_0 + \frac{k\nu_1}{a_0} \sin \nu_0 = 0 \tag{7.5.65}$$

$$b_1' - 2a_0 b_0 a_1 \sin \gamma_0 + (\mu - a_0^2 \sin \gamma_0)b_1 - a_0^2 b_0 \gamma_1 \cos \gamma_0 = 0 \tag{7.5.66}$$

$$\tfrac{1}{2}\gamma_1' - \nu_1' - 2a_0 a_1(2 + \cos \gamma_0) - 6b_0 b_1 + a_0^2 \gamma_1 \sin \gamma_0 = 0 \tag{7.5.67}$$

Equations (7.5.64) through (7.5.67) are a set of coupled linear ordinary-differential equations with constant coefficients. Thus the perturbations have an exponential form. That is,

$$(a_1', b_1', \gamma_1', \nu_1') = \lambda(a_1, b_1, \gamma_1, \nu_1) \tag{7.5.68}$$

Substituting (7.5.68) into (7.5.64) through (7.5.67) leads to a system of four linear algebraic homogeneous equations. For a nontrivial solution the determinant of the coefficient matrix must be zero. This leads to a fourth-order equation for λ. If none of the roots of this equation has a positive definite real part, the motion is stable; otherwise it is unstable. The various possibilities are discussed when $\mu = 0$.

When $b_0 = 0$, substituting (7.5.68) into (7.5.64) through (7.5.67) and using (7.5.56), we obtain

$$\lambda a_1 - k\nu_1 \cos \nu_0 = 0 \tag{7.5.69}$$

$$\lambda \nu_1 + \left(6a_0 + \frac{k}{a_0^2} \cos \nu_0\right) a_1 = 0 \tag{7.5.70}$$

$$(\lambda - a_0^2 \sin \gamma_0)b_1 = 0 \tag{7.5.71}$$

$$[\sigma - a_0^2(2 + \cos \gamma_0)]\, b_1 = 0 \tag{7.5.72}$$

We note that the equations uncouple in such a way that the exponents for a_1 and ν_1 can differ from those of b_1. From (7.5.54) and (7.5.55) we see that γ is indeterminate when $b = 0$; however any small perturbation in b fixes the value of γ as (7.5.71) and (7.5.72) indicate. Here, therefore, we must treat γ_0 as an unknown.

It follows from (7.5.69) and (7.5.70) that

$$\lambda^2 = -k \cos \nu_0 (6a_0 + k a_0^{-2} \cos \nu_0) \tag{7.5.73}$$

and hence that λ is pure imaginary for all a_0 when $\nu_0 = 0$ and for all $a < a_C$ when $\nu_0 = \pi$, where $a_C = (k/6)^{1/3}$. It follows from (7.5.71) and (7.5.72) that

$$a_0^2 \cos \gamma_0 = \sigma - 2a_0^2 \quad \text{and} \quad \lambda = a_0^2 \sin \gamma_0 \tag{7.5.74}$$

Eliminating γ_0 from (7.5.74) and using (7.5.56) to eliminate σ, we obtain

$$\lambda^2 = a_0^4 - (\sigma - 2a_0^2)^2 = k \cos \nu_0 (2a_0 - ka_0^{-2} \cos \nu_0) \tag{7.5.75}$$

Thus we find that the exponent of b_1 is pure imaginary for all a_0 when $\nu_0 = \pi$.

Summarizing the stability of the planar motion, we find that when $\nu_0 = 0$ the predicted motion is stable for all $\sigma < \sigma_A$ and unstable for all $\sigma > \sigma_A$. For $\nu_0 = \pi$ we find that the motion is stable for all $a < a_C$ and unstable for all $a > a_C$.

When $b_0 \neq 0$ and $\mu = 0$, (7.5.64) through (7.5.68) give

$$\lambda a_1 - a_0 b_0^2 \gamma_1 - k\nu_1 \cos \nu_0 = 0 \tag{7.5.76}$$

$$(6a_0 + ka_0^{-2} \cos \nu_0) a_1 + 2b_0 b_1 + \lambda \nu_1 = 0 \tag{7.5.77}$$

$$\lambda b_1 + a_0^2 b_0 \gamma_1 = 0 \tag{7.5.78}$$

$$-2a_0 a_1 - 6b_0 b_1 + \tfrac{1}{2} \lambda \gamma_1 - \lambda \nu_1 = 0 \tag{7.5.79}$$

For a nontrivial solution the determinant of the coefficient matrix of (7.5.76) through (7.5.79) must vanish. That is,

$$\lambda^4 + 16a_0^4 \lambda^2 + 4ka_0^2 \cos \nu_0 (16a_0^3 - 5k \cos \nu_0 - \tfrac{3}{2} k^2 a_0^{-3}) = 0 \tag{7.5.80}$$

Hence

$$\lambda^2 = -8a_0^4 \mp [64a_0^8 + 4ka_0^2 \cos \nu_0 \, (\tfrac{3}{2} k^2 a_0^{-3} + 5k \cos \nu_0 - 16a_0^3)]^{1/2} \tag{7.6.81}$$

The motion is unstable if λ^2 is positive definite or complex. Thus for stability

$$\cos \nu_0 (\tfrac{3}{2} k^2 a_0^{-3} + 5k \cos \nu_0 - 16a_0^3) \leqslant 0 \tag{7.5.82}$$

$$64a_0^8 + 4ka_0^2 \cos \nu_0 (\tfrac{3}{2} k^2 a_0^{-3} + 5k \cos \nu_0 - 16a_0^3) \geqslant 0 \tag{7.5.83}$$

When $\nu_0 = 0$ we find that (7.5.82) and (7.5.83) are satisfied if $a \geqslant a_A = (k/2)^{1/3}$. When $\nu_0 = \pi$, (7.5.82) is satisfied for all $a \leqslant a_D = (3k/16)^{1/3}$, while (7.5.83) is satisfied for all $a > a_E \approx 0.564(k)^{1/3}$. Thus $a_D - a_E \approx 0.008(k)^{1/3}$ and a_D and a_E are very close.

The results of the stability analysis are shown in Figure 7-12. Summarizing the stability of the nonplanar motion, we find that when $\nu_0 = 0$ the predicted motion is stable for all $a \geqslant a_A$ and no real value for b exists for all $a < a_A$. Thus a stable out-of-plane motion develops when a is greater than a_A. Recalling that a_A^3 is proportional to k, we note that this result is in agreement with the experi-

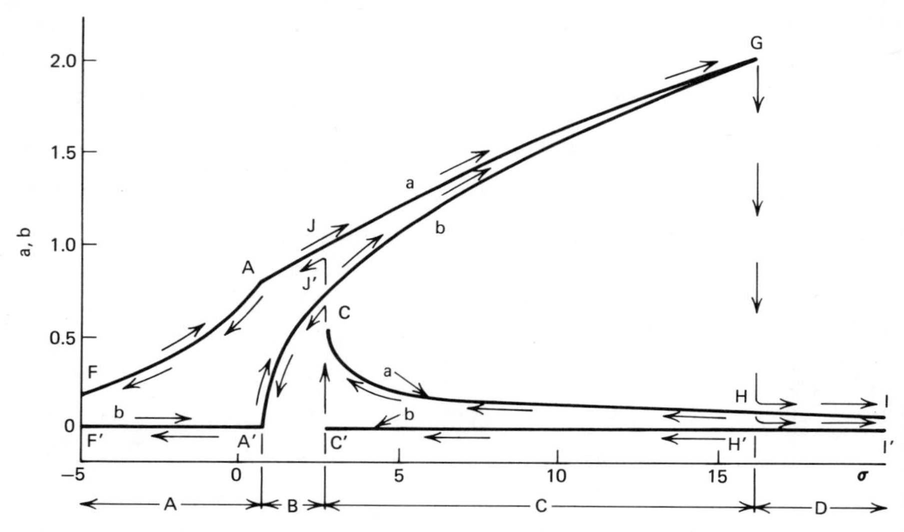

Figure 7-13. Frequency–response curves for a string under the action of linear damping.

ments of Murthy and Ramakrishna (1965). When $\nu_0 = \pi$ the nonplanar motion is stable only in the small interval $a_E \leqslant a \leqslant a_D$.

We are now in a position to describe the predicted motion as the frequency of the excitation is varied very slowly back and forth through resonance, while the amplitude of the excitation is held constant. Since damping, no matter how small, causes the branches of the frequency–response curves to terminate or close for some finite value of positive σ, we calculated the frequency–response curves using (7.5.52) through (7.5.55) for $k = 1$ and $\mu = 0.25$. The stable portions of these curves are shown in Figure 7-13. The following discussion refers to this figure.

Starting with a negative value of σ, we expect to initiate planar motion. Then as σ increases, the amplitudes of the two components of the motion move from F toward A and from F' toward A'. When $\sigma = \sigma_A$ the amplitude of the out-of-plane component begins to grow, and as σ increases further the two amplitudes increase through J and J' and coalesce. At point G, where the two amplitudes are equal, the motion becomes unstable. Any further increase in σ causes the amplitudes to jump spontaneously down to H and H'. During this jump transients develop and eventually decay (see Section 4.4), and the motion changes from nonplanar back to planar.

Reversing the procedure, we start with a large value of σ. Again we expect to initiate planar motion. As σ decreases the amplitudes of the two components move from I through H toward C and from I' through H' toward C'. When $\sigma = \sigma_C$ any further decrease in σ causes the amplitudes to jump spontaneously up to J and J'. During this jump transients develop and eventually decay, and the motion changes from planar to nonplanar. As σ decreases further both amplitudes decrease. When $\sigma = \sigma_A$ the amplitude of the out-of-plane component is zero, and the motion is planar again. The motion remains planar as σ continues to decrease.

We note that the jump occurs at different values of σ for the two processes. For increasing σ the motion first smoothly changes from planar to nonplanar and then jumps from nonplanar back to planar. For decreasing σ the motion first jumps from planar to nonplanar and then smoothly changes back to planar. The jump produces a 180-degree change in the phase between response and excitation when there is no damping.

When σ is either less than σ_A or greater than its value at the downward jump σ_G, the motion must be planar. When σ is greater than σ_A and less than σ_C, the motion must be nonplanar. And when σ is greater than σ_C and less than σ_G, the motion can be either planar or nonplanar. A finite disturbance is required to make the motion jump when σ is between σ_C and σ_G.

Next we describe the predicted motion as the amplitude of the excitation is varied very slowly up and down when σ is slightly above the resonant value. Referring to Figure 7-14 we note that as k increases from zero, initially only

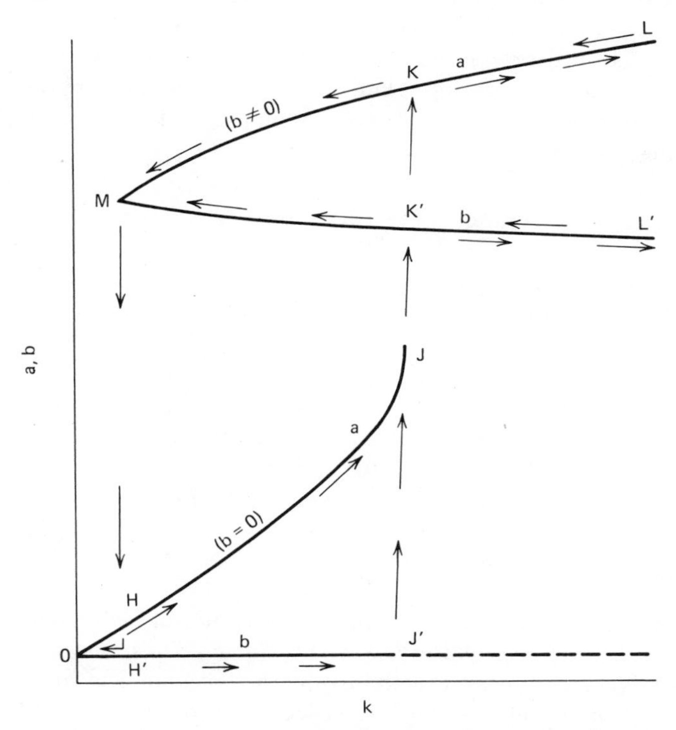

Figure 7-14. Response curves for a string under the action of linear damping.

planar motion is possible. The amplitude of the in-plane component increases linearly with k initially (in agreement with the linear theory) through H toward J. When $k = k_J$ the in-plane motion becomes unstable and any further increase in k causes the amplitudes to jump spontaneously from J up to K and from J' up to K'. During the jump the motion changes from planar to nonplanar. As k continues to increase the amplitude of the in-plane component increases from K toward L, while the amplitude of the out-of-plane component decreases from K' toward L'. If k is increased substantially the out-of-plane component vanishes smoothly and the motion becomes planar again. [This can be seen for the undamped case by considering (7.5.61) and (7.5.62); for $\sigma = 8, b = 0$ when $a = \sqrt{8}$ and $k = 2(8)^{3/2}$.] Reversing this procedure, we find that, as k decreases from a large value, the in-plane component of the motion decreases from L through K toward M, while the out-of-plane component increases from L' through K' toward M. At point M, where the two amplitudes are equal, the nonplanar motion becomes unstable. Any further decrease in k causes the amplitudes to jump spontaneously from M to H and H', and the motion becomes planar again.

Again the jumps produce a 180-degree change in the phase between the excitation and the response in the undamped case. In the region between the jumps the motion can be either planar or nonplanar, and a finite disturbance is required to make the motion change from one branch to the other.

We note that, had we run this experiment when σ is negative, no out-of-plane component would have developed in the predicted motion for any value of k.

For steady-state nonplanar undamped motions it follows from (7.5.48) and the fact that $\gamma = \pi$ that

$$\alpha = \sigma T_1 - \nu \quad \text{and} \quad \beta = \sigma T_1 + \frac{\pi}{2} - \nu \tag{7.5.84}$$

where ν is either zero or π. Thus the displacement of a point on the string can be written in the following dimensionless form:

$$\mathbf{\Delta} = [a \cos (\Omega t - \nu)\mathbf{j} - b \sin (\Omega t - \nu)\mathbf{k}] \sin \frac{s\pi x}{l} \tag{7.5.85}$$

It follows that the trajectory of a point is an ellipse having a semimajor axis $a \sin (s\pi x/l)$ along the x-axis. As the trajectory becomes circular the motion becomes unstable and spontaneously reverts to planar motion.

7.6. Plates

In this section we consider the nonlinear vibrations of plates neglecting transverse shear deformations and rotary inertia. These effects were included in the studies of Wu and Vinson (1969a, b); Singh, Das, and Sundararajan (1971); Kanaka Raju and Venkateswara Rao (1976a); Kozhemyakina and Morgaevskii

(1976); and Sathyamoorthy (1977). Yu and Lai (1966) determined the influence of transverse shear on the nonlinear vibrations of sandwich plates. Thickness-shear flexibility was included in the analyses of Ambartsumyan (1976); Wu and Vinson (1971); and Singh, Sundararajan, and Das (1974). For simplicity we consider only homogeneous plates. Nonuniform plates were analyzed by Huang and Meng (1972); Huang (1974); Ramachandran (1975); Ramachandran and Reddy (1975); Huang, Woo, and Walker (1976); and Kanaka Raju (1977).

7.6.1. GOVERNING EQUATIONS

We consider the nonlinear vibrations of a plate that is initially flat. A comprehensive treatment of the linear problem is given in the monograph of Leissa (1969). We introduce a Cartesian coordinate system xyz, with the x- and y-axes lying in the undisturbed position of the middle surface of the plate, and let u, v, and w denote the displacements of the middle surface in the x-, y-, and z-directions, respectively. Then the equations of motion are

$$\frac{\partial N_1}{\partial x} + \frac{\partial N_{12}}{\partial y} = \rho h \frac{\partial^2 u}{\partial t^2} \tag{7.6.1}$$

$$\frac{\partial N_{12}}{\partial x} + \frac{\partial N_2}{\partial y} = \rho h \frac{\partial^2 y}{\partial t^2} \tag{7.6.2}$$

$$\frac{\partial^2 M_1}{\partial x^2} + \frac{\partial^2 M_2}{\partial y^2} + 2\frac{\partial^2 M_{12}}{\partial x \partial y} + \frac{\partial}{\partial x}\left(N_1 \frac{\partial w}{\partial x}\right) + \frac{\partial}{\partial y}\left(N_2 \frac{\partial w}{\partial y}\right) + \frac{\partial}{\partial x}\left(N_{12}\frac{\partial w}{\partial y}\right) + \frac{\partial}{\partial y}\left(N_{12}\frac{\partial w}{\partial x}\right) = \rho h \frac{\partial^2 w}{\partial t^2} + f \tag{7.6.3}$$

where f is the transverse load and ρ and h are the density and thickness of the plate, respectively. Here N_1, N_2, and N_{12} are the forces per unit of length, characteristic of the state of stress in the plane of the plate; M_1 and M_2 are the bending moments parallel to the x- and y-axes; and M_{12} is the twisting moment.

Here we consider isotropic linear elastic materials. Studies dealing with nonlinear elastic materials are mentioned in the introduction to this chapter. Eringen (1955) examined the nonlinear oscillations of viscoelastic plates with hereditary damping included in the stress–strain relations. A number of investigators analyzed the nonlinear oscillations of anisotropic plates. Yu (1962), Yu and Lai (1966), and Shahin (1974) studied the nonlinear vibrations of sandwich plates, while Alwar and Adimurthy (1975) studied the response of sandwich panels to pulse and shock excitations. Yu (1963) investigated the nonlinear vibration of layered plates and shells. Hassert and Nowinski (1962), Wu and Vinson (1969b), Sathyamoorthy and Pandalai (1970), and Ramachandran (1974c) treated a rectangular plate with special rectangular orthotropy. Nowinski (1963b) and

Pal (1973) analyzed orthotropic circular plates; Nowinski and Ismail (1965), Vendhan and Dhoopar (1973), and Karmakar (1975) analyzed orthotropic triangular plates; while Sathyamoorthy and Pandalai (1973c) analyzed rectilinearly orthotropic skew plates. Huang (1972a, 1973) and Huang and Woo (1973) used a Ritz-Kantorovich method to analyze cylindrically anisotropic circular and annular plates, while Venkateswara Rao, Kanaka Raju, and Raju (1976) used a finite-element method to analyze orthotropic circular plates. Mayberry and Bert (1969) experimentally investigated the free vibrations of various laminated anisotropic rectangular plates and compared their result with an orthotropic plate analysis. Wu and Vinson (1971) extended this analysis by including thickness-shear flexibility. Bennett (1971) and Chandra and Basava Raju (1975b) investigated, respectively, the forced and free vibrations of angle ply-laminated rectangular plates, while Chandra and Basava Raju (1975a) and Chandra (1976) investigated the free vibrations of cross ply-laminated rectangular plates. Bennett (1972), Bert (1973), and Schmidt (1973) investigated the nonlinear oscillations of an arbitrarily laminated rectangular plate.

For a Hookean isotropic material the load-displacement equations are (Timoshenko, 1940)

$$N_1 = Eh(1 - \nu^2)^{-1} (u_x + \tfrac{1}{2} w_x^2 + \nu v_y + \tfrac{1}{2} \nu w_y^2) + N_1^{(i)} \tag{7.6.4}$$

$$N_2 = Eh(1 - \nu^2)^{-1} (v_y + \tfrac{1}{2} w_y^2 + \nu u_x + \tfrac{1}{2} \nu w_x^2) + N_2^{(i)} \tag{7.6.5}$$

$$N_{12} = Gh(u_y + v_x + w_x w_y) + N_{12}^{(i)} \tag{7.6.6}$$

$$M_1 = -D(w_{xx} + \nu w_{yy}) \tag{7.6.7}$$

$$M_2 = -D(w_{yy} + \nu w_{xx}) \tag{7.6.8}$$

$$M_{12} = -D(1 - \nu) w_{xy} \tag{7.6.9}$$

where E is Young's modulus, G is the shear modulus, ν is Poisson's ratio, $D = \frac{1}{12} Eh^3 (1 - \nu^2)^{-1}$ is the flexural rigidity, and the terms with superscript i stand for the in-plane applied edge loads. The nonlinear terms in (7.6.4) through (7.6.6) are due to the fact that an element of the plate is not only sheared and elongated but also rotated about the x- and y-axes. Substituting (7.6.4) through (7.6.9) into (7.6.1) through (7.6.3) leads to the following equations (Chu and Herrmann, 1956):

$$u_{xx} + w_x w_{xx} + \nu(v_{xy} + w_y w_{xy}) + \tfrac{1}{2}(1 - \nu)(u_{yy} + v_{xy} + w_x w_{yy} + w_y w_{xy}) = c_p^{-2} u_{tt} \tag{7.6.10}$$

$$v_{yy} + w_y w_{yy} + \nu(u_{xy} + w_x w_{xy}) + \tfrac{1}{2}(1 - \nu)(u_{xy} + v_{xx} + w_x w_{xy} + w_y w_{xx}) = c_p^{-2} v_{tt} \tag{7.6.11}$$

$$\tfrac{1}{12}h^2\nabla^4 w - N_1^{(i)}w_{xx} - 2N_{12}^{(i)}w_{xy} - N_2^{(i)}w_{yy} - u_x w_{xx} - \tfrac{1}{2}w_x^2 w_{xx}$$
$$- v_y w_{yy} - \tfrac{1}{2}w_y^2 w_{yy} - \nu(v_y w_{xx} + \tfrac{1}{2}w_y^2 w_{xx} + u_x w_{yy} + \tfrac{1}{2}w_x^2 w_{yy})$$
$$- (1-\nu)(u_y w_{xy} + v_x w_{xy} + w_x w_y w_{xy}) = c_p^{-2}(w_x u_{tt} + w_y v_{tt}$$
$$+ w_{tt}) + f \tag{7.6.12}$$

where $c_p^2 = E/[\rho(1-\nu^2)]$. Equations (7.6.10) through (7.6.12) are usually called the *dynamic analog of the von Karman equations* because they reduce to von Karman's equations (1910) in the absence of the time derivatives.

Equations (7.6.10) through (7.6.12) can be derived alternatively by using a variational principle. The normal and shearing strains of the middle surface are related to the displacements by

$$\epsilon_1 = u_x + \tfrac{1}{2}w_x^2, \quad \epsilon_2 = v_y + \tfrac{1}{2}w_y^2$$
$$\epsilon_{12} = u_y + v_x + w_x w_y \tag{7.6.13}$$

The first and second invariants of the middle surface are defined by

$$e_1 = \epsilon_1 + \epsilon_2, \quad e_2 = \epsilon_1\epsilon_2 - \tfrac{1}{4}\epsilon_{12}^2 \tag{7.6.14}$$

Then the sum of the membrane and bending energies in a linear Hookean elastic plate may be written in the form

$$V = \tfrac{1}{2}D\iint_A \left\{(\nabla^2 w)^2 + \frac{12}{h^2}e_1^2 - 2(1-\nu)\left[\frac{12}{h^2}e_2 + w_{xx}w_{yy} - w_{xy}^2\right]\right\} dx\,dy \tag{7.6.15}$$

The kinetic energy of the plate is

$$T = \tfrac{1}{2}\rho h \iint_A (u_t^2 + v_t^2 + w_t^2)\,dx\,dy \tag{7.6.16}$$

and the Lagrangian is given by

$$L = T - V \tag{7.6.17}$$

According to Hamilton's principle

$$\delta\int_{t_1}^{t_2} L\,dt = \delta\int_{t_1}^{t_2}(T-V)\,dt = 0 \tag{7.6.18}$$

where δ indicates the variation of the integral. Substituting (7.6.15) and (7.6.16) into (7.6.18) and performing the variations, we obtain (7.6.10) through (7.6.12) as the Euler–Lagrange equations of this variational problem.

To determine an approximation to the static counterparts of (7.6.10) through (7.6.12), Berger (1955) neglected the second invariant e_2 in (7.6.15) and obtained the so-called Berger's equations from the variational problem. Nash and Modeer (1959) followed Berger and neglected e_2 in (7.6.15) for the dynamic problem and obtained the following equations from the variational problem:

$$e_x = c_p^{-2} u_{tt} \quad \text{and} \quad e_y = c_p^{-2} v_{tt} \tag{7.6.19}$$

$$D\nabla^4 w - \rho h c_p^2 \left[\frac{\partial}{\partial x}(e w_x) + \frac{\partial}{\partial y}(e w_y) \right] + \rho h w_{tt} = 0 \tag{7.6.20}$$

These equations are usually referred to as the dynamic analog of Berger's equations. Although no complete explanation of this approximation is offered, the static stresses and deflections obtained by Berger (1955) for both rectangular and circular plates agree well with those obtained by using (7.6.10) through (7.6.12). Neglecting the in-plane inertia, Nash and Modeer (1959) and Wah (1963a) showed that application of (7.6.19) and (7.6.20) to simply-supported plates leads to results that are in excellent agreement with those obtained from (7.6.10) through (7.6.12).

Lee, Blotter, and Yen (1971) found that the errors introduced by applying the Berger hypothesis to a clamped circular plate depend on Poisson's ratio and the ratio of the radius to the thickness of the plate. Moreover they found that the error is minimized when ν increases. Comparing (7.6.19) and (7.6.20) with (7.6.10) through (7.6.12), we find that the former set of equations can be obtained from the latter by letting $\nu \to 1$ while keeping D and c_p fixed. Using the simplified equations, Wah (1963a) found that the ratio of the nonlinear and linear fundamental frequencies of simply-supported rectangular plates does not depend on the aspect ratio AR of the plate, in contrast with the results obtained by Chu and Herrmann (1956) from the von Karman equations. A comparison of the solutions of Wah (1963a) and Chu and Herrmann (1956) shows that the error introduced by the Berger approximation is minimized for $AR \approx 0.5$ for all a/h, where a is the amplitude of the motion. The error increases as AR deviates from 0.5 and as $a/h \to 1$. Nowinski and Ohnabe (1972) pointed out that the Berger hypothesis may lead to grave inaccuracies and even meaningless results for plates without the in-plane restraint. Huang and Al-Khattat (1977) showed that for radially restrained circular plates, solutions based on the Berger hypothesis are accurate at low amplitudes of vibration but that the accuracy decreases as the amplitude increases. Moreover they found that the Berger hypothesis is entirely unsuitable for plates with movable edges.

As shown in Sections 7.1, 7.2, and 7.5 for beams and strings, the in-plane inertia can be neglected if the transverse modes under consideration are widely separated from the fundamental in-plane mode. In this case one can neglect u_{tt} and v_{tt} in (7.6.10) through (7.6.12) and in (7.6.19) and (7.6.20). In the case of the von Karman equations, investigators followed two alternatives: a small group used (7.6.10) through (7.6.12), while the other group worked with (7.6.1) through (7.6.3) and (7.6.7) through (7.6.9) rewritten as follows. When $u_{tt} = v_{tt} = 0$, (7.6.1) and (7.6.2) are satisfied exactly if we introduce the stress function F defined by

$$N_1 = N_1^{(i)} + F_{yy}, \quad N_2 = N_2^{(i)} + F_{xx}, \quad N_{12} = N_{12}^{(i)} - F_{xy} \tag{7.6.21}$$

Crawford and Atluri (1975) examined the effect of initial stresses on the nonlinear vibration of simply-supported rectangular plates. Substituting (7.6.21) into (7.6.3) and using (7.6.7) through (7.6.9) to eliminate M_1, M_2, and M_{12}, we obtain

$$D\nabla^4 w + \rho h w_{tt} = -(F_{yy} + N_1^{(i)}) w_{xx} - (F_{xx} + N_2^{(i)}) w_{yy} + 2(F_{xy} + N_{12}^{(i)}) w_{xy} + f \tag{7.6.22}$$

The compatibility equations for the in-plane strains provides

$$\nabla^4 F = Eh(w_{xy}^2 - w_{xx} w_{yy}) \tag{7.6.23}$$

Equations (7.6.22) and (7.6.23) govern the nonlinear transverse motion of the plate if the in-plane inertia is negligible.

One can obtain the equations governing the nonlinear vibration of membranes from (7.6.22) and (7.6.23) by setting $D = 0$. Eringen (1951) analyzed the nonlinear axisymmetric vibrations of circular membranes using the initial membrane strain as a perturbation parameter. Yen and Lee (1975) used the Lindstedt-Poincaré technique to determine the nonlinear vibrations of a circular membrane including the effects of the longitudinal inertia. They found that their solution is invalid when the linear fundamental in-plane frequency is twice the fundamental transverse frequency; this occurs when the initial strain is about 0.4882. Then they obtained an expansion valid for this internal resonance case; however they did not indicate what material these results can be applied to. We note that for most metals the initial strain cannot be made to exceed about 0.05. Chobotov and Binder (1964) used a combination of a Ritz-Galérkin procedure and a perturbation technique to determine the effect of sinusoidal excitations on the nonlinear response of circular membranes.

A number of investigators studied the dynamics of spinning disks because of their use as sawblades and turbine wheels. The importance of flexural waves in causing spinning-disk failures was recognized a long time ago by Richards (1872, p. 135). Such waves have been observed in practice and studied experimentally (Campbell, 1924; von Freudenreich, 1925; Tobias and Arnold, 1957; Tobias, 1958; Williams and Tobias, 1963; Krauter and Bulkeley, 1970) either as free or forced motions in disks that have been spun in air. Nowinski (1964) and Nowinski and Woodall (1964) studied the nonlinear transverse vibrations of spinning disks.

Tobias (1957) and Efstathiades (1971) studied the nonlinear vibration of imperfect spinning disks, and Schaeffer and McElman (1965) analyzed the nonlinear dynamic response of a membrane in a freestream. Advani (1967a, b) found exact solutions for the motion of free-spinning membranes in the form of forward and backward waves traveling circumferentially around the membrane and found the latter waves that travel as fast as the membrane spins forward (stationary). Advani and Bulkeley (1969) obtained an exact nonlinear vibration mode in addition to the above-mentioned traveling waves. Advani and Bhattacharjie (1969) found an uncoupled representation of the equations governing the nonlinear axisymmetric motion of free-spinning membranes. Bulkeley (1973) examined the stability of the aforementioned exact solutions.

Chu (1961a) derived the counterpart of (7.6.22) and (7.6.23) for circular cylindrical shells. In the body of this chapter we do not discuss the nonlinear vibration of shells. Cylindrical shells were studied experimentally by Olson (1965) and Matsuzaki and Kobayashi (1969b) and theoretically by Chu (1961a); Evensen (1963); Nowinski (1963a); Goodier and McIvor (1964); McIvor (1966); Bieniek, Fan, and Lackman (1966); Evensen and Fulton (1967); Mayers and Wrenn (1967); McIvor and Lovell (1968); Dowell and Ventres (1968); Lovell and McIvor (1969); Matsuzaki and Kobayashi (1969a); Atluri (1972); Mente (1973); Ginsberg (1974); Ahmad and Lee (1975), Chen and Babcock (1975); Nguyen (1975); and Radwan and Genin (1976). Spherical shells were studied by McIvor and Sonstegard (1966) and Grossman, Koplik, and Yu (1969); hemispherical shells were studied by Jordan (1973). Conical shells were studied by Sun and Lu (1968); square shells were studied by Chauhan and Ashwell (1969); and axisymmetric shells and solids were studied by Nagarajan and Popov (1975a, b). Shells of revolution were studied by Stricklin, Martinez, Tillerson, Hong, and Haisler (1971); Lehner and Batterman (1974); Belytschko and Hsieh (1974); and Kanaka Raju and Venkateswara Rao (1976b). Shallow shells were studied by Alekseeva (1969); Leissa and Kadi (1971); El-Zaouk and Dym (1973); and Singh, Sundararajan, and Das (1974). Curved panels were studied by Hayes and Miles (1956); Greenspon (1960); Cummings (1964); and Rehfield, Girl, and Sparrow (1974). Homogeneous and layered plates and shells were studied by Yu (1963). Cylindrical panels were studied by Wu and Witmer (1975), Vol'mir and Kulterbaev (1975), and Ramachandran and Murthy (1976). Shahinpoor (1973) analyzed the combined radial-axial large-amplitude oscillations of hyperelastic cylindrical tubes, while Thurman and Mote (1969) analyzed the nonlinear oscillation of a cylinder containing flowing fluid. The postbuckling of thin elastic shells was examined by Lange and Newell (1971).

Neglecting the in-plane inertia in (7.6.19) gives

$$e = u_x + v_y + \tfrac{1}{2} w_x^2 + \tfrac{1}{2} w_y^2 = e(t) \tag{7.6.24}$$

according to (7.6.13) and (7.6.14). Integrating (7.6.24) over the area A of the plate we have

$$Ae(t) = \iint_A (u_x + v_y)\, dx\, dy + \tfrac{1}{2} \iint_A (w_x^2 + w_y^2)\, dx\, dy \tag{7.6.25}$$

If we impose the condition that u and v vanish at the boundaries, we have

$$Ae(t) = \tfrac{1}{2} \iint_A (w_x^2 + w_y^2)\, dx\, dy \tag{7.6.26}$$

When the edges are either simply supported or clamped, integrating (7.6.26) by parts leads to

$$Ae = -\tfrac{1}{2} \iint_A w \nabla^2 w\, dx\, dy \tag{7.6.27}$$

Since $e = e(t)$, (7.6.20) can be rewritten as

$$D\nabla^4 w - \rho h e c_p^2 (w_{xx} + w_{yy}) + \rho h w_{tt} = 0 \tag{7.6.28}$$

where e is given by (7.6.26) or (7.6.27), depending on the boundary conditions.

It is clear that the Berger hypothesis led to a great simplification in the governing equations from two coupled fourth-order equations to a single fourth-order equation. Besides Nash and Modeer (1959), Wah (1963a), Lee, Blotter, and Yen (1971) and Huang and Al-Khattat (1977), a number of investigators used Berger's hypothesis in their analyses of the nonlinear vibration of plates. Wah (1963b), Srinivasan (1965), Pal (1970), and Ramachandran (1977) analyzed circular plates, while Chiang and Chen (1972) analyzed circular plates with concentric rigid masses. Wu and Vinson (1969a), Singh, Das, and Sundararajan (1971), and Mei (1973b) treated rectangular plates, while Ramachandran and Reddy (1972) treated rectangular plates with cutouts. Sathyamoorthy (1977) treated skew plates, while Gajendar (1967), Mack and McQueary (1967), Kishor and Rao (1974), and Sircar (1974) treated plates resting on elastic foundations; and Ramachandran (1974b) treated restrained circular plates. Wu and Vinson (1969b, 1971) and Mayberry and Bert (1969) studied rectangular orthotropic skew plates, and Bert (1973) studied arbitrarily laminated rectangular plates.

7.6.2. FORCED SYMMETRIC RESPONSE OF CIRCULAR PLATES

In the remainder of this section we discuss the nonlinear forced symmetric response of circular plates. We use the dynamic analog of the von Karman equations because the simplified Burger equation may be inaccurate and it is similar to that of beams which is discussed in detail in Sections 7.3 and 7.4. Besides Chu and Herrmann (1956), a number of investigators used the von Karman equations in their analyses of the nonlinear vibrations of plates. Rectangular plates were analyzed by Smith, Malme, and Gogos (1961); Yamaki (1961); Eisley (1964b); Murthy and Sherbourne (1972); Bayles, Lowery, and Boyd (1973); Donaldson (1973b); Vendhan (1975a); Crawford and Atluri (1975); and Venkateswara Rao, Raju, and Kanaka Raju (1976a). Rectangular sandwich plates were analyzed by Yu (1962), and rectangular anisotropic plates were analyzed by Sathyamoorthy and Pandalai (1970), Bennett (1971), Chandra and Basava Raju (1975a, b), and Chandra (1976). Circular plates were treated by Yamaki (1961), Nowinski (1962), Bulkeley (1963); Crose and Ang (1969); Farnsworth and Evan-Iwanowski (1970); Huang and Sandman (1971); Huang (1972a, b, 1973, 1974); Kung and Pao (1972); Ramachandran (1974a); and Sridhar, Mook, and Nayfeh (1975, 1978); annular plates were treated by Sandman and Huang (1971); Huang and Woo (1973); Goldberg and Koening (1975); and Huang, Woo, and Walker (1976); and ring sector plates were treated by Chisyaki and Takahashi (1972). Triangular plates were treated by Vendhan and Dhoopar (1973), Vendhan and Das (1975), and Vendhan (1975b); elliptic plates were treated by Lobitz, Nayfeh, and Mook (1977); and skew plates were treated by Sathyamoorthy and Pandalai (1972b, 1973a, c, d).

In the present discussion we employ (7.6.22) and (7.6.23), where $f = f(r, t)$ is a known function of time—an external excitation. Parametric excitations of plates, beams, and shells are discussed in Section 5.4.4 in connection with the effect of follower forces on the lateral vibration of columns. In some problems f may be a function of w; such is the case for plates on elastic foundations and plates interacting with an airstream (flutter problem). In the latter case as the plate deforms, it modifies the airflow which in turn exerts pressure and shear perturbations on the plate. These pressure and shear perturbations are functions of the plate displacement as well as the airstream velocity, density, and viscosity. When the dynamic pressure $(\frac{1}{2}\rho_f V_f^2)$ of the flow is low, the response of the plate is random, with dominant frequency components near the lower natural frequencies. When the dynamic pressure reaches a critical value, the linear response of the plate grows exponentially with time and the motion is termed unstable. If the linear frequency of the combined air/plate system corresponding to the unstable mode is zero, this (static) instability is called *divergence*; and if the linear frequency of the system corresponding to the unstable mode is different from zero, this dynamic instability is called *flutter*. For more details on flutter we refer the reader to the book of Dowell (1975).

As the critical dynamic pressure is exceeded, the linear theory indicates that the plate's motion grows exponentially with time. However as the deflection increases, nonlinear effects (midplane stretching) come into play and generally restrain the motion to a bounded limit cycle, especially for supersonic airflows.

Besides the book of Dowell (1975), a number of investigators treated nonlinear flutter problems. Dimantha and Roorda (1969) examined the domain of asymptotic stability of nonlinear conservative systems. Flat plates were examined by Kobayashi (1962a, b); Bolotin (1963); Librescu (1965); Dugundji (1966); Dowell (1966, 1967a, 1973), Morino (1969); Ventres and Dowell (1970); Eastep and McIntosh (1971); and Kuo, Morino, and Dugundji (1972). Curved plates were examined by Dowell (1969, 1970b). Cylindrical shells were studied by Olson and Fung (1966) and Gordon and Atluri (1974), while shallow shells were studied by Gajl (1974). Kirchman and Greenspon (1957) and Lin (1962) studied the response of a panel to periodic and randomly varying loads. Interaction of panel flutter with external excitations was analyzed by Dowell (1970a); and Kuo, Morino, and Dugundji (1973). Ginsberg (1975) and Nayfeh and Kelly (1978) analyzed the nonlinear interaction of an acoustic fluid and a flat plate under-harmonic excitation.

The symmetric responses of uniform circular plates can be obtained by rewriting (7.6.22) and (7.6.23) in the absence of applied in-plane loads in the following polar form:

$$\rho h w_{tt} + D\nabla^4 w = \frac{1}{r}\frac{\partial}{\partial r}(F_r w_r) - 2\mu w_t + f(r, t) \tag{7.6.29}$$

$$\nabla^4 F = -\frac{Eh}{2r}\frac{\partial}{\partial r}(w_r^2) \tag{7.6.30}$$

where μ is the damping coefficient and

$$\nabla^4 = \left(\frac{\partial^2}{\partial r^2} + \frac{1}{r}\frac{\partial}{\partial r}\right)^2 \tag{7.6.31}$$

The relationships among F, the radial displacement u, and the deflection w are given by

$$Eh(u_r + \tfrac{1}{2} w_r^2) = \frac{1}{r}F_r - \nu F_{rr} \tag{7.6.32}$$

and

$$Eh\frac{u}{r} = F_{rr} - \frac{\nu}{r} F_r \tag{7.6.33}$$

It is convenient to rewrite (7.6.29) through (7.6.33) in terms of dimensionless variables. These variables are defined as follows:

$$r = Rr^*, \quad t = R^2\sqrt{\frac{\rho h}{D}}\, t^*, \quad w = \frac{h^2}{R} w^*$$

$$u = \frac{h^4}{R^3} u^*, \quad \mu = \frac{24(1 - \nu^2)}{R^4} \sqrt{\rho h^5 D}\ \mu^*$$

$$f = \frac{Dh^2}{R^5} f^*, \quad \text{and} \quad F = \frac{Eh^5}{R^2} F^*$$

where R is the radius of the plate. Substituting these definitions into (7.6.29) through (7.6.33) and dropping the asterisks we obtain

$$w_{tt} + \nabla^4 w = \epsilon \left[\frac{1}{r} \frac{\partial}{\partial r} (F_r w_r) - 2\mu w_t \right] + f \tag{7.6.34}$$

$$\nabla^4 F = -\frac{1}{2r} \frac{\partial}{\partial r} (w_r^2) \tag{7.6.35}$$

$$u_r + \tfrac{1}{2} w_r^2 = \frac{1}{r} F_r - \nu F_{rr} \tag{7.6.36}$$

$$\frac{u}{r} = F_{rr} - \frac{\nu}{r} F_r \tag{7.6.37}$$

where $\epsilon = 12(1 - \nu^2) h^2/R^2$. All the dimensionless variables are $O(1)$ as $\epsilon \to 0$.

The boundary conditions for clamped edges are

$$w = 0, \quad u = 0, \quad \text{and } w_r = 0 \quad \text{at } r = 1 \tag{7.6.38}$$

Equations (7.6.36) and (7.6.37) can be combined to yield the following equation for F:

$$\tfrac{1}{2} w_r^2 = \frac{1}{r} F_r - F_{rr} - rF_{rrr} \tag{7.6.39}$$

which is a first integral of (7.6.35). Then it follows from (7.6.37) and (7.6.38) that the boundary condition for F is

$$F_{rr} - \nu F_r = 0 \quad \text{at } r = 1 \tag{7.6.40}$$

In addition, w and F are required to be finite at $r = 0$.

We seek approximate solutions which are uniformly valid for small ϵ. As in the case of beams, we use the method of multiple scales. We assume

$$w(r, t; \epsilon) = \sum_{m=1}^{\infty} \psi_m(t; \epsilon)\, \phi_m(r) \tag{7.6.41}$$

where the ϕ_m are the linear, free-oscillation modes. Thus the ϕ_m are the solutions of the following eigenvalue problem:

$$\nabla^4 \phi_m - \omega_m^2 \phi_m = 0 \tag{7.6.42}$$

$$\phi_m(1) = 0, \quad \phi'_m(1) = 0, \quad \text{and } \phi_m(0) < \infty \tag{7.6.43}$$

The eignevalues ω_m are the natural frequencies of the plate.

Following the procedure used in Section 7.1, one can show that the ϕ_m are orthogonal with respect to the weighting function r. The amplitude of each mode is chosen such that

$$\int_0^1 r\phi_n \phi_m \, dr = \delta_{nm} \tag{7.6.44}$$

Next we obtain the solution of the eigenvalue problem defined above. We rewrite (7.6.42) in the following convenient form:

$$\left(\frac{d^2}{dr^2} + \frac{1}{r}\frac{d}{dr} - \omega_m^2\right)\left(\frac{d^2}{dr^2} + \frac{1}{r}\frac{d}{dr} + \omega_m^2\right)\phi_m = 0 \tag{7.6.45}$$

Thus we can obtain the four linearly independent solutions of (7.6.45) from the following two equations:

$$\left(\frac{d^2}{dr^2} + \frac{1}{r}\frac{d}{dr} - \omega_m^2\right)\phi_m = 0 \tag{7.6.46}$$

$$\left(\frac{d^2}{dr^2} + \frac{1}{r}\frac{d}{dr} + \omega_m^2\right)\phi_m = 0 \tag{7.6.47}$$

From (7.6.46) we obtain

$$\phi_m^{(1)} = E_1 I_0(\kappa_m r) + E_2 K_0(\kappa_m r) \tag{7.6.48}$$

and from (7.6.47) we obtain

$$\phi_m^{(2)} = E_3 J_0(\kappa_m r) + E_4 Y_0(\kappa_m r) \tag{7.6.49}$$

where the E_m are constants of integration and $\kappa_m^2 = \omega_m$. The complete solution is

$$\phi_m = \phi_m^{(1)} + \phi_m^{(2)} \tag{7.6.50}$$

The condition that $\phi_m(0)$ be bounded demands that $E_2 = E_4 = 0$ because both K_0 and Y_0 have logarithmic singularities at the origin. Thus it follows from (7.6.43) that

$$\phi_m = C_m \left[J_0(\kappa_m r) I_0(\kappa_m) - J_0(\kappa_m) I_0(\kappa_m r)\right] \tag{7.6.51}$$

where the κ_m are the roots of

$$I_0(\kappa) J'_0(\kappa) - J_0(\kappa) I'_0(\kappa) = 0 \tag{7.6.52}$$

and the C_m are obtained from (7.6.44).

The first five natural frequencies obtained from (7.6.52) are $\omega_1 = 10.2158$, $\omega_2 = 39.7710$, $\omega_3 = 89.1040$, $\omega_4 = 158.1830$, and $\omega_5 = 247.0050$. We note that

$$\omega_1 + 2\omega_2 = 89.7578 \approx \omega_3 \tag{7.6.53}$$

and introduce a detuning parameter σ_1 according to

$$\omega_1 + 2\omega_2 = \omega_3 + \epsilon\sigma_1 \tag{7.6.54}$$

Hence there is an internal resonance involving three modes.

Equations (7.6.39) and (7.6.40) suggest that it may be more convenient to solve for F_r instead of F. Thus we put

$$G = F_r \tag{7.6.55}$$

Substituting (7.6.55) into (7.6.40) we obtain the following boundary condition for G:

$$G_r - \nu G = 0 \quad \text{at } r = 1 \tag{7.6.56}$$

Substituting (7.6.55) and (7.6.41) into (7.6.39) we obtain

$$r^2 G_{rr} + rG_r - G = -\tfrac{1}{2} r \left(\sum_{m=1}^{\infty} \psi_m \phi_m' \right)^2 \tag{7.6.57}$$

The function G can be represented by an expansion in terms of a complete set of orthogonal eigenfunctions. Because

$$\left(r^2 \frac{d^2}{dr^2} + r \frac{d}{dr} - 1 \right) J_1(\zeta_m r) = -\zeta_m^2 r^2 J_1(\zeta_m r)$$

it is convenient to express G as follows:

$$G(r, t) = \sum_{m=1}^{\infty} \eta_m(t) J_1(\zeta_m r) \tag{7.6.58}$$

where the ζ_m are chosen such that (7.6.56) is satisfied. That is, the ζ_m are the roots of

$$\zeta J_0(\zeta) - (1 + \nu) J_1(\zeta) = 0 \tag{7.6.59}$$

For $\nu = \frac{1}{3}$, the first 12 roots of (7.6.59) are

$$\zeta_1 = 1.545, \quad \zeta_2 = 5.266, \quad \zeta_3 = 9.497, \quad \zeta_4 = 11.68$$
$$\zeta_5 = 14.84, \quad \zeta_6 = 18.00, \quad \zeta_7 = 21.15, \quad \zeta_8 = 24.30$$
$$\zeta_9 = 30.59, \quad \zeta_{10} = 33.74, \quad \zeta_{11} = 36.88, \quad \zeta_{12} = 40.03$$

To obtain the functions $\eta_m(t)$, we substitute (7.6.58) into (7.6.57), multiply by $r^{-1}J_1(\zeta_n r)$, and integrate from $r = 0$ to $r = 1$. The result is

$$\eta_n(t) = \sum_{q=1}^{\infty} \sum_{p=1}^{\infty} S_{npq} \psi_q(t) \psi_p(t) \tag{7.6.60}$$

where

$$S_{npq} = [(\zeta_n^2 - 1 + \nu^2) J_1^2(\zeta_n)]^{-1} \int_0^1 \phi_p' \phi_q' J_1(\zeta_n r)\, dr \tag{7.6.61}$$

Using (7.6.60) we can now rewrite (7.6.58) in the following form:

$$G(r, t) = \sum_{m=1}^{\infty} \sum_{n=1}^{\infty} \sum_{p=1}^{\infty} S_{mnp} \psi_n(t) \psi_p(t) J_1(\zeta_m r) \tag{7.6.62}$$

To obtain the equations governing the ψ_n, we substitute (7.6.62) and (7.6.41) into (7.6.34), multiply by $r\phi_n$, and integrate from $r = 0$ to $r = 1$. The result is

$$\ddot{\psi}_n + \omega_n^2 \psi_n = \epsilon \left[\sum_{m=1}^{\infty} \sum_{p=1}^{\infty} \sum_{q=1}^{\infty} \Gamma_{nmpq} \psi_m \psi_p \psi_q - 2\mu_n \dot{\psi}_n \right] + f_n(t) \tag{7.6.63}$$

where

$$f_n(t) = \int_0^1 r\phi_n f(r, t)\, dr \quad \text{and} \quad \mu_n = \int_0^1 \mu r \phi_n\, dr \tag{7.6.64}$$

To obtain the expression for the Γ_{nmpq}, we consider

$$\int_0^1 \frac{\partial}{\partial r}(Gw_r) \phi_n\, dr = Gw_r \phi_n \Big|_0^1 - \int_0^1 Gw_r \phi_n'\, dr \tag{7.6.65}$$

The first term vanishes as a result of the boundary conditions and the symmetry of the deflection. Substituting (7.6.41) and (7.6.62) into (7.6.65) leads to

$$\int_0^1 \frac{\partial}{\partial r}(Gw_r) \phi_n\, dr =$$

$$- \sum_{m=1}^{\infty} \sum_{p=1}^{\infty} \sum_{q=1}^{\infty} \left[\sum_{k=1}^{\infty} S_{kpq} \int_0^1 J_1(\zeta_k r) \phi_n' \phi_m'\, dr \right] \psi_m \psi_p \psi_q$$

Using (7.6.61) leads to

$$\Gamma_{nmpq} = \sum_{k=1}^{\infty} \frac{\int_0^1 \phi_p' \phi_q' J_1(\zeta_k r)\, dr \int_0^1 \phi_n' \phi_m' J_1(\zeta_k r)\, dr}{(\zeta_k^2 - 1 + \nu^2) J_1^2(\zeta_k)} \tag{7.6.66}$$

Equations (7.6.63) and (7.1.38) have precisely the same form. Though the definitions of Γ_{nmpq} in (7.1.41) and (7.6.66) are different, (7.1.38) also applies to plates. Finally we note that the internal resonance for plates involves three modes, while internal resonances for beams involve either two or four modes. Next we consider several cases of primary resonances. Bauer (1968) considered the nonlinear response of a plate to a pulse excitation.

In this case we put

$$f_n(t) = 2\epsilon k_n \cos(\Omega t) \tag{7.6.67}$$

where

$$\Omega = \omega_s + \epsilon\sigma_2 \tag{7.6.68}$$

Following the method of multiple scales we let

$$\psi_n(t; \epsilon) = \psi_{n0}(T_0, T_1) + \epsilon\psi_{n1}(T_0, T_1) + \cdots \tag{7.6.69}$$

in (7.6.63), equate coefficients of like powers of ϵ, and obtain the equations describing the ψ_{n0} and ψ_{n1}. The solution of the equations governing the ψ_{n0} can be written as

$$\psi_{n0} = A_n(T_1) \exp(i\omega_n T_0) + cc \tag{7.6.70}$$

Using (7.6.70) we find that secular terms are eliminated from the ψ_{n1} if

$$-2i\omega_1(A_1' + \mu_1 A_1) + \delta_{1s} k_1 \exp(i\sigma_2 T_1) + Q_1 \bar{A}_2^2 A_3 \exp(-i\sigma_1 T_1) + A_1 \sum_{n=1}^{\infty} \alpha_{1n} A_n \bar{A}_n = 0 \tag{7.6.71}$$

$$-2i\omega_2(A_2' + \mu_2 A_2) + \delta_{2s} k_2 \exp(i\sigma_2 T_1) + Q_2 \bar{A}_1 \bar{A}_2 A_3 \exp(-i\sigma_1 T_1) + A_2 \sum_{n=1}^{\infty} \alpha_{2n} A_n \bar{A}_n = 0 \tag{7.6.72}$$

$$-2i\omega_3(A_3' + \mu_3 A_3) + \delta_{3s} k_3 \exp(i\sigma_2 T_1) + Q_3 A_1 A_2^2 \exp(i\sigma_1 T_1) + A_3 \sum_{n=1}^{\infty} \alpha_{3n} A_n \bar{A}_n = 0 \tag{7.6.73}$$

and for $m > 3$

$$-2i\omega_m(A_m' + \mu_m A_m) + \delta_{ms} k_m \exp(i\sigma_2 T_1) + A_m \sum_{n=1}^{\infty} \alpha_{mn} A_n \bar{A}_n = 0 \tag{7.6.74}$$

where

$$\alpha_{nj} = \alpha_{jn} = \begin{cases} 2(2\Gamma_{njnj} + \Gamma_{nnjj}) & \text{for } n \neq j \\ 3\Gamma_{nnnn} \end{cases} \tag{7.6.75}$$

$$Q_1 = Q_3 = 2\Gamma_{1223} + \Gamma_{1322} \quad \text{and} \quad Q_2 = 2Q_1 \tag{7.6.76}$$

Henceforth we let

$$Q_1 = 8Q \tag{7.6.77}$$

It is convenient to introduce the polar notation

$$A_n = \tfrac{1}{2} a_n \exp(i\beta_n) \tag{7.6.78}$$

Substituting (7.6.78) into (7.6.71) through (7.6.74) and separating the result into real and imaginary parts, we obtain

$$\omega_1(a_1' + \mu_1 a_1) + Q a_2^2 a_3 \sin\gamma_1 - \delta_{1s} k_1 \sin\gamma_2 = 0 \tag{7.6.79}$$

$$\omega_1 a_1 \beta_1' + Q a_2^2 a_3 \cos\gamma_1 + \tfrac{1}{8} a_1 \sum_{n=1}^{\infty} \alpha_{1n} a_n^2 + \delta_{1s} k_1 \cos\gamma_2 = 0 \tag{7.6.80}$$

$$\omega_2(a_2' + \mu_2 a_2) + 2Q a_1 a_2 a_3 \sin\gamma_1 - \delta_{2s} k_2 \sin\gamma_2 = 0 \tag{7.6.81}$$

$$\omega_2 a_2 \beta_2' + 2Q a_1 a_2 a_3 \cos\gamma_1 + \tfrac{1}{8} a_2 \sum_{n=1}^{\infty} \alpha_{2n} a_n^2 + \delta_{2s} k_2 \cos\gamma_2 = 0 \tag{7.6.82}$$

$$\omega_3(a_3' + \mu_3 a_3) - Q a_1 a_2^2 \sin\gamma_1 - \delta_{3s} k_3 \sin\gamma_2 = 0 \tag{7.6.83}$$

$$\omega_3 a_3 \beta_3' + Q a_1 a_2^2 \cos\gamma_1 + \tfrac{1}{8} a_3 \sum_{n=1}^{\infty} \alpha_{3n} a_n^2 + \delta_{3s} k_3 \cos\gamma_2 = 0 \tag{7.6.84}$$

$$\omega_m(a_m' + \mu_m a_m) - \delta_{ms} k_m \sin\gamma_2 = 0 \tag{7.6.85}$$

$$\omega_m a_m \beta_m' + \tfrac{1}{8} a_m \sum_{n=1}^{\infty} \alpha_{mn} a_n^2 + \delta_{ms} k_m \cos\gamma_2 = 0 \tag{7.6.86}$$

where

$$\gamma_1 = \sigma_1 T_1 + \beta_1 + 2\beta_2 - \beta_3 \quad \text{and} \quad \gamma_2 = \sigma_2 T_1 - \beta_s \tag{7.6.87}$$

There are four cases of interest: $s = 1, 2, 3$, and any number greater than 3. These cases are discussed separately.

Case I, $s \geqslant 4$. It follows from (7.6.85) that for all m, except $m = 1, 2, 3$, and s,

$$a_m' + \mu_m a_m = 0$$

Thus the amplitudes decay. We can also establish that the first three modes decay. To do this, we consider the steady-state solutions and suppose that a_2 and a_3 are not zero. Then from (7.6.81) and (7.6.83) it follows that

$$\frac{a_2^2}{a_3^2} = -\frac{2\mu_3 \omega_3}{\mu_2 \omega_2} \tag{7.6.88}$$

This is an impossible relationship; thus we conclude that a_2 and a_3 must be zero. Then it follows immediately from (7.6.79) that $a_1 = 0$.

Thus only the sth mode is excited and the solution is the same as that for the Duffing equation for a hardening spring, which is discussed in Section 4.1. In the steady state the deflection has the form

$$w(r, t) = a_s \cos(\Omega t - \gamma_2)\, \phi_s(r) + O(\epsilon) \tag{7.6.89}$$

Case II, $s = 1$. In this case we find immediately that the a_m for $m \geqslant 4$ decay. We can also establish that the second and third modes decay. To do this, we consider the steady-state solutions and suppose that a_2 and a_3 are not zero. Again we obtain (7.6.88); thus we conclude that a_2 and a_3 are zero. Only the first mode is excited, and the solution is the same as that for the Duffing equation for a hardening spring. In the steady state the deflection has the form

$$w(r, t) = a_1 \cos(\Omega t - \gamma_2)\, \phi_1(r) + O(\epsilon) \tag{7.6.90}$$

Case II, $s = 2$. Proceeding in a similar way we find that only the second mode is excited. Thus in the steady state

$$w(r, t) = a_2 \cos(\Omega t - \gamma_2)\, \phi_2(r) + O(\epsilon) \tag{7.6.91}$$

Case IV, $s = 3$. In this case we find that the a_m for $m \geqslant 4$ decay. Now, however, if we try to establish that a_1 and a_2 decay, we find that

$$\left(\frac{a_1}{a_2}\right)^2 = \frac{\mu_2 \omega_2}{2\mu_1 \omega_1} \tag{7.6.92}$$

In contrast with (7.6.88), this relationship is possible. Thus there are two possibilities: either a_1 and a_2 are zero or neither one is zero. These possibilities are discussed next.

When a_1 and a_2 are zero, only the third mode is excited. This case is the same as those discussed above, and the steady-state deflection has the form

$$w(r,t) = a_3 \cos(\Omega t - \gamma_2)\,\phi_3(r) + O(\epsilon) \tag{7.6.93}$$

When a_1 and a_2 differ from zero, we must solve (7.6.79) through (7.6.84) numerically. For the steady-state solution γ_1' and γ_2' are zero, and it follows from (7.6.87) that

$$\beta_3' = \sigma_2 \quad \text{and} \quad \beta_1' + 2\beta_2' = \sigma_2 - \sigma_1 \tag{7.6.94}$$

Then from (7.6.80), (7.6.82), and (7.6.84) we obtain

$$\sigma_2 - \sigma_1 + \tfrac{1}{8}\left[\left(\frac{\alpha_{11}}{\omega_1} + \frac{2\alpha_{21}}{\omega_2}\right)a_1^2 + \left(\frac{\alpha_{12}}{\omega_1} + \frac{2\alpha_{22}}{\omega_2}\right)a_2^2 + \left(\frac{\alpha_{13}}{\omega_1} + \frac{2\alpha_{23}}{\omega_2}\right)a_3^2\right] + Q\left(\frac{a_2^2}{\omega_1 a_1} + \frac{4a_1}{\omega_2}\right)a_3 \cos\gamma_1 = 0 \tag{7.6.95}$$

$$\omega_3\sigma_2 a_3 + \tfrac{1}{8}a_3(\alpha_{31}a_1^2 + \alpha_{32}a_2^2 + \alpha_{33}a_3^2) + Qa_1a_2^2 \cos\gamma_1 + k_3 \cos\gamma_2 = 0 \tag{7.6.96}$$

and from (7.6.79), (7.6.81), and (7.6.83) we obtain

$$\omega_1\mu_1 a_1 + Qa_2^2 a_3 \sin\gamma_1 = 0 \tag{7.6.97}$$

$$\omega_2\mu_2 a_2 + 2Qa_1a_2a_3 \sin\gamma_1 = 0 \tag{7.6.98}$$

$$\omega_3\mu_3 a_3 - Qa_1a_2^2 \sin\gamma_1 - k_3 \sin\gamma_2 = 0 \tag{7.6.99}$$

We can solve (7.6.95) through (7.6.99) for a_1, a_2, a_3, γ_1, and γ_2. Then we can solve (7.6.80) and (7.6.82) for β_1' and β_2', but β_1 and β_2 can only be determined to

TABLE 7-3. Variations of the Coefficients Appearing in Equations (7.6.95) through (7.6.99) with the Number of Terms in Equation (7.6.66)

Coefficients	Values		
	2 terms	10 terms	11 terms
$\alpha_{11} = 3\Gamma_{1111}$	-162.22	-162.22	-162.22
$\alpha_{12} = 2(2\Gamma_{1212} + \Gamma_{1122}) = \alpha_{21}$	-848.34	-883.80	-883.80
$\alpha_{13} = 2(2\Gamma_{1313} + \Gamma_{1133}) = \alpha_{31}$	-1428.6	-1644.8	-1644.8
$\alpha_{22} = 3\Gamma_{2222}$	-4991.2	-5552.1	-5552.1
$\alpha_{23} = 2(2\Gamma_{3232} + \Gamma_{2233}) = \alpha_{32}$	-10333	-14220	-14220
$\alpha_{33} = 3\Gamma_{3333}$	-27914	-34401	-34401
$Q_1 = 2\Gamma_{1223} + \Gamma_{1322} = Q_3$	-331.04	-566.77	-566.77
$Q_2 = 2Q_1$	-662.08	-1113.5	-1113.5

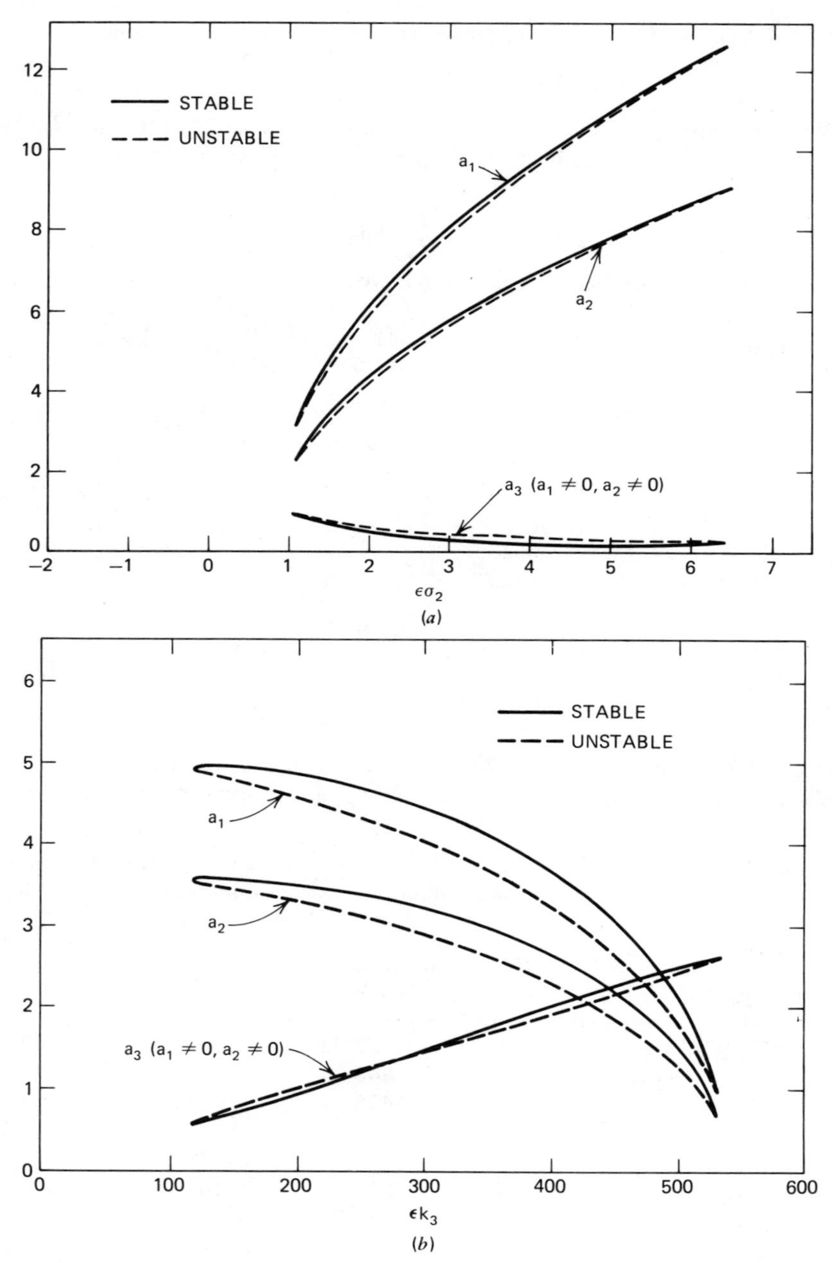

Figure 7-15. Amplitudes of the modes for a circular plate as a function of (*a*) the frequency and (*b*) the amplitude of the excitation; $\Omega \simeq \omega_2$.

within arbitrary constants, τ_1 and τ_2, which depend on the initial conditions. Equation (7.6.87) yields β_3.

The phases for the first and second modes (given here by the constants τ_1 and τ_2) depend on the initial conditions. Had there been a two-frequency excitation, each near one of the natural frequencies, then τ_1 and τ_2 could have been determined, as in the cases when only one mode is excited. And if there had been a three-term excitation, each near a different natural frequency, then in general no steady-state solution could have been found.

The nonlinearity adjusts the frequencies of the three strongly excited modes so that the third mode is tuned exactly to the frequency of the excitation and the frequencies of the first three modes satisfy the condition of commensurability exactly. To see this, we consider the following:

$$\omega_1 + \epsilon\beta_1' + 2(\omega_2 + \epsilon\beta_2') = \omega_1 + 2\omega_2 + \epsilon(\sigma_2 - \sigma_1) = \omega_3 + \epsilon\sigma_2 = \omega_3 + \epsilon\beta_3' = \Omega \tag{7.6.100}$$

The steady-state deflection is

$$w(r, t) = a_1 \cos [(\omega_1 + \epsilon\beta_1') t + \tau_1] \phi_1(r) + a_2 \cos [(\omega_2 + \epsilon\beta_2') t + \tau_2] \phi_2(r) + a_3 \cos (\Omega t - \gamma_2) \phi_3(r) + O(\epsilon) \tag{7.6.101}$$

The coefficients needed to solve (7.6.95) through (7.6.99) are given in Table 7-3. The number of terms refers to the summation of the series in (7.6.66), and the three values show the convergence of this series.

The amplitudes a_1, a_2, and a_3 are plotted as functions of the frequency of the excitation in Figure 7-15*a* and as functions of the amplitude of the excitation in Figure 7-15*b*. As before, we note that when a_1 and a_2 are not zero, there is a wide range of the parameters of the excitation for which a_1 is larger than a_2 and for which a_2 is considerably larger than a_3. Moreover we note that when the amplitude of the excitation is small, the three-mode solution is impossible. Thus these results reduce to the familiar solution of the linearized problem as the amplitude of the excitation decreases.

Exercises

7.1. Nonlinear longitudinal forced vibrations in a nonuniform bar in the presence of viscous damping are given by

$$\rho(x) A(x) \frac{\partial^2 u}{\partial t^2} = \frac{\partial}{\partial x} [A(x)\sigma] - 2\mu \frac{\partial u}{\partial t} + F(x, t)$$

where

$$\sigma = E(x) e [1 + E_1(x) e + E_2(x) e^2 + \cdots], \quad e = \frac{\partial u}{\partial x}$$

For fixed ends,

$$u(0, t) = u(1, t) = 0$$

(a) Eliminate σ from these equations and obtain

$$\rho(x)A(x)\frac{\partial^2 u}{\partial t^2} = \frac{\partial}{\partial x}\left[E(x)A(x)\frac{\partial u}{\partial x}\right]$$

$$+\frac{\partial}{\partial x}\left[E(x)A(x)E_1(x)\left(\frac{\partial u}{\partial x}\right)^2 + E(x)A(x)E_2(x)\left(\frac{\partial u}{\partial x}\right)^3\right]$$

$$-2\mu\frac{\partial u}{\partial t} + F(x, t)$$

(b) Show that the linear undamped mode shapes and frequencies of this system are given by the eigenvalue problem

$$\frac{d}{dx}\left[E(x)A(x)\frac{d\phi}{dx}\right] + \omega^2\rho(x)A(x)\phi = 0$$

$$\phi(0) = \phi(1) = 0$$

where ϕ is the mode shape corresponding to the frequency ω.

(c) Show that the eigenfrequencies are real.

(d) If $\phi_n(x)$ and $\phi_m(x)$ are the eigenmodes corresponding respectively to the eigenfrequencies ω_n and ω_m, show that they are orthogonal and can be made orthonormal so that

$$\int_0^1 \rho(x)A(x)\phi_n(x)\phi_m(x)\,dx = \delta_{mn}$$

(e) Seek a solution to the nonlinear, damped, forced system in the form

$$u = \sum_{n=1}^{\infty} \psi_n(t)\phi_n(x)$$

Substitute into the governing equation, use the orthornormality condition, and obtain

$$\ddot{\psi}_n + \omega_n^2\psi_n = -2\mu_n\dot{\psi}_n + N_n(\psi_1, \psi_2, \ldots, \psi_m, \ldots) + f_n(t)$$

where the damping is assumed to be modal,

$$f_n(t) = \int_0^1 F(x, t)\phi_n(x)\,dx$$

$$N_n = \sum_{p,q} \psi_p \psi_q \int_0^1 \phi_n \frac{d}{dx}\left[EAE_1 \frac{d\phi_p}{dx}\frac{d\phi_q}{dx}\right] dx + \sum_{p,q,r} \psi_p \psi_q \psi_r$$

$$\cdot \int_0^1 \phi_n \frac{d}{dx}\left[EAE_2 \frac{d\phi_p}{dx}\frac{d\phi_q}{dx}\frac{d\phi_r}{dx}\right] dx$$

(f) Let

$$f_n(t) = K_n \cos \Omega t$$

and determine the response of the system to this excitation. Bojadziev and Lardner (1973) analyzed the response of a nonlinear elastic rod to a monofrequency excitation.

(g) Let

$$f_n(t) = K_{1n} \cos \Omega_1 t + K_{2n} \cos (\Omega_2 t + \tau_n)$$

and determine the response of the system to this excitation.

7.2. Consider the response of a nonlinear elastic rod to an external excitation. The problem is governed by

$$\rho \frac{\partial^2 u}{\partial t^2} = \frac{\partial \sigma}{\partial x} - 2\rho\mu \frac{\partial u}{\partial t} + \rho F(x, t)$$

$$\sigma = Ee(1 + E_1 e + E_2 e^2 + \cdots), \quad e = \frac{\partial u}{\partial x}$$

$$u(0, t) = 0, \quad u(1, t) + h \frac{\partial u(1, t)}{\partial x} = 0$$

(a) Eliminate σ to obtain

$$\frac{\partial^2 u}{\partial t^2} - c^2 \frac{\partial^2 u}{\partial x^2} = 2c^2 E_1 \frac{\partial u}{\partial x}\frac{\partial^2 u}{\partial x^2} + 3c^2 E_2 \left(\frac{\partial u}{\partial x}\right)^2 \frac{\partial^2 u}{\partial x^2} - 2\mu \frac{\partial u}{\partial t} + F(x, t)$$

(b) Show that the linear, undamped eigenmodes and eigenfrequencies are $\sin k_n x$, where $\omega_n = ck_n$ and the k_n are the roots of

$$\tan k = -hk$$

Determine the lowest five eigenvalues for the case $h = 1$.

(c) Let

$$u = \sqrt{2} \sum_{n=1}^{N} \psi_n(t) \sin k_n x$$

in the governing equation, use orthogonality of eigenmodes, and obtain

$$\ddot{\psi}_n + \omega_n^2 \psi_n = -2\mu_n \dot{\psi}_n + c^2 \sum_{p,q} \Gamma_{npq} \psi_p \psi_q + c^2 \sum_{p,q,r} \Gamma_{npqr} \psi_p \psi_q \psi_r + f_n(t)$$

where the damping is assumed to be modal and

$$\Gamma_{npq} = -4\sqrt{2}\, E_1 k_p k_q^2 \int_0^1 \sin k_n x \sin k_q x \cos k_p x \, dx$$

$$\Gamma_{npqr} = -12 E_2 k_p k_q^2 k_r \int_0^1 \sin k_n x \sin k_q x \cos k_p x \cos k_r x \, dx$$

$$f_n(t) = \sqrt{2} \int_0^1 F(x, t) \sin k_n x \, dx$$

(d) Determine the steady-state response of the system when $f_n(t) = K_n \cos \Omega t$.
(e) Determine the long time response of the system when

$$f_n(t) = K_{1n} \cos \Omega_1 t + K_{2n} \cos (\Omega_2 t + \tau_n).$$

7.3. Consider the preceding exercise with the new boundary conditions

$$u(0, t) - h_1 \frac{\partial u}{\partial x}(0, t) = 0, \quad u(1, t) + h_2 \frac{\partial u}{\partial x}(1, t) = 0$$

(a) Show that the undamped, linear eigenmodes and eigenfrequencies are

$$\phi_m = c_m \left[\sin k_m x + h_1 k_m \cos k_m x\right], \quad \omega_m = c k_m$$

where the k_n are the solutions of

$$(1 - h_1 h_2 k^2) \tan k + (h_2 + h_1) k = 0$$

and the c_m are chosen so that the ϕ's are orthonormal.

(b) Show that there are no values for h_1 and h_2 such that one frequency is twice another frequency.

(c) Show that the second frequency is approximately three times the first frequency when $h_1 = 1$ and $h_2 = 1.51883$ (Bojadziev and Lardner, 1974; Bojadziev, 1976).

(d) If $f_n(t) = K_n \cos \Omega t$, determine an approximate solution for the case in part (c) when (i) $\Omega \approx \omega_1$ and (ii) $\Omega \approx \omega_2$.

7.4. The nonlinear forced planar transverse vibrations of a fixed wire are governed by

$$w_{tt} - c_0^2 w_{xx} = \frac{c_1^2}{2l} w_{xx} \int_0^l w_x^2 \, dx - 2\hat{\mu} \frac{\partial w}{\partial t} + F(x, t)$$

$$w(0, t) = w(l, t) = 0$$

where $\hat{\mu}$ is a damping coefficient and $F(x, t)$ is an external excitation.

(a) Show that the natural frequencies and corresponding mode shapes are

$$\omega_n = \frac{n\pi c_0}{l}, \quad \phi_n = \sqrt{2} \sin \frac{n\pi x}{l}$$

(b) Seek a solution in the form

$$w(x, t) = \sqrt{2} \sum_{n=1}^{\infty} \psi_n(t) \sin \frac{n\pi x}{l}$$

assume modal damping, and obtain the following coupled equations for the ψ_n:

$$\ddot{\psi}_n + \omega_n^2 \psi_n = -\frac{c_1^2 n^2 \pi^4}{2l^4} \psi_n \sum_{m=1}^{\infty} m^2 \psi_m^2 - 2\hat{\mu}_n \dot{\psi}_n + f_n(t)$$

(c) When $f_n(t) = \delta_{1n} K_1 \cos \Omega t$, show that $\psi_n \to 0$ as $t \to \infty$ for all $n \geqslant 2$. Then determine the steady-state response when

(i) $\Omega \approx \omega_1$
(ii) $\Omega \approx 3\omega_1$
(iii) $\Omega \approx \frac{1}{3}\omega_1$

(d) When $f_1(t) = K_1 \cos \Omega t, f_3(t) = K_3 \cos(\Omega t + \tau)$, and all other f_n are zero, determine the steady-state response and its stability when $\Omega \approx 3\omega_1$.

(e) When $f_n(t) = K_n \cos(\Omega t + \tau_n)$, do you think that all the free-oscillation modes are excited when Ω is approximately near one of the ω_n?

7.5. Consider the problem of Exercise 7.4 but subject to different boundary conditions. Thus consider the response of the system

$$w_{tt} - c_0^2 w_{xx} = \frac{c_1^2}{2l} w_{xx} \int_0^l w_x^2 \, dx - 2\hat{\mu} \frac{\partial w}{\partial t} + F(x, t)$$

$$w(0, t) - h_1 w_x(0, t) = 0, \quad w(1, t) + h_2 w_x(1, t) = 0$$

(a) Show that the linear undamped natural frequencies and modes are $\omega_n = c_0 k_n$, $\phi_n = \delta_n[\sin k_n x + h_1 k_n \cos k_n x]$, where δ_n is a normalizing constant and the k_n are the roots of

$$(1 - h_1 h_2 k^2) \tan k + (h_2 + h_1) k = 0$$

(b) Show that $\omega_2 \approx 3\omega_1$ if $h_1 = 1$ and $h_2 = 1.52$.

(c) Determine the response of the system when $F(x, t) = \hat{F}(x) \cos \Omega t$ and $\Omega \approx \omega_1$ or ω_2.

7.6. When the tension in the string of Exercise 7.4 is $T_0[1+p(t)]$, its nonlinear transverse vibrations are governed by

$$w_{tt} - c_0^2[1+p(t)]w_{xx} = \frac{c_1^2}{2l} w_{xx} \int_0^l w_x^2\, dx - 2\hat{\mu} w_t$$

$$w(0) = w(l) = 0$$

where $c_0^2 = T_0/\rho$, $c_1^2 = EA/\rho$, ρ is the density per unit length, E is the modulus of elasticity, A is the cross-sectional area of the string, and $\hat{\mu}$ is a damping coefficient.

(a) Assume modal damping, seek a solution in the form

$$w = \sqrt{2} \sum_{n=1}^{\infty} u_n(t) \sin \frac{n\pi x}{l}$$

and obtain

$$\ddot{u}_n + \omega_n^2[1+p(t)]u_n = -2\hat{\mu}_n \dot{u}_n - \alpha n^2 u_n \sum_{m=1}^{\infty} m^2 u_m^2$$

where $\omega_n = n\pi c_0/l$ and $\alpha = c_1^2 \pi^4/2l^4$.

(b) When $p(t) = 2\epsilon \cos \Omega t$, where $\Omega = 2\omega_1 + \epsilon\sigma$, seek an asymptotic expansion in the form

$$u_n = \epsilon^{1/2} A_n(T_1) \exp(i\omega_n T_0) + cc + O(\epsilon^{3/2})$$

where

$$\hat{\mu}_n = \epsilon \mu_n$$

$$2i\omega_1(A_1' + \mu_1 A_1) + \omega_1^2 \bar{A}_1 \exp(i\sigma T_1) + 3\alpha A_1^2 \bar{A}_1 + 2\alpha A_1 \sum_{m=2}^{\infty} m^2 A_m \bar{A}_m = 0$$

$$2i\omega_n(A_n' + \mu_n A_n) + 2\alpha n^2 A_n \sum_{m \neq n}^{\infty} m^2 A_m \bar{A}_m + 3\alpha n^4 A_n^2 \bar{A}_n = 0$$

(c) As $T_1 \to \infty$, show that $A_n \to 0$ for $n \geqslant 2$ and hence

$$2i\omega_1(A_1' + \mu_1 A_1) + \omega_1^2 \bar{A}_1 \exp(i\sigma T_1) + 3\alpha A_1^2 \bar{A}_1 = 0$$

Express A_1 in the polar form $\frac{1}{2}a_1 \exp(i\beta_1)$ and determine the frequency-response equation. Plot a_1 as a function of σ.

(d) When $\alpha = 0$, solve for A_1 and determine the condition for unbounded motions. This solution explains the observation of Melde (1859).

(e) When $p(t) = 2\epsilon[p_1 \cos(\Omega_1 t + \theta_1) + p_2 \cos(\Omega_2 T + \theta_2)]$, where $\Omega_1 = 2\omega_1 + \epsilon\sigma_1$ and $\Omega_2 = 2\omega_2 + \epsilon\sigma_2$, show that

$$u_n = \epsilon^{1/2} A_n(T_1) \exp(i\omega_n T_0) + cc + O(\epsilon^{3/2})$$

where

$$2i\omega_1(A_1' + \mu_1 A_1) + \omega_1^2 p_1 \bar{A}_1 \exp(i\sigma_1 T_1 + i\theta_1) + 3\alpha A_1^2 \bar{A}_1 + 2\alpha A_1 \sum_{m=2}^{\infty} m^2 A_m \bar{A}_m = 0$$

$$2i\omega_2(A_2' + \mu_2 A_2) + \omega_2^2 p_2 \bar{A}_2 \exp(i\sigma_2 T_1 + i\theta_2) + 48\alpha A_2^2 \bar{A}_2 + 8\alpha A_2 \sum_{m \neq 2}^{\infty} m^2 A_m \bar{A}_m = 0$$

$$2i\omega_m(A_m' + \mu_m A_m) + 3\alpha n^4 A_n^2 \bar{A}_n + 2\alpha n^2 A_n \sum_{m \neq n}^{\infty} m^2 A_m \bar{A}_m = 0$$

(f) As $T_1 \to \infty$, show that $A_n \to 0$ for $n \geqslant 3$. Express A_1 and A_2 in polar form and determine the steady-state amplitudes and their stability.

7.7. Consider the forced excitation of a hinged-hinged beam (Section 7.3).

(a) Let $f_1 = 2\epsilon k_{11} \cos \Omega_1 t$, $f_2 = 2\epsilon k_{21} \cos(\Omega_2 t + \theta)$, and $P_n = 0$ for $n \geqslant 3$. When $\Omega_1 = \pi^2 + \epsilon\sigma_1$ and $\Omega_2 = 4\pi^2 + \epsilon\sigma_2$, show that

$$u_n = A_n(T_1) \exp(in^2\pi^2 T_0) + cc + O(\epsilon)$$

$$2i\pi^2(A_1' + \mu_1 A_1) + \tfrac{3}{2}\pi^4 A_1^2 \bar{A}_1 + \pi^4 A_1 \sum_{m=2}^{\infty} m^2 A_m \bar{A}_m - k_{11} \exp(i\sigma_1 T_1) = 0$$

$$8i\pi^2(A_2' + \mu_2 A_2) + 24\pi^4 A_2^2 \bar{A}_2 + 4\pi^4 A_2 \sum_{m \neq 2}^{\infty} m^2 A_m \bar{A}_m - k_{21} \exp(i\sigma_2 T_1 + i\theta) = 0$$

and

$$2in^2(A_n' + \mu_n A_n) + \tfrac{3}{2} n^4 \pi^2 A_n^2 \bar{A}_n + n^2\pi^2 A_n \sum_{m \neq n}^{\infty} m^2 A_m \bar{A}_m = 0$$

for $n \geqslant 3$.

(b) As $T_1 \to \infty$, show that $A_n \to 0$ for $n \geqslant 3$ and that the response consists of the lowest two modes.

(c) Express A_1 and A_2 in polar form and determine the equations describing the steady-state solution. Must the detunings be related for a steady-state solution? Plot a_1 and a_2 as functions of the detunings.

7.8. Consider the forced response of hinged-hinged beam (Section 7.3). When $f_n = 2K_n \cos(\Omega t + \theta_n)$, where $\Omega = 2\pi^2 + \epsilon\sigma$. Note that $\omega_2 - 2\omega_1 \approx \Omega$ in this case.

(a) Show that

$$u_n = A_n(T_1) \exp(in^2\pi^2 T_0) + \Lambda_n \exp[i(\Omega T_0 + \theta_n)] + cc + O(\epsilon)$$

where

$$\Lambda_n = \frac{K_n}{n^4\pi^4 - \Omega^2}$$

$$2i(A_1' + \mu_1 A_1) + \tfrac{3}{2}\pi^2 A_1(A_1\bar{A}_1 + 2\Lambda_1^2) + \pi^2 A_1 \sum_{m=2}^{\infty} m^2(A_m\bar{A}_m + \Lambda_m^2)$$
$$+ 4\pi^2 A_2\bar{A}_1\Lambda_2 \exp[-i(\sigma T_1 + \theta_2)] = 0$$

$$8i\pi^2(A_2' + \mu_2 A_2) + 24\pi^4 A_2(A_2\bar{A}_2 + 2\Lambda_2^2) + 4\pi^4 A_2 \sum_{m\neq 2}^{\infty} m^2(A_m\bar{A}_m + \Lambda_m^2)$$
$$+ 2\pi^4 A_1^2\Lambda_2 \exp[i(\sigma T_1 + \theta_2)] = 0$$

and

$$2in^2(A_n' + \mu_n A_n) + \tfrac{3}{2}n^4\pi^2 A_n(A_n\bar{A}_n + 2\Lambda_n^2)$$
$$+ n^2\pi^2 A_n \sum_{m\neq n}^{\infty} m^2(A_m\bar{A}_m + \Lambda_m^2) = 0$$

for $n \geqslant 3$.

(b) As $T_1 \to \infty$, show that $A_n \to 0$ for $n \geqslant 3$. Thus the free-oscillation part of any mode that is not involved in a resonance with the forced excitation decays with time. Then express A_n in polar form and determine the equations describing the steady-state solution. Plot a_1 and a_2 as functions of σ. Are there any jumps?

7.9. Consider the forced response of a hinged-hinged beam (Section 7.3) when

$$f_n = 2K_{n1}\cos(\Omega_1 t + \theta_{n1}) + 2K_{n2}\cos(\Omega_2 t + \theta_{n2})$$

where

$$3\Omega_1 = \pi^2 + \epsilon\sigma_1 \quad \text{and} \quad \Omega_2 = 3\pi^2 + \epsilon\sigma_2$$

Note that $\omega_1 \approx 3\Omega_1 \approx \frac{1}{3}\Omega_2$ and $\omega_3 \approx 3\Omega_2$ in this case.

(a) Show that

$$u_n = A_n \exp(in^2\pi^2 T_0) + \Lambda_{n1}\exp(i\Omega_1 T_0 + i\theta_{n1})$$
$$+ \Lambda_{n2}\exp(i\Omega_2 T_0 + i\theta_{n2}) + cc + O(\epsilon)$$

where

$$\Lambda_{nm} = \frac{K_{nm}}{n^4\pi^4 - \Omega_m^2}$$

$$2i\pi^2(A_1' + \mu_1 A_1) + \tfrac{3}{2}\pi^4 A_1(A_1\bar{A}_1 + 2\Lambda_{11}^2 + 2\Lambda_{12}^2)$$

$$+ \pi^4 A_1 \sum_{m=2}^{\infty} m^2(A_m\bar{A}_m + \Lambda_{m1}^2 + \Lambda_{m2}^2)$$

$$+\tfrac{1}{2}\pi^4\Lambda_{11}\sum_{m=1}^{\infty} m^2\Lambda_{m1}^2 \exp\left[i(\sigma_1 T_1 + 2\theta_{m1} + \theta_{11})\right]$$

$$+\tfrac{3}{2}\pi^4\bar{A}_1^2\Lambda_{12} \exp\left[i(\sigma_2 T_1 + \theta_{12})\right] = 0$$

$$18i\pi^2(A_3' + \mu_3 A_3) + \tfrac{243}{2}\pi^4 A_3(A_3\bar{A}_3 + 2\Lambda_{31}^2 + 2\Lambda_{32}^2)$$

$$+ 9\pi^4 A_3 \sum_{m\neq 3}^{\infty} m^2(A_m\bar{A}_m + \Lambda_{m1}^2 + \Lambda_{m2}^2)$$

$$+\tfrac{9}{2}\pi^4\Lambda_{32}\sum_{m=1}^{\infty} m^2\Lambda_{m2}^2 \exp\left[i(3\sigma_2 T_1 + 2\theta_{m2} + \theta_{31})\right] = 0$$

$$2in^2\pi^2(A_n' + \mu_n A_n) + \tfrac{3}{2}n^4\pi^4 A_n(A_n\bar{A}_n + 2\Lambda_{n1}^2 + 2\Lambda_{n2}^2)$$

$$+ n^2\pi^4 A_n \sum_{m\neq n}^{\infty} m^2(A_m\bar{A}_m + \Lambda_{m1}^2 + \Lambda_{m2}^2) = 0$$

for $n \neq 1$ and 3.

(b) As $T_1 \to \infty$, show that $A_n \to 0$ for $n \neq 1$ and 3. Then express A_1 and A_3 in polar form and determine the steady-state motions and their stability.

7.10. Consider the forced response of a hinged-hinged beam (Section 7.3) when

$$f_n = 2\sum_{m=1}^{3} K_{nm} \cos(\Omega_m t + \theta_{nm})$$

where

$$\Omega_1 + \Omega_2 + \Omega_3 = \pi^2 + \epsilon\sigma.$$

(a) Show that

$$u_n = A_n(T_1)\exp(in^2\pi^2 T_0) + \sum_{m=1}^{3}\Lambda_{nm}\exp(i\Omega_m T_0) + cc + O(\epsilon)$$

where

$$\Lambda_{nm} = K_{nm}\frac{\exp(i\theta_{nm})}{n^4\pi^4 - \Omega_m^2}$$

$$2i\pi^2(A_1' + \mu_1 A_1) + \tfrac{3}{2}\pi^4 A_1\left(A_1\bar{A}_1 + 2\sum_{m=1}^{3}|\Lambda_{1m}|^2\right)$$

$$+ \pi^4 A_1 \sum_{m=2}^{\infty} m^2\left(A_m\bar{A}_m + \sum_{k=1}^{3}|\Lambda_{mk}|^2\right) + \pi^4\sum_{m=1}^{\infty} m^2(\Lambda_{m1}\Lambda_{m2}\Lambda_{13}$$

$$+ \Lambda_{m1}\Lambda_{m3}\Lambda_{12} + \Lambda_{m2}\Lambda_{m3}\Lambda_{11})\exp(i\sigma T_1) = 0$$

$$2in^2\pi^2(A_n' + \mu_n A_n) + \tfrac{3}{2}n^4\pi^4\left(A_n\bar{A}_n + 2\sum_{m=1}^{3}|\Lambda_{nm}|^2\right)$$

$$+ n^2\pi^4 A_n \sum_{m\neq n}^{\infty} m^2\left(A_m\bar{A}_m + \sum_{k=1}^{3}|\Lambda_{mk}|^2\right) = 0$$

for $n \geqslant 2$.

(b) Show that as $T_1 \to \infty$, $A_n \to 0$ for $n \geqslant 2$. Then express A_1 in the polar form $\frac{1}{2}a_1 \exp(i\beta_1)$ and determine the equations describing a_1 and β_1. Obtain the frequency-response equation. Plot a_1 as a function of σ. Indicate the jumps.

7.11. Consider the transverse vibrations of a hinged-hinged, uniform beam under the combined influence of parametric and external excitations as in (7.1.38).

(a) Show that (7.1.38) simplifies to

$$\ddot{u}_n + n^4\pi^4 u_n = -2\epsilon\mu_n\dot{u}_n - \tfrac{1}{2}\epsilon n^2\pi^4 u_n \sum_{m=1}^{\infty} m^2 u_m^2 - \epsilon n^2\pi^2 p u_n + f_n(t)$$

(b) When $f(x,t) \equiv 0$ and $p(t) = 2p_1 \cos\Omega_1 t + 2p_2 \cos(\Omega_2 t + \theta)$, where $\Omega_1 = 2\omega_1 + \epsilon\sigma_1$ and $\Omega_2 = 2\omega_2 + \epsilon\sigma_2$, show that

$$u_n = A_n(T_1)\exp(in^2\pi^2 T_0) + cc + O(\epsilon)$$

where

$$2i(A_1' + \mu_1 A_1) + p_1\bar{A}_1 \exp(i\sigma_1 T_1) + \tfrac{3}{2}\pi^2 A_1^2\bar{A}_1 + \pi^2 A_1 \sum_{m=2}^{\infty} m^2 A_m\bar{A}_m = 0$$

$$8i(A_2' + \mu_2 A_2) + 4p_2\bar{A}_2 \exp(i\sigma_2 T_1 + i\theta) + 24\pi^2 A_2^2\bar{A}_2$$

$$+ 4\pi^2 A_2 \sum_{m\neq 2}^{\infty} m^2 A_m\bar{A}_m = 0$$

$$2in^2(A_n' + \mu_n A_n) + \tfrac{3}{2}n^4\pi^2 A_n^2\bar{A}_n + n^2\pi^2 A_n \sum_{m\neq n}^{\infty} m^2 A_m\bar{A}_m = 0$$

(c) As $T_1 \to \infty$, show that $A_n \to 0$ for $n \geqslant 3$. Then express A_1 and A_2 in polar form and determine the steady-state motions and their stability.

(d) When $f_n(t) = 2K_n \cos(\Omega_1 t + \theta_n)$ and $p(t) = 2\epsilon\cos(\Omega_2 t)$, where $\Omega_2 - \Omega_1 = \omega_1 + \epsilon\sigma$ and no other resonances exist, show that

$$u_n = A_n(T_1)\exp(in^2\pi^2 T_0) + \Lambda_n \exp(i\Omega_1 T_0 + i\theta_n) + cc + O(\epsilon)$$

where

$$\Lambda_n = \frac{K_n}{n^4\pi^4 - \Omega_1^2}$$

$$2i(A_1' + \mu_1 A_1) + \Lambda_1 \exp(i\sigma T_1 - i\theta_1) + \tfrac{3}{2}\pi^2 A_1(A_1\bar{A}_1 + 2\Lambda_1^2)$$

$$+ \pi^2 A_1 \sum_{m=2}^{\infty} m^2(A_m\bar{A}_m + \Lambda_m^2) = 0$$

$$2in^2(A_n' + \mu_n A_n) + \tfrac{3}{2}n^4\pi^2 A_n(A_n\bar{A}_n + 2\Lambda_n^2) + n^2\pi^2 A_n \sum_{m \neq n}^{\infty} m^2(A_m\bar{A}_m + \Lambda_m^2) = 0$$

(f) As $T_1 \to \infty$, show that $A_n \to 0$ for $n \geqslant 2$ and hence

$$2i(A_1' + \mu_1 A_1) + \tfrac{3}{2}\pi^2 A_1(A_1\bar{A}_1 + 2\Lambda_1^2) + \pi^2 A_1 \sum_{m=2}^{\infty} m^2\Lambda_m^2 + \Lambda_1 \exp(i\sigma T_1 - i\theta_1) = 0$$

7.12. Consider the response of a uniform beam hinged at one end and clamped at the other (Section 7.4). Let $f_n(t) = 2K_n \cos \Omega t$, where Ω is near $2\omega_1$. Show that after a long time the deflection is given by

$$w(x, t) = a_1\phi_1(x) \cos [\tfrac{1}{2}\Omega t - \tfrac{1}{4}(\gamma_1 + \gamma_2)] + a_2\phi_2(x) \cos [\tfrac{3}{2}\Omega t + \tfrac{1}{4}(\gamma_1 - 3\gamma_2)] + 2\left[\sum_{n=1}^{\infty} \Lambda_n\phi_n(x)\right] \cos (\Omega t)$$

The $\phi_n(x)$ are the linear free-oscillation modes and a_1, a_2, γ_1, and γ_2 are solutions of

$$-\omega_1\mu_1 a_1 + \tfrac{1}{8}Q_1 a_1^2 a_2 \sin\gamma_1 + \tfrac{1}{2}H_{12}a_2 \sin\gamma_2 = 0$$

$$-\omega_2\mu_2 a_2 - \tfrac{1}{8}Q_2 a_1^3 \sin\gamma_1 + \tfrac{1}{2}H_{12}a_1 \sin\gamma_2 = 0$$

$$\omega_1\sigma_2 a_1 + \tfrac{1}{2}Q_1 a_1^2 a_2 \cos\gamma_1 + \tfrac{1}{2}a_1 \sum_{n=1}^{2} \alpha_{1n}a_n^2 + 4H_{11}a_1 + 2H_{12}a_2 \cos\gamma_2 = 0$$

$$\omega_2\left(\frac{3\sigma_2}{4} - \sigma_1\right)a_2 + \tfrac{1}{8}Q_2 a_1^3 \cos\gamma_1 + \tfrac{1}{8}a_2 \sum_{n=1}^{2} \alpha_{2n}a_n^2 + H_{22}a_2 + \tfrac{1}{2}H_{12}a_1 \cos\gamma_2 = 0$$

where

$$\epsilon\sigma_1 = \omega_2 - 3\omega_1, \quad \epsilon\sigma_2 = 2\Omega - 4\omega_1, \quad \Lambda_n = \frac{K_n}{\omega_n^2 - \Omega^2}$$

and the Q_n, α_{nm}, and H_{nm} are defined by (7.4.15), (7.4.16), and (7.4.47). Discuss the various possible solutions.

7.13. Consider the response of a uniform beam hinged at one end and clamped at the other (Section 7.4). Let $f_n = 2K_n \cos \Omega t$, where Ω is near $5\omega_1$. Show that after a long time the deflection is given by

$$w(x, t) = a_1\phi_1(x) \cos [\tfrac{1}{5}\Omega t - \tfrac{1}{5}(\gamma_1 + \gamma_2)] + a_2\phi_2(x) \cos [\tfrac{3}{5}\Omega t + \tfrac{1}{5}(2\gamma_1 - 3\gamma_2)] + 2\left[\sum_{n=1}^{\infty} \Lambda_n\phi_n(x)\right] \cos (\Omega t)$$

The $\phi_n(x)$ are the linear free-oscillation modes and a_1, a_2, γ_1, and γ_2 are the solutions of

$$-\omega_1\mu_1 a_1 + \tfrac{1}{8}Q_1 a_1^2 a_2 \sin\gamma_1 + H_{12}a_1 a_2 \sin\gamma_2 = 0$$

$$-\omega_2\mu_2 a_2 - \tfrac{1}{8}Q_2 a_1^3 \sin\gamma_1 + \tfrac{1}{2}H_{12}a_1^2 \sin\gamma_2 = 0$$

$$\tfrac{1}{5}\omega_1\sigma_2 a_1 + \tfrac{1}{8}Q_1 a_1^2 a_2 \cos\gamma_1 + \tfrac{1}{8}a_1 \sum_{n=1}^{2} \alpha_{1n} a_n^2 + H_{11}a_1 + H_{12}a_1 a_2 \cos\gamma_2 = 0$$

$$\omega_2\left(\frac{3\sigma_2}{5} - \sigma_1\right) a_2 + \tfrac{1}{8}Q_2 a_1^3 \cos\gamma_1 + \tfrac{1}{8}a_2 \sum_{n=1}^{2} \alpha_{2n} a_n^2 + H_{22}a_2 + \tfrac{1}{2}H_{12}a_1^2 \cos\gamma_2 = 0$$

where

$$\epsilon\sigma_1 = \omega_2 - 3\omega_1, \quad \epsilon\sigma_2 = \Omega - 5\omega_1, \quad \Lambda_n = \frac{K_n}{\omega_n^2 - \Omega^2}$$

and the Q_n, α_{nm}, and H_{nm} are defined by (7.4.15), (7.4.16), and (7.4.47). Discuss the various possible solutions.

7.14. Consider the nonlinear vibrations of clamped-clamped beams. Tseng and Dugundji (1970) studied analytically and experimentally the response of such beams to harmonic excitations.

(a) Show that the linear mode shapes are

$$\phi_n = C_n[(\cos\eta_n x - \cosh\eta_n x)(\sin\eta_n - \sinh\eta_n) - (\sin\eta_n x - \sinh\eta_n x)(\cos\eta_n - \cosh\eta_n)]$$

the natural frequencies are $\omega_n = \eta_n^2$ and the η_n are the roots of

$$\cos\eta_n \cosh\eta_n = 1$$

(b) Show that the first four frequencies are 22.37, 61.67, 120.91, and 199.85 and that

$$\omega_1 + \omega_2 + \omega_3 = \omega_4 + \epsilon\sigma_1$$

where $\epsilon\sigma_1 = 5.10$.

(c) Let $f_n = 2\epsilon\delta_{ns}k_n \cos\Omega t$ and $\Omega = \omega_s + \epsilon\sigma_2$ in (7.1.21), where $s = 1$ or 2 or 3 or 4. Show that for the clamped-clamped beam, the equations describing the A_n are

$$-2i\omega_1(A_1' + \mu_1 A_1) + A_1 \sum_{m=1}^{\infty} \alpha_{1m} A_m \bar{A}_m + 8QA_4\bar{A}_3\bar{A}_2 \exp(-i\sigma_1 T_1) + \delta_{1s}k_1 \exp(i\sigma_2 T_1) = 0$$

$$-2i\omega_2(A_2' + \mu_2 A_2) + A_2 \sum_{m=1}^{\infty} \alpha_{2m} A_m \bar{A}_m + 8QA_4\bar{A}_3\bar{A}_1 \exp(-i\sigma_1 T_1) + \delta_{2s}k_2 \exp(i\sigma_2 T_1) = 0$$

$$-2i\omega_3(A_3' + \mu_3 A_3) + A_3 \sum_{m=1}^{\infty} \alpha_{3m} A_m \bar{A}_m + 8QA_4\bar{A}_2\bar{A}_1 \exp(-i\sigma_1 T_1) + \delta_{3s}k_3 \exp(i\sigma_2 T_1) = 0$$

$$-2i\omega_4(A_4' + \mu_4 A_4) + A_4 \sum_{m=1}^{\infty} \alpha_{4m} A_m \bar{A}_m + 8QA_3A_2A_1 \exp(i\sigma_1 T_1) + \delta_{4s} k_4 \exp(i\sigma_2 T_1) = 0$$

$$-2i\omega_n(A_n' + \mu_n A_n) + A_n \sum_{m=1}^{\infty} \alpha_{nm} A_m \bar{A}_m = 0 \quad \text{for } n \geqslant 5$$

where $8Q = 2(\Gamma_{1234} + \Gamma_{1324} + \Gamma_{1423})$.

(d) Show that $A_n \to 0$ as $T_1 \to \infty$ for $n \geqslant 5$.

(e) When $s = 1$ show that in the steady-state case $\omega_2\mu_2 a_2^2 + \omega_4\mu_4 a_4^2 = 0$. Hence $a_2 = a_3 = a_4 = 0$, and only the first mode is excited and the solution is the same as that of the Duffing equation.

(f) When $s = 2$ show that in the steady-state response $a_1 = a_3 = a_4 = 0$ and only the second mode is excited and the solution is the same as that of the Duffing equation.

(g) When $s = 3$ show that $a_1 = a_2 = a_4 = 0$ and only the third mode is excited.

(h) When $s = 4$ show that there are two possibilities: (i) $a_1 = a_2 = a_3 = 0$ and $a_4 \neq 0$ and the solution is the same as that of the Duffing equation, and (ii) $a_n \neq 0$ for $n = 1, 2, 3,$ and 4 and determine the equations describing the steady-state response in this case.

7.15. Consider the response of a uniform beam clamped at both ends (Exercise 7.14). Let

$$f_n(t) = 2K_n \cos \Omega t$$

where $\Omega = \omega_3 - \omega_1 - \omega_2 + \epsilon\sigma_2$.

Show that after a long time the deflection is given by

$$w(x, t) = \sum_{n=1}^{4} a_n \phi_n(x) \cos[(\omega_n + \epsilon\beta_n')t + \tau_n] + 2\left[\sum_{n=1}^{\infty} \Lambda_n \phi_n(x)\right] \cos \Omega t$$

The $\phi_n(x)$ are the linear free-oscillation modes and the a_n and β_n' are solutions of

$$-\omega_1\mu_1 a_1 + Qa_2a_3a_4 \sin\gamma_1 + Fa_3a_2 \sin\gamma_2 = 0$$

$$-\omega_2\mu_2 a_2 + Qa_1a_3a_4 \sin\gamma_1 + Fa_1a_3 \sin\gamma_2 = 0$$

$$-\omega_3\mu_3 a_3 + Qa_1a_2a_4 \sin\gamma_1 - Fa_2a_1 \sin\gamma_2 = 0$$

$$-\omega_4\mu_4 a_4 - Qa_1a_2a_3 \sin\gamma_1 = 0$$

$$\omega_1 a_1 \beta_1' + Qa_2a_3a_4 \cos\gamma_1 + \tfrac{1}{8}a_1 \sum_{n=1}^{4} \alpha_{1n} a_n^2 + H_{11}a_1 + Fa_3a_2 \cos\gamma_2 = 0$$

$$\omega_2 a_2 \beta_2' + Qa_1a_3a_4 \cos\gamma_1 + \tfrac{1}{8}a_2 \sum_{n=1}^{4} \alpha_{2n} a_n^2 + H_{22}a_2 + Fa_1a_3 \cos\gamma_2 = 0$$

$$\omega_3 a_3 \beta_3' + Qa_1a_2a_4 \cos\gamma_1 + \tfrac{1}{8}a_3 \sum_{n=1}^{4} \alpha_{3n} a_n^2 + H_{33}a_3 + Fa_1a_2 \cos\gamma_2 = 0$$

$$\omega_4 a_4 \beta_4' + Qa_1a_2a_3 \cos\gamma_1 + \tfrac{1}{8}a_4 \sum_{n=1}^{4} \alpha_{4n} a_n^2 + H_{44}a_4 = 0$$

where

$$\gamma_1 = \beta_4 - \beta_3 - \beta_2 - \beta_1 - \sigma_1 T_1$$

$$\gamma_2 = \beta_3 - \beta_2 - \beta_1 - \sigma_2 T_1$$

$$F = 8 \sum_{n=1}^{\infty} (\Gamma_{123n} + \Gamma_{231n} + \Gamma_{132n})\Lambda_n$$

$$H_{pq} = \sum_{i=1}^{\infty} \sum_{j=1}^{\infty} (\Gamma_{pqij} + 2\Gamma_{piqj})\Lambda_i\Lambda_j$$

There are two possibilities: either a_1, a_2, a_3, and a_4 are zero or none is zero. For the latter case, combine the last four equations into two, making it possible to solve for $a_1, a_2, a_3, a_4, \gamma_1$, and γ_2. Then $\beta_1', \beta_2', \beta_3'$, and β_4' can be found. But β_1, β_2, β_3, and β_4 can only be determined to within the arbitrary constants τ_1, τ_2, τ_3, and τ_4, which depend on the initial conditions.

Show how the nonlinearity adjusts the frequencies so that the conditions for the internal and external resonances are satisfied exactly.

7.16. Consider the response of a uniform beam clamped at both ends (Exercise 7.14). Let $f_n(t) = 2K_n \cos \Omega t$, where $2\Omega = \omega_4 - \omega_1 + \epsilon\sigma_2$. Show that after a long time the deflection is given by

$$w(x, t) = \sum_{n=1}^{4} a_n \phi_n(x) \cos[(\omega_n + \epsilon\beta_n')t + \tau_n] + \left[\sum_{n=1}^{\infty} \Lambda_n \phi_n(x)\right] \cos \Omega t$$

The $\phi_n(x)$ are the linear free-oscillation modes and the a_n and the β_n' are the solutions of

$$-\omega_1\mu_1 a_1 + Q a_2 a_3 a_4 \sin\gamma_1 + \tfrac{1}{2} H_{14} a_4 \sin\gamma_2 = 0$$

$$-\omega_2\mu_2 a_2 + Q a_1 a_3 a_4 \sin\gamma_1 = 0$$

$$-\omega_3\mu_3 a_3 + Q a_1 a_2 a_4 \sin\gamma_1 = 0$$

$$-\omega_4\mu_4 a_4 - Q a_1 a_2 a_3 \sin\gamma_1 - \tfrac{1}{2} H_{14} a_1 \sin\gamma_2 = 0$$

$$\omega_1 a_1 \beta_1' + Q a_2 a_3 a_4 \cos\gamma_1 + \tfrac{1}{8} a_1 \sum_{n=1}^{4} \alpha_{1n} a_n^2 + H_{11} a_1 + \tfrac{1}{2} H_{14} a_4 \cos\gamma_2 = 0$$

$$\omega_2 a_2 \beta_2' + Q a_1 a_3 a_4 \cos\gamma_1 + \tfrac{1}{8} a_2 \sum_{n=1}^{4} \alpha_{2n} a_n^2 + H_{22} a_2 = 0$$

$$\omega_3 a_3 \beta_3' + Q a_1 a_2 a_4 \cos\gamma_1 + \tfrac{1}{8} a_3 \sum_{n=1}^{4} \alpha_{3n} a_n^2 + H_{33} a_3 = 0$$

$$\omega_4 a_4 \beta_4' + Q a_1 a_2 a_3 \cos\gamma_1 + \tfrac{1}{8} a_4 \sum_{n=1}^{4} \alpha_{4n} a_n^2 + H_{44} a_4 + \tfrac{1}{2} H_{14} a_1 \cos\gamma_2 = 0$$

where

$$\gamma_1 = \beta_4 - \beta_3 - \beta_2 - \beta_1 - \sigma_1 T_1, \quad \gamma_2 = \beta_4 - \beta_1 - \sigma_2 T_1$$

$$\Lambda_n = \frac{K_n}{\omega_n^2 - \Omega^2}$$

Discuss the various possible solutions. For the nontrivial cases show how the equations can be combined to eliminate all the β_n' and make it possible to solve for the a_n, γ_1, and γ_2. After this is done, one can solve for the β_n'. When can the β_n be completely determined and when can they only be found to within arbitrary constants, τ_n, which depend on the initial conditions? Obtain the τ_n for the case when the β_n can be completely determined. Show that the nonlinearity adjusts the frequencies so that the conditions for internal and external resonances are satisfied exactly.

7.17. Consider the coupling of longitudinal and transverse oscillations of a beam (Section 7.2) when $g_1 = 0$ in (7.2.21) through (7.2.27).

(a) When $f_{22} = 0$, show that in the steady state

$$\lambda_1 \nu_1 a_1^2 + \omega_2 \mu_2 b_2^2 = 0$$

Hence $a_1 = b_2 = 0$ and the only nonzero mode is the third transverse mode. Show that the response to this order is linear and that one needs to carry out the expansion one more order to include the nonlinear effects.

(b) When $f_{33} = 0$, show that in the steady state $a_1 = b_3 = 0$ and that the only nonzero mode is the second transverse mode.

(c) When f_{22} and f_{33} are different from zero, show that the steady-state equations can be combined to give

$$\tan \gamma_I = \frac{\nu_1}{\sigma_2 + \sigma_3 - \sigma_I}$$

$$b_2^2 \left[\omega_2 \mu_2 + \frac{k^2 \sin^2 \gamma_I}{\lambda_1 \nu_1} b_3^2\right]^2 + b_2^2 \left[\omega_2 \sigma_2 - \frac{k^2 \sin 2\gamma_I}{2\lambda_1 \nu_1} b_3^2\right]^2 = f_{22}^2$$

$$b_3^2 \left[\omega_3 \mu_3 + \frac{k^2 \sin^2 \gamma_I}{\lambda_1 \nu_1} b_2^2\right]^2 + b_3^2 \left[\omega_3 \sigma_3 - \frac{k^2 \sin 2\gamma_I}{2\lambda_1 \nu_1} b_2^2\right]^2 = f_{33}^2$$

Do you expect to find stable solutions of these equations?

7.18. Expanding the normal displacement for a circular cylindrical shell in terms of normal modes with time-dependent coefficients, one obtains the following equations for the time variations (Dowell and Ventres, 1968; Atluri, 1972):

$$\ddot{u}_m + \omega_m^2 u_m = -2\epsilon \mu_m \dot{u}_m + \epsilon \sum_{p,q} \alpha_{mpq} u_p u_q + f_m(t) + O(u^3)$$

(a) When $f_m(t) = \epsilon \delta_{mn} k_n \cos \Omega t$, where $\Omega = \omega_n + \epsilon \sigma_n$ for $n = 1$ or 2 or 3 and $\omega_3 = \omega_1 + \omega_2 + \epsilon \sigma$ and no other resonances exist, show that

$$u_m = \epsilon A_m(T_1) \exp(i\omega_m T_0) + cc + O(\epsilon^2)$$

where

$$2i\omega_1(A_1' + \mu_1 A_1) = (\alpha_{123} + \alpha_{132}) A_3 \bar{A}_2 \exp(i\sigma T_1) + \delta_{1n} k_n \exp(i\sigma_n T_1)$$

$$2i\omega_2(A_2' + \mu_2 A_2) = (\alpha_{213} + \alpha_{231}) A_3 \bar{A}_1 \exp(i\sigma T_1) + \delta_{2n} k_n \exp(i\sigma_n T_1)$$

$$2i\omega_3(A_3' + \mu_3 A_3) = (\alpha_{312} + \alpha_{321}) A_1 A_2 \exp(-i\sigma T_1) + \delta_{3n} k_n \exp(i\sigma_n T_1)$$

$$A_n' + \mu_n A_n = 0 \quad \text{for } n \geqslant 4$$

(b) When $n = 1$ or 2, show that the steady-state consists of the first or second mode only.

(c) When $n = 3$, show that the steady-state response contains the third mode when $k_3 \leqslant k_c$, where k_c is a critical value that depends on the σ's and μ's. When $k > k_c$, show that the third mode saturates and the first and the second modes are excited. Determine k_c and the steady-state amplitudes and phases.

7.19. Consider small but finite oscillations of a taut string, Section 7.5. When the effects of linear damping are included, the equations governing the motion have the following form [(7.5.52) to (7.5.55)]:

$$a' + \mu a + ab^2 \sin \gamma - k \sin \nu = 0$$

$$\sigma - \nu' - 3a^2 - b^2(2 + \cos \gamma) + \frac{k}{a} \cos \nu = 0$$

$$b' + \mu b - a^2 b \sin \gamma = 0$$

$$b[\sigma - \nu' + \tfrac{1}{2}\gamma' - 3b^2 - a^2(2 + \cos \gamma)] = 0$$

(a) Show that when b is zero, the frequency-response equation has the following form:

$$\sigma = 3a^2 \pm \left(\frac{k^2}{a^2} - \mu^2\right)^{1/2}$$

in agreement with (4.1.19).

(b) Show that infinitesimal perturbations of this steady-state solution are governed by

$$\begin{aligned} (\lambda + \mu)\,\Delta a - (k \cos \nu)\,\Delta \nu &= 0 \\ \left(6a + \frac{k}{a^2}\cos \nu\right)\Delta a + \left(\lambda + \frac{k}{a}\sin \nu\right)\Delta \nu &= 0 \end{aligned} \tag{1}$$

$$\begin{aligned} (\lambda + \mu - a^2 \sin \gamma)\,\Delta b &= 0 \\ (\sigma - 2a^2 - a^2 \cos \gamma) b &= 0 \end{aligned} \tag{2}$$

where λ is an eigenvalue of the perturbation. Again we see that the exponent for Δa and $\Delta \nu$ can differ from the exponent for Δb.

(c) From (1), show that

$$\lambda = -\mu \pm [(\sigma - 3a^2)(9a^2 - \sigma)]^{1/2}$$

Then show that the points where the small perturbations in a and ν grow are located between the two points on the frequency–response curve having vertical tangents.

(d) From (2), show that

$$\lambda = -\mu \pm [(3a^2 - \sigma)(\sigma - a^2)]^{1/2}$$

Then show that when σ is slowly increasing and the point representing the motion is moving along the upper branch, the out-of-plane component begins to grow when $a = a_A$, where a_A is a root of

$$4\mu^2 a^8 - 4k^2 a^6 + k^4 = 0$$

What are the other points on the frequency–response curve at which the out-of-plane component begins to grow.

(e) For small values of μ/k, show that

$$a_A = (\tfrac{1}{2}k)^{1/3} + \frac{\mu^2}{12k} + O(\mu^3)$$

Compare this result with that in Section 7.5.2. Thus the effect of the damping is to raise the amplitude at which the out-of-plane component begins to grow.

(f) Describe the response of the string as the frequency of the excitation is slowly varied up and down through resonance while the amplitude of the excitation is held constant.

7.20. The nonlinear response of a relief valve fixed at one end and connected at the other end by an end mass and a nonlinear spring can be described by the following dimensionless mathematical problem (Dokainish and Elmadany, 1978):

$$u_{tt} = u_{xx}$$

$$u = 0 \quad \text{at } x = 0$$

$$u_{xx} + \alpha_1 u_x + \alpha_2 u + \alpha_3 u^3 = F_0 + F_1 \cos(\Omega t) \quad \text{at } x = 1$$

where the α_n and F_n are constants. The response of systems having nonlinear constraints was analyzed by Porter and Billet (1965).

(a) When $F_1 = 0$, show that

$$u = bx$$

where b is given by

$$\alpha_1 b + \alpha_2 b + \alpha_3 b^3 = F_0$$

(b) When $F_1 \neq 0$, let $u = bx + v(x, t)$. Show that v is governed by

$$v_{tt} = v_{xx}$$

$$v = 0 \quad \text{at } x = 0$$

$$v_{xx} + \alpha_1 v_x + (\alpha_2 + 3b^2\alpha_3)v + 3b\alpha_3 v^2 + \alpha_3 v^3 = F_1 \cos \Omega t \quad \text{at } x = 1$$

(c) Show that the linear solution is

$$v = \sum_{m=1}^{\infty} a_m \sin \omega_m x \cos(\omega_m t + \beta_m) + \Lambda \sin \Omega x \cos \Omega t$$

where the ω_m are the roots of

$$(\alpha_2 + 3b^2\alpha_3 - \omega^2)\tan \omega + \alpha_1 \omega = 0$$

$$\Lambda = F_1[\alpha_1 \Omega \cos \Omega + (a_2 + 3b^2\alpha_3 - \Omega^2)\sin \Omega]^{-1}$$

Thus a primary resonance occurs whenever $\Omega \approx \omega_n$ for any n.

(d) Use the method of multiple scales to determine a uniform solution when $\Omega \approx \omega_1$.

(e) Use the method of multiple scales to determine uniform solutions when $\Omega \approx 2\omega_1$, $\Omega \approx \frac{1}{2}\omega_1$, and $\Omega \approx \omega_1 + \omega_2$.

7.21. Consider the symmetric response of a uniform circular plate clamped along its edge (Section 7.6.2). Let $f_n(t) = 2K_n \cos \Omega T_0$, where $\Omega \approx 3\omega_1$. Show that the steady-state deflection is given by

$$w(r, t) = a_1 \phi_1(r) \cos\left(\tfrac{1}{3}\Omega t - \tfrac{1}{3}\gamma\right) + \left[2 \sum_{n=1}^{\infty} \Lambda_n \phi_n(r)\right] \cos \Omega t + O(\epsilon)$$

The $\phi_n(r)$ are the linear free-oscillation modes and a_1 and γ are the solutions of

$$a_1(\omega_1 \mu_1 - F a_1 \sin \gamma) = 0$$

$$a_1(\omega_1 \sigma + 3H_{11} + \tfrac{9}{8}\alpha_{11} a_1^2 + 3F a_1 \cos \gamma) = 0$$

where

$$\epsilon\sigma = \Omega - 3\omega_1, \quad F = \tfrac{3}{4} \sum_{n=1}^{\infty} \Gamma_{111n} \Lambda_n, \quad \Lambda_n = \frac{K_n}{\omega_n^2 - \Omega^2}$$

and the Γ_{nmpq}, the H_{nm}, and the α_{nm} are defined by (7.6.66), (7.4.47), and (7.6.75).

Note that the solution for a_1 and γ was discussed in Section 4.1.4.

7.22. Consider the symmetric response of a uniform circular plate clamped along its edge. Let

$$f_n(t) = 2K_n \cos \Omega T_0$$

where $\omega_1 \approx 3\Omega$. Show that the steady-state deflection is given by

$$w(r, t) = a_1 \phi_1(r) \cos(3\Omega t - \gamma) + \left[2 \sum_{n=1}^{\infty} \Lambda_n \phi_n(r)\right] \cos(\Omega t) + O(\epsilon)$$

The $\phi_n(r)$ are the linear free-oscillation modes and a_1 and γ are the solutions of

$$\omega_1 \mu_1 a_1 - F \sin \gamma = 0$$

$$\omega_1 \sigma a_1 + H_{11} a_1 + \tfrac{1}{8}\alpha_{11} a_1^3 + F \cos \gamma = 0$$

where

$$\epsilon\sigma = 3\Omega - \omega_1, \quad F = \sum_{m=1}^{\infty} \sum_{p=1}^{\infty} \sum_{q=1}^{\infty} \Gamma_{1mpq} \Lambda_m \Lambda_p \Lambda_q, \quad \Lambda_n = \frac{K_n}{\omega_n^2 - \Omega^2}$$

and the Γ_{nmpq}, the α_{1n}, and the H_{nm} are defined by (7.6.66), (7.6.75), and (7.4.47).

Note that the solution for a_1 and γ was discussed in Section 4.1.3.

Show that the nonlinearity adjusts the frequencies so that the conditions for internal and external resonances are exactly satisfied.

7.23. Consider the symmetric response of a uniform circular plate clamped along its edges (Section 7.6.2). Let $f_n(t) = 2K_n \cos \Omega T_0$, where 3Ω is near ω_2. Show that the steady-state deflection is given by

$$w(r, t) = a_2 \phi_2(r) \cos(3\Omega t - \gamma) + 2\left[\sum_{n=1}^{\infty} \Lambda_n \phi_n(r)\right] \cos \Omega t + O(\epsilon)$$

The $\phi_n(r)$ are the linear free-oscillation modes and a_2 and γ are the solutions of

$$\omega_2\mu_2 a_2 - F\sin\gamma = 0$$

$$\omega_2\sigma a_2 + \tfrac{1}{8}\alpha_{22}a_{22}^3 + H_{22}a_2 + F\cos\gamma = 0$$

where $\epsilon\sigma = 3\Omega - \omega_2$, $F = \Sigma_{m=1}^{\infty}\,\Sigma_{p=1}^{\infty}\,\Sigma_{q=1}^{\infty}\,\Gamma_{2mpq}\Lambda_m\Lambda_p\Lambda_q$, $\Lambda_n = K_n/(\omega_n^2 - \Omega^2)$, and the α_{nm}, the H_{nm}, and the Γ_{nmpq} are defined by (7.6.75), (7.4.47), and (7.6.66).

Show that the nonlinearity adjusts the frequencies so that the conditions for internal and external resonances are exactly satisfied.

7.24. Consider the symmetric response of a uniform circular plate clamped along its edge (Section 7.6.2). Let $f_n(t) = 2K_n\cos\Omega T_0$, where 2Ω is near $\omega_1 + \omega_2$. Show that after a long time the deflection is given by

$$w(r,t) = \sum_{n=1}^{3} a_n\phi_n(r)\cos\left[(\omega_n + \epsilon\beta_n')t + \tau_n\right] + 2\left[\sum_{n=1}^{\infty}\Lambda_n\phi_n(r)\right]\cos\Omega t + O(\epsilon)$$

The $\phi_n(r)$ are the linear free-oscillation modes, and a_1, a_2, a_3, β_1', β_2', and β_3' are solutions of

$$\omega_1\mu_1 a_1 + Qa_2^2 a_3\sin\gamma_1 + \tfrac{1}{2}H_{12}a_2\sin\gamma_2 = 0$$

$$\omega_2\mu_2 a_2 + 2Qa_1a_2a_3\sin\gamma_1 + \tfrac{1}{2}H_{12}a_1\sin\gamma_2 + \tfrac{1}{2}H_{23}a_3\sin(\gamma_1 - \gamma_2) = 0$$

$$\omega_3\mu_3 a_3 - Qa_1a_2^2\sin\gamma_1 - \tfrac{1}{2}H_{23}a_2\sin(\gamma_1 - \gamma_2) = 0$$

$$\omega_1 a_1\beta_1' + Qa_2^2 a_3\cos\gamma_1 + \tfrac{1}{8}a_1\sum_{n=1}^{3}\alpha_{1n}a_n^2 + H_{11}a_1 + \tfrac{1}{2}H_{12}a_2\cos\gamma_2 = 0$$

$$\omega_2 a_2\beta_2' + 2Qa_1a_2a_3\cos\gamma_1 + \tfrac{1}{8}a_2\sum_{n=1}^{3}\alpha_{2n}a_n^2 + H_{22}a_2$$
$$+ \tfrac{1}{2}H_{12}a_1\cos\gamma_2 + \tfrac{1}{2}H_{23}a_3\cos(\gamma_1 - \gamma_2) = 0$$

$$\omega_3 a_3\beta_3' + Qa_1a_2^2\cos\gamma_1 + \tfrac{1}{8}a_3\sum_{n=1}^{3}\alpha_{3n}a_n^2 + H_{33}a_3 + \tfrac{1}{2}H_{23}a_2\cos(\gamma_1 - \gamma_2) = 0$$

where

$$\Lambda_n = \frac{K_n}{\omega_n^2 - \Omega^2}$$

$$-\gamma_1 = \beta_3 - 2\beta_2 - \beta_1 - \sigma_1 T_1, \quad -\gamma_2 = \sigma_2 T_1 - \beta_1 - \beta_2$$

$$\epsilon\sigma_1 = \omega_1 + 2\omega_2 - \omega_3, \quad \epsilon\sigma_2 = 2\Omega - \omega_1 - \omega_2$$

and the Γ_{nmpq}, the α_{nm}, the H_{nm}, and Q are defined by (7.6.66), (7.6.75), (7.4.47), and (7.6.77).

There are two possibilities: either a_1, a_2, and a_3 are zero or none is zero. For the latter, show that the last three equations can be combined to yield

$$\sigma_2 + Q\left(\frac{a_2^2 a_3}{\omega_1 a_1} + \frac{2a_1 a_3}{\omega_2}\right)\cos\gamma_1 + \tfrac{1}{8}\sum_{n=1}^{\infty}\left(\frac{\alpha_{1n}}{\omega_1} + \frac{\alpha_{2n}}{\omega_2}\right)a_n^2 + \frac{H_{11}}{\omega_1} + \frac{H_{22}}{\omega_2}$$

$$+ \tfrac{1}{2}H_{12}\left(\frac{a_2}{\omega_1 a_1} + \frac{a_1}{\omega_2 a_2}\right)\cos\gamma_2 + \tfrac{1}{2}H_{23}\left(\frac{a_3}{\omega_2 a_2}\right)\cos(\gamma_1 - \gamma_2) = 0$$

and

$$\sigma_1 + \sigma_2 + Q\left(\frac{a_1 a_2^2}{\omega_3 a_3} - \frac{2a_1 a_3}{\omega_2}\right)\cos\gamma_1 - \tfrac{1}{8}\sum_{n=1}^{3}\left(\frac{\alpha_{2n}}{\omega_2} - \frac{\alpha_{3n}}{\omega_3}\right)a_n^2 + \frac{H_{33}}{\omega_3} - \frac{H_{22}}{\omega_2}$$

$$- \tfrac{1}{2}H_{12}\left(\frac{a_1}{\omega_2 a_2}\right)\cos\gamma_2 + \tfrac{1}{2}H_{23}\left(\frac{a_2}{\omega_3 a_3} - \frac{a_3}{\omega_2 a_2}\right)\cos(\gamma_1 - \gamma_2) = 0$$

Thus one can solve for a_1, a_2, a_3, γ_1, and γ_2. Then β_1', β_2', and β_3' can be obtained to within the arbitrary constants τ_1, τ_2, and τ_3, which depend on the initial conditions.

Show that the nonlinearity adjusts the frequencies of the three modes so that the conditions for internal and external resonances are satisfied exactly.

7.25. Consider the symmetric response of a uniform circular plate clamped along its edge (Section 7.6.2). Let $f_n(t) = 2K_n \cos \Omega T_0$, where 2Ω is near $\omega_3 + \omega_N$ for $N \geqslant 4$. Show that after a long time the deflection is given by

$$w(r,t) = \sum_{n=1}^{3} a_n\phi_n(r)\cos[(\omega_n + \epsilon\beta_n')t + \tau_n]$$

$$+ a_N\phi_N(r)\cos[(\omega_N + \epsilon\beta_N')t + \tau_N] + 2\left[\sum_{n=1}^{\infty}\Lambda_n\phi_n(r)\right]\cos\Omega t + O(\epsilon)$$

The $\phi_n(r)$ are the linear free-oscillation modes and $a_1, a_2, a_3, a_N, \beta_1', \beta_2', \beta_3'$, and β_N' are the solutions of

$$\omega_1\mu_1 a_1 + Qa_2^2 a_3 \sin\gamma_1 = 0$$

$$a_2(\omega_2\mu_2 + 2Qa_1a_3\sin\gamma_1) = 0$$

$$\omega_3\mu_3 a_3 - Qa_1a_2^2\sin\gamma_1 + \tfrac{1}{2}H_{3N}a_N\sin\gamma_2 = 0$$

$$\omega_N\mu_N a_n + \tfrac{1}{2}H_{3N}a_3\sin\gamma_2 = 0$$

$$\omega_1 a_1\beta_1' + Qa_2^2 a_3\cos\gamma_1 + \tfrac{1}{8}a_1\sum_{n=1}^{3}\alpha_{1n}a_n^2 + \tfrac{1}{8}\alpha_{1N}a_1a_N^2 + H_{11}a_1 = 0$$

$$a_2(\omega_2\beta_2' + 2Qa_1a_3\cos\gamma_1 + \tfrac{1}{8}\sum_{n=1}^{3}\alpha_{2n}a_n^2 + \tfrac{1}{8}\alpha_{2N}a_N^2 + H_{22}) = 0$$

$$\omega_3 a_3 \beta_3' + Qa_1 a_2^2 \cos \gamma_1 + \tfrac{1}{2} H_{3N} a_N \cos \gamma_2 + \tfrac{1}{8} a_3 \sum_{n=1}^{3} \alpha_{3n} a_n^2 + \tfrac{1}{8} \alpha_{3N} a_3 a_N^2 + H_{33} a_3 = 0$$

$$\omega_N a_n \beta_N' + \tfrac{1}{8} a_N \sum_{n=1}^{3} \alpha_{Nn} a_n^2 + \tfrac{1}{8} \alpha_{NN} a_N^3 + H_{NN} a_N + \tfrac{1}{2} H_{3N} a_3 \cos \gamma_2 = 0$$

where

$$\gamma_1 = \beta_3 - 2\beta_2 - \beta_1 - \sigma_1 T_1, \quad \gamma_2 = \sigma_2 T_1 - \beta_3 - \beta_N$$

$$\Lambda_n = \frac{K_n}{\omega_n^2 - \Omega^2}$$

$$-\epsilon\sigma_1 = \omega_3 - 2\omega_2 - \omega_1, \quad \epsilon\sigma_2 = 2\Omega - \omega_3 - \omega_N$$

and the Γ_{nmpq}, the α_{nm}, the H_{nm}, and Q are defined by (7.6.66), (7.6.75), (7.4.47), and (7.6.77).

Discuss the various possible solutions. In the general case show how the number of equations can be reduced to six, yielding $a_1, a_2, a_3, a_N, \gamma_1$, and γ_2. Show how the nonlinearity adjusts the frequencies so that the conditions for internal and external resonances are satisfied exactly.

7.26. Consider the symmetric response of a uniform circular plate clamped along its edge (Section 7.6.2). Let $f_n(t) = 2K_n \cos \Omega t$, where Ω is near $\omega_1 + \omega_2 + \omega_3$. Show that after a long time the solution is given by

$$w(r, t) = \sum_{n=1}^{3} a_n \phi_n(r) \cos [(\omega_n + \epsilon\beta_n')t + \tau_n] + 2\left[\sum_{n=1}^{\infty} \Lambda_n \phi_n(r)\right] \cos \Omega t + O(\epsilon)$$

where the $\phi_n(r)$ are the linear free-oscillation modes and $a_1, a_2, a_2, \beta_1', \beta_2'$, and β_3' are solutions of

$$\omega_1 \mu_1 a_1 + Qa_2^2 a_3 \sin \gamma_1 + Fa_2 a_3 \sin \gamma_2 = 0$$

$$\omega_2 \mu_2 a_2 + 2Qa_1 a_2 a_3 \sin \gamma_1 + Fa_1 a_3 \sin \gamma_2 = 0$$

$$\omega_3 \mu_3 a_2 - Qa_1 a_2^2 \sin \gamma_1 + Fa_1 a_2 \sin \gamma_2 = 0$$

$$\omega_1 a_1 \beta_1' + Qa_2^2 a_3 \cos \gamma_1 + \tfrac{1}{8} a_1 \sum_{n=1}^{3} \alpha_{1n} a_n^2 + H_{11} a_1 + Fa_2 a_3 \cos \gamma_2 = 0$$

$$\omega_2 a_2 \beta_2' + 2Qa_1 a_2 a_3 \cos \gamma_1 + \tfrac{1}{8} a_2 \sum_{n=1}^{3} \alpha_{2n} a_n^2 + H_{22} a_2 + Fa_1 a_3 \cos \gamma_2 = 0$$

$$\omega_3 a_3 \beta_3' + Qa_1 a_2^2 \cos \gamma_1 + \tfrac{1}{8} a_3 \sum_{n=1}^{3} \alpha_{3n} a_n^2 + H_{33} a_3 + Fa_1 a_2 \cos \gamma_2 = 0$$

where

$$F = 2 \sum_{n=1}^{\infty} (\Gamma_{123n} + \Gamma_{231n} + \Gamma_{132n})\Lambda_n, \quad \Lambda_n = \frac{K_n}{\omega_n^2 - \Omega^2}$$

$$\gamma_1 = \beta_3 - 2\beta_2 - \beta_1 - \sigma_1 T_1, \quad \gamma_2 = \sigma_2 T_1 - \beta_1 - \beta_2 - \beta_3$$

$$\epsilon\sigma_1 = \omega_1 + 2\omega_2 - \omega_3, \quad \epsilon\sigma_2 = \Omega - \omega_1 - \omega_2 - \omega_3$$

and the Γ_{nmpq}, the α_{nm}, the H_{nm}, and Q are defined by (7.6.66), (7.6.75), (7.4.47), and (7.6.77).

Discuss the various possible solutions. For the general case show how the number of equations can be reduced to five, yielding a_1, a_2, a_3, γ_1, and γ_2. Show how the nonlinearity adjusts the frequencies so that the conditions for internal and external resonances are satisfied exactly.

7.27. Consider the symmetric response of a uniform circular plate clamped along its edge (Section 7.6.2). Let $f_n(t) = 2\epsilon k_n \cos(\Omega_n T_0 + \tau)$, where Ω_1 is near ω_1, Ω_2 is near ω_2, and Ω_3 is near ω_3, and $k_n = 0$ for $n \geqslant 4$.

Show that after a long time the deflection is given by

$$w(r, t) = \sum_{n=1}^{3} a_n \phi_n(r) \cos(\Omega_n t - \gamma_n) + O(\epsilon)$$

The $\phi_n(r)$ are the linear free-oscillation modes, and $a_1, a_2, a_3, \gamma_1, \gamma_2$, and γ_3 are the solutions

$$\omega_1 a_1' + \omega_1 \mu_1 a_1 + Q a_2^2 a_3 \sin\gamma_I - k_1 \sin\gamma_1 = 0$$

$$\omega_2 a_2' + \omega_2 \mu_2 a_2 + 2Q a_1 a_2 a_3 \sin\gamma_I - k_2 \sin\gamma_2 = 0$$

$$\omega_3 a_3' + \omega_3 \mu_3 a_3 - Q a_2^2 a_1 \sin\gamma_I - k_3 \sin\gamma_3 = 0$$

$$-\omega_1 a_1 \gamma_1' + a_1\left(\omega_1 \sigma_1 + \tfrac{1}{8} \sum_{n=1}^{3} \alpha_{1n} a_n^2\right) + Q a_2^2 a_3 \cos\gamma_I + k_1 \cos\gamma_1 = 0$$

$$-\omega_2 a_2 \gamma_2' + a_2\left(\omega_2 \sigma_2 + \tfrac{1}{8} \sum_{n=1}^{3} \alpha_{2n} a_n^2\right) + 2Q a_1 a_2 a_3 \cos\gamma_I + k_2 \cos\gamma_2 = 0$$

$$-\omega_3 a_3 \gamma_3' + a_3\left(\omega_3 \dot{\sigma}_2 + \tfrac{1}{8} \sum_{n=1}^{3} \alpha_{3n} a_n^2\right) + Q a_1 a_2^2 \cos\gamma_I + k_3 \cos\gamma_3 = 0$$

where

$$\epsilon\sigma_I = \omega_3 - 2\omega_2 - \omega_1, \quad \epsilon\sigma_n = \Omega_n - \omega_n$$

$$-\gamma_I = (\sigma_3 - 2\sigma_2 - \sigma_1 - \sigma_I)T_1 - \gamma_3 + 2\gamma_2 + \gamma_1$$

and the α_{nm} and Q are defined by (7.6.75) and (7.6.77).

Discuss the various possible solutions when

(a) k_1 is zero but k_2 and k_3 are not
(b) k_2 is zero but k_1 and k_3 are not
(c) k_3 is zero but k_1 and k_2 are not
(d) none is zero

Show that when a steady-state solution exists, the nonlinearity adjusts the frequencies so that the conditions for internal and external resonances are satisfied exactly.

7.28. In the Berger approximation, the nonlinear forced vibrations of a clamped plate are given by (Section 7.6.1)

$$D\nabla^4 w - \rho h e c_p^2 \nabla^2 w + \rho h w_{tt} = -2\mu w_t + f(\mathbf{r}, t)$$

$$e = -\frac{1}{2A} \iint_A w \nabla^2 w \, dx \, dy$$

(a) For the symmetric response of a clamped circular plate, introduce the dimensionless variables defined in Section 7.6.2 and rewrite the governing equation as

$$\nabla^4 w - \alpha \nabla^2 w \int_0^1 r \left(\frac{\partial w}{\partial r}\right)^2 dr + w_{tt} = -2\alpha\mu w_t + f(r, t)$$

where $\alpha = 12h^2/R^2$.

(b) Show that the response can be expressed in the form

$$w(r, t) = \sum_{n=1}^{\infty} \psi_n(t) \phi_n(r)$$

where the ϕ_n are defined by (7.6.51).

(c) Substitute in the dimensionless equation, use the orthonormality of the ϕ_n, assume modal damping, and show that the equations describing the ψ_n can be written as

$$\ddot{\psi}_n + \omega_n^2 \psi_n = \alpha \sum_{m=1}^{\infty} \sum_{p=1}^{\infty} \sum_{q=1}^{\infty} \Gamma_{nmpq} \psi_m \psi_p \psi_q - 2\alpha\mu_n \dot{\psi}_n + f_n(t)$$

where the ω_n are defined by (7.6.42) and (7.6.43) and

$$\Gamma_{nmpq} = -\left[\int_0^1 r\phi_n' \phi_m' \, dr\right]\left[\int_0^1 r\phi_p' \phi_q' \, dr\right]$$

(d) Compare these equations with (7.6.63). Can you conclude anything about the validity of the Berger approximation when ν is varied?

7.29. Consider the forced response of a simply-supported rectangular plate of sides a and b by using the Berger approximation (7.6.26) and (7.6.28).

(a) Seek a solution in the form

$$w(x, y, t) = \sum_{n=1}^{\infty} \sum_{m=1}^{\infty} \psi_{mn}(t) \sin \frac{n\pi x}{a} \sin \frac{m\pi y}{b}$$

Show that

$$e = \frac{\pi^2}{8} \sum_{n=1}^{\infty} \sum_{m=1}^{\infty} \left(\frac{n^2}{a^2} + \frac{m^2}{b^2}\right) \psi_{mn}^2$$

Then show that

$$\ddot{\psi}_{mn} + \omega_{mn}^2 \psi_{mn} = \alpha \left(\frac{n^2}{a^2} + \frac{m^2}{b^2}\right) \sum_{p=1}^{\infty} \sum_{q=1}^{\infty} \left(\frac{p^2}{a^2} + \frac{q^2}{b^2}\right) \psi_{pq}^2 - 2\mu_{mn}\dot{\psi}_{mn} + f_{mn}(t)$$

where $\alpha = \rho h \pi^4 c_p^2 / 8D$ and $\omega_{mn}^2 = (D/\rho h)[(n^2/a^2) + (m^2/b^2)]^2$. Note that a modal damping has been added as well as an external excitation.

(b) When $f_{11} = k_{11} \cos \Omega t$, where $\Omega \approx \omega_{11}$ and all other $f_{mn} = 0$, show that the steady-state response for small but finite amplitudes consists of one mode and the frequency-response equation is similar to that of the Duffing equation.

(c) When $a = b$ and $f_{12} = k_{12} \cos \Omega t$, where $\Omega \approx \omega_{12}$ and all other $f_{mn} = 0$, show that the steady-state response consists of either the 12-mode alone or both the 12- and the 21-modes.

7.30. The equations governing large-amplitude motions of a spinning membrane are (Nowinski, 1964)

$$\rho h(w_{tt} + r\Omega^2 w_r + \tfrac{1}{2} r^2 \Omega^2 \nabla^2 w) = w_{rr}\left(\frac{1}{r} F_r + \frac{1}{r^2} F_{\theta\theta}\right) + \left(\frac{1}{r} w_r + \frac{1}{r^2} w_{\theta\theta}\right) F_{rr} - 2 \frac{\partial}{\partial r}\left(\frac{1}{r} w_\theta\right) \frac{\partial}{\partial r}\left(\frac{1}{r} F_\theta\right)$$

$$\nabla^4 F = 2\rho h(1 - \nu)\Omega^2 + Eh\left[-\left(\frac{1}{r} w_r + \frac{1}{r^2} w_{\theta\theta}\right) w_{rr} + \frac{1}{r^2} w_{r\theta}^2 - \frac{2}{r^3} w_{r\theta} w_\theta + \frac{1}{r^4} w_\theta^2\right]$$

where Ω is the angular velocity of the membrane and the stress resultants N_1, N_2, and N_{12} are related to F by

$$N_1 = \frac{1}{r} F_r + \frac{1}{r^2} F_{\theta\theta} - \tfrac{1}{2} \rho h \Omega^2 r^2$$

$$N_2 = F_{rr} - \tfrac{1}{2} \rho h \Omega^2 r^2, \quad N_{12} = -\frac{\partial}{\partial r}\left(\frac{1}{r} F_\theta\right)$$

(a) Show that the above equations are satisfied exactly by (Advani, 1967a, 1969a)

$$w = Ar^2 \sin 2(\theta \pm ct)$$

$$F = Br^4 \cos 4(\theta \pm ct) + \tfrac{1}{32}[2hEA^2 + (1 - \nu)\rho h\Omega^2]r^4 - Cr^2$$

where the phase velocity c is related to A and B by

$$c^2 - \tfrac{1}{8}(5 - \nu)\Omega^2 = \frac{E}{4\rho}A^2 - \frac{12B}{\rho h}$$

(b) For a free-spinning membrane show that

$$B = 0, \quad 2C = [\tfrac{1}{4}hEA^2 - \tfrac{1}{8}(3 + \nu)\rho h\Omega^2]R^2$$

where R is the radius of the membrane.

(c) For stationary waves show that $c = \Omega$ and hence (Advani, 1967a) $\Omega^2 = 2EA^2/\rho(3 + \nu)$.

(d) Seek a solution to the original equations in the form

$$w = A(r/R)^2 \cos(2\theta)\,\psi(t)$$

and show that (Nowinski, 1964; Advani and Bulkeley, 1969)

$$F = \tfrac{1}{16}EhA^2\left[\left(\frac{r}{R}\right)^4 - 2\left(\frac{r}{R}\right)^2\right]\psi^2 + \tfrac{1}{32}\rho h(1 - \nu)\Omega^2 r^4 + \tfrac{1}{16}\rho h(3 + \nu)\Omega^2 R^2 r^2$$

Then show that ψ is governed by

$$\ddot{\psi} + \alpha_1\psi + \alpha_3\psi^3 = 0$$

and determine α_1 and α_3.

CHAPTER 8

Traveling Waves

In contrast with the preceding chapter, which deals with standing waves, this chapter deals with traveling waves. To analyze nonlinear waves we need to distinguish between the two main classes of waves–dispersive and nondispersive waves. In this book we base the distinction on the types of solutions rather than on the types of governing equations. To this end we seek an oscillatory solution of the linearized problem proportional to $\exp[i(\mathbf{k}\cdot\mathbf{r}-\omega t)]$, where $\mathbf{k}$ and ω are constants which may be complex numbers. Then the dispersion relation $\omega=\Omega(\mathbf{k})$ connecting the wavenumber and the frequency is used to distinguish between the two classes of waves. Waves are called *dispersive* if the phase speed $\omega/k=\Omega(\mathbf{k})/k$ is not independent of k; otherwise they are called *nondispersive.*

As an example of nondispersive waves, we consider linear longitudinal waves along a uniform elastic bar. They are governed by

$$\frac{\partial^2 u}{\partial t^2}-c_0^2\frac{\partial^2 u}{\partial x^2}=0,\quad c_0=\sqrt{E/\rho}$$

where E is Young's modulus, ρ is the density of the bar, $u(x,t)$ is the longitudinal displacement, and c_0 is the linear wave speed. Seeking oscillatory solutions of the form

$$u(x,t)=a\exp[i(kx-\omega t)]$$

leads to the dispersion relation

$$\omega=\pm c_0 k$$

Since the phase speed $\omega/k=\pm c_0$ is independent of the wavenumber k, longitudinal waves along a uniform elastic bar are nondispersive.

As an example of dispersive waves, we consider linear transverse waves along a uniform elastic bar. They are governed by the equation

$$\frac{\partial^2 w}{\partial t^2}+\frac{EI}{\rho A}\frac{\partial^4 w}{\partial x^4}=0$$

where $w(x, t)$ is the transverse displacement and A and I are the cross-sectional area and moment of inertia. Seeking oscillatory solutions of the form

$$w(x, t) = a \exp [i(kx - \omega t)]$$

yields the dispersion relation

$$\omega = \pm\sqrt{EI/\rho A}\, k^2$$

Since the phase speed ω/k is a function of k, transverse waves along a uniform elastic bar are dispersive.

In the next section we take up nonlinear nondispersive waves in connection with longitudinal waves along a bar. Then we take up nonlinear dispersive waves in connection with transverse waves along a bar. For a more complete discussion and applications to other physical systems, we refer the reader to the books of Courant and Friedrichs (1948), Jeffrey and Taniuti (1966), Cristescu (1967), Whitham (1974), Leibovich and Seebass (1974), Beyer (1974), and Karpman (1975) and the review articles of Fleishman (1963), Thurston and Shapiro (1967), Lick (1970), Hayes (1971), Blackstock (1972), Jeffrey and Kakutani (1972), Phillips (1974), Rudenko, Soluyan, and Khokhlov (1974), and Seymour and Mortell (1975a).

8.1 Longitudinal Waves Along a Bar

We consider the behavior of nonlinear waves in a bar whose properties vary slowly along its length. We introduce dimensionless quantities by using a reference length L, a reference density ρ_r, a reference Young's modulus E_r, and a reference cross-sectional area A_r. We let $u(x, t)$ denote the longitudinal displacement at time t of the section of the bar that was at the position x in the initial state. In terms of these dimensionless quantities, the equation of motion is (e.g., Cristescu, 1967)

$$\rho A \frac{\partial^2 u}{\partial t^2} = \frac{\partial}{\partial x}(A\sigma) \tag{8.1.1}$$

where σ is the dimensionless longitudinal stress. The stress is related to the strain by the constitutive relation $\sigma = \sigma(e)$, where the strain $e = \partial u/\partial x$. In this book we consider only constitutive equations that can be written in the finite form $\sigma = h(e)$, where h is usually a monotonically increasing function of e as shown in Figure 8-1. Hence (8.1.1) can be rewritten as

$$\frac{\partial^2 u}{\partial t^2} = \frac{1}{\rho A} \frac{\partial}{\partial x} \left[A\sigma \left(\frac{\partial u}{\partial x} \right) \right] \tag{8.1.2}$$

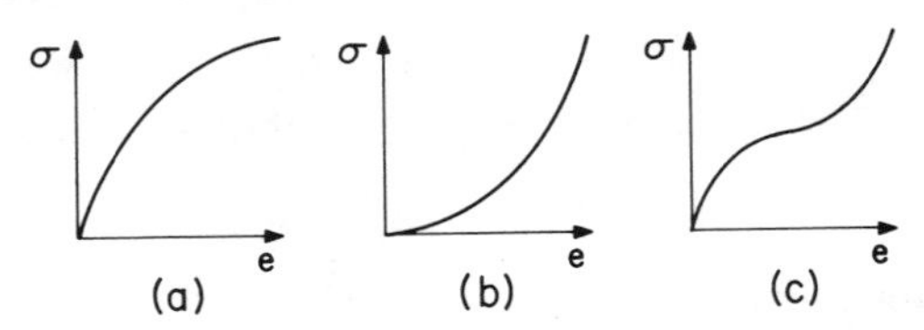

Figure 8-1. Stress–strain curves: (*a*) work-hardening material; (*b*) some rubbers, soils, and some metals; (*c*) changing concavity.

It is assumed that σ is an analytic function of e so that it can be expanded in a Taylor series according to

$$\sigma = Ee(1 + E_1 e + E_2 e^2 + \cdots) \tag{8.1.3}$$

where E and the E_n are known functions of x. Substituting for σ from (8.1.3) into (8.1.1), letting $e = \partial u/\partial x$, neglecting cubic and higher-order terms, and rearranging, we obtain

$$\frac{\partial^2 u}{\partial x^2} - \frac{1}{c_0^2}\frac{\partial^2 u}{\partial t^2} = -(\ln EA)'\frac{\partial u}{\partial x}\left(1 + E_1\frac{\partial u}{\partial x}\right) - E_1'\left(\frac{\partial u}{\partial x}\right)^2 - 2E_1\frac{\partial u}{\partial x}\frac{\partial^2 u}{\partial x^2} \tag{8.1.4}$$

where a prime denotes differentiation with respect to the argument and $c_0^2 = E(x)/\rho(x)$. We note that time has been made dimensionless by using $L\sqrt{\rho_r/E_r}$.

In the next section we discuss the significance of characteristics for hyperbolic equations and obtain an exact solution for simple waves. In Section 8.1.2 we apply the method of multiple scales to simple waves along a uniform bar. In Sections 8.1.3 and 8.1.4 we discuss the problem of shock fitting by using the velocity and displacement, respectively. In Section 8.1.5 we discuss the effect of heterogeneity. In Section 8.1.6 we discuss the case of finite uniform bars, while in Section 8.1.7 we discuss finite heterogeneous bars.

8.1.1. HYPERBOLIC EQUATIONS AND CHARACTERISTICS

In this section we discuss the significance of characteristics for hyperbolic equations and obtain an exact solution for the case of uniform bars. Letting A and ρ be constant, we rewrite (8.1.2) as

$$\frac{\partial^2 u}{\partial t^2} = c^2(e)\frac{\partial^2 u}{\partial x^2}, \quad c^2 = \frac{1}{\rho}\frac{d\sigma}{de} \tag{8.1.5}$$

where $c(e)$ is shown below to be the nonlinear velocity of propagation of the wave. In what follows we assume that $c^2(e) > 0$. Equation (8.1.5) is usually called a *quasi-linear equation* because it is linear in the highest derivatives. To analyze the behavior of the solutions of the second-order equation (8.1.5), we first replace it by a system of two first-order equations. To this end we let $v = \partial u/\partial t$, where v is the particle velocity. If u is twice continuously differentiable, then

$$\frac{\partial e}{\partial t} - \frac{\partial v}{\partial x} = 0 \tag{8.1.6}$$

Moreover (8.1.5) can be rewritten in terms of e and v as

$$\frac{\partial v}{\partial t} - c^2(e)\frac{\partial e}{\partial x} = 0 \tag{8.1.7}$$

Characteristics for a Second-Order System. We note that in each of (8.1.6) and (8.1.7) the combination of $\partial e/\partial t$ and $\partial e/\partial x$ is different from the combination of $\partial v/\partial t$ and $\partial v/\partial x$. That is, in a tx-plane each equation involves differentiation in more than one direction; namely, the t- and x-directions. The question arises whether (8.1.6) and (8.1.7) can be transformed into a system of equations such that each equation involves differentiation in one direction only. To this end we form a linear combination of (8.1.6) and (8.1.7) by adding p_1 times the former equation to p_2 times the latter equation. The result is

$$p_1 \frac{\partial e}{\partial t} - p_2 c^2 \frac{\partial e}{\partial x} + p_2 \frac{\partial v}{\partial t} - p_1 \frac{\partial v}{\partial x} = 0 \tag{8.1.8}$$

The combination of $\partial e/\partial t$ and $\partial e/\partial x$ will be the same as the combination of $\partial v/\partial t$ and $\partial v/\partial x$ in (8.1.8) if

$$\frac{p_1}{p_2 c^2} = \frac{p_2}{p_1} \tag{8.1.9}$$

or

$$p_1 = \pm c p_2 \tag{8.1.10}$$

Hence (8.1.8) assumes the so-called *characteristic*, or *canonical* form

$$c\left(\frac{\partial e}{\partial t} - c\frac{\partial e}{\partial x}\right) + \left(\frac{\partial v}{\partial t} - c\frac{\partial v}{\partial x}\right) = 0 \tag{8.1.11}$$

$$c\left(\frac{\partial e}{\partial t} + c\frac{\partial e}{\partial x}\right) - \left(\frac{\partial v}{\partial t} + c\frac{\partial v}{\partial x}\right) = 0 \tag{8.1.12}$$

We note that (8.1.11) involves differentiation in the single direction defined by

$$\frac{dx}{dt} = -c \tag{8.1.13}$$

while (8.1.12) involves differentiation in the single direction defined by

$$\frac{dx}{dt} = c \tag{8.1.14}$$

Thus c is the velocity of propagation. Equations (8.1.13) and (8.1.14) define two families of *characteristics* C_- and C_+ along which small disturbances propagate. Since $c = c(e)$ varies from point to point, the characteristics are curves with variable slopes. These slopes are not known a priori but need to be determined simultaneously with the unknown e. The second-order system of equations (8.1.6) and (8.1.7) is of the hyperbolic type because it has two linearly independent characteristics. In general a system of n first-order partial-differential equations is called *hyperbolic* if it has a full set of n linearly independent characteristics.

Equations (8.1.11) and (8.1.12) can be rewritten as

$$
\begin{aligned}
dv + c\,de = 0 \quad &\text{along } C_-\!: \frac{dx}{dt} = -c \\
dv - c\,de = 0 \quad &\text{along } C_+\!: \frac{dx}{dt} = c
\end{aligned}
\tag{8.1.15}
$$

Since $c = c(e)$, (8.1.15) can be integrated to yield

$$
\begin{aligned}
v + \int c\,de = J_- \quad &\text{along } C_- \\
v - \int c\,de = J_+ \quad &\text{along } C_+
\end{aligned}
\tag{8.1.16}
$$

where J_- and J_+ are called the *Riemann invariants*; they are constants along the characteristics C_- and C_+, respectively.

If the constitutive equation has the form

$$
\sigma = Ee(1 + E_1 e) \tag{8.1.17}
$$

then

$$
c^2 = \frac{E}{\rho}(1 + 2E_1 e) = c_0^2(1 + 2E_1 e) \tag{8.1.18}
$$

with c_0 being unity for a uniform bar. Hence (8.1.13), (8.1.14), and (8.1.16) become

$$
\begin{aligned}
\frac{dx}{dt} &= -(1 + 2E_1 e)^{1/2} \\
\frac{dx}{dt} &= (1 + 2E_1 e)^{1/2}
\end{aligned}
\tag{8.1.19}
$$

$$
\begin{aligned}
v + \frac{1}{3E_1}\left[(1 + 2E_1 e)^{3/2} - 1\right] &= J_- \\
v - \frac{1}{3E_1}\left[(1 + 2E_1 e)^{3/2} - 1\right] &= J_+
\end{aligned}
\tag{8.1.20}
$$

where the limits of integration are chosen so that $J_- = 0$ and $J_+ = 0$ when $v = 0$ and $e = 0$.

The characteristics of (8.1.6) and (8.1.7) can be obtained by the following alternate method. The characteristics are curves in the tx-plane, at whose intersection e and v are continuous but their derivatives may be discontinuous. If $t = t(s)$ and $x = x(s)$ represent one of these curves, then

$$\begin{aligned} \frac{dv}{ds} &= \frac{\partial v}{\partial x}\frac{dx}{ds} + \frac{\partial v}{\partial t}\frac{dt}{ds} \\ \frac{de}{ds} &= \frac{\partial e}{\partial x}\frac{dx}{ds} + \frac{\partial e}{\partial t}\frac{dt}{ds} \end{aligned} \tag{8.1.21}$$

or

$$\begin{aligned} dv &= \frac{\partial v}{\partial x}dx + \frac{\partial v}{\partial t}dt \\ de &= \frac{\partial e}{\partial x}dx + \frac{\partial e}{\partial t}dt \end{aligned} \tag{8.1.22}$$

Solving (8.1.6), (8.1.7), and (8.1.22) we obtain

$$\begin{aligned} &\frac{\partial e}{\partial t} = \frac{\partial v}{\partial x} = \frac{dv\,dx - c^2\,de\,dt}{dx^2 - c^2\,dt^2} \\ &\frac{\partial e}{\partial x} = \frac{de\,dx - dv\,dt}{dx^2 - c^2\,dt^2}, \quad \frac{\partial v}{\partial t} = c^2\frac{\partial e}{\partial x} \end{aligned} \tag{8.1.23}$$

Since the characteristics are the curves in the tx-plane at whose intersection the derivatives of e and v may be discontinuous, it follows from (8.1.23) that they are given by

$$dx^2 = c^2\,dt^2$$

or

$$\frac{dx}{dt} = \pm c(e) \tag{8.1.24}$$

in agreement with (8.1.13) and (8.1.14). In order that the derivatives of e and v be finite along these characteristics, it follows from (8.1.23) that the following differential relations must be satisfied along these characteristics (consistency condition).

$$dv = \pm c\,de \tag{8.1.25}$$

in agreement with (8.1.15).

Characteristics of an nth-Order System. Next we show how one can generalize the method of finding the characteristics to the nth-order quasi-linear system

$$\frac{\partial \mathbf{v}}{\partial t} + [B] \frac{\partial \mathbf{v}}{\partial x} + \mathbf{d} = 0 \tag{8.1.26}$$

where $\mathbf{v}$ and $\mathbf{d}$ are column vectors with n components and $[B]$ is an $n \times n$ matrix. Since the system is quasi-linear, $[B]$ and $\mathbf{d}$ are in general functions of x, t, and the components of $\mathbf{v}$ but not its derivatives.

To classify the system (8.1.26) and determine its characteristics, we attempt to transform it into a new system such that each equation involves differentiation in one direction only. To this end we form a linear combination of the equations in (8.1.26) by multiplying (8.1.26) by the horizontal vector $\mathbf{p}_k$ having n components. The result is

$$\mathbf{p}_k \frac{\partial \mathbf{v}}{\partial t} + \mathbf{p}_k [B] \frac{\partial \mathbf{v}}{\partial x} + \mathbf{p}_k \mathbf{d} = 0, \quad k = 1, 2, \ldots, n \tag{8.1.27}$$

We choose the $\mathbf{p}_k$ such that

$$\mathbf{p}_k [B] = c_k \mathbf{p}_k, k = 1, 2, \ldots, n \tag{8.1.28}$$

Thus the $\mathbf{p}_k$ are the characteristic vectors of the matrix $[B]$ corresponding to its characteristic values c_k which are solutions of

$$| [B] - c [I] | = 0 \tag{8.1.29}$$

With (8.1.28), (8.1.27) can be rewritten as

$$\mathbf{p}_k \left[\frac{\partial \mathbf{v}}{\partial t} + c_k \frac{\partial \mathbf{v}}{\partial x} + \mathbf{d} \right] = 0, \quad k = 1, 2, \ldots, n \tag{8.1.30}$$

If all the c_k are real and different from zero and all the $\mathbf{p}_k$ are linearly independent, then (8.1.30) is the *characteristic*, or *canonical* form of (8.1.26) because each equation involves differentiation in a single direction defined by one of the following ordinary-differential equations:

$$\frac{dx}{dt} = c_k, \quad k = 1, 2, \ldots, n \tag{8.1.31}$$

If these conditions are satisfied, the system of equations (8.1.26) is called *hyperbolic*. If the characteristic values c_k are complex, the system is called *elliptic*. Thus the type of the quasi-linear system (8.1.26) depends on the characteristic values and the characteristic vectors of the matrix $[B]$.

As an example we rewrite (8.1.6) and (8.1.7) in the matrix form

$$\frac{\partial}{\partial t} \begin{bmatrix} e \\ v \end{bmatrix} + \begin{bmatrix} 0 & -1 \\ -c^2 & 0 \end{bmatrix} \frac{\partial}{\partial x} \begin{bmatrix} e \\ v \end{bmatrix} = 0 \tag{8.1.32}$$

Hence the characteristic velocities are the roots of

$$\begin{vmatrix} -\lambda & -1 \\ -c^2 & -\lambda \end{vmatrix} = 0 \tag{8.1.33}$$

or

$$\lambda = \pm c \tag{8.1.34}$$

The corresponding left characteristic vectors are $(-c, 1)$ and $(c, 1)$. Therefore (8.1.32) can be reduced to the canonical form

$$\begin{aligned} \frac{\partial v}{\partial t} + c\frac{\partial v}{\partial x} - c\left(\frac{\partial e}{\partial t} + c\frac{\partial e}{\partial x}\right) = 0 \\ \frac{\partial v}{\partial t} - c\frac{\partial v}{\partial x} + c\left(\frac{\partial e}{\partial t} - c\frac{\partial e}{\partial x}\right) = 0 \end{aligned} \tag{8.1.35}$$

in agreement with (8.1.11) and (8.1.12).

Significance of Characteristics. To complete the problem formulation, we need to supplement the second-order equation (8.1.5) or its equivalent first-order equations (8.1.6) and (8.1.7) by initial and boundary conditions. We consider first the Cauchy initial-value problem. Thus we assume that at $t = 0$, the distributions of e and v are known along the x-coordinate, and it is required to determine e and v for $t > 0$. A set of C_+ and C_- characteristics, originating from the x-axis, exists in the tx-plane as shown in Figure 8-2. It follows from (8.1.16) that the values of J_- and J_+ and hence the values of e and v at any point P are determined only by the values of these quantities at the initial time $t = 0$; that is, by the values of e and v at the points A and B, where AP is the C_+ characteristic originating from A while BP is the C_- characteristic originating from B.

We note that the intersection point P of the characteristics AP and BP depends on their paths. Moreover the paths of these characteristics depend on the values of e at the intermediate points between A and P and B and P. But the value of e at an intermediate point N_1 is determined not only by the initial value at A but

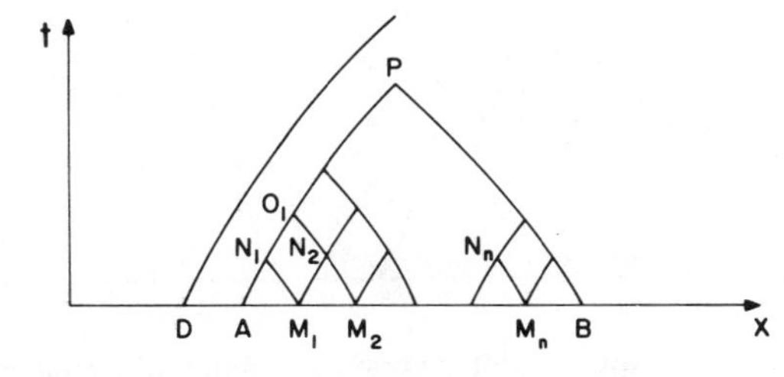

Figure 8-2. Construction of the characteristic network for a Cauchy boundary-value problem.

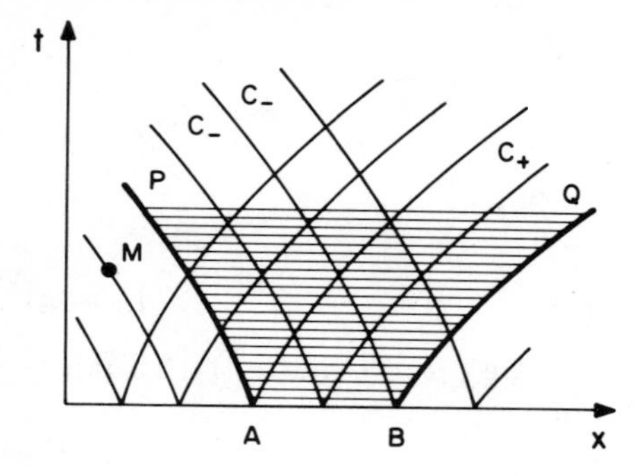

Figure 8-3. Diagram illustrating the region of influence.

also by the initial value at M_1. Thus the values of e and v at P are completely determined by the initial values along the segment AB of the x-axis and are absolutely independent of the initial values outside this segment. A slight change in the initial values at the point D will not have any effect on those at P owing to the fact that the disturbance created by this change will not be able to arrive at the point P at time t_p. Hence AB is called the *domain of dependence* of point P and APB is called the *domain of determinacy* of the segment AB. Similarly the initial values along the segment AB can influence the values of e and v at those points in the so-called *domain of influence* bounded by the C_- characteristic originating from the point A, the segment AB, and the C_+ characteristic originating from the point B as shown in Figure 8-3. There is no influence of the initial conditions along the segment AB on the values of e and v at points outside this domain such as M because a disturbance created by a change in these initial conditions cannot reach the point M at time t_M.

We consider next a mixed boundary- and initial-value problem by considering the bounded domain shown in Figure 8-4. This problem is often encountered in dynamics when various boundary conditions are imposed at the edges of the

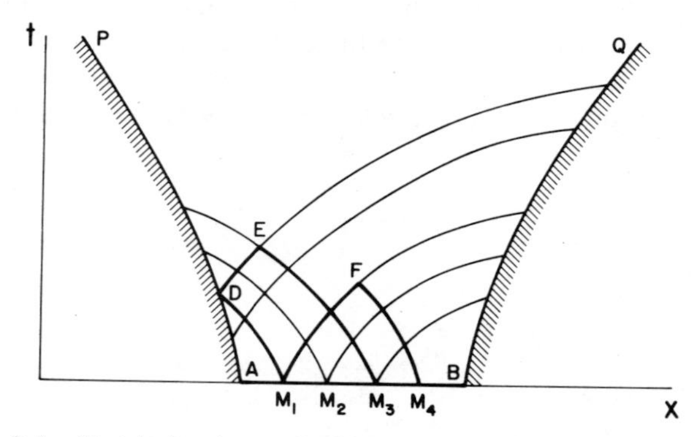

Figure 8-4. Sketch showing a mixed boundary- and initial-value problem.

body. For example, the ends of a bar can be fixed, the stress can be specified, or a law of motion can be specified such as when two bars are joined together.

The segment AB is called *spacelike* in the sense that the characteristics corresponding to increasing time originating from one of its points lie on one side of this segment. On the other hand, the segment AP is called *timelike* in the sense that the characteristics corresponding to increasing time originating from one of its points lie on opposite sides of this segment. Similarly the segment BQ is timelike.

For example, let e and v be specified along the segment AB at $t = 0$ and let v be specified along the segments AP and BQ for all t. We note that points, such as F lying at the intersection of a C_+ and a C_- characteristic originating from points inside the initial segment AB, do not differ from points in an unbounded bar. Thus the values of e and v at such points are determined completely by the initial values along the segment AB. On the other hand, a point on the segment AP such as D is reached only from the past by the C_- characteristic originating at the point M_1 on the initial segment AB, and it is not reached from the past by the C_+ characteristic. Hence the value of e at D is a function of the initial values at M_1 and the value of v at D. Since the intersection of the C_- characteristic with the segment AP depends on the initial values along the segment AM_1 and the boundary values at AD, AM_1D is the domain of dependence of the point D. The values of e and v at the point D are carried in the future by the C_+ characteristic originating from it. Thus the values of e and v at the point E depend on the initial values along the segment AM_3 as well as the boundary values along the segment AD. Therefore the states at points such as D and E depend on the initial as well as the boundary values, in contrast with the states at points such as F which depend on the initial values only.

Simple Waves. We consider the case of an initially undeformed, semiinfinite bar at rest. If the origin of the x-axis is at the left end of the bar, then the initial conditions are

$$e(x, 0) = v(x, 0) = \sigma(x, 0) = 0 \quad \text{for } x > 0 \tag{8.1.36}$$

while the boundary condition at $x = 0$ is taken in the form of a specified e, v, or σ for $t \geqslant 0$. For definiteness we let

$$v(0, t) = \epsilon\phi(t) \quad \text{for } t \geqslant 0 \tag{8.1.37}$$

where ϵ is a dimensionless parameter which is the order of the ratio of the amplitude of the particle velocity to the linearized speed of sound.

Figure 8-5 shows that the C_- characteristic reaching any point in the tx-plane starts at the x-axis, where $e = v = 0$. Substituting these conditions into the first

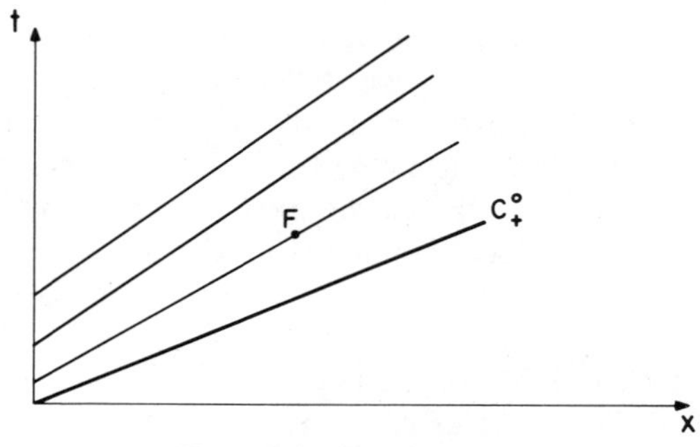

Figure 8-5. Simple waves.

of equations (8.1.16) yields $J_- = 0$ and hence

$$v = -\Psi(e) \quad \text{on all } C_-, \text{ where } \Psi(e) = \int_0^e c\, de \tag{8.1.38}$$

Since every point in the tx-plane is reached by a C_- characteristic, $v = -\Psi(e)$ in the whole tx-plane. Hence the second of equations (8.1.16) gives

$$v = \tfrac{1}{2}J_+ \quad \text{and } \Psi(e) = -\tfrac{1}{2}J_+ \text{ on } C_+ \tag{8.1.39}$$

Consequently $c(e)$ is constant on C_+. To satisfy the boundary condition (8.1.37), we find it convenient to rewrite (8.1.14) as

$$\frac{dt}{dx} = \frac{1}{c(e)} \text{ on } C_+ \tag{8.1.40}$$

Since $c(e)$ is constant on C_+, the C_+ characteristics are straight lines, and (8.1.40) can be integrated to give

$$t - \frac{x}{c(e)} = \xi \tag{8.1.41}$$

where ξ is a constant on C_+. Thus

$$v(x, t) = f\left(t - \frac{x}{c(e)}\right) \tag{8.1.42}$$

We note that all C_+ characteristics below C_+^0 corresponding to $\xi = 0$ start at the x-axis, and hence $v = e = 0$ below C_+^0. Using this fact and the boundary condi-

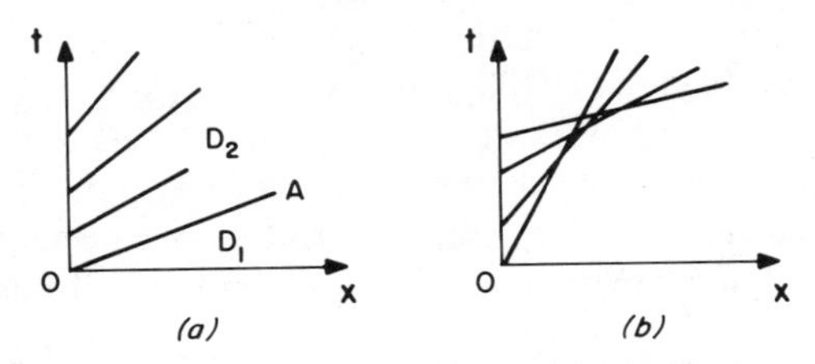

Figure 8-6. Characteristic fields showing (*a*) a divergent family of characteristic lines and (*b*) a convergent bundle of characteristic lines and the possibility of forming shock waves.

tion (8.1.37), we find that $f(t) = \epsilon\phi(t)$, and hence

$$v(x, t) = \epsilon\phi\left(t - \frac{x}{c(e)}\right) \quad \text{for } \xi \geqslant 0$$
$$e = \Psi^{-1}(-v) \tag{8.1.43}$$

where Ψ^{-1} is the inverse of Ψ.

When σ has the special form (8.1.17), c is given by (8.1.18), and hence (8.1.43) becomes

$$v(x, t) = \epsilon\phi[t - x(1 - 3\epsilon E_1\phi)^{-1/3}] \tag{8.1.44}$$

When $c(e)$ decreases with e corresponding to the stress–strain curve shown in Figure 8-1*a*, the characteristic lines diverge as shown in Figure 8-6. On the other hand, when $c(e)$ increases with e corresponding to the stress–strain curve shown in Figure 8-1*b*, the characteristic lines converge as in Figure 8-6 and a shock wave forms as the envelope of these characteristics. Shock fitting is discussed in Sections 8.1.3 and 8.1.4. In the case of the stress–strain curve shown in Figure 8-1*c*, the characteristic lines first diverge and then converge.

8.1.2. THE METHOD OF MULTIPLE SCALES

In the preceding section we obtained a solution for simple waves which is an exact solution of the governing equations and boundary and initial values in the absence of reflections, heterogeneities such as stratifications, dissipative mechanisms, and rate dependence. A number of techniques can be used to determine approximate solutions to these general problems. These include the method of strained coordinates with its variant the method of renormalization, the analytic method of characteristics (Lin, 1954; Fox, 1955), the method of multiple scales, the method of averaging, and geometrical acoustics. Nayfeh and Kluwick (1976) compared three of these techniques: the method of renormalization, the analytic method of characteristics, and the method of multiple scales.

Nayfeh (1977) compared the method of multiple scales and the method of strained coordinates. Nayfeh (1973b, Exercise 3.30), Kluwick (1974), Van Dyke (1975), and Nayfeh and Kluwick (1976) showed that the results of the method of renormalization depend on whether u or $\partial u/\partial t$ is normalized. Nayfeh and Kluwick (1976) showed that this arbitrariness does not occur when using either the analytic method of characteristics or the method of multiple scales. However special consideration must be given to determining u from $\partial u/\partial t$ or $\partial u/\partial x$ obtained by the analytic method of characteristics. No such consideration is needed when using the method of multiple scales. Therefore in this book we use the method of multiple scales. It has been applied to problems of this type by Luke (1966); Kakutani, Ono, Taniuti, and Wei (1968); Kakutani and Ono (1969); Asano and Taniuti (1969, 1970); Choquet-Bruhat (1969); Whitham (1970); Nariboli (1970); Keller and Kogelman (1970); Kakutani and Kawahara (1970); Nariboli and Sedov (1970); Asano (1970; Chikwendu and Kevorkian (1972); Nariboli and Lin (1973); Davy and Ames (1973); Taniuti (1974); Kakutani (1974); Kakutani and Matsuuchi (1975); Nayfeh (1975a, b); Nayfeh and Kluwick (1976); Nozaki and Taniuti (1976); Asano, Nishihara, Nozaki, and Taniuti (1976); Mortell (1977); Nayfeh (1977); and Lardner (1977). Fink, Hall, and Hausrath (1974) and Eckhaus (1975) gave some rigor to some of the results obtained by the above authors.

To gain confidence in the results of the method of multiple scales, we start by its application to a nontrivial problem, which has an exact solution for comparison. Thus we consider the problem of simple waves governed by

$$\frac{\partial^2 u}{\partial x^2} - \frac{\partial^2 u}{\partial t^2} = -2E_1 \frac{\partial u}{\partial x} \frac{\partial^2 u}{\partial x^2} \tag{8.1.45}$$

subject to the initial conditions

$$\frac{\partial u}{\partial t}(x, 0) = \frac{\partial u}{\partial x}(x, 0) = 0 \tag{8.1.46}$$

and the boundary condition

$$v(0, t) = \frac{\partial u}{\partial t}(0, t) = \epsilon\phi(t) \tag{8.1.47}$$

where ϵ is a small, dimensionless quantity which is the order of the ratio of the amplitude of the particle velocity v to the linearized speed of sound.

Seeking a straightforward expansion for this problem in the form

$$u(x, t) = \epsilon u_1(x, t) + \epsilon^2 u_2(x, t) + \cdots \tag{8.1.48}$$

one obtains the following expression for the velocity:

$$v = \frac{\partial u}{\partial t} = \epsilon\phi(s_1) - \epsilon^2 E_1 x \phi(s_1)\phi'(s_1) + \cdots \tag{8.1.49}$$

where $s_1 = t - x$ is the linearized right-running characteristic. Equation (8.1.49) shows that $v \to \infty$ as $x \to \infty$, contrary to the exact solution (8.1.44). Moreover, (8.1.49) is not valid when $x \geqslant O(\epsilon^{-1})$ because the second term becomes the same order or larger than the first term, contrary to the assumption under which it is derived.

To determine the source of nonuniformity, we expand the exact solution (8.1.44) for small ϵ. Expanding the radical in (8.1.44) we have

$$(1 - 3\epsilon E_1 \phi)^{-1/3} = 1 + \epsilon E_1 \phi + 2\epsilon^2 E_1^2 \phi^2 + \cdots \tag{8.1.50}$$

which is uniform when $|\phi| < \infty$. Then (8.1.44) becomes

$$v = \epsilon\phi(t - x - \epsilon E_1 x\phi - 2\epsilon^2 E_1^2 x\phi^2 + \cdots) \tag{8.1.51}$$

because $c_0 = 1$ as a result of the nondimensionalization, which is bounded for large x. We note that the argument in (8.1.51) is an approximation to the nonlinear right-running characteristics and that this approximation did not lead to a nonuniformity.

Expanding (8.1.51) for small ϵ and keeping terms to $O(\epsilon^2)$ yield (8.1.49), which is not valid for large x. Thus this is the step that led to the nonuniformity. We note that approximating the Taylor series expansion of (8.1.51) by only two terms is justified when the combination ϵx and not only ϵ is small. This observation is the basic idea underlying the method of multiple scales. Thus to obtain a uniformly valid expansion for large x, one has to consider ϵx as a single entity which we denote as x_1, which is obviously a slow scale. Hence we introduce this slow scale in addition to the fast scales $x_0 = x$ and $t_0 = t$ and consider e to be a function of x_0, t_0, and x_1. It turns out that it is more convenient to use the linearized characteristics $s_1 = t - x$ and $s_2 = t + x$ as the fast scales. With these new variables we seek an expansion in the form

$$u(x, t; \epsilon) = \epsilon u_1(s_1, s_2, x_1) + \epsilon^2 u_2(s_1, s_2, x_1) + \cdots \tag{8.1.52}$$

In terms of the new independent variables, the time and space derivatives become

$$\begin{aligned} \frac{\partial}{\partial t} &= \frac{\partial}{\partial s_2} + \frac{\partial}{\partial s_1} \\ \frac{\partial^2}{\partial t^2} &= \left(\frac{\partial}{\partial s_2} + \frac{\partial}{\partial s_1}\right)^2 \end{aligned} \tag{8.1.53}$$

$$\begin{aligned} \frac{\partial}{\partial x} &= \frac{\partial}{\partial s_2} - \frac{\partial}{\partial s_1} + \epsilon \frac{\partial}{\partial x_1} \\ \frac{\partial^2}{\partial x^2} &= \left(\frac{\partial}{\partial s_2} - \frac{\partial}{\partial s_1}\right)^2 + 2\epsilon \frac{\partial}{\partial x_1}\left(\frac{\partial}{\partial s_2} - \frac{\partial}{\partial s_1}\right) + \epsilon^2 \frac{\partial^2}{\partial x_1^2} \end{aligned} \tag{8.1.54}$$

Substituting (8.1.52) through (8.1.54) into (8.1.45) and equating coefficients of like powers of ϵ, we obtain

$$\frac{\partial^2 u_1}{\partial s_1\,\partial s_2} = 0 \tag{8.1.55}$$

$$4\frac{\partial^2 u_2}{\partial s_1\,\partial s_2} = 2\frac{\partial}{\partial x_1}\left(\frac{\partial u_1}{\partial s_2} - \frac{\partial u_1}{\partial s_1}\right) + 2E_1\left(\frac{\partial u_1}{\partial s_2} - \frac{\partial u_1}{\partial s_1}\right)\left(\frac{\partial}{\partial s_2} - \frac{\partial}{\partial s_1}\right)^2 u_1 \tag{8.1.56}$$

The solution of (8.1.55) is taken in the form of a right-running wave. That is,

$$u_1 = f(s_1, x_1) \tag{8.1.57}$$

Hence (8.1.56) becomes

$$4\frac{\partial^2 u_2}{\partial s_1\,\partial s_2} = -2\frac{\partial^2 f}{\partial s_1\,\partial x_1} - 2E_1\frac{\partial f}{\partial s_1}\frac{\partial^2 f}{\partial s_1^2} \tag{8.1.58}$$

For a uniformly valid expansion, u_2/u_1 must be bounded for all s_2. This condition demands the vanishing of the right-hand side of (8.1.58). That is,

$$\frac{\partial^2 f}{\partial s_1\,\partial x_1} + E_1\frac{\partial f}{\partial s_1}\frac{\partial^2 f}{\partial s_1^2} = 0 \tag{8.1.59}$$

The general solution of (8.1.59) is

$$f(s_1, x_1) = F(\xi) + \tfrac{1}{2}E_1 x_1 F'^2(\xi) \tag{8.1.60}$$

where

$$s_1 = \xi + E_1 x_1 F'(\xi) \tag{8.1.61}$$

This can be shown as follows. Differentiating (8.1.61) with respect to s_1 and x_1 yields

$$\begin{aligned} 1 &= \frac{\partial \xi}{\partial s_1} + E_1 x_1 F''\frac{\partial \xi}{\partial s_1} \\ 0 &= \frac{\partial \xi}{\partial x_1} + E_1 F' + E_1 x_1 F''\frac{\partial \xi}{\partial x_1} \end{aligned} \tag{8.1.62}$$

Hence

$$\begin{aligned} \frac{\partial \xi}{\partial s_1} &= \frac{1}{1 + E_1 x_1 F''} \\ \frac{\partial \xi}{\partial x_1} &= -\frac{E_1 F'}{1 + E_1 x_1 F''} \end{aligned} \tag{8.1.63}$$

Differentiating (8.1.60) with respect to s_1 yields

$$\frac{\partial f}{\partial s_1} = F'[1 + E_1 x_1 F''] \frac{\partial \xi}{\partial s_1} = F' \tag{8.1.64}$$

on account of (8.1.62). Differentiating (8.1.64) with respect to s_1 and x_1 we have

$$\frac{\partial^2 f}{\partial s_1^2} = F'' \frac{\partial \xi}{\partial s_1}, \quad \frac{\partial^2 f}{\partial s_1 \, \partial x_1} = F'' \frac{\partial \xi}{\partial x_1} \tag{8.1.65}$$

Hence

$$\frac{\partial^2 f}{\partial s_1 \, \partial x_1} + E_1 \frac{\partial f}{\partial s_1} \frac{\partial^2 f}{\partial s_1^2} = F'' \left[\frac{\partial \xi}{\partial x_1} + E_1 F' \frac{\partial \xi}{\partial s_1} \right]$$

which is zero, on account of (8.1.63).

Combining (8.1.52), (8.1.57), (8.1.60), and (8.1.61) and using the definitions of x_1 and s_1, we obtain

$$u = \epsilon [F(\xi) + \tfrac{1}{2} \epsilon E_1 x F'^2(\xi)] + \cdots \tag{8.1.66}$$

where

$$t - x = \xi + \epsilon E_1 x F'(\xi) + \cdots \tag{8.1.67}$$

Since $v = \partial u / \partial t = \partial u / \partial s_1$, it follows from (8.1.64) and (8.1.66) that

$$v = \epsilon F'(\xi) + \cdots \tag{8.1.68}$$

Since $\xi = t$ at $x = 0$ according to (8.1.67), imposing the boundary condition $v(0, t) = \epsilon \phi(t)$ on (8.1.68) yields

$$F'(\xi) = \phi(\xi)$$

Hence the first approximation to the solution of (8.1.45) through (8.1.47) is

$$v(x, t; \epsilon) = \epsilon \phi(\xi) + \cdots \tag{8.1.69}$$

where

$$t - x = \xi + \epsilon E_1 x \phi(\xi) + \cdots \tag{8.1.70}$$

which is in agreement with (8.1.51) to $O(\epsilon)$. Thus the method of multiple scales yields a uniform approximation to the exact solution. Since

$$e = \frac{\partial u}{\partial x} \quad \text{and} \quad \frac{\partial u}{\partial x} = -\frac{\partial u}{\partial s_1} + O(\epsilon)$$

it follows from (8.1.64) and (8.1.66) that the first approximation to the strain is given by

$$e = -\epsilon\phi(\xi) + \cdots \tag{8.1.71}$$

where ξ is defined by (8.1.70).

Equations (8.1.69) and (8.1.70) show that the velocity wave propagates such that v remains constant along curves with constant ξ; these curves are characteristics of the nonlinear equation (8.1.45) as shown in the preceding section. Thus each characteristic curve represents a moving wavelet, and information is carried by that wavelet. Thus to the first approximation the solution to our problem is the same as the linearized solution with the linearized characteristic being replaced by the characteristic calculated by including the first-order nonlinear term (Whitham, 1952).

Since the speed of propagation along the characteristic curve ξ = constant is given by

$$\left(\frac{dt}{dx}\right)^{-1} \approx 1 - E_1 e \tag{8.1.72}$$

from (8.1.70) and (8.1.71), different values of e propagate with different velocities. Hence any finite-amplitude wave is distorted by the nonlinearity as it propagates along the bar. To show this distortion quantitatively, we consider next an initially sinusoidal wave.

When $F(t) = \cos \omega t$, the first approximations to the displacement and strain become

$$\begin{aligned} \hat{u} &= \frac{u}{\epsilon} = \cos \hat{\xi} + \tfrac{1}{2}\hat{\epsilon}\hat{x} \sin^2 \hat{\xi} \\ \hat{e} &= \frac{e}{\epsilon\omega} = \sin \hat{\xi} \end{aligned} \tag{8.1.73}$$

where

$$\begin{gathered} \hat{t} - \hat{x} = \hat{\xi} - \hat{\epsilon}\hat{x} \sin \hat{\xi} \\ \hat{x} = \omega x, \quad \hat{t} = \omega t, \quad \hat{\epsilon} = \epsilon\omega E_1, \quad \hat{\xi} = \omega\xi \end{gathered} \tag{8.1.74}$$

Figure 8-7 shows the development of an initially sinusoidal velocity wave as it propagates along a nonlinear elastic bar. It shows the typical nonlinear distortion of the wave as it propagates. Figure 8-8 shows that when the nonlinear elastic coefficient E_1 is negative (i.e., $\hat{\epsilon} < 0$), the distortion has the opposite tendency to that shown in Figure 8-7. Figure 8-9 shows the nonlinear distortion of the displacement.

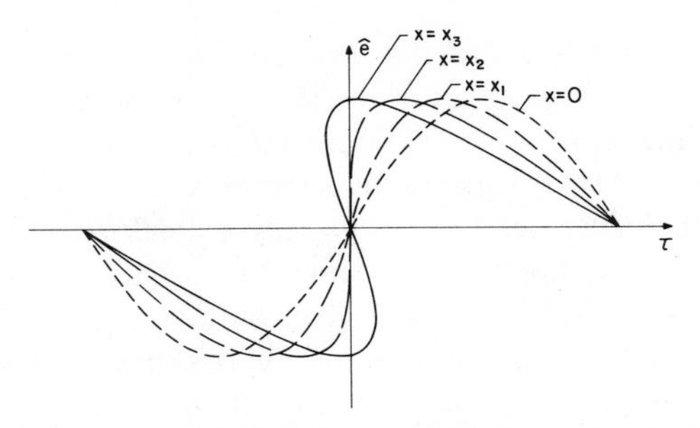

Figure 8-7. Sketch showing the steepening of an initially sinusoidal strain wave for $E_1 > 0$.

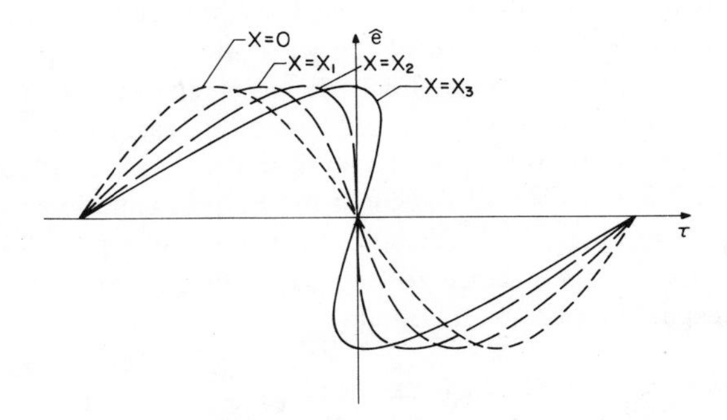

Figure 8-8. Sketch showing the steepening of an initially sinusoidal strain wave for $E_1 < 0$.

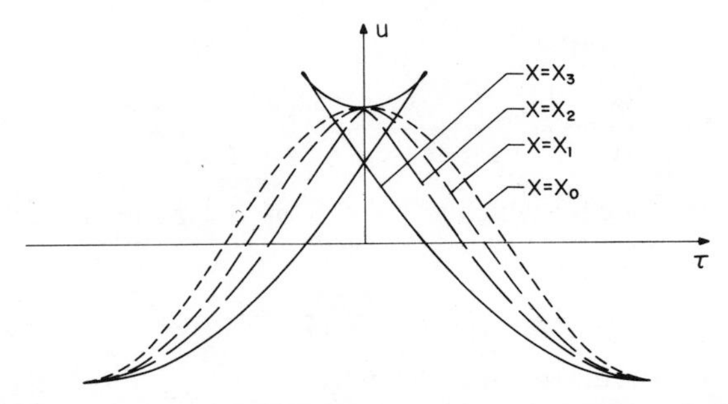

Figure 8-9. Sketch showing the nonlinear distortion of the displacement of an initially sinusoidal wave.

Figures 8-7 through 8-9 show that the continuous-wave solution given by (8.1.69) through (8.1.71) becomes multivalued at some distance $x > x_B$ due to the steepening of the waveform. Therefore the continuous wave solution is not valid beyond that distance because velocity or strain disturbances cannot be multivalued, either in space or time. Physically a shock wave begins to form and viscous and heat conduction effects cannot be neglected. This constitutes a major limitation of the lossless model. To continue the solution further, one includes viscous effects in the analysis and approximates the wave propagation by a *Burgers' equation* (Burgers, 1948). Alternatively one can represent the solution to the physical problem by a single-valued solution consisting of the continuous wave solution (8.1.69) through (8.1.71) together with discontinuous shocks (e.g., Blackstock, 1966; Whitham, 1974, Sections 2.3 and 2.4). However the representation of the solution in the shock-free regions by a simple right-running wave demands the shocks to be weak so that the reflected waves can be neglected. Since the representation of the wave motion by a continuous waveform together with discontinuous shocks is easier analytically than the representation of the wave motion by a Burgers' equation, especially for complicated problems, we limit our discussion in this book to "weak shock theory" and refer the reader to Whitham (1974) for details of the Burgers equation approach.

The breaking distance x_B is the distance x at which the profile of v or e first develops an infinite slope. Hence to determine this distance we differentiate (8.1.69) with respect to x and t and obtain

$$\frac{\partial v}{\partial x} = \epsilon\phi' \frac{\partial \xi}{\partial x}, \quad \frac{\partial v}{\partial t} = \epsilon\phi' \frac{\partial \xi}{\partial t} \tag{8.1.75}$$

Differentiating (8.1.70) with respect to x and t yields

$$\frac{\partial \xi}{\partial x} = -\frac{1 + \epsilon E_1 \phi}{1 + \epsilon E_1 x\phi'}, \quad \frac{\partial \xi}{\partial t} = \frac{1}{1 + \epsilon E_1 x\phi'} \tag{8.1.76}$$

Combining (8.1.75) and (8.1.76) we have

$$\frac{\partial v}{\partial x} = -\epsilon\phi' \frac{1 + \epsilon E_1 \phi}{1 + \epsilon E_1 x\phi'}, \quad \frac{\partial v}{\partial t} = \frac{\epsilon\phi'}{1 + \epsilon E_1 x\phi'} \tag{8.1.77}$$

Since any characteristic for which $\epsilon E_1 \phi' < 0$, $\partial v/\partial x$ and $\partial v/\partial t$ become infinite at

$$x = -\frac{1}{\epsilon E_1 \phi'}$$

the breaking first occurs on the characteristic $\xi = \xi_B$ for which $\epsilon E_1 \phi'(\xi) < 0$ and $|\phi'(\xi)|$ is a maximum. Therefore the breaking distance is given by

$$x_B = \frac{1}{|\epsilon E_1| \phi'_m} \tag{8.1.78}$$

where ϕ'_m is the maximum of $|\phi'|$.

In the preceding analysis the material is assumed to be elastic and the energy can be dissipated only by shocks. In the presence of linear viscous damping the wave propagation is governed by

$$\frac{\partial^2 u}{\partial x^2} - \frac{\partial^2 u}{\partial t^2} = -2E_1 \frac{\partial u}{\partial x} \frac{\partial^2 u}{\partial x^2} + 2\epsilon\mu \frac{\partial u}{\partial t} \tag{8.1.79}$$

where $2\epsilon\mu$ is a constant dimensionless damping coefficient. Following a procedure similar to the one above, we find that (Nayfeh, 1975a)

$$\begin{aligned} v &= \epsilon \exp(-\epsilon\mu x)\phi(\xi) \\ t - x &= \xi + E_1\mu^{-1}[1 - \exp(-\epsilon\mu x)]\,\phi(\xi) \end{aligned} \tag{8.1.80}$$

which was first obtained by Varley and Rogers (1967). As in the inviscid case, the solution (8.1.80) becomes multivalued and develops a shock at a finite distance x_B due to the steepening of the waveform, unless the viscous coefficient μ equals or exceeds a critical value. To determine x_B and the critical viscous coefficient, we differentiate (8.1.80) with respect to either x or t and find the minimum distance for which either $\partial v/\partial x$ or $\partial v/\partial t$ is infinite. The result is

$$\frac{\mu}{|E_1|\phi'_m} = 1 - \exp(-\epsilon\mu x_B)$$

Hence

$$x_B = -\frac{\ln(1 - \mu/|E_1|\phi'_m)}{\epsilon\mu} \tag{8.1.81}$$

Therefore a shock forms at a finite distance x_B unless μ equals or exceeds the critical value

$$\mu_c = |E_1|\phi'_m \tag{8.1.82}$$

Thus although the viscous dissipation may be small, it has a cumulative effect of attenuating the wave which can prevent shock formation when $\mu \geqslant \mu_c$.

This cumulative effect is not limited to viscous damping; it can be caused by all other forms of energy loss such as dissipation. Most of the studies that account for the effect of internal dissipation consider the transmitting material to be predominantly linearly elastic with a small amount of nonlinear elasticity and a small amount of linear viscoelasticity. A number of authors, including Varley and Cumberbatch (1966), Varley and Rogers (1967), and Seymour and Varley (1970), studied the high-frequency limit in which the constitutive equation is approximated by

$$\frac{\partial \sigma}{\partial t} = E(1 + 2E_1 e)\frac{\partial e}{\partial t} - \tau e \tag{8.1.83}$$

where τ is the ratio of the input to the attenuation time of the medium. Nariboli and Sedov (1970) and Nariboli and Lin (1973) studied the low-frequency limit in which the constitutive equation of the material is

$$\sigma = E(e + E_1 e^2) + 2\mu \frac{\partial e}{\partial t} \tag{8.1.84}$$

which is the constitutive equation for a Voigt material. We note that low-frequency damping structures the shock but does not prevent its formation. The magnitude of μ determines the thickness of the shock (see Exercise 8.8).

8.1.3. SHOCK FITTING BY USING VELOCITY

In this section we represent an approximate solution to the problem of wave propagation along an infinite bar by a continuous right-running wave solution together with discontinuous shocks. The representation of the continuous wave solution by a simple (right-running) wave demands the shocks to be weak so that reflected waves are negligible.

Shocks can be fitted into the continuous-wave solution (8.1.69) through (8.1.71) by two alternate ways. One of these ways utilizes the shock velocity (e.g., Whitham, 1974; Blackstock 1966). If the continuous-wave solution gives birth to a shock at time t_B at a distance x_B from the origin, the time t_s of arrival of the shock at a distance x from the origin is

$$t_s = t_B + \int_{x_B}^{x} v_s^{-1}(x)\, dx \tag{8.1.85}$$

where $v_s(x)$ is the shock speed. But for a weak shock (Whitham, 1974, Section 2.5),

$$v_s = \tfrac{1}{2}(c_a + c_b) \tag{8.1.86}$$

where c_a and c_b are the speeds of propagation ahead and behind the shock, respectively. But the speeds of propagation can be obtained from (8.1.70) as

$$c_a = 1 - \epsilon E_1 \phi(\xi_a), \qquad c_b = 1 - \epsilon E_1 \phi(\xi_b) \tag{8.1.87}$$

where ξ_a and ξ_b are the values of the characteristics ahead and behind the shock, respectively. Hence

$$v_s = 1 - \tfrac{1}{2}\epsilon E_1 [\phi(\xi_a) + \phi(\xi_b)] \tag{8.1.88}$$

and

$$v_s^{-1} = 1 + \tfrac{1}{2}\epsilon E_1 [\phi(\xi_a) + \phi(\xi_b)] = 1 + \tfrac{1}{2} E_1 (v_a + v_b) \tag{8.1.89}$$

In terms of the retarded time $s_{1s} = t_s - x$, (8.1.89) can be rewritten as

$$\frac{ds_{1s}}{dx} = \frac{1}{v_s(x)} - 1 = \tfrac{1}{2} E_1 (v_a + v_b) \tag{8.1.90}$$

To determine the path and strength of each shock, we follow Blackstock (1966) and combine (8.1.69) and (8.1.70) in principle to obtain

$$s_1 = \phi^{-1}\left(\frac{v}{\epsilon}\right) + E_1 x v \tag{8.1.91}$$

where ϕ^{-1} stands for the inverse function. Applying (8.1.91) just ahead and just behind the shock gives

$$\begin{aligned} s_{1s} &= \phi^{-1}\left(\frac{v_a}{\epsilon}\right) + E_1 x v_a \\ s_{1s} &= \phi^{-1}\left(\frac{v_b}{\epsilon}\right) + E_1 x v_b \end{aligned} \tag{8.1.92}$$

Equations (8.1.90) and (8.1.92) can be solved simultaneously for the three unknowns s_{1s}, v_a, and v_b.

Therefore (8.1.90) through (8.1.92) are sufficient for the determination of the entire wave motion. Equation (8.1.91) determines the waveform in the shock-free regions, while (8.1.90) and (8.1.92) determine the strength and path of each shock. However explicit solutions of these equations exist only for special cases (Varley and Rogers, 1967; Seymour and Varley, 1970; Whitham, 1974). For general waveforms they need to be solved numerically.

One of the special cases for which an explicit solution can be obtained is that of the sawtooth waveform shown in Figure 8-10. Mathematically the boundary condition for the velocity can be expressed as

$$v(0, t) = -\frac{v_0}{T}(t - 2nT), \qquad n = 0, \pm 1, \pm 2, \ldots \tag{8.1.93}$$

where v_0 is the amplitude of the disturbance and $2T$ is the period. Many investigators have shown that nonlinear waves tend to approach a stable waveform of sawtooth shape with rounded peaks. The thicknesses of the shocks relative to the wavelength and the amount of rounding depend on the amplitude of the waves and the strength of the dissipative mechanisms. For high amplitudes and a small amount of dissipation, the waveform is closely approximated by a

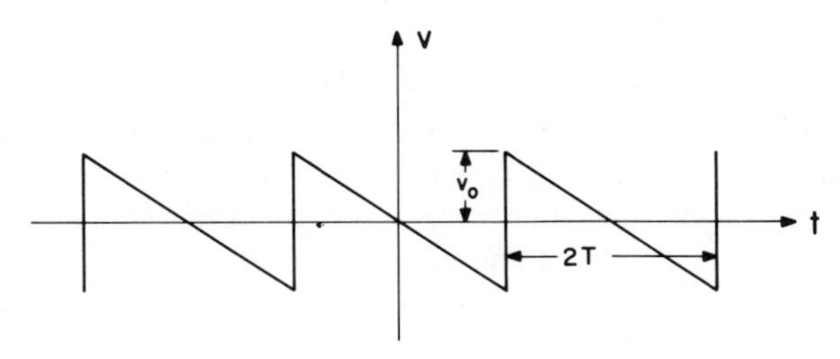

Figure 8-10. Sawtooth waveform.

sharp sawtooth form with shocks of zero thickness similar to the one shown in Figure 8-10.

Using the boundary condition (8.1.93) in (8.1.69) and using the expression (8.1.70) for ξ, we find that

$$v(x, t) = -\frac{v_0}{T}(\xi - 2nT) \tag{8.1.94}$$

Hence

$$\phi^{-1}\left(\frac{v_b}{\epsilon}\right) = \xi_b = -\frac{v_b T}{v_0} + 2nT \tag{8.1.95}$$

Eliminating $\phi^{-1}(v_b/\epsilon)$ from (8.1.92) and (8.1.95) and recalling that $s_{1s} = (2n - 1)T$ we have

$$v_b = \frac{v_0 T}{T - E_1 v_0 x} \tag{8.1.96}$$

As $x \to \infty$,

$$v_b \to -\frac{T}{E_1 x} \tag{8.1.97}$$

Whitham (1952) showed that any periodic disturbance decays according to (8.1.97) at large distances (but not so large that dissipation over the entire waveform dominates the decay). Thus the main effect of a shock is to attenuate the amplitude of the wave. Once a shock forms, it begins to move faster than the wavelets ahead of it, and hence the wavelets are caught by, and coalesce into, the shock. On the other hand, the wavelets behind the shock move faster than it, and hence they catch up with, and coalesce into, the shock.

For general waveforms, Pestorius and Blackstock (1973) devised an algorithm for the solution of (8.1.90) through (8.1.92) by casting them in the form of difference equations. The wave is allowed to propagate a small distance Δx, and the original time scale is incremented by $E_1 v \, \Delta x$ because differentiating (8.1.70) with respect to x at constant ξ yields

$$\frac{ds_1}{dx} = E_1 v \tag{8.1.98}$$

The new scale is scanned for multivaluedness. If none is found, the wave is allowed to propagate by another small distance Δx. The process is repeated until multivaluedness is found, signifying the presence of shock waves. The locations of the shocks are then determined by using (8.1.90) and (8.1.92). Their computer code monitors the motions of the shocks and accounts for their merging. The results of this algorithm for sound waves are in full agreement with experimental results.

8.1.4. SHOCK FITTING BY USING DISPLACEMENT

The other technique of fitting shocks into the continuous-wave solution utilizes the fact that the displacement is continuous across the shocks (Nayfeh and Kluwick, 1976). For other wave problems one can utilize the fact that the potential function or the stream function is continuous across the shocks. Thus for the problem of waves propagating along a bar, the shock waves are given by

$$u(\xi_a, x) = u(\xi_b, x), \qquad t(\xi_a, x) = t(\xi_b, x) \tag{8.1.99}$$

Substituting for u and t from (8.1.66) and (8.1.70) into (8.1.99) we have

$$F(\xi_b) - F(\xi_a) = \tfrac{1}{2}\epsilon E_1 x\,[F'^2(\xi_a) - F'^2(\xi_b)] \tag{8.1.100}$$

$$s_{1s} = \xi_a + \epsilon E_1 x F'(\xi_a) \tag{8.1.101}$$

$$s_{1s} = \xi_b + \epsilon E_1 x F'(\xi_b) \tag{8.1.102}$$

Equations (8.1.100) through (8.1.102) constitute three equations for the three unknowns s_{1s}, ξ_a, and ξ_b. Therefore (8.1.66) and (8.1.70) determine the solution in the shock-free regions, while (8.1.100) through (8.1.102) determine the strength and path of each shock.

Eliminating s_{1s} from (8.1.101) and (8.1.102) yields

$$\xi_b - \xi_a = \epsilon E_1 x\,[F'(\xi_a) - F'(\xi_b)] \tag{8.1.103}$$

Hence (8.1.100) and (8.1.103) constitute a system of two equations for the unknowns ξ_a and ξ_b. Eliminating x from (8.1.100) and (8.1.103) gives

$$F(\xi_b) - F(\xi_a) = \tfrac{1}{2}(\xi_b - \xi_a)[F'(\xi_b) + F'(\xi_a)] \tag{8.1.104}$$

which is essentially the equal area rule (Landau, 1965; Whitham, 1974, Section 2.8). That is, the multivalued parts of the continuous-wave profile are replaced by a discontinuity which cuts off lobes of equal area as shown in Figure 8-11.

Since $\epsilon f \approx u$, we can rewrite the solvability condition (8.1.59) in the form

$$\frac{\partial}{\partial s_1}\left[\frac{\partial}{\partial x_1} + \tfrac{1}{2}E_1\frac{\partial u}{\partial s_1}\frac{\partial}{\partial s_1}\right]u = 0 \tag{8.1.105}$$

Equation (8.1.105) shows the interesting property of equidisplacement lines. The slope of the line u = constant is the arithmetic mean of the slope of the

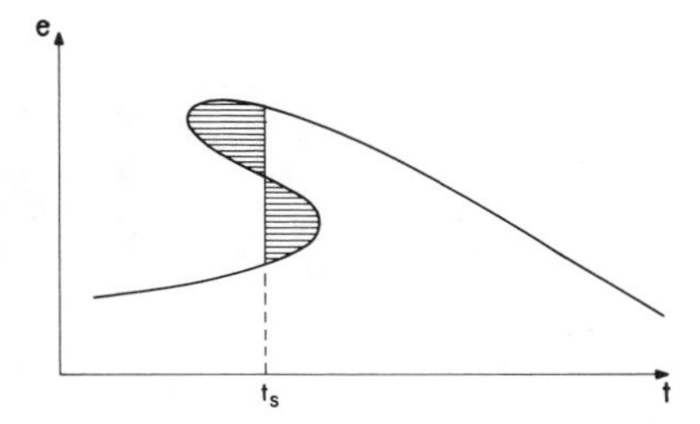

Figure 8-11. Equal-area rule.

characteristic in the unperturbed bar and the slope of the characteristic at the position under consideration. For a shock running into the unperturbed bar, this is one form of the bisector rule which states that in the tx-plane the shock curve bisects the angle between the characteristics that meet on the shock.

8.1.5. EFFECT OF HETEROGENEITY

In this section we consider the propagation of simple right-running waves along a nonuniform bar. The simple-wave assumption demands the reflections to be small and hence the heterogeneity to vary slowly with position. Thus we consider the solutions of (8.1.4), (8.1.46), and (8.1.47) with $c_0(x)$, $E_1(x)$, $E(x)$, and $A(x)$ being slowly varying functions of x. Nonlinear wave propagation in media with slowly varying properties was studied by Seymour and Varley (1967), Butler and Gribben (1968), Asano and Taniuti (1969, 1970), Asano (1970), Varley and Cumberbatch (1970), Kakutani (1971a, b), and Nayfeh (1975a, b). Considerable work has been done on nonlinear waves propagating in a stratified atmosphere. For example, Randall (1968) and Hayes and Runyan (1970) studied the problem of sonic boom, while Einaudi (1969), Romanova (1970), and Bois (1976) studied finite-amplitude waves in an isothermal atmosphere. Varley, Venkataraman, and Cumberbatch (1971) studied the propagation of large-amplitude tsunamis across a basin of changing depth. Seymour and Mortell (1975b) and Mortell and Seymour (1976) studied the problem of wave propagation in nonlinear laminated materials. For more Russian literature see Romanova (1970).

To determine an approximate solution to (8.1.4), (8.1.46), and (8.1.47) following Nayfeh (1975a), we express the slow variation of the bar properties by assuming that they are functions of the slow scale x_1 and seek an expansion by using the method of multiple scales in the form

$$u(x, t; \epsilon) = \epsilon u_1(s_1, s_2, x_1) + \epsilon^2 u_2(s_1, s_2, x_1) + \cdots \tag{8.1.106}$$

where

$$s_1 = t - \int \frac{dx}{c_0}, \qquad s_2 = t + \int \frac{dx}{c_0}, \qquad x_1 = \epsilon x \tag{8.1.107}$$

In terms of these new variables, the space and time derivatives become

$$\frac{\partial}{\partial t} = \frac{\partial}{\partial s_2} + \frac{\partial}{\partial s_1} \tag{8.1.108}$$

$$\frac{\partial^2}{\partial t^2} = \left(\frac{\partial}{\partial s_2} + \frac{\partial}{\partial s_1}\right)^2 \tag{8.1.109}$$

$$\frac{\partial}{\partial x} = \frac{1}{c_0}\left(\frac{\partial}{\partial s_2} - \frac{\partial}{\partial s_1}\right) + \epsilon \frac{\partial}{\partial x_1} \tag{8.1.110}$$

$$\frac{\partial^2}{\partial x^2} = \frac{1}{c_0^2}\left(\frac{\partial}{\partial s} - \frac{\partial}{\partial s_1}\right)^2 + \frac{2\epsilon}{c_0}\frac{\partial}{\partial x_1}\left(\frac{\partial}{\partial s_2} - \frac{\partial}{\partial s_1}\right) - \frac{\epsilon c_0'}{c_0^2}\left(\frac{\partial}{\partial s_2} - \frac{\partial}{\partial s_1}\right) + \epsilon^2 \frac{\partial^2}{\partial x_1^2} \tag{8.1.111}$$

Substituting (8.1.106) into (8.1.4), using (8.1.107) through (8.1.111), and equating coefficients of like powers of ϵ, we obtain

$$\frac{\partial^2 u_1}{\partial s_1 \partial s_2} = 0 \tag{8.1.112}$$

$$\frac{4}{c_0^2}\frac{\partial^2 u_2}{\partial s_1 \partial s_2} = \frac{2}{c_0}\frac{\partial}{\partial x_1}\left(\frac{\partial u_1}{\partial s_2} - \frac{\partial u_1}{\partial s_1}\right) - \frac{c_0'}{c_0^2}\left(\frac{\partial u_1}{\partial s_2} - \frac{\partial u_1}{\partial s_1}\right) + \frac{1}{c_0}(\ln EA)'\left(\frac{\partial u_1}{\partial s_2} - \frac{\partial u_1}{\partial s_1}\right) + \frac{2E_1}{c_0^3}\left(\frac{\partial u_1}{\partial s_2} - \frac{\partial u_1}{\partial s_1}\right)\left(\frac{\partial}{\partial s_2} - \frac{\partial}{\partial s_1}\right)^2 u_1 \tag{8.1.113}$$

The right-running wave solution of (8.1.112) has the form

$$u_1 = f(s_1, x_1) \tag{8.1.114}$$

Then (8.1.113) becomes

$$\frac{4}{c_0^2}\frac{\partial^2 u_2}{\partial s_1 \partial s_2} = -\frac{2}{c_0}\frac{\partial^2 f}{\partial x_1 \partial s_1} + \frac{c_0'}{c_0^2}\frac{\partial f}{\partial s_1} - \frac{1}{c_0}(\ln EA)'\frac{\partial f}{\partial s_1} - \frac{2E_1}{c_0^3}\frac{\partial f}{\partial s_1}\frac{\partial^2 f}{\partial s_1^2} \tag{8.1.115}$$

In order that u_2/u_1 be bounded for all s_2, the right-hand side of (8.1.115) must vanish. That is,

$$\frac{\partial^2 f}{\partial x_1 \partial s_1} - \frac{c_0'}{2c_0}\frac{\partial f}{\partial s_1} + \tfrac{1}{2}(\ln EA)'\frac{\partial f}{\partial s_1} + \frac{E_1}{c_0^2}\frac{\partial f}{\partial s_1}\frac{\partial^2 f}{\partial s_1^2} = 0 \tag{8.1.116}$$

To determine the solution of (8.1.116), we first transform it into an equation with constant coefficients. To this end we let

$$f = Q(x_1)F(s_1, z), \qquad z = Z(x_1) \tag{8.1.117}$$

in (8.1.116) and obtain

$$QZ'\left[\frac{\partial^2 F}{\partial z\, \partial s_1} + \frac{E_1 Q}{c_0^2 Z'}\frac{\partial F}{\partial s_1}\frac{\partial^2 F}{\partial s_1^2}\right] + \left[Q' - \frac{c_0'}{2c_0}Q + \tfrac{1}{2}Q(\ln EA)'\right]\frac{\partial F}{\partial s_1} = 0 \tag{8.1.118}$$

To render (8.1.118) an equation with constant coefficients, we choose Q and Z such that

$$Q' - \frac{c_0'}{2c_0}Q + \tfrac{1}{2}Q(\ln EA)' = 0 \qquad Z' = E_1 c_0^{-2} Q \tag{8.1.119}$$

so that (8.1.118) becomes

$$\frac{\partial^2 F}{\partial z\,\partial s_1} + \frac{\partial F}{\partial s_1}\frac{\partial^2 F}{\partial s_1^2} = 0 \tag{8.1.120}$$

Equations (8.1.119) are satisfied by

$$Q = (c_0/EA)^{1/2}$$
$$Z = \int E_1 c_0^{-1} (c_0 EA)^{-1/2}\, dx_1 \tag{8.1.121}$$

while from Section 8.1.2 it follows that (8.1.120) has the exact solution

$$F = h(s_1 - zh') + \tfrac{1}{2} zh'^2 (s_1 - zh') \tag{8.1.122}$$

Using the boundary condition (8.1.47) one finds that $h' = \phi$ and hence that

$$u = \epsilon \sqrt{\frac{c_0}{EA}}\, h(\xi) + \tfrac{1}{2}\epsilon^2 \left[\int_0^x \frac{E_1}{c_0\sqrt{c_0 EA}}\, dx\right] \sqrt{\frac{c_0}{EA}}\, \phi^2(\xi) + \cdots \tag{8.1.123}$$

$$v = \epsilon \sqrt{\frac{c_0}{EA}}\, \phi(\xi) + \cdots \tag{8.1.124}$$

where the reference quantities are taken at $x = 0$ and

$$s_1 = \xi + \epsilon\phi(\xi) \int_0^x \frac{E_1}{c_0\sqrt{c_0 EA}}\, dx + \cdots \tag{8.1.125}$$

which is the same as the nonlinear geometrical acoustic result (Seymour and Mortell, 1975). As in the case of uniform bars, the velocity given by (8.1.124) along nonuniform bars becomes multivalued at some distance x_B due to the steepening of the waveform. The distance x_B corresponds to the distance at which $\partial v/\partial x$ is infinite, and hence it corresponds to the distance at which $\partial\xi/\partial x$ first becomes infinite. To determine this distance we differentiate (8.1.125) with respect to x and obtain

$$\frac{\partial \xi}{\partial x} = -\frac{\dfrac{1}{c_0} + \epsilon \dfrac{E_1}{c_0\sqrt{c_0 EA}}\phi}{1 + \epsilon\phi' \displaystyle\int_0^x \frac{E_1}{c_0\sqrt{c_0 EA}}\, dx}$$

Hence the breaking first occurs on the characteristic $\xi = \xi_B$ for which $\epsilon\phi' E_1 < 0$ and $|\phi'(\xi)|$ is a maximum. Therefore the breaking distance is given by

$$1 + \epsilon\phi_m' \int_0^{x_B} \frac{E_1}{c_0\sqrt{c_0 EA}} dx = 0 \tag{8.1.126}$$

where ϕ_m' is the value of $|\phi'|$ on the characteristic ξ_B on which $|\phi'|$ is a maximum. Equation (8.1.126) shows that decreasing the speed of sound c_0 in the direction of propagation leads to a smaller x_B and hence an earlier formation of the shocks due to the increase in the rate of progress of finite-amplitude distortion. Equation (8.1.126) shows also that similar effects take place if the cross-sectional area of the rod or Young's modulus decreases in the direction of propagation.

8.1.6. FINITE UNIFORM BARS

In the last four sections we discussed simple, right-running waves along a semi-infinite bar. In this section we discuss the nonlinear behavior of waves in finite, uniform bars. In this case the key technical problem is the interaction of the right- and left-running waves as the Riemann invariants (8.1.16) show.

When a right-running wave impinges on an end of a bar, part of it is radiated and the other part is reflected as a left-running wave. The ratio of the reflected part to the radiated part depends on the boundary conditions. For example, at a fixed end the wave is totally reflected. Hence the analysis of waves in finite bars demands the inclusion of both right- and left-running waves. It should be noted that even in semiinfinite systems more than one type of progressive wave might travel in the same direction and nonlinearly interact. Collins (1967), Davison (1968), and Nair and Nemat-Nasser (1971) studied the interaction of longitudinal and shear waves; Parker and Varley (1968) studied the interaction between simple stretching and deflection waves in elastic membranes and strings; and Advani (1969b) studied the interaction of flexural and longitudinal waves in an infinite plate.

Oppositely traveling waves in a finite bar were studied by Mortell and Varley (1970), Mortell and Seymour (1972, 1973b), and Cekirge and Varley (1973) by using an approximation based on the analytic method of characteristics; Lardner (1975, 1976), by using the method of averaging; and Mortell (1977) and Nayfeh (1977), by using the method of multiple scales. In the method of averaging, the solution is expressed in terms of all the modes in an eigenfunction expansion, in contrast with the generalized version of the method of multiple scales, which treats the waveform. It should be noted that application of the derivative-expansion version of the method of multiple scales leads to an infinite set of ordinary-differential equations describing the evolution of the amplitudes of the harmonics of the initial disturbance (Keller and Kogelman, 1970). Since the generalized version of the method of multiple scales leads to two uncoupled first-order partial-differential equations, it is better suited to this problem than either the method of averaging or the derivative-expansion version of the method of multiple scales. Therefore we restrict our discussion to the generalized method of multiple scales.

In the closely related problem of periodic oscillations produced in a closed-end gas-filled tube by the motion of a piston at one end, Saenger and Hudson (1960) found experimentally that in a narrow frequency band around each resonant frequency shock waves form, traveling backward and forward and being repeatedly reflected from the closed end. Moreover near the fundamental frequency one shock forms, near the second frequency two shocks appear, and near the nth frequency n shocks form. This problem has been studied analytically by Betchov (1958), Chester (1964), Collins (1971), Seymour and Mortell (1973b), and Keller (1975, 1976). Sturtevant (1973, 1974) studied shock focusing and nonlinear resonances in a shock tube. Van Wijngaarden (1968), Collins (1971), Mortell (1971a), Mortell and Seymour (1973a), Seymour and Mortell (1973a), Jimenez (1973), and Keller (1977a, b, c) studied periodic oscillations near resonance produced by a piston in an open pipe. Chu and Ying (1963) and Rehm (1968) analyzed thermally driven nonlinear oscillations in a pipe, while Chu (1963) and Mortell (1971b) studied self-sustained thermally driven oscillations in a gas-filled pipe.

If a semiinfinite bar is disturbed at its left end, then the disturbance propagates to the right, and it is nonlinearly distorted as it travels. On the other hand, when a finite bar is given an initial disturbance, a standing wave results which is nonlinearly distorted with time. To allow the investigation of the nonlinear distortion with either space or time, we introduce a slow time scale $T_1 = \epsilon t$ and a long scale $x_1 = \epsilon x$. Moreover we seek an expansion in the form

$$u(x, t; \epsilon) = \epsilon u_1(s_1, s_2, x_1, T_1) + \epsilon^2 u_2(s_1, s_2, x_1, T_1) + \cdots$$
$$s_1 = t - x, \qquad s_2 = t + x \tag{8.1.127}$$

The space and time derivatives are transformed according to

$$\frac{\partial}{\partial x} = \frac{\partial}{\partial s_2} - \frac{\partial}{\partial s_1} + \epsilon \frac{\partial}{\partial x_1}$$
$$\frac{\partial}{\partial t} = \frac{\partial}{\partial s_2} + \frac{\partial}{\partial s_1} + \epsilon \frac{\partial}{\partial T_1} \tag{8.1.128}$$

Substituting (8.1.127) and (8.1.128) into (8.1.45) and equating the coefficients of ϵ and ϵ^2 yields

$$\frac{\partial^2 u_1}{\partial s_1 \partial s_2} = 0 \tag{8.1.129}$$

$$4\frac{\partial^2 u_2}{\partial s_1 \partial s_2} = 2\frac{\partial}{\partial x_1}\left(\frac{\partial u_1}{\partial s_2} - \frac{\partial u_1}{\partial s_1}\right) - 2\frac{\partial}{\partial T_1}\left(\frac{\partial u_1}{\partial s_2} + \frac{\partial u_1}{\partial s_1}\right) + 2E_1\left(\frac{\partial u_1}{\partial s_2} - \frac{\partial u_1}{\partial s_1}\right)\left(\frac{\partial}{\partial s_2} - \frac{\partial}{\partial s_1}\right)^2 u_1 \quad (8.1.130)$$

The general solution of (8.1.129) is

$$u_1 = f(s_1, x_1, T_1) + g(s_2, x_1, T_1) \quad (8.1.131)$$

where the equations describing f and g are obtained by imposing the solvability condition at the next level of approximation.

Substituting for u_1 in (8.1.130) yields

$$4\frac{\partial^2 u_2}{\partial s_1 \partial s_2} = -2\frac{\partial^2 f}{\partial s_1 \partial T_1} - 2\frac{\partial^2 f}{\partial s_1 \partial x_1} - 2\frac{\partial^2 g}{\partial s_2 \partial T_1} + 2\frac{\partial^2 g}{\partial s_2 \partial x_1} - 2E_1\left[\frac{\partial f}{\partial s_1}\frac{\partial^2 f}{\partial s_1^2} + \frac{\partial f}{\partial s_1}\frac{\partial^2 g}{\partial s_2^2} - \frac{\partial g}{\partial s_2}\frac{\partial^2 g}{\partial s_2^2} - \frac{\partial g}{\partial s_2}\frac{\partial^2 f}{\partial s_1^2}\right] \quad (8.1.132)$$

The particular solution of (8.1.132) is

$$2u_2 = -\left[\frac{\partial f}{\partial T_1} + \frac{\partial f}{\partial x_1} + \tfrac{1}{2}E_1\left(\frac{\partial f}{\partial s_1}\right)^2\right]s_2 - \left[\frac{\partial g}{\partial T_1} - \frac{\partial g}{\partial x_1} - \tfrac{1}{2}E_1\left(\frac{\partial g}{\partial s_2}\right)^2\right]s_1 - E_1 f\frac{\partial g}{\partial s_2} + E_1 g\frac{\partial f}{\partial s_1} \quad (8.1.133)$$

Thus if f and g and their first derivatives are bounded, the terms proportional to s_1 and s_2 are secular terms. For a uniformly valid expansion these terms must vanish. That is,

$$\frac{\partial f}{\partial T_1} + \frac{\partial f}{\partial x_1} + \tfrac{1}{2}E_1\left(\frac{\partial f}{\partial s_1}\right)^2 = 0$$
$$\frac{\partial g}{\partial T_1} - \frac{\partial g}{\partial x_1} - \tfrac{1}{2}E_1\left(\frac{\partial g}{\partial s_2}\right)^2 = 0 \quad (8.1.134)$$

Inspection of the right-hand side of (8.1.132) shows that secular terms will not appear in the particular solution of u_2 if

$$\frac{\partial^2 f}{\partial s_1 \partial T_1} + \frac{\partial^2 f}{\partial s_1 \partial x_1} + E_1\frac{\partial f}{\partial s_1}\frac{\partial^2 f}{\partial s_1^2} = 0$$
$$\frac{\partial^2 g}{\partial s_2 \partial T_1} - \frac{\partial^2 g}{\partial s_2 \partial x_1} - E_1\frac{\partial g}{\partial s_2}\frac{\partial^2 g}{\partial s_2^2} = 0 \quad (8.1.135)$$

We note that the conditions (8.1.135) are more general than the conditions (8.1.134).

To investigate the nonlinear distortion of waves, we find it convenient to determine the distortion as a function of T_1. Thus we assume that f and g are independent of x_1. Then, as in Section 8.1.2, the solutions of (8.1.135) are

$$\begin{aligned} f(s_1, T_1) &= F(\xi) + \tfrac{1}{2} E_1 T_1 F'^2(\xi) \\ g(s_2, T_1) &= G(\eta) - \tfrac{1}{2} E_1 T_1 G'^2(\eta) \end{aligned} \tag{8.1.136}$$

where

$$\begin{aligned} s_1 &= \xi + E_1 T_1 F'(\xi) \\ s_2 &= \eta - E_1 T_1 G'(\eta) \end{aligned} \tag{8.1.137}$$

Therefore to the first approximation

$$u = \epsilon F(\xi) + \epsilon G(\eta) + \tfrac{1}{2} \epsilon^2 E_1 t F'^2(\xi) - \tfrac{1}{2} \epsilon^2 E_1 t G'^2(\eta) \tag{8.1.138}$$

where

$$\begin{aligned} t - x &= \xi + \epsilon E_1 t F'(\xi) \\ t + x &= \eta - \epsilon E_1 t G'(\eta) \end{aligned} \tag{8.1.139}$$

Since the strain e is $\partial u/\partial x$ and the velocity v is $\partial u/\partial t$, it follows from (8.1.138) and (8.1.139) that

$$\begin{aligned} e &= \epsilon G'(\eta) - \epsilon F'(\xi) \\ v &= \epsilon G'(\eta) + \epsilon F'(\xi) \end{aligned} \tag{8.1.140}$$

If the bar is rigidly fixed at its ends and if it is of length 1, then the boundary conditions are

$$v(0, t) = v(1, t) = 0 \tag{8.1.141}$$

Using the boundary condition $v(0, t) = 0$ in (8.1.140) yields $F'(\xi) = -G'(\eta)$. Hence $x = 0$ corresponds to $\xi = \eta$ according to (8.1.139) and

$$F'(\xi) = -G'(\xi) \tag{8.1.142}$$

When $x = 1$, it follows from (8.1.139) that

$$\begin{aligned} \xi &= -1 + t + \epsilon E_1 t G'(\xi) \\ \eta &= 1 + t + \epsilon E_1 t G'(\eta) \end{aligned} \tag{8.1.143}$$

and hence, using the boundary condition $v(1, t) = 0$ in (8.1.140) yields

$$G'(\xi + 2) = G'(\xi), \qquad \eta = \xi + 2 \quad \text{at } x = 1 \tag{8.1.144}$$

or G' is a periodic function of period 2. Consequently (8.1.140) becomes

$$\begin{aligned} e &= \epsilon G'(\eta) + \epsilon G'(\xi) \\ v &= \epsilon G'(\eta) - \epsilon G'(\xi) \end{aligned} \tag{8.1.145}$$

where G' is a periodic function of period 2 and

$$
\begin{aligned}
t - x &= \xi - \epsilon E_1 t G'(\xi) \\
t + x &= \eta - \epsilon E_1 t G'(\eta)
\end{aligned}
\tag{8.1.146}
$$

This is essentially the solution obtained by Mortell and Varley (1970) on replacing t by x and Lardner (1975) by using the method of averaging.

So far we have not utilized the initial conditions. Let us consider the case

$$
e(x, 0) = \epsilon\psi(x), \qquad v(x, 0) = \epsilon\phi(x) \qquad \text{for } 0 \leqslant x \leqslant 1 \tag{8.1.147}
$$

Since $t = 0$ corresponds to $\xi = -x$ and $\eta = x$ from (8.1.146), (8.1.145) and (8.1.147) give

$$
\begin{aligned}
G'(x) + G'(-x) &= \psi(x) \\
G'(x) - G'(-x) &= \phi(x)
\end{aligned}
\tag{8.1.148}
$$

Hence

$$
\begin{aligned}
G'(x) &= \tfrac{1}{2}\,[\psi(x) + \phi(x)] \\
G'(-x) &= \tfrac{1}{2}\,[\psi(x) - \phi(x)]
\end{aligned}
\tag{8.1.149}
$$

We note that (8.1.139) and (8.1.140) show that the oppositely traveling waves do not interact in the body of the bar. This fact was predicted in gas columns experimentally by Saenger and Hudson (1960) and analytically by Kruskal and Zabusky (1964). However (8.1.142) shows that the two waves interact at the boundary.

As time progresses, the solution (8.1.145) and (8.1.146) becomes multivalued indicating the formation of a shock wave and the present continuous solution ceases to be valid. As in Section 8.1.2, the time at which the shock forms corresponds to the time at which $\partial v/\partial x$ and $\partial v/\partial t$ first become infinite. Thus differentiating (8.1.146) with respect to t yields

$$
\left[\frac{\partial \xi}{\partial t}, \frac{\partial \eta}{\partial t}\right] = \left[\frac{1 + \epsilon E_1 G'(\xi)}{1 - \epsilon E_1 t G''(\xi)}, \frac{1 + \epsilon E_1 G'(\eta)}{1 - \epsilon E_1 t G'(\eta)}\right] \tag{8.1.150}
$$

Therefore the shock forms at $t = t_s$, where

$$
t_s = [\epsilon |E_1| G''_m]^{-1} \tag{8.1.151}
$$

where G''_m is the maximum of $|G''(\xi)|$. Substituting for t_s in (8.1.146) yields the following shock location:

$$
x_s = \pm t_s [1 + \epsilon E_1 G'(\xi_m)] \mp \xi_m \tag{8.1.152}
$$

where ξ_m denotes the value of ξ at which $|G''(\xi)|$ is a maximum. The positive and negative signs apply to the ξ and η families of characteristics, respectively. By suitably adjusting ξ_m by multiples of 2, one can always find an x_s in the interval [0, 1].

The presence of the shock results in the dissipation of the disturbance and hence the nonexistence of finite-amplitude, periodic standing waves in a finite bar, in agreement with the theorem of Lax (1964).

8.1.7. FINITE HETEROGENEOUS BARS

For a bar whose properties vary slowly with position, it is convenient to use the long scale $x_1 = \epsilon x$ to determine the nonlinear distortion of a standing wave because this scale has to be included in the analysis to account for the effects of heterogeneity. Thus we seek an expansion in the form of (8.1.106) and (8.1.107). Substituting the result into (8.1.4) and equating the coefficients of ϵ and ϵ^2 to zero leads to (8.1.112) and (8.1.113)

To analyze standing waves, we need to include both right- and left-running-wave solutions for (8.1.112). Thus instead of (8.1.114), we write the solution of (8.1.112) as

$$u_1 = f(s_1, x_1) + g(s_2, x_1) \tag{8.1.153}$$

Then (8.1.113) becomes

$$\begin{aligned}\frac{4}{c_0^2}\frac{\partial^2 u_2}{\partial s_1 \partial s_2} = &-\frac{2}{c_0}\frac{\partial^2 f}{\partial s_1 \partial x_1} + \frac{c_0'}{c_0^2}\frac{\partial f}{\partial s_1} - \frac{1}{c_0}(\ln EA)'\frac{\partial f}{\partial s_1} - \frac{2E_1}{c_0^3}\frac{\partial f}{\partial s_1}\frac{\partial^2 f}{\partial s_1^2} \\ &+ \frac{2}{c_0}\frac{\partial^2 g}{\partial s_2 \partial x_1} - \frac{c_0'}{c_0^2}\frac{\partial g}{\partial s_2} + \frac{1}{c_0}(\ln EA)'\frac{\partial g}{\partial s_2} + \frac{2E_1}{c_0^3}\frac{\partial g}{\partial s_2}\frac{\partial^2 g}{\partial s_2^2} \\ &- \frac{2E_1}{c_0^3}\frac{\partial f}{\partial s_1}\frac{\partial^2 g}{\partial s_2^2} + \frac{2E_1}{c_0^3}\frac{\partial^2 f}{\partial s_1^2}\frac{\partial g}{\partial s_2}\end{aligned} \tag{8.1.154}$$

If f and g and their first derivatives are bounded, then secular terms will not appear in the particular solution of (8.1.154) if

$$\frac{\partial^2 f}{\partial s_1 \partial x_1} - \frac{c_0'}{2c_0}\frac{\partial f}{\partial s_1} + \tfrac{1}{2}(\ln EA)'\frac{\partial f}{\partial s_1} + \frac{E_1}{c_0^2}\frac{\partial f}{\partial s_1}\frac{\partial^2 f}{\partial s_1^2} = 0 \tag{8.1.155}$$

$$\frac{\partial^2 g}{\partial s_2 \partial x_1} - \frac{c_0'}{2c_0}\frac{\partial g}{\partial s_2} + \tfrac{1}{2}(\ln EA)'\frac{\partial g}{\partial s_2} + \frac{E_1}{c_0^2}\frac{\partial g}{\partial s_2}\frac{\partial^2 g}{\partial s_2^2} = 0 \tag{8.1.156}$$

As in Section 8.1.5, the solutions of (8.1.155) and (8.1.156) are

$$f = Q[F(s_1 - zF') + \tfrac{1}{2}zF'^2(s_1 - zF')] \tag{8.1.157}$$

$$g = Q[G(s_2 - zG') + \tfrac{1}{2}zG'^2(s_2 - zG')] \tag{8.1.158}$$

where Q and z are defined in (8.1.121). Therefore to the first approximation

$$\begin{aligned} e &= \epsilon Q[G'(\eta) - F'(\xi)] \\ v &= \epsilon Q[G'(\eta) + F'(\xi)] \end{aligned} \tag{8.1.159}$$

where

$$\xi = t - \int_0^x \frac{dx}{c_0} - \epsilon z_0 F'(\xi)$$
$$\eta = t + \int_1^x \frac{dx}{c_0} - \epsilon z_1 G'(\eta) \tag{8.1.160}$$

and

$$z_0 = \int_0^x \frac{E_1}{c_0\sqrt{c_0 EA}} dx, \qquad z_1 = \int_1^x \frac{E_1}{c_0\sqrt{c_0 EA}} dx \tag{8.1.161}$$

Note that the characteristic curves have been parameterized so that $\xi = t$ when $x = 0$ and $\eta = t$ when $x = 1$.

Next we consider the case of fixed ends. That is,

$$v(0, t) = v(1, t) = 0 \tag{8.1.162}$$

At $x = 0$, it follows from (8.1.160) that $\xi = t$ and

$$\eta = \xi - P - \epsilon z_1(0) G'(\eta) \tag{8.1.163}$$

where

$$P = \int_0^1 c_0^{-1}\, dx \tag{8.1.164}$$

Since the condition $v(0, t) = 0$ demands that

$$F'(\xi) + G'(\eta) = 0 \qquad \text{at } x = 0 \tag{8.1.165}$$

it follows from (8.1.163) that

$$F'[R + P + \epsilon z_1(0) G'(R)] + G'(R) = 0 \tag{8.1.166}$$

for all $R \geqslant 0$. Similarly at $x = 1$ it follows from (8.1.160) that $\eta = t$ and

$$\xi = \eta - P - \epsilon z_0(1) F'(\xi) \tag{8.1.167}$$

But the condition $v(1, t) = 0$ demands that

$$F'(\xi) + G'(\eta) = 0 \qquad \text{at } x = 1 \tag{8.1.168}$$

Therefore

$$F'(\xi) + G'[\xi + P + \epsilon z_0(1) F'(\xi)] = 0 \tag{8.1.169}$$

for all $\xi \geqslant 0$.

To eliminate G from (8.1.166) and (8.1.169), we let

$$R = \xi + P + \epsilon z_0(1)F'(\xi) \tag{8.1.170}$$

Then (8.1.166) and (8.1.169) become

$$F'[\xi + 2P + \epsilon z_0(1)F'(\xi) - \epsilon z_0(1)G'(R)] + G'(R) = 0 \tag{8.1.171}$$

$$F'(\xi) + G'(R) = 0 \tag{8.1.172}$$

where use has been made of the fact that $z_1(0) = -z_0(1)$. Eliminating $G'(R)$ from (8.1.171) and (8.1.172) yields

$$F'[\xi + \chi(\xi)] = F'(\xi) \tag{8.1.173}$$

where

$$\chi(\xi) = 2[P + \epsilon z_0(1)F'(\xi)] \tag{8.1.174}$$

Similarly eliminating F' from (8.1.166) and (8.1.169) yields

$$G'[\eta + \Gamma(\eta)] = G'(\eta) \tag{8.1.175}$$

where

$$\Gamma(\eta) = 2[P - z_0(1)G'(\eta)] \tag{8.1.176}$$

We note that $F'(\xi)$ is given by the nonlinear functional equation (8.1.173); this is essentially the result in Mortell and Seymour (1972). Although $F'(\xi)$ is not periodic with period $2P$, it repeats at a sequence of times separated by the time interval $\chi = 2[P + \epsilon z_0(1)F'(\xi)]$, and zeros of $F'(\xi)$ are separated by the time interval $2P$.

To solve the functional equation (8.1.73), we follow Mortell and Seymour (1972) and note that according to (8.1.160) and (8.1.161), the $\hat{\xi}$-wavelet which arrives at $x = 1$ at $t = 0$ is given by

$$\hat{\xi} = -P - \epsilon z_0(1)F'(\hat{\xi}) \tag{8.1.177}$$

and it left the end $x = 0$ at the time $t = \frac{1}{2}\chi_0 = -\hat{\xi}$. Moreover this $\hat{\xi}$-wavelet will be reflected from the end $x = 1$ at $t = 0$ as the wavelet $\eta = 0$, and it arrives at the end $x = 0$ at $t = \frac{1}{2}\chi_1$, where

$$\chi_1 = 2P - 2\epsilon z_0(1)G'(0) \tag{8.1.178}$$

Therefore $t = \frac{1}{2}(\chi_1 + \chi_0)$ represents the period of the $\hat{\xi}$-wavelet that arrives at $x = 1$ at $t = 0$. Since $F'(\xi)$ repeats itself at a sequence of times separated by the interval $\chi(\xi)$,

$$\chi_1 + \chi_0 = \chi(\hat{\xi}) = \chi(-\tfrac{1}{2}\chi_0) \tag{8.1.179}$$

Since the end $x = 0$ is fixed, it follows from (8.1.159) that

$$F'(\hat{\xi}) + G'(0) = 0 \tag{8.1.180}$$

and hence from (8.1.177) that

$$\chi_0 = -2\hat{\xi} = 2P - 2\epsilon z_0(1)G'(0) = \chi_1 \tag{8.1.181}$$

Therefore if $F'(\xi)$ is continuous and specified at $x = 0$ during the interval $-\frac{1}{2}\chi_0 \leqslant \xi \leqslant \frac{1}{2}\chi_0$ as

$$F'(\xi) = h(\xi) \qquad \text{for } -\tfrac{1}{2}\chi_0 \leqslant \xi \leqslant \tfrac{1}{2}\chi_0 \tag{8.1.182}$$

then $F'(\xi)$ can be determined for all $\xi \geqslant -\frac{1}{2}\chi_0$ by a repeated application of equations (8.1.173) and (8.1.174). Thus

$$F'(\xi_1) = h(\alpha) \qquad \text{for } \tfrac{1}{2}\chi_0 \leqslant \xi_1 \leqslant \tfrac{3}{2}\chi_0 \tag{8.1.183}$$

where

$$\xi_1 = \alpha + \chi(\alpha) = \alpha + 2P + 2\epsilon z_0(1)h(\alpha) \tag{8.1.184}$$

Moreover

$$F'(\xi_2) = F'(\xi_1) \qquad \text{for } \tfrac{3}{2}\chi_0 \leqslant \xi_2 \leqslant \tfrac{5}{2}\chi_0 \tag{8.1.185}$$

where

$$\xi_2 = \xi_1 + \chi(\xi_1) \tag{8.1.186}$$

It follows from (8.1.183) and (8.1.185) that $F'(\xi_1) = h(\alpha)$ and hence from (8.1.174) that $\chi(\xi_1) = \chi(\alpha)$. Therefore (8.1.185) and (8.1.186) can be rewritten as

$$F'(\xi_2) = h(\alpha) \qquad \text{for } \tfrac{3}{2}\chi_0 \leqslant \xi_2 \leqslant \tfrac{5}{2}\chi_0 \tag{8.1.187}$$

where

$$\xi_2 = \alpha + 2\chi(\alpha) = \alpha + 4[P + \epsilon z_0(1)h(\alpha)] \tag{8.1.188}$$

Repeating the above procedure, we can write the solution of the functional equations (8.1.173) and (8.1.174) as

$$F'(\xi_n) = h(\alpha) \qquad \text{for } (n - \tfrac{1}{2})\chi_0 \leqslant \xi_n \leqslant (n + \tfrac{1}{2})\chi_0 \tag{8.1.189}$$

where ξ_n is related to α by

$$\xi_n = \alpha + 2nP + 2\epsilon n z_0(1)h(\alpha) \tag{8.1.190}$$

Following a similar procedure we can write the solution of the functional equations (8.1.175) and (8.1.176) as

$$G'(\eta_n) = k(\beta) \qquad \text{for } (n - \tfrac{1}{2})\Gamma_0 \leqslant \eta_n \leqslant (n + \tfrac{1}{2})\Gamma_0 \tag{8.1.191}$$

where

$$\eta_n = \beta + 2nP - 2\epsilon n z_0(1)k(\beta) \tag{8.1.192}$$

$$\Gamma_0 = 2P + 2\epsilon z_0(1)F'(0) \tag{8.1.193}$$

Substituting for $F'(\xi)$ and $G'(\eta)$ from (8.1.189) through (8.1.192) into (8.1.159) and (8.1.160), we have

$$\begin{aligned} e &= \epsilon Q\,[k(\beta) - h(\alpha)] \\ v &= \epsilon Q\,[k(\beta) + h(\alpha)] \end{aligned} \tag{8.1.194}$$

where α and β are related implicitly to ξ and η by

$$\xi = \alpha + 2nP + 2\epsilon n z_0(1)h(\alpha) \tag{8.1.195}$$

in the range

$$\begin{gathered} (n - \tfrac{1}{2})\chi_0 \leqslant \xi \leqslant (n + \tfrac{1}{2})\chi_0 \qquad \text{for } n = 0, 1, 2, \ldots \\ \eta = \beta + 2mP - 2\epsilon m z_0(1)k(\beta) \end{gathered} \tag{8.1.196}$$

in the range

$$(m - \tfrac{1}{2})\Gamma_0 \leqslant \eta \leqslant (m + \tfrac{1}{2})\Gamma_0, \qquad m = 0, 1, 2, \ldots$$

To complete the solution we need to relate $h(\alpha)$ and $k(\beta)$ to the initial conditions (8.1.147). It follows from (8.1.147) and (8.1.194) that

$$\begin{aligned} K(x) - H(x) &= Q^{-1}(x)\phi(x) \\ K(x) + H(x) &= Q^{-1}(x)\psi(x) \end{aligned} \tag{8.1.197}$$

Hence

$$\begin{aligned} H(x) &= \tfrac{1}{2}Q^{-1}(x)[\psi(x) - \phi(x)] \\ K(x) &= \tfrac{1}{2}Q^{-1}(x)[\psi(x) + \phi(x)] \end{aligned} \tag{8.1.198}$$

Thus the problem reduces to relating the functions h and k to the functions H and K.

When $t = 0$, it follows from (8.1.160) and (8.1.195) that $n = 0$, and hence $\xi = \alpha$, and α is related to x by

$$\alpha = -\int_0^x \frac{dx}{c_0} - \epsilon z_0(x)h(\alpha) \tag{8.1.199}$$

Moreover the initial condition requires that

$$h(\alpha) = H(x) \qquad \text{for } 0 \leqslant x \leqslant 1 \qquad \text{at } t = 0 \tag{8.1.200}$$

Therefore

$$h(\alpha) = H(-R) \qquad \text{for } -1 \leqslant R \leqslant 0, \qquad -\tfrac{1}{2}\chi_0 \leqslant \alpha \leqslant 0 \tag{8.1.201}$$

where

$$\alpha = \int_0^R \frac{dx}{c_0} + \epsilon z_0(R)H(-R) \tag{8.1.202}$$

When $t = 0$, it follows from (8.1.160) and (8.1.196) that $m = 0$, and hence $\eta = \beta$ and

$$\beta = \int_1^x \frac{dx}{c_0} - \epsilon z_1(x)k(\beta) \tag{8.1.203}$$

The initial condition requires that

$$k(\beta) = K(x) \qquad \text{for } 0 \leqslant x \leqslant 1 \qquad \text{at } t = 0 \tag{8.1.204}$$

Letting $x = S + 1$ in (8.1.204) yields

$$k(\beta) = K(S+1) \qquad \text{for } -1 \leqslant S \leqslant 0, \qquad -\tfrac{1}{2}\Gamma_0 \leqslant \beta \leqslant 0 \tag{8.1.205}$$

where

$$\beta = \int_0^S \frac{dx}{c_0} - \epsilon z_1(S+1)K(S+1) \tag{8.1.206}$$

To determine $h(\alpha)$ for $0 \leqslant \alpha \leqslant \frac{1}{2}\chi_0$, we rewrite (8.1.166) as

$$h[r + P + \epsilon z_1(0)k(r)] + k(r) = 0 \tag{8.1.207}$$

for $r \geqslant 0$. Substituting for k from (8.1.205) into (8.1.207) yields

$$h(\alpha) = -K(R) \qquad \text{for } 0 \leqslant R \leqslant 1, \qquad 0 \leqslant \alpha \leqslant \tfrac{1}{2}\chi_0 \tag{8.1.208}$$

where

$$\alpha = r + P + \epsilon z_1(0)K(R) \tag{8.1.209}$$

and from (8.1.206)

$$r = \int_1^R \frac{dR}{c_0(R)} - \epsilon z_1(R)K(R) \tag{8.1.210}$$

We note that

$$\int_1^R \frac{dR}{c_0(R)} = \int_1^0 \frac{dR}{c_0(R)} + \int_0^R \frac{dR}{c_0(R)} = -P + \int_0^R \frac{dR}{c_0(R)}$$

according to (8.1.164) and that

$$z_1(R) = z_1(0) + z_0(R)$$

according to (8.1.161). Therefore on eliminating r from (8.1.209) and (8.1.210), we have

$$\alpha = \int_0^R \frac{dR}{c_0} - \epsilon z_0(R)K(R) \tag{8.1.211}$$

Following a similar procedure we find that

$$k(\beta) = -H(1 - S), \quad 0 \leqslant S \leqslant 1, \quad 0 \leqslant \beta \leqslant \tfrac{1}{2}\Gamma_0 \tag{8.1.212}$$

where

$$\beta = \int_0^S \frac{dx}{c_0} + \epsilon z_0(S)H(1 - S) \tag{8.1.213}$$

$$z_0(S) = \int_0^S \Lambda(S - 1)\, dS, \quad \Lambda = \frac{E_1}{c_0\sqrt{c_0 EA}}$$

When the bar is uniform, $c_0 = E = A = 1$, E_1 is a constant, and (8.1.159) and (8.1.160) reduce to

$$\begin{aligned} e &= \epsilon G'(\eta) - \epsilon F'(\xi) \\ v &= \epsilon G'(\eta) + \epsilon F'(\xi) \end{aligned} \tag{8.1.214}$$

where

$$\begin{aligned} \xi &= t - x - \epsilon E_1 x F'(\xi) \\ \eta &= t + x - 1 - \epsilon E_1 (x - 1) G'(\eta) \end{aligned} \tag{8.1.215}$$

This is essentially the solution obtained by Mortell and Varley (1970) by using the analytic method of characteristics. If we introduce a new set of characteristic variables α and β according to

$$\begin{aligned} \xi &= \alpha[1 + \epsilon E_1 F'(\xi)] \\ \eta &= (\beta - 1)[1 - \epsilon E_1 G'(\eta)] \end{aligned} \tag{8.1.216}$$

then

$$\begin{aligned} \xi &= \alpha[1 + \epsilon E_1 F'(\alpha)] + \cdots \\ \eta &= (\beta - 1)[1 - \epsilon E_1 G'(\beta - 1)] + \cdots \end{aligned} \tag{8.1.217}$$

Therefore (8.1.214) and (8.1.215) become

$$\begin{aligned} e &= \epsilon \tilde{G}'(\beta) - \epsilon F'(\alpha) + \cdots \\ v &= \epsilon \tilde{G}'(\beta) + \epsilon F'(\alpha) + \cdots \end{aligned} \tag{8.1.218}$$

where

$$\begin{aligned} t - x &= \alpha + \epsilon E_1 t F'(\alpha) + \cdots \\ t + x &= \beta - \epsilon E_1 t \tilde{G}'(\beta) + \cdots \end{aligned} \tag{8.1.219}$$

with $\tilde{G}'(\beta) = G'(\beta - 1)$.

Since (8.1.218) and (8.1.219) are identical in form to (8.1.139) and (8.1.140), then apart from second-order quantities in ϵ, (8.1.218) and (8.1.219) are the same as the solution of the preceding section. Comparing the algebra involved in satisfying the initial and boundary conditions by the forms of the solution in this and in the preceding section, we conclude that the form of the solution in the preceding section is more convenient to work with in the case of uniform bars. For nonuniform bars it appears that one does not have any choice but to work with the form of solution in this section.

We note that slow heterogeneous effects are not cumulative from cycle to cycle, and hence although they may accelerate or delay the appearance of shocks they do not prevent their formation. This is in contrast with dissipation and radiation effects, which are cumulative from cycle to cycle and can thus prevent shock formation.

8.2. Transverse Waves Along a Beam on an Elastic Foundation

In the rest of this chapter we consider dispersive waves. As an example of dispersive waves we discuss transverse waves along a uniform, infinite beam that rests on a nonlinear elastic foundation. That is, we consider solutions of the equation

$$EI \frac{\partial^4 \tilde{w}}{\partial \tilde{x}^4} + \rho A \frac{\partial^2 \tilde{w}}{\partial \tilde{t}^2} + \alpha_1 \tilde{w} + \alpha_3 \tilde{w}^3 = 0 \tag{8.2.1}$$

where E is the modulus of elasticity; A and I are the area and moment of inertia of the cross section, respectively; ρ is the density of the beam per unit length; and the α_n are the elastic coefficients of the foundation. In what follows α_1 will be taken positive while α_3 may be positive or negative depending on the foundation. Using a characteristic transverse deflection w_c, we introduce dimensionless quantities defined by

$$w = \frac{\tilde{w}}{w_c}, \quad x = \tilde{x} \left(\frac{\alpha_1}{EI} \right)^{1/4}, \quad t = \tilde{t} \left(\frac{\alpha_1}{\rho A} \right)^{1/2}$$

Then (8.2.1) becomes

$$\frac{\partial^4 w}{\partial x^4} + \frac{\partial^2 w}{\partial t^2} + w + \epsilon^2 \alpha w^3 = 0 \tag{8.2.2}$$

where $\epsilon^2 = w_c^2 \, |\alpha_3|/\alpha_1$, and $\alpha = 1$ when $\alpha_3 > 0$ and $\alpha = -1$ when $\alpha_3 < 0$. In what follows ϵ is assumed to be small but finite.

In the next section we derive two alternate partial-differential equations describing the spatial and temporal evolution of the amplitude and phase of a wavepacket. We use these differential equations in Section 8.2.2 to determine an approximate expression for monochromatic waves and then determine their stability. In Section 8.2.3 we discuss modulations of the envelope of stable periodic waves, while in Section 8.2.4 we discuss modulations of the envelope of unstable periodic waves. The problem of wave-wave interaction (in this case harmonic resonance) is taken up in Section 8.2.5, where we derive two coupled nonlinear parabolic complex equations for the interaction of two wavepackets. The final section of this chapter deals with stationary forms for the envelopes of the interacting waves.

8.2.1. A NONLINEAR PARABOLIC EQUATION

In this section we consider the propagation of a group of waves centered around the wavenumber k and the frequency ω. To accomplish this, we use the method of multiple scales and introduce the slow time scales $T_1 = \epsilon t$ and $T_2 = \epsilon^2 t$ in addition to the original time scale $T_0 = t$. Moreover we introduce the long scales $X_1 = \epsilon x$ and $X_2 = \epsilon^2 x$ in addition to the original space scale $X_0 = x$. Hence the time and space derivatives become

$$\begin{aligned} \frac{\partial}{\partial t} &= \frac{\partial}{\partial T_0} + \epsilon \frac{\partial}{\partial T_1} + \epsilon^2 \frac{\partial}{\partial T_2} \\ \frac{\partial}{\partial x} &= \frac{\partial}{\partial X_0} + \epsilon \frac{\partial}{\partial X_1} + \epsilon^2 \frac{\partial}{\partial X_2} \end{aligned} \tag{8.2.3}$$

Then we seek a second-order solution in the form

$$w(x, t; \epsilon) = \sum_{n=0}^{2} \epsilon^n w_n(X_0, X_1, X_2, T_0, T_1, T_2) + O(\epsilon^3) \tag{8.2.4}$$

Substituting (8.2.4) into (8.2.2), using (8.2.3), and equating coefficients of like powers of ϵ, we obtain

$$\frac{\partial^4 w_0}{\partial X_0^4} + \frac{\partial^2 w_0}{\partial T_0^2} + w_0 = 0 \tag{8.2.5}$$

$$\frac{\partial^4 w_1}{\partial X_0^4} + \frac{\partial^2 w_1}{\partial T_0^2} + w_1 = -4 \frac{\partial^4 w_0}{\partial X_0^3 \, \partial X_1} - 2 \frac{\partial^2 w_0}{\partial T_0 \, \partial T_1} \tag{8.2.6}$$

$$\frac{\partial^4 w_2}{\partial X_0^4} + \frac{\partial^2 w_2}{\partial T_0^2} + w_2 = -4\frac{\partial^4 w_1}{\partial X_0^3\,\partial X_1} - 2\frac{\partial^2 w_1}{\partial T_0\,\partial T_1} - 4\frac{\partial^4 w_0}{\partial X_0^3\,\partial X_2} - 6\frac{\partial^4 w_0}{\partial X_0^2\,\partial X_1^2}$$

$$- 2\frac{\partial^2 w_0}{\partial T_0\,\partial T_2} - \frac{\partial^2 w_0}{\partial T_1^2} - \alpha w_0^3 \quad (8.2.7)$$

To analyze the propagation of a wavepacket centered around the wavenumber k and the frequency ω, we take the solution of (8.2.5) in the form

$$w_0 = A(X_1, X_2, T_1, T_2) \exp\,[i(kX_0 - \omega T_0)] + cc \quad (8.2.8)$$

Substituting (8.2.8) into (8.2.5) leads to the dispersion relation

$$\omega^2 = k^4 + 1 \quad (8.2.9)$$

The complex function A is unknown at this level of approximation; it is determined at the next levels of approximation by imposing the solvability conditions. If we express A in the polar form

$$A = \tfrac{1}{2}a(X_1, X_2, T_1, T_2) \exp\,[i\beta(X_1, X_2, T_1, T_2)] \quad (8.2.10)$$

then a and β are the amplitude and the phase of the wave. If A is a function of T_1 and T_2 only, then (8.2.8) describes the propagation of a uniform wave whose wavenumber is k and whose frequency is shifted from ω by the nonlinearity. On the other hand, if A is a function of X_1 and X_2 only, then (8.2.8) represents a uniform wave whose frequency is ω and whose wavenumber is shifted from k by the nonlinearity. Equation (8.2.8) describes the propagation of a wavepacket only when A is a function of both space and time.

Substituting for w_0 from (8.2.8) into (8.2.6) yields

$$\frac{\partial^4 w_1}{\partial X_0^4} + \frac{\partial^2 w_1}{\partial T_0^2} + w_1 = \left(4ik^3\frac{\partial A}{\partial X_1} + 2i\omega\frac{\partial A}{\partial T_1}\right) \exp\,[i(kX_0 - \omega T_0)] + cc \quad (8.2.11)$$

Since ω and k are related by the dispersion relation (8.2.9), the particular solution of (8.2.11) contains secular terms in either X_0 or T_0 or both which lead to a nonuniform expansion for either long times or large distances or both. For a uniform expansion we require the vanishing of the right-hand side of (8.2.11); this solvability condition leads to the following equation for A:

$$2k^3\frac{\partial A}{\partial X_1} + \omega\frac{\partial A}{\partial T_1} = 0 \quad (8.2.12)$$

Then (8.2.11) becomes the same as (8.2.5), and consequently $w_1 = 0$, without loss of generality.

Substituting for w_0 from (8.2.8) into (8.2.7) and recalling that $w_1 = 0$, we have

$$\frac{\partial^4 w_2}{\partial X_0^4} + \frac{\partial^2 w_2}{\partial T_0^2} + w_2 = \left[4ik^3 \frac{\partial A}{\partial X_2} + 6k^2 \frac{\partial^2 A}{\partial X_1^2} + 2i\omega \frac{\partial A}{\partial T_2} - \frac{\partial^2 A}{\partial T_1^2}\right]$$

$$\cdot \exp\left[i(kX_0 - \omega T_0)\right] - \alpha A^3 \exp\left[3i(kX_0 - \omega T_0)\right] - 3\alpha A^2 \bar{A}$$

$$\cdot \exp\left[i(kX_0 - \omega T_0)\right] + cc \tag{8.2.13}$$

Eliminating the terms that produce secular terms in (8.2.13) leads to

$$4ik^3 \frac{\partial A}{\partial X_2} + 2i\omega \frac{\partial A}{\partial T_2} + 6k^2 \frac{\partial^2 A}{\partial X_1^2} - \frac{\partial^2 A}{\partial T_1^2} - 3\alpha A^2 \bar{A} = 0 \tag{8.2.14}$$

Then the solution of (8.2.13) becomes

$$w_2 = \frac{\alpha A^3}{8(1 - 9k^4)} \exp\left[3i(kX_0 - \omega T_0)\right] + cc \tag{8.2.15}$$

where use has been made of (8.2.9).

Equations (8.2.12) and (8.2.14) describe the evolution of the complex amplitude A with the slow and long scales. These equations can be combined to yield one of two alternate single partial-differential equations for A. This is accomplished by eliminating either $\partial A/\partial T_1$ or $\partial A/\partial X_1$ from (8.2.12) and (8.2.14). Solving for $\partial A/\partial T_1$ from (8.2.12) yields

$$\frac{\partial A}{\partial T_1} = -\frac{2k^3}{\omega} \frac{\partial A}{\partial X_1}$$

Hence

$$\frac{\partial^2 A}{\partial T_1^2} = -\frac{2k^3}{\omega} \frac{\partial^2 A}{\partial X_1 \, \partial T_1} = -\frac{2k^3}{\omega} \frac{\partial}{\partial X_1}\left(\frac{\partial A}{\partial T_1}\right) = \frac{4k^6}{\omega^2} \frac{\partial^2 A}{\partial X_1^2} \tag{8.2.16}$$

Combining (8.2.14) and (8.2.16) yields

$$2i\omega \frac{\partial A}{\partial T_2} + 4ik^3 \frac{\partial A}{\partial X_2} + \left(6k^2 - \frac{4k^6}{\omega^2}\right) \frac{\partial^2 A}{\partial X_1^2} - 3\alpha A^2 \bar{A} = 0 \tag{8.2.17}$$

Equation (8.2.17) can be put in a general form if one uses the dispersion relation (8.2.9). To do this, we differentiate (8.2.9) with respect to k and obtain

$$\omega\omega' = 2k^3 \tag{8.2.18}$$

where $\omega' = d\omega/dk$ is the group velocity. Differentiating (8.2.18) with respect to k yields

$$\omega\omega'' = -\omega'^2 + 6k^2$$

which on eliminating ω' by using (8.2.18) becomes

$$\omega\omega'' = 6k^2 - \frac{4k^6}{\omega^2} \tag{8.2.19}$$

where $\omega'' = d^2\omega/dk^2$. Using (8.2.18) and (8.2.19) in (8.2.17), we rewrite the latter in the form

$$2i\omega \frac{\partial A}{\partial T_2} + 2i\omega\omega' \frac{\partial A}{\partial X_2} + \omega\omega'' \frac{\partial^2 A}{\partial X_1^2} - 3\alpha A^2 \bar{A} = 0 \tag{8.2.20}$$

Expressing X_1, X_2, and T_2 in terms of the original x and t variables, we rewrite (8.2.20) in the final form

$$\frac{\partial A}{\partial t} + \omega' \frac{\partial A}{\partial x} - \tfrac{1}{2} i\omega'' \frac{\partial^2 A}{\partial x^2} + \tfrac{3}{2} i\epsilon^2 \alpha\omega^{-1} A^2 \bar{A} = 0 \tag{8.2.21}$$

Had we eliminated $\partial A/\partial X_1$ from (8.2.12) and (8.2.14), we would have obtained the following alternate partial-differential equation for A:

$$\frac{\partial A}{\partial x} + k' \frac{\partial A}{\partial t} + \tfrac{1}{2} ik'' \frac{\partial^2 A}{\partial t^2} + \tfrac{3}{4} i\epsilon^2 \alpha k^{-3} A^2 \bar{A} = 0 \tag{8.2.22}$$

where $k' = dk/d\omega$ and $k'' = d^2k/d\omega^2$. Equations (8.2.21) and (8.2.22) are nonlinear parabolic equations of the Schrödinger equation type.

Although (8.2.21) and (8.2.22) were derived for a wavepacket propagating transversly along a beam, they describe the propagation of wavepackets in any dispersive, lossless medium if one replaces ω', ω'', k', and k'' by using the dispersion relation and α by the nonlinear interaction coefficient appropriate to the specific medium. Equations similar to (8.2.21) were derived to describe the propagation of two-dimensional wavepackets in plasmas by Washimi and Taniuti (1966), Taniuti and Washimi (1968), Asano, Taniuti, and Yajima (1969), Watanabe (1969), and Kakutani and Sugimoto (1974); on the surface of a fluid of finite depth by Davey (1972), Hasimoto and Ono (1971), and Nayfeh and Saric (1972b); on the interface of two fluids of infinite depth by Nayfeh (1976a); on the surface of a cylindrical column of fluid by Kakutani, Inoue, and Kan (1974); in ducts by Nayfeh (1975c); in nonlinear optics by Karpman (1975) and Asfar and Nayfeh (1976), for example; for torsional waves in an elastic rod by Hirao and Sugimoto (1977); for a two-dimensional Klein-Gordon equation by Watanabe and Taniuti (1977); for general nonlinear partial- and integropartial-differential equations by Taniuti and Yajima (1969, 1973), Kadomtsev and Karpman (1971), Kakutani and Sugimoto (1974), Asano (1974), and Inoue and Matsumoto (1974); and for the nonlinear stability for a wave system in plane Poiseuille flow by Stewartson and Stuart (1971) and DiPrima,

Eckhaus, and Segel (1971). Davey and Stewartson (1974) and Kakutani and Michihiro (1976) derived two coupled nonlinear partial-differential equations for the evolution of three-dimensional wavepackets on water of finite depth.

We use (8.2.21) and (8.2.22) in the next section to study monochromatic waves (uniform wave trains) and determine their stability. The modulation of the envelopes of these uniform waves is discussed in Sections 8.2.3 and 8.2.4.

Solutions of evolution equations of the type (8.2.21) and (8.2.22) were obtained by using an inverse scattering method by Gardner, Green, Kruskal, and Miura (1967); Lax (1968); Lamb (1971, 1973); Zakharov and Shabat (1972); Wadati (1973); Ablowitz, Kaup, Newell, and Segur (1974); Zakharov and Manakov (1973); Kato (1974); Kaup (1976); and Asano and Kato (1977). Steady-state solutions of the Schrödinger equation generally represent wave trains which can be expressed in terms of the Jacobian elliptic functions. They include a bright and a dark envelope soliton, a phase jump, and a plane wave with constant amplitude as special cases (e.g., Lighthill, 1965; Taniuti and Washimi, 1968; Hasimoto and Ono, 1971; Hasegawa and Tappert, 1973). Some of these steady-state solutions are discussed next.

8.2.2. MONOCHROMATIC WAVES AND THEIR STABILITY

If the wave has a fixed, single wavenumber, then we can use (8.2.21) to determine the nonlinear frequency shift. Since k is fixed, A must be independent of x and (8.2.21) reduces to

$$\frac{\partial A}{\partial t} + \tfrac{3}{2} i\epsilon^2 \alpha\omega^{-1} A^2 \bar{A} = 0 \tag{8.2.23}$$

whose solution is

$$A = \tfrac{1}{2} a_0 \exp\left(-\tfrac{3}{8} i\epsilon^2 a_0^2 \alpha\omega^{-1} t + i\beta_0\right) \tag{8.2.24}$$

where a_0 and β_0 are constants. Substituting for A in (8.2.8), using (8.2.4), and recalling that $w_1 = 0$, we obtain

$$w = a_0 \cos(kx - \hat{\omega} t + \beta_0) + O(\epsilon^2) \tag{8.2.25}$$

where

$$\hat{\omega} = \omega + \tfrac{3}{8} \epsilon^2 a_0^2 \alpha\omega^{-1} + O(\epsilon^3) \tag{8.2.26}$$

Thus the effect of the nonlinearity is to produce a frequency shift.

If the wave has a fixed, single frequency ω, then A is independent of t. In this case it is more convenient to use (8.2.22) because the resulting equation is a first-order equation, compared with a second-order equation resulting from the use of (8.2.21). Thus monochromatic waves with fixed frequency are given by dropping the time derivatives in (8.2.22) and obtaining

$$\frac{\partial A}{\partial x} + \tfrac{3}{4} i\epsilon^2 \alpha k^{-3} A^2 \bar{A} = 0 \tag{8.2.27}$$

The solution of (8.2.27) is

$$A = \tfrac{1}{2} a_0 \exp\left[-\tfrac{3}{16} i\epsilon^2 a_0^2 \alpha k^{-3} x + i\beta_0\right] \tag{8.2.28}$$

where a_0 and β_0 are constants. Substituting for A in (8.28), using (8.24), and recalling that $w_1 = 0$, we obtain

$$w = a_0 \cos(\hat{k}x - \omega t + \beta_0) + O(\epsilon^2) \tag{8.2.29}$$

where

$$\hat{k} = k - \tfrac{3}{16} \epsilon^2 a_0^2 \alpha k^{-3} + O(\epsilon^3) \tag{8.2.30}$$

Thus the effect of the nonlinearity is to produce a wavenumber shift.

The above-obtained monochromatic wave solutions may be unstable. To determine the stability of, say, the solutions given by (8.2.25) and (8.2.26), we first express A in the polar form

$$A(x, t) = \tfrac{1}{2} a(x, t) \exp[i\beta(x, t)] \tag{8.2.31}$$

Substituting (8.2.31) into (8.2.21) and separating real and imaginary parts we have

$$\frac{\partial a}{\partial t} + \omega' \frac{\partial a}{\partial x} + \omega'' \left[\frac{\partial a}{\partial x}\frac{\partial \beta}{\partial x} + \tfrac{1}{2} a \frac{\partial^2 \beta}{\partial x^2}\right] = 0 \tag{8.2.32}$$

$$a \frac{\partial \beta}{\partial t} + \omega' a \frac{\partial \beta}{\partial x} - \tfrac{1}{2}\omega'' \left[\frac{\partial^2 a}{\partial x^2} - a\left(\frac{\partial \beta}{\partial x}\right)^2\right] + \tfrac{3}{8}\epsilon^2 \alpha \omega^{-1} a^3 = 0 \tag{8.2.33}$$

Next we perturb a and β from the monochromatic wave solution of (8.2.24). That is, we let

$$\begin{aligned} a &= a_0 + a_1(x, t) \\ \beta &= -\tfrac{3}{8}\epsilon^2 a_0^2 \alpha \omega^{-1} t + \beta_0 + \beta_1(x, t) \end{aligned} \tag{8.2.34}$$

where a_1 and β_1 are assumed to be infinitesimal. Substituting (8.2.34) into (8.2.32) and (8.2.33) and keeping only linear terms in the perturbation quantities, we obtain

$$\begin{aligned} &\frac{\partial a_1}{\partial t} + \omega' \frac{\partial a_1}{\partial x} + \tfrac{1}{2}\omega'' a_0 \frac{\partial^2 \beta_1}{\partial x^2} = 0 \\ &a_0 \frac{\partial \beta_1}{\partial t} + \omega' a_0 \frac{\partial \beta_1}{\partial x} - \tfrac{1}{2}\omega'' \frac{\partial^2 a_1}{\partial x^2} + \tfrac{3}{4}\epsilon^2 a_0^2 \alpha \omega^{-1} a_1 = 0 \end{aligned} \tag{8.2.35}$$

Since (8.2.35) have constant coefficients, one can represent their solutions in the form

$$\begin{aligned} a_1 &= a_{10} \exp[i(Kx - \Omega t)] \\ \beta_1 &= \beta_{10} \exp[i(Kx - \Omega t)] \end{aligned} \tag{8.2.36}$$

where a_{10}, β_{10}, K, and Ω are constants. Substituting (8.2.36) into (8.2.35) yields

$$\begin{aligned} i(\omega' K - \Omega) a_{10} - \tfrac{1}{2}\omega'' a_0 K^2 \beta_{10} &= 0 \\ (\tfrac{1}{2}\omega'' K^2 + \tfrac{3}{4}\epsilon^2 a_0^2 \alpha\omega^{-1}) a_{10} + i a_0 (\omega' K - \Omega)\beta_{10} &= 0 \end{aligned} \tag{8.2.37}$$

For a nontrivial solution the determinant of the coefficient matrix of (8.2.37) must vanish. That is,

$$(\Omega - \omega' K)^2 = \tfrac{1}{4}\omega''^2 K^4 \left(1 + \frac{3\epsilon^2 a_0^2 \alpha}{2K^2 \omega\omega''}\right) \tag{8.2.38}$$

For a given K, Ω is always real, and hence a_1 and β_1 are bounded according to (8.2.36) if, and only if, $\alpha\omega'' > 0$. Otherwise Ω will be complex for values of $K < \sqrt{\frac{3}{2}}\, \epsilon a_0 \, |\alpha/\omega\omega''|^{1/2}$, and a_1 and β_1 will be unbounded. Hence the monochromatic wave solution given by (8.2.25) and (8.2.26) is stable only if $\alpha\omega'' > 0$. Combining (8.2.9) and (8.2.19) gives

$$\omega'' = \frac{2k^6 + 6k^2}{\omega^3}$$

Hence $\omega'' > 0$ and monochromatic waves are stable or unstable depending on whether α and hence α_3 is positive or negative.

However the linearized equations (8.2.35) are valid only for short times. The solution for long times is obtained in the next section for the case of stable waves and in Section 8.2.4 for the case of unstable waves.

8.2.3. MODULATIONAL STABILITY

In this section we consider the nonlinear modulation of the periodic-wave solution (8.2.25) and (8.2.26) when $\alpha > 0$. To do this we introduce the new independent variables

$$\xi = x - \omega' t, \quad \tau = \omega'' t \tag{8.2.39}$$

in (8.2.32) and (8.2.33) and obtain

$$\frac{\partial a}{\partial \tau} + \frac{\partial a}{\partial \xi}\frac{\partial \beta}{\partial \xi} + \tfrac{1}{2} a \frac{\partial^2 \beta}{\partial \xi^2} = 0 \tag{8.2.40}$$

$$\frac{\partial \beta}{\partial \tau} - \frac{1}{2a}\frac{\partial^2 a}{\partial \xi^2} + \tfrac{1}{2}\left(\frac{\partial \beta}{\partial \xi}\right)^2 + \chi a^2 = 0 \tag{8.2.41}$$

where $\chi = \frac{3}{8}\epsilon^2 \alpha(\omega\omega'')^{-1}$.

Equations (8.2.40) and (8.2.41) are satisfied by periodic- and solitary-wave (soliton) solutions. These solutions can be obtained by considering stationary-wave solutions. Thus we seek a solution for (8.2.40) and (8.2.41) in the form

$$a = a_0 + a_1(\xi - v\tau), \quad \beta = -\chi a_0^2 \tau + \beta_1(\xi - v\tau) \tag{8.2.42}$$

where a_0 and v are constants. Substituting (8.2.42) into (8.2.40) and (8.2.41) gives

$$-va_1' + a_1'\beta_1' + \tfrac{1}{2}a\beta_1'' = 0 \tag{8.2.43}$$

$$-v\beta_1' - \frac{a_1''}{2a} + \tfrac{1}{2}\beta_1'^2 + \chi(a^2 - a_0^2) = 0 \tag{8.2.44}$$

Multiplying (8.2.43) by $2(a_0 + a_1)$ and integrating the resulting equation yield

$$a^2\beta_1' = va^2 + c_1 \tag{8.2.45}$$

where c_1 is a constant of integration. Since $\beta_1' = 0$ when $a_1 = 0$ according to (8.2.42), $c_1 = -va_0^2$. Hence

$$\beta_1' = \frac{v(a^2 - a_0^2)}{a^2} \tag{8.2.46}$$

which when substituted into (8.2.44) gives

$$a'' = 2\chi a(a^2 - a_0^2) - v^2\left(a - \frac{a_0^4}{a^3}\right) \tag{8.2.47}$$

A first integral of (8.2.47) is

$$a'^2 = \chi a^4 - (2\chi a_0^2 + v^2)a^2 - \frac{v^2 a_0^4}{a^2} + c_2 \tag{8.2.48}$$

where c_2 is a constant of integration. Letting $\zeta = a^2$, we rewrite (8.2.48) as

$$\tfrac{1}{4}\zeta'^2 = \chi\zeta^3 - (2\chi\zeta_0 + v^2)\zeta^2 + c_2\zeta - v^2\zeta_0^2 \tag{8.2.49}$$

In the special case in which $a \to a_0$ and $a' \to 0$ as $\xi \to \infty$, it follows from (8.2.48) that

$$c_2 = \chi a_0^4 + 2v^2 a_0^2$$

and hence (8.2.49) becomes

$$\tfrac{1}{4}\zeta'^2 = \chi\zeta^3 - (2\chi\zeta_0 + v^2)\zeta^2 + (\chi\zeta_0^2 + 2v^2\zeta_0)\zeta - v^2\zeta_0^2 \tag{8.2.50}$$

Factoring the right-hand side of (8.2.50) we have

$$\tfrac{1}{4}\zeta'^2 = (\chi\zeta - v^2)(\zeta_0 - \zeta)^2 \tag{8.2.51}$$

For a soliton, $v^2 \leqslant \chi\zeta$. Hence $v \leqslant \sqrt{\chi\zeta_{\min}}$, where $\zeta_{\min}$ is the minimum of a^2.

Equation (8.2.51) can be solved by letting

$$\zeta = \zeta_0(1 - b\,\mathrm{sech}^2\,\theta), \quad b = \left(1 - \frac{v^2}{\chi\zeta_0}\right) \tag{8.2.52}$$

so that

$$d\zeta = -2b\zeta_0\,\mathrm{sech}^2\,\theta\,\tanh\theta\,d\theta$$

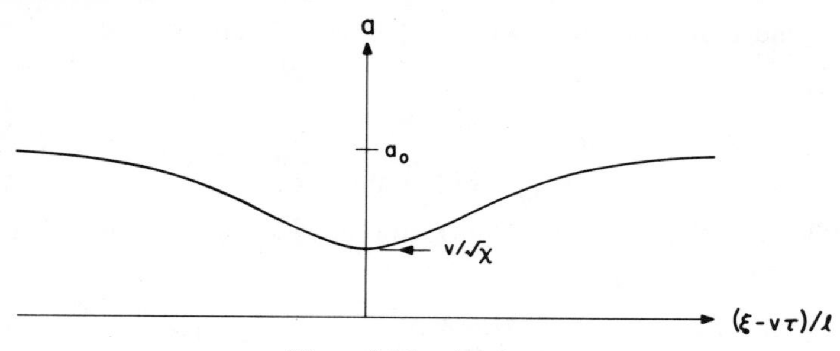

Figure 8-12. A soliton.

Hence (8.2.51) can be integrated to give

$$\theta = \sqrt{\chi\zeta_0 - v^2}\,(\xi - v\tau) \tag{8.2.53}$$

Therefore the soliton, a wave consisting of a single dip of constant shape and speed as shown in Figure 8-12, is given by

$$a^2 = a_0^2 - \left(a_0^2 - \frac{v^2}{\chi}\right)\operatorname{sech}^2\frac{\xi - v\tau}{l} \tag{8.2.54}$$

where

$$l^{-1} = \sqrt{\chi a_0^2 - v^2}, \quad v \leqslant a_{\min}\sqrt{\chi} \tag{8.2.55}$$

We note that the modulation (soliton) of the envelope of the uniform-wave train consists of a hole that moves with the speed v. The phase jump across the hole is given by

$$\Delta\beta = \int_{-\infty}^{\infty} \beta_1'(\xi)\,d\xi = \int_{-\infty}^{\infty} \frac{v(a^2 - a_0^2)}{a^2}\,d\xi = -2\arctan\frac{\sqrt{a_0^2\chi - v^2}}{v\sqrt{\chi}} \tag{8.2.56}$$

If $a_{\min} = 0$, $v = 0$ and (8.2.54) reduces to

$$a^2 = a_0^2\left(1 - \operatorname{sech}^2\frac{\xi - v\tau}{l}\right) = a_0^2\tanh^2\frac{\xi - v\tau}{l}$$

or

$$a = a_0\tanh\frac{\xi - v\tau}{l} \tag{8.2.57}$$

Moreover (8.2.56) becomes

$$\Delta\beta = -\pi \tag{8.2.58}$$

In the general case a does not tend to a_0 and a' does not tend to zero as $\xi \to \infty$, and the roots ζ_n of the right-hand side of (8.2.49) are distinct. In terms of these roots (8.2.49) can be rewritten as

$$\tfrac{1}{4}\zeta'^2 = \chi(\zeta - \zeta_1)(\zeta - \zeta_2)(\zeta - \zeta_3) \tag{8.2.59}$$

which can be put in the standard form of an equation for a Jacobian elliptic function as follows. Letting $\zeta = \zeta_2 - (\zeta_2 - \zeta_1)u^2$, we can rewrite (8.2.59) as

$$u'^2 = K^2(1 - u^2)(1 - \nu^2 + \nu^2 u^2) \tag{8.2.60}$$

where

$$K = \sqrt{(\zeta_3 - \zeta_1)\chi}, \quad \nu = \left(\frac{\zeta_2 - \zeta_1}{\zeta_3 - \zeta_1}\right)^{1/2}$$

Hence $u = \text{cn}[K(\xi - v\tau), \nu]$ and

$$a^2 = \zeta_2 - (\zeta_2 - \zeta_1)\,\text{cn}^2[K(\xi - v\tau), \nu] \tag{8.2.61}$$

which Korteweg and de Vries named *cnoidal waves.*

8.2.4. MODULATIONAL INSTABILITY

In this section we consider the modulation of the envelope of the unstable periodic solution of Section 8.2.2 (sometimes termed self-modulation); that is, we consider solutions of (8.2.40) and (8.2.41) when α and hence χ are negative.

First we investigate periodic and solitary solutions of (8.2.40) and (8.2.41). Thus we seek a solution in the form

$$a = a_0 + a_1(\xi - v\tau), \quad \beta = -\chi a_0^2 \tau + \beta_1(\xi - v\tau) \tag{8.2.62}$$

where a_0 and v are constants. Substituting (8.2.62) into (8.2.40) and (8.2.41) yields

$$-va' + a'\beta_1' + \tfrac{1}{2}a\beta_1'' = 0 \tag{8.2.63}$$

$$-\frac{a''}{2a} - v\beta_1' + \tfrac{1}{2}\beta_1'^2 - \hat{\chi}(a^2 - a_0^2) = 0 \tag{8.2.64}$$

where $\hat{\chi} = -\chi$.

Equation (8.2.63) can be solved for β_1'. The result is

$$\beta_1' = \frac{v(a^2 - c_1)}{a^2}$$

We shall assume in the following that $\beta_1' = 0$ when $a = a_0$ so that $c_1 = a_0^2$, and hence

$$\beta_1' = \frac{v(a^2 - a_0^2)}{a^2} \tag{8.2.65}$$

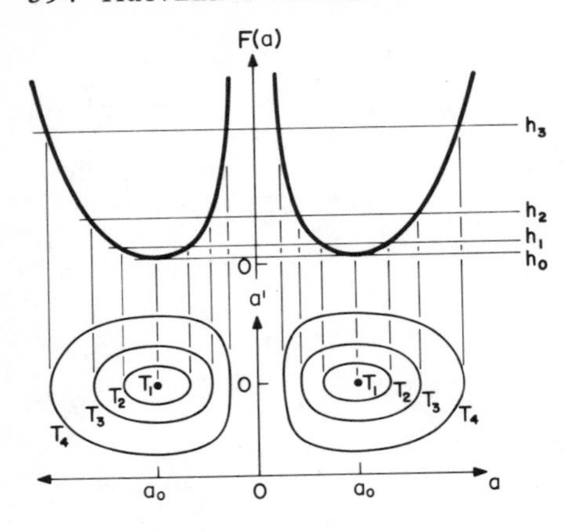

Figure 8-13. Phase plane for the case of modulational instability when $v \neq 0$.

Then (8.2.64) becomes

$$a'' = -2\hat{\chi}a(a^2 - a_0^2) - v^2 \left(a - \frac{a_0^4}{a^3}\right) \tag{8.2.66}$$

Multiplying (8.2.66) by a' and integrating the resulting equation gives

$$\tfrac{1}{2}a'^2 = -\tfrac{1}{2}\hat{\chi}a^2(a^2 - 2a_0^2) - \tfrac{1}{2}v^2\left(a^2 + \frac{a_0^4}{a^2}\right) + h = -F(a) + h \tag{8.2.67}$$

where h is a constant of integration.

Figure 8-13 shows the character of the solutions in the phase plane for the case $v \neq 0$. In this case there are two centers at $a = \pm a_0$ when $h = h_0$, and there is no modulation. But when $h > h_0$ the modulation of the envelope is oscillatory about $a = \pm a_0$.

Figure 8-14 shows the character of the solutions in the phase plane for the case $v = 0$. In this case there are two centers at $a = \pm a_0$ and a saddle point at $a = 0$. When $h = h_0$, there is no modulation of the wave motion. When $h_0 < h < h_2$ such as $h = h_1$ the modulation of the envelope is oscillatory about $a = \pm a_0$. The separatrix corresponds to $h = h_2$, and it corresponds to the following solution of the soliton type

$$a = \sqrt{2}a_0 \operatorname{sech}\left[\sqrt{2\hat{\chi}}a_0(\xi - v\tau - \xi_0 + v\tau_0)\right] \tag{8.2.68}$$

where ξ_0 and τ_0 are the initial values of ξ and τ_0. When $h > h_2$, the modulation of the envelope is periodic, and a in this case passes through zero.

Nonstationary modulations of the envelopes of periodic waves with constant amplitudes were treated numerically by Karpman and Kruskal (1969) and analytically by Zakharov and Shabat (1972). They considered solutions of (8.2.40) and (8.2.41) when $\chi < 0$ in the form

$$a = a_0 + a_1(\xi, \tau), \quad \beta = -\chi a_0^2 \tau + \beta_1(\xi, \tau) \tag{8.2.69}$$

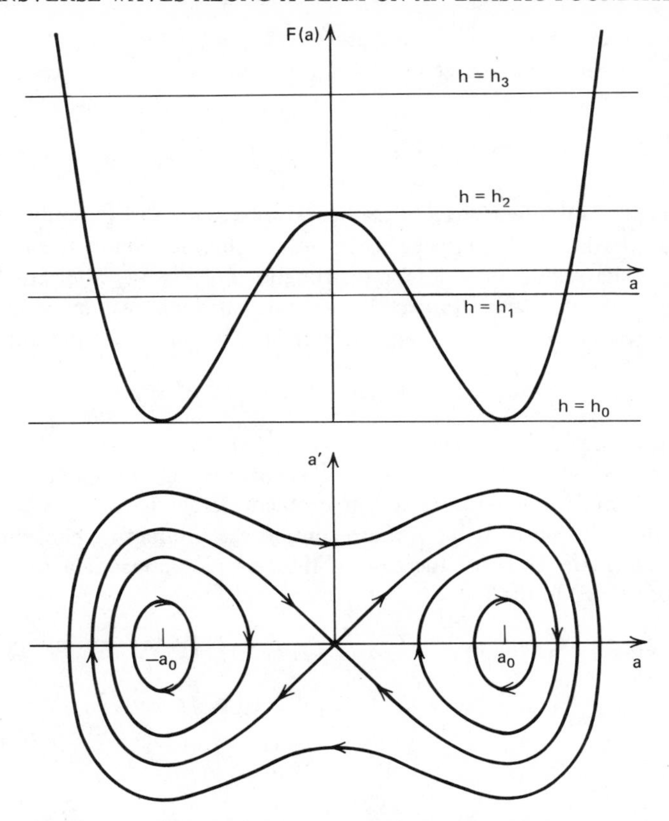

Figure 8-14. Phase plane for the case of modulational instability when $v = 0$.

where a_0 is a constant subject to the initial conditions

$$a_1(\xi, 0) = a_0 f(\xi), \quad \beta_1(\xi, 0) = 0 \tag{8.2.70}$$

where $f(\xi) \to 0$ as $|\xi| \to \infty$. They found that when the initial periodic wave is unstable, some perturbations cause it to decay very rapidly into a series of solitons, one of which is given by (8.2.68).

8.2.5. NONLINEAR COUPLED PARABOLIC EQUATIONS

Although the solutions obtained in the preceding two sections are valid for a wide range of wavenumbers, they are not valid when $k^2 \approx \frac{1}{3}$ because $w_2 \to \infty$ as $k^2 \to \frac{1}{3}$ according to (8.2.15). The term $\epsilon^2 w_2$ becomes the same order as w_0 when $k^2 - \frac{1}{3} = O(\epsilon^2)$ and dominates it when $k^2 - \frac{1}{3} < O(\epsilon^2)$, contrary to the implicit assumption used in deriving the expansion. To understand the reason for the expansion breaking down and hence find a method for determining a uniform expansion when $k^2 \approx \frac{1}{3}$, we use the dispersion relation (8.2.9) to write down the linearized phase speed of the waves. That is,

$$c = \frac{\omega}{k} = \left(k^2 + \frac{1}{k^2}\right)^{1/2} \tag{8.2.71}$$

Equation (8.2.71) shows that a fundamental wave with a wavenumber $k = 1/\sqrt{n}$, $n \geqslant 2$, has the same phase speed as its nth harmonic. Hence the fundamental and its nth harmonic may interact strongly, and we say that there exists a condition of *harmonic resonance.* In general, harmonic resonance may exist if (ω, k) and $(n\omega, nk)$ satisfy the same dispersion relation. In the present case

$$\omega^2 = k^4 + 1 \quad \text{and} \quad n^2\omega^2 = n^4k^4 + 1 \tag{8.2.72}$$

hold simultaneously. Eliminating ω from (8.2.72) yields $k^2 = 1/n$.

Since the wave with wavenumber $k^2 \approx \frac{1}{3}$ is harmonically resonant with its third harmonic, a uniform expansion to our problem demands that the zeroth-order solution, the solution of (8.2.5), must contain the interacting harmonics rather than one of them. Thus for the case of third-harmonic resonance, we take the solution of (8.2.5) in the form

$$w_0 = A_1(X_1, X_2, T_1, T_2) \exp\left[i(k_1X_0 - \omega_1T_0)\right] + A_3(X_1, X_2, T_1, T_2) \cdot \exp\left[i(k_3X_0 - \omega_3T_0)\right] + cc \tag{8.2.73}$$

where

$$\omega_n^2 = k_n^4 + 1$$

$$\omega_3 \approx 3\omega_1 \quad \text{and} \quad k_3 \approx 3k_1 \tag{8.2.74}$$

Substituting w_0 into (8.2.6) yields

$$\frac{\partial^4 w_1}{\partial X_0^4} + \frac{\partial^2 w_1}{\partial T_0^2} + w_1 = \sum_{n=1,3}\left(4ik_n^3 \frac{\partial A_n}{\partial X_1} + 2i\omega_n \frac{\partial A_n}{\partial T_1}\right) \cdot \exp\left[i(k_nX_0 - \omega_nT_0)\right] + cc \tag{8.2.75}$$

Eliminating terms that produce secular terms in (8.2.75) we have

$$2k_n^3 \frac{\partial A_n}{\partial X_1} + \omega_n \frac{\partial A_n}{\partial T_1} = 0 \quad \text{for } n = 1 \text{ and } 3 \tag{8.2.76}$$

Then the solution of (8.2.75) can be taken as $w_1 = 0$, without loss of generality.

Substituting w_0 into (8.2.7) and recalling that $w_1 = 0$, we obtain

$$\frac{\partial^4 w_2}{\partial X_0^4} + \frac{\partial^2 w_2}{\partial T_0^2} + w_2 = \sum_{n=1,3} \left[4ik_n^3 \frac{\partial A_n}{\partial X_2} + 6k_n^2 \frac{\partial^2 A_n}{\partial X_1^2} + 2i\omega_n \frac{\partial A_n}{\partial T_2} - \frac{\partial^2 A_n}{\partial T_1^2} \right]$$
$$\cdot \exp\left[i(k_n X_0 - \omega_n T_0)\right] - 3\alpha(A_1\bar{A}_1 + 2A_3\bar{A}_3)A_1 \exp\left[i(k_1 X_0 - \omega_1 T_0)\right]$$
$$- 3\alpha(2A_1\bar{A}_1 + A_3\bar{A}_3)A_3 \exp\left[i(k_3 X_0 - \omega_3 T_0)\right] - \alpha A_1^3$$
$$\cdot \exp\left[3i(k_1 X_0 - \omega_1 T_0)\right] - \alpha A_3^3 \exp\left[3i(k_3 X_0 - \omega_3 T_0)\right] - 3\alpha A_1^2 A_3$$
$$\cdot \exp\left[i(k_3 + 2k_1)X_0 - i(\omega_3 + 2\omega_1)T_0\right] - 3\alpha A_1 A_3^2 \exp\left[i(2k_3 + k_1)X_0\right.$$
$$\left. - i(2\omega_3 + \omega_1)T_0\right] - 3\alpha A_3 \bar{A}_1^2 \exp\left[i(k_3 - 2k_1)X_0 - i(\omega_3 - 2\omega_1)T_0\right]$$
$$- 3\alpha A_3^2 \bar{A}_1 \exp\left[i(2k_3 - k_1)X_0 - i(2\omega_3 - \omega_1)T_0\right] + cc \tag{8.2.77}$$

To express the nearness of the resonance, we introduce detuning parameters σ_1 and σ_2 defined by

$$k_3 - 3k_1 = \epsilon^2 \sigma_1, \quad \omega_3 - 3\omega_1 = \epsilon^2 \sigma_2 \tag{8.2.78}$$

Then we express the near-resonance terms in (8.2.77) as

$$3k_1 X_0 - 3\omega_1 T_0 = k_3 X_0 - \omega_3 T_0 - \epsilon^2 \sigma_1 X_0 + \epsilon^2 \sigma_2 T_0 = k_3 X_0 - \omega_3 T_0 - \Gamma(X_2, T_2) \tag{8.2.79}$$

$$(k_3 - 2k_1)X_0 - (\omega_3 - 2\omega_1)T_0 = k_1 X_0 - \omega_1 T_0 + \Gamma(X_2, T_2) \tag{8.2.80}$$

where

$$\Gamma = \sigma_1 X_2 - \sigma_2 T_2 \tag{8.2.81}$$

Using (8.2.79) and (8.2.80) in eliminating the terms that produce secular terms in (8.2.77), we have

$$4ik_1^3 \frac{\partial A_1}{\partial X_2} + 6k_1^2 \frac{\partial^2 A_1}{\partial X_1^2} + 2i\omega_1 \frac{\partial A_1}{\partial T_2} - \frac{\partial^2 A_1}{\partial T_1^2} = 3\alpha(A_1\bar{A}_1 + 2A_3\bar{A}_3)A_1 + 3\alpha A_3 \bar{A}_1^2 \exp(i\Gamma) \tag{8.2.82}$$

$$4ik_3^3 \frac{\partial A_3}{\partial X_2} + 6k_3^2 \frac{\partial^2 A_3}{\partial X_1^2} + 2i\omega_3 \frac{\partial A_3}{\partial T_2} - \frac{\partial^2 A_3}{\partial T_1^2} = 3\alpha(2A_1\bar{A}_1 + A_3\bar{A}_3)A_3 + \alpha A_1^3 \exp(-i\Gamma) \tag{8.2.83}$$

Equations (8.2.76), (8.2.82), and (8.2.83) constitute a system of four complex partial-differential equations describing the interaction of two wavepackets centered around (ω_1, k_1) and (ω_3, k_3). As in the case of the propagation of one wavepacket, these interaction equations can be combined into a pair of coupled partial-differential equations by eliminating either the T_1 or the X_1 derivatives. First we eliminate the T_1 derivatives. From (8.2.76) it follows that

$$\frac{\partial A_n}{\partial T_1} = -\frac{2k_n^3}{\omega_n}\frac{\partial A_n}{\partial X_1}, \quad \frac{\partial^2 A_n}{\partial T_1^2} = \frac{4k_n^6}{\omega_n^2}\frac{\partial^2 A_n}{\partial X_1^2}$$

Then eliminating the T_1 derivatives, we rewrite (8.2.82) and (8.2.83) as

$$2i\omega_1 \frac{\partial A_1}{\partial T_2} + 4ik_1^3 \frac{\partial A_1}{\partial X_2} + \left(6k_1^2 - \frac{4k_1^6}{\omega_1^2}\right)\frac{\partial^2 A_1}{\partial X_1^2} = 3\alpha(A_1\bar{A}_1 + 2A_3\bar{A}_3)A_1 + 3\alpha A_3 \bar{A}_1^2 \exp(i\Gamma) \tag{8.2.84}$$

$$2i\omega_3 \frac{\partial A_3}{\partial T_2} + 4ik_3^3 \frac{\partial A_3}{\partial X_2} + \left(6k_3^2 - \frac{4k_3^6}{\omega_3^2}\right)\frac{\partial^2 A_3}{\partial X_1^2} = 3\alpha(2A_1\bar{A}_1 + A_3\bar{A}_3)A_3 + \alpha A_1^3 \exp(-i\Gamma) \tag{8.2.85}$$

Using (8.2.18) and (8.2.19) with ω and k replaced respectively by ω_n and k_n, we can rewrite (8.2.84) and (8.2.85) in terms of the original variables as

$$\frac{\partial A_1}{\partial t} + \omega_1' \frac{\partial A_1}{\partial x} - \tfrac{1}{2} i\omega_1'' \frac{\partial^2 A_1}{\partial x^2} = -\tfrac{3}{2} i\epsilon^2 \alpha\omega_1^{-1}(A_1\bar{A}_1 + 2A_3\bar{A}_3)A_1 - \tfrac{3}{2} i\epsilon^2\alpha\omega_1^{-1} A_3\bar{A}_1^2 \exp(i\Gamma) \tag{8.2.86}$$

$$\frac{\partial A_3}{\partial t} + \omega_3' \frac{\partial A_3}{\partial x} - \tfrac{1}{2} i\omega_3'' \frac{\partial^2 A_3}{\partial x^2} = -\tfrac{3}{2} i\epsilon^2 \alpha\omega_3^{-1}(2A_1\bar{A}_1 + A_3\bar{A}_3)A_3 - \tfrac{1}{2} i\epsilon^2\alpha\omega_3^{-1} A_1^3 \exp(-i\Gamma) \tag{8.2.87}$$

We note that the terms proportional to $\exp(\pm i\Gamma)$ are the result of the harmonic resonance. Thus in the absence of harmonic resonance, the equations describing the interaction of any two wavepackets centered at (k_1, ω_1) and (k_3, ω_3) are

$$\frac{\partial A_1}{\partial t} + \omega_1' \frac{\partial A_1}{\partial x} - \tfrac{1}{2} i\omega_1'' \frac{\partial^2 A_1}{\partial x^2} = -\tfrac{3}{2} i\epsilon^2 \alpha\omega_1^{-1}(A_1\bar{A}_1 + 2A_3\bar{A}_3)A_1 \tag{8.2.88}$$

$$\frac{\partial A_3}{\partial t} + \omega_3' \frac{\partial A_3}{\partial x} - \tfrac{1}{2} i\omega_3'' \frac{\partial^2 A_3}{\partial x^2} = -\tfrac{3}{2} i\epsilon^2 \alpha\omega_3^{-1}(2A_1\bar{A}_1 + A_3\bar{A}_3)A_3 \tag{8.2.89}$$

Inoue (1977) derived equations similar to (8.2.88) and (8.2.89) for the interaction of two wavepackets in an isotropic dielectric material when the group velocities ω_1' and ω_3' are the same.

Had we eliminated the X_1 rather than the T_1 derivatives, we would have obtained the alternate equations

$$\frac{\partial A_1}{\partial x} + k_1' \frac{\partial A_1}{\partial t} + \tfrac{1}{2} i k_1'' \frac{\partial^2 A_1}{\partial t^2} = -\tfrac{3}{4} i \epsilon^2 \alpha k_1^{-3} (A_1 \bar{A}_1 + 2 A_3 \bar{A}_3) A_1 - \tfrac{3}{4} i \epsilon^2 \alpha k_1^{-3} A_3 \bar{A}_1^2 \exp(i\Gamma) \quad (8.2.90)$$

$$\frac{\partial A_3}{\partial x} + k_3' \frac{\partial A_3}{\partial t} + \tfrac{1}{2} i k_3'' \frac{\partial^2 A_3}{\partial t^2} = -\tfrac{3}{4} i \epsilon^2 \alpha k_3^{-3} (2 A_1 \bar{A}_1 + A_3 \bar{A}_3) A_3 - \tfrac{1}{4} i \epsilon^2 \alpha k_3^{-3} A_1^3 \exp(-i\Gamma) \quad (8.2.91)$$

Putting $A_n = \frac{1}{2} a_n \exp(i\beta_n)$ and separating real and imaginary parts in (8.2.90) and (8.2.91), we obtain

$$\frac{\partial a_1}{\partial x} + k_1' \frac{\partial a_1}{\partial t} - k_1'' \left[\frac{\partial a_1}{\partial t} \frac{\partial \beta_1}{\partial t} + \tfrac{1}{2} a_1 \frac{\partial^2 \beta_1}{\partial t^2} \right] = \tfrac{3}{16} \epsilon^2 \alpha k_1^{-3} a_3 a_1^2 \sin\gamma \quad (8.2.92)$$

$$\frac{\partial a_3}{\partial x} + k_3' \frac{\partial a_3}{\partial t} - k_3'' \left[\frac{\partial a_3}{\partial t} \frac{\partial \beta_3}{\partial t} + \tfrac{1}{2} a_3 \frac{\partial^2 \beta_3}{\partial t^2} \right] = -\tfrac{1}{16} \epsilon^2 \alpha k_3^{-3} a_1^3 \sin\gamma \quad (8.2.93)$$

$$a_1 \frac{\partial \beta_1}{\partial x} + k_1' a_1 \frac{\partial \beta_1}{\partial t} + \tfrac{1}{2} k_1'' \left[\frac{\partial^2 a_1}{\partial t^2} - a_1 \left(\frac{\partial \beta_1}{\partial t} \right)^2 \right] = -\tfrac{3}{16} \epsilon^2 \alpha k_1^{-3} (a_1^2 + 2a_3^2) a_1 - \tfrac{3}{16} \epsilon^2 \alpha k_1^{-3} a_3 a_1^2 \cos\gamma \quad (8.2.94)$$

$$a_3 \frac{\partial \beta_3}{\partial x} + k_3' a_3 \frac{\partial \beta_3}{\partial t} + \tfrac{1}{2} k_3'' \left[\frac{\partial^2 a_3}{\partial t^2} - a_3 \left(\frac{\partial \beta_3}{\partial t} \right)^2 \right] = -\tfrac{3}{16} \epsilon^2 \alpha k_3^{-3} (2a_1^2 + a_3^2) a_3 - \tfrac{1}{16} \epsilon^2 \alpha k_3^{-3} a_1^3 \cos\gamma \quad (8.2.95)$$

where

$$\gamma = \beta_3 - 3\beta_1 + (k_3 - 3k_1) x - (\omega_3 - 3\omega_1) t \quad (8.2.96)$$

Equations (8.2.92) through (8.2.96) describe the interaction of two wavepackets centered at (k_1, ω_1) and (k_3, ω_3), where $\omega_3 \approx 3\omega_1$ and $k_3 \approx 3k_1$. Since there are no solutions available yet for these equations subject to general initial conditions, we discuss stationary-wave solutions in the next section.

8.2.6. INTERACTION OF STATIONARY WAVES

In this section we consider stationary-wave solutions of (8.2.92) through (8.2.96) of the form

$$a_n = a_n(x - vt), \quad \beta_n = \beta_n(x - vt) \quad (8.2.97)$$

Substituting this form of the solution into (8.2.92) through (8.2.95) yields

$$(1 - k_1'v)a_1' - k_1''v^2(a_1'\beta_1' + \tfrac{1}{2}a_1\beta_1'') = \tfrac{3}{16}\epsilon^2\alpha k_1^{-3}a_3a_1^2 \sin\gamma \quad (8.2.98)$$

$$(1 - k_3'v)a_3' - k_3''v^2(a_3'\beta_3' + \tfrac{1}{2}a_3\beta_3'') = -\tfrac{1}{16}\epsilon^2\alpha k_3^{-3}a_1^3 \sin\gamma \quad (8.2.99)$$

$$(1 - k_1'v)a_1\beta_1' + \tfrac{1}{2}k_1''v^2(a_1'' - a_1\beta_1'^2) = -\tfrac{3}{16}\epsilon^2\alpha k_1^{-3}(a_1^2 + 2a_3^2)a_1 - \tfrac{3}{16}\epsilon^2\alpha k_1^{-3}a_3a_1^2 \cos\gamma \quad (8.2.100)$$

$$(1 - vk_3')a_3\beta_3' + \tfrac{1}{2}k_3''v^2(a_3'' - a_3\beta_3'^2) = -\tfrac{3}{16}\epsilon^2\alpha k_3^{-3}(2a_1^2 + a_3^2)a_3 - \tfrac{1}{16}\epsilon^2\alpha k_3^{-3}a_1^3 \cos\gamma \quad (8.2.101)$$

For stationary-wave solutions to exist, (8.2.98) through (8.2.101) must be functions of $\xi = x - vt$ only. Hence it follows from (8.2.96) that stationary-wave solutions are possible when

$$v = \frac{\omega_3 - 3\omega_1}{k_3 - 3k_1} \quad (8.2.102)$$

so that

$$\gamma = \beta_3 - 3\beta_1 + (k_3 - 3k_1)(x - vt) \quad (8.2.103)$$

In the perfect resonance case (i.e., $k_3 = 3k_1$ and $\omega_3 = 3\omega_1$) stationary waves exist for all speeds of propagation v. In general one needs to solve numerically (8.2.98) through (8.2.101) and (8.2.103) subject to initial conditions to determine the a_n and the β_n.

For the special case of $v = 0$, the problem governing stationary-wave solutions reduces to

$$a_1' = \tfrac{3}{16}\epsilon^2\alpha k_1^{-3}a_3a_1^2 \sin\gamma \quad (8.2.104)$$

$$a_3' = -\tfrac{1}{16}\epsilon^2\alpha k_3^{-3}a_1^3 \sin\gamma \quad (8.2.105)$$

$$a_1\beta_1' = -\tfrac{3}{16}\epsilon^2\alpha k_1^{-3}(a_1^2 + 2a_3^2)a_1 - \tfrac{3}{16}\epsilon^2\alpha k_1^{-3}a_3a_1^2 \cos\gamma \quad (8.2.106)$$

$$a_3\beta_3' = -\tfrac{3}{16}\epsilon^2\alpha k_3^{-3}(2a_1^2 + a_3^2)a_3 - \tfrac{1}{16}\epsilon^2\alpha k_3^{-3}a_1^3 \cos\gamma \quad (8.2.107)$$

$$\gamma = \beta_3 - 3\beta_1 + \sigma_1 x, \quad \epsilon^2\sigma_1 = k_3 - 3k_1 \quad (8.2.108)$$

Equations (8.2.104) through (8.2.108) have the same form as (6.3.23) through (6.3.27). It follows from Section 6.3 that these equations have two classes of solutions. The first class consists of constant a_1 and a_3 and the β_n are adjusted such that $k_3 + \beta_3' = 3(k_1 + \beta_1')$; thus the wave motion is periodic with constant amplitudes. The second class of solutions consists of periodic a_n and β_n so that the wave motion is aperiodic. It should be noted that the periodic-wave motion is unstable because any small disturbance would lead to an aperiodic motion. We note also that (8.2.104) through (8.2.108) do not possess solitary-wave solutions, in contrast with the case of quadratic nonlinearity, such as in the interaction of capillary-gravity waves (Simmons, 1969; McGoldrick, 1970b; Nayfeh, 1973c).

McGoldrick (1970a, 1972) and Kim and Hanratty (1971) demonstrated experimentally the interaction of two capillary-gravity waves on the surface of deep water when the phase speeds are approximately the same. The problem of second-harmonic resonance was studied for the case of deep water by Wilton (1915), Simmons (1969), and McGoldrick (1970b); for the case of finite-depth water by Nayfeh (1970b); and for the case of an airstream adjacent to finite-depth water by Nayfeh (1973c). Nayfeh (1970c, 1971a) and Lekoudis, Nayfeh, and Saric (1977) studied third-harmonic resonances in deep water.

Franken, Hill, Peters, and Weinreich (1961) performed an experiment on second-harmonic generation in the field of nonlinear optics. Armstrong, Bloembergen, Ducuing, and Pershan (1962) considered the case of second-harmonic resonance in optics; while Matsumoto, Sugimoto, and Inoue (1975) studied second-harmonic resonances in a cold collisionless plasma.

May (1960) and Meitzler (1961) demonstrated experimentally that a longitudinal wave can be coupled with a flexural wave at the particular frequency at which these waves propagate with the same phase speed. Sugimoto and Hirao (1977) studied second-harmonic resonances in the interaction of torsional and longitudinal modes.

Nishikawa, Hojo, Mima, and Ikezi (1974) and Kawahara, Sugimoto, and Kakutani (1975) treated the interaction of the envelopes of short and long waves.

Three or more waves can interact in systems governed by higher dimensional problems. Some of these interactions were studied by McGoldrick (1965); Benney and Saffman (1966); Benney (1967); Davidson (1967, 1969, 1972); Benney and Newell (1967a, b, 1969); Hoult (1968); Newell (1968); Sagdeev and Galeev (1969); Tsytovich (1970); Stenflo (1973); Larsson and Stenflo (1973); and Inoue (1975).

Exercises

8.1. Which of the following equations describes dispersive waves?

(a) $\dfrac{\partial u}{\partial t} + \dfrac{\partial u}{\partial x} = 0$

(b) $\dfrac{\partial u}{\partial t} + \dfrac{\partial u}{\partial x} - \dfrac{\partial^2 u}{\partial x^2} = 0$

(c) $\dfrac{\partial u}{\partial t} + \dfrac{\partial u}{\partial x} + \dfrac{\partial^3 u}{\partial x^3} = 0$

(d) $\dfrac{\partial^2 u}{\partial t^2} - \dfrac{\partial^2 u}{\partial x^2} + u = 0$

8.2. Consider right-running waves along a uniform semiinfinite bar. The problem is governed by

$$\frac{\partial^2 u}{\partial x^2} - \frac{\partial^2 u}{\partial t^2} = -2E_1 \frac{\partial u}{\partial x}\frac{\partial^2 u}{\partial x^2}$$

$$u(0, t) = \epsilon\phi(t)$$

$$\frac{\partial u}{\partial t}(x, 0) = \frac{\partial u}{\partial x}(x, 0) = 0$$

(a) Show that the straightforward expansion is

$$u = \epsilon\phi(s_1) - \tfrac{1}{2}\epsilon^2 E_1 x\phi'^2(s_1) + \cdots$$

$$s_1 = t - x$$

(b) Show that the strain e is

$$e = -\epsilon\phi'(s_1) - \tfrac{1}{2}\epsilon^2 E_1[\phi'^2(s_1) - 2x\phi'(s_1)\,\phi''(s_1)] + \cdots$$

Hence it is not valid for distances $O(\epsilon^{-1})$ or larger.

(c) Use the method of renormalization and let

$$s_1 = \xi + \epsilon\xi_1(x, \xi) + \cdots$$

in the expansion for e. Expand for small ϵ and show that the nonuniform term disappears if $\xi_1 = E_1 x\phi'(\xi)$. Therefore a uniform expansion is

$$e = -\epsilon\phi'(\xi) + \cdots$$

$$s_1 = \xi + \epsilon E_1 x\phi'(\xi) + \cdots$$

in agreement with (8.1.71) and (8.1.70) obtained by using the method of multiple scales.

(d) Show that if ξ_1 is chosen to eliminate the secular term in u rather than e, the resulting expansion is

$$u = \epsilon\phi(\xi) + \cdots$$

$$s_1 = \xi + \tfrac{1}{2}\epsilon E_1 x\phi'(\xi) + \cdots$$

Substitute this solution into e and show that

$$e = -\epsilon\phi'(\xi) - \tfrac{1}{2}\epsilon^2 E_1[\phi'^2(\xi) - x\phi'(\xi)\,\phi''(\xi)] + \cdots$$

which is nonuniform for large x (see Nayfeh and Kluwick, 1976).

8.3. The problem of simple waves propagating in a duct carrying a uniform mean flow of Mach number M is given by

$$\phi_{tt} - \phi_{xx} = -2\phi_x\phi_{xt} + (1-\gamma)\,[\phi_t + \tfrac{1}{2}\phi_x^2 - \tfrac{1}{2}M^2]\,\phi_{xx} - \phi_x^2\phi_{xx}$$

$$u(0,t) = \phi_x(0,t) = M + \epsilon f(t)$$

$$u = 0 \text{ upstream}$$

(a) Show that the straightforward expansion for the solution of this problem is (Nayfeh and Kaiser, 1975)

$$u = M + \epsilon f(\xi) + \tfrac{1}{2}\epsilon^2(\gamma+1)(1+M)^{-2}\, x\, f(\xi)\, f'(\xi) + O(\epsilon^3)$$

where $\xi = t - (1+M)^{-1}\,x$. This expansion is not uniform for large x.

(b) Let $\xi = s + \epsilon s_1(s,x) + \cdots$, use the method of renormalization, and obtain the following uniform expansion (Nayfeh and Kaiser, 1975):

$$u = M + \epsilon f(s) + \cdots$$

where

$$t - (1+M)^{-1}\,x = s - \tfrac{1}{2}\epsilon(\gamma+1)(1+M)^{-2}\,x\,f(s) + \cdots$$

8.4. Consider the equation

$$\frac{\partial^2 u}{\partial t^2} - \frac{\partial^2 u}{\partial x^2} + \epsilon\left(\frac{\partial u}{\partial t}\right)^3 = 0$$

(a) Use the method of multiple scales to show that right-running waves can be approximated by (Chikwendu and Kevorkian, 1972)

$$u = f(s_1, T_1) + O(\epsilon)$$

where $s_1 = x - t$ and f satisfies the equation

$$2\frac{\partial^2 f}{\partial s_1\,\partial T_1} + \left(\frac{\partial f}{\partial s_1}\right)^3 = 0$$

(b) Show that this equation has the integral

$$\frac{\partial f}{\partial s_1} = [T_1 + F(s_1)]^{-1/2}$$

(c) If $u(x,0) = a\sin\omega x$ and $\partial u(x,0)/\partial t = -a\omega\cos\omega x$, show that

$$F(s_1) = \frac{\sec^2 \omega s_1}{a^2\omega^2}$$

and then

$$u = \frac{1}{\omega\sqrt{T_1}}\arcsin\left[\left(\frac{a^2\omega^2 T_1}{1 + a^2\omega^2 T_1}\right)^{1/2}\sin\omega s_1\right]$$

8.5. Consider the equation

$$\frac{\partial^2 u}{\partial t^2} - \frac{\partial^2 u}{\partial x^2} + \epsilon \left[\beta \frac{\partial u}{\partial t} - \alpha \left(\frac{\partial u}{\partial t} \right)^3 \right] = 0$$

This equation was proposed by Myerscough (1973) as a model for wind-induced oscillations of overhead power lines and treated by Keller and Kogelman (1970), Chikwendu and Kevorkian (1972), and Lardner (1977).

(a) Use the method of multiple scales to show that right-running waves can be approximated by (Chikwendu and Kevorkian, 1972)

$$u = f(s_1, T_1) + O(\epsilon)$$

where $s_1 = x - t$ and

$$2 \frac{\partial^2 f}{\partial s_1 \, \partial T_1} - \beta \frac{\partial f}{\partial s_1} + \alpha \left(\frac{\partial f}{\partial s_1} \right)^3 = 0$$

(b) Show that this equation has the integral

$$\frac{\partial f}{\partial s_1} = \left[\frac{\alpha}{\beta} + F(s_1)\, e^{-\beta T_1} \right]^{-1/2}$$

(c) If $u(x, 0) = a \sin \omega x$ and $\partial u(x, 0)/\partial t = -a\omega \cos \omega x$, show that

$$F(s_1) = \frac{\sec^2 \omega s_1}{a^2 \omega^2} - \frac{\alpha}{\beta}$$

$$\frac{\partial f}{\partial s_1} = \frac{a\omega \cos \omega s_1}{\left[\frac{\alpha}{\beta} a^2 \omega^2 (1 - e^{-\beta T_1}) \cos^2 \omega s_1 + e^{-\beta T_1} \right]^{1/2}}$$

$$u = \frac{1}{\omega} \left(\frac{\beta/\alpha}{1 - e^{-\beta T_1}} \right)^{1/2} \arcsin \left\{ \left[\frac{\frac{\alpha}{\beta} a^2 \omega^2 (1 - e^{-\beta T_1})}{e^{-\beta T_1} + \frac{\alpha}{\beta} a^2 \omega^2 (1 - e^{-\beta T_1})} \right]^{1/2} \sin \omega s_1 \right\}$$

8.6. In the presence of viscous damping, waves propagating along a uniform, initially undeformed nonlinear elastic bar are governed by

$$\frac{\partial^2 u}{\partial x^2} - \frac{1}{c^2} \frac{\partial^2 u}{\partial t^2} = -2E_1 \frac{\partial u}{\partial x} \frac{\partial^2 u}{\partial x^2} + 2\epsilon\mu \frac{\partial u}{\partial t}$$

(a) Show that right-running waves are approximated by

$$e = \epsilon f(s_1, x_1, T_1) + \cdots$$

where $s_1 = t - (x/c)$, $x_1 = \epsilon x$, $T_1 = \epsilon t$, and f satisfies

$$\frac{\partial f}{\partial x_1} + \frac{1}{c} \frac{\partial f}{\partial T_1} + \mu c f + \frac{E_1}{c} f \frac{\partial f}{\partial s_1} = 0$$

(b) For waves that are "nonlinearly distorted with distance," show that

$$e = \epsilon \exp(-\epsilon\mu cx)\, \phi(\xi) + \cdots$$

where

$$s_1 = \xi + E_1 \mu^{-1} c^{-2} [1 - \exp(-\epsilon\mu cx)]\, \phi(\xi) + \cdots$$

Hint: Let $f = Q(x_1)\, h(s_1, z)$, where $z = Z(x_1)$ and choose Q and Z so that h is given by

$$\frac{\partial h}{\partial z} + \frac{E_1}{c^2} h \frac{\partial h}{\partial s_1} = 0$$

(c) For waves that are "nonlinearly distorted with time," show that

$$e = \epsilon \exp(-\epsilon\mu c^2 t)\, \phi(\xi) + \cdots$$

where

$$s_1 = \xi + E_1 \mu^{-1} c^{-2} [1 - \exp(-\epsilon\mu c^2 t)]\, \phi(\xi) + \cdots$$

(d) Under what conditions can oppositely traveling waves be approximated by

$$e = \epsilon f(s_1, x_1, T_1) + \epsilon g(s_2, x_1, T_1) + \cdots$$

where f is described in (a) while g satisfies the equation

$$\frac{\partial g}{\partial x_1} - \frac{1}{c} \frac{\partial g}{\partial T_1} + \frac{E_1}{c} g \frac{\partial g}{\partial x_1} - \mu c g = 0$$

8.7. For high frequencies the constitutive equation for a homogeneous viscoelastic material can be approximated by

$$\frac{\partial \sigma}{\partial t} = E(1 + 2E_1 e) \frac{\partial e}{\partial t} - 2\tau e$$

where E, E_1, and τ are constants.

(a) Show that waves propagating along a uniform bar made of this material are governed by the equations

$$\rho \frac{\partial^2 u}{\partial t^2} = \frac{\partial \sigma}{\partial x}$$

$$\frac{\partial \sigma}{\partial t} = E\left(1 + 2E_1 \frac{\partial u}{\partial x}\right) \frac{\partial^2 u}{\partial x\, \partial t} - 2\hat{\tau} \frac{\partial u}{\partial x}$$

(b) For small- but finite-amplitude right-running waves, let $\hat{\tau} = \epsilon\tau$ and seek an expansion in the form

$$u = \epsilon u_1(s_1, s_2, x_1, T_1) + \epsilon^2 u_2(s_1, s_2, x_1, T_1) + \cdots$$

$$\sigma = \epsilon \sigma_1(s_1, s_2, x_1, T_1) + \epsilon^2 \sigma_2(s_1, s_2, x_1, T_1) + \cdots$$

where

$$s_1 = t - \frac{x}{c}, \quad s_2 = t + \frac{x}{c}, \quad x_1 = \epsilon x, \quad T_1 = \epsilon t, \quad c^2 = \frac{E}{\rho}$$

For right-running waves take $u_1 = f(s_1, x_1, T_1)$ and show that $\sigma_1 = -\rho c(\partial f/\partial s_1)$

(c) Then show that the equations describing u_2 and σ_2 are

$$\rho c \left(\frac{\partial}{\partial s_2} + \frac{\partial}{\partial s_1}\right)^2 u_2 - \left(\frac{\partial}{\partial s_2} - \frac{\partial}{\partial s_1}\right) \sigma_2 = -2\rho c \frac{\partial^2 f}{\partial s_1 \, \partial T_1} - \rho c^2 \frac{\partial^2 f}{\partial s_1 \, \partial x_1}$$

$$\left(\frac{\partial}{\partial s_2} + \frac{\partial}{\partial s_1}\right) \sigma_2 - \frac{E}{c}\left(\frac{\partial^2}{\partial s_2^2} - \frac{\partial^2}{\partial s_1^2}\right) u_2 = E \frac{\partial^2 f}{\partial s_1 \, \partial x_1} + \frac{2EE_1}{c^2} \frac{\partial f}{\partial s_1} \frac{\partial^2 f}{\partial s_1^2} + \frac{2\tau}{c} \frac{\partial f}{\partial s_1}$$

Subtract these equations to obtain

$$2 \frac{\partial}{\partial s_2} \left[\rho c \left(\frac{\partial}{\partial s_2} + \frac{\partial}{\partial s_1}\right) u_2 - \sigma_2\right] = -2\rho c \frac{\partial^2 f}{\partial s_1 \, \partial T_1} - 2\rho c^2 \frac{\partial^2 f}{\partial s_1 \, \partial x_1} - 2\rho E_1 \frac{\partial f}{\partial s_1} \frac{\partial^2 f}{\partial s_1^2} - \frac{2\tau}{c} \frac{\partial f}{\partial s_1}$$

Hence show that the solvability condition is

$$\frac{\partial^2 f}{\partial s_1 \, \partial x_1} + \frac{1}{c} \frac{\partial^2 f}{\partial s_1 \, \partial T_1} + \frac{E_1}{c^2} \frac{\partial f}{\partial s_1} \frac{\partial^2 f}{\partial s_1^2} + \frac{\tau}{\rho c^3} \frac{\partial f}{\partial s_1} = 0$$

(d) Under what conditions can oppositely traveling waves be approximated by

$$u = \epsilon f(s_1, x_1, T_1) + \epsilon g(s_2, x_1, T_1) + \cdots$$

$$\sigma = -\rho c \epsilon [f(s_1, x_1, T_1) - g(s_2, x_1, T_1)] + \cdots$$

where f is described by the above equation and g is described by

$$\frac{\partial^2 g}{\partial s_2 \, \partial x_1} - \frac{1}{c} \frac{\partial^2 g}{\partial s_2 \, \partial T_1} + \frac{E_1}{c^2} \frac{\partial g}{\partial s_2} \frac{\partial^2 g}{\partial s_2^2} + \frac{\tau}{\rho c^3} \frac{\partial g}{\partial s_2} = 0$$

(e) When the waves are "nonlinearly distorted with distance" only, show that the equation describing f becomes

$$\frac{\partial F}{\partial x_1} + \frac{E_1}{c^2} F \frac{\partial F}{\partial s_1} + \frac{\tau}{\rho c^2} F = 0$$

where $F = \partial f / \partial s_1$. Show that

$$\sigma = -\epsilon \rho c \exp\left(-\frac{\epsilon \tau}{\rho c^2} x\right) \psi(\xi)$$

where

$$s_1 = t - \frac{x}{c} = \xi + \frac{E_1 \rho}{\tau} \left[1 - \exp\left(-\frac{\epsilon \tau x}{\rho c^2}\right)\right] \psi(\xi)$$

Hint: Let $F = Q(x_1)\, H(s_1, z)$, $z = Z(x_1)$, and choose Q and Z such that H satisfies $(\partial H/\partial z) + (E_1/c^2)\, H(\partial H/\partial s_1) = 0$

8.8. For low frequencies the constitutive equation for a viscoelastic material can be approximated by the Voigt model

$$\sigma = E(e + E_1 e^2) + 2\hat{\mu}\,\frac{\partial e}{\partial t}$$

(a) Show that nonlinear waves in a uniform bar made of this material are governed by the equation

$$\frac{\partial^2 u}{\partial x^2} - \frac{1}{c^2}\frac{\partial^2 u}{\partial t^2} = -2E_1 \frac{\partial u}{\partial x}\frac{\partial^2 u}{\partial x^2} - 2\frac{\hat{\mu}}{E}\frac{\partial^3 u}{\partial x^2\, \partial t}$$

(b) To determine an approximation to small but finite-amplitude progressive waves in this rod, set $\hat{\mu} = \epsilon\mu E$ and obtain

$$u = \epsilon f(s_1, x_1, T_1) + \cdots$$

where f satisfies the following equation:

$$\frac{\partial^2 f}{\partial s_1\, \partial x_1} + \frac{1}{c}\frac{\partial^2 f}{\partial s_1\, \partial T_1} + \frac{E_1}{c^2}\frac{\partial f}{\partial s_1}\frac{\partial^2 f}{\partial s_1^2} - \frac{\mu}{c}\frac{\partial^3 f}{\partial s_1^3} = 0$$

where $s_1 = t - (x/c)$, $x_1 = \epsilon x$, and $T_1 = \epsilon t$. Letting $F = \partial f/\partial s_1$, rewrite this equation as the generalized Burgers equation

$$\frac{\partial F}{\partial x_1} + \frac{1}{c}\frac{\partial F}{\partial T_1} + \frac{E_1}{c^2} F \frac{\partial F}{\partial s_1} - \frac{\mu}{c}\frac{\partial^2 F}{\partial s_1^2} = 0$$

If the distortion is a function of T_1, F is governed by

$$\frac{\partial F}{\partial T_1} + \frac{E_1}{c} F \frac{\partial F}{\partial s_1} - \mu \frac{\partial^2 F}{\partial s_1^2} = 0$$

which was obtained by Nariboli and Sedov (1970) and Nariboli and Lin (1973) using the method of multiple scales and by Lardner (1976) using the method of averaging.

(c) Under what conditions can an approximate solution to the case of oppositely traveling waves along this bar be given by

$$u = \epsilon f(s_1, x_1, T_1) + \epsilon g(s_1, x_1, T_1) + \cdots$$

where the equation describing f is given in (b) above, while g is governed by

$$\frac{\partial^2 g}{\partial s_2\, \partial x_1} - \frac{1}{c}\frac{\partial^2 g}{\partial s_2\, \partial T_1} + \frac{E_1}{c^2}\frac{\partial g}{\partial s_2}\frac{\partial^2 g}{\partial s_2^2} + \frac{\mu}{c}\frac{\partial^3 g}{\partial s_2^3} = 0$$

Thus the waves do not interact in the body of the material although they may interact at the boundaries (Mortell, 1977; Nayfeh, 1977).

8.9. Consider Burgers' equation

$$\frac{\partial u}{\partial t} + u \frac{\partial u}{\partial x} = \tfrac{1}{2} \delta \frac{\partial^2 u}{\partial x^2}$$

Show that the change of variables

$$\psi = \exp\left(-\frac{1}{\delta}\int u\, dx\right)$$

$$u = -\frac{\delta}{\psi}\frac{\partial \psi}{\partial x}$$

transforms this equation into the heat equation (Hopf, 1950; Cole, 1951)

$$\frac{\partial \psi}{\partial t} = \tfrac{1}{2} \delta \frac{\partial^2 \psi}{\partial x^2}$$

8.10. Seek stationary solutions of Burgers' equation

$$\frac{\partial u}{\partial t} + u \frac{\partial u}{\partial x} = \tfrac{1}{2} \delta \frac{\partial^2 u}{\partial x^2}$$

in the form

$$u = f(x - ct)$$

Derive a nonlinear ordinary equation for f. Advani (1969b) used this approach to study waves propagating in an infinite plate, while Yen and Tang (1970, 1976) used it to study waves generated by a steady moving load on an elastic string and plate, respectively.

8.11. Consider the Burgers equation

$$\frac{\partial u}{\partial t} + u \frac{\partial u}{\partial x} = \nu \frac{\partial^2 u}{\partial x^2}, \quad \nu > 0$$

(a) Show that it has a steady solution of the form

$$u_0 = -u_\infty \tanh\left(\frac{u_\infty x}{2\nu}\right), \quad u_\infty > 0$$

where u_∞ is a constant. Show that this solution is a shocklike solution.

(b) Let $u = u_0 + v$, where $|v| << |u_0|$ and show that

$$\frac{\partial v}{\partial t} + u_0 \frac{\partial v}{\partial x} + \frac{du_0}{dx} v = \nu \frac{\partial^2 v}{\partial x^2}$$

(c) Use the variational equation to study the stability of the above steady solution (Jeffery and Kakutani, 1970).

8.12. Consider the Korteweg-de Vries equation

$$\frac{\partial u}{\partial t} + u \frac{\partial u}{\partial x} = \epsilon \frac{\partial^3 u}{\partial x^3}, \quad \epsilon > 0$$

(a) Show that it has the steady solution

$$u_0 = u_\infty \left[1 - 3\,\mathrm{sech}^2 \left(\frac{u_\infty}{4\epsilon}\right)^{1/2} x\right], \quad u_\infty > 0$$

(b) Let $u = u_0 + v$, where $|v| < |u_0|$, and investigate the stability of this steady solution (Jeffery and Kakutani, 1970).

8.13. Consider the sine-Gordon equation

$$\psi_{xx} - \psi_{tt} = \sin\psi$$

Let $\psi = \pi + \phi(s)$, where $s = x - vt$ and $v < 1$. Show that ϕ is governed by (Sabata, 1974)

$$\phi'' + (1 - v^2)^{-1} \sin\phi = 0$$

Investigate the solutions of this pendulum equation and discuss the solutions of the original equation.

8.14. Waves propagating along a uniform nonlinear elastic bar are governed by the equation

$$\frac{\partial^2 u}{\partial x^2} - \frac{1}{c^2}\frac{\partial^2 u}{\partial t^2} = -2E_1 \frac{\partial u}{\partial x}\frac{\partial^2 u}{\partial x^2}$$

(a) Show that its characteristics are given by

$$\left(1 + 2E_1 \frac{\partial u}{\partial x}\right)\left(\frac{dt}{dx}\right)^2 = \frac{1}{c^2}$$

Hence

$$\frac{dt}{dx} = \pm\frac{1}{c}\left(1 - E_1 \frac{\partial u}{\partial x}\right) + \text{higher-order terms}$$

(b) Change the independent variables x and t to the nonlinear characteristics ξ and η. Hence show that the equations of the characteristics can be rewritten as

$$\frac{\partial t}{\partial \eta} = \left(\frac{1}{c} - \frac{E_1 e}{c}\right)\frac{\partial x}{\partial \eta} = c_1 \frac{\partial x}{\partial \eta}$$

$$\frac{\partial t}{\partial \xi} = \left(-\frac{1}{c} + \frac{E_1 e}{c}\right)\frac{\partial x}{\partial \xi} = c_2 \frac{\partial x}{\partial \xi}$$

where $e = \partial u/\partial x$.

(c) Use the chain rule to transform the derivatives to

$$\frac{\partial}{\partial x} = -\frac{1}{c_2 - c_1}\left[\frac{c_1}{x_\xi}\frac{\partial}{\partial \xi} - \frac{c_2}{x_\eta}\frac{\partial}{\partial \eta}\right]$$

$$\frac{\partial}{\partial t} = \frac{1}{c_2 - c_1}\left[\frac{1}{x_\xi}\frac{\partial}{\partial \xi} - \frac{1}{x_\eta}\frac{\partial}{\partial \eta}\right]$$

where $x_\xi = \partial x/\partial \xi$ and $x_\eta = \partial x/\partial \eta$. Then rewrite the set of equations as

$$x_\xi(e_\eta + c_2 v_\eta) - x_\eta(e_\xi + c_1 v_\xi) = 0$$

$$x_\xi[v_\eta + c^2 c_2(1 + 2E_1 e)\, e_\eta] - x_\eta[v_\xi + c^2 c_1(1 + 2E_1 e)\, e_\xi] = 0$$

where $v = \partial u/\partial t$.

(d) Show that a straightforward expansion for this problem is (Lin, 1954)

$$e = \epsilon f(\xi) + \epsilon g(\eta) + \cdots$$

$$v = -\epsilon c f(\xi) + \epsilon c g(\eta) + \cdots$$

$$t = \tfrac{1}{2}(\eta + \xi) - \tfrac{1}{4}\epsilon E_1 \left[f(\eta - \xi) + g(\xi - \eta) + \int_\eta^\xi f(s)\, ds + \int_\xi^\eta g(s)\, ds\right] + \cdots$$

$$x = \tfrac{1}{2}c(\eta - \xi) + \tfrac{1}{4}\epsilon c E_1 \left[f(\eta - \xi) - g(\xi - \eta) - \int_\eta^\xi f(s)\, ds + \int_\xi^\eta g(s)\, ds\right] + \cdots$$

where the parametrization was fixed by imposing the conditions $x(\xi, \xi) = 0$ and $t(\xi, \xi) = \xi$. Then show that

$$t - \frac{x}{c} = \xi - \tfrac{1}{2}\epsilon E_1 \left[\frac{2x}{c} f(\xi) + \int_\xi^\eta g(s)\, ds\right] + \cdots$$

$$t + \frac{x}{c} = \eta + \tfrac{1}{2}\epsilon E_1 \left[\frac{2x}{c} g(\eta) - \int_\eta^\xi f(s)\, ds\right] + \cdots$$

Under what conditions can the integral terms be neglected? Show that except for the presence of the integral terms in the expressions for ξ and η, this solution is in full agreement with that obtained by using the method of multiple scales.

8.15. Consider the problem of forced excitations of a nonlinear finite elastic bar. The problem is governed by

$$\frac{\partial^2 u}{\partial x^2} - \frac{\partial^2 u}{\partial t^2} = -2E_1 \frac{\partial u}{\partial x}\frac{\partial^2 u}{\partial x^2}$$

$$u(0, t) = 0, \quad u(1, t) = \epsilon p \cos \omega t$$

(a) Seek a straightforward expansion in the form

$$u = \epsilon u_1(x, t) + \epsilon^2 u_2(x, t) + \cdots$$

(b) Show that

$$u_1 = p \cos \omega t \sin \omega x\, (\sin \omega)^{-1}$$

(c) Show that

$$u_2 = \frac{E_1 p^2 \omega^2}{8 \sin^2 \omega}\left[\frac{1}{\omega}(x \sin 2\omega - \sin 2\omega x) - \left(x \cos 2\omega x - \frac{\cos 2\omega \sin 2\omega x}{\sin 2\omega}\right)\cos 2\omega t\right]$$

(d) Show that the straightforward expansion breaks down when $\omega \approx n\pi$; these are the resonant frequencies of the bar and the excitation is a primary resonance.

(e) Show also that the straightforward expansion breaks down when $\omega \approx (n + \frac{1}{2})\pi$; the bar is subharmonically excited.

(f) Determine an expansion valid near $\omega \approx n\pi$, where n is an integer; Chester (1964) and Collins (1971) obtained uniform expansions for gas-filled columns.

(g) Determine an expansion valid near $\omega \approx (n + \frac{1}{2})\pi$, where n is an integer.

8.16. Consider one-dimensional wave propagation in an inviscid isentropic gas. If x is the position of a fluid particle in the undisturbed state and if $\eta(x, t)$ is its displacement from the position x,

(a) Show that the motion is governed by

Conservation of mass

$$\rho_0 = \left(1 + \frac{\partial \eta}{\partial x}\right)\rho$$

Conservation of momentum

$$\rho_0 \frac{\partial^2 \eta}{\partial t^2} = -\frac{\partial p}{\partial x}$$

Equation of state

$$\frac{p}{p_0} = \left(\frac{\rho}{\rho_0}\right)^{\gamma}$$

where ρ_0 and p_0 are the density and pressure in the undisturbed state, ρ and p are the instantaneous density and pressure of the gas, and γ is the gas specific heat ratio.

(b) Combine these equations to arrive at the following Lagrangian form of the wave equation:

$$\frac{\partial^2 \eta}{\partial t^2} = c_0^2 \left(1 + \frac{\partial \eta}{\partial x}\right)^{-(\gamma+1)} \frac{\partial^2 \eta}{\partial x^2}$$

where $c_0 = (\gamma p_0/\rho_0)^{1/2}$ is the speed of sound in the undisturbed fluid.

8.17. Consider the Lagrangian form of the wave equation in an inviscid isentropic gas

$$\frac{\partial^2 \eta}{\partial t^2} = c_0^2 \left(1 + \frac{\partial \eta}{\partial x}\right)^{-(\gamma+1)} \frac{\partial^2 \eta}{\partial x^2}$$

(a) Assume that the particle velocity $u = \partial\eta/\partial t$ is a function of the local fluid density. That is,

$$u = f\left(\frac{\partial \eta}{\partial x}\right)$$

Substitute this form of solution in the wave equation and obtain

$$f'^2 = c_0^2 \left(1 + \frac{\partial \eta}{\partial x}\right)^{-(\gamma+1)}$$

(b) Since in the undisturbed state $u = \partial\eta/\partial x = 0$, show that

$$u = \pm \frac{2c_0}{1-\gamma} \left[1 - \left(1 + \frac{\partial \eta}{\partial x}\right)^{(1-\gamma)/2}\right]$$

(c) Show that the exact characteristics of the wave equation are given by

$$\frac{dt}{dx} = \pm \frac{1}{c}, \quad c = c_0 \left(1 + \frac{\partial \eta}{\partial x}\right)^{-(\gamma+1)/2}$$

(d) Show that

$$c = c_0 \left(1 \mp \frac{1-\gamma}{2c_0} u\right)^{(\gamma+1)/(\gamma-1)}$$

(e) Therefore the exact solution for right-running waves is (Earnshaw, 1860)

$$u = F\left(t - \frac{x}{c}\right)$$

or

$$u = F\left[t - \frac{x}{c_0}\left(1 + \frac{\gamma-1}{2c_0} u\right)^{-(\gamma+1)/(\gamma-1)}\right]$$

(f) For small but finite waves, show that u can be approximated by

$$u = F\left[t - \frac{x}{c_0}\left(1 - \frac{\gamma+1}{2c_0} u\right)\right]$$

(g) Compare this approximate solution with a solution obtained by using the method of multiple scales.

8.18. An approximate solution for right-running waves in an inviscid isentropic gas is given in Exercise 8.17 as

$$u = F\left[t - \frac{x}{c_0} + \frac{\gamma+1}{2c_0^2} xu\right]$$

(a) If $u(0, t) = u_0 \sin \omega t$, show that

$$u = u_0 \sin \left[\omega t - \frac{\omega x}{c_0} + \frac{\gamma+1}{2c_0^2} \omega x u\right] = u_0 \sin \xi$$

(b) Show that u can be represented by the following Fourier series (Fubini-Ghiron, 1935):

$$\frac{u}{u_0} = \frac{4c_0^2}{(\gamma+1)\,\omega u_0 x} \sum_{n=1}^{\infty} \frac{1}{n} J_n \left[\frac{n(\gamma+1)\,\omega u_0 x}{2c_0^2}\right] \sin \left[n\left(\omega t - \frac{\omega x}{c_0}\right)\right]$$

$$\text{Hint: } \frac{1}{\pi}\int_0^{2\pi} \sin\xi \sin ns\, ds = -\frac{1}{n}\sin\xi\cos ns \Big|_0^{2\pi} + \frac{1}{n\pi}\int_0^{2\pi} \cos ns \cos\xi\, d\xi$$

$$\sigma\cos\xi\, d\xi = d\xi - ds$$

$$\frac{1}{\pi}\int_0^{\pi} \cos(n\sigma\sin\xi - n\xi)\, d\xi = J_n(n\sigma)$$

8.19. Consider one-dimensional wave propagation in an inviscid isentropic gas.

(a) Show that the governing equations in Eulerian form are

Conservation of mass

$$\frac{1}{\rho}\frac{\partial\rho}{\partial t} + \frac{u}{\rho}\frac{\partial\rho}{\partial x} = -\frac{\partial u}{\partial x}$$

Conservation of momentum

$$\frac{\partial u}{\partial t} + u\frac{\partial u}{\partial x} = -\frac{1}{\rho}\frac{\partial p}{\partial x}$$

Equation of state

$$\frac{p}{p_0} = \left(\frac{\rho}{\rho_0}\right)^{\gamma}$$

where ρ, p, and u are the gas density, pressure, and velocity, respectively.

(b) Add $\pm c(c^2 = \partial p/\partial\rho)$ times the mass equation to the momentum equation and obtain

$$\frac{\partial J_1}{\partial t} = -(u + c)\frac{\partial J_1}{\partial x}$$

$$\frac{\partial J_2}{\partial t} = -(u - c)\frac{\partial J_2}{\partial x}$$

where

$$J_1 = u + \int_{\rho_0}^{\rho} \frac{c}{\rho}\, d\rho = u + \frac{2}{\gamma - 1}[c - c_0]$$

$$J_2 = u - \int_{\rho_0}^{\rho} \frac{c}{\rho}\, d\rho = u - \frac{2}{\gamma - 1}[c - c_0]$$

(c) Show that J_1 is invariant along the curve $dx/dt = u + c$, while J_2 is invariant along the curve $dx/dt = u - c$. These are the Riemann invariants (Riemann, 1858).

(d) Show that simple waves are given by (Whitham, 1974, Section 6.8)

$$u = F\left[t - \frac{x}{c_0 + \frac{\gamma+1}{2} u}\right]$$

8.20. Consider linear inviscid acoustic waves propagating in a hardwalled duct.

(a) Show that the problem is governed by

$$\nabla^2 p = \frac{1}{c^2} \frac{\partial^2 p}{\partial t^2}$$

$$\frac{\partial p}{\partial n} = 0 \quad \text{on } \Gamma$$

where p is the pressure, c is the speed of sound, Γ is the duct surface, and $\partial p/\partial n$ is the normal derivative.

(b) Seek a solution in the form

$$p = \psi(y, z) \exp\left[i(kx - \omega t)\right]$$

and obtain

$$\frac{\partial^2 \psi}{\partial y^2} + \frac{\partial^2 \psi}{\partial z^2} + \gamma^2 \psi = 0$$

$$\frac{\partial \psi}{\partial n} = 0 \quad \text{on } \Gamma$$

where

$$\omega^2 = c^2 (k^2 + \gamma^2)$$

(c) Investigate whether the waves are dispersive or nondispersive for a rectangular duct. Are there any harmonic resonances?

(d) Which waves are dispersive in a cylindrical duct? Are there any harmonic resonances?

8.21. Consider the problem of linear waves propagating on the surface of an inviscid liquid of finite depth. Since the liquid is inviscid, its motion can be expressed in terms of a potential function $\phi(x, y, t)$ with the velocity vector $\mathbf{v} = \nabla\phi$. Then the governing equations and boundary conditions are

$$\nabla^2 \phi = 0$$

$$\frac{\partial \phi}{\partial y} = 0 \quad \text{at } y = -h$$

$$\frac{\partial \eta}{\partial t} - \frac{\partial \phi}{\partial y} = 0 \quad \text{at } y = 0$$

$$\eta + \phi_t - W\eta_{xx} = 0 \quad \text{at } y = 0$$

where $\eta(x, t)$ is the elevation of the interface above its undisturbed position, h is the dimensionless depth of the liquid, and W is the reciprocal of the Weber number.

(a) Show that a solution of this problem can be expressed as

$$\eta = \eta_0 \exp\,[i(kx - \omega t)]$$

$$\phi = -\frac{i\omega\eta_0}{k}\frac{\cosh\,[k(y+h)]}{\sinh kh}\exp\,[i(kx - \omega t)]$$

where η_0 is a constant and ω and k satisfy the dispersion relationship

$$\omega^2 = k \tanh\,(kh)\,(1 + Wk^2)$$

(b) For a small surface tension, $W \approx 0$ and

$$\omega^2 = k \tanh\,(kh)$$

Show that the waves are nondispersive when $h \to 0$ (i.e., shallow water).

(c) For deep water and nonnegligible surface tension, show that the dispersion relation becomes

$$\omega^2 = k(1 + Wk^2)$$

Determine the conditions under which both (ω, k) and $(n\omega, nk)$ for any integer n satisfy this dispersion relation; these are the conditions for the existence of harmonic resonance.

8.22. Consider waves governed by the Klein-Gordon equation

$$\frac{\partial^2 u}{\partial t^2} - \frac{\partial^2 u}{\partial x^2} - \gamma u = \epsilon\alpha u^3$$

(a) Use the method of multiple scales to show that an approximate solution to a wavepacket is

$$u = A(x, t) \exp\,[i(kx - \omega t)] + \cdots$$

where $\omega^2 = k^2 - \gamma$ and A is governed by either

$$\frac{\partial A}{\partial t} + \omega' \frac{\partial A}{\partial x} - \tfrac{1}{2} i\omega'' \frac{\partial^2 A}{\partial x^2} = \frac{3i\epsilon\alpha}{2\omega} A^2 \bar{A}$$

or

$$\frac{\partial A}{\partial x} + k' \frac{\partial A}{\partial t} + \tfrac{1}{2} ik'' \frac{\partial^2 A}{\partial t^2} = \frac{3i\epsilon\alpha}{2k} A^2 \bar{A}$$

(b) When ω is real, the motion is stable (i.e., it is bounded for all t). Then show that monochromatic waves with fixed wavenumbers are given by

$$u = a \cos\,(kx - \hat{\omega} t) + \cdots$$

where a is a constant and

$$\hat{\omega} = \omega - \frac{3\epsilon\alpha a^2}{8\omega} + \cdots$$

(c) Investigate the stability of the preceding monochromatic wave solution.

(d) Show that monochromatic waves with fixed frequency are given by

$$u = a \cos (\hat{k}x - \omega t) + \cdots$$

where a is a constant and

$$\hat{k} = k + \frac{3\epsilon\alpha a^2}{8k} + \cdots$$

(e) Investigate the stability of the preceding monochromatic wave solution.

(f) When $\gamma > 0$, determine the critical wavenumber k_c separating stability from instability as a function of the disturbance amplitude (Nayfeh, 1976a).

References

The italic numbers that follow each reference indicate the pages on which the reference is cited.

Ablowitz, M. J., B. Funk, and A. C. Newell (1973). Semi-resonant interactions and frequency dividers. *Stud. Appl. Math.*, **52,** 51–88. *163*

Ablowitz, M. J., D. J. Kaup, A. C. Newell, and H. Segur (1974). The inverse scattering transform Fourier analysis for nonlinear problems. *Stud. Appl. Math.*, **53,** 249–315. *588*

Adams, O. E. and R. M. Evan-Iwanowski (1973). Stationary and nonstationary responses of a reinforced cylindrical shell near parametric resonance. *Int. J. Non-Linear Mech.*, **8,** 656–673. *269*

Adimurthy, N. K. See Alwar and Adimurthy.

Adler, L. and M. A. Breazeale (1970). Generation of fractional harmonics in a resonant ultrasonic wave system. *J. Acoust. Soc. Am.*, **48,** 1077–1083. *367, 416*

Advani, S. H. (1967a). Large amplitude asymmetric waves in a spinning membrane. *J. Appl. Mech.*, **34,** 1044–1045. *506, 543*

Advani, S. H. (1967b). Stationary waves in a thin spinning disk. *Int. J. Mech. Sci.*, **9,** 307–313. *506*

Advani, S. H. (1969a). Nonlinear transverse vibrations and waves in spinning membrane discs. *Int. J. Non-Linear Mech.*, **4,** 123–127. *543*

Advani, S. H. (1969b). Wave propagation in an infinite plate. *J. Inst. Math. Appl.*, **5,** 271–282. *571, 608*

Advani, S. H. and A. Bhattacharjie (1969). Large amplitude axisymmetric transverse vibrations of spinning membrane disks. *J. Sound Vib.*, **9,** 59–64. *506*

Advani, S. H. and P. Z. Bulkeley (1969). Nonlinear transverse vibrations and waves in spinning membrane discs. *Int. J. Non-Linear Mech.*, **4,** 123–127. *506, 543*

Agrawal, B. N. (1975). Nonstationary and nonlinear vibration analysis and its application. *Shock Vib. Dig.*, **7,** 77–87. *402*

Agrawal, B. J., and R. M. Evan-Iwanowski (1973). Resonances in nonstationary, nonlinear, multi-degree-of freedom systems. *AIAA J.*, **11,** 907–912. *219, 417, 418*

Ahmad, J., and H. C. Lee (1975). Nonlinear acoustic response of a cylindrical shell. *J. Franklin Inst.*, **299,** 171–189. *506*

Ahuja, R., and R. C. Duffield (1975). Parametric instability of variable cross-section beams resting on an elastic foundation. *J. Sound Vib.*, **39,** 159–174. *267*

Aks, S., and R. Carhart (1970). Renormalized perturbation theory for the weakly nonlinear oscillator. *J. Math. Phys.*, **11,** 214–222. *163*

Aks, S. See Varga and Aks.

Akulenko, L. D. (1968a). On the analysis of resonances in nonlinear systems. *PMM*, **32,** 1124–1129. *417*

Akulenko, L. D. (1968b). On the oscillatory and rotational resonant motions. *PMM*, **32,** 298–305. *418*

Alekseeva, N. K. (1969). On the question of the dynamic stability of shallow shells. *Prikl. Mekh.*, **5**, 60–68. *506*

Alfriend, K. T. (1970). The stability of the triangular Lagrangian points for commensurability of order two. *Celestial Mech.*, **1**, 315–359. *378, 379*

Alfriend, K. T. (1971a). Stability and motion about L_4 at three-to-one commensurability. *Celestial Mech.*, **4**, 60–77. *378*

Alfriend, K. T. (1971b). Stability and motion in two degree of freedom Hamiltonian systems for three-to-one commensurability. *Int. J. Non-Linear Mech.*, **6**, 563–578. *396*

Alfriend, K. T. (1971c). Stability and motion in two-degree-of-freedom Hamiltonian systems for two-to-one commensurability. *Celestial Mech.*, **3**, 247–265. *379, 396*

Alfriend, K. T. (1977). Comment on "analytical solution for planar librations of a gravity stabilized satellite." *J. Spacecraft*, **14**, 767–768. *257*

Alfriend, K. T., and R. H. Rand (1969). Stability of the triangular points in the elliptic restricted problem of three bodies. *AIAA J.*, **7**, 1024–1028. *331, 379*

Alfriend, K. T., and D. L. Richardson (1973). Third and fourth order resonances in Hamiltonian systems. *Celestial Mech.*, **7**, 408–420. *396*

Ali, S. M. J. See Singh and Ali.

Ali Hasan, S., and A. D. Barr (1974). Nonlinear and parametric vibration of thin-walled beams of equal angle-section. *J. Sound Vib.*, **32**, 25–47. *267, 269*

Al-Khattat, I. M. See Huang and Al-Khattat.

Alwar, R. S., and N. K. Adimurthy (1975). Non-linear dynamic response of sandwich panels under pulse and shock type excitations. *J. Sound Vib.*, **39**, 43–54. *501*

Ambartsumyan, S. A. (1976). *Theory of Anisotropic Plates.* Technomic, Stamford. *501*

Ambartsumyan, S. A., and A. A. Khachaturian (1960). On the stability of vibrations of anisotropic plates. *Izv. Akad. Nauk*, **1**, 113–122. *268*

Ames, W. E. See Davy and Ames.

Ames, W. F. See Lee and Ames.

Anand, G. V. (1966). Nonlinear resonance in stretched strings with viscous damping. *J. Acoust. Soc. Am.*, **40**, 1517–1528. *486*

Anand, G. V. (1969a). Large-amplitude damped free vibrations of a stretched string. *J. Acoust. Soc. Am.*, **45**, 1089–1096. *486*

Anand, G. V. (1969b). Stability of nonlinear oscillations of stretched strings. *J. Acoust. Soc. Am.*, **46**, 667–677. *486*

Anand, G. V. (1972). Natural modes of a coupled nonlinear system. *Int. J. Non-Linear Mech.*, **7**, 81–91. *365*

Anand, G. V. (1973). Negative resistence mode of forced oscillations of a string. *J. Acoust. Soc. Am.*, **54**, 692–698. *486*

Anderson, G. L. (1974a). An asymptotic analysis of the response of nonlinear damped systems subjected to step function excitation. *J. Sound Vib.*, **34**, 424–440. *162*

Anderson, G. L. (1974b). Application of ultraspherical polynomials to non-linear, nonconservative systems subjected to step function excitation. *J. Sound Vib.*, **32**, 101–108. *162*

Anderson, G. L. (1975a). A modified perturbation method for treating nonlinear systems under harmonic excitation. *J. Sound Vib.*, **40**, 219–225. *163*

Anderson, G. L. (1975b). On the extensional and flexural vibrations of rotating bars. *Int. J. Non-Linear Mech.*, **10**, 223–236. *448*

Anderson, G. L. (1975c). The method of averaging applied to a damped, non-linear system under harmonic excitation. *J. Sound Vib.*, **40**, 219–225. *163*

Anderson, W. J. See Hyer, Anderson, and Scott; Kovac, Anderson, and Scott.

Andronov, A., and M. A. Leontovich (1927). On the vibrations of systems with periodically varying parameters. *Zh. Russk. Fiz.-Khim. Obshch.*, **59**, 429–443. *259, 267*

Andronov, A., and A. Vitt (1930). Zur Theorie des Mitnehmens von van der Pol. *Arch. Elektrotech.*, **24**, 99–110. *206*

Andronov, A., A. Vitt, and S. Khaikin (1966). *Theory of Oscillators.* Addison-Wesley, Reading. *vii, 117, 144, 259, 262*

Ang, A. H. S. See Crose and Ang.

Angelopoulos, T. See Argyris, Dunne, and Angelopoulos.

Anh, L. S. (1973). Mechanical relaxation oscillations. *Mech. Solids*, **8**, 37–40. *144*

Ansari, K. A. (1974). Nonlinear flexural vibrations of a rotating Myklestad beam. *AIAA J.*, **12**, 98–99. *448*

Ansari, K. A. (1975). Nonlinear vibrations of a rotating pretwisted blade. *Comput. Struct.*, **5**, 101–118. *448*

Antipov, A. A. See Popov, Antipov, and Krzhechkovskii.

Aravamudan, K. S., and P. N. Murthy (1973). Non-linear vibration of beams with time-dependent boundary conditions. *Int. J. Non-Linear Mech.*, **8**, 195–212. *454*

Argyris, J. H., and D. W. Scharpf (1969). Finite elements in time and space. *Nucl. Eng. Des.*, **10**, 456–464. *445*

Argyris, J. H., P. C. Dunne, and T. Angelopoulos (1973). Nonlinear oscillations using the finite element technique. *Comput. Meth. Appl. Mech. Eng.*, **2**, 203–250. *50*

Armstrong, J. A., N. Bloembergen, J. Ducuing, and P. S. Pershan (1962). Interactions between light waves in a nonliner dielectric. *Phys. Rev.*, **127**, 1918–1939. *601*

Arnold, F. R. (1955). Steady-state behavior of systems provided with nonlinear dynamic vibration absorbers. *J. Appl. Mech.*, **22**, 487–402. *417*

Arnold, R. N. See Tobias and Arnold.

Arnold, V. I. (1963). Small denominators and the problems of stability of motion in the classical and celestial mechanics. *Usp. Math. Nauk*, **18**, 91–192. *395*

Arscott, F. (1964). *Periodic Differential Equations.* Pergamon, New York. *259*

Arya, J. C., G. N. Bojadziev, and A. S. Farooqui (1975). Response of a nonlinear vibrator under the influence of a time-dependent external force. *IEEE Trans.*, **AC-20**, 232–234. *119, 156, 162, 255*

Arya, J. C., and R. W. Lardner (1974). The effect of kinematical nonlinearities on the vibration frequencies of a stretched string. *Utilitas Math.*, **6**, 307–320. *445, 489*

Asano, N. (1970). Reductive perturbation method for nonlinear wave propagation in inhomogeneous media. III. *J. Phys. Soc. Japan*, **29**, 220–224. *556, 568*

Asano, N. (1974). Modulation for nonlinear waves in dissipative or unstable media. *J. Phys. Soc. Japan*, **36**, 861–868. *587*

Asano, N., and Y. Kato (1977). Spectrum method for a general evolution equation. *Prog. Theor. Phys.*, **58**, 161–174. *588*

Asano, N., K. Nishihara, K. Nozaki, and T. Taniuti (1976). Weak thermonuclear reaction wave in high-density plasma. *J. Phys. Soc. Japan*, **41**, 1774–1777. *556*

Asano, N., and T. Taniuti (1969). Reductive perturbation method for nonlinear wave propagation in inhomogeneous media. I. *J. Phys. Soc. Japan*, **27**, 1059–1062. *556, 568*

Asano, N., and T. Taniuti (1970). Reductive perturbation method for nonlinear wave propagation in inhomogeneous media. II. *J. Phys. Soc. Japan*, **29**, 209–214. *556, 568*

Asano, N., T. Taniuti, and N. Yajima (1969). Perturbation method for a nonlinear wave modulation. II. *J. Math. Phys.*, **10**, 2020–2024. *587*

Asfar, O. R., and A. H. Nayfeh (1976). Nonlinear modulation of TM waves in a circular waveguide. *J. Appl. Phys.*, **47**, 88–91. *587*

Asfar, O. E. See Nayfeh and Asfar.

Ashwell, D. G., and A. P. Chauhan (1973). A study of 1/2-subharmonic oscillations by the method of harmonic balance. *J. Sound Vib.*, **27**, 313–324. *199*

Ashwell, D. G. See Chauhan and Ashwell.

Asmis, K. G., and W. K. Tso (1972). Combination resonance in a nonlinear two-degree-of freedom system. *J. Appl. Mech.*, **E39**, 832–834. *418*

Asmis, K. G. See Tso and Asmis.

Atkinson, C. P. (1956). Electronic analog computer solutions of nonlinear vibratory systems of two degrees of freedom. *J. Appl. Mech.*, **23**, 629–634. *417*

Atkinson, C. P. (1957). Superharmonic oscillators as solutions to Duffing's equation as solved by an electronic differential analysis. *J. Appl. Mech.*, **24**, 520–525. *175*

Atkinson, C. P. (1962). On the superposition method for determining frequencies of nonlinear systems. *ASME Proc. 4th Natl. Congr. Appl. Mech.*, 57–62. *365*

Atkinson, C. P. See Rosenberg and Atkinson.

Atluri, S. (1972). A perturbation analysis of nonlinear free flexural vibrations of a circular cylindrical shell. *Int. J. Solids Struct.*, **8**, 549–569. *506, 533*

Atluri, S. (1973). Nonlinear vibrations of a hinged beam including nonlinear inertia effects. *J. Appl. Mech.*, **40**, 121–126. *445*

Atluri, S. See Crawford and Atluri; Gordon and Atluri.

Atta, E., O. A. Kandil, D. T. Mook, and A. H. Nayfeh (1977). Unsteady aerodynamic loads on arbitrary wings including wing-tip and leading-edge separation. *AIAA Paper No. 77-156. 98*

Babcock, C. D. See Chen and Babcock.

Bailey, H. R., and L. Cesari (1958). Boundedness of solutions of linear differential systems with periodic coefficients. *Arch. Ration. Mech. Anal.*, **1**, 246–271. *304*

Bainov, D. D. (1966). Periodic oscillations of a non-self contained quasilinear system with n degrees of freedom in the presence of resonance frequencies. *Izv. Mat. Inst. Bulg. Akad. Nauk*, **9**, 191–199. *417*

Bainov, D. D. See Plotnikova and Bainov.

Baker, N. H., D. W. Moore, and E. A. Spiegel (1971). Aperiodic behaviour of a non-linear oscillator. *Q. J. Mech. Appl. Math.*, **24**, 391–422. *93, 152*

Balachandra, M. (1973). Periodic solutions of singularly perturbed equations arising from gyroscopic systems. *SIAM J. Appl. Math.*, **24**, 562–573. *144, 396*

Balachandra, M. (1975). Some new basic results for singularly perturbed ordinary differential equations. *J. Math. Anal. Appl.*, **49**, 302–316. *144*

Balachandra, M., and P. R. Sethna (1975). A generalization of the method of averaging for systems with two time scales. *Arch. Rat. Mech. Anal.*, **58**, 261–283. *137*

Balachandra, M. See Sethna and Balachandra.

Balmer, H. A. See Witmer, Balmer, Leech, and Pian.

Barba, P. M. See Kane and Barba.

Barbera, F. J. (1969). An analytical technique for studying the anamolous roll behavior of reentry vehicles. *J. Spacecraft*, **6**, 1279–1284. *228*

Barkham, P. G. D. See Soudack and Barkhma.

Barr, A. D. See Ali Hasan and Barr; Ibrahim and Barr.

Basava Raju, B. See Chandra and Basava Raju.

Bastin, H., and M. Delchambre (1975). Duffing oscillator submitted to forced vibrations of different frequencies. *Bull. Cl. Sci. Acad. R. Belg.*, **61**, 281–292. *163*

Batchelor, D. B. (1976). Parametric resonance of systems with time-varying dissipation. *Appl. Phys. Lett.*, **29**, 280–281. *349*

Batterman, S. C. See Lehner and Batterman.

Bauer, H. F. (1968). Nonlinear response of elastic plates to pulse excitations. *J. Appl. Mech.*, **35**, 47–52. *514*

Bauer, H. F. (1970). Pure subharmonic vibrations in a nonlinear two-degree-of-freedom system. In *Dev. Theo. Appl. Mech.*, **8**, 29–49. *418*

Baum, H. R. (1972). On the weakly damped harmonic oscillator. *Q. Appl. Math.*, **30**, 573–576. *155, 156*

Baxa, P. A. See Broucke and Baxa.

Bayles, D. J., R. L. Lowery, and D. E. Boyd (1973). Nonlinear vibrations of rectangular plates. *ASCE J. Struct. Div.*, **99**, 853–864. *445, 508*

Beal, T. R. (1965). Dynamic stability of a flexible missile under constant and pulsating thrusts. *AIAA J.*, **3**, 486–494. *296*

Beilin, E. A., and G. U. Dzhanelidze (1952). Survey of work on the dynamic stability of elastic systems. *PMM*, **16**, 635–648. *260*

Beletskii, V. V. (1960). The libration of a satellite. In *Artificial Earth Satellites* (L. V. Kurnosova, Ed.). Plenum, New York, pp. 18–45. *396*

Beletskii, V. V. (1968). Libration boundaries of a tri-axial satellite in a gravitational field. *PMM*, **31**, 1107–1111. *396*

Beliaev, N. M. (1924). Stability of prismatic rods, subject to variable longitudinal forces. Collection of Papers: *Eng. Construct. Struct. Mech., Put'*, Leningrad, pp. 149–167. *259, 267*

Bellman, R. (1953). *Stability Theory* of *Differential Equations*. McGraw-Hill, New York. *259*

Belytschko, T., and B. J. Hsieh (1974). Nonlinear transient analysis of shells and solids of revolution by convected elements. *AIAA J.*, **12**, 1032–1035. *506*

Benedetti, G. A. (1974). Dynamic stability of a beam loaded by a sequence of moving particles. *J. Appl. Mech.*, **41**, 1069–1071. *269*

Bennett, A. (1966). Analytical determination of characteristic exponents. *Prog. Astronaut. Aeronaut.*, **17**, 101–113. *379*

Bennett, J. A. (1971). Nonlinear vibration of simply supported angle ply laminated plates. *AIAA J.*, **9**, 1997–2003. *502, 508*

Bennett, J. A. (1972). Some approximations in the nonlinear vibrations of unsymmetrically laminated plates. *AIAA J.*, **10**, 1145–1146. *502*

Bennett, J. A. (1973). Ultraharmonic motion of a viscously damped nonlinear beam. *AIAA J.*, **11**, 710–715. *463*

Bennett, J. A., and J. G. Eisley (1970). A multiple-degree-of-freedom approach to nonlinear beam vibrations. *AIAA J.*, **8**, 734–739. *460*

Bennett, J. A., and R. L. Rinkel (1972). Ultraharmonic vibrations of nonlinear beams. *AIAA J.*, **10,** 715–716. *460*

Bennett, J. A. See Eisley and Bennett.

Benney, D. J. (1967). The asymptotic behavior of nonlinear dispersive waves. *J. Math. Phys.*, **46,** 115–132. *601*

Benney, D. J., and A. C. Newell (1967a). Sequential time closures for interacting random waves. *J. Math. and Phys.*, **46,** 363–393. *601*

Benney, D. J., and A. C. Newell (1967b). The propagation of nonlinear wave envelopes. *J. Math. and Phys.*, **46,** 133–139. *601*

Benney, D. J., and A. C. Newell (1969). Random wave closures. *Stud. Appl. Math.*, **48,** 29–53. *601*

Benney, D. J., and P. G. Saffman (1966). Nonlinear interactions of random waves in a dispersive medium. *Proc. Roy. Soc.* (London), **A289,** 301–320. *601*

Berezovskii, A. A., and L. F. Shulenzhko (1963). Nonlinear formulation of the problem of a parametric instability of plates. *Dop. Akad. Nauk*, **8,** 889–993. *268*

Berg, G. V., and D. A. DaPeppo (1960). Dynamic analysis of elastoplastic structures. *Proc. ASCE*, **86,** 35–58. *98*

Berger, H. M. (1955). A new approach to the analysis of large deflections of plates. *J. Appl. Mech.*, **22,** 465–472. *504*

Bernard, J. E. See McIvor and Bernard.

Bernstein, S. A. (1947). *The Foundations of the Dynamics of Structures.* Stroiizdat, Moscow. *266*

Berry, M. V. (1978). Regular and irregular motion. In *Nonlinear Dynamics* (S. Jorne, Ed.). AIP, New York, *56, 388*

Bert, C. W. (1961). Analysis of a nonlinear torsional coupling having dynamic links. *Int. J. Mech. Sci.*, **3,** 80–90. *395*

Bert, C. W. (1973). Nonlinear vibration of a rectangular plate arbitrarily laminated of anisotropic material. *J. Appl. Mech.*, **40,** 452–458. *502, 507*

Bert, C. W., and C. A. Fisher (1972). Stability analysis of a pinned-end beam undergoing nonlinear free vibration. *J. Sound Vib.*, **22,** 129–131. *460*

Bert, C. W. See Mayberry and Bert; Ray and Bert.

Beshai, M. E., and M. A. Dokainish (1975). Transient response of a forced nonlinear system. *J. Sound Vib.*, **41,** 53–62. *119, 163*

Betchov, R. (1958). Nonlinear oscillations of a column of gas. *Phys. Fluids*, **1,** 205–212. *572*

Beth, H. J. E. (1913). The oscillations about a position of equilibrium where a simple linear relation exists between the frequencies of the principal vibrations. *Phil. Mag.*, **26,** 268–324. *379*

Beyer, R. T. (1974). *Nonlinear Acoustics*. Naval Ship Systems Command, Washington, D.C. *545*

Bhansali, S., and V. R. Thiruvenkatachar (1975). Free oscillations of a coupled nonlinear system with two degrees of freedom. *ZAMM*, **56,** 276–279. *396*

Bhattacharjie, A. See Advani and Bhattacharjie.

Bieniek, M. P., T. C. Fan, and L. M. Lackman (1966). Dynamic stability of cylindrical shells. *AIAA J.*, **4,** 495–500. *269, 506*

Billet, R. A. See Porter and Billet.

Binder, R. O. See Chobotov and Binder.

Bishop, R. E. D., and A. G. Parkinson (1965). Second order vibration of flexible shafts. *Phil. Trans. Roy. Soc.*, **259,** 1–31. *269*

Black, H. F., and A. J. McTernan (1968). Vibration of a rotating asymmetric shaft supported in asymmetric bearings. *J. Mech. Eng. Sci.*, **10,** 252–261. *269, 331*

Blackstock, D. T. (1966). Connection between the Fay and Fubini Solutions for plane sound waves of finite amplitude. *J. Acoust. Soc. Am.*, **39,** 1019–1026. *562, 564, 565*

Blackstock, D. T. (1972). Nonlinear acoustics (Theoretical). In *American Institute of Physics Handbook* (D. Bray, Ed.) McGraw-Hill, New York, Chapt. 3n, pp. 3-183–3-205. *545*

Blackstock, D. T. See Pestorius and Blackstock.

Blanton, J. N. See Junkins, Jacobson, and Blanton.

Blaquière, A. (1966). *Nonlinear System Analysis*. Academic, New York. *vii, 87, 110, 259, 262, 379, 388, 417, 423*

Bloch, F. (1928). Über die Quantenmechanik der Elektronen in Kristallgittern. *Z. Phys.*, **52,** 555–600. *273*

Bloembergen, N. See Armstrong, Bloembergen, Ducuing, and Pershan.

Blotter, P. T. See Lee, Blotter, and Yen.

Bodger, W. K. (1967). Deceleration of an unbalanced rotor through a critical speed. *Trans. ASME*, **B89,** 582–586. *219*

Bodner, V. A. (1938). The stability of plates subjected to longitudinal periodic forces. *PMM*, **2,** 87–104. *268*

Bogdanoff, J. L. (1962). Influence on the behavior of a linear dynamical system of some imposed rapid motions of small amplitude. *J. Acoust. Soc. Am.*, **34,** 1055–1062. *302*

Bogdanoff, J. L., and S. J. Citron (1965). Experiments with an inverted pendulum subject to random parametric excitation. *J. Acoust. Soc. Am.*, **38,** 447–452. *261, 302*

Bogoliubov, N. N., and Y. A. Mitropolsky (1961). *Asymptotic Methods in the Theory of Nonlinear Oscillations*. Gordon and Breach, New York. *vii, 219, 259, 261, 339, 388, 417*

Bogoliubov, N. N. See Krylov and Bogoliubov.

Bohn, M. P., and G. Herrmann (1974). The dynamic behavior of articulated pipes conveying fluid with periodic flow rate. *J. Appl. Mech.*, **41,** 55–62. *269*

Bois, P. A. (1976). Propagation linéaire et non linéaire d'ondes atmosphériques. *J. Mec.*, **15,** 781–811. *568*

Bojadziev, G. N. (1976). Non-linear vibrating systems in resonance governed by hyperbolic differential equations. *Int. J. Non-Linear Mech.*, **11,** 347–354. *522*

Bojadziev, G. N., and R. W. Lardner (1973). Monofrequent oscillations in mechanical systems governed by second order hyperbolic differential equations with small nonlinearities. *Int. J. Non-Linear Mech.*, **8,** 289–302. *521, 522*

Bojadziev, G. N., and R. W. Lardner (1974). Second order hyperbolic equations with small nonlinearities in the case of internal resonance. *Int. J. Non-Linear Mech.*, **9,** 397–407. *522*

Bojadziev, G. N. See Arya, Bojadziev, and Farooqui.

Bolotin, V. V. (1951). On the transverse vibrations of rods excited by periodic longitudinal forces. Collection of Papers: *Trans. Vib. Crit. Velocities*, **1,** 46–77. *267, 268*

Bolotin, V. V. (1953). On the parametric excitation of transverse vibrations. Collection of Papers: *Trans. Vib. Crit. Velocity*, **2,** 5–44. *266*

Bolotin, V. V. (1963). *Nonconservative Problems of the Theory of Elastic Stability.* Pergamon, New York. *100, 259, 266, 509*

Bolotin, V. V. (1964). *The Dynamic Stability of Elastic Systems.* Holden-Day, San Francisco. *259, 266, 267, 268, 269, 277, 284, 285, 286, 338, 339, 444*

Bond, V. R. (1974). The uniform, regular differential equations of the KS transformed perturbed two-body problem. *Celestial Mech.*, **10,** 303–318. *41*

Bondar, N. G. (1953). Dynamic stability and vibrations of hingeless parabolic arches. *Inzh. Sb. Akad. Nauk SSSR*, **13,** 87–102. *268*

Bondarenko, G. V. (1936). *The Hill Differential Equation and Its Uses in Engineering Vibration Problems.* Akademiia Nauk SSSR, Moscow. *259, 267, 283, 295*

Bonthron, R. J. See Messal and Bonthron.

Borges, C. A., L. Cesari, and D. A. Sanchez (1974). Functional analysis and the method of harmonic balance. *Q. Appl. Math.*, **32,** 457–464. *61*

Boston, J. R. (1971). Response of a nonlinear form of the Mathieu equation. *J. Acoust. Soc. Am.*, **49,** 299–305. *338*

Bouc, R. See Van Dooren and Bouc.

Boyd, D. E. See Bayles, Lowery, and Boyd.

Brachkovskii, B. Z. (1942). On the dynamic stability of elastic systems. *PMM*, **6,** 87–88. *266*

Bradley, J. W. See Murphy and Bradley.

Brauer, F., and J. A. Nohel (1969). *The Qualitative Theory of Ordinary Differential Equations*. Benjamin, New York. *173*

Breakwell, J. V., and R. Pringle (1966a). Nonlinear resonances affecting gravity-gradient stability. In *Astrodynamics* (M. Lunc, Ed.), Gauthier-Villars, Paris, pp. 305–325. *396*

Breakwell, J. V., and R. Pringle, Jr. (1966b). Resonances affecting motion near the earth-moon equilateral libration points. *AIAA Prog. Astronaut. Aeronaut.*, **17,** 55–74. *378*

Breakwell, J. V. See Hitzl and Breakwell.

Breazeale, M. A. See Adler and Breazeale.

Brereton, R. C. See Modi and Brereton.

Brillouin, L. (1897). Théorie d'un alternateur auto-excitateur. *Eclairage Elect.*, **11,** 49–59. *262*

Brillouin, N. (1956). *Wave Propagation in Periodic Structures*. Dover, New York. *258*

Brockett, R. W. (1970). *Finite Dimensional Linear Systems*, Wiley, New York. *283, 284*

Brocklehurst, A. See Christopher and Brocklehurst.

Broucke, R., and P. A. Baxa (1973). Periodic solutions of a spring-pendulum system. *Celestial Mech.*, **8,** 261–267. *370, 379*

Bublik, B. M., and V. I. Merkulov (1960). On the dynamic stability of a thin elastic shell filled with a liquid. *PMM*, **24,** 941–947. *269*

Budynas, R. See Poli and Budynas.

Bulkeley, P. Z. (1963). An axisymmetric nonlinear vibration of circular plates. *J. Appl. Mech.*, **30,** 630–631. *508*

Bulkeley, P. Z. (1973). Stability of transverse waves in a spinning membrane disk. *J. Appl. Mech.*, **40,** 133–136. *506*

Bulkeley, P. Z. See Advani and Bulkeley; Krauter and Bulkeley.

Burgers, J. M. (1948). A mathematical model illustrating the theory of turbulence. *Adv. Appl. Mech.*, **1,** 171–199. *562*

Burgreen, D. (1951). Free vibrations of a pin-ended column with constant distance between ends. *J. Appl. Mech.*, **18,** 135–139. *460*

Burnashev, I. A. (1954). On the dynamic stability of the plane bending form of a beam. *Dokl. Akad. Nauk Uzb. SS*, **R3**, 7-12. *267*

Busby, H. R., Jr., and V. I. Weingarten (1972). Non-linear response of a beam to periodic loading. *Int. J. Non-Linear Mech.*, **7**, 289-303. *446, 460*

Butenin, N. (1965). *Elements of the Theory of Nonlinear Oscillations*. Blaisdel, New York. *vii, 117, 375, 388, 395, 417*

Butler, D. S., and R. J. Gribben (1968). Relativistic formulation for nonlinear waves in a nonuniform plasma. *J. Plasma Phys.*, **2**, 257-281. *568*

Bykov, Ya. V., and T. Chinkaraev (1972). The periodic oscillations of nonlinear systems at resonance: Part 1. *Diff. Eqs.*, **5**, 1146-1159. *163*

Campbell, W. (1924). The protection of steam-turbine disk wheels from axial vibration. *Trans. ASME*, **46**, 31-160. *505*

Cap. F. F. (1974). Averaging method for the solution of non-linear differential equations with periodic non-harmonic solutions. *Int. J. Non-Linear Mech.*, **9**, 441-450. *119, 158*

Cap, F. F., and H. Lashinsky (1973). On an equation related to nonlinear saturation of convection phenomena. *Zagadnienia Drgan Nieliniowych*, **14**, 519-528. *156*

Caprioli, L. (1954). Su un modello meccanico per le oscillazioni di rilassamento. *Rend. Accad. Sci., Ist. Bologna*, **1**, 152-168. *144*

Carhart, R. See Aks and Carhart.

Carrier, G. F. (1945). On the non-linear vibration problem of the elastic string. *Q. Appl. Math.*, **3**, 157-165. *486*

Carrier, G. F. (1949). A note on the vibrating string. *Q. Appl. Math.*, **7**, 97-101. *486*

Carter, W. J., and F. C. Liu (1961). Steady-state behavior of nonlinear dynamic vibration absorber. *J. Appl. Mech.*, **28**, 67-70. *417*

Cartwright, M. L., and J. E. Littlewood (1947). On nonlinear differential equations of the second order. *Ann. Math.*, **2**, 472-494. *144, 179*

Caughey, T. K. (1954). The existence and stability of ultraharmonics and subharmonics in forced nonlinear oscillations. *J. Appl. Mech.*, **21**, 327-335. *163*

Caughey, T. K. (1960a). Forced oscillations of a semi-infinite rod exhibiting weak bilinear hysteresis. *J. Appl. Mech.*, **27**, 644-648. *100, 236*

Caughey, T. K. (1960b). Random excitation of a system with bilinear hysteresis. *J. Appl. Mech.*, **27**, 649-652. *98*

Caughey, T. K. (1960c). Sinusoidal excitation of a system with bilinear hysteresis. *J. Appl. Mech.*, **27**, 640-643. *98*

Caughey, T. K. (1963). Equivalent linearization techniques. *J. Acoust. Soc. Am.*, **35**, 1706-1711. *87*

Caughey, T. K., and A. H. Gray (1964). On the almost sure stability of linear dynamic systems with stochastic coefficients. *J. Appl. Mech.*, **31**, 365-372. *267, 284*

Caughey, T. K., and A. Vijayaraghavan (1976). Stability analysis of the periodic solution of a piecewise-linear non-linear dynamic system. *Int. J. Non-Linear Mech.*, **11**, 127-134. *98*

Caughey, T. K. See Dickerson and Caughey; Tso and Caughey.

Cekirge, H. M., and E. Varley (1973). Large amplitude waves in bounded media. I. Reflection and transmission of large amplitude shockless pulses at an interface. *Phil. Trans. Roy. Soc. London*, **A273**, 261-313. *571*

Cesari, L. (1940). Sulla stabilità delle soluzioni dei sistemi di equazioni differenziali lineari a coefficienti periodici. *Atti Accad. Italia, Mem. Cl. Fis. Mat. Nat.*, **11**, 633-692. *304*

Cesari, L. (1971). *Asymptotic Behavior and Stability Problems in Ordinary Differential Equations*, 3rd ed., Springer-Verlag, New York. *260, 304*

Cesari, L., and J. K. Hale (1954). Second order linear differential systems with periodic L-integrable coefficients. *Riv. Mat. Univ. Parma*, **5**, 55–61. *304*

Cesari, L. See Bailey and Cesari. Borges, Cesari, and Sanchez.

Chandra, R. (1976). Large deflection vibration of cross-ply laminated plates with certain edge conditions. *J. Sound Vib.*, **47**, 509–514. *502, 508*

Chandra, R., and B. Basava Raju (1975a). Large amplitude flexural vibration of cross ply laminated composite plates. *Fibre Sci. Tech.*, **8**, 243–263. *502, 508*

Chandra, R., and B. Basava Raju (1975b). Large deflection vibration of angle ply laminated plates. *J. Sound Vib.*, **40**, 393–408. *502, 508*

Chao, L. L., and D. L. Sikarskie (1974). A new type of nonlinear approximation with application to the Duffing equation. *Int. J. Non-Linear Mech.*, **9**, 179–191. *163*

Chauhan, A. P., and D. G. Ashwell (1969). Small and large amplitude free vibrations of square shallow shells. *Int. J. Mech. Sci.*, **11**, 337–349. *506*

Chauhan, A. P. See Ashwell and Chauhan.

Chelomei, V. N. (1939). *The Dynamic Stability of Elements of Aircraft Structures.* Aeroflot, Moscow. *259, 266, 267, 268*

Chen, S. S. (1971). Dynamic stability of a tube conveying fluid. *ASCE J. Eng. Mech. Div.*, 97, 1469–1485. *269*

Chen, J. C., and C. D. Babcock (1975). Nonlinear vibration of cylindrical shells. *AIAA J.*, **13**, 868–876. *506*

Chen, S. S. H. See Chiang and Chen.

Chen, Y. S. See Jong and Chen.

Cheng, W.-H. See Hsu and Cheng.

Cherry, T. M. (1924). Integrals developable about a singular point of a Hamiltonian system of differential equations. *Proc. Camb. Phil. Soc.*, **22**, 510–533. *395*

Cheshankov, B. I. (1971). On subharmonic oscillations of a pendulum. *PMM*, **35**, 343–348. *261*

Cheshankov, B. I. (1974a). Two-frequency resonance oscillations in conservative systems. *PMM*, **38**, 700–708. *379*

Cheshankov, B. I. (1974b). Multifrequency oscillations under external excitations in the case of resonance of mixed second and third rank. *ZAMM*, **54**, 362–366. *418*

Cheshankov, B. I. (1975). An asymptotic method in the field of nonlinear oscillations. *ZAMM*, **55**, 665–670. *163*

Chester, W. (1964). Resonant oscillations in closed tubes. *J. Fluid Mech.*, **18**, 44–64. *572, 611*

Chester, W. (1975). The forced oscillations of a simple pendulum. *J. Inst. Math. Appl.*, **15**, 189–306. *261, 338*

Chhatpar, C. K. See Dugundji and Chhatpar.

Chiang, D. C., and S. S. H. Chen (1972). Large amplitude vibration of a circular plate with concentric rigid mass. *J. Appl. Mech.*, **39**, 577–583. *507*

Chikwendu, S. C., and J. Kevorkian (1972). A perturbation method for hyperbolic equations with small nonlinearities. *SIAM J. Appl. Math.*, **22**, 235–258. *556, 601, 604*

Chinkaraev, T. See Bykov and Chinkaraev.

Chisyaki, T., and K. Takahashi (1972). Nonlinear vibration of ring sector plates. *Proc. Japan Soc. Civ. Eng.*, **204**, 1–13. *508*

Chobotov, V. A., and R. O. Binder (1964). Nonlinear response of a circular membrane to sinusoidal acoustic excitation. *J. Acoust. Soc. Am.*, **36,** 59–71. *505*

Choquet-Bruhat, Y. (1969). Ondes asymptotiques et approachées pour des systèmes d'equations aux derivées partielles non lineaires. *J. Math. Pures Appl.*, **48,** 117–158. *556*

Chou, C. C., and S. C. Sinha (1975). Application of a weighted mean square method of linearization to oscillations governed by Liénard's equations. *J. Appl. Mech.*, **42,** 496–498. *87*

Christopher, P. A. T., and A. Brocklehurst (1974). A generalized form of an approximate solution to a strongly non-linear, second-order differential equation. *Int. J. Control,* **19,** 831–839. *119*

Chu, B. T. (1963). Analysis of a self-sustained thermally driven nonlinear vibration. *Phys. Fluids*, **6,** 1638–1644. *572*

Chu, B. T., and S. J. Ying (1963). Thermally driven nonlinear oscillations in a pipe with traveling shock waves. *Phys. Fluids*, **6,** 1625–1637. *572*

Chu, H. N. (1961a). Influence of large amplitudes on flexural vibrations of a thin cylindrical shell. *J. Aero. Sci.*, **28,** 602–609. *506*

Chu, H.-N. (1961b). Influence of transverse shear on nonlinear vibrations of sandwich beams with honeycomb cores. *J. Aero. Sci.*, **28,** 405–410. *448*

Chu, H.-N., and G. Herrmann (1956). Influence of large amplitudes on free flexural vibrations of rectangular elastic plates. *J. Appl. Mech.*, **23,** 532–540. *502, 504, 508*

Citron, S. J. See Bogdanoff and Citron.

Clare, T. A. (1971). Resonance instability for finned configurations having nonlinear aerodynamic properties. *J. Spacecraft,* **8,** 278–283. *228*

Clark, L. G. See McQuery and Clark.

Cochran, J. E. (1972). Effects of gravity-gradient torque on the rotational motion of a triaxial satellite in a precessing elliptic orbit. *Celestial Mech.*, **6,** 127–150. *396*

Cochran, J. E. (1977). Nonlinear resonances in the attitude motion of dual-spin spacecraft. *J. Spacecraft*, **14,** 562–572. *396*

Coddington, E. A., and N. Levinson, (1955). *Theory of Ordinary Differential Equations.* McGraw-Hill, New York. *108, 144, 259*

Cohen, H. (1955). On subharmonic synchronization of nearly-linear systems. *Appl. Math.*, **13,** 102–105. *214*

Cole, J. D. (1951). On a quasilinear parabolic equation occurring in aerodynamics. *Q. Appl. Math.*, **9,** 225–236. *608*

Cole, J. D. (1968). *Perturbation Methods in Applied Mathematics*. Blaisdell, Waltham. *144*

Collins, W. D. (1967). The propagation and interaction of one-dimensional nonlinear waves in an incompressible isotropic elastic half-space. *Q. J. Mech. Appl. Math.*, **20,** 429–452. *571*

Collins, W. D. (1971). Forced oscillations of systems governed by one-dimensional nonlinear wave equations. *Q. J. Mech. Appl. Math.*, **24,** 129–153. *572, 611*

Corbeiller, P. L. (1931). Les systèmes auto-entretenues et les oscillations de relaxation. In *Librairie Scientifique*, Hermann, Paris. *144*

Courant, R., and K. O. Friedrichs (1948). *Supersonic Flow and Shock Waves.* Interscience, New York. *545*

Craig, R. R. See Kana and Craig.

Crandall, S. H. (1974). Nonlinearities in structural dynamics. *Shock Vib. Dig.*, **6,** 2–14. *444*

Crawford, J., and S. Atluri (1975). Nonlinear vibrations of a flat plate with initial stresses. *J. Sound Vib.*, **43,** 117–129. *505, 508*

Crespo da Silva, M. R. M. (1970). Attitude stability of a gravity-stabilized gyrostat satellite. *Celestial Mech.*, **2,** 147–165. *396*

Crespo, da Silva, M. R. M. (1972a). Non-linear resonant attitude motions in gravity-stabilized gyrostat satellites. *Int. J. Non-Linear Mech.*, **7,** 621–641. *396*

Crespo da Silva, M. R. M. (1972b). On the dynamical equivalence between two types of vehicles with rotors. *J. Br. Interplanetary Soc.*, **25,** 177–181. *396*

Crespo da Silva, M. R. M. (1974). A transformation approach of dynamical systems. *Int. J. Non-Linear Mech.*, **9,** 241–249. *379*

Crespo da Silva, M. R. M. (1977). The effect of a slow spin on the nonlinear resonant motions of a dynamical system. *J. Sound Vib.*, **52,** 201–209. *370*

Cristescu, N. (1967). *Dynamic Plasticity.* North-Holland, Amsterdam. *545*

Croll, J. G. A. (1975). Kinematically coupled non-linear vibrations. *J. Sound Vib.*, **40,** 77–85. *304*

Crose, J. G., and A. H. S. Ang (1969). Nonlinear analysis method for circular plates. *J. Eng. Mech. Div.*, **95,** 979–999. *508*

Cumberbatch, E. See Varley and Cumberbatch; Varley, Venkataraman, and Cumberbatch.

Cummings, B. E. (1964). Large-amplitude vibration and response of curved panels. *AIAA J.* **2,** 709–716. *506*

Cunningham, W. J. (1958). *Introduction to Nonlinear Analysis.* McGraw-Hill, New York. *259, 277*

Dallos, P. J. (1966). On the generation of odd-fractional subharmonics. *J. Acoust. Soc. Am.*, **40,** 1381–1391. *367, 416*

Dallos, P. J., and C. O. Linnell (1966a). Even-order subharmonics in the peripheral auditory system. *J. Acoust. Soc. Am.*, **40,** 561–564. *367, 416*

Dallos, P. J., and C. O. Linnell (1966b). Subharmonic components in cochlear-microphonic potentials. *J. Acoust. Soc. Am.*, **40,** 4–11. *367, 416*

Danby, J. M. A. (1964). Stability of the triangular points in the elliptic restricted problem of three bodies. *Astron. J.*, **69,** 165–172. *331, 379*

DaPeppo, D. A. See Berg and DaPeppo.

Das, Y. C. See Singh, Das, and Sundararajan; Singh, Sundararajan, and Das; Vendhan and Das.

Dasarathy, B. V. (1975). Equivalent linear models for non-linear nonautonomous systems. *J. Sound Vib.*, **42,** 447–452. *87*

Dasarathy, B. V., and P. Srinivasan (1969). On the study of a third-order mechanical oscillator. *J. Sound Vib.*, **9,** 49–52. *151*

Dasarathy, B. V. See Rangacharyulu, Srinivasan, and Dasarathy; Srirangarajan and Dasarathy.

Davey, A. (1972). The propagation of a weak nonlinear wave. *J. Fluid Mech.*, **53,** 769–781. *587*

Davey, A., and K. Stewartson (1974). On three dimensional packets of surface waves. *Proc. Roy. Soc. London*, **A338,** 101–110. *588*

Davidenkov, N. N. (1938). Energy dissipation in vibrations. *J. Tech. Phys.*, **8,** 483–499. *100*

Davidson, R. C. (1967). The evolution of wave correlations in uniformly turbulent, weakly nonlinear systems. *J. Plasma Phys.*, **1,** 341–359. *601*

Davidson, R. C. (1969). General weak turbulence theory of resonant four-wave processes. *Phys. Fluids*, **12,** 149–161. *601*

Davidson, R. C. (1972). *Methods in Nonlinear Plasma Theory.* Academic, New York. *601*

Davis, H. (1962). *Introduction to Nonlinear Differential and Integral Equations.* Dover, New York. *vii*

Davison, L. (1968). Perturbation theory of nonlinear elastic wave propagation. *Int. J. Solids Struct.*, **4**, 301–322. *571*

Davy, D. T., and W. E. Ames (1973). An asymptotic solution of an initial value problem for a nonlinear viscoelastic rod. *Int. J. Non-Linear Mech.*, **8**, 59–71. *449*

Davydov, I. S. (1970). Parametric vibrations in a single stage spur gear transmission. *Russ. Eng. J.*, **50**, 36. *269*

Delchambre, M. See Bastin and Delchambre.

DeNeef, P., and H. Lashinsky (1973). Van der Pol model for unstable waves on a beam-plasma system. *Phys. Rev. Lett.*, **31**, 1039–1041. *107*

Den Hartog, J. P. (1947). *Mechanical Vibrations*. McGraw-Hill, New York. *vii, 259*

Deprit, A. (1969). Canonical transformations depending on a small parameter. *Celestial Mech.*, **1**, 12–30. *378, 396*

Deprit, A., and A. Deprit-Bartholomé (1967). Stability of the triangular Lagrangian points. *Astron. J.*, **72**, 173–179. *378*

Deprit-Bartholomé, A. See Deprit and Deprit-Bartholomé.

Dewan, E. M., and H. Lashinsky (1969). Asynchronous quenching of the van der Pol oscillator. *IEEE Trans.*, **AC-14**, 212–214. *210*

Dhoopar, B. L. See Vendhan and Dhoopar.

Dickerson, J. R., and T. K. Caughey (1969). Stability of continuous dynamic systems with parametric excitation. *J. Appl. Mech.*, **36**, 212–216. *284*

Dimantha, P. C., and J. Roorda (1969). On the domain of asymptotic stability of nonlinear nonconservative systems. *Appl. Sci. Res.*, **29**, 272–288. *509*

Dimentberg, F. M. (1961). *The Flexural Vibration of Rotating Shafts*. Butterworth, London. *269*

Dimentberg, F. M. (1976). Response of a nonlinearly damped oscillator to combined periodic parametric and random external excitation. *Int. J. Non-Linear Mech.*, **1**, 83–87. *356*

DiPrima, R. C., W. Eckhaus, and L. A. Segel (1971). Nonlinear wavenumber interaction in near-critical two-dimensional flows. *J. Fluid Mech.*, **49**, 705–744. *588*

Dobias, I. (1974). Response of a nonlinear mechanical system with one degree of freedom to mixed harmonic and stochastic input. *Zagadnienia Drgan Nieliniowych*, **15**, 233–246. *163*

Dokainish, M. A., and M. Elmadany (1978). On the nonlinear response of a relief valve. *ASME J. Mech. Des.*, **100**, 675–680. *445, 535*

Dokainish, M. A. See Beshai and Dokainish.

Donaldson, B. K. (1973a). A new approach to the forced vibration of flat skin-stringer-frame structures. *J. Sound Vib.*, **30**, 419–435. *447*

Donaldson, B. K. (1973b). A new approach to the forced vibration of thin plates. *J. Sound Vib.*, **30**, 397–417. *508*

Dorodnicyn, A. A. (1947). Asymptotic solution of van der Pol's equation. *PMM*, **11**, 313–328. *144, 146*

Dougherty, C. B. See Cole and Dougherty.

Dowell, E. H. (1966). Nonlinear oscillations of a fluttering plate. *AIAA J.*, **4**, 1267–1275. *509*

Dowell, E. H. (1967a). Nonlinear oscillations of a fluttering plate II. *AIAA J.*, **5**, 1856–1862. *509*

Dowell, E. H. (1967b). On the nonlinear flexural vibration of rings. *AIAA J.*, **5**, 1508–1509. *447*

Dowell, E. H. (1969). Nonlinear flutter of curved plates. *AIAA J.*, **7**, 424–431. *509*

Dowell, E. H. (1970a). Noise or flutter or both? *J. Sound Vib.*, **11**, 159–180. *509*

Dowell, E. H. (1970b). Nonlinear flutter of curved plates. II. *AIAA J.*, **8**, 259–261. *509*

Dowell, E. H. (1973). Theoretical vibration and flutter studies of point supported panels. *J. Spacecraft*, **10**, 389–395. *509*

Dowell, E. H. (1975). *Aeroelasticity of Plates and Shells*. Noordhoff, Leyden. *107, 508, 509*

Dowell, E. H., and C. S. Ventres (1968). Model equations for the nonlinear flexural vibrations of a cylindrical shell. *Int. J. Solids Struct.*, **4**, 975–991. *506, 533*

Dowell, E. H. See Ventres and Dowell.

Drew, J. H. (1974). Stability of certain periodic solutions of a forced system with hysteresis. *Int. J. Non-Linear Mech.*, **9**, 121–125. *98, 236*

Ducuing, J. See Armstrong, Bloembergen, Ducuing, and Pershan.

Duffield, R. C., and N. Willems (1972). Parametric resonance of stiffened rectangular plates. *J. Appl. Mech.*, **39**, 217–226. *268*

Duffield, R. C. See Ahuja and Duffield.

Dugundji, J. (1966). Theoretical considerations of panel flutter at high supersonic Mach numbers. *AIAA J.*, **4**, 1257–1266; *Errata and Addenda*, **7**, 1663. *509*

Dugundji, J., and C. K. Chhatpar (1970). Dynamic stability of a pendulum under parametric excitation. *Rev. Roum. Sci. Tech. Mech. Appl.*, **15**, 741–763. *261, 338*

Dugundji, J., and V. Mukhopadhyay (1973). Lateral bending-torsion vibrations of a thin beam under parametric excitation. *J. Appl. Mech.*, **40**, 693–698. *267, 306*

Dugundji, J. See Kuo, Morino, and Dugundji; Tseng and Dugundji.

Dunne, P. C. See Argyris, Dunne, and Angelopoulos.

Dym, C. L. See El-Zaouk and Dym.

Dysthe, K. B., and O. T. Gudmestad (1975). On resonance and stability of conservative systems. *J. Math. Phys.*, **16**, 56–64. *424*

Dzhanelidze, G. U. (1953). Theorems on the separation of variables in problems of the dynamic stability of rods. *Tr. Leningr. Inst. Inzh. Vodn. Transp.*, **20**, 193–198. *266*

Dzhanelidze, G. U. (1955). The general problem of dynamic stability of elastic systems. *Usp. Math. Nauk*, **10**, 202–203. *266*

Dzhanelidze, G. U., and M. A. Radstig (1940). Dynamic stability of rings subject to normal periodic forces. *PMM*, **5**, 55–60. *268*

Dzhanelidze, G. U. See Beilin and Dzhanelidze.

Dzygadlo, Z. (1965). Parametric self-excited vibration of a simple supported plate in supersonic flow. *Proc. Vib. Prob.*, **4**, 381–394. *268*

Dzygadlo, Z., and A. Krzyanowski (1972). Self-excited and forced vibrations of an aeroelastic system. Subject to a follower force. *Proc. Vib. Prob.*, **13**, 259–280. *268*

Dzygadlo, Z., and A. Wielgus (1974). Parametric and parametric self-excited vibrations of rectangular multi-span planes in supersonic flow. *Proc. Vib. Prob.*, **15**, 167–178. *269*

Earnshaw, S. (1860). On the mathematical theory of sound. *Phil. Trans. Roy. Soc. London*, **150**, 133–148. *612*

Eastep, F. E., and S. C. McIntosh (1971). Analysis of nonlinear panel flutter and response under random excitation or nonlinear aerodynamic loading. *AIAA J.*, **9**, 411–418. *509*

Eckhaus, W. (1975). New approach to the asymptotic theory of nonlinear oscillations and wave propagation. *J. Math. Anal. Appl.*, **49**, 575–611. *417, 556*

Eckhaus, W. See DiPrima, Eckhaus, and Segel.

Efstathiades, G. J. (1968). Subharmonic instability in non-linear two-degree-of-freedom systems. *Int. J. Mech. Sci.*, **10**, 829–847. *418*

Efstathiades, G. J. (1969). Subharmonic instability and coupled motions in non-linear vibration isolating suspensions. *J. Sound Vib.*, **10**, 81–102. *418*

Efstathiades, G. J. (1971). A new approach to the large-deflection vibrations of imperfect circular disks using Galerkin's procedure. *J. Sound Vib.*, **16**, 231–253. *506*

Efstathiades, G. J. (1974). Combination tones in single mode motion of a class of nonlinear systems with two-degrees-of-freedom. *J. Sound Vib.*, **34**, 379–397. *185*

Efstathiades, G. J., and C. J. H. Williams (1967). Vibration isolation using nonlinear springs. *Int. J. Mech. Sci.*, **9**, 27–44. *417*

Ehrich, F. F. (1971). Sum and difference frequencies in vibration of high speed rotating machinery. *J. Eng. Ind.*, Paper No. 71-Vibr.-103. *269*

Einaudi, F. (1969). Singular perturbation analysis of acoustic-gravity waves. *Phys. Fluids*, **12**, 752–756. *568*

Einaudi, F. (1975). On the convergence of an iterative method for studying nonlinear oscillations without damping. *SIAM J. Appl. Math.*, **29**, 1–11. *50*

Einaudi, R. (1936). Sulle configurazioni di equilibrio instabili di una piastra sollecitata da sforzi tangentiali pulsante. *Atti Accad. Gioenia*, **20**, 1–5; *Mem.*, **5**, 1–20 (1937). *268*

Eisley, J. G. (1964a). Large amplitude vibration of buckled beams and rectangular plates. *AIAA J.*, **2**, 2207–2209. *447*

Eisley, J. G. (1964b). Nonlinear vibration of beams and rectangular plates. *ZAMP*, **15**, 167–175. *460, 508*

Eisley, J. G. (1966). Nonlinear deformation of elastic beams, rings and strings. *Appl. Mech. Surveys*, 285–290. *444*

Eisley, J. G., and J. A. Bennett (1970). Stability of large amplitude forced motion of a simply supported beam. *Int. J. Non-Linear Mech.*, **5**, 645–657. *460*

Eisley, J. G. See Bennett and Eisley; Ho, Scott, and Eisley; Min and Eisley.

Elachi, C. (1976). Waves in active and passive periodic structures: a review. *Proc. IEEE*, **64**, 1666–1698. *258*

Eller, A. I. (1972). Driven non-linear oscillations of a string. *J. Acoust. Soc. Am.*, **51**, 960–966. *485*

Eller, A. I. (1973). Fractional-harmonic frequency pairs in nonlinear systems. *J. Acoust. Soc. Am.*, **53**, 758–765. *367, 402, 416*

Eller, A. I. (1974). Subharmonic response of bubbles to underwater sound. *J. Acoust. Soc. Am.*, **55**, 871–873. *199*

Eller, A. I., and H. G. Flynn (1969). Generation of subharmonics of order one-half by bubbles in a sound field. *J. Acoust. Soc. Am.*, **46**, 722–727. *199*

Elmadany, M. See Dokainish and Elmadany.

Elmaraghy, R., and B. Tabarrok (1975). On the dynamic stability of an axially oscillating beam. *J. Franklin Inst.*, **300**, 25–39. *267*

El Owaidy, H. (1974). On perturbations of a class of self-excited oscillators. *Polytech. Mech. Eng.*, **18**, 61–70. *129*

El-Zaouk, B. R., and C. L. Dym (1973). Nonlinear vibrations of orthotropic doubly-curved shallow shells. *J. Sound Vib.*, **31**, 89–103. *506*

Eminhizer, C. R., R. H. G. Helleman, and E. W. Montroll (1976). On a convergent nonlinear perturbation theory without small denominators or secular terms. *J. Math. Phys.*, **17**, 121–140. *56, 163, 388*

Ericsson, L. E. See Price and Ericsson.

Eringen, A. C. (1951). On the nonlinear vibration of circular membranes. *Proc. 1st U.S. Nat. Congr. Appl. Mech.*, 139–145. *505*

Eringen, A. C. (1952). On the nonlinear vibration of elastic bars. *Q. Appl. Math.*, **9**, 361–369. *448*

Eringen, C. (1955). On the nonlinear oscillations of viscoelastic plates. *J. Appl. Mech.*, **22**, 563–567. *445, 501*

Erugin, N. P. (1966). *Linear Systems of Ordinary Differential Equations with Periodic and Quasi-Periodic Coefficients*. Academic, New York. *259*

Euler, J. A. See Stanisic and Euler.

Evan-Iwanowski, R. M. (1965). On the parametric response of structures. *Appl. Mech. Rev.*, **18**, 699–702. *260, 267*

Evan-Iwanowski, R. M. (1969). Nonstationary vibrations of mechanical systems. *Appl. Mech. Rev.*, **22**, 213–219. *219*

Evan-Iwanowski, R. M. (1976). *Resonance Oscillations in Mechanical Systems*. Elsevier, New York. *vii, 219, 260, 267, 268, 402, 417, 418*

Evan-Iwanowski, R. M., W. F. Sanford, and T. Kehagioglou (1970). Nonstationary parametric response of nonlinear column. *Dev. Theor. Appl. Mech.*, **5**, 715–744. *268*

Evan-Iwanowski, R. M. See Adams and Evan-Iwanowski; Agrawal and Evan-Iwanowski; Evensen and Evan-Iwanowski; Farnsworth and Evan-Iwanowski; Mozer and Evan-Iwanowski; Rajac and Evan-Iwanowski; Somerset and Evan-Iwanowski; Stevens and Evan-Iwanowski; Vijayaraghavan and Evan-Iwanowski.

Evensen, D. A. (1963). Some observations on the nonlinear vibrations of thin cylindrical shells. *AIAA J.*, **1**, 2857–2858. *506*

Evensen, D. A. (1966). A theoretical and experimental study of non-linear flexural vibrations of thin circular rings. *J. Appl. Mech.*, **33**, 553–560. *447*

Evensen, D. A. (1968). Nonlinear vibrations of beams with various boundary conditions. *AIAA J.*, **6**, 370–372. *460, 469*

Evensen, D. A., and R. E. Fulton (1967). Some studies on the nonlinear dynamic response of shell-type structures. *Int. Conf. Dyn. Stab. Struct.*, 237–254. *506*

Evensen, H. A., and R. M. Evan-Iwanowski (1966). Effects of longitudinal inertia upon the parametric response of elastic columns. *J. Appl. Mech.*, **33**, 141–148. *267, 268*

Fan, T. C. See Bieniek, Fan, and Lackman.

Faraday, M. (1831). On a peculiar class of acoustical figures and on certain forms assumed by a group of particles upon vibrating elastic surfaces. *Phil. Trans. Roy. Soc.* (London), 299–318. *20, 258*

Farnsworth, C. E., and R. M. Evan-Iwanowski (1970). Resonance response of nonlinear circular plates subjected to uniform static load. *J. Appl. Mech.*, **37**, 1043–1049. *508*

Farooqui, A. S. See Arya, Bojadziev, and Farooqui.

Farrenkopf, R. L. See Scher and Farrenkopf.

Fedorchenko, A. M. (1958). On the motion of a heavy nonsymmetrical top with a vibrating support. *Ukr. Matem.*, **10**, 209–218. *331*

Fedorchenko, A. M. (1961). On a dynamical method for increasing stability of a rapidly rotating symmetrical gyroscope. *PMM*, **25**, 938–940. *331*

Federhofer, K. (1954). Die durch pulsierende Axialkräfte gedrückte Kreiszylinderschale. Österr. *Akad. Wiss., Math. Naturw. Kl.*, **163**, 41–54. *269*

Federhofer, K. (1959). Nonlinear bending vibrations of circular rings. *Ing. Arch.*, **28**, 53–58. *447*

Fiala, V. See Tondl, Fiala, and Šklíba.

Filippov, A. P. (1956). *Vibrations of Elastic Systems*. Kiev, Akad. Nauk Ukraine, USSR. *219*

Finizio, N. J. (1974). Stability of columns subjected to periodic axial forces of impulsive type. *Q. Appl. Math.*, **31**, 455–465. *267*

Fink, J. P., W. S. Hall, and A. R. Hausrath (1974). Convergent two-time method for periodic differential equations. *J. Diff. Eqs.*, **15**, 459–498. *556*

Fisher, C. A. See Bert and Fisher.

Fjeld, M. (1968). Asynchronous quenching of nonlinear systems exhibiting limit cycles. *IEEE Trans.*, **AC-13**, 201–202. *210*

Flatto, L., and N. Levinson (1955). Periodic solutions of singularly perturbed systems. *J. Rat. Mech. Anal.*, **4**, 943–950. *144*

Fleishman, B. A. (1963). Wave propagation in non-simple media. *Nonlinear Diff. Eqs. Non-Linear Mech.*, 211–217. *545*

Fletcher, J. E. See Struble and Fletcher.

Fletcher, W. H. W. See Keen and Fletcher.

Floquet, G. (1883). Sur les équations différentielles linéaires à coefficients périodiques. *Ann. Sci. École Norm. Sup.*, **12**, 47–89. *273*

Flügge, W. (1967). *Viscoelasticity*. Blaisdell, Waltham, Mass. *100*

Flynn, H. G. See Eller and Flynn.

Ford, J. (1975). The statistical mechanics of classical analytic dynamics. In *Fundamental Problems in Statistical Mechanics*, Vol. 3 (E. D. G. Cohen, Ed.). North-Holland, Amsterdam, pp. 215–255. *388*

Ford, J. See Walker and Ford.

Fost, R. See Oden, Key, and Fost.

Fox, P. A. (1955). Perturbation theory of wave propagation based on the method of characteristics. *J. Math. and Phys.*, **34**, 133–151. *555*

Franken, P. A., A. E. Hill, C. W. Peters, and G. Weinreich (1961). Generation of optical harmonics. *Phys. Rev. Lett.*, **7**, 118–119. *601*

Fried, I. (1969). Finite element analysis of time-dependent phenomena. *AIAA J.*, **7**, 1170–1173. *445*

Friedmann, P. P. (1977). Recent developments in rotary-wing aeroelasticity. *J. Aircraft*, **14**, 1027–1041. *269*

Friedmann, P., C. E. Hammond, and T.-H. Woo (1977). Efficient numerical treatment of periodic systems with application to stability problems. *Int. J. Num. Meth. Eng.*, **11**, 1117–1136. *269, 283*

Friedmann, P., and L. J. Silverthorn (1974). Aeroelastic stability of periodic systems with application to rotor blade flutter. *AIAA J.*, **12**, 1559–1565. *269, 283*

Friedmann, P., and L. J. Silverthorn (1975). Aeroelastic stability of coupled flap-lag motion of hingeless helicopter blades at arbitrary advance ratios. *J. Sound Vib.*, **39**, 409–428. *269*

Friedmann, P., and P. Tong (1973). Non-linear flap-lag dynamics of hingeless helicopter blades in hover and in forward flight. *J. Sound Vib.*, **30**, 9–31. *269*

Friedrichs, K. O. (1965). *Advanced Ordinary Differential Equations*. Gordon and Breach, New York. *vii*

Friedrichs, K. O. See Courant and Friedrichs.

Froude, W. (1863). Remarks on Mr. Scott Russell's paper on rolling. *Trans. Inst. Naval Arch.*, **4**, 232–275. *28, 373*

Fu, C. C. (1974). Stable harmonic and subharmonic oscillations of a system with piecewise linear restoring forces. *J. Appl. Mech.*, **41**, 273–277. *179*

Fu, F. C. L., and S. Nemat-Nasser (1972a). On the stability of steady-state response of certain nonlinear dynamic systems subjected to harmonic excitations. *Ing.-Arch.*, **41**, 407–420. *284, 322*

Fu, F. C. L., and S. Nemat-Nasser (1972b). Stability of solution of systems of linear differential equations with harmonic coefficients. *AIAA J.*, **10**, 30–36. *284, 322*

Fu, F. C. L., and S. Nemat-Nasser (1975). Response and stability of linear dynamic systems with many degrees of freedom subjected to nonconservative and harmonic forces. *J. Appl. Mech.*, **42**, 458–463. *322*

Fubini-Ghiron, S. (1935). Anomalie nella propagzione di onde acustici che die grande ampiezza. *Alta Frequenza*, **4**, 530–581. *612*

Fugiwara, N. See Sugiyama, Fugiwara, and Sekiya.

Fujii, H. (1976). Librations of a parametrically resonant dual-spin satellite with an energy damper. *J. Spacecraft*, **13**, 416–423. *396*

Fulton, R. E. See Evensen and Fulton.

Fung, Y. C. (1963). Some recent contributions to panel flutter research. *AIAA J.*, **1**, 898–909. *107*

Fung, Y. C. See Olson and Fung.

Funk, B. See Ablowitz, Funk, and Newell.

Furuike, D. M. See Iwan and Furuike.

Furukawa, S. See Maezawa and Furukawa.

Gajendar, N. (1967). Large amplitude vibrations of plates on elastic foundations. *Int. J. Non-Linear Mech.*, **2**, 163–172. *507*

Gajl, B. (1974). Analysis of non-linear self excited limit cycle oscillations of shallow shells in a non-linear supersonic flow. *Mech. Teor. Stosow*, **12**, 17–34. *509*

Galeev, A. A. See Sagdeev and Galeev.

Gambill, R. A. (1954). Stability criteria for linear differential systems with periodic coefficients. *Riv. Math. Univ. Parma*, **5**, 169–181. *304*

Gambill, R. A. (1955). Criteria for parametric instability for linear differential systems with periodic coefficients. *Riv. Math. Univ. Parma*, **6**, 37–43. *304*

Gambill, R. A., and J. K. Hale (1956). Subharmonic and ultraharmonic solutions for weakly nonlinear systems. *J. Rat. Mech. Anal.*, **5**, 353–394. *179*

Gardner, C. S., J. M. Greene, M. D. Kruskal, and R. M. Miura (1967). Method for solving the Korteweg-de Vries equation. *Phys. Rev. Lett.*, **19**, 1095–1097. *588*

Garfinkel, B. (1966). Formal solution in the problem of small divisors. *Astron. J.*, **71**, 657–669. *395*

Gastev, V. A. (1949). Transverse vibrations and stability of rods subjected to periodic repeated pulses. *Tr. Leningrad Inst. Aviat., Prob. 1. 267*

Gaylord, E. W. (1959). Natural frequencies of two nonlinear systems compared with the pendulum. *J. Appl. Mech.*, **26**, 145–146. *85*

Gebman, J. R., and D. L. Mingori (1976). Perturbation solution for the flat spin recovery of a dual-spin spacecraft. *AIAA J.*, **14**, 859–867. *396*

Genin, J., and J. S. Maybee (1969). Stability in the three dimensional whirling problem. *Int. J. Non-Linear Mech.*, **4**, 205–215. *396*

Genin, J., and J. S. Maybee (1970). External and material damped three-dimensional rotor system. *Int. J. Non-Linear Mech.*, **5**, 287–297. *396*

Genin, J. See Radwan and Genin.

George, J. H., R. W. Gunderson, and H. Hahn (1975). Sustained small oscillations in nonlinear control systems. *AIAA J.*, **13**, 1251–1252. *129*

Ghobarah, A. A. (1972). Dynamic stability of monosymmetrical thin-walled structures. *J. Appl. Mech.*, **39**, 1055–1059. *269*

Ghobarah, A. A. and W. K. Tso (1972). Parametric stability of thin-walled beams of open section. *J. Appl. Mech.*, **39**, 201–206. *267, 268*

Gilchrist, A. O. (1961). The free oscillations of conservative quasilinear systems with two degrees of freedom. *Int. J. Mech. Sci.*, **3**, 286–311. *388*

Ginsberg, J. H. (1972a). An equivalence principle for systems with secondary generalized coordinates. *J. Sound Vib.*, **23**, 55–61. *417*

Ginsberg, J. H. (1972b). The effects of damping on a nonlinear system with two degrees of freedom. *Int. J. Non-Linear Mech.*, **7**, 323–336. *417*

Ginsberg, J. H. (1973). The dynamic stability of a pipe conveying a pulsatile flow. *Int. J. Eng. Sci.*, **11**, 1013–1024. *269*

Ginsberg, J. H. (1974). Nonlinear axisymmetric free vibrations in simply supported cylindrical shells. *J. Appl. Mech.*, **41**, 310–311. *506*

Ginsberg, J. H. (1975). Multi-dimensional non-linear acoustic wave propagation, Part II. The nonlinear interaction of an acoustic fluid and plate under harmonic excitation. *J. Sound Vib.*, **40**, 359–379. *509*

Girl, J. See Rehfield, Girl, and Sparrow.

Goeoskokov, E. G. (1966). *Nectachonarnie Kolebaniia Mekanicheskiv System*. Akad. Nauk Ukr. SSR, Kiev. *227*

Gogos, C. M. See Smith, Malme, and Gogos.

Golay, M. J. E. (1964). Normalized equations of the regenerative oscillator-noise, phase-locking, and pulling. *Proc. IEEE*, **52**, 1311–1330. *255*

Goldberg, J., and H. A. Koenig (1975). The non-linear dynamic response of an impulsively loaded circular plate with a central hole. *Int. J. Solids Struct.*, **11**, 985–997. *508*

Gol'denblat, I. I. (1947). *Contemporary Problems of Vibrations and Stability of Engineering Structures*. Stroiizdat, Moscow. *267, 268*

Goloskokov, E. G. See Ovcharova and Goloskokov.

Goodier, J. N., and I. K. McIvor (1964). The elastic cylindrical shell under nearly uniform radial impulse. *J. Appl. Mech.*, **31**, 259–266. *506*

Goodman, L. E., and J. H. Klumpp. (1956). Analysis of slip damping with reference to turbine-blade vibration. *J. Appl. Mech.*, **23**, 421–429. *98*

Goodstein, R. (1959). A perturbation solution of the equations of motion of a gyroscope. *J. Appl. Mech.*, **26**, 349–352. *396*

Gordon, J. T., and S. Atluri (1974). Nonlinear flutter of a cylindrical shell. *Dev. Theor. Appl. Mech.*, **7**, 285–307. *509*

Gorelik, G., and A. Witt (1933). Swing of an elastic pendulum as an example of two parametrically bound linear vibration systems. *J. Tech. Phys.* (USSR), **3**, 244-307. *370*

Graffi, D. (1942). Le oscillazioni di rilassamento. *Alta Frequenza*, **11**, 80–98. *144*

Grammel, G. (1952). Zur Stabilität erzwungener Schwingungen elastischer Körper mit Geschwindigkeit-proportionaler Dämpfung. *Ing.-Arch.*, **20**, 170–183. *268*

Grammel, R. (1953). Oscillations non linéaires avec une infinité de degrés de liberté. Biezeno Anniversary Volume of *Appl. Mech.* Haarlem, Antwerpen, pp. 108–118. *444*

Gray A. H. See Caughey and Gray.

Grebenikov, E. A. (1964). On the stability of the Lagrangian triangular solutions of elliptic restricted three-body problem. *Astron. Zh.*, **41**, 451–459. *331*, *379*

Greene, J. M. See Gardner, Greene, Kruskal, and Miura.

Greenspon, J. E. (1960). A simplified expression for the period of nonlinear oscillations of curved and flat panels. *J. Aero. Sci.*, **27**, 138–139. *506*

Greenspon, J. E. See Kirchman and Greenspon.

Gribben, R. J. See Butler and Gribben.

Grossley, F. R. E. (1952). The free oscillation of the centrifugal pendulum with wide angles. *J. Appl. Mech.*, **19**, 315–319. *388*

Grossman, P. L., B. Koplik, and Y.-Y. Yu (1969). Nonlinear vibrations of shallow spherical shells. *J. Appl. Mech.*, **36**, 451–458. *506*

Grybos, R. (1972). Parametric vibrations of a simple gear transmission. *Rozpr. Inz.*, **20**, 3–17. *269*, *308*

Gudmestad, O. T. See Dysthe and Gudmestad.

Gumpert, W. (1974). On the existence of non-relaxation cycles for friction self-vibration. *Zagadnienia Drgan Nieliniowych*, 311–314. *130*

Gunderson, H., H. Rigas, and F. S. A. van Vleck (1974). Technique for determining stability regions for the damped Mathieu equation. *SIAM J. Appl. Math.*, **26**, 345–349. *300*

Gunderson, R. W. See George, Gunderson, and Hahn.

Gurtin, M. E. See Pasley and Gurtin.

Gustavson, F. (1966). On constructing formal integrals of a Hamiltonian system near an equilibrium point. *Astron. J.*, **71**, 670–686. *395*

Gyozo, L. (1974). The elimination of self-excited vibration by Lanchester-type damper. *Zagadnienia Drgan Nieliniowych*, 299–309. *130*

Haacke, W. (1951). Bemerkungen zur Stabilisierung eines physikalischen Pendels. I. *ZAMM*, **31**, 161–169. *261*

Habakow, H. (1972). A perturbation procedure for weakly coupled oscillators. *Int. J. Non-Linear Mech.*, **7**, 125–137. *379*, *388*

Hablani, H. B., and S. K. Shrivastava (1977). Analytical solution for planar librations of a gravity stabilized satellite. *J. Spacecraft*, **14**, 126–128. *257*

Hagedorn, P. (1968). Aum Instabilitatsbereich erster ordnung der Mathieugleichung mit quadratischer Dampfung. *ZAMM*, **48**, T256–T260. *339*

Hagedorn, P. (1969). Kombinationsresonanz und Instabilitätsberichte zweiter Art bei parametererregten Schwingungen mit nichtlinearer Dämpfung. *Ing.-Arch.*, **38**, 80–96. *339*

Hagedorn, P. (1970a). Die Mathieugleichung mit nichtlinearen Dämpfungsund Rückstellgliedern. *ZAMM*, **50**, 321–324. *339*

Hagedorn, P. (1970b). Über Kombinationsresonanz bei parametererregten Systemen mit Coulombscher Dämpfung. *ZAMM*, **50**, T228–T231. *339*

Hagedorn, P. (1978). *Nichtlineare Schwingungen*. Akademische Verlagsgesellschaft, Wiesbaden, W. Germany. *vii*

Hahn, H. See George, Gunderson and Hahn.

Hahn, W. (1963). *Theory and Application of Liapunov's Direct Method*. Prentice-Hall, Englewood Cliffs. *284*

Hahn, W. (1967). *Stability of Motion*. Springer-Verlag, New York. *172*

Haight, E. C., and W. W. King (1971). Stability of parametrically excited vibrations of an elastic rod. *Dev. Theor. Appl. Mech.*, **5**, 677–714. *449*

Haight, E. C., and W. W. King (1972). Stability of nonlinear oscillations of an elastic rod. *J. Acoust. Soc. Am.*, **52**, 899–911. *448*

Haisler, W. E. See Stricklin, Martinez, Tillerson, Hong, and Haisler.

Hale, J. K. (1954). On boundedness of the solution of linear differential systems with periodic coefficients. *Riv. Mat. Univ. Parma*, **5**, 133–167. *304*

Hale, J. K. (1957). On a class of linear differential equations with periodic coefficients. *Ill. J. Math.*, **1**, 98–104. *304*

Hale, J. K. (1958). Linear systems of first and second order differential equations with periodic coefficients. *Ill. J. Math.*, **2**, 586–591. *304*

Hale, J. K. (1963). *Oscillations in Nonlinear Systems*. McGraw-Hill, New York. *144, 259, 304*

Hale, J. K. (1969). *Ordinary Differential Equations*. Wiley, New York. *137*

Hale, J. K., and G. Seifert (1961). Bounded and almost periodic solutions of singularly perturbed equations. *J. Math. Anal. Appl.*, **3**, 18–24. *144*

Hale, J. K. See Cesari and Hale; Gambill and Hale.

Hall, W. S. See Fink, Hall, and Hausrath.

Hamer, K., and M. A. Smith (1972). Stability of Hill's equation with three independent parameters. *J. Appl. Mech.*, **39**, 276–278. *295*

Hammond, C. E. (1974). An application of Floquet theory to the prediction of mechanical instability. *J. Am. Helicopter Soc.*, **19**, 14–23. *269*

Hammond, C. E. See Friedmann, Hammond, and Woo.

Hanratty, T. J. See Kim and Hanratty.

Hanson, R. D. See Townsend and Hanson.

Harrison, H. (1948). Plane and circular motion of a string. *J. Acoust. Soc. Am.*, **20**, 874–875. *485*

Harrison, J. A. See Mingori and Harrison.

Hasegawa, A., and F. Tappert (1973). Transmission of stationary nonlinear optical pulses in dispersive dielectric fibers. II. Normal dispersion. *Appl. Phys. Lett.*, **23**, 171–175. *588*

Hasimoto, H., and H. Ono (1971). Nonlinear modulation of gravity waves. *J. Phys. Soc. Japan*, **33**, 805–811. *587, 588*

Hassert, J. E., and J. L. Nowinski (1962). Nonlinear transverse vibrations of a flat rectangular orthotropic plate supported by stiff ribs. *Proc. 5th Int. Sym. Space Tech. Sci.*, 561–570. *501*

Hausrath, A. R. See Fink, Hall, and Hausrath.

Hayashi, C. (1953a). *Forced Oscillations in Nonlinear Systems*. Nippon, Osaka, Japan. *259*

Hayashi, C. (1953b). Subharmonic oscillations in nonlinear systems. *J. Appl. Phys.*, **24**, 521–529. *179*

Hayashi, C. (1964). *Nonlinear Oscillations in Physical Systems*. McGraw-Hill, New York. *259*

Hayashi, C., and M. Kuramitsu (1974). Self-excited oscillations in a system with two degrees of freedom. *Mem. Fac. Eng. Kyoto Univ.*, **36**, 87–104. *375*

Hayashi, S. See Yamamoto and Hayashi.

Hayes, W. D. (1953). On the equation for a damped pendulum under constant torque. *ZAMP*, **4**, 398–401. *149*

Hayes, W. D. (1971). Sonic Boom. *Ann. Rev. Fluid Mech.*, **3**, 269–290. *545*

Hayes, W. D. and J. W. Miles (1956). The free oscillations of a buckled panel. *Q. Appl. Math.*, **14**, 19–26. *506*

Hayes, W. D., and H. L. Runyan (1970). Sonic-boom propagation through a stratified atmosphere. *J. Acoust. Soc. Am.*, **51**, 695–701. *568*

Heidebrecht, A. C. See Tso, Pollner, and Heidebrecht.

Heinbockel, J. H., and R. A. Struble (1963). Resonant oscillations of an extensible pendulum. *ZAMP*, **14**, 262–269. *370, 379*

Heinbockel, J. H. See Struble and Heinbockel.

Helfenstein, H. (1950). Über eine spezielle Lamesche Differentialgleichung, mit Anwendung auf eine approximative Resonanzformel der Duffingschen schwingungsgleichung. *162*

Helleman, R. (1978). Variational solutions of nonintegrable systems, recurrent solutions of arbitrary period. In *Nonlinear Dynamics* (S. Jorna, Ed.). AIP, New York. *56, 388*

Helleman, R. H. G., and E. W. Montroll (1974). On a nonlinear perturbation theory without secular terms. *Physica*, **74**, 22–74. *56, 388*

Helleman, R. H. G. See Eminhizer, Helleman, and Montroll; Montroll and Helleman.

Hemp, G. W. (1972). Dynamic analysis of the runaway escapement mechanism. *Shock Vib. Bull.*, **42**, 125–133. *97*

Hemp, G. W., and P. R. Sethna (1964). The effect of high-frequency support oscillation on the motion of a spherical pendulum. *J. Appl. Mech.*, **31**, 351–354. *261, 369*

Hemp, G. W., and P. R. Sethna (1968). On dynamical systems with high frequency parametric excitation. *Int. J. Non-Linear Mech.*, **3**, 351–365. *302*

Hemp, G. W. See Sethna and Hemp.

Henrard, J. (1970). Periodic orbits emanating from a resonant equilibrium. *Celestial Mech.*, **1**, 437–466. *378, 396*

Henry, R. F. (1962). Coupled periodic motion in free undamped nonlinear two-degree-of-freedom systems. *J. Mech. Eng. Sci.*, **4**, 149–155. *388*

Henry, R. F., and S. A. Tobias (1959). Instability and steady-state coupled motions in vibration isolating suspensions. *Int. J. Mech. Sci.*, **1**, 19–29. *388*

Henry, R. F., and S. A. Tobias (1961). Modes at rest and their stability in coupled nonlinear systems. *J. Mech. Eng. Sci.*, **3**, 163–173. *388*

Herrmann, G. (1967). Stability of equilibrium of elastic systems subjected to nonconservative forces. *Appl. Mech. Rev.*, **20**, 103–108. *100*

Herrmann, G., and I. C. Jong (1965). On the destabilizing effect of damping in nonconservative elastic systems. *J. Appl. Mech.*, **32**, 592–597. *100*

Herrmann, G., and I. C. Jong (1966). On nonconservative stability problems of elastic systems with slight damping. *J. Appl. Mech.*, **33**, 125–133. *100*

Herrmann, G. See Bohn and Herrmann; Chu and Herrmann; Nemat-Nasser and Herrmann; Nemat-Nasser, Prasad, and Herrmann.

Hill, A. E. See Franken, Hill, Peters, and Weinreich.

Hill, G. W. (1886). On the part of the lunar perigee which is a function of the mean motions of the sun and moon. *Acta Math.*, **8**, 1–36. *273, 284, 295*

Hirano, I., and K. Matsukura (1968). Behavior of a vibrating system passing through the resonance. *Mitsubishi Denki*, **42**, 1511. *219*

Hirao, M., and N. Sugimoto (1977). Nonlinear modulation of torsional waves in elastic rod. *J. Phys. Soc. Japan*, **42**, 2056–2064. *587*

Hirao, M. See Sugimoto and Hirao.

Hirsch, P. (1930). Das Pendel mit oszillierendem Aufhangepunkt. *ZAMM*, **10**, 41–52. *261, 338*

Hitzl, D. L. (1969). Nonlinear attitude motion near resonance. *AIAA J.*, **7**, 1039–1047. *396*

Hitzl, D. L. (1971). Resonant attitude instabilities for asymmetric satellite in a circular orbit. AIAA Paper No. 71-88. *396*

Hitzl, D. L., and J. V. Breakwell (1971). Resonant and non-resonant gravity-gradient perturbations of a tumbling tri-axial satellite. *Celestial Mech.*, **3**, 346–383. *396*

Ho, C.-H., R. A. Scott, and J. G. Eisley (1975). Non-planar, non-linear oscillations of a beam–I. Forced motions. *Int. J. Non-Linear Mech.*, **10**, 113–127. *449*

Ho, C.-H., R. A. Scott, and J. G. Eisley (1976). Non-planar, non-linear oscillations of a beam–II. Free motions. *J. Sound Vib.*, **47**, 333–339. *449*

Ho, F. H., and J. L. Lai (1970). Parametric shimmy of a nosegear. *J. Aircraft*, **7**, 373–375. *269*

Hoff, N. J. (1951). The dynamics of the buckling of elastic columns. *J. Appl. Mech.*, **18**, 68–74. *455*

Hohenemser, K. H., and S. K. Yin (1972). Some applications of the method of multiblade coordinates. *J. Am. Helicopter Soc.*, **17**, 3–12. *269*

Hohenemser, K. H. See Peters and Hohenemser.

Hojo, H. See Nishikawa, Hojo, Mima, and Ikezi.

Holzer, S. M. (1974). Degree of stability of equilibrium. *J. Struct. Mech.*, **3**, 61–75. *423*

Hong, J. H. See Stricklin, Martinez, and Tillerson; Hong and Haisler.

Hoogstraten, H. W., and B. Kaper (1975). An asymptotic theory for a class of weakly nonlinear oscillations. *Arch. Rat. Mech. Anal.*, **58**, 239–260. *388*

Hopf, E. (1950). The partial differential equation $u_t + uu_x = \mu u_{xx}$. *Comm. Pure Appl. Math.*, **3**, 201–230. *608*

Hori, G. I. (1966). Theory of general perturbations with unspecified canonical variables. *Publ. Astron. Soc. Japan*, **18**, 287–296. *395*

Hori, G. I. (1967). Nonlinear coupling of two harmonic oscillations. *Publ. Astron. Soc. Japan*, **19**, 229–241. *388*

Horvay, G., and S. W. Yuan (1947). Stability of rotor blade flapping motion when the hinges are tilted. *J. Aero. Sci.*, **14**, 583–593. *269*

Houben, H. (1970). Einfluss des Antriebs in Schwingungssystemen mit Parametererregung. *ZAMM*, **50**, T231–T234. *269*

Hoult, D. P. (1968). Euler-Lagrange relationship for random dispersive waves. *Phys. Fluids*, **11**, 2082–2086. *601*

Hovanessian, S. A. (1959). Analysis of a nonlinear mechanical system with three degrees of freedom. *J. Appl. Mech.*, **26**, 546–548. *417*

Hsieh, B. J. See Belytschko and Hsieh.

Hsieh, D. (1975). Variational methods and nonlinear forced oscillations. *J. Math. Phys.*, **16**, 275–280. *163*

Hsu, C. S. (1959). On simple subharmonics. *Q. Appl. Math.*, **17**, 102–105. *179*

Hsu, C. S. (1960). On the application of elliptic functions in nonlinear forced oscillations. *Q. Appl. Math.*, **17**, 393–407. *162*

Hsu, C. S. (1961). On a restricted class of coupled Hill's equations and some applications. *J. Appl. Mech.*, **28**, 551–556. *304*

Hsu, C. S. (1963). On the parametric excitation of a dynamic system having multiple degrees of freedom. *J. Appl. Mech.*, **30**, 367–372. *308*

Hsu, C. S. (1965). Further results on parametric excitation of a dynamic system. *J. Appl. Mech.*, **32**, 373–377. *308, 309*

Hsu, C. S. (1972). Impulsive parametric excitation: theory. *J. Appl. Mech.*, **39**, 551–558. *283*

Hsu, C. S. (1974a). On approximating a general linear periodic system. *J. Math. Anal. Appl.*, **45**, 234–251. *283*

Hsu, C. S. (1974b). On parametric excitation and snap-through stability problems of shells. In *Thin-Shell Structures*. Prentice-Hall, Englewood Cliffs, pp. 103–131. *269*

Hsu, C. S. (1974c). Some simple exact periodic responses for a nonlinear system under parametric excitation. *J. Appl. Mech.*, **41**, 1135–1137. *338*

Hsu, C. S. (1975a). Limit cycle oscillations of parametrically excited second-order nonlinear systems. *J. Appl. Mech.*, **42**, 176–182. *339*

Hsu, C. S. (1975b). The response of a parametrically excited hanging string in fluid. *J. Sound Vib.*, **39**, 305–316. *268, 486*

Hsu, C. S., and W.-H. Cheng (1973). Applications of the theory of impulsive parametric excitation and new treatments of general parametric excitation problems. *J. Appl. Mech.*, **40**, 78–86. *264, 283*

Hsu, C. S., and W.-H. Cheng (1974). Steady-state response of a dynamical system under combined parametric and forcing excitations. *J. Appl. Mech.*, **41**, 371–378. *339, 356*

Hsu, C. S., and T. H. Lee (1971). A stability study of continuous systems under parametric excitation via Liapunov's direct method. In *IUTAM Symposium*, *Herrenalb*, *1969*, Springer-Verlag, New York, pp. 112–118. *284*

Hsu, C. S. See Lee and Hsu; Troger and Hsu.

Hu, K.-K., and P. G. Kirmser (1971). On the nonlinear vibrations of free-free beams. *J. Appl. Mech.*, **38**, 461–466. *444*

Huang, C.-L. (1972a). Nonlinear axisymmetric flexural vibration equations of a cylindrically anistropic plate. *AIAA J.*, **10**, 1378–1379. *502, 508*

Huang, C.-L. (1972b). Nonlinear responses for a circular plate subjected to a dynamic ring load. *Dev. Theor. Appl. Mech.*, **6**, 489–514. *508*

Huang, C.-L. (1973). Finite amplitude vibrations of an orthotropic circular plate with an isotropic core. *Int. J. Non-Linear Mech.*, **8**, 445–457. *446, 502, 508*

Huang, C.-L. (1974). Nonlinear oscillations of an annulus with variable thickness. *Dev. Theor. Appl. Mech.*, **7**, 271–284. *501, 508*

Huang, C.-L., and I. M. Al-Khattat (1977). Finite amplitude vibrations of a circular plate. *Int. J. Non-Linear Mech.*, **12**, 297–306. *504, 507, 508*

Huang, C.-L., and Y. J. Meng. (1972). Non-linear oscillations of a nonuniform fixed circular plate. *Int. J. Non-Linear Mech.*, **7**, 557–569. *446, 501*

Huang, C.-L., and B. E. Sandman (1971). Large amplitude vibrations of a rigidly clamped circular plate. *Int. J. Non-Linear Mech.*, **6**, 451–468. *446, 508*

Huang, C.-L., and H. K. Woo (1973). Large oscillations of an orthotropic annulus. *Dev. Mech.*, **7**, 1027–1039. *502, 508*

Huang, C.-L. H. K. Woo, and H. S. Walker (1976). Nonlinear flexural oscillations of a partially tapered annular plate. *Int. J. Non-Linear Mech.*, **11**, 89-97. *446, 501, 508*

Huang, C.-L. See Sandman and Huang.

Huang, T. C. (1954). Subharmonic oscillations in nonlinear systems of two degrees of freedom. *J. Appl. Mech.*, **21**, 95–101. *418*

Huang, T. C. (1955). Harmonic oscillations of nonlinear two-degree-freedom systems. *J. Appl. Mech.*, **22**, 107–110. *417*

Huber, H., and H. Strehlow (1976). Hingeless rotor dynamics in high speed flight. *Vertica*, **1**, 39–53. *269*

Hübner, W. (1965). Die Wechselwirkung zwichen Schwingen und Antrieb bei erzwungenen Schwingungen. *Ing.-Arch.*, **34**, 411–422. *227*

Hudson, G. E. See Saenger and Hudson.

Hull, E. H. (1961). Shaft whirling as influenced by stiffness asymmetry. *Trans. Ser. B ASME*, **83**, 219–226. *269*

Hunt, G. W. See Thompson and Hunt.

Hunter, J. H. See Phelps and Hunter.

Husid, R. See Jennings and Husid.

Hwang, C., and W. S. Pi (1972). Nonlinear acoustic response analysis of plates using the finite element method. *AIAA J.*, **10**, 276–281. *446*

Hyer, M. W., W. J. Anderson, and R. A. Scott (1976). Nonlinear vibrations of three-layer beams with viscoelastic cores. I. Theory. *J. Sound Vib.*, **46**, 121–136. *449*

Iakubovich, V. A. (1959). The method of small parameters for systems with periodic coefficients. *PMM*, **23**, 17–43. *308*

Ibragimov, K. Kh. (1975). Nonlinear vibrations of viscoelastic systems with two degrees of freedom. *Izv. Akad. Nauk Uzb. SSR, Ser. Tekh Nauk*, **No. 4**, 36–39. *160*

Ibrahim, R. A., and A. D. S. Barr (1975a). Autoparametric resonance in a structure containing a liquid, Part I: Two-mode interaction. *J. Sound Vib.*, **42**, 159–179. *269*

Ibrahim, R. A., and A. D. S. Barr (1975b). Autoparametric resonance in a structure containing a liquid, Part II: Three-mode interaction. *J. Sound Vib.*, **42**, 181–200. *269*

Ikezi, H. See Nishikawa, Hojo, Mima, and Ikezi.

Infante, E. F., and R. H. Plaut (1969). Stability of a column subjected to a time-dependent axial load. *AIAA J.*, **7**, 766–768. *267*

Inoue, Y. (1975). Resonant four-wave interaction in a dispersive medium. *J. Phys. Soc. Japan*, **39**, 1092–1099. *601*

Inoue, Y. (1977). Nonlinear interaction of dispersive waves with equal group velocity. *J. Phys. Soc. Japan*, **43**, 243–249. *598*

Inoue, Y., and Y. Matsumoto (1974). Nonlinear wave modulation in dispersive media. *J. Phys. Soc. Japan*, **36**, 1446–1455. *587*

Inoue, Y. See Kakutani, Inoue, and Kan; Matsumoto, Sugimoto, and Inoue.

Isakovitch, M. A. (1957). Scattering of sound waves on small irregularities in a wave guide. *Akust. Zh.*, **3**, 37–45. *362*

Ishida, Y. See Yamamoto and Ishida; Yamamoto, Ishida, and Kawasumi.

Ishihara, K. See Iwatsubo, Sugiyama, and Ishihara.

Ismail, I. A. See Nowinski and Ismail.

Issid, N. T. See Paidoussis and Issid.

Ivanova, N. I., B. I. Morgunov, and I. E. Troyanovskii (1974). Calculation of nonlinear vibrations of a viscoelastic bar. *Polyn. Mech.* (USSR), **10,** 481–483. *449*

Iwan, W. D. (1965). The steady-state response of a two-degree-of-freedom bilinear hysteretic system. *J. Appl. Mech.*, **32,** 151–156. *98, 236*

Iwan, W. D. (1966). A distributed element model for hysteresis and its steady-state dynamic response. *J. Appl. Mech.*, **33,** 893–900. *98*

Iwan, W. D. (1967). On a class of models for the yielding behavior of continuous and composite systems. *J. Appl. Mech.*, **34,** 612–617. *98*

Iwan, W. D. (1968a). Response of multi-degree-of-freedom yielding systems. *ASCE J. Eng. Mech. Div.*, **94,** 421–437. *98*

Iwan, W. D. (1968b). Steady-state dynamic response of a limited slip system. *J. Appl. Mech.*, **35,** 322–326. *98*

Iwan, W. D. (1969). On defining equivalent systems for certain ordinary non-linear differential equations. *Int. J. Non-Linear Mech.*, **4,** 325–334. *87*

Iwan, W. D. (1970). On the steady-state response of a one-dimensional yielding continuum. *J. Appl. Mech.*, **37,** 720–727. *98*

Iwan, W. D. (1973). A generalization of the concept of equivalent linearization. *Int. J. Non-Linear Mech.*, **8,** 279–287. *87*

Iwan, W. D., and D. M. Furuike (1973). The transient and steady-state reponse of a hereditary system. *Int. J. Non-Linear Mech.*, **3,** 394–406. *98*

Iwan, W. D., and R. K. Miller (1977). The steady-state response of systems with spatially localized nonlinearity. *Int. J. Non-Linear Mech.*, **12,** 165–173. *87*

Iwan, W. D. See Patula and Iwan.

Iwatsubo, T., H. Kanki, and R. Kawai (1972). Vibration of asymmetric rotor through critical speed with limited power supply. *J. Mech. Eng. Sci.*, **14,** 184–194. *227*

Iwatsubo, T., M. Saigo, and Y. Sugiyama (1973). Parametric instability of clamped-clamped and clamped-simply supported columns under periodic axial load. *J. Sound Vib.*, **30,** 65–77. *266, 267, 306*

Iwatsubo, T., Y. Sugiyama, and K. Ishihara (1972). Stability and nonstationary vibration of columns under periodic loads. *J. Sound Vib.*, **23,** 245–257. *266, 267*

Iwatsubo, T., Y. Sugiyama, and S. Ogino (1974). Simple and combination resonances of columns under periodic axial loads. *J. Sound Vib.*, **33,** 211–222. *266, 267*

Iwatsubo, T., A. Tomita, and R. Kawai (1973). Vibrations of asymmetric rotors supported by asymmetric bearings. *Ing.-Arch.*, **42,** 416–432. *269, 331*

Jacobsen, L. S. (1952). Dynamic behavior of simplified structures up to the point of collapse. In *Symposium on Earthquake and Blast effects on Structures*, Los Angeles, pp. 112–113. *98*

Jacobson, I. D. See Junkins, Jacobson, and Blanton.

Jacobson, M. J. and R. C. W. van der Heyde (1972). Acoustic fatigue design information for honeycomb panels with fiber-reinforced facings. *J. Aircraft,* **9,** 31–42. *472*

Jain, P. C., and V. Srinivasan (1975). A review of self-excited vibrations in oil film journal bearings. *Wear*, **31,** 219–225. *107*

Janssens, P., R. van Dooren, and M. Melchambre (1969). Sur les oscillateurs nonlinéaires du type de Duffing à deux degrés libertè. In *CBRM Colloque Equations Différentielles Non Linéaires*, Mons, pp. 171–184. *417*

Javan, A. (1957). Theory of a three-level maser. *Phys. Rev.*, **107,** 1579–1589.

Jeffery, A., and T. Kakutani (1970). Stability of the Burgers shock wave and the Korteweg-de Vries soliton. *Indiana Univ. Math. J.*, **20,** 463–468. *608, 609*

Jeffrey, A., and T. Kakutani (1972). Weak nonlinear dispersive waves: a discussion centered around the Korteweg-de Vries equation. *SIAM Rev.*, **14,** 582–643. *545*

Jeffrey, A., and T. Taniuti (1966). *Nonlinear Wave Propagation*, Academic, New York. *545*

Jennings, P. C. (1964). Periodic response of a general yielding structure. *ASCE J. Eng. Mech. Div.*, **90,** 131–166. *98*

Jennings, P. C., and R. Husid (1968). Collapse of yielding structures during earthquakes. *ASCE J. Eng. Mech. Div.*, **94,** 1045–1065. *98*

Jimenez, J. (1973). Nonlinear gas oscillations in pipes, Part I. Theory. *J. Fluid Mech.*, **59,** 23–46. *572*

Johansen, K. F., and T. R. Kane (1969). A simple description of the motion of a spherical pendulum. *J. Appl. Mech.*, **36,** 408–411. *369*

Jong, I. C. (1969). On stability of a circulatory system with bilinear hysteresis damping. *J. Appl. Mech.*, **36,** 76–82. *98, 339*

Jong, I. C., and Y. S. Chen (1971). Wave propagation in a rod with distributed-yielding hysteresis. *Int. J. Non-Linear Mech.*, **6,** 511–527. *98*

Jong, I. C. See Herrmann and Jong.

Jordan, P. F. (1973). Nonlinear dynamics of hemispherical shells. *AIAA J.*, **11,** 1117–1122. *506*

Jorna, S. (1978). *Nonlinear Dynamics*. AIP, New York. *388*

Joshi, S. G., H. R. Srirangarajan, and P. Srinivasan (1976). Stability of third-order systems. *J. Sound Vib.*, **48,** 578–581. *152*

Junkins, J. L., I. D. Jacobson, and J. N. Blanton (1973). A nonlinear oscillator analog of rigid body dynamics. *Celestial Mech.*, **7,** 398–407. *396*

Kadi, A. S. See Leissa and Kadi.

Kadomtsev, B. B., and V. I. Karpman (1971). Nonlinear waves. *Sov. Phys. Uspekhi*, **14,** 40–60. *587*

Kadyrbekov, T. (1975). Nonlinear vibrations of a viscoelastic beam. *Polym. Mech. (USSR)*, **9,** 832–835. *449*

Kahn, M. E. See Kane and Kahn.

Kaiser, J. E. See Nayfeh and Kaiser.

Kakutani, T. (1971a). Effect of an uneven bottom on gravity waves. *J. Phys. Soc. Japan*, **30,** 272–276. *568*

Kakutani, T. (1971b). Weak nonlinear magneto-acoustic waves in an inhomogeneous plasma. *J. Phys. Soc. Japan*, **31,** 1246–1248. *568*

Kakutani, T. (1974). Plasma waves in the long wave approximation. *Prog. Theor. Phys. Suppl.*, **55,** 97–119. *556*

Kakutani, T., Y. Inoue, and T. Kan (1974). Nonlinear capillary waves on the surface of liquid column. *Phys. Soc. Japan*, **37,** 529–538. *587*

Kakutani, T., and T. Kawahara (1970). Weak ion-acoustic shock waves. *J. Phys. Soc. Japan*, **29,** 1068–1073. *556*

Kakutani, T. and K. Matsuuchi (1975). Effect of viscosity on long gravity waves. *J. Phys. Soc. Japan*, **39**, 237–246. *556*

Kakutani, T., and K. Michihiro (1976). Nonlinear modulation of stationary water waves. *J. Phys. Soc. Japan*, **41**, 1792–1799. *588*

Kakutani, T., and H. Ono (1969). Weak nonlinear hydromagnetic waves in a cold collision-free plasma. *J. Phys. Soc. Japan*, **26**, 1305–1318. *556*

Kakutani, T., H. Ono, T. Taniuti, and C.-C Wei (1968). Reductive perturbation method in nonlinear wave propagation. II. *J. Phys. Soc. Japan*, **24**, 1159–1166. *556*

Kakutani, T., and N. Sugimoto (1974). Krylov-Bogoliubov-Mitropolsky method for nonlinear wave modulation. *Phys. Fluids*, **17**, 1617–1625. *587*

Kakutani, T. See Jeffrey and Kakutani; Kawahara, Sugimoto, and Kakutani.

Kamel, A. A. (1969). Expansion formulae in canonical transformations depending on a small parameter. *Celestial Mech.*, **1**, 190–199. *378, 396*

Kamel, A. A. (1970). Perturbation method in the theory of nonlinear oscillations. *Celestial Mech.*, **3**, 90–106. *396*

Kamel, A. A. (1971). Lie transforms and the Hamiltonization of non-Hamiltonian systems. *Celestial Mech.*, **4**, 397–405. *396*

Kamel, A. A. See Nayfeh and Kamel.

Kan, T. See Kakutani, Inoue, and Kan.

Kana, D. D., and R. R. Craig (1968). Parametric oscillations of a longitudinally excited cylindrical shell containing liquid. *J. Spacecraft*, **5**, 13–21. *269*

Kanaka Raju, K. (1977). Large amplitude vibrations of circular plates with varying thickness. *J. Sound Vib.*, **50**, 399–403. *446, 501*

Kanaka Raju, K., and G. Venkateswara Rao (1976a). Axisymmetric vibrations of circular plates including the effects of geometric nonlinearity, shear deformation and rotary inertia. *J. Sound Vib.*, **47**, 179–184. *446, 500*

Kanaka Raju, K., and G. Venkateswara Rao (1976b). Large amplitude asymmetric vibrations of some thin shells of revolution. *J. Sound Vib.*, **44**, 327–333. *506*

Kanaka Raju, K. See Raju, Venkateswara Rao, and Kanaka Raju; Venkateswara Rao, Kanaka Raju, and Raju; Venkateswara Rao, Raju, and Kanaka Raju.

Kandil, O. A. See Atta, Kandil, Mook, and Nayfeh; Thrasher, Mook, Kandil, and Nayfeh.

Kane, T. R. (1966). Attitude stability of earthpointing satellites. *AIAA J.*, **3**, 726–731. *396*

Kane, T. R., and P. M. Barba (1966). Attitude stability of a spinning satellite in an elliptic orbit. *J. Appl. Mech.*, **33**, 402–405. *331*

Kane, T. R., and M. E. Kahn (1968). On a class of two-degree-of-freedom oscillations. *J. Appl. Mech.*, **35**, 547–552. *370, 379*

Kane, T. R., and D. L. Mingori (1965). Effect of a rotor on the attitude stability of a satellite in a circular orbit. *AAIA J.*, **3**, 936–940. *331, 396*

Kane, T. R., and D. J. Shippy (1963). Attitude stability of a spinning unsymmetrical satellite in a circular orbit. *J. Astro. Sci.*, **10**, 114–119. *331*

Kane, T. R., and D. Sobala (1963). A new method for attitude stabilization. *AIAA J.*, **1**, 1365–1367. *283, 331*

Kane, T. R. See Johansen and Kane.

Kaneta, K. See Tanabashi and Kaneta

Kanki, H. See Iwatsubo, Kanki, and Kawai.

Kaper, B. (1976). Perturbed nonlinear oscillations. *SIAM J. Appl. Math.*, **31**, 519–546. *138*

Kaper, B. See Hoogstraten and Kaper.

Kapitza, P. L. (1951). Dynamic stability of a pendulum whose point of suspension oscillates. *Zh. Eksper Teor. Fiz.*, **21**, 588–597. *261*

Karasudhi, P. G., Tan and S. L. Lee (1974). Vibration of frame foundation with bilinear hysteresis for rotating machinery. *J. Eng. Ind.*, **96**, 1010–1014. *98*

Karimov, A. U. (1974). On a method of studying nonlinear oscillations of viscoelastic systems. *PMM*, **38**, 901–994. *100*

Karmakar, B. M. (1975). Amplitude-frequency characteristic of nonlinear vibrations of a thin anistropic right-angled triangular plate resting on elastic foundation. *J. Ind. Inst. Sci.*, **57**, 204–214. *502*

Karpasiuk, V. K. (1973). Determination of the frequency of the approximate solution of Hill's equation. *PMM*, **37**, 365–366. *295*

Karpman, V. I. (1975). *Nonlinear Waves in Dispersive Media*. Pergamon, New York. *545, 587*

Karpman, V. I., and E. M. Kruskal (1969). Modulated waves in nonlinear dispersive media. *Sov. Phys. (JETP)*, **28**, 277–281. *594*

Karpman, V. I. See Kadomtsev and Karpman.

Katayama, T. See Sugiyama, Katayama, and Sekiya.

Kato, Y. (1974). Inverse scattering method for the initial value problem of the nonlinear equation. *Prog. Theor. Phys. Suppl.*, **55**, 247–283. *588*

Kato, Y. See Asano and Kato.

Kauderer, H. (1958). *Nichtlineare Mechanik*. Springer-Verlag, Berlin. *vii, 259, 261, 338*

Kaup, D. K. (1976). The three-wave interaction—a nondispersive phenomenon. *Stud. Appl. Math.*, **55**, 9–44. *588*

Kaup, D. J. See Ablowitz, Kaup, Newell, and Segur.

Kawahara, T., N. Sugimoto, and T. Kakutani (1975). Nonlinear interaction between short and long capillary-gravity waves. *J. Phys. Soc. Japan*, **39**, 1379–1386. *601*

Kawahara, T. See Kakutani and Kawahara.

Kawai, R. See Iwatsubo, Kanki, and Kawai; Iwatsubo, Tomita, and Kawai.

Kawasumi, J. See Yamamoto, Ishida, and Kawasumi.

Keen, B. E. and W. H. W. Fletcher (1970). Suppression of a plasma instability by the method of "asynchronous quenching." *Phys. Rev. Lett.*, **24**, 130–134. *107*

Kehagioglou, T. See Evan-Iwanowski, Sanford, and Kehagioglou.

Kellenberger, W. (1955). Forced, double-frequency, flexural vibrations in a rotating horizontal, cylindrical shaft. *Brown Boveri Rev.*, **42**, 79. *269*

Keller, J. B., and S. Kogelman (1970). Asymptotic solutions of initial value problems for nonlinear partial differential equations. *SIAM J. Appl. Math.*, **18**, 748–758. *556, 572, 604*

Keller, J. J. (1975). Subharmonic nonlinear acoustic resonances in closed tubes. *ZAMP*, **26**, 395. *572*

Keller, J. J. (1976). Third order resonances in closed tubes. *ZAMP*, **27**, 303–323. *572*

Keller, J. J. (1977a). Nonlinear acoustic resonances in shock tubes with varying cross-sectional area. *ZAMP*, **28**, 107–122. *572*

Keller, J. J. (1977b). Resonant oscillations in open tubes. *ZAMP*, **28**, 237–263. *572*

Keller, J. J. (1977c). Subharmonic nonlinear acoustic resonances in open tubes—Part I: Theory. *ZAMP*, **28**, 419–431. *572*

Kelly, S. G. See Nayfeh and Kelly.

Kelzon, A. S., and V. I. Yakovlev (1974). Experimental investigation of self-excited vibrations of a high speed rotor. *Mashinovedenie*, **5**, 21–28. *129*

Kevorkian, J. (1971). Passage through resonance for a one-dimensional oscillator with slowly varying frequency. *SIAM J. Appl. Math.*, **20**, 364–373. *219*

Kevorkian, J. (1974). On a model for reentry roll-resonance. *SIAM J. Appl. Math.*, **26**, 638–669. *228*

Kevorkian, J. See Chikwendu and Kevorkian.

Key, J. E. See Oden, Key and Fost.

Khachaturian, A. A. See Ambartsumyan and Khachaturian.

Khaikin, S. See Andronov, Vitt, and Khaikin.

Khalilov, Z. I. (1942). The dynamic stability of a plate under the action of periodic longitudinal forces. *Tr. Azerb. Gos. Univ. Ser. Mar.*, **1**, 28–32. *268*

Kharlamov, S. A. See Novikov and Kharlamov.

Khazin, L. G. (1971). On the stability of Hamiltonian systems in the presence of resonances. *PMM*, **35**, 423–431. *396*

Khazin, L. G., and F. Kh. Tsel'man (1970). On the nonlinear interaction of resonant oscillators. *Dokl. Akad. Nauk SSSR*, **193**, 317–319. *379*

Khokhlov, R. V. See Rudenko, Soluyan, and Khokhlov.

Kim, Y. Y., and T. J. Hanratty (1971). Weak quadratic interactions of two-dimensional waves. *J. Fluid Mech.*, **50**, 107–132. *601*

King, W. W., and C. C. Lin (1974). Applications of Bolotin's method to vibrations of plates. *AIAA J.*, **12**, 399–400. *268*

King, W. W. See King and Haight.

Kinney, W. D., and R. M. Rosenberg (1965). On steady-state forced vibrations in strongly nonlinear systems having two degrees of freedom. In *Les Vibrations Frocées dans Les Systèmes Nonlinéaires.* Proc. CNRS Int. Coll. **148,** pp. 351–373. *417*

Kirchman, E. J., and J. E. Greenspon (1957). Nonlinear response of aircraft panels in acoustic noise. *J. Acoust. Soc. Am.*, **29**, 854–857. *509*

Kirmser, P. G. See Hu and Kirmser.

Kishor, B., and J. S. Rao (1974). Non-linear vibration analysis of a rectangular plate on a viscoelastic foundation. *Aeronautical Q.*, **25**, 37–46. *507*

Klotter, K. (1953a). A comprehensive stability for forced vibrations in nonlinear systems. *J. Appl. Mech.*, **20**, 9–12. *163*

Klotter, K. (1953b). Steady-state vibrations in systems having arbitrary restoring and arbitrary damping forces. In *Proc. Symp. Nonlinear Circuit Anal.*, New York, pp. 234–257. *163*

Klotter, K. (1954). Steady state oscillations in nonlinear multiloop circuits. *Trans. Inst. Radio Eng.*, **CT-1**, 13–18. *417*

Klotter, K. (1955). Free oscillations of systems having quadratic damping and arbitrary restoring forces. *J. Appl. Mech.*, **22**, 493–499. *119, 129, 154*

Klotter, K., and G. Kotowski (1943a). Über die Gleichung $y'' + (\lambda + \gamma_1 \cos x + \gamma_2 \cos 2x)y = 0$. *ZAMM*, **23**, 149–155. *295*

Klotter, K., and G. Kotowski (1943b). Über die stabilität der Lösungen Hillscher differential Gleichungen mit drei unabhängigen Parametern. *ZAMM*, **23**, 149–155. *295*

Klotter, K., and E. Kreyszig (1957). Über eine besondere Klasse selbsterregter Schwingungen. *Ing.-Arch.*, **25**, 389–403. *129*

Klotter, K., and E. Kreyszig (1960). On a special class of self-sustained oscillations. *J. Appl. Mech.*, **27**, 568–574. *129*

Klotter, K., and E. Pinney (1953). A comprehensive stability criterion for forced vibrations in nonlinear systems. *J. Appl. Mech.*, **20**, 9–12. *163*

Klumpp, J. H. See Goodman and Klumpp.

Kluwick, A. (1974). Gleichmassiggültige Lösungen and kumulative Effekte bei Wellenaubreitungsvorgängen. *J. Mecanique*, **13**, 131–157. *556*

Kluwick, A. See Nayfeh and Kluwick.

Kobayashi, S. (1962a). Flutter of simply-supported rectangular panels in a supersonic flow—two-dimensional panel flutter. I. Simply-supported panel. II. Clamped panel. *Trans. Japan Soc. Aero Space Sci.*, **5**, 79–118. *509*

Kobayashi, S. (1962b). Two-dimensional panel flutter. II. Clamped panel. *Trans. Japan Soc. Aero Space Sci.*, **5**, 103–118. *509*

Kobayashi, S. See Matsuzaki and Kobayashi.

Kochin, N. E. (1934). On the torsional vibrations of crankshafts. *PMM*, **2**, 3–28. *259*

Koenig, H. A. See Goldberg and Koenig.

Kogelman, S. See Keller and Kogelman.

Kohnkin, V. M. See Tarantovich and Kohnkin.

Kohut, J., and M. V. Mathews (1971). Study of motion of a bowed violin string. *J. Acoust. Soc. Am.*, **49**, 532–537. *485*

Koltunov, M. A. See Maiboroda, Koltunov, and Morgunov.

Kononenko, V. O. (1964). Forced oscillations of a nonlinear quasi-harmonical system. *Proc. Eleventh Int. Congr. Appl. Mech.*, pp. 173–177. *163*

Kononenko, V. O. (1969). *Vibrating Systems with a Limiting Power Supply*. Iliffe, London. *227, 260, 360*

Kononenko, V. O., and S. S. Korablev (1959). An experimental investigation of the resonance phenomena with a centrifugally excited alternating force. *Tr. Mosk. Tekhn. Inst. Lekh. Prom*, No. **14**. *224*

Kononenko, V. O., and P. S. Koval'chuk (1973a). Dynamic interaction of mechanisms generating oscillations in nonlinear systems. *Mech. Solids* (USSR), **8**, 48–56. *227*

Kononenko, V. O., and P. S. Koval'chuk (1973b). Interaction of self oscillation in mechanical oscillating systems. *PMM*, 3–9. *375, 379*

Koplik, B. See Grossman, Koplik, and Yu.

Korablev, S. S. See Kononenko and Korablev.

Korolev, V. I., and L. V. Postnikov (1973). On the theory of locking of a self-excited oscillator at fractional frequencies. *Radiophys. Quantum Electron.*, **13**, 966–970. *214*

Kotowski, G. See Klotter and Kotowski.

Kovac, E. J., W. J. Anderson, and R. A. Scott (1971). Forced nonlinear vibrations of a damped sandwich beam. *J. Sound Vib.*, **17**, 25–39. *449*

Koval'chuk, P. S. See Kononenko and Koval'chuk.

Kozhemyakina, I. F., and A. B. Morgaevskii (1976). Effect of rotary inertia and shearing force on nonlinear oscillations of plates under moving loads. *Prikl. Mekh.*, **12**, 75–80. *500*

Krauter, A. I., and P. Z. Bulkeley (1970). The effect of central clamping on transverse vibrations of spinning membrane disks. *J. Appl. Mech.*, **37**, 1037–1042. *505*

Kravchuk, A. S., B. I. Morgunov, and I. E. Troyanovskii (1974). Forced nonlinear vibrations of a viscoelastic body. *Polym. Mech.*, **10**, 589–593. *100, 445*

Kreyszig, E. See Klotter and Kreyszig.

Krogdahl, W. S. (1960). Numerical solutions of van der Pol's equation. *ZAMP*, **11**, 59–63. *144*

Kronauer, R. E., and S. A. Musa (1966a). Exchange of energy between oscillations in weakly nonlinear conservative systems. *J. Appl. Mech.*, **33**, 451–452. *424*

Kronauer, R. E., and S. Musa (1966b). Necessary conditions for subharmonic and superharmonic synchronization in weakly nonlinear systems. *Q. Appl. Math.*, **24**, 153–160. *179*

Kronauer, R. E. See Musa and Kronauer; Subramanian and Kronauer.

Kruschul, M. J. (1960). On almost periodic solutions of quasi-linear systems with resonances. On the theory of self-excited rotors. *Isv. AN SSR Mech. Masch.*, **1**, 90. *402, 418*

Kruskal, E. M. See Karpman and Kruskal.

Kruskal, M. D., and N. J. Zabusky (1964). Stroboscopic-perturbation procedure for treating a class of nonlinear wave equations. *J. Math. Phys.*, **5**, 231–244. *486, 575*

Kruskal, M. D. See Gardner, Greene, Kruskal, and Miura.

Krylov, N. M., and N. N. Bogoliubov (1935). Calculations of the vibration of frame construction with the consideration of normal forces with the help of the methods of nonlinear mechanics. In *Investigation of Vibration of Structures ONTI Kharkov*, Kiev, pp. 5–24. *266, 267*

Krylov, N. N., and N. N. Bogoliubov (1947). *Introduction to Nonlinear Mechanics.* Princeton University, Princeton. *vii, 137, 259, 266*

Krzyanowski, A. See Dzygadlo and Krzyanowski.

Krzhnechkovskii, P. G. See Popov, Antipov, and Krzhnechkovskii.

Kucharski, W. (1950). Beiträge zur Theorie der durch gleichförmigen schub beanspruchten Platte. *Ing.-Arch.*, **18**, 385–408. *268*

Kuczynski, W. A. See Sissingh and Kuczynski.

Kulterbacv, K. P. See Vol'mir and Kulterbacv.

Kung, G. C., and Y. H. Pao (1972). Nonlinear flexural vibrations of a clamped circular plate. *J. Appl. Mech.*, **39**, 1050–1054. *508*

Kuo, C.-C., L. Morino, and J. Dugundji (1972). Perturbation and harmonic balance methods for nonlinear panel flutter. *AIAA J.*, **10**, 1479–1484. *509*

Kuo, C.-C., L. Morino, and J. Dugundji (1973). Nonlinear interaction of panel flutter with harmonic forcing excitation. *AIAA J.*, **11**, 419–420. *509*

Kuo, J. K. See Rosenberg and Kuo.

Kuramitsu, M. See Hayashi and Kuramitsu.

Kuroda, M. (1974). Periodic solutions of the free vibration of a two degree-of-freedom nonlinear spring-mass system. *JSME*, **17**, 59–58. *388*

Kuzmak, G. E. (1959). Asymptotic solutions of nonlinear second order differential equations with variable coefficients. *PMM*, **23**, 730–744. *138*

Kyner, W. T. (1969). *Lectures on Nonlinear Resonance.* NASA PM-81, pp. 255–299. *395*

Lackman, L. M. See Bieniek, Fan, and Lackman.

Lai, J. L. See Ho and Lai; Yu and Lai.

Lamb, G. L. (1971). Analytical descriptions of ultrashort optical pulse propagation in a resonant medium. *Res. Mod. Phys.*, **43**, 99–124. *588*

Lamb, G. L. (1973). Phase variation in coherent-optical-pulse propagation. *Phys. Rev. Lett.*, **31**, 196–199. *588*

Lamb, H. (1932). *Hydrodynamics.* Cambridge University, London. *370*

Lamb, W. E. (1960). Quantum mechanical amplifiers. In *Lectures in Theoretical Physics* 2 (W. E. Brittin and B. W. Downs, Eds.), Wiley, New York. *163*

Lamb, W. E. (1964). Theory of an optical maser. *Phys. Rev.*, **A134**, 1429–1450. *107*

Landau, L. D. (1965). On shock waves at long distances from the place of their origin. In *Collected Papers of L. D. Landau.* Pergamon, New York. *567*

Lansdowne, D. L., and A. C. Soudack (1971). Extension of Krylov-Bogoliubov equivalent linearization principle to coupled nonlinear systems. *Int. J. Control*, **13**, 1151–1159. *388*

Lange, C. G., and A. C. Newell (1971). The post-buckling problem for thin elastic shells. *SIAM J. Appl. Math.*, **21**, 605–629. *506*

Lardner, R. W. (1975). The formation of shock waves in Krylov-Bogoliubov solutions of hyperbolic partial differential equations. *J. Sound Vib.*, **39**, 489–502. *571, 575*

Lardner, R. W. (1976). The development of plane shock waves in nonlinear viscoelastic media. *Proc. Roy. Soc. London*, **A347**, 329–344. *571, 607*

Lardner, R. W. (1977). Asymptotic solutions of nonlinear wave equations using the methods of averaging and two-timing. *Q. Appl. Math.*, **35**, 225–238. *556, 604*

Lardner, R. W. See Arya and Lardner; Bojadziev and Lardner.

Larsson, J., and L. Stenflo (1973). Three-wave interactions in magnetized plasmas. *Beitr. Plasma Phys.*, **13**, 169–181. *601*

LaSalle, J. P. (1949). Relaxation oscillations. *Q. Appl. Math.*, **7**, 1–19. *144*

LaSalle, J. P., and S. Lefschetz (1961). *Stability by Liapunov's Direct Method.* Academic, New York. *172, 284*

Lashinsky, H. (1969). Periodic pulling and the transition to turbulence in a system with discrete modes. In *Turbulence of Fluids and Plasmas* (J. Fox, Ed.), Wiley, New York, pp. 29–46. *107*

Lashinsky, H. See Cap and Lashinsky; DeNeff and Lashinsky; Dewan and Lashinsky.

Laura, P. A. A. See Susemihl and Laura.

Lauterborn, W. (1970). Resonanzkurven von Gasblasen in Flüssigkeiten. *Acustica*, **23**, 73–81. *196, 199*

Lax, P. D. (1964). Development of singularities of solutions of nonlinear hyperbolic partical differential equations. *J. Math. Phys.*, **5**, 611–613. *575*

Lax, P. D. (1968). Integrals of nonlinear equations of evolution and solitary waves. *Comm. Pure Appl. Math.*, **21**, 467–490. *588*

Lazan, B. J. (1968). *Damping of Materials and Members in Structural Mechanics.* Pergamon, New York. *95, 100, 449*

Lazarev, V. (1937). Parametrical excitation of combination oscillations. *Tech. Phys. USSR*, **4**, 885. *308*

Lee, E. W. (1957). Non-linear forced vibration of a stretched string. *Br. J. Appl. Phys.*, **8**, 411–413. *485, 486*

Lee H. C. See Ahmad and Lee.

Lee, S. L. See Karasudhi, Tan, and Lee.

Lee, S. Y., and W. F. Ames (1973). A class of general solutions to the nonlinear dynamic equations of elastic strings. *J. Appl. Mech.*, **40**, 1035–1039. *486*

Lee, T-C. (1976). A study of coupled Mathieu equations by use of infinite determinants. *J. Appl. Mech.*, **43**, 349–352. *284*

Lee, T. H., and C. S. Hsu (1972). Liapunov stability criteria for continuous systems under parametric excitation. *J. Appl. Mech.*, **39**, 244–250. *284*

Lee, T. H. See Hsu and Lee.

Lee, T. W., P. T. Blotter, and D. H. Y. Yen (1971). On the nonlinear vibrations of a clamped circular plate. *Dev. Mech.*, **6**, 907–920. *504, 507*

Lee, T. W. See Yen and Lee.

Leech, J. W. See Morino, Leech, and Witmer; Witmer, Balmer, Leech, and Pian.

Lefschetz, S. (1956). Linear and nonlinear oscillations. In *Modern Mathematics for the Engineer.* (E. F. Beckenback, Ed.), McGraw-Hill, New York, pp. 7–30. *181*

Lefschetz, S. (1959). *Differential Equations: Geometric Theory.* Wiley, New York. *vii*

Lefschetz, S. See LaSalle and Lefschetz.

Lehner, J. R., and S. C. Batterman (1974). Non-linear static and dynamic deformations of shells of revolution. *Int. J. Non-Linear Mech.*, **9**, 501–519. *506*

Lei, M. M. See Raville, Ueng, and Lei.

Leibovich, S., and A. R. Seebass (1974). *Nonlinear Waves.* Cornell University Press, Ithaca, New York. *545*

Leimanis, E. (1965). *The General Problem of the Motion of Coupled Rigid Bodies About a Fixed Point.* Springer-Verlag, New York. *395*

Leipholz, H. E. (1964). Über den Einfluss der Dämpfung bei nichtkonservativen Stabilitätsproblemen elastischer Stäbe. *Ing.-Arch.*, **33**, 308–321. *100*

Leipholz, H. E. (1970). *Stability Theory.* Academic, New York. *173*

Leissa, A. W. (1969). *Vibration of Plates.* NASA-Sp-160. *501*

Leissa, A. W., and A. S. Kadi (1971). Curvature effects on shallow shell vibrations. *J. Sound Vib.*, **16**, 173–187. *506*

Leissa, A. W. See Simons and Leissa.

Lekoudis, S. G., A. H. Nayfeh, and W. S. Saric (1977). Third-order resonant wave interactions. *Phys. Fluids*, **20**, 1793–1795. *601*

Lekoudis, S. G. See Saric, Nayfeh, and Lekoudis.

Leontovic, A. M. (1962). On the stability of the Lagrange periodic solutions of the restricted problem of three bodies. *Sov. Math. Dokl.*, **3**, 425–428. *378*

Leontovich, M. A. See Andronov and Leontovich.

Levenson, M. E. (1949). Harmonic and subharmonic response for the Duffing equation. *J. Appl. Phys.*, **20**, 1045–1051. *179*

Levenson, M. E. (1968). A numerical determination of ultrasubharmonic response for Duffing's equation. *Q. Appl. Math.*, **26**, 456–461. *179*

Levinson, N., and O. Smith (1942). A general equation for relaxation oscillations. *Duke Math. J.*, **9**, 382–403. *144*

Levinson, N. See Coddington and Levinson; Flatto and Levinson.

Lewis, F. M. (1932). Vibration during acceleration through a critical speed. *Trans. ASME*, **54**, 253–261. *219*

Liapunov, A. M. (1966). *Stability of Motion.* Academic, New York. *45, 284*

Librescu, L. (1965). Aeroelastic stability of orthotropic heterogeneous thin panels in the vicinity of the flutter critical boundary. *J. Mechanique*, **4**, 51–76. *509*

Lick, W. (1970). Nonlinear wave propagation in fluids. *Am. Rev. Fluid Mech.*, **2**, 113–136. *545*

Liénard, A. (1928). Etude des oscillations entretenues. *Rev. Gén. d'Elect.*, **23**, 901–946. *118*

Lighthill, M. J. (1965). Contributions to the theory of waves in nonlinear dispersive systems. *J. Inst. Math. Appl.*, **1**, 269–306. *588*

Likins, P. W. (1965). Stability of a symmetrical satellite in attitudes fixed in an orbiting reference frame. *J. Astron. Sci.*, **12**, 18–24. *396*

Likins, P. W. (1967). Attitude stability criteria for dual-spin spacecraft. *J. Spacecraft,* **4**, 1638–1643. *396*

Likins, P. W., G.-T. Tseng, and D. L. Mingori (1971). Stable limit cycles due to nonlinear damping in dual-spin spacecraft. *J. Spacecraft*, **8**, 568–574. *396*

Likins, P. W., and G. M. Wrout (1969). Bounds on the librations of parametrically resonant satellites. *AIAA J.*, **7**, 1134–1139. *396*

Likins, P. W. See Lindh and Likins; Mingori, Tseng, and Likins; Nishinaga and Likins; Schneider and Likins; White and Likins.

Lin, C. C. (1954). On a perturbation method based on the method of characteristics. *J. Math. Phys.*, **33**, 117–134. *555*

Lin, C. C. See King and Lin.

Lin, W. C. See Nariboli and Lin.

Lin, Y. K. (1962). Response of a nonlinear flat panel to periodic and randomly varying loadings. *J. Aeronaut. Sci.*, **29**, 1029–1034. *509*

Lindh, K. G., and P. W. Likins (1970). Infinite determinant methods for stability analysis of periodic-coefficient differential equations. *AIAA J.*, **8**, 680–686. *284, 331*

Linnell, C. O. See Dallos and Linnell.

Lion, P. M. (1966). Stability of linear periodic systems. *J. Franklin Inst.*, **28**, 27–40. *308*

Littlewood, J. E. See Cartwright and Littlewood.

Liu, F. C. See Carter and Liu.

Liu, J. J. F. (1974). Satellite motion about an oblate earth. *AIAA J.*, **12**, 1511–1516. *396*

Lobitz, D. W., A. H. Nayfeh, and D. T. Mook (1977). Nonlinear analysis of vibrations of irregular plates, *J. Sound Vib.*, **49**, 203–217. *446, 447, 508*

Lobitz, D. W. See Nayfeh, Mook, and Lobitz.

Longcope, D. B. See Roderman, Longcope, and Shampine.

Lou, C. L., and D. L. Sikarskie (1975). Nonlinear vibration of beams using a form-function approximation. *J. Appl. Mech.*, **42**, 209–214. *460*

Loud, W. S. (1955). On periodic solutions of Duffing's equation with damping. *J. Math. Phys.*, **34**, 173–178. *163*

Loud, W. S. (1957). Behavior of certain forced nonlinear systems of second order under large forcing. *Duke Math. J.*, **24**, 235–247. *162*

Loud, W. S. (1965). Verzweigung periodischer Lösungen gewisser nichtlinearer Differentialgleichungen. *ZAMM*, **45**, T54–T55. *163*

Loud, W. S. (1968). Branching phenomena for periodic solutions of nonautonomous piecewise linear systems. *Int. J. Non-Linear Mech.*, **3**, 273–293. *162*

Loud, W. S. (1969). Nonsymmetric periodic solutions of certain second order nonlinear differential equations. *J. Diff. Eqs.*, **5**, 352–368. *162*

Loud, W. S. (1972). Subharmonic solutions of second order equations arising near harmonic solutions. *J. Diff. Eqs.*, **11**, 628–660. *179*

Loud, W. S., and P. R. Sethna (1966). Some explicit estimates for domains of attractions. *J. Diff. Eqs.*, **2,** 158–172. *173, 417*

Lovell, E. G., and I. K. McIvor (1969). Nonlinear response of a cylindrical shell to an impulsive pressure. *J. Appl. Mech.*, **36,** 277–284. *506*

Lovell, E. G. See McIvor and Lovell.

Lowenstern, E. R. (1932). The stabilizing effect of imposed oscillations of high frequency on a dynamical system. *Phil. Mag.*, **13,** 458–486. *302*

Lowery, R. L. See Bayles, Lowery, and Boyd.

Lu, S. Y. See Sun and Lu.

Lu, Y.-C. (1976). *Singularity Theory and an Introduction to Catastrophe Theory.* Springer-Verlag, New York. *171*

Lubkin, S., and J. J. Stoker (1943). Stability of columns and strings under periodically varying forces. *Q. Appl. Mech.*, **1,** 215–236. *267*

Ludeke, C. A., and W. S. Wagner (1968). The generalized Duffing equation with large damping. *Int. J. Non-Linear Mech.*, **3,** 383–395. *119*

Luke, J. C. (1966). A perturbation method for nonlinear dispersive wave problems. *Proc. Roy. Soc. London*, **A292,** 403–412. *556*

Luk'ianov, L. G. (1969). On the stability in the first approximation of the triangular Lagrangian solutions of the elliptic restricted three-body problem. *Bull. Inst. Teor. Astron. Akad. Nauk* (USSR), **11,** 693–704. *331, 379*

Luukkala, M. (1967). Fine structure of fractional harmonic phonons. *Phys. Lett.*, **25A,** 76–77. *367, 416*

McDonald, P. H. (1955). Nonlinear dynamic coupling in a beam vibration. *J. Appl. Mech.*, **22,** 573–578. *460*

McElman, J. A. See Schaeffer and McElman.

McFarlane, A. G. J. See Quazi and McFarlane.

McGoldrick, L. F. (1965). Resonant interactions among capillary-gravity waves. *J. Fluid Mech.*, **21,** 305–331. *601*

McGoldrick, L. F. (1970a). An experiment on second-order capillary gravity resonant wave interactions. *J. Fluid Mech.*, **40,** 251–271. *601*

McGoldrick, L. F. (1970b). On Wilton's ripples: a special case of resonant interactions. *J. Fluid Mech.*, **42,** 193–200. *601*

McGoldrick, L. F. (1972). On the rippling of small waves: a harmonic nonlinear nearly resonant interaction. *J. Fluid Mech.*, **52,** 725–751. *601*

McIntosh, S. C. See Eastep and McIntosh.

McIvor, I. K. (1964). Dynamic stability of axially vibrating columns. *ASCE J. Eng. Mech. Div.*, **89,** 191–210. *445, 455*

McIvor, I. K. (1966). The elastic cylindrical shell under radial impulse. *J. Appl. Mech.*, **33,** 831–837. *506*

McIvor, I. K., and J. E. Bernard (1973). The dynamic response of columns under short duration axial loads. *J. Appl. Mech.*, **40,** 688–692. *455*

McIvor, I. K., and E. G. Lovell (1968). Dynamic response of finite-length cylindrical shells to nearly uniform radial impulse. *AIAA J.*, **6,** 2346–2351. *506*

McIvor, I. K., and D. A. Sonstegard (1966). Axisymmetric response of a closed spherical shell to a nearly uniform radial impulse. *J. Acoust. Soc. Am.*, **40,** 1540–1547. *506*

McIvor, I. K. See Goodier and McIvor; Lovell and McIvor.

McLachlan, N. W. (1947). *Theory and Application of Mathieu Functions.* Oxford University Press, New York. *259, 283*

McLachlan, N. W. (1950). *Ordinary Nonlinear Differential Equations in Engineering and Physical Sciences.* Clarendon Press, Oxford. *vii, 259*

McQueary, C. E., and L. G. Clark (1967). Periodic oscillations of a class of nonautonomous nonlinear elastic continua. *Int. J. Non-Linear Mech.*, **2,** 331–342. *444*

McQueary, C. E. See Mack and McQueary.

McTernan, A. J. See Black and McTernan.

Mack, L. R. and C. E. McQueary (1967). Oscillations of a circular membrane on a nonlinear elastic foundation. *J. Acoust. Soc. Am.*, **42,** 60–65. *507*

Mackie, A. G. (1968). The nonlinear oscillations of a string. *Q. Appl. Math.*, **26,** 468–469. *486*

Maezawa, S., and S. Furukawa (1973). Superharmonic resonance in piecewise-linear system. (Effect of damping and stability problem.) *Bull. JSME*, **16,** 931–941. *175*

Magnus, W., and S. Winkler (1966). *Hill's Equation.* Wiley, New York. *259, 283, 295*

Mahaffey, R. A. (1976). A Harmonic oscillator description of plasma oscillations. *Phys. Fluids*, **19,** 1387–1391. *60, 61, 62*

Mahalingam, S. (1975). The response of vibrating systems with Coulomb and linear damping inserts. *J. Sound Vib.*, **41,** 311–320. *96*

Maiboroda, V. P., M. A. Koltunov, B. I. Morgunov (1972). Vibrations of a shock absorbing device on a nonlinear viscoelastic base. *Polym. Mech.*, **8,** 635–641. *100*

Maiboroda, V. P., and B. I. Morgunov (1972). Calculation of the nonlinear viscoelastic oscillations of an antivibration layer. *Polym. Mech.*, **8,** 282–287. *100*

Maiboroda, V. P., B. I. Morgunov, and M. A. Koltunov (1972). Forced vibrations of a viscoelastic damping layer. *Polym. Mech.*, **8,** 106–113.

Makushin, V. M. (1947). The dynamic stability of the deformed state of elastic rods. *Tr. Mosk. Vissh Tek. Uchil.*, 61–84. *267*

Malkin, I. G. (1944). Stability in the case of constantly acting disturbances. *PMM*, **8,** 241–245. *272*

Malkin, I. G. (1956). *Some Problems in the Theory of Nonlinear Oscillations. GITTL*, Moscow. *vii, 259, 261, 367*

Malkina, R. L. (1953). Stability of curved arches subjected to longitudinal periodic forces. *Inzh. Sb.*, **14,** 123–130, *268*

Malme, C. I. See Smith, Malme, and Gogos.

Manakov, S. V. See Zakharov and Manakov.

Mandlestam, L., and N. Papalexi (1934). On the establishment of vibrations according to a resonance of the nth form. *J. Tech. Phys.*, **4,** 67–77. *262, 338*

Mansour, W. M. (1972). Quenching of limit cycles of a van der Pol oscillator. *J. Sound Vib.*, **25,** 395–405. *210, 375*

Markeev, A. P. (1967a). Resonance effects and stability of stationary rotation of a satellite. *Kosm. Issled.*, **5,** 365–375. *396*

Markeev, A. P. (1967b). The rotational motion of a dynamically symmetric satellite in an elliptical orbit. *Kosm. Issled.*, **5,** 530–539. *331*

Markeev, A. P. (1968). Stability of a canonical system with two degrees of freedom in the presence of resonance. *PMM*, **32,** 766–772. *378, 395, 424*

Markeev, A. P. (1969a). On the stability of a non-autonomous Hamiltonian system with two degrees of freedom. *PMM*, **33,** 563–569. *378, 379, 395*

Markeev, A. P. (1969b). On the stability of the triangular libration points in the circular bounded three-body problem. *PMM*, **33**, 112–116. *379*

Markeev, A. P. (1970). On the stability of triangular libration points in the elliptic restricted three-body problem. *PMM*, **34**, 227–232. *331, 379*

Markeev, A. P. (1972). Stability of the triangular Lagrangian solutions of the restricted three-body problem in the three dimensional circular case. *Sov. Astron.*, 682–686. *378*

Markov, A. N. (1949). The dynamic stability of anisotropic cylindrical shells. *PMM*, **13**, 145–150. *269*

Marples, V. (1965). Transition of a rotating shaft through a critical speed. *Proc. Inst. Mech. Eng.*, **180**, 8. *219*

Marshall, L. R. and Morrow, T. B. (1975). Wave induced instability of semi-submersible oil drilling platforms. In *ASME Fluid Mechanics in the Petroleum Industry*, Winter Annual Meeting. *374, 379, 402*

Marshall, L. R. See Mook, Marshall, and Nayfeh; Nayfeh, Mook, and Marshall.

Martinez, J. E. See Stricklin, Martinez; Tillerson, Hong, and Haisler.

Mathews, M. V. See Kohut and Mathews.

Mathieu, E. (1868). Mémoire sur le mouvement vibratoire d'une membrane de forme elliptique. *J. Math.*, **13**, 137–203. *274, 284*

Matkowsky, B. J., E. H. Rogers, and L. A. Rubenfeld (1971). Generation and stability of subharmonic and modulated subharmonic oscillations in nonlinear systems. *Q. Appl. Math.*, **30**, 329–336. *214*

Matowsky, B. J. See Reiss and Matowsky.

Matsukura, K. See Hirano and Matsukura.

Matsumoto, Y., N. Sugimoto, and Y. Inoue (1975). The second-harmonic resonance for nonlinear hydromagnetic waves. *J. Plasma Phys.*, **14**, 53–64. *601*

Matsumoto, Y. See Inoue and Matsumoto.

Matsuuchi, K. See Kakutani and Matsuuchi.

Matsuzaki, Y. and S. Kobayashi (1969a). An analytical study of the nonlinear flexural vibration of thin circular cylindrical shells. *J. Japan Soc. Aero Space Sci.*, **17**, 308–315. *506*

Matsuzaki, Y., and S. Kobayashi (1969b). A theoretical and experimental study of the nonlinear flexural vibration of thin circular cylindrical shells with clamped ends. *Japan Soc. Aero Space Sci.*, **12**, 55–62. *506*

May, J. E. (1960). Wire-type dispersive ultrasonic delay lines. *IRE Trans. Ultrason. Eng.*, UE-7, 44–53. *601*

Maybee, J. S. See Genin and Maybee.

Mayberry, B. L., and C. W. Bert (1969). Experimental investigation of nonlinear vibrations of laminated anisotropic panels. *Shock Vib. Bull.*, **39**, 191–199. *502, 507*

Mayers, J., and B. G. Wrenn (1967). On the nonlinear free vibrations of thin cylindrical shells. *Dev. Mech.*, **4**, 819. *506*

Mayers, J. See Wrenn and Mayers.

Mei, C. (1972). Nonlinear vibration of beams by matrix displacement method. *AIAA J.*, **10**, 355–357. *446, 469*

Mei, C. (1973a). Finite element analysis of nonlinear vibration of beam columns. *AIAA J.*, **11**, 115–117. *446, 469*

Mei, C. (1973b). Finite element displacement method for large amplitude free flexural vibrations of beams and plates. *Comput. Struct.*, **3**, 163–174. *446, 507*

Meirovitch, L. (1970). *Methods of Analytical Dynamics*. McGraw-Hill, New York. *260, 373*

Meirovitch, L., and F. B. Wallace (1967). Attitude instability regions of a spinning unsymmetrical satellite in a circular orbit. *J. Astronaut. Sci*., **14,** 123–133. *284, 331*

Meirovitch, L. See Ryland and Meirovitch; Wallace and Meirovitch.

Meissner, E. (1932). *Graphische Analysis vermittelst des Linienbildes einer Funktion.* Verlag der Schweiz-Bauzeitung, Zürich. *82, 126*

Meitzler, A. H. (1961). Mode coupling occurring in the propagation of elastic pulses in wires. *J. Acoust. Soc. Am*., **33,** 435–445. *601*

Melchambre, M. See Janssens van Dooren and Melchambre.

Melde, F. (1859). Über Erregung stehender Wellen eines fadenformigen Körpers. *Ann. Phys. Chem*., **109,** 193–215. *20, 258, 524*

Mendelson, K. S. (1970). Perturbation theory for damped nonlinear oscillations. *J. Math. Phys*., **11,** 3413–3415. *119, 156*

Meng, Y. J. See Huang and Meng.

Mente, L. J. (1973). Dynamic nonlinear response of cylindrical shells to asymmetric pressure loading. *AIAA J*., **11,** 793–800. *506*

Merkulov, V. I. See Bublik and Merkulov.

Messal, E. E., and R. J. Bonthron (1972). Subharmonic rotor instability due to elastic asymmetry. *J. Eng. Ind*., **94,** 185–192. *269*

Mettler, E. (1940). Biegeschwingungen eines Stabes unter pulsierender Axiallast. *Mitt. Forsch.-Anst. GHH-Konzern*, **8,** 1–12. *267*

Mettler, E. (1941). Biegeschwingungen eines Stabes mit kleiner Vorkrümmung, exzentrisch angreifender pulsierender Axiallast und statischer Querbelastung. *Forsch. Geb. Stahl*, **4,** 1–23. *268*

Mettler, E. (1947). Eine Theorie der Stabilität der elastischen Bewegung. *Ing.-Arch*., **16,** 135–146, *267*

Mettler, E. (1949). Allgemeine Theorie der Stabilität erzwungener Schwingungen elastischer Körper. *Ing.-Arch*, **17,** 418–449. *266, 267, 308*

Mettler, E. (1951). Zum Problem der Stabilität erzwungener Schwingungen elastischer Körper. *ZAMM*, **31,** 263–264. *266*

Mettler, E. (1955). Nichtlineare Schwingungen und kinetische Instabilität bei Saiten und Stäben. *Ing.-Arch*., **23,** 354–364. *268*

Mettler, E. (1962). Dynamic buckling. In *Handbook of Engineering Mechanics* (W. Flügge, Ed.). McGraw-Hill, New York. *260, 267*

Mettler, E. (1963). Kleine Schwingungen und Methode der säkularen Störungen. *ZAMM*, **43,** T81–T85. *379*

Mettler, E. (1967). Stability and vibration problems of mechanical systems under harmonic excitation. In *Dynamic Stability of Structures* (G. Herrmann, Ed.). Pergamon, New York, pp. 169–188. *260, 267, 308, 417, 447*

Mettler, E. (1968). Combination resonances in mechanical systems under harmonic excitations. In *Proc. 4th Conf. Nonlinear Oscillations*, Prague, pp. 51–70. *370, 379*

Mettler, E. (1975). Über höhere Näherungen in der Theorie des elastischen Pendels mit innerer Resonanz. *ZAMM*, **55,** 49–62. *370, 379*

Mettler, E., and F. Weidenhammer (1956). Der axial pulsierend belastete Stab mit Endmasse. *ZAMM*, **36,** 284–287. *267, 268*

Mettler, E., and F. Weidenhammer (1962). Zum Problem des kinetischen Durchschlagens schwach gekrümmter Stäbe. *Ing.-Arch.*, **31**, 421–432. *402, 433*

Meyer, R. E. (1976). Adiabatic variation. V. Nonlinear near-periodic oscillator. *ZAMP*, **27**, 181–195. *138*

Michihiro, K. See Kakutani and Michihiro.

Mikhlin, I. V. (1974). Resonance modes of near-conservative nonlinear systems. *PMM*, **38**, 425–429. *417*

Miles, J. W. (1962). Stability of forced oscillations of a spherical pendulum. *Q. Appl. Math.*, **20**, 21–32. *369*

Miles, J. W. (1965). Stability of forced oscillations of a vibrating string. *J. Acoust. Soc. Am.*, **38**, 855–861. *486*

Miles, J. W. See Hayes and Miles.

Miller, R. K. (1977). The steady-state response of systems with hardening hysteresis. *ASME J. Mech. Des., Paper No. 77-DET-71. 98*

Miller, R. K. See Iwan and Miller.

Milner, P. M. See Pengilley and Milner.

Mima, K. See Nishikawa, Hojo, Mima, and Ikezi.

Min, G.-B and J. G. Eisley (1972). Nonlinear vibration of buckled beams. *J. Eng. Ind.*, **94**, 637–646. *447*

Mingori, D. L. (1969). Effects of energy dissipation on the attitude stability of dual-spin satellites. *AIAA J.*, **7**, 20–27. *283, 331, 396*

Mingori, D. L. (1973). Stability of whirling shafts with internal and external damping. *Int. J. Non-Linear Mech.*, **8**, 155–159. *396*

Mingori, D. L., and J. A. Harrison (1974). Circularly constrained particle motion in spinning and coning bodies. *AIAA J.*, **12**, 1553–1558. *158*

Mingori, D. L., G. T. Tseng, and P. W. Likins (1972). Constant and variable amplitude limit cycles in dual-spin spacecraft. *AIAA J.*, **9**, 825–830. *396*

Mingori, D. L. See Gebman and Mingori; Kane and Mingori; Likins, Tseng, and Mingori.

Minorsky, N. (1947). *Non-Linear Mechanics.* J. W. Edwards, Ann Arbor. *vii, 259*

Minorsky, N. (1962). *Nonlinear Oscillations.* Van Nostrand, Princeton. *vii, 259, 262, 370, 375*

Minorsky, N. (1967). Comments on asynchronous quenching. *IEEE Trans.*, **AC-12**, 225–227. *210*

Mishra, A. K., and M. C. Singh (1974). The normal modes of non-linear symmetric systems by group representation theory. *Int. J. Non-Linear Mech.*, **9**, 463–480. *365*

Mishra, A. K., and M. C. Singh (1976). Non-linear forced vibrations of symmetric systems by group representation theory. *J. Tech. Phys.*, **17**, 171–181. *163*

Mitropolsky, Y. A. (1965). *Problems of the Asymptotic Theory of Nonstationary Vibrations.* Daniel Davey, New York. *vii, 137, 162, 219*

Mitropolsky, Y. A. See Bogoliubov and Mitropolsky.

Miura, R. M. See Gardner, Greene, Kruskal, and Miura.

Modeer, J. R. See Nash and Modeer.

Modi, V. J., and R. C. Brereton (1969a). Periodic solutions associated with the gravity-gradient-oriented systems. Part 1. Analytic and numerical determination. *AIAA J.*, **7**, 1217–1224. *396*

Modi, V. J., and R. C. Brereton (1969b). Periodic solutions associated with the gravity-gradient-oriented systems. Part II. Stability analysis. *AIAA J.*, **7**, 1465–1468. *396*

Moeller, T. L. See Iwan and Moeller.

Mojaddidy, Z., D. T. Mook, and A. H. Nayfeh (1977). Nonlinear analysis of the aperiodic responses of beams. *Proc. Sixth Canadian Congress Appl. Mech.*, **6**, 387–388. *185, 193, 460*

Momot, I. P. See Samoilenko and Momot.

Month, L. A., and R. H. Rand (1977). The stability of bifurcating periodic solutions in a two-degree-of-freedom nonlinear system. *J. Appl. Mech.*, **44**, 782–784. *388*

Montroll, E. W., and R. H. G. Helleman (1976). On a nonlinear perturbation theory without secular terms. In *Topics in Statistical Mechanics and Biophysics* (R. A. Piccirelli, Ed.). AIP, New York, pp. 75–110. *56, 388*

Montroll, E. W. See Eminhizer, Helleman, and Montroll; Helleman and Montroll.

Moody, L. M. (1967). The parametric response of imperfect columns. *Proc. 10th Midwestern Mech. Conf. Dev.*, **4**, 329–346. *266, 267*

Mook, D. T., L. R. Marshall, and A. H. Nayfeh (1974). Subharmonic and superharmonic resonances in the pitch and roll modes of ship motions. *J. Hydronautics*, **8**, 32–40. *374, 402*

Mook, D. T. See Atta, Kandil, Mook, and Nayfeh; Lobitz, Nayfeh, and Mook; Mojaddidy, Mook, and Nayfeh; Nayfeh and Mook; Nayfeh, Mook, and Lobitz; Nayfeh, Mook, and Marshall; Nayfeh, Mook, and Sridhar; Sridhar, Mook, and Nayfeh; Sridhar, Nayfeh, and Mook; Tezak, Mook, and Nayfeh; Thrasher, Mook, Kandil, and Nayfeh.

Moore, D. W. See Baker, Moore, and Spiegel.

Moran, T. J. (1970). Transient motions in dynamical systems with high frequency parametric excitation. *Int. J. Non-Linear Mech.*, **6**, 633–644. *302*

Moran, T. J. See Sethna and Moran.

Morgaevskii, A. B. See Kozhemyakina and Morgaevskii.

Morgunov, B. I. See Ivanova, Morgunov, and Troyanovskii; Kravchuk, Morgunov, and Troyanovskii; Maiboroda, Koltunov, and Morgunov; Maiboroda and Morgunov; Maiboroda, Morgunov, and Koltunov.

Morino, L. (1969). A perturbation method for treating nonlinear panel flutter problems. *AIAA J.*, **7**, 405–411. *509*

Morino, L., J. W. Leech, and E. A. Witmer (1971). An improved numerical calculation technique of large elastic-plastic deformations of thin shells. *J. Appl. Mech.*, **38**, 423–436. *445*

Morino, L. See Kuo, Morino, and Dugundji.

Morris, N. F. (1965). The dynamic stability of beam columns with a fixed distance between supports. *J. Franklin Inst.*, **280**, 163–173. *460*

Morrow, T. B. See Marshall and Morrow.

Mortell, M. P. (1971a). Resonant oscillations: a regular perturbation approach. *J. Math. Phys.*, **12**, 1069–1075. *572*

Mortell, M. P. (1971b). Resonant thermal acoustic oscillations. *Int. J. Eng. Sci.*, **9**, 175–192. *572*

Mortell, M. P. (1977). The evolution of nonlinear standing waves in bounded media. *ZAMP*, **28**, 33–46. *556, 571, 607*

Mortell, M. P., and B. R. Seymour (1972). Pulse propagation in a nonlinear viscoelastic rod of finite length. *SIAM J. Appl. Math.*, **22**, 209–224. *571, 575*

Mortell, M. P., and B. R. Seymour (1973a). Standing waves in an open pipe: a nonlinear initial-boundary value problem. *ZAMP*, **24**, 473–487. *572*

Mortell, M. P., and B. R. Seymour (1973b). The evolution of a self-sustained oscillation in a nonlinear continuous system. *J. Appl. Mech.*, **40**, 53–63. *571*

Mortell, M. P., and B. R. Seymour (1976). Wave propagation in a nonlinear laminated material: a derivation of geometrical acoustics. *Q. J. Mech. Appl. Math.*, **29**, 457–466. *568*

Mortell, M. P., and E. Varley (1970). Finite amplitude waves in bounded media: nonlinear free vibrations of an elastic panel. *Proc. Roy. Soc. London*, **A318**, 169–196. *571, 575, 582*

Mortell, M. P. See Seymour and Mortell.

Moseenkov, B. I. (1957). Transverse oscillations of a beam rigid with respect to two axis in a transient regime. *PMM*, **3**, 155–168. *219*

Moser, J. (1958). New aspects in the theory of stability of Hamiltonian systems. *Comm. Pure Appl. Math.*, **9**, 81–114. *395*

Moser, J. (1965). Combination tones for the Duffing equation. *Comm. Pure Appl. Math.*, **18**, 167–181. *163*

Moser, J. (1968). Lectures on Hamiltonian systems. *Memoirs Am. Math. Soc.*, **81**, 1–60. *395*

Moser, J. (1973). *Stable and Random Motions in Dynamical Systems.* Princeton University Press, Princeton. *388*

Moser, J. (1978). Nearly integrable and integrable systems. In *Nonlinear Dynamics* (S. Jorna, Ed.). AIP, New York. *388*

Mostaghel, N., and J. L. Sackman (1970). On the stability of Hill's equation. *J. Franklin Inst.*, **289**, 147–154. *295*

Mote, C. D. (1968). Dynamic stability of an axially moving band. *J. Franklin Inst.*, **285**, 329–346. *269*

Mote, C. D. See Thurman and Mote.

Movlyankulov, Kh. (1974). Investigation of nonlinear oscillation of viscoelastic systems by the averaging method. *Polym. Mech.* (USSR), **10**, 415–418. *100*

Mozer, D. T., and R. M. Evan-Iwanowski (1972). Parametrically excited column with hysteretic material properties. *Shock Vib. Bull.*, **42**, 153–160. *100, 268*

Mukhopadyay, V. See Dugundji and Mukhopadyay.

Mumford, W. W. (1960). Some notes on the history of parametric transducers. *Proc. IRE*, **48**, 848–853. *262*

Murphy, C. H. (1957). The prediction of nonlinear pitching and yawing motion of symmetric missiles. *J. Aeronaut. Sci.*, **24**, 473–479. *228*

Murphy, C. H. (1963). Quasi-linear analysis of the nonlinear motion of a nonspinning symmetric missile. *ZAMP*, **14**, 630–643. *228*

Murphy, C. H. (1965). Angular motion of a reentering symmetric missile. *AIAA J.*, **3**, 1275–1282. *228*

Murphy, C. H. (1968). Nonlinear motion of a symmetric missile acted on by a double-valued static moment. *AIAA J.*, **6**, 713–717. *228, 442*

Murphy, C. H. (1971a). Nonlinear motion of a missile with slight configurational asymmetries. *J. Spacecraft*, **8**, 259–263. *228, 442*

Murphy, C. H. (1971b). Response of an asymmetric missile to spin varying through resonance. *AIAA J.*, **9**, 2197–2201. *228, 442*

Murphy, C. H. (1973). Subharmonic behavior of a slightly asymmetric missile. *AIAA J.*, **11**, 884–885. *228, 443*

Murphy, C. H., and J. W. Bradley (1975). Nonlinear limit motions of a slightly asymmetric re-entry vehicle. *AIAA J.*, **13**, 851–857. *228*

Murthy, D. N. S., and A. N. Sherbourne (1972). Free flexural vibrations of damped plates. *J. Appl. Mech.*, **39**, 298–300. *508*

Murthy, G. S. S., and B. S. Ramakrishna (1965). Nonlinear character of resonance in stretched strings. *J. Acoust. Soc. Am.*, **38,** 461–471. *485, 486, 498*

Murthy, K. A. V. See Verma and Murthy.

Murthy, P. A. K. See Ramachandran and Murthy.

Murthy, P. N. See Aravamudan and Murthy.

Musa, S. A., and R. E. Kronauer (1968). Sub- and super-harmonic synchronization in weakly nonlinear systems: integral constraints and duality. *Q. Appl. Math.*, **25,** 399–414. *424*

Musa, S. A. See Kronauer and Musa.

Myerscough, C. J. (1973). A simple model of the growth of wind-induced oscillations in overhead lines. *J. Sound Vib.*, **28,** 699–713. *604*

Nagarajan, S., and E. P. Popov (1975a). Nonlinear dynamic analysis of axisymmetric shells. *Int. J. Numer. Meth. Eng.*, **9,** 535–550. *506*

Nagarajan, S., and E. P. Popov (1975b). Nonlinear finite element-dynamic analysis of axisymmetric solids. *Earthquake Eng. Struct. Dyn.*, **3,** 385–399. *506*

Nagasaka, I. See Yamamoto, Yasuda, and Nagasaka.

Nagoh, T. See Yamamoto, Yasuda, and Nagoh.

Naguleswaran, S., and C. J. H. Williams (1968). Lateral vibration of band-saw blades, pulley belts and the like. *Int. J. Mech. Sci.*, **10,** 239–250. *269*

Nair, S., and S. Nemat-Nasser (1971). On finite amplitude waves in heterogeneous elastic solids. *Int. J. Eng. Sci.*, **9,** 1087–1105. *571*

Nakamura, T. See Yamamoto, Yasuda, and Nakamura.

Nakamura, Y. (1971). Suppression and excitation of electron oscillation in a beam-plasma system. *J. Phys. Soc. Japan*, **31,** 273–279. *107*

Nakao, Y. See Yamamoto and Nakao.

Nambudiripad, K. B. M., and V. V. Neis (1976). Determination of mechanical response of nonlinear viscoelastic solids based on frechet expansion. *Int. J. Non-Linear Mech.*, **11,** 135–145. *100, 445*

Narashimha, R. (1968). Nonlinear vibration of an elastic string. *J. Sound Vib.*, **8,** 134–146. *486*

Nariboli, G. A. (1970). Nonlinear longitudinal dispersive waves in elastic rods. *J. Math. Phys. Sci.*, **4,** 64–73. *556*

Nariboli, G. A., and W. C. Lin (1973). A new type of Burgers' equation. *ZAMM,* **53,** 505–510. *556, 564, 607*

Nariboli, G. A., and A. Sedov (1970). Burgers' Korteweg-de Vries equation for viscoelastic rods and plates. *J. Math. Anal. Appl.*, **32,** 661–677. *556, 564, 607*

Nash, W. A., and J. R. Modeer (1959). Certain approximate analyses of the nonlinear behavior of plates and shallow shells. *Proc. Symp. Theor. Thin Elast. Shells*, 331–354. *504, 507*

Naumov, K. A. (1946). The stability of prismatic rods allowing for damping. *Tr. Mosk Inst. Inzh. Transp.*, **69,** 132–141. *268*

Nayfeh, A. H. (1967). The van der Pol oscillator with delayed amplitude limiting. *Proc. IEEE*, **55,** 111–112. *151*

Nayfeh, A. H. (1968). Forced oscillations of the van der Pol oscillator with delayed amplitude limiting. *IEEE Trans.*, **CT-15,** 192–200. *147, 152, 210, 255*

Nayfeh, A. H. (1969). A multiple time scaling analysis of reentry roll dynamics. *AIAA J.*, **7,** 2155–2157. *138*

Nayfeh, A. H. (1970a). Characteristic exponents for the triangular points in the elliptic restricted problem of three bodies. *AIAA J.*, **8,** 1916–1917. *331, 379*

Nayfeh, A. H. (1970b). Finite amplitude surface waves in a liquid layer. *J. Fluid Mech.*, **40,** 671–684. *601*

Nayfeh, A. H. (1970c). Triple-and quintuple-dimpled wave profiles in deep water. *Phys. Fluids*, **13,** 545–550. *601*

Nayfeh, A. H. (1971a). Third-harmonic resonance in the interaction of capillary and gravity waves. *J. Fluid Mech.*, **48,** 385–395. *601*

Nayfeh, A. H. (1971b). Two-to-one resonances near the equilateral libration points. *AIAA J.*, **9,** 23–27. *150, 378, 379, 424*

Nayfeh, A. H. (1972). Characteristic exponents and stability of Hill's equation. *J. Appl. Mech.*, **39,** 1156–1158. *295*

Nayfeh, A. H. (1973a). Nonlinear transverse vibrations of beams with properties that vary along the length. *J. Acoust. Soc. Am.*, **53,** 766–770. *445, 447, 448, 453, 469*

Nayfeh, A. H. (1973b). *Perturbation Methods*. Wiley, New York. *51, 62, 119, 158, 260, 284, 370, 379, 438, 556*

Nayfeh, A. H. (1973c). Second-harmonic resonance in the interaction of an air stream with capillary-gravity waves. *J. Fluid Mech.*, **59,** 803–816. *150, 601*

Nayfeh, A. H. (1974). Sound waves in two-dimensional ducts with sinusoidal walls. *J. Acoust. Soc. Am.*, **56,** 768–770. *362*

Nayfeh, A. H. (1975a). Finite-amplitude longitudinal waves in nonuniform bars. *J. Sound Vib.*, **42,** 357–361. *556, 568*

Nayfeh, A. H. (1975b). Finite-amplitude plane waves in ducts with varying properties. *J. Acoust. Soc. Am.*, **57,** 1413–1415. *556, 568*

Nayfeh, A. H. (1975c). Nonlinear propagation of a wave packet in a hard-walled circular duct. *J. Acoust. Soc. Am.*, **57,** 803–809. *587*

Nayfeh, A. H. (1976a). Nonlinear propagation of wave-packets on fluid interfaces. *J. Appl. Mech.*, **43,** 584–588. *587, 616*

Nayfeh, A. H. (1976b). Numerical-perturbation methods in nonlinear mechanics. In *Numerical Methods in Geo-Mechanics* 1. ASCE, New York, pp. 168–182. *446*

Nayfeh, A. H. (1977). Perturbation methods and nonlinear hyperbolic waves. *J. Sound Vib.*, **54,** 605–609. *556, 571, 607*

Nayfeh, A. H., and O. R. Asfar (1974). Parallel plate wave guide with sinusoidally perturbed boundaries. *J. Appl. Phys.*, **45,** 4797–4800. *363*

Nayfeh, A. H., and J. E. Kaiser (1975). Nonlinear propagation of arbitrary plane waves in ducts carrying plug flows. *AIAA Paper No. 75-846. 603*

Nayfeh, A. H., and A. A. Kamel (1970a). Stability of the triangular points in the elliptic restricted problem of three bodies. *AIAA J.*, **8,** 221–223. *331, 379*

Nayfeh, A. H., and A. A. Kamel (1970b). Three-to-one resonances near the equilateral libration points. *AIAA J.*, **8,** 2245–2251. *378, 424*

Nayfeh, A. H., and S. G. Kelly (1978). Nonlinear interactions of acoustic fields with plates under harmonic excitations. *J. Sound Vib.*, **60,** 371–377. *509*

Nayfeh, A. H., and A. Kluwick (1976). A comparison of three perturbation methods for non-linear hyperbolic waves. *J. Sound Vib.*, **48,** 293–299. *555, 556, 567, 602*

Nayfeh, A. H., and D. T. Mook (1977). Parametric excitations of linear systems having many degrees of freedom. *J. Acoust. Soc. Am.*, **62,** 375–381. *265, 267, 310, 311, 319*

Nayfeh, A. H., D. T. Mook, and D. W. Lobitz (1974). Numerical-perturbation method for the nonlinear analysis of structural vibrations. *AIAA J.*, **12,** 1222–1228. *446, 447, 469*

Nayfeh, A. H., D. T. Mook, and L. R. Marshall (1973). Nonlinear coupling of pitch and roll modes in ship motions. *J. Hydronautics*, **7**, 145–152. *367, 374, 379, 402, 406*

Nayfeh, A. H., D. T. Mook, and L. R. Marshall (1974). Perturbation-energy approach for the development of the nonlinear equations of ship motion. *J. Hydronautics.*, **8**, 130–136. *371*

Nayfeh, A. H., D. T. Mook, and S. Sridhar (1974). Nonlinear analysis of the forced response of structural elements. *J. Acoust. Soc. Am.*, **55**, 281–291. *469*

Nayfeh, A. H., and S. Nemat-nasser (1972). Elastic waves in inhomogeneous elastic media. *J. Appl. Mech.*, **39**, 696–702. *348*

Nayfeh, A. H., and W. S. Saric (1971a). Nonlinear Kelvin-Helmholtz instability. *J. Fluid Mech.*, **46**, 209–231. *107*

Nayfeh, A. H., and W. S. Saric (1971b). Nonlinear resonances in the motion of rolling reentry bodies. *AIAA Paper No. 71-47. 228, 442*

Nayfeh, A. H., and W. S. Saric (1972a). An analysis of asymmetric rolling bodies with nonlinear aerodynamics. *AIAA J.*, **10**, 1004–1011. *228, 443*

Nayfeh, A. H., and W. S. Saric (1972b). Nonlinear waves in a Kelvin-Helmholtz flow. *J. Fluid Mech.*, **55**, 311–327. *587*

Nayfeh, A. H., and W. S. Saric (1973). Nonlinear acoustic response of a spherical bubble. *J. Sound Vib.*, **30**, 445–453. *199*

Nayfeh, A. H. See Asfar and Nayfeh; Atta, Kandil, Mook, and Nayfeh; Lekoudis, Nayfeh, and Saric; Lobitz, Nayfeh, and Mook; Mojaddidy, Mook, and Nayfeh; Mook, Marshall, and Nayfeh; Saric, Nayfeh, and Lekoudis; Sridhar, Mook, and Nayfeh; Sridhar, Nayfeh, and Mook; Tezak, Mook and Nayfeh; Thrasher, Mook, Kandil, and Nayfeh.

Neis, V. V. See Nambudiripad and Neis.

Nemat-Nasser, S. (1967). On the stability of the equilibrium of nonconservative continuous systems with slight damping. *J. Appl. Mech.*, **34**, 344–348. *100*

Nemat-Nasser, S. (1970). Thermoelastic stability under general loads. *Appl. Mech. Rev.*, **23**, 615–624. *100*

Nemat-Nasser, S., and G. Herrmann (1966). Some general considerations concerning the destabilizing effect in nonconservative systems. *ZAMP*, **17**, 305–313. *100*

Nemat-Nasser, S., S. Prasad, and G. Herrmann (1966). Destabilizing effect of velocity-dependent forces in nonconservative continuous systems. *AIAA J.*, **4**, 1276–1280. *100*

Nemat-Nasser, S. See Fu and Nemat-Nasser; Nair and Nemat-Nasser; Nayfeh and Nemat-Nasser.

Neppiras, E. A. (1969). Subharmonic and other low-frequency emission from bubbles in sound-irradiated liquids. *J. Acoust. Soc. Am.*, **46**, 587–601. *199*

Ness, D. J. (1967). Small oscillations of a stabilized, inverted pendulum. *Am. J. Phys.*, **35**, 964–967. *261, 338*

Ness, D. J. (1971). Resonance classification in a cubic system. *J. Appl. Mech.*, **38**, 585–590. *163, 339*

Newell, A. C. (1968). The closure problem in a system of random gravity waves. *Rev. Geophys.*, **6**, 1–31. *601*

Newell, A. C. See Ablowitz, Kaup, Newell, and Segur; Ablowitz, Funk, and Newell; Benney and Newell; Lange and Newell.

Newland, D. E. (1965). On the methods of Galerkin, Ritz and Krylov-Bogoliubov in the theory of nonlinear vibrations. *Int. J. Mech. Sci.*, **7**, 159–172. *163, 388*

Nguyen, K. T. (1975). Applying the relaxation method to the solution of nonlinear dynamic problems for cylindrical shells. *PMM*, **11**, 29–33. *506*

Nguyen, van D. (1975a). Interaction between forced and self-excited oscillations in multidimensional systems. *ZAMM*, **55**, 683–684. *375*

Nguyen, van D. (1975b). Interaction between parametric and forced oscillations in multidimensional systems. *J. Tech. Phys.*, **16**, 213–225. *339, 356*

Nguyen, van D. (1976a). Parametric resonance of 4th order in a nonlinear vibrating system under the influence of frictions. *J. Tech. Phys.*, **17**, 435–440. *339*

Nguyen, van D. (1976b). Some properties of the generalized van der Pol equation. *J. Tech. Phys.*, **17**, 183–190. *129*

Nishihara, K. See Asano, Nishihara, Nozaki, and Taniuti.

Nishikawa, K., H. Hojo, K. Mima, and H. Ikezi (1974). Coupled nonlinear electron-plasma and ion acoustic waves. *Phys. Rev. Lett.*, **33**, 148–151. *601*

Nishikawa, Y., and P. Y. Willems (1969). A method for stability investigation of a periodic dynamic system with many degrees of freedom. *J. Franklin Inst.*, **287**, 143–157. *331*

Nishinaga, R. G., and P. W. Likins (1974). Exploitation of nonlinear resonance in gravity-stabilized satellites. *Int. J. Non-Linear Mech.*, **9**, 23–43. *396*

Nocilla, S., and R. Riganti (1974). Forced vibrations of a class of nonlinear systems with one degree of freedom. *Mech. Res. Comm.*, **1**, 197–202. *163*

Nohel, J. A. See Brauer and Nohel.

Novikov, L. Z., and S. A. Kharlamov (1973). Behavior of a pendulum with an elastic suspension and a vibrating base. *Mech. Solids* (USSR), **8**, 1–9. *370*

Novosiolov, V. S. (1976). On examination of non-linear oscillations by methods of Hamiltonian mechanics. *Vestn. Leningr. Univ. Mat. Mekh. Astron.*, **N.4**, 123–125. *424*

Nowinski, J. L. (1962). Non-linear transverse vibrations of circular elastic plates built-in at the boundary. *Proc. 4th U.S. Nat. Congr. Appl. Mech.*, 325–334. *508*

Nowinski, J. L. (1963a). Nonlinear transverse vibrations of orthotropic cylindrical shells. *AIAA J.*, **1**, 617–620. *506*

Nowinski, J. L. (1963b). Non-linear vibrations of elastic circular plates exhibiting rectilinear orthotropy. *ZAMP*, **14**, 112–124. *501*

Nowinski, J. L. (1964). Nonlinear transverse vibrations of a spinning disc. *J. Appl. Mech.*, **31**, 72–78. *505, 542, 543*

Nowinski, J. L., and I. A. Ismail (1965). Large oscillations of an anisotropic-triangular plate. *J. Franklin Inst.*, **280**, 417–424. *502*

Nowinski, J. L., and H. Ohnabe (1972). On certain inconsistencies in Berger equations for large deflections of plastic plates. *Int. J. Mech. Sci.*, **14**, 165–170. *504*

Nowinski, J. L., and S. R. Woodall (1964). Finite vibrations of free rotating anisotropic membrane. *J. Acoust. Soc. Am.*, **36**, 2113–2118. *505*

Nowinski, J. L. See Hassert and Nowinski.

Nozaki, K., and T. Taniuti (1976). A note on the reductive perturbation method. *J. Phys. Soc. Japan*, **40**, 572–576. *556*

Nozaki, K. See Asano, Nishihara, Nozaki, and Taniuti.

Nustrov, V. S. (1974). On a case of resonance for non-linear systems. *PMM*, **38**, 936–939. *388*

Obi, C. (1950). Subharmonic solution of nonlinear differential equations of the second order. *J. London Math. Soc.*, **25**, 217–226. *179*

Oden, J. T., J. E. Key, and R. Fost (1974). A note on the analysis of nonlinear dynamics of elastic membranes by the finite element method. *Comput. Struct.*, **4**, 445–452. *445*

Ogino, S. See Iwatsubo, Sugiyama, and Ogino.

Ohnabe, H. See Nowinski and Ohnabe.

Olson, M. D. (1965). Some experimental observations on the nonlinear vibrations of cylindrical shells. *AIAA J.*, **3**, 1775–1777. *506*

Olson, M. D., and Y. C. Fung (1966). Comparing theory and experiment for the supersonic flutter of circular cylindrical shells. *AIAA J.*, **5**, 1849–1856. *509*

Olsson, M. G. (1976). Why does a mass on a spring sometimes misbehave? *Am. J. Phys.*, **44**, 1211–1212. *370*

Oniashvili, O. D. (1951). *Certain Dynamic Problems of the Theory of Shells.* Friedman, West Newton, Mass. *269*

Ono, H. See Hasimoto and Ono; Kakutani and Ono.

Oplinger, D. W. (1960). Frequency response of a nonlinear stretched string. *J. Acoust. Soc. Am.*, **32**, 1529–1538. *486*

Osinski, Z. (1962). Longitudinal, torsional and bending vibrations of a uniform bar with nonlinear internal friction and relaxation. *Nonlinear Vib. Probl.*, **4**, 159–166. *445*

Ottl, D. (1975). Transient motion of an oscillator with a damping force of constant magnitude. *ZAMM*, **55**, 58–60. *149*

Ovcharova, D. K., and E. G. Goloskokov (1975). Forced oscillations of a rotor on a roller bearing. *Prikl. Mekh.*, **11**, 95–100. *163*

Paidoussis, M. P., and N. T. Issid (1974). Dynamic stability of pipes conveying fluid. *J. Sound Vib.*, **33**, 267–294. *269*

Paidoussis, M. P., and N. T. Issid (1976). Experiments on parametric resonance of pipes containing pulsatile flow. *J. Appl. Mech.*, **43**, 198–202. *269*

Paidoussis, M. P., and C. Sundararajan (1975). Parametric and combination resonances of a pipe conveying pulsating fluid. *J. Appl. Mech.*, **42**, 780–784. *269*

Pal, M. C. (1970). Large amplitude vibration of circular plates subjected to aerodynamic heating. *Int. J. Solids Struct.*, **6**, 301–313. *507*

Pal, M. C. (1973). Static and dynamic non-linear behaviour of heated orthotropic circular plates. *Int. J. Nonlinear Mech.*, **8**, 489–504. *502*

Paltov, N. P. See Popov and Paltov.

Pandalai, K. A. V. See Sathyamoorthy and Pandalai.

Pao, Y. H. See Kung and Pao.

Papalexi, N. See Mandlestam and Papalexi.

Parekh, C. J. See Zienkiewicz and Parekh.

Paria, G. (1968). Time-hardening and time-softening elastic materials. *Indian J. Mech. Math.*, **4**, 117–125. *138*

Parker, D. F., and E. Varley (1968). The interaction of finite amplitude deflection and stretching waves in elastic membranes and strings. *Q. J. Mech. Appl. Math.*, **21**, 329–352. *571*

Parkinson, A. G. See Bishop and Parkinson.

Pasley, P. R., and M. E. Gurtin (1960). The vibration response of linear undamped system resting on a nonlinear spring. *J. Appl. Mech.*, **27**, 272–274. *445*

Patula, E. J., and W. D. Iwan (1972). On the validity of equation difference minimization techniques. *Int. J. Non-Linear Mech.*, **7**, 1–17. *87*

Paulling, J. R., and R. M. Rosenberg (1959). On unstable ship motions resulting from nonlinear coupling. *J. Ship Res.*, **3**, 36–46. *374, 379*

Pengilley, C. J., and P. M. Milner (1967). On asynchronous quenching. *IEEE Trans.*, **AC-12**, 224–225. *210*

Pershan, P. S. See Armstrong, Bloembergen, Ducuing, and Pershan.

Pestorius, F. M., and D. T. Blackstock (1973). Nonlinear distortion in the propagation of intense acoustic noise. In *Interagency Symposium on University Research on Transportation Noise*, Stanford University, pp. 565–577. *566*

Peters, C. W. See Franken, Hill, Peters, and Weinreich.

Peters, D. A., and K. H. Hohenemser (1971). Application of the Floguet transition matrix to problems of lifting rotor stability. *J. Am. Helicopter Soc.*, **16**, 25–33. *269, 283*

Phelps, F. M., and J. H. Hunter (1965). An analytical solution of the inverted pendulum. *Am. J. Phys.*, **33**, 285–295. *261*

Phelps, F. M., and J. H. Hunter (1966). Reply to Joshi's comments on a damping term in the equations of motion of the inverted pendulum. *Am. J. Phys.*, **34**, 533–535. *261*

Phillips, O. M. (1974). Nonlinear dispersive waves. *Ann. Rev. Fluid Mech.*, **6**, 93–110. *545*

Pi, W. S. See Hwang and Pi.

Pian, T. H. H. See Witmer, Balmer, Leech, and Pian.

Pinney, E. See Klotter and Pinney.

Pipes, L. A. (1953). Analysis of a nonlinear dynamic vibration absorber. *J. Appl. Mech.*, **20**, 515–518. *417*

Pisarenko, G. S. (1948). Forced transverse vibrations of clamped cantilevers allowing for hysteresis losses. *Inzh. Sb.*, **5**, 108–132. *98*

Piszczek, K. (1955). Longitudinal and transverse vibrations of a rod subjected to an axial pulsating force, taking into account nonlinear members into consideration. *Arch. Mech.*, **7**, 345–362. *268*

Piszczek, K. (1961). Second kind resonance region for a load whose direction follows the deformation of the body. *Rozpr. Inz.*, **9**, 155. *268, 308, 320*

Plakhtienko, N. P. (1975). Using Duffing's model in the dynamic analysis of vibratory systems. *Prikl. Mekh.*, **11**, 128–132. *163*

Platus, D. H. (1969). Angle-of-attack convergence and windward-meridian rotation rate of rolling re-entry vehicles. *AIAA J.*, **12**, 2324–2332. *228*

Plaut, R. H. See Infante and Plaut.

Plotnikova, G. V. (1962). On the construction of periodic solutions of a nonautonomous quasi-linear system with one degree of freedom near resonance in the case of double roots of the equation of fundamental amplitudes. *PMM*, **26**, 1123–1133. *163*

Plotnikova, G. V. (1963a). On the construction of periodic solutions of a nonautonomous quasi-linear system with two degrees of freedom. *PMM*, **27**, 544–550. *417*

Plotnikova, G. V. (1963b). On the stability of periodic solutions of nonautonomous quasi-linear systems with one degree of freedom. *PMM*, **27**, 241–254. *163*

Plotnikova, G. V. (1965). On the stability of periodic solutions of non-self-contained quasi-linear systems with two degrees of freedom. *PMM*, **29**, 1273–1281. *418*

Poli, C., and R. Budynas (1971). On the stability of motion of two mechanical systems. *Bull. Mech. Eng. Educ.*, **10**, 325–336. *396*

Pollner, E. See Tso, Pollner, and Heidebrecht.

Ponomarev, A. T. See Vol'mir and Ponomarev.

Ponzo, P. J., and N. Wax (1974). Weakly coupled harmonic oscillators. *SIAM J. Appl. Math.*, **26**, 508–527. *418*

Popelar, C. H. (1972). Dynamic stability of thin-walled column. *ASCE J. Eng. Mech. Div.*, **98**, 657–677. *267*

Popov, E. P. See Nagarajan and Popov.

Popov, I. P., and N. P. Paltov (1960). *Approximate Methods for Analyzing Nonlinear Automatic Systems.* State Press for Physics and Mathematical Literature, Moscow. *119, 156*

Popov, V. G., A. A. Antipov, and P. G. Krzhechkovskii (1973). Parametric vibrations of three-layered cylindrical shells. *Strength Mater.*, **5**, 803–805. *269*

Porter, B., and R. A. Billet (1965). Harmonic and subharmonic vibration of a continuous system having non-linear constraint. *Int. J. Mech. Sci.*, **7**, 431–439. *445, 535*

Postnikov, L. V. (1974). The problem of finding and investigating quasiharmonic oscillations in weakly nonlinear systems. *Radiophys. Quant. Electron.*, **14**, 1332–1337. *388*

Postnikov, L. V. See Korolev and Postnikov.

Prasad, S. See Nemat-Nasser, Prasad, and Herrmann.

Price, D. A., and L. E. Ericsson (1970). A new treatment of roll-pitch coupling for ballistic reentry vehicles. *AIAA J.*, **8**, 1608–1615. *228*

Pringle, R. (1964). Bounds on the librations of a symmetrical satellite. *AIAA J.*, **2**, 908–912. *396*

Pringle, R. (1968). Exploitation of nonlinear resonance in damping an elastic dumbbell satellite. *AIAA J.*, **6**, 1217–1222. *396*

Pringle, R. (1969). Stability of the force-free motions of a dual-spin spacecraft. *AIAA J.*, **7**, 1054–1063. *396*

Pringle, R. (1973). Satellite vibration-rotation motions studied via canonical transformations. *Celestial Mech.*, **7**, 495–518. *396*

Pringle, R. See Breakwell and Pringle.

Proskuriakov, A. P. (1946). Characteristic values of solutions of second-order differential equations with periodic coefficients. *PMM*, **10**, 545–558. *284*

Proskuriakov, A. P. (1960a). On a property of periodic solutions of quasi-linear autonomous system with several degrees of freedom. *PMM*, **24**, 1098–1104. *388*

Proskuriakov, A. P. (1960b). Periodic oscillations of quasilinear autonomous systems with two degrees of freedom. *PMM*, **24**, 1671–1680. *388*

Proskuriakov, A. P. (1962). On the construction of periodic solutions of quasi-linear autonomous systems with several degrees of freedom. *PMM*, **26**, 519–528. *388*

Proskuriakov, A. P. (1965). Stability of single-frequency periodic solutions of quasi-linear self-contained systems with two degrees of freedom. *PMM*, **29**, 1106–1113. *388*

Proskuriakov, A. P. (1968). The structure of the periodic solutions of a quasi-linear self-contained system with several degrees of freedom in the case of differing, but partly noncommensurate frequencies. *PMM*, **32**, 566–570. *388*

Proskuriakov, A. P. (1971). Influence of resistence forces on existence of subharmonic oscillations of quasi-linear systems. *PMM*, **35**, 111–113. *179, 257*

Prosperetti, A. (1976). Subharmonics and ultraharmonics in the forced oscillations of weakly nonlinear systems. *Am. J. Phys.*, **44**, 548–554. *179*

Pun, L. (1973). Initial conditioned solutions of a second-order nonlinear conservative differential equation with a periodically varying coefficient. *J. Franklin Inst.*, **294**, 193–216. *339*

Quazi, A. S., and A. G. J. McFarlane (1967). The controlled transition of a rotating shaft through its critical speed. *Int. J. Contr.*, **6**, 301–315. *219*

Radstig, M. A. See Dzhanelidze and Radstig.

Radwan, H. R., and J. Genin (1976). Nonlinear vibrations of thin cylinders. *J. Appl. Mech.*, **43**, 370–372. *506*

Rajac, T. J., and R. M. Evan-Iwanowski (1976). Interaction of a motor with limited power with dissipative (hysteretic) foundation. In *Proc. 8th SECTAM*, pp. 349–361. *100, 227*

Raju, B. B. See Chandra and Raju.

Raju, I. S., G. Venkateswara Rao, and K. Kanaka Raju (1976a). Effect of longitudinal or inplane deformation and inertia on the large amplitude flexural vibrations of slender beams and thin plates. *J. Sound Vib.*, **49**, 415–422.

Raju, I. S., G. Venkatsewara Rao, and K. Kanaka Raju (1976b). Large amplitude free vibrations of tapered beams. *AIAA J.*, **14**, 280–282. *447*

Raju, I. S. See Venkateswara Rao, Kanaka Raju, and Raju; Venkateswara Rao, Raju, and Kanaka Raju.

Ramachandran, J. (1974a). Large amplitude vibrations of circular plates with mixed boundary conditions. *Comput. Struct.*, **4**, 871–877. *508*

Ramachandran, J. (1974b). Large-amplitude vibrations of elastically restrained circular plates. *J. Acoust. Soc. Am.*, **55**, 880–882. *507*

Ramachandran, J. (1974c). Nonlinear vibrations of elastically restrained rectangular orthropic plates. *Nucl. Eng. Des.*, **30**, 402–407. *501*

Ramachandran, J. (1975). Non-linear vibrations of circular plates with linearly varying thickness. *J. Sound Vib.*, **38**, 225–232. *501*

Ramachandran, J. (1977). Frequency analysis of plates vibrating at large amplitudes. *J. Sound Vib.*, **51**, 1–5. *507*

Ramachandran, J., and P. A. K. Murthy (1976). Non-linear vibrations of a shallow cylindrical panel on an elastic foundation. *J. Sound Vib.*, **47**, 495–500. *506*

Ramachandran, J., and D. V. Reddy (1972). Nonlinear vibrations of rectangular plates with cutouts. *AIAA J.*, **10**, 1709–1710. *507*

Ramachandran, J., and D. V. Reddy (1975). Nonlinear vibrations of rectangular plates with linearly varying thickness. *Appl. Sci. Res.*, **31**, 52–66 and 67–80. *501*

Ramakrishna, B. S. See Murthy and Ramakrishna.

Raman, C. V. (1912). Experimental investigations on the maintenance of vibrations. *Proc. Indian Assoc. Cultivation Sci. Bull.*, **6**, 405. *258*

Rand, D. A. See Holmes and Rand.

Rand, R. H. (1969). On the stability of Hill's equation with four independent parameters. *J. Appl. Mech.*, **36**, 885–888. *295*

Rand, R. H. (1974). A direct method for non-linear normal modes. *Int. J. Non-Linear Mech.*, **9**, 363–368. *365*

Rand, R. H., and S. F. Tseng (1969). On the stability of a differential equation with application to the vibrations of a particle in the plane. *J. Appl. Mech.*, **36**, 311–313. *295*

Rand, R. H., and R. P. Vito (1972). Nonlinear vibrations of two-degree-of-freedom systems with repeated linearized natural frequencies. *J. Appl. Mech.*, **39**, 296–297. *388*

Rand, R. H. See Alfriend and Rand; Month and Rand.

Randall, D. G. (1968). Sonic bang intensities in a stratified, still atmosphere. *J. Sound Vib.*, **8**, 196–214. *568*

Rangacharyulu, M. A. V., and P. A. Srinivasan (1973). Note on the period of oscillation of non-linear systems. *J. Sound Vib.*, **29**, 207–214. *396*

Rangacharyulu, M. A. V., P. A. Srinivasan, and B. V. Dasarathy (1974). Transient response of coupled non-linear non-conservative systems. *J. Sound Vib.*, **37**, 467–473. *388*

Rao, C. K. (1975). Nonlinear torsional vibrations of thin-walled beams of open section. *J. Appl. Mech.*, **42**, 240–241. *448*

Rao, J. S. See Kishor and Rao.

Rashkovich, D. See Stevanovich and Rashkovich.

Rasmussen, M. L. (1970). Uniformly valid approximations for non-linear oscillations with small damping. *Int. J. Non-Linear Mech.*, **5**, 687–696. *119*

Rasmussen, M. L. (1973). Non-linear oscillations with small speed-dependent damping. *Dev. Mech.*, **7**, 437–448. *119*

Rasmussen, M. L. (1977). On the damping decrement for non-linear oscillations. *Int. J. Non-Linear Mech.*, **12**, 81–90. *119*

Rauscher, M. (1938). Steady oscillations of systems with nonlinear and unsymmetrical elasticity. *J. Appl. Mech.*, **5**, A169–A177. *241*

Ray, J. D., and C. W. Bert (1969). Nonlinear vibrations of a beam with pinned ends. *J. Eng. Ind.*, **91**, 997–1104. *453, 460*

Reddy, D. V. See Ramachandran and Reddy.

Rehfield, L. W. (1973). Nonlinear free vibrations of elastic structures. *Int. J. Solids Struct.*, **9**, 581–590. *460*

Rehfield, L. W. (1974a). Large amplitude forced vibrations of elastic structures. *AIAA J.*, **12**, 388–390. *460*

Rehfield, L. W. (1974b). Nonlinear flexural oscillations of shallow arches. *AIAA J.*, **12**, 91–93. *447*

Rehfield, L. W. (1975). A simple, approximate method for analyzing nonlinear free vibrations of elastic structures. *J. Appl. Mech.*, **42**, 509–510. *460*

Rehfield, L. W., J. Girl, and C. A. Sparrow (1974). Edge restraint effect on vibration of curved panels. *AIAA J.*, **12**, 239–241. *506*

Rehm, R. G. (1968). Radiative energy addition behind a shock wave. *Phys. Fluids*, **11**, 1872–1883. *572*

Reiss, E. L., and B. J. Matkowsky (1971). Nonlinear dynamic buckling of a compressed elastic column. *Q. Appl. Math.*, **29**, 245–260. *460*

Reuter, G. E. H. (1949). Subharmonics in a nonlinear system with unsymmetrical restoring force. *Q. J. Mech. Appl. Math.*, **2**, 198–207. *179*

Rhodes, J. E., Jr. (1971). Parametric self-excitation of a belt into transverse vibration. *J. Appl. Mech.*, **37**, 1055–1060. *269*

Richards, J. (1872). *Wood-Working Machines.* Spon, London. *505*

Richardson, D. L. See Alfriend and Richardson.

Riead, H. D. (1974). Nonlinear response using normal modes. *AIAA Paper No. 74-138.*

Riemann, B. (1858). Über die Fortpflanzung ebener Luftwellen von erdlicher Schwingungsweite. *Göttingen Abhandlungen*, **8**, 43–65. *612*

Riganti, R. See Nocilla and Riganti.

Rigas, H. See Gunderson, Rigas, and van Vleck.

Rinkel, R. L. See Bennett and Rinkel.

Robb, A. M. (1952). *Theory of Naval Architecture*. Charles Griffin, London, Chapt. 18. *29*

Roberson, R. E. (1952). Synthesis of a nonlinear dynamic vibration absorber. *J. Franklin Inst.*, **254**, 205–220. *417*

Roderman, R. D., B. Longcope, and L. F. Shampine (1976). Response of a string to an accelerating mass. *J. Appl. Mech.*, **43**, 675–680. *486*

Rodgers, L. C., and W. H. Warner (1973). Dynamic stability of out-of-plane motion of curved elastic rods. *SIAM J. Appl. Math.*, **24**, 36–43. *449*

Rogers, E. H. See Matkowsky, Rogers, and Rubenfeld.

Rogers, T. G. See Varley and Rogers.

Romanova, N. N. (1970). The vertical propagation of short acoustic waves in the real atmosphere. *Izv. Atm. Ocean Phys.*, **6**, 134–145. *568*

Roorda, J. See Dimantha and Roorda.

Roseau, M. (1966). *Vibrations non Linéaires et Théorie de la Stabilité.* Springer-Verlag, New York. *173*

Roseau, M. (1969). Sur une classe de systèmes dynamiques soumis à des excitations périodiques de lonque période. *C.R. Acad. Sci. Paris*, **258**, 409–412. *137*

Rosen, G. (1965). Formal equivalence of the nonlinear string and one-dimensional fluid flow. *Q. Appl. Math.*, **23**, 286–287. *486*

Rosenberg, R. M. (1955). Discussion of harmonic oscillations of nonlinear two-degree of freedom systems. *J. Appl. Mech.*, **22**, 602–603. *417*

Rosenberg, R. M. (1958). On the periodic solution of the forced oscillator equation. *Q. Appl. Math.*, **15**, 341–354. *179*

Rosenberg, R. M. (1960). Normal modes of nonlinear dual mode systems. *J. Appl. Mech.*, **27**, 263–268. *365*

Rosenberg, R. M. (1961). On normal vibration of a general class of nonlinear dual mode systems. *J. Appl. Mech.*, **28**, 275–283. *365*

Rosenberg, R. M. (1962). The normal modes of nonlinear *n*-degree-of-freedom systems. *J. Appl. Mech.*, **29**, 7–14. *365*

Rosenberg, R. M. (1964). On normal mode vibration. *Proc. Camb. Phil. Soc.*, **60**, 595–611. *365*

Rosenberg, R. M. (1966). On nonlinear vibrations of systems with many degrees of freedom. *Adv. Appl. Mech.*, **9**, 155–241. *365*

Rosenberg, R. M. (1968). Normal modes of non-linear dual mode systems. *J. Appl. Mech.*, **27**, 263–268. *365*

Rosenberg, R. M., and C. P. Atkinson (1959). On the natural modes and their stability in nonlinear two-degree-of-freedom systems. *J. Appl. Mech.*, **26**, 377–385. *365*

Rosenberg, R. M., and J. K. Kuo (1964). Nonsimilar normal mode vibrations of nonlinear systems having two degrees of freedom. *J. Appl. Mech.*, **31**, 283–290. *365*

Rosenberg, R. M. See Kinney and Rosenberg; Paulling and Rosenberg; Yang and Rosenberg.

Rubenfeld, L. A. (1973). The stability surfaces of a Hill's equation with several small parameters. *J. Appl. Mech.*, **40**, 1107–1109. *295*

Rubenfeld, L. A. (1974). Nonlinear dynamic buckling of a compressed elastic column. *Q. Appl. Math.*, **32**, 163–171. *460*

Rubenfeld, L. A. (1977). The passage of weakly coupled nonlinear oscillators through internal resonance. *Stud. Appl. Math.*, **57**, 77–92. *138, 438*

Rubenfeld, L. A. See Matkowsky, Rogers, and Rubenfeld.

Rudenko, O. V., S. I. Soluyan, and V. Khokhlov (1974). Problems in the theory of nonlinear acoustics. *Sov. Phys. Acoust.*, **20,** 271–275. *545*

Runyan, H. L. See Hayes and Runyan.

Ryland, G., and L. Meirovitch (1977). Stability boundaries of a swinging spring with oscillating support. *J. Sound Vib.*, **51,** 547–560. *261, 360, 370*

Sabata, H. (1974). A pair of exactly soluble nonlinear oscillators. *Int. J. Non-Linear Mech.*, **9,** 435–439. *609*

Sackman, J. L. See Mostaghel and Sackman.

Saenger, R. A., and G. E. Hudson (1960). Periodic shock waves in resonating gas columns. *J. Acoust. Soc. Am.*, **32,** 961–970. *572, 575*

Safar, M. H. (1970). The exploitation of the subharmonic pressure waves from pulsating gas bubbles in an acoustic field in liquids. *J. Phys. D: Appl. Phys.*, **3,** 635–636. *199*

Saffman, P. G. See Benney and Saffman.

Sagdeev, R. Z., and A. A. Galeev (1969). *Nonlinear Plasma Theory.* Benjamin, New York. *601*

Saigo, M. See Iwatsubo, Saigo, and Sugiyama.

St. Hilaire, A. O. (1976). Analytical prediction of the non-linear response of a self-excited structure. *J. Sound Vib.*, **47,** 185–205. *129*

Saito, A. See Yamamoto and Saito.

Saito, H., K. Sato, and T. Yutani (1976). Non-linear forced vibrations of a beam carrying concentrated mass under gravity. *J. Sound Vib.*, **46,** 515–525. *447*

Saito, H. See Tsuchiya and Saito.

Salant, R. F. (1973). Acoustic propagation in waveguides with sinusoidal walls. *J. Acoust. Soc. Am.*, **53,** 504–507. *362*

Salion, V. E. (1956). The dynamic stability of a curved arch under the action of periodic moments. *Collection of Papers: Problems of Stability and Strength*, Akad. Nauk Ukr. SSR, Kiev, pp. 123–127. *268*

Samoilenko, A. M., and I. P. Momot. (1974). Quasi-periodic oscillations in the resonance region of second order nonlinear systems. *Mat. Fiz.*, **77,** 140–152. *163*

Samuels, J. S. (1959). On propagation of waves in slightly rough ducts. *J. Acoust. Soc. Am.*, **31,** 319–325. *362*

Sanchez, D. A. See Borges, Cesari, and Sanchez.

Sandman, B. E., and C.-L. Huang (1971). Finite amplitude oscillations of a thin elastic annulus. *Dev. Mech.*, **6,** 921–934. *508*

Sandman, B. E. See Huang and Sandman.

Sanford, W. F. See Evan-Iwanowski, Sanford, and Kehagioglou.

Sankar, T. S. See Srinivasan and Sankar.

Saric, W. S., A. H. Nayfeh, and S. G. Lekoudis (1976). Experiments on the stability of liquid films adjacent to supersonic boundary layers. *J. Fluid Mech.*, **77,** 63–80. *107*

Saric, W. S. See Nayfeh and Saric; Lekoudis, Nayfeh, and Saric.

Sathyamoorthy, M. (1977). Shear and rotary inertia effects on large amplitude vibration of skew plates. *J. Sound Vib.*, **52,** 155–163. *501, 507*

Sathyamoorthy, M., and K. A. V. Pandalai (1970). Nonlinear flexural vibrations of orthotropic rectangular plates. *J. Aeronaut. Soc. India*, **4,** 264–266. *501, 508*

Sathyamoorthy, M. and K. A. V. Pandalai (1971). Nonlinear flexural vibrations of oval rings. *J. Aeronaut. Soc. India*, **23,** 1–12. *447*

Sathyamoorthy, M. and K. A. V. Pandalai (1972a). Large amplitude vibrations of certain deformable bodies. Part I–discs, membranes and rings. *J. Aeronaut. Soc. India*, **24**, 409–414. *444*

Sathyamoorthy, M., and K. A. V. Pandalai (1972b). Nonlinear flexural vibration of orthotropic skew plates. *J. Sound Vib.*, **24**, 115–120. *508*

Sathyamoorthy, M., and K. A. V. Pandalai (1973a). Large amplitude flexural vibration of simply supported skew plates. *AIAA J.*, **11**, 1279–1282. *508*

Sathyamoorthy, M., and K. A. V. Pandalai (1973b). Large amplitude vibrations of certain deformable bodies. Part II–plates and shells. *J. Aeronaut. Soc. India*, **25**, 1–10. *444*

Sathyamoorthy, M., and K. A. V. Pandalai (1973c). Nonlinear vibrations of elastic skew plates exhibiting rectilinear orthotropy. *J. Franklin Inst.*, **296**, 359–369. *502, 508*

Sathyamoorthy, M., and K. A. V. Pandalai (1973d). Vibration of simply supported clamped skew plates at large amplitudes. *J. Sound Vib.*, **27**, 37–46. *508*

Sathyamoorthy, M. See Pandalai and Sathyamoorthy.

Sato, K. See Saito, Sato, and Yutani.

Schaeffer, H. G., and J. A. McElman (1965). Dynamic response of a nonlinear membrane in supersonic flow. *AIAA J.*, **3**, 543–544. *506*

Scharpf, D. W. See Argyris and Scharpf.

Scheifele, G. See Stiefel and Scheifele.

Scher, M. P., and R. L. Farrenkopf (1974). Dynamic trap states of dual spin spacecraft. *AIAA J.*, **12**, 1721–1725. *396*

Schmidt, G. (1961a). Instabilität gedämpfter rheolinearer Schwingungen. *Math. Nach.*, **23**, 301–318. *268*

Schmidt, G. (1961b). Über die Biegeschwingungen des gelenkig gelagerten axial pulsierend belasteten Stabes. *Math. Nach.*, **23**, 76–132. *268*

Schmidt, G. (1961c). Über die Querschwingungen schwach vergekrümmter Stabe bei exzentrisch angreifender pulsierender Längsbelastung. *Math. Nach.*, **23**, 292–300. *268*

Schmidt, G. (1964). Zur Stabilität der Längs- und Querschwingungen eines längs-pulsierend belasteten Stabes. *Math. Nach.*, **27**, 341–351. *267*

Schmidt, G. (1974). *Parametererregte Schwingungen.* VEB Deutcher Verlag der Wissenschaften, Berlin. *259, 268*

Schmidt, G., and F. Weidenhammer (1961). Instabilitäten gedämpten rheolinearer Schwingungen. *Math. Nach.*, **23**, 301–318. *268, 308, 320*

Schmidt, R. (1963). Untersuchungen des dynamischen Stabilitätsverhaltens von dünnwandigen Trägern mit beliebiger, offener Profilform unter der Wirkung von harmonisch pulsierend Längskraften und Biegemomenten. *Wiss. Z. Tech. Hoch. Otto von Guernicke Magdeburg*, **7**, 49–67. *268*

Schmidt, R. (1973). Nonlinear vibration of a rectangular plate arbitrarily laminated of anisotropic material. *J. Appl. Mech.*, **40**, 1148–1149. *502*

Schneider, C. C., and P. W. Likins (1973). Nutation dampers vs precession dampers for asymmetric spinning spacecraft. *J. Spacecraft*, **10**, 218–222. *396*

Schneider, K. R. (1972). Methods for the approximate computation of the periodic solutions of systems of nonlinear periodic differential equations. *Computing*, **10**, 63–82. *339*

Schwarz, H. R. (1970). Über periodische Lösungen einer nichtlinearen Differentialgleichung mit periodischen Koeffizienten. *ZAMM*, **50**, T23–T25. *339*

Scott, R. A. See Hyer, Anderson, and Scott; Ho, Scott, and Eisley; Kovac, Anderson, and Scott.

Sedov, A. See Nariboli and Sedov.

Seebass, A. R. See Leibovich and Seebass.

Segel, L. A. See DiPrima, Eckhaus, and Segel.

Segur, H. See Ablowitz, Kaup, Newell, and Segur.

Seifert, G. See Hale and Seifert.

Sekiya, T. See Sugiyama, Fugiwara, and Sekiya; Sugiyama, Katayama, and Sekiya.

Sethna, P. R. (1954). Steady state motion of one and two degrees of freedom vibrating systems with a nonlinear restoring force. *Proc. Second U.S. Nat. Congr. Appl. Mech.*, pp. 69–78. *163, 417*

Sethna, P. R. (1960). Steady-state undamped vibrations of a class of nonlinear discrete systems. *J. Appl. Mech.*, **27**, 187–195. *417*

Sethna, P. R. (1962). Free vibrations of beams with nonlinear viscoelastic material properties. *Proc. Fourth U.S. Nat. Congr. Appl. Mech.*, pp. 1103–1112. *445, 449*

Sethna, P. R. (1963a). Coupling in certain classes of weakly nonlinear vibrating systems. In *Nonlinear Differential Equations and Nonlinear Mechanics.* (J. P. LaSalle and S. Lefschetz, Eds.) Academic, New York, pp. 58–70. *317*

Sethna, P. R. (1963b). Transients in certain autonomous multiple-degree-of-freedom nonlinear vibrating systems. *J. Appl. Mech.*, **30**, 44–50. *388*

Sethna, P. R. (1965). Vibrations of dynamical systems with quadratic nonlinearities. *J. Appl. Mech.*, **32**, 576–582. *268, 370, 402*

Sethna, P. R. (1967a). An extension of the method of averaging. *Q. Appl. Math.*, **25**, 205-211. *137*

Sethna, P. R. (1967b). Domains of attraction of some quasiperiodic solutions. In *Differential Equations and Differential Systems*, Academic, New York, 323–332. *173*

Sethna, P. R. (1969). Systems with fast and slow times. *Proc. 5th Conf. on Nonlinear Oscillations*, Kiev. *137*

Sethna, P. R. (1973). Method of averaging for systems bounded for positive time. *J. Math. Anal. Appl.*, **41**, 621–631. *137, 261, 272*

Sethna, P. R., and M. Balachandra (1974a). Some asymptotic results for systems with multiple time scales. *SIAM J. Appl. Math.*, **27**, 611–625. *137*

Sethna, P. R., and M. Balachandra (1974b). Transients in high spin gyroscopic systems. *J. Appl. Mech.*, **41**, 787–792. *396*

Sethna, P. R., and M. B. Balachandra (1976). On nonlinear gyroscopic systems. In *Mechanics Today.* (S. Nemat-Nasser, Ed.) Pergamon, New York, pp. 191–242. *395*

Sethna, P. R., and G. W. Hemp (1965). Nonlinear oscillations of a gyroscopic pendulum with an oscillating point of suspension. In *Les Vibrations Forcées dans les Systèmes Non-Linéaires.* Centre National de la Recherche Scientifique, Paris, pp. 375–391. *261, 338*

Sethna, P. R., and T. J. Moran (1968). Some nonlocal results for weakly nonlinear dynamical systems. *Q. Appl. Math.*, **26**, 175–185. *417*

Sethna, P. R. See Balachandra and Sethna; Hemp and Sethna; Loud and Sethna.

Sevin, E. (1960). On the elastic bending of columns due to dynamic axial forces including the effect of axial inertia. *J. Appl. Mech.*, **27**, 125–131. *455*

Sevin, E. (1961). On the parametric excitation of a pendulum-type vibration absorber. *J. Appl. Mech.*, **28**, 330–334. *370, 379*

Seymour, B. R., and M. P. Mortell (1973a). Nonlinear resonant oscillations in open tubes. *J. Fluid Mech.*, **60**, 733–749. *572*

Seymour, B. R., and M. P. Mortell (1973b). Resonant acoustic oscillations with damping: small rate theory. *J. Fluid Mech.*, **58**, 353–373. *572*

Seymour, B. R., and M. P. Mortell (1975a). Nonlinear geometrical acoustics. In *Mechanics Today* **2**, Pergamon, London, pp. 251–312. *545*

Seymour, B. R., and M. P. Mortell (1975b). Propagation of pulses and weak shocks in nonlinear laminated composites. *J. Appl. Mech.*, **42**, 832–836. *568*

Seymour, B. R., and E. Varley (1967). High frequency, pulsatile flow in a tapering viscoelastic tube of varying stiffness. In *Proceedings of the First International Conference on Hemorheology.* Pergamon, New York, pp. 131–142. *568*

Seymour, B. R., and E. Varley (1970). High frequency, periodic disturbances in dissipative systems. I. Small amplitude, finite rate theory. *Proc. Roy. Soc. London*, **A314**, 387–415. *563, 565*

Seymour, B. R. See Mortell and Seymour.

Shabat, A. B. See Zakharov and Shabat.

Shafey, N. A. See Sun and Shafey.

Shahin, R. M. (1974). Non-linear vibrations of multi-layer orthotropic sandwich plates. *J. Sound Vib.*, **36**, 361–374. *501*

Shahinpoor, M. (1973). Combined radial-axial large amplitude oscillations of hyperelastic cylindrical tubes. *J. Math. Phys. Sci.*, **7**, 111–128. *506*

Shampine, L. F. See Roderman, Longcope, and Shampine.

Shapiro, M. J. See Thurston and Shapiro.

Shen, C.-N. (1959). Stability of forced oscillations with nonlinear second-order terms. *J. Appl. Mech.*, **26**, 499–502. *163*

Sherbourne, A. N. See Murthy and Sherbourne.

Shibata, A. See Shiga, Shibata, and Takahashi.

Shiga, T., A. Shibata, and J. Takahashi (1974). Experimental study on dynamic properties of reinforced concrete shear walls. *Proc. Fifth World Conf. Earthquake Eng.*, Rome, Vol. 1, pp. 1157–1166. *98*

Shih, L. Y. (1971). Three-dimensional nonlinear vibration of a traveling string. *Int. J. Non-Linear Mech.*, **6**, 427.

Shippy, D. J. See Kane and Shippy.

Shkenev, Y. S. (1960). Nonlinear vibrations of circular rings. *Inzh. Sb.*, **28**, 82–86. *447*

Shrivastava, S. K. See Hablani and Shrivastava.

Shtokalo, I. Z. (1961). *Linear Differential Equations with Variable Coefficients.* Gordon and Breach, New York. *259*

Shulenzhko, L. F. See Berezovskii and Shulenzhko.

Sikarskie, D. L. See Chao and Sikarskie; Lou and Sikarskie.

Siljak, D. D. (1969). *Nonlinear Systems.* Wiley, New York. *vii, 92*

Silverthorn, L. J. See Friedmann and Silverthorn.

Simanov, S. N. (1952). On the theory of quasiharmonic oscillations. *PMM*, **16**, 129–146. *308*

Simmons, W. F. (1969). A variational method for weak resonant wave interations. *Proc. Roy. Soc. London*, **A309**, 551–575. *150, 601*

Simons, D. A., and A. W. Leissa (1971). Vibrations of rectangular cantilever plates subjected to inplane acceleration loads. *J. Sound Vib.*, **17**, 407–422. *268*

Singh, M. C. See Mishra and Singh.

Singh, P. N., and S. M. J. Ali (1975). Non-linear vibration of a moderately thick shallow clamped arch. *J. Sound Vib.*, **41**, 275–282. *447*

Singh, P. N., Y. C. Das, and V. Sundararajan (1971). Large-amplitude vibration of rectangular plates. *J. Sound Vib.*, **17**, 235–240. *500, 507*

Singh, P. N., V. Sundararajan, and Y. C. Das (1974). Large-amplitude vibration of some moderately thick structural elements. *J. Sound Vib.*, **36**, 375–387. *501, 506*

Sinha, S. C. See Chou and Sinha.

Sinitsyn, S. B. (1976). Oscillations of a cylindrical shell made of nonlinearly elastic materials in passing through a half-resonance. *Sov. Appl. Mech.*, **10**, 1016–1019. *444*

Sircar, R. (1974). Vibration of rectilinear plates on elastic foundation at large-amplitude. *Bull. Acad. Pol. Sci. Ser. Sci. Tech.*, **22**, 293–299. *507*

Sissingh, G. J. (1968). Dynamics of rotors operating at high advance ratios. *J. Am. Helicopter Soc.*, **13**, 56–63. *269*

Sissingh, G. J., and W. A. Kuczynski (1970). Investigations on the effect of blade torsion on the dynamics of the flapping motion. *J. Am. Helicopter Soc.*, **15**, 2–9. *269*

Skalak, R., and M. I. Yarymovych (1960). Subharmonic oscillations of a pendulum. *J. Appl. Mech.*, **27**, 159–164. *261, 338*

Šklíba, J. See Tondl, Fiala, and Šklíba.

Smirnov, A. F. (1947). *The Static and Dynamic Stability of Structures.* Transzeldorizdat, Moscow. *266*

Smith, D. M. (1933). The motion of a rotor carried by a flexible shaft in flexible bearings. *Proc. Roy. Soc. London*, **142**, 92. *269, 331*

Smith, M. A. See Hamer and Smith.

Smith, O. See Levinson and Smith.

Smith, P. W., C. T. Malme, and C. M. Gogos (1961). Non-linear response of a simple clamped panel. *J. Acoust. Soc. Am.*, **33**, 1475–1480. *508*

Sobala, D. See Kane and Sobala.

Sobolev, V. A. (1954). The dynamic stability of deformation of a strip in excentric compression and pure bending. *Inzh. Sb.*, **19**, 65–72. *267*

Soluyan, S. I. See Rodenko, Soluyan, and Khokhlov.

Somerset, J. H. (1967). Transition mechanisms attendent to large-amplitude parametric vibrations of rectangular plates. *J. Eng. Ind. Trans. ASME*, **89**, 619–625. *268*

Somerset, J. H., and R. M. Evan-Iwanowski (1965). Experiments on parametric instability of columns. *Dev. Appl. Mech.*, **2**, 503–525. *267*

Somerset, J. H., and R. M. Evan-Iwanowski (1967). Influence of nonlinear inertia on the parametric response of rectangular plates. *Int. J. Non-Linear Mech.*, **2**, 217–232. *268*

Sommerfeld, A. (1904). *VDI Zeitschr.*, **48**, 631–636. *224*

Sonstegard, D. A. See McIvor and Sonstegard.

Soudack, A. C. and P. G. D. Barkham (1971). Further results on non-linear non-autonomous second-order differential equations. *Int. J. Control*, **12**, 763–767. *119*

Soudack, A. C. See Lansdowne and Soudack.

Sparrow, C. A. See Rehfield, Girl, and Sparrow.

Spiegel, E. A. See Baker, Moore, and Spiegel.

Sridhar, S., D. T. Mook, and A. H. Nayfeh (1975). Non-linear resonances in the forced responses of plates. Part I: symmetric responses of circular plates. *J. Sound Vib.*, **41**, 359–373. *418, 469, 508*

Sridhar, S., D. T. Mook, and A. H. Nayfeh (1978). Non-linear resonances in the forced response of plates. Part II: asymmetric responses of circular plates. *J. Sound Vib.*, **59**, 159–170. *508*

Sridhar, S., A. H. Nayfeh, and D. T. Mook (1975). Non-linear resonances in a class of multi-degree-of-freedom systems. *J. Acoust. Soc. Am.*, **58**, 113–123. *469*

Sridhar, S. See Nayfeh, Mook and Sridhar.

Srinivasan, A. V. (1965). Large amplitude-free oscillations of beams and plates. *AIAA J.*, **3**, 1951–1953. *460, 507*

Srinivasan, A. V. (1966a). Non-linear vibrations of beams and plates. *Int. Non-Linear Mech.*, **1**, 179–191. *460*

Srinivasan, A. V. (1966b). Steady-state response of beams supported on non-linear springs. *AIAA J.*, **4**, 1863–1864.

Srinivasan, A. V. (1967). Dynamic stability of beam columns. *AIAA J.*, **5**, 1685–1686. *460*

Srinivasan, P., and T. S. Sankar (1974). Autoparametric self-excitation of a pendulum type elastic oscillator. *J. Sound Vib.*, **35**, 549–557. *370, 379*

Srinivasan, P. See Dasarathy and Srinivasan; Jain and Srinivasan; Joshi, Srirangarajan, and Srinivasan; Rangacharyulu and Srinivasan; Rangacharyulu, Srinivasan, and Dasarathy; Srirangarajan and Srinivasan.

Srirangarajan, H. R., and B. V. Dasarathy (1975). Study of third-order non-linear systems–variation of parameters approach. *J. Sound Vib.*, **40**, 173–178. *151*

Srirangarajan, H. R., and P. Srinivasan (1973). The transient response of certain third-order non-linear systems. *J. Sound Vib.*, **29**, 215–226. *151, 162, 255, 417*

Srirangarajan, H. R., and P. Srinivasan (1974). Application of ultraspherical polynominals to forced oscillations of a third-order non-linear system. *J. Sound Vib.*, **36**, 513–519. *151, 255*

Srirangarajan, H. R., and P. Scrinivasan (1976). The pulse response of non-linear systems. *J. Sound Vib.*, **44**, 369–377.

Srirangarajan, H. R., P. Srinivasan, and B. V. Dasarathy (1974). Analysis of two degrees of freedom systems through weighted mean square linearization approach. *J. Sound Vib.*, **36**, 119–131. *87*

Srirangarajan, H. R. See Joshi, Srirangarajan, and Srinivasan.

Stanišić, M. M., and J. A. Euler (1973). General perturbational solution of a harmonically forced nonlinear oscillator equation. *Int. J. Non-Linear Mech.*, **8**, 523–538. *163*

Starzhinskii, V. M. See Yakubovich and Starzhinskii.

Steele, C. R. (1967). Nonlinear effects in the problem of the beam on a foundation with a moving load. *Int. J. Solids Struct.*, **3**, 565–585. *445*

Stenflo, L. (1973). Three-wave interaction in cold magnetized plasmas. *Planet. Space Sci.*, **21**, 391–397. *601*

Stenflo, L. See Larsson and Stenflo.

Stephenson, A. (1906). On a class of forced oscillations. *Q. J. Math.*, **37**, 353–360. *258*

Stephenson, A. (1908). On a new type of dynamical stability. *Mem. Proc. Manch. Lit. Phil. Soc.*, **52**, 1907–1908. *259, 261, 267*

Stevanovich, K., and D. Rashkovich (1974). Many frequency vibration in one frequency regime of nonlinear systems with several degrees of freedom. *Zagadnienia Drgan Nieliniowych*, **15**, 201–220. *418*

Stevens, K. K. (1966). On the parametric excitation of a viscoelastic column. *AIAA J.*, **4**, 2111–2116. *268, 308*

Stevens, K. K., and R. M. Evan-Iwanowski (1969). Parametric resonance of viscoelastic columns. *Int. J. Solids Struct.*, **5**, 755–765. *268*

Stewartson, K., and J. T. Stuart (1971). A nonlinear instability theory for a wave system in plane Poiseuille flow. *J. Fluid Mech.*, **48**, 529–545. *587*

Stewartson, K. See Davey and Stewartson.

Stiefel, E. L., and G. Scheifele (1971). *Linear and Regular Celestial Mechanics.* Springer-Verlag, New York. *41*

Stoker, J. J. (1950). *Nonlinear Vibrations*. Wiley, New York. *vii, 144, 146, 179, 259, 261*

Stoker, J. J. See Lubkin and Stoker.

Strehlow, H. See Huber and Strehlow.

Stricklin, J. A., J. E. Martinez, J. R. Tillerson, J. H. Hong, and W. E. Haisler (1971). Nonlinear dynamic analysis of shells of revolution by matrix displacement method. *AIAA J.*, **9**, 629–636. *445, 506*

Stuart, J. T. See Stewartson and Stuart.

Struble, R. A. (1962). *Nonlinear Differential Equations*. McGraw-Hill, New York. *vii, 108, 173, 210, 259*

Struble, R. A. (1963). On the subharmonic oscillations of a pendulum. *J. Appl. Mech.*, **30**, 301–303. *261, 338*

Struble, R. A., and J. E. Fletcher (1962). General perturbations solution to the Mathieu-Hill equation. *SIAM J. Appl. Math.*, **10**, 314–328. *295*

Struble, R. A., and J. H. Heinbockel (1963). Resonant oscillations of a beam-pendulum system. *J. Appl. Mech.*, **30**, 181–188. *370, 379*

Struble, R. A. See Heinbockel and Struble.

Strutt, M. J. O. (1887). On the maintenance of vibrations by forces of double frequency and on the propagation of waves through a medium endowed with a periodic structure. *Phil. Mag.*, **24**, 145–159. *258*

Strutt, M. J. O. (1928). Eigenschwingungin einer saite mit sinus-fermiger. *Massenverteilung Ann. Phys.*, **85**, 129–136. *277*

Strutt, M. J. O. See van der Pol and Strutt.

Sturtevant, B. (1973). Studies of shock focusing and nonlinear resonance in shock tubes. In *Shock Tube Research*, Stanford University Press, Stanford, pp. 23–34. *572*

Sturtevant, B. (1974). Nonlinear gas oscillations in pipes. Part 2. Experiment. *J. Fluid Mech.*, **63**, 97–120. *572*

Subramanian, R., and R. E. Kronauer (1972). Escape from a potential well (Part I). *Q. Appl. Math.*, **29**, 459–491. *423*

Subramanian, R., and R. E. Kronauer (1973). Escape from a potential well (Part II). *Q. Appl. Math.*, **30**, 127–142. *423*

Subramanian, R. See Tiwari and Subramanian.

Sugimoto, N., and M. Hirao (1977). Nonlinear mode coupling of elastic waves. *J. Acoust. Soc. Am.*, **62**, 23–32. *601*

Sugimoto, N. See Hirao and Sugimoto; Kakutani and Sugimoto; Kawahara, Sugimoto, and Kakutani; Matsumoto, Sugimoto, and Inoue.

Sugiyama, Y., N. Fugiwara, and T. Sekiya (1970). Studies on nonconservative problems of instability of columns by means of an analogue computer. *Proc. 18th Japan Nat. Congr. Appl. Mech.* 1968, pp. 113–126. *266, 267, 306*

Sugiyama, Y., T. Katayama, and T. Sekiya (1971). Studies on nonconservative problems of instability of columns by difference method. *Proc. 19th Japan Nat. Congr. Appl. Mech.* 1969, pp. 23–31. *266, 267*

Sugiyama, Y. See Iwatsubo, Saigo, and Sugiyama; Iwatsubo, Sugiyama, and Ishihara; Iwatsubo, Sugiyama, and Ogino.

Sun, C. L., and S. Y. Lu (1968). Nonlinear dynamic behavior of heated conical and cylindrical shells. *Nucl. Eng. Des.*, **7**, 113–122. *506*

Sun, C. T., and N. A. Shafey (1976). Influence of physical non-linearity on the dynamic behaviour of composite plates. *J. Sound Vib.*, **46**, 225–232. *444*

Sundararajan, C. See Paidoussis and Sundararajan; Singh, Sundararajan, and Das; Singh, Das, and Sundararajan.

Susemihl, E. A., and P. A. A. Laura (1975). Analysis of non-linear vibrating systems by the collocation technique. *J. Sound Vib.*, **41**, 256–258. *50*

Szebehely, V. (1967). *Theory of Orbits.* Academic, New York. *337*

Szego, G. P. (1966). Liapunov second method. *Appl. Mech. Rev.*, **19**, 833–838. *172*

Szemplinska-Stupnicka, W. (1968). Higher-harmonic oscillations in heteronomous nonlinear systems with one degree of freedom. *Int. J. Non-Linear Mech.*, **3**, 17–30. *175*

Szemplinska-Stupnicka, W. (1969). On the phenomenon of the combination type resonance in non-linear two-degree-of-freedom systems. *Int. J. Non-Linear Mech.*, **4**, 335–359. *418*

Szemplinska-Stupnicka, W. (1970). On the asymptotic, averaging and W. Ritz method in the theory of steady-state vibrations of nonlinear systems with many degrees of freedom. *Arch. Mech.*, **22**, 161–181. *417*

Szemplinska-Stupnicka, W. (1972). On the averaging and W. Ritz methods in the theory of non-linear resonances in vibrating systems with multiple degrees of freedom. *Arch. Mech.*, **24**, 67–88. *417, 418*

Szemplinska-Stupnicka, W. (1973). On the stability limit of non-linear resonances in multiple-degree-of freedom vibrating systems. *Arch. Mech.*, **25**, 501–511. *418*

Szemplinska-Stupnicka, W. (1974). On the methods of treating secondary resonances in nonlinear multi-degree-of-freedom vibrating systems. *Bull. Acad. Pol. Sci., Ser. Sci. Tech.*, **22**, 39–48. *418*

Szemplinska-Stupnicka, W. (1975). A study of main and secondary resonances in nonlinear multi-degree-of-freedom vibrating systems. *Int. J. Non-Linear Mech.*, **10**, 289–304. *418*

Szulkin, P. (1960). On forced vibrations in a nonlinear system with many degrees of freedom. *Bull. Acad. Pol. Sci., Ser. Sci. Tech.*, **5**, 8. *417*

Tabarrok, B. See Elmaraghy and Tabarrok.

Takahashi, K. See Chisyaki and Takashashi; Shiga, Shibata, and Takahashi.

Tan, G. See Karasudhi, Tan, and Lee.

Tanabashi, R., and K. Kaneta (1962). On the relation between the restoring force characteristics of structures and the pattern of earthquake ground motion. *Proc. Japan Nat. Symp. Earthquake Eng.*, Tokyo. *98*

Tang, S.-C., and D. H. Y. Yen (1970). A note on the nonlinear response of an elastic beam on a foundation to a moving load. *Int. J. Solids Struct.*, **6**, 1451-1461. *445*

Tang, S.-C. See Yen and Tang.

Tani, J. (1974). Dynamic instability of truncated conical shells under periodic axial loads. *Int. J. Solids Struct.*, **10**, 169–176. *269*

Tani, J. (1976). Influence of deformations prior to instability on the dynamic instability of conical shells under periodic axial load. *J. Appl. Mech.*, **43**, 87–91. *269*

Taniuti, T. (1974). Reductive perturbation method and far fields of wave equations. *Prog. Theor. Phys. Suppl.*, **55**, 1–35. *556*

Taniuti, T., and H. Washimi (1968). Self-trapping and instability of hydromagnetic waves along the magnetic field in a cold plasma. *Phys. Rev. Lett.*, **21**, 209–212. *587, 588*

Taniuti, T., and C.-C. Wei (1968). Reductive perturbation method in nonlinear wave propagation. *J. Phys. Soc. Japan*, **24**, 941–946.

Taniuti, T., and N. Yajima (1969). Perturbation method for a nonlinear wave modulation. I. *J. Math. Phys.*, **10**, 1369–1372. *587*

Taniuti, T., and N. Yajima (1973). Perturbation method for a nonlinear wave modulation. III. *J. Math. Phys.*, **14**, 1389–1397. *587*

Taniuti, T. See Asano, Nishihara, Nozaki, and Taniuti; Asano and Taniuti; Asano, Taniuti, and Yajima; Jeffrey and Taniuti; Kakutani, Ono, Taniuti, and Wei; Nozaki and Taniuti; Washimi and Taniuti; Watanabe and Taniuti.

Tappert, F. See Hasegawa and Tappert.

Tarantovich, T. M., and V. M. Kohnkin (1975). Dynamics of a two-frequency self-excited oscillator with "stiff" excitation. *Radiophys. Quantum Electron.*, **16**, 165–168. *129*

Tezak, E. G., D. T. Mook, and A. H. Nayfeh (1978). Nonlinear analysis of the lateral response of columns to periodic loads. *Trans. ASME J. Mech. Des.*, **100**, 651–659. *268, 426*

Theodorchik, K. F. (1948). *Self-Excited Oscillations* (GITTL), State Publ. House Thoer. Tech. Lit., Moscow. *374*

Thiruvenkatachar, V. R. See Bhansali and Thiruvenkatachar.

Thompson, J. M. T., and G. W. Hunt (1973). *A General Theory of Elastic Stability*. Wiley, New York. *424*

Thomson, W. T. (1957). Analog computer for nonlinear system with hysteresis. *J. Appl. Mech.*, **24**, 245–247. *98*

Thorne, C. J. (1961). Some gyroscopic oscillations. *J. Appl. Mech.*, **28**, 57–66. *396*

Thrasher, D. F., D. T. Mook, O. A. Kandil, and A. H. Nayfeh (1977). Application of the vortex-lattice concept to general, unsteady lifting surface problems. *AIAA Paper No. 77-1157. 99*

Thurman, A. L., and C. D. Mote (1969). Nonlinear oscillation of a cylinder containing flowing fluid. *J. Eng. Ind.*, **91**, 1147–1155. *506*

Thurston, G. A, (1973). Floquet theory and Newton's method. *J. Appl. Mech.*, **40**, 1091–1096. *283*

Thurston, G. A., and M. J. Shapiro (1967). Interpretation of ultrasonic experiments on finite-amplitude waves. *J. Acoust. Soc. Am.*, **41**, 1112–1125. *545*

Tillerson, J. R. (1975). Selecting solution procedures for nonlinear structural dynamics. *Shock Vib. Dig.*, **7**, 2–13. *445*

Tillerson, J. R. See Stricklin, Martinez, Tillerson, Hong, and Haisler.

Timoshenko, S. (1940). *Theory of Plates and Shells.* McGraw-Hill, New York. *502*

Timoshenko, S. (1955). *Vibration Problems in Engineering*. Van Nostrand, Princeton. *vii, 259*

Tissot, M. A. (1852). Thesis de mécanique. *J. Math. Pures Appl.*, **17**, 88–116. *369*

Tiwari, R. N., and R. Subramanian (1976). Subharmonic and superharmonic synchronization in weakly non-linear systems. *J. Sound Vib.*, **47**, 501–508. *185*

Tobias, S. A. (1957). Free undamped non-linear vibrations of imperfect circular disks. *Proc. Inst. Mech. Eng.*, **171**, 691–701. *506*

Tobias, S. A. (1958). Non-linear forced vibrations of circular disks. An experimental investigation. *Engineering*, **186**, 51–56. *505*

Tobias, S. A. (1959). Design of small isolator units for the suppression of low frequency vibration. *J. Mech. Eng. Sci.*, **3**, 280–292. *387*

Tobias, S. A., and R. N. Arnold (1957). The influence of dynamical imperfection on the vibration of rotating disks. *Proc. Inst. Mech. Eng.*, **171**, 669–690. *505*

Tobias, S. A. See Henry and Tobias; Williams and Tobias.

Tomáš, A., and A. Tondl (1967). On the existence of subultraharmonic resonance of a single-mass nonlinear system excited by periodic force. *Nonlinear Vib. Prob.*, **8**, 11–22. *179, 185*

Tomita, A. See Iwatsubo, Tomita, and Kawai.

Tondl, A. (1963a). On the combination resonance of a nonlinear system with two degrees of freedom. *Rev. Mech. Appl.*, **8**, 573–588. *418*

Tondl, A. (1963b). On the internal resonance of a nonlinear system with two degrees of freedom. *Zag. Drgan Nil*, **5**, 207–222. *418*

Tondl, A. (1964). An analysis of resonance vibrations of nonlinear systems with two degrees of freedom. *Rozpr. CSAV Rada*, **TV 8**, 1–82. *418*

Tondl, A. (1965). *Some Problems of Rotor Dynamics*. Chapman and Hall, London. *269, 331, 395, 418*

Tondl, A. (1966). Über die innere Resonanz eines nichtlinearen Einmassensystems mit zwei Freiheitsgraden. *Z. Schwingungs Schwachstromtechnik*, **20**, 249–257. *402*

Tondl, A. (1968). Über die Stabilität selbsterregter Schwingungen mit Trockenreibung. *ZAMM*, **48**, T289–T291. *129, 152*

Tondl, A. (1970a). *Domains of Attraction for Nonlinear Systems.* National Research Institute for Machine Design, Bečhovice, Czechoslovakia. *163, 173*

Tondl, A. (1970b). *Self-Excited Vibrations.* National Research Institute for Machine Design, Bečhovice, Czechoslovakia. *129, 144, 375*

Tondl, A. (1972). *Resonance Vibrations of Nonlinear Systems Excited by a Periodic Nonharmonic Force.* National Research Institute for Machine Design, Bečhovice, Czechoslovakia. *185, 417*

Tondl, A. (1973a). *Analysis of Stability of Steady-Locally Stable Solutions for not Fully Determined Disturbances.* National Research Institute for Machine Design, Bečhovice, Czechoslovakia. *163, 173*

Tondl, A. (1973b). Some properties of nonlinear system characteristics and their application to damping identification. *Acta Tech. ČSAV*, **2**, 166–179. *119, 163*

Tondl, A. (1974). Notes on the solution of forced oscillations of a third-order non-linear system. *J. Sound Vib.*, **37**, 273–279. *152, 255*

Tondl, A. (1975a). Quenching of self-excited vibrations: equilibrium aspects. *J. Sound Vib.*, **42**, 251–260. *210, 375*

Tondl, A. (1975b). Quenching of self-excited vibrations: one- and two-frequency vibrations. *J. Sound Vib.*, **42**, 261–271. *210, 375*

Tondl, A. (1976a). Additional note on a third-order system. *J. Sound Vib.*, **47**, 133–135. *255*

Tondl, A. (1976b). *On the Interaction Between Self-Excited and Forced Vibrations.* National Research Institute for Machine Design, Bečhovice, Czechoslovakia. *129, 210*

Tondl, A. (1976c). Parametric vibration of a nonlinear system. *Ing.-Arch.*, **45**, 317–324. *339*

Tondl, A. (1976d). Quenching of self-excited vibrations: effect of dry friction. *J. Sound Vib.*, **45**, 285–294. *210*

Tondl, A., V. Fiala, and J. Skliba (1970). *Domains of Attraction for Non-Linear Systems.* National Research Institute for Machine Design, Bečhovice, Czechoslovakia. *110*

Tondl, A. See Tomáš and Tondl.

Tong, P. See Friedmann and Tong.

Townsend, W. H., and R. D. Hanson (1974). Hysteresis loops for reinforced concrete beam-column connections. *Proc. Fifth World Conf. Earthquake Eng.*, Rome, Vol. 1, pp. 1131–1134. *98*

Troger, H. (1975). Notes on pendulum with oscillating suspension point. *ZAMM*, **55**, 68–69. *261, 338*

Troger, H., and C. S. Hsu (1977). Response of a nonlinear system under combined parametric and forcing excitation. *Appl. Mech.*, **44.** *339, 358*

Troyanovskii, I. E. See Ivanova, Morgunov, and Troyanovskii; Kravchuk, Morgunov, and Troyanovskii.

Tsel'man, F. Kh. (1970). On pumping transfer of energy between non-linearly-coupled oscillators in third-order resonance *PMM*, **34**, 916–922. *379, 424*

Tsel'man, F. Kh. (1971). On the oscillations of a system of coupled oscillators with one third-order resonance. *PMM,* **35**, 1038–1044. *379, 396, 424*

Tsel'man, F. Kh. See Kahzin and Tsel'man.

Tseng, G. T. See Likins, Tseng, and Mingori; Mingori, Tseng, and Likins.

Tseng, S. F. See Rand and Tseng.

Tseng, W. Y., and J. Dugundji (1970). Nonlinear vibrations of a beam under harmonic excitation. *J. Appl. Mech.*, **37**, 292–297. *530*

Tseng, W.-Y., and J. Dugundji (1971). Nonlinear vibrations of a buckled beam under harmonic excitation. *J. Appl. Mech.*, **38**, 467–476. *433, 447*

Tso, W. K. (1968). Parametric torsional stability of a bar under axial excitation. *J. Appl. Mech.*, **35**, 13–19. *268*

Tso, W. K., and K. G. Asmis (1970). Parametric excitation of a pendulum with bilinear hysteresis. *J. Appl. Mech.*, **37**, 1061–1068. *98, 261, 339*

Tso, W. K., and K. G. Asmis (1974). Multiple parametric resonance in a nonlinear two degree of freedom system. *Int. J. Non-Linear Mech.*, **9**, 269–277. *339, 424*

Tso, W. K., and T. K. Caughey (1965). Parametric excitation of a non-linear system. *J. Appl. Mech.*, **32**, 899–902. *268, 338, 354*

Tso, W. K., E. Pollner, and A. C. Heidebrecht (1974). Cyclic loading on externally reinforced masonary walls. *Proc. Fifth World Conf. Earthquake Eng.*, Rome, Vol. 1, pp. 1177–1186. *98*

Tso, W. K. See Asmis and Tso; Ghobarah and Tso.

Tsuchiya, K. (1975). Thermally induced nutational body motion of a spinning spacecraft with flexible appendages. *AIAA J.*, **13**, 448–453. *442*

Tsuchiya, K. (1977). Thermally induced vibrations of a flexible appendage attached to a spacecraft. *AIAA J.*, **15**, 505–510. *442*

Tsuchiya, K. and H. Saito (1974). Dynamics of a spin-stabilized satellite having flexible appendages. *AIAA J.*, **12**, 490–495. *442*

Tsytovich, V. N. (1970). *Nonlinear Effects in Plasma.* Plenum, New York. *601*

Urabe, M. (1963). Numerical study of periodic solutions of the van der Pol equation. In *Nonlinear Differential Equations and Nonlinear Mechanics.* (J. P. LaSalle and S. Lefschetzy, Eds.) Academic, New York, pp. 184–192. *144, 146*

Valeev, K. G. (1960a). On Hill's method in the theory of linear differential equations with periodic coefficients. *PMM*, **24**, 1493–1505. *284, 308, 313*

Valeev, K. G. (1960b). On the solution and characteristic exponents of solutions of some systems of linear differential equations with periodic coefficients. *PMM*, **24**, 877–902. *308, 313*

Valeev, K. G. (1961). On Hill's method in the theory of linear differential equations with periodic coefficients. Determination of the characteristic exponents. *PMM*, **25**, 460–466. *284, 308, 313*

Valeev, K. G. (1963). On the danger of combination resonances. *PMM*, **27**, 1745–1759. *268, 308, 320*

Van Dao, N. (1973). Parametric vibrations of mechanical systems with several degrees of freedom under the action of electromagnetic force. *Proc. Vib. Probl.*, **14**, 85–94. *308*

van der Burgh, A. H. P. (1968). On the asymptotic solutions of the differential equations for the elastic pendulum. *J. Mec.*, **7**, 507–520. *370, 379*

van der Burgh, A. H. P. (1975). On the higher order asymptotic approximations for the solutions of the equations of motion of an elastic pendulum. *J. Sound Vib.*, **42**, 463–475. *370, 379*

van der Burgh, A. H. P. (1976). On the asymptotic approximations of the solutions of a system of two nonlinearly coupled harmonic oscillators. *J. Sound Vib.*, **49**, 93–103. *370, 379*

van der Heyde, R. C. W. See Jacobson and van der Heyde.

van der Mark, J. See van der Pol and van der Mark.

van der Pol, B. (1922). On oscillation hysteresis in a simple triode generator. *Phil. Mag.*, **43**, 700–719. *105, 142*

van der Pol, B., and J. van der Mark (1928). Le battement du coeur considéré comme oscillation de relaxation. *Onde Électrique*, **7**, 365–392. *143*

van der Pol, F., and M. J. O. Strutt (1928). On the stability of the solutions of Mathieu's equation. *Phil. Mag.*, **5**, 18–38. *277*

van der Werff, T. J. (1973). A new weighting function for solving non-linear oscillation problems. *Int. J. Mech. Sci.*, **15**, 913–920. *87*

van der Werff, J. (1975). A note on least squares, Kryloff and Bogoliuboff and nonlinear vibrations. *S. Afr. Mech. Eng.*, **25**, 325–327. *87*

van Dooren, R. (1971a). Combination tones of summed type in a non-linear damped vibratory system with two degrees of freedom. *Int. J. Non-Linear Mech.*, **6**, 237–254. *402*

van Dooren, R. (1971b). Recherche numérique d'oscillations composées du type additif dans un système oscillant non linéaire amorti à deux degrés de liberté. *Acad. R. Belg., Cl. Sci.*, **57**, 524–544. *402*

van Dooren, R. (1972a). An analytical method for certain weakly non-linear periodic differential systems. *Acad. Roy. Belg., Cl. Sci.*, **58**, 605–621. *417*

van Dooren, R. (1972b). Forced oscillations in coupled Duffing equations by an analytical method of varying amplitudes and phase angles. *Koninklijke Vlaamse Academie van België*, Vol. 34, No. 2. *417*

van Dooren, R. (1973a). An analytical method for certain highly non-linear periodic differential equations. *Funkcialaj Ekvacioj*, **16**, 169–180. *163*

van Dooren, R. (1973b). Differential tones in a damped mechanical system with quadratic and cubic nonlinearities. *Int. J. Non-Linear Mech.*, **8**, 575–583. *402*

van Dooren, R. (1973c). Numerical computation of forced oscillations in coupled Duffing equations. *Num. Math.*, **20**, 300–311. *417*

van Dooren, R. (1975). Two mode subharmonic vibrations of order 1/9 of a non-linear beam forced by a two mode harmonic load. *J. Sound Vib.*, **41**, 133–142. *418*

van Dooren, R., and R. Bouc (1975). Two mode subharmonic and harmonic vibrations of a non-linear beam forced by a two mode harmonic load. *Int. J. Non-Linear Mech.*, **10**, 271–280. *418*

van Dooren, R. See Janssens, van Dooren, and Melchambre.

van Dyke, M. (1975). *Perturbation Methods in Fluid Mechanics.* Parabolic Press, Stanford. *556*

van Vleck, F. S. A. See Gunderson, Rigas, and van Vleck.

van Wijngaarden, L. (1968). On the oscillations near and at resonance in open pipes. *J. Eng. Math.*, **2,** 225–240. *572*

Varga, B. B., and S. Ø. Aks. (1974). Multiple time scales and the ϕ^4 model of quantum field theory. *J. Math. Phys.*, **15,** 149–154. *163, 388*

Varley, E., and E. Cumberbatch (1966). Nonlinear, high frequency sound waves. *J. Inst. Math. Appl.*, **2,** 133–143. *563*

Varley, E., and E. Cumberbatch (1970). Large amplitude waves in a stratified media: Acoustic pulses. *J. Fluid Mech.*, **43,** 513–537. *568*

Varley, E., and T. G. Rogers (1967). The propagation of high frequency, finite acceleration pulses and shocks in viscoelastic materials. *Proc. Roy. Soc. London*, **A296,** 498–518. *563, 565*

Varley, E., R. Venkataraman, and E. Cumberbatch (1971). The propagation of large amplitude tsunamis across a basin of changing depth. Part 1. Off-shore behavior. *J. Fluid Mech.*, **49,** 775–801. *568*

Varley, E. See Cekirge and Varley; Mortell and Varley; Parker and Varley; Seymour and Varley.

Vendhan, C. P. (1975a). Modal equations for the nonlinear flexural vibrations of plates. *AIAA J.*, **13,** 1092–1094. *508*

Vendhan, C. P. (1975b). A study of Berger equations applied to non-linear vibrations of elastic plates. *Int. J. Mech. Sci.*, **17,** 461–468. *508*

Vendhan, C. P., and Y. C. Das (1975). Application of Rayleigh-Ritz and Galerkin methods to non-linear vibration of plates. *J. Sound Vib.*, **39,** 147–157. *508*

Vendhan, C. P., and B. L. Dhoopar (1973). Nonlinear vibration of orthotropic triangular plates. *AIAA J.*, **11,** 704–709. *502, 508*

Venkataraman, R. See Varley, Venkataraman, and Cumberbatch.

Venkateswara Rao, G., K. Kanaka Raju, and I. S. Raju (1976). Finite element formulation for the large amplitude free vibrations of beams and orthotropic circular plates. *Comput. Struct.*, **6,** 169–172. *446, 502*

Venkateswara Rao, G., I. S. Raju, and K. Kanaka Raju (1976a). A finite element formulation for large amplitude flexural vibrations of thin rectangular plates. *Comput. Struct.*, **6,** 163–167. *446, 508*

Venkateswara Rao, G., I. S. Raju, and K. Kanaka Raju (1976b). Nonlinear vibrations of beams considering shear deformation and rotary inertia. *AIAA J.*, **14,** 685–687. *448*

Venkateswara Rao, G. See Kanaka Raju and Venkateswara Rao; Raju, Venkateswara Rao, and Kanaka Raju.

Ventres, C. S., and E. H. Dowell (1970). Comparison of theory and experiment for nonlinear flutter of loaded plates. *AIAA J.*, **8,** 2022–2030. *509*

Ventres, C. S. See Dowell and Ventres.

Verma, M. K., and K. A. V. Murthy (1974). Non-linear vibrations of nonuniform beams with concentrated masses. *J. Sound Vib.*, **33,** 1–12. *447*

Vijayaraghavan, A., and R. M. Evan-Iwanowski (1967). Parametric instability of circular cylindrical shells. *J. Appl. Mech.*, **34**, 985–990. *269*

Vijayaraghavan, A. See Caughey and Vijayaraghavan.

Vinson, J. R. See Wu and Vinson.

Vito, R. (1973). On nonsimilar normal mode vibration in nonlinear two-degree-of-freedom systems. *J. Appl. Mech.*, **40**, 1119–1120. *365*

Vito, R. (1974). On the stability of vibrations of a particle in a plane constrained by identical non-linear springs. *Int. J. Non-Linear Mech.*, **9**, 325–330. *84*

Vito, R. See Rand and Vito.

Vitt, A. See Andronov and Vitt; Andronov, Vitt, and Khaikin.

Vol'mir, A. S., and Kh. P. Kulterbacv (1975). Nonlinear oscillations of cylindrical panels under wind action. *Sov. Appl. Mech.*, **10**, 259–263. *506*

Vol'mir, A. S., and A. T. Ponomarev (1973). Non-linear parametric vibration of closed cylindrical shells. *Izv. Akad. Nauk Arm. SSR Mekh.*, **26**, 44–50. *269*

Volosov, V. M. (1961). The method of averaging. *Trans. Sov. Math. Dokl.*, **2**, 221–224. *137*

von Freudenreich, J. (1925). Vibration of steam turbine discs. *Engineering*, **119**, 2–4 and 31–34. *505*

von Kármán, T. (1910). Festigkeitsprobleme in Maschinenbau. *Encyklopädie der Mathematischen Wissenschaften*, Vol. 3, (P. R. Halmos, Ed.), American Mathematical Society, pp. 211–385. *503*

von Kármán, T. (1940). The engineer grapples with nonlinear problems. *Bull. Am. Math. Soc.*, **46**, 615–683. *181*

Wadati, M. (1973). The modified Korteweg-de Vries equation. *J. Phys. Soc. Japan*, **34**, 1289–1296. *588*

Wagner, H. (1965). Large amplitude free vibrations of a beam. *J. Appl. Mech.*, **32**, 887–892. *460*

Wagner, W. S. See Ludeke and Wagner.

Wah, T. (1963a). Large amplitude flexural vibration of rectangular plates. *Int. J. Mech. Sci.*, **5**, 425–438. *504, 507*

Wah, T. (1963b). Vibration of circular plates at large amplitudes. *ASCE J. Eng. Mech. Div.*, **89**, 1–15. *507*

Walker, G. H., and J. Ford (1969). Amplitude instability and ergodic behavior for conservative nonlinear oscillator systems. *Phys. Rev.*, **188**, 416–432. *388*

Walker, H. S. See Huang, Woo, and Walker.

Wallace, F. B., and L. Meirovitch (1967). Attitude instability regions of a spinning symmetric satellite in an elliptic orbit. *AIAA J.*, **5**, 1642–1650. *331*

Wallace, F. B. See Meirovitch and Wallace.

Warncke, E. (1973). Self-excited vibrations of a rigid rotor running in a sliding bearing. *ZAMM*, **53**, 60–62. *129*

Warner, W. H. See Rodgers and Warner.

Washimi, H., and T. Taniuti (1966). Propagation of ion-acoustic solitary waves of small amplitude. *Phys. Rev. Letts.*, **17**, 996–998. *587*

Washimi, H. See Taniuti and Washimi.

Watanabe, K., and T. Taniuti (1977). Reductive perturbation method for wave-modulation in multi-dimensional space. *J. Phys. Soc. Japan*, **42**, 1396–1403. *587*

Watanabe, T. (1969). A nonlinear theory of two-stream instability. *J. Phys. Soc. Japan.*, **27**, 1341–1350. *587*

Watanabe, T. (1978). Forced vibration of continuous system with non-linear boundary conditions, *ASME J. Mech. Des.*, **100,** 487–491. *445*

Watson, G. N. See Whittaker and Watson.

Wax, N. See Ponzo and Wax.

Wehrli, C. (1963). Über kritische Drehzahlen unter pulsierender Torsion. *Ing.-Arch.*, **33,** 73–84. *269*

Wei, C.-C. See Kakutani, Ono, Taniuti, and Wei; Taniuti and Wei.

Weidenhammer, F. (1951). Der eingespannte, axial-pulsierend belastete Stab als Stabilitätsproblem. *Ing.-Arch.*, **19,** 162–191. *267, 268*

Weidenhammer, G. (1952). Nichtlineare Biegeschwingungen des axial-pulsierend belasteten Stabes. *Ing.-Arch.*, **20,** 315–330. *268*

Weidenhammer, F. (1956). Das Stabilitätsverhalten der nichtlinearen Biegeschwingungen des axial-pulsierend belasteten Stabes. *Ing.-Arch.*, **24,** 53–68. *268, 338*

Weidenhammer, F. See Mettler and Weidenhammer; Schmidt and Weidenhammer.

Weingarten, V. I. See Busby and Weingarten.

Weinreich, G. See Franken, Hill, Peters, and Weinreich.

Wenzke, W. (1963). Die dynamische Stabilität der axial-pulsierend belasteten Kreiszylinderschale. *Wiss. Tech. Hoch.*, **7,** 93–124. *269*

Westbrook, D. R. (1976). Small strain non-linear dynamics of plates. *J. Sound Vib.*, **44,** 75–82.

White, E. W., and P. W. Likins (1969). The influence of gravity torque on dual-spin satellite attitude stability. *J. Astronaut. Sci.*, **16,** 32–37. *396*

Whitham, G. B. (1952). The flow pattern of a supersonic projectile. *Comm. Pure Appl. Math.*, **5,** 301–348. *560, 566*

Whitham, G. B. (1970). Two-timing, variational principles and waves. *J. Fluid Mech.*, **44,** 373–395. *556*

Whitham, G. B. (1974). *Linear and Nonlinear Waves.* Wiley, New York. *545, 562, 564, 565, 567, 574, 614*

Whittaker, E. T. (1961). *A. Treatise on the Analytical Dynamics of Particles and Rigid Bodies.* Cambridge University Press. *369, 373*

Whittaker, E. T., and G. N. Watson (1962). *A Course of Modern Analysis.* Cambridge University Press. *259, 284, 285*

Wielgus, A. See Dzygadlo and Wielgus.

Willems, N. See Duffield and Willems.

Willems, P. Y. See Nishikawa and Willems.

Williams, C. J. H. (1966). The stability of nodal patterns in disk vibration. *Int. J. Mech. Sci.*, **8,** 421–432. *417, 441*

Williams, C. J. H., and S. A. Tobias (1963). Forced undamped nonlinear vibrations of imperfect circular discs. *J. Mech. Eng. Sci.*, **5,** 325–335. *505*

Williams, C. J. H. See Efstathiades and Williams; Naguleswaran and Williams.

Wilton, J. R. (1915). On ripples. *Phil. Mag.*, **29,** 688–700. *601*

Winkler, S. See Magnus and Winkler.

Witmer, E. A., H. A. Balmer, J. W. Leech, and T. H. H. Pian (1963). Large dynamic deformations of beams, rings, plates, and shells. *AIAA J.*, **1,** 1848–1857. *445*

Witmer, E. A. See Morino, Leech, and Witmer; Wu and Witmer.

Witt, A. See Gorelik and Witt.

Woinowsky-Krieger, S. (1942). Stability of rings. *Ing.-Arch.*, **13**, 90. *268*

Woinowsky-Krieger, S. (1950). The effect of an axial force on the vibration of hinged bars. *J. Appl. Mech.*, **17**, 35–36. *460*

Witmer, E. A. See Morino, Leech, and Witmer.

Woo, H. K. See Huang and Woo; Huang, Woo, and Walker.

Woo, T.-H. See Friedmann, Hammond, and Woo.

Woodall, S. R. (1966). On the large amplitude oscillations of a thin elastic beam. *Int. J. Non-Linear Mech.*, **1**, 217–238. *460*

Woodall, S. R. See Nowinski and Woodall.

Wrenn, B. G. and J. Mayers (1970). Nonlinear beam vibration with variable axial boundary restraint. *AIAA J.*, **8**, 1718–1720. *448, 453, 460*

Wrenn, B. G. See Mayers and Wrenn.

Wrout, G. N. See Likins and Wrout.

Wu, C.-I., and J. R. Vinson (1969a). Influences of large amplitudes, transverse shear deformation, and rotatory inertia on lateral vibrations of transversely isotropic plates. J. *Appl. Mech.*, **36**, 254–260. *500, 507*

Wu, C.-I., and J. R. Vinson (1969b). On the nonlinear oscillations of plates composed of composite materials. *J. Composite Mat.*, **3**, 548–561. *500, 501, 507*

Wu, C.-I., and J. R. Vinson (1971). Non-linear oscillations of laminated especially orthotropic plates with clamped and simply supported edges. *J. Acoust. Soc. Am.*, **49**, 1561–1567. *501, 502, 507*

Wu, R. W.-H., and E. A. Witmer (1971). Finite-element analysis of large elastic-plastic transient deformations of simple structures. *AIAA J.*, **9**, 1719–1724. *445*

Wu, R. W.-H., and E. A. Witmer (1973). Nonlinear transient responses of structures by the spatial finite-element method. *AIAA J.*, **11**, 1110–1117. *445*

Wu, R. W.-H., and E. A. Witmer (1975). Analytical and experimental studies of nonlinear transient responses of stiffened cylindrical panels. *AIAA J.*, **13**, 1171–1178. *506*

Yajima, N. See Asano, Taniuti, and Yajima; Taniuti and Yajima.

Yakovlev, V. I. See Kelzon and Yakovlev.

Yakubovich, V. A. (1958). On the dynamic stability of elastic systems. *Dokl. Akad. Nauk SSSR*, **121**, 602–605. *308*

Yakubovich, V. A., and V. M. Starzhinskii (1975). *Linear Differential Equations with Periodic Coefficients.* Wiley, New York. *259, 284*

Yamaki, N. (1961). Influence of large amplitudes on flexural vibrations of elastic plates. *ZAMM*, **41**, 501–510. *508*

Yamamoto, T. (1957). On the vibrations of a rotating shaft. *Mem. Fac. Eng. Nagoya Univ.*, **9**, 19–115. *367, 396*

Yamamoto, T. (1960). Response curves at the critical speeds of subharmonic and summed and differential harmonic oscillations. *Bull. JSME*, **3**, 397–403. *367, 396, 414*

Yamamoto, T. (1961a). On subharmonic and summed and differential harmonic oscillations of rotating shaft. *Bull. JSME*, **4**, 51–58. *396, 402*

Yamamoto, T. (1961b). Summed and differential harmonic oscillation in nonlinear vibratory systems. *Bull. JSME*, **4**, 658–664. *402, 418*

Yamamoto, T., and S. Hayashi (1963). On the response curves and stability of summed and differential harmonic oscillation. *Bull. JSME*, **6**, 420–429. *402, 418*

Yamamoto, T., and S. Hayashi (1964). Combination tones of differential type in nonlinear vibratory system. *Bull. JSME*, **7**, 690–698. *203, 402*

Yamamoto, T., and Y. Ishida (1974). The particular vibration phenomena due to ball bearings at the major critical speed. *Bull. JSME*, **17**, 59–67. *396*

Yamamoto, T., and Y. Ishida (1977). Theoretical discussions on vibrations of a rotating shaft with nonlinear spring characteristics. *Ing.-Arch.*, **46**, 125–135. *396*

Yamamoto, T., Y. Ishida, and J. Kawasumi (1975). Oscillations of a rotating shaft with symmetrical nonlinear spring characteristics. *Bull. JSME*, **18**, 965–975. *396*

Yamamoto, T., Y. Ishida, and J. Kawasumi (1977). The particular vibration phenomena due to ball bearings at the major critical speed. *Bull. JSME*, **20**, 33–40. *396*

Yamamoto, T., and Y. Nakao (1963). Combination tones of differential type in nonlinear vibratory systems. *Bull. JSME*, **6**, 682–689. *203, 402*

Yamamoto, T., and A. Saito (1970). On the vibrations of summed and differential types under parametric excitation. *Mem. Fac. Eng. Nagoya Univ.*, **22**, 54–123. *306, 308, 311*

Yamamoto, T., and K. Yasuda (1974). Occurrence of the summed and differential harmonic oscillations in a nonlinear multi-degree-of freedom vibratory system. *J. Appl. Mech.*, **41**, 781–786. *418*

Yamamoto, T., and K. Yasuda (1977). On the internal resonance in a nonlinear two-degree-of-freedom system. *Bull. JSME*, **20**, 169–175. *418*

Yamamoto, T., K. Yasuda, and I. Nagasaka (1976). Ultra-subharmonic oscillations in a nonlinear vibratory system. *Bull. JSME*, **19**, 1442–1447. *179*

Yamamoto, T., K. Yasuda, and I. Nagasaka (1977). On the internal resonance in a nonlinear two-degree-of-freedom system. *Bull. JSME*, **20**, 1093–1100. *418*

Yamamoto, T., K. Yasuda, and T. Nagoh (1975). Super-division harmonic oscillations in a nonlinear multi-degree-of-freedom system. *Bull. JSME*, **18**, 1082–1089.

Yamamoto, T., K. Yasuda, and T. Nagoh (1977). Super-division harmonic oscillation caused by nonlinearity of the fourth order. *Bull. JSME*, **20**, 24–32.

Yamamoto, T., K. Yasuda, and T. Nakamura (1974a). Combination oscillations in a nonlinear vibratory system with one degree of freedom. *Bull. JSME*, **17**, 560–568. *185*

Yamamoto, T., K. Yasuda, and T. Nakamura (1974b). Sub-combination tones in a nonlinear vibratory system. *Acta Tech. Čsav.*, **2**, 143–161. *185*

Yamamoto, T., K. Yasuda, and T. Nakamura (1974c). Sub-combination tones in a nonlinear vibratory system (caused by symmetrical nonlinearity). *Bull. JSME*, **17**, 1426–1437. *185*

Yanagiwara, H. (1960). A periodic solution of van der Pol's equation with a damping coefficient $\lambda=20$. *Fac. Sci. Hiroshima Univ.*, **A24**, 201–217. *144*

Yang, T. L., and R. M. Rosenberg (1967). On the vibrations of a particle in the plane. *Int. J. Non-Linear Mech.*, **2**, 1–25. *295, 388*

Yao, J. C. (1963). Dynamic stability of cylindrical shells under static and periodic axial and radial loads. *AIAA J.*, **1**, 1391–1396. *269*

Yao, J. C. (1965). Nonlinear elastic buckling and parametric excitation of a cylinder under axial loads. *J. Appl. Mech.*, **32**, 109–115. *269*

Yarymovych, M. I. See Skalak and Yarymovych.

Yasuda, K. See Yamamoto and Yasuda; Yamamoto, Yasuda, and Nagasaka; Yamamoto, Yasuda, and Nagoh; Yamamoto, Yasuda, and Nakamura.

Yen, D. H. Y. (1974). On the normal modes of nonlinear dual-mass systems. *Int. J. Non-Linear Mech.*, **9**, 45–53. *365*

Yen, D. H. Y., and T. W. Lee (1975). On the non-linear vibrations of a circular membrane. *Int. J. Non-Linear Mech.*, **10**, 47–62. *505*

Yen, D. H. Y., and S.-C. Tang (1970). On the nonlinear response of an elastic string to a moving load. *Int. J. Non-Linear Mech.*, **5**, 465–474. *608*

Yen, D. H. Y., and S.-C. Tang (1976). Nonlinear waves generated by a steadily moving line load on an elastic plate. *Int. J. Solids Struct.*, **12**, 467–477. *608*

Yen, D. H. Y. See Lee, Blotter, and Yen; Tang and Yen.

Yin, S. K. See Hohenemser and Yin.

Ying, S. J. See Chu and Ying.

Yu, Y.-Y. (1962). Nonlinear flexural vibrations of sandwich plates. *J. Acoust. Soc. Am.*, **34**, 1176–1183. *501, 508*

Yu, Y.-Y. (1963). Application of variation equation of motion to the nonlinear vibration analysis of homogeneous and layered plates and shells. *J. Appl. Mech.*, **30**, 79–86. *501, 506*

Yu, Y. Y. and J. L. Lai (1966). Influence of transverse shear and edge condition on nonlinear vibration and dynamic buckling of homogeneous and sandwich plates. *J. Appl. Mech.*, **33**, 934–936. *501*

Yu, Y.-Y. See Grossman, Koplik, and Yu.

Yuan, S. W. See Horvay and Yuan.

Yutani, T. See Saito, Sato, and Yutani.

Zabusky, N. J. (1962). Exact solution for the vibrations of a nonlinear continuous model string. *J. Math. Phys.*, **3**, 1028–1039. *486*

Zabusky, N. J. See Kruskal and Zabusky.

Zahradka, J. (1975). Subharmonic oscillations in a non-linear system with cardan shafts. *Maschinenbautechnik*, **24**, 218–220, 226.

Zakharov, V. E., and S. V. Mankov (1973). Resonant interaction of wave packets in nonlinear media. *Sov. Phys.-JETP Letts.*, **18**, 243–245. *588*

Zakharov, V. E., and A. B. Shabat (1972). Exact theory of two-dimensional self focusing and one-dimensional self modulation of waves in nonlinear media. *Sov. Phys.*, **34**, 62–69. *588, 594*

Ziegler, H. (1952). Die Stabilitätskriterien der Elastomechanik. *Ing.-Arch.* **20**, 49–56. *100*

Ziegler, H. (1953). Linear elastic stability. *ZAMP*, **4**, 89–121. *100*

Ziegler, H. (1956). On the concept of elastic stability. *Adv. Appl. Mech.*, **4**, 351–403. *100*

Ziegler, H. (1968). *Principles of Structural Stability*. Blaisdell, Waltham, Mass. *100*

Zienkiewicz, O. C., and C. J. Parekh (1970). Transient field problems; two-dimensional and three-dimensional analysis by isoparametric finite element. *Int. J. Num. Meth. Eng.*, **2**, 61–71. *445*

INDEX